DIVINE ACTION AND NATURAL SELECTION

SCIENCE, FAITH AND EVOLUTION

DIVINE ACTION AND NATURAL SELECTION

SCIENCE, FAITH AND EVOLUTION

EDITORS

JOSEPH SECKBACH

The Hebrew University of Jerusalem, Israel

RICHARD GORDON

University of Manitoba, Canada

World Scientific

NEW JERSEY • LONDON • SINGAPORE • BEIJING • SHANGHAI • HONG KONG • TAIPEI • CHENNAI

Published by

World Scientific Publishing Co. Pte. Ltd.
5 Toh Tuck Link, Singapore 596224
USA office: 27 Warren Street, Suite 401-402, Hackensack, NJ 07601
UK office: 57 Shelton Street, Covent Garden, London WC2H 9HE

Library of Congress Cataloging-in-Publication Data
Divine action and natural selection : science, faith, and evolution /
editors, Joseph Seckbach, Richard Gordon.
p. cm.
Includes bibliographical references and index.
ISBN-13: 978-981-283-433-1 (hardcover : alk. paper)
ISBN-10: 981-283-433-8 (hardcover : alk. paper)
ISBN-13: 978-981-283-434-8 (pbk. : alk. paper)
ISBN-10: 981-283-434-6 (pbk. : alk. paper)
1. Evolution (Biology) 2. Natural selection. 3. Religion and science.
I. Seckbach, J. (Joseph) II. Gordon, Richard, 1943–
QH366.2.D57 2008
576.8--dc22

2008042460

British Library Cataloguing-in-Publication Data
A catalogue record for this book is available from the British Library.

Printed by FuIsland Offset Printing (S) Pte Ltd, Singapore

This volume is dedicated to a close colleague and prominent scientist **Professor Aharon Oren** (The Hebrew University of Jerusalem), the steady assistant and author of chapters in books of the *Cellular Origins, Life in Extreme Habitats and Astrobiology* series (http://www.springer.com/series/5775). His assistance is very valuable and remarkable. Professor Oren has been a faithful advisor. I wish him and his family the best blessing.

This volume is also dedicated to our respective grandchildren: **Zur**, **Harel**, **Dor**, **Even**, **Elyakim**, **Naveh**, **Aharon**, **Bitya**, **Oriyah** and **Ruth** [J.S.], and **Evan**, **Ethan**, **Emma**, **Alex**, **Nick** and **Ethan** [R.G.].

Contents

Part 1. Background in Theology, Philosophy and Science

Part 2. Towards Harmony Between Science and Religion

Part 3. Is a Purely Scientific Approach to the Origin of Life in the Universe Sufficient?

Acknowledgements

I wish to thank my wife **Fern L. Seckbach** for her patience, cooperation and advice in many ways in this volume. Thanks are due to our faithful authors of this volume who had lots of patience to wait for the responses of the reviewers and of the editors' replies. Special appreciation is due to Professor **Julian Chela-Flores** (ICTP, Trieste, IT) and to many colleagues for so much advice and many hints around this volume. I appreciate the special effort of some dialoguers who went far from their routine tasks and fully discussed the specific chapters in this volume.

I acknowledge the efforts our peer reviewers and external referees for evaluating the manuscripts of this volume. Their names are listed in an alphabetical order (with special bold printed names to those who were outstandingly active in evaluating several manuscripts): Amnon Albeck, Doron Aurbach, Oktar Babuna, **Peter Bahn**, John Baumgardner, **Michael Behe**, Valentino Braitenberg, **Taner Edis**, George Ellis, Benjamin Fain, Raphael Falk, **Jerom Yehuda Gellman**, Arnie Gotfryd, Mordechai Halperin, Oren Harman, Arie Issar, Kamaluddin, Charles H. Lineweaver, Michel Morange, Edward Oakes, John Oller, Robert Pollack, Jin Qusheng, **Fred Rainey**, **Fuz Rana**, Florence Raulin Cerceau, Werner Reisser, Juan Roederer, Mary Beth Saffo, Gerald Schroeder, Tahir Shah, **Josef Svoboda** and **Nakdimon Umiel**.

The process of reviewing was not an easy task since a reviewer from one group of evolutionary school could not always accept the opposite view. We did our best to judge a chapter on the basis of its scientific or philosophical/theological value and improve the English (US) language. We hope that the readers of this volume will evaluate and judge the chapters by their own understanding and personal tendencies.

Gratitude is due to the assistance of Professor **Wladyslaw Altermann** and Dr. **Michaela Frei** from the Ludwig Maximilians University in Munich, Germany. They welcomed and hosted me in their Department during the generosity of my DAAD-2006 (German Academic Exchange Program) fellowship.

Joseph Seckbach
Hebrew University of Jerusalem
seckbach@huji.ac.il
January 2008

Further Acknowledgements

I would like to thank Joseph Seckbach for inviting me to contribute a chapter and then, when I proposed adding dialogue between the authors and others, took me up on it. This has been a strange journey for a theoretical biologist, but one that may be rewarded with better science. I thank all of our authors and guests who have tried to be civil or more with one another while discussing such contentious issues. The resulting dialogue often seems to end in mid-thought, with conflicts, logical inconsistencies, rebuttals, and even agreements achieved, as if suspended for another day. Some potential dialogues never got started, so coverage is incomplete and uneven. We allowed half a year for it, and it is clear that dialogue profitably could have gone on for another full year. Thus we regard this book as but the start of what we hope will eventually be a fruitful dialogue between, and perhaps eventually a consensus on, science and religion.

Richard Gordon
University of Manitoba
gordonr@cc.umanitoba.ca
July 2008

List of Authors and Dialoguers and Their Addresses and E-mail Addresses

Altan, Cem Sedat [Dialogue]
President of the Science Research Foundation, Turkey
seda_aral@yahoo.com

Aurbach, Doron [Author: Chapter 33, Dialogue]
Department of Chemistry, Bar Ilan University, Ramat Gan 52900 Israel
aurbach@mail.biu.ac.il

Babuna, Oktar [Author: Chapter 16, Dialogue]
Istanbul University Medical School, Istanbul University Cerrahpasa Medical School, Department of Neurosurgery, Istanbul, Turkey
University of California San Francisco Medical School, Department of Neurosurgery, California, USA, and New York University Medical School, Department of Neurosurgery, New York, USA
oktar_babuna@yahoo.com

Bahn, Peter R. [Author: Chapter 14, Dialogue]
Bahn Biotechnology Company, 10415 E. Boyd Rd., Mt. Vernon, Illinois, 62864 USA
pbahn@mvn.net

Ball, Philip [Dialogue]
Nature, 4-6 Crinan Street, London N1 9XW, UK
p.ball@nature.com

Barbalet, Tom [Author: Chapter 6, Dialogue]
Las Vegas NV USA
tom@nobleape.com

Barr, Stephen M. [Author: Chapter 23, Dialogue]
Bartol Research Institute, Department of Physics and Astronomy, University of Delaware, Newark, DE, 19716 USA
smbarr@bartol.udel.edu

Baumgardner, John [Author: Chapter 41, Dialogue]
Institute for Creation Research; 10946 Woodside Avenue North, Santee, California 92071 USA
jrbaumgardner@cox.net

Bazant, Pavel [Dialogue]
Faculty of Mathematics & Physics, Univerzita Karlova v Praze, Praha
Czech Republic
pbazant@gmail.com

Behe, Michael J. [Author: Chapter 34, Dialogue]
Department of Biological Sciences, Lehigh University Beth-Lehem, Pennsylvania 18015 USA
mjb1@lehigh.edu

Beloussov, Lev V. [Dialogue]
Department of Embryology, Moscow State University
Moscow State University, Moscow 119899, Russia
morphogenesis@yandex.ru

Berger, Dov [Author: Chapter 11, Dialogue]
Committee on Interdisciplinary Studies, Bar Ilan University, Ramat Gan 52900, Israel
berger1@macam.ac.il

Bienfait, Frits [Author: Chapter19]
Oudegracht 285 Bis, 3511 Pa Utrecht, The Netherlands
bienfait@ision.nl

Chela-Flores, Julian [Author: Chapters 4, 45, Dialogue]
The Abdus Salam International Centre for Theoretical Physics, Strada Costiera 11; 34136 Trieste, Italy and Instituto De Estudios Avanzados, Caracas 1015A, Venezuela
chelaf@ictp.itc

Coyne, George V. [Author: Chapter 2]
Vatican Observatory, V-00120 Vatican City State
gcoyne@specola.va

Damer, Bruce [Author: Chapter5]
Digital Space Commons, 343 Soquel Avenue, # 70 Santa Cruz CA 95062-2305, USA
bdamer@digitalspace.com

De Duve, Christian [Author: Chapter 1]
Rockefeller University, 1230 York Avenue, New York NY 10065, USA CDE
cdeduve@mail.rockefeller.edu

Edis, Taner [Author: Chapter 26, Dialogue]
Truman State University, Department of Physics. Kirksville MO 63501,USA
edis@truman.edu

Fain, Benjamin [Author: Chapter21]
School of Chemistry, Tel Aviv University, 69978 Tel Aviv, Israel
fain@post.tau.ac.il

Furstenberg, Harry [Author: Forward]
Department of Mathematics, Hebrew University of Jerusalem, Israel
harry@math.huji.ac.il

Gellman, Jerome Yehuda [Author: Chapter 22]
Ben-Gurion University of the Negev, Beer Sheva, Israel,
gellman@bgumail.bgu.ac.il

Gordon, Richard [Editor, Author: Chapters 12, 17, Dialogue]
Department of Radiology, University of Manitoba
Room GA216, HSC, 820 Sherbrook Street, Winnipeg, Manitoba R3A 1R9
Canada
gordonr@cc.umanitoba.ca

Gotfryd, Arnie [Author: Chapter 44]
3727 Chesswood Drive Toronto, ON, M3J 2P6, Canada
arnie@gotfryd.com

Grand, Steve [Dialogue]
The Open University, Walton Hall, Milton Keynes, MK7 6AA UK
steve@cyberlife-research.net

Grandpierre, Attila [Author: Chapter 28, Dialogue]
Konkoly Observatory of the Hungarian Academy of Sciences, H-1525 Budapest, P. O. Box 67, Hungary
grandp@ella.hu

Gündoğdu, Cihat [Author: Chapter 37, Dialogue]
Marmara University School of Medicine, Istanbul Department of Family Medicine, Haydarpasa Numune Educational and Research State Hospital, Istanbul, Turkey
CGundogdu@scienceresearchfoundation.org

Halperin, Mordechai [Author: Chapter 18]
Israeli Ministry of Health, 4 Man Street, Givaat Mordechai, Jerusalem, Israel
medical.ethics@moh.health.gov.il

Harman, Oren Solomon [Author: Chapter 9, Dialogue]
Program in Science, Technology and Societybar Ilan University, Israel
harman@vms.huji.ac.il

Issar, Arie S. [Author: Chapter 42, Dialogue]
Ben-Gurion University of the Negev, Sede Boker Campus, Israel
issar@bgu.ac.il

Leibowitz, Elia M. [Author: Chapter 36]
School of Physics and Astronomy, Tel Aviv University,
Tel Aviv 69978, Israel
elia@wise.tau.ac.il

Lineweaver, Charles [Author: Chapter 10, Dialogue]
Planetary Science Institute, Research School of Astronomy and Astrophysics and Research School of Earth Sciences, The Australian National University, Canberra, Act, 0200 Australia
charley@mso.anu.edu.au

McGrew, Stephen P. [Author: Chapter 7, Dialogue]
New Light Industries, Ltd., 9713 W. Sunset Hwy.
Spokane WA 99204 USA
stevem@nli-ltd.com

McMullin, Ernan [Author: Chapter 8, Dialogue]
Department of Philosophy, University of Notre Dame,
Notre Dame IN 46556 USA
mcmullin.1@nd.edu

Muravnik, Galina L. [Author: Chapter 30]
Orthodox Lyceum of Spiritual Culture Named After Venerable Seraphim of Sarov 3 Parkovaja St., 34/1, 105043 Moscow, Russia
ar-paramir@yandex.ru

Negrón-Mendoza, Alicia [Author: Chapter 43]
Instituto De Ciencias Nucleares, Unam, A.P. 70-543, México, D.F. Mexico
negron@nucleares.unam.mx

Negrotti, Massimo [Dialogue]
Faculty of Sociology, University of Urbino, Via Saffi,
15 I-61029, Urbino, Italy
massimo.negrotti@libero.it

Oakes, Edward T. [Author: Chapter 3]
University of St. Mary of the Lake. Mundelein Seminary, 1000 East Maple Avenue, Mundelein, Illinois 60060 USA
eoakes@usml.edu

Prothero, Donald R. [Author: Chapter 24]
Department of Geology, Occidental College, Los Angeles,
CA 90041 USA
prothero@oxy.edu

Ramos-Amézquita, Sandra [Author: Chapter 43]
Facultad De Ciencias Políticas Y Sociales, Unam, México,
sra.sandra@gmail.com

Ramos-Bernal, Sergio [Author: Chapter 43]
Instituto De Ciencias Nucleares, Unam, A.P. 70-543, México, D.F. Mexico
ramos@nucleares.unam.mx

Raulin Cerceau, Florence [Author: Chapter 13, Dialogue]
Centre Alexandre Koyré, (Cnrs-Ehess-Mnhn- Umr 8560) Case Postale 25, Muséum National D'histoire Naturelle 57 Rue Cuvier – 75005 Paris – France
raulin@mnhn.fr

Roederer, Juan G. [Author: Chapter 39, Dialogue]
Geophysical Institute, University of Alaska-Fairbanks, Fairbanks, AK 99775, USA
jgr@gi.alaska.edu

Roth, Hermann Josef [Author: Chapter 31]
Universitätsbibliothek, Naturhistorischer Veren, Nußalle 15a,D - 53115 Bonn, Germany
Hermannjroth@aol.com

Sabath, Karol (Deceased) **[Author: Chapter 32]**
Museum of Evolution, Institute of Paleobiology, Polish Academy of Sciences, Twarda 51/55, 00-818 Warsaw, Poland

Schrader, Malcolm E. [Author: Chapter 25]
Department of Inorganic and Analytical Chemistry, Institute of Chemistry, The Hebrew University of Jerusalem 91904, Jerusalem, Israel
schrader@vms.huji.ac.il

Schroeder, Gerald [Author: Chapter 35, Dialogue]
Aish Hatorah College of Jewish Studies, Hahish 5, Jerusalem, 93223 Israel
gs@geraldschroeder.com

Seckbach, Joseph [Editor]
Hebrew University of Jerusalem
P.O.Box 1132, Efrat, 90435 Israel
seckbach@huji.ac.il

Shah, K. Tahir [Author: Chapter 20]
Faculty of Engineering University of Trieste, Trieste, Italy
shah@ictp.it

Shostak, Seth [Author: Chapter 40]
Seti Institute 515 N. Whisman Road, Mountain View, CA 94043, USA
seth@seti.org

Slifkin, Natan [Author: Chapter 29]
2a Nachal Raziel, Ramat Bet Shemesh 99632, Israel
zoorabbi@zootorah.com

Spetner, Lee M. [Author: Chapter 38, Dialogue]
Redoxia Israel Ltd., 27 Hakablan St., Jerusalem 93874, Israel
LSpetner@alum.mit.edu

Stenger, Victor J. [Dialogue]
Department of Philosophy, University of Colorado, Boulder, CO 80301 USA
vic.stenger@comcast.net

Svoboda, Josef [Author: Chapter 27, Dialogue]
Department of Biology, University of Toronto at Mississauga Mississauga, Ontario, L5L 1C6, Canada
josef.svoboda@utoronto.ca

Tuszynski, Jack A. [Dialogue]
Department of Physics, University of Alberta and Division of Experimental Oncology, Cross Cancer Institute, Room 3336, Cross Cancer Institute, 11560 University Avenue, Edmonton, AB T6G 1Z2, Canada
jtus@phys.ualberta.ca

Valderrama-Negrón, Ninel [Author: Chapter 43]
Facultad De Filosofía Y Letras, Unam, México
ninelvn@gmail.com

Yahya, Harun (Pen Name For Adnan Oktar) **[Author: Chapter 15, Dialogue]**
Academy of Fine Arts at Mimar Sinan University of Istanbul and, Department of Philosophy at Istanbul University, Turkey
info@harunyahya.com
seda_aral@yahoo.com

Biography

Hillel (Harry) Furstenberg is an Israeli mathematician. He is known for his application of probability theory and ergodic theory methods to other areas including number theory combinatorics, and Lie group theory. He was born in Berlin in 1935, and soon immigrated to the US. He concluded his BA and MSc studies at Yeshiva University, New York in 1955. He obtained his Ph.D. under Salomon Bochner at Princeton University in 1958. After several years at the University of Minnesota he became a professor of Mathematics at the Hebrew University of Jerusalem in 1965 and has taught at Bar Ilan University, Israel as well. He is a member of the Israel Academy of Sciences and Humanities and U.S. National Academy of Sciences. He is the recipient of several prizes: The Israel Prize in Mathematics, the Rothschild Prize, the Harvey Prize, the Emet Prize, and the Wolf Prize for Mathematics.

Foreword

HILLEL (HARRY) FURSTENBERG

Department of Mathematics
Hebrew University of Jerusalem, Israel
harry@math.huji.ac.il

1. Darwinism — A Two-pronged Challenge to Religion

In a discussion of magic and the prohibition in Judaism of practicing magic, the Talmud points out that the Hebrew word for witchcraft can be seen as an acronym for "encroachment on the heavenly domain", or, in other words, curtailing the sphere of Divine intervention. It is fair to say that the Rabbis of the Talmud would find contemporary science also encroaching on the heavenly domain. The conflict may be inevitable. The task of science is to demystify phenomena, finding "everyday" laws governing as much of our experience as possible, whereas religion — as also poetry — seeks the miraculous in the mundane, heightening the "everyday" experience.

The classic example of this clash of interests — though by no means the earliest — is the development of Newtonian physics and its implication for celestial motion. In this context, Laplace, the author of a treatise on celestial mechanics, is famously reported to have responded to Napoleon's query as to how he could have omitted God's name from his work, "I have no need for this hypothesis." Newton himself, a staunch believer, still had a need for this hypothesis, particularly so since he wasn't able to explain all the known astronomical phenomena, whereas Laplace's improved analysis did away with the gaps in Newtonian astronomy. The spokesmen for Christianity at the time criticized Newton's theory of gravitation and laws of motion vigorously, arguing that he "took from God that direct action on his works so constantly ascribed to him in

Scripture, and transferred it to material mechanism", or that he "substituted gravitation for Providence" (p. 16 in: White, AD: *A History of the Warfare of Science with Theology in Christendom.* New York: D. Appleton and Co.; 1896.)

Providence implies perpetual supervision, and is more in line with Aristotle's physics whereby a persistent force is needed to maintain motion. Newtonian theory not only dispenses with this need, but in some sense obligates God to abide by laws capable of formulation by human hand.

This clash doesn't spare Jewish thinkers who, following Nachmanides, emphasize the ongoing character of Divine Providence. On the other hand it wouldn't disturb Maimonides, for whom immutable law is the clear indicator of Divine intelligence built into the Creation. Indeed religious thought has come to terms with the Newtonian challenge, with the idea of God functioning outside the realm of time harmonizing with timeless laws of nature, and not subject to whimsy. Newton's explanation of the miracle of the clockwork-like movement of heavenly bodies is the prototype of scientific understanding, basing the observed phenomenon on universal laws of motion and gravity that can be confirmed at every level of experience.

The Darwinian challenge, at the heart of what probably remains the most contentious of science/religion debates, is of a more complex nature, the role of the Creator being usurped not by a universal deterministic law, but by a combination of what Darwin himself called "blind chance and necessity." The role of chance in evolutionary theory is seen by many as directly implying the lack of any governance, and a phenomenon enabled by chance events surely has "no need" for the implicit intelligence of a Creator. The component of necessity in Darwinian Theory appears in the fact that in a highly competitive environment, the underperformer will not pass its genes to successive generations. This is a law of simple logic — or common sense. There is some irony in the idea that chance events which are compatible with Divine intervention (though not giving testimony to such intervention) and may have been thought to be repudiated by the strict determinism of Newtonian mechanics, are fully restored in Darwinism, but, nonetheless "encroach on the heavenly domain." Once again the Church responded vigorously: "The principle of natural selection is absolutely incompatible with the word of God" (p. 70 in White); "Christians have a right to protest the arraying of probabilities against the clear evidence of the Scriptures"; "to ignore design as manifested in God's creation is to dethrone God" (p. 79 in White). In 1996, however, the Church presented the very constructive comments of the late Pope John Paul II: "New knowledge has led us to realize that the theory of evolution is no longer a mere hypothesis.... The convergence, neither sought nor fabricated, of the results of work that was conducted independently is in itself a significant argument in favor of this theory" (Paul II, J: Address to the Pontifical Academy of Sciences, October 22, 1996. In: *Evolutionary and Molecular Biology: Scientific Perspectives on Divine Action.* Edited by Russell RJ, Stoeger WR, Ayala FJ: University of Notre Dame Press; 1999:2-9).

We have two ingredients appearing in the Darwinian scientific upheaval that do not play a role in the Newtonian revolution, and these will be at the focus of many of the studies in the present volume. These features, related to one another in a natural way, are the place of chance in scientific theory, and the evidence of design in scientific phenomena. Darwinism welcomes the former and the latter lies beyond its frontiers. Religion is ambivalent about chance, but insists on the significance of design.

2. Blind Chance and Random Process

In Darwin's theory chance is called upon to account for variability among members of a species. The mechanism for this is made precise in contemporary evolutionary theory, with variability stemming from "random" mutations within genes. Many mutations will be detrimental to the carrier of the affected gene, but some mutations, if not directly beneficial, may lead to a version of the organism better adapted to its environment. With this pinpointing of the place of chance, there is a subtle evolution of the notion of "blind chance" to the more respectable idea of "random process", an established notion in contemporary science. One shouldn't fail to see the oxymoronic character of this expression; randomness referring to inexplicability of a phenomenon, and "process" implying an understandable mechanism driving the phenomenon.

Nevertheless the notion has been very fruitful, and random processes with precise mathematical definitions serve as important models for physical and chemical phenomena (nuclear fission, diffusion). As a consequence a physical law involving random behavior may well be granted the same status as a deterministic law. This is particularly the case when a large number of chance events are involved and one encounters a "Law of Large Numbers" which guarantees a high probability to certain outcomes. This latter circumstance, however, is not present in Darwin's evolutionary theory. The chains of events called upon in the Darwinistic paradigm are enormously complex, and the probabilities involved defy estimation. Conceivably at some point the theory will be sufficiently developed so that the adaptive tendency of organisms appears as a rigorous outcome of a random process. In the present state of the theory, the adaptive tendency observed in the bio-world appears as an unlikely — but possible — outcome of chance events.

In the absence of Laws of Large Numbers virtually ensuring the occurrence of certain fortuitous events with some frequency, such events must be seen as "accidents", ostensibly not subject to scientific analysis. Accidents do happen, but no more can be said. To what (or whom) an accident is ascribed depends very much on one's philosophical agenda, and there is no objective guide in this matter. Anatole France is quoted as saying with some cynicism that "chance is the pseudonym of God when He didn't want to sign". In the *Bible*, the two are hardly distinguished and chance is most clearly regarded as the expression of Divine intervention. An extreme formulation is the prayer in Genesis (Chapter 24) by Abraham's servant Eliezer asking God to "chance before him" and lead him to a suitable wife for Abraham's son Isaac.

When the religiously inclined person attributes a chance event to Providence, he is regarding the event not as Divine whim, but as manifestation of Divine intelligence. It is seen as part of some larger scheme whose rationale escapes the human observer, but is confidently believed to be present. There is no real clash here between our state of ignorance and the intelligent workings of an omniscient Deity.

Probability theory uses human ignorance systematically to create a useful scientific discipline. This in itself is something of a miracle.

There remains however — at least on the psychological level — a clash between our perception of chance events and the presence of intelligence.

The conception of God as Supreme Designer does not square with assigning to Him the role of Random Processor. It is thus the stochastic element in Darwinism that remains most objectionable to the religious mind. But independent of the religious ramifications, the aspect of design and purposiveness in the bio-world on the one hand, and the aspect of randomness on the other, is a feature of Darwinism that, at least superficially, creates a tension within evolutionary theory. It is to this tension that the contemporary notion of Intelligent Design is addressed.

3. Design as Paradigm

The inescapable appearance of design in the bio-world has been an essential feature of biological science since its beginnings. The teleological nature of explanation in biology with its emphasis on function tacitly posits a Designer in the background. And while the Designer hypothesis is no longer in vogue, the appearance of design (or "designoid" as Dawkins refers to this) remains a basic feature in the study of the animate world (Dawkins, R: *Climbing Mount Improbable*. New York: W.W. Norton & Co.; 1996; Dawkins, R: *The Ancestor's Tale: A Pilgrimage to the Dawn of Evolution*; 2004). The phenomenon of adaptation is a sine qua non in the domain of living matter, and adaptation is essentially synonymous with design, the phenomenon that living creatures are endowed with properties enabling them to survive and reproduce. Explaining the phenomenon of adaptation is the central problem of evolutionary theory.

The scenario that would be welcomed by proponents of Intelligent Design would envision the development of a scientific discipline that might be called "Design Processes" which would achieve recognition alongside the theory of Random (Stochastic) Processes. Such a theory would provide an alternative mechanism for evolution, which while not invoking supernatural forces, would accord more harmoniously with theological assumptions. It is difficult in the present state of scientific development to give credence to such a scenario. Yet it shouldn't be ruled out, and its possibility sets limits on how compelling are the consequences of the contemporary Darwinian paradigm.

Withal, alongside the phenomena suggesting design in the bio-world are features which exhibit a lack of design, and in fact are suggestive of the kind of randomness encountered in Darwinism. These features are often overlooked by the proponents of Intelligent Design, and deserve more extensive treatment. The outstanding example is the appearance of "homologous" structural similarities in different species where in some of these species no recognizable purpose is served by the organs in question (e.g., vestigial organs). There are even disadvantageous homologies. An example of this is one of the cranial nerves which runs from the brain to the larynx passing close to the heart. In fish this is a direct route, but in giraffes this path involves an unexplainably long detour. Phenomena of this type fit well with a paradigm of small random modifications of organisms, and place a limit on Intelligent Design.

These phenomena also challenge the widely held view (shared by the rabbis) of a sharp delineation of the animal world (and plant world for that matter) into distinct species traceable to Creation. Orthodox Darwinism goes beyond this with the belief that a single progenitor represents the origin of all animal life on earth. But the more basic message of Darwinism is the affinity of all life forms.

The religious individual holding to his belief in an Almighty God, though resisting forming a visual picture of God, nevertheless maintains some mental model of God's activity (e.g., as king, as warrior, as judge, etc.) This model is infringed upon by the laws science uncovers, which we naturally, if not justifiably, perceive as the world running itself. This took place centuries ago with regard to the physical world, and may now be inevitable in our perception of the animate world. But while changing the details of one's mental model, this need not diminish devoutness, no small part of which is constituted by human appreciation for the wonders of God's creation. In this context it is most instructive to read Darwin's concluding lines in the "*Origin of the Species*":

"Thus, from the war of nature, from famine and death, the most exalted object which we are capable of conceiving, namely, the production of the higher animals follows. There is grandeur in this view of life, with its several powers, having been originally breathed into a few forms or into one; and that, whilst this planet has gone cycling on according to the fixed law of gravity, from so simple a beginning endless forms most beautiful and most wonderful have been and are being evolved."

Biographies

Professor Joseph Seckbach is the initiator and chief editor of **Cellular Origins, Life in Extreme Habitats and Astrobiology (COLE)** series. www.springer.com/sereis/5775. He is the author of several chapters in this series. Dr. Seckbach earned his Ph.D. from the University of Chicago, Chicago, IL (1965). Recently he spent three months in Ludwig Maximilians University in Munich with a DAAD fellowship from the German service of exchange academicians, where several forward steps of this volume have been performed.

Among his publications are books, scientific articles concerning plant ferritin (phytoferritin), cellular evolution, acidothermophilic algae (mainly on *Cyanidium caldarium*), and life in extreme environments. He also edited and translated several popular books. Dr. Seckbach is the co-author (with R. Ikan) of the Chemistry Lexicon (1991, 1999) and other volumes, such as the *Proceedings of Endocytobiology VII Conference* (Freiburg, Germany, 1998) and the *Proceedings of Algae and Extreme Environments Meeting* (Trebon, Czech Republic, 2000) http://www.schweizerbart.de/pubs/books/bo/novahedwig-051012300-desc.ht). His recent interest is in the field of enigmatic microorganisms and life in extreme environments.

Professor Julian Chela-Flores was born in Caracas, República Bolivariana de Venezuela and studied in the University of London, England, where he obtained his Ph.D. in quantum mechanics (1969). He was a researcher at the Venezuelan Institute for Scientific Research (IVIC) and Professor at Simon Bolivar University (USB), Caracas until his retirement in 1990. During his USB tenure he was Dean of Research for six years. He is a Fellow of The Latin American Academy of Sciences, The Academy of Sciences of the Developing World, the Academy of Creative Endeavors (Moscow) and a Corresponding Member of the Venezuelan “Academia de Fisica, Matematicas y Ciencias Naturales”. His current positions are Staff Associate of the Abdus Salam International Center for Theoretical Physics (ICTP), Trieste, Research Associate, Dublin Institute for Advanced Studies (DIAS) and Profesor-Titular, Institute of Advanced Studies (IDEA), Caracas. His particular area of expertise is astrobiology, in which he is the author of numerous papers. He organized a series of Conferences on Chemical Evolution and the Origin of Life from 1992 till 2003. In 2001 he published the book: *The New Science of Astrobiology From Genesis of the Living Cell to Evolution of Intelligent Behavior in the Universe*.

Preface 1
Where Did We Come From?

JOSEPH SECKBACH[1] AND JULIAN CHELA-FLORES[2]

[1]*Hebrew University of Jerusalém, P.O.Box 1132, Efrat, 90435, Israel*
[2]*The Abdus Salam International Centre for Theoretical Physics, Strada Costiera 11; 34014 Trieste, Italy and Instituto de Estudios Avanzados, Caracas 1015A, Republica Bolivariana de Venezuela*

This book appears 200 years after Darwin's birth and 150 after the publication of the principle of natural selection (co-discovered by Alfred Russel Wallace). It focuses on the implications of his landmark contribution to the study of evolution. His seminal contribution has played a significant role in the dialogue between faith and reason. Evolution is currently a subject of much religion-inspired discussions while, at the same time, significant related progress has been achieved from the scientific point of view. Darwin's classical scientific masterpiece *The Origin of Species* is especially relevant for the educational system, as well as for the scientific community. This issue is not new. Science and religion have been the focus of debate for centuries. Evolution of life on Earth is revisited in this volume to encourage a dialogue on the old question: *Where did we come from?* There are multiple sources that deal with these evolutionary issues. One can find relevant information in journals, magazines, a number of books, as well as in the 'Internet ocean of information'.

In spite of the abundance of previous publications, there is room for the present book. It attempts to encourage a constructive dialogue between two approaches to the origin and evolution of life on Earth. Firstly, the book discusses a scientific theory that is strongly and convincingly supported by observation and experiments. Secondly, this volume contains an approach to life on Earth within the domain of theology. The following chapters address the ongoing dialogue between Darwinian theory and Intelligent Design (ID), the latter being a new term that refers to divine action by a theistic designer of all life on Earth. Within the theological domain ID views God's creation and guidance as influencing all living creatures, as expressed in the book Genesis, and other Books of the *Bible.* Within the scientific domain Darwinism assumes that natural selection is the main mechanism for evolution.

Both sides of the dialogue have proponents, as well as opponents that frequently express some degree of skepticism. Furthermore, currently there are debates as how evolution should be inserted in the high school and college science curricula. In the USA and Europe there are arguments at government level on the relevance of evolution in education. However, there is some opposition to teaching Darwinian evolution, or at least there are requests that ID should also be introduced into science classes. In this context there has been an official decision. A court case involved the U.S. District Judge John E. Jones of Pennsylvania (USA), whose verdict ruled that teaching ID would violate the separation of church and state. This decision has influenced the selection of topics that should be taught in those schools.

Chapters in this volume cover both approaches to the question of the origin and evolution of life on Earth. But we emphasize that the frontiers of science and the humanities should be clearly outlined. Prominent authors have written these contributions. We have gathered together over 40 chapters. The majority of these contributions come from USA and Israeli scholars, while manuscripts arrived from many other countries. All the manuscripts in this volume have been peer-reviewed, or sent out to external referees.

Some contributors suggest definite answers to what they consider to be a debate between science and religion, while others encourage a dialogue, rather than a debate. Indeed, these discussions are not new to Christianity, Islam or, to a lesser extent, to Judaism. We should hope that the dialogue between science and faith should not be sterile in either direction, but rather that it should lead to a deeper understanding of these issues: *What is the domain of divine action, and where are the boundaries of experiments and rational hypotheses?*

Our personal view is that a dialogue rather than a debate in faith and reason is to be encouraged for the reasons we shall enumerate. Firstly, it is not relevant for science to address the question of divine action, because science and religion have their own specific frontiers. Secondly, controversy arising from discussing faith and reason should be paid due attention, especially when it concerns the proposal of teaching both science and creationism in the same school as part of training in science (cf., section on "Background in theology, philosophy and science" in this volume). The origin and evolution of life can be considered either within philosophy, or within science. ID does not necessarily have to reject a scientific approach to the origin of life, since astrobiology lies outside the domain of religion. In the following pages it is advocated that observant believers may well accept the scientific approach of evolution (cf., for instance, the chapters in this book by Halperin and by Chela-Flores).

Finally, there is the question of whether the book of Genesis should be interpreted literarily. One should remember that the Holy Books can be interpreted allegorically, as advocated by Augustine of Hippo in his classic masterpiece *The City of God*. The six days of creation do not have to be necessarily interpreted as six periods of 24 hours each, but rather Genesis can be interpreted in terms of eras consisting of long periods of time, in agreement with well-established geologic measurements, or astronomical observations. Or, the attractive story for youngsters of Noah's ark and the flood could be interpreted just as a moral lesson for human beings to behave properly. To take literarily the first sections of Genesis, or specifically the first Books of the *Bible* (*Old Testament*), is not the aim of the *Torah* (*Pentateuch*). One should conclude that "creation" (in its religious interpretation) and the scientific approach to Darwinism are not necessarily contradictory, but are opposite sides of the same coin.

All manuscripts and dialogues in this volume are solely the views of their authors and do not necessarily reflect the opinion of either the editors or publisher. We present both interpretations of the origin and evolution of life on Earth with the intention of encouraging a dialogue. Consequently, we expect all the authors of this book to be entitled to their own opinions, and hopefully not to accept all the editors' views, since in that case there would be no room for a dialogue. Likewise, the editors do not necessarily agree with every idea expressed in this volume. Furthermore, the editors are not responsible for the views of the contributors. It is our hope and expectation that careful readers of the present volume will be exposed to various aspects and issues related to the origin and evolution of life on Earth, and will draw their own conclusions. We also expect some readers to find some views expressed in this book that might be different from their own. If this were the case, we believe that the effort invested in the production of this book has been worthwhile.

Biography

Dr. Richard Gordon is a theoretical biologist whose endeavors range from AIDS prevention to breast cancer imaging on the medical side and from the effects of microgravity on amphibian embryos to the delights of diatom motility and morphogenesis on the basic science side. He inadvertently wrote the first paper on diatom nanotechnology. He has a B.Sc. in Mathematics from the University of Chicago (1963) and a Ph.D. in Chemical Physics from the University of Oregon (1967), and is now a Professor of Radiology at the University of Manitoba, where he has also held appointments in Botany, Computer Science, Electrical & Computer Engineering, Pathology, Physics and Zoology, perhaps foretold by his 1981 Rh Institute Grant for Outstanding Contributions to Scholarship and Research in the Interdisciplinary Category. His forays into how people treat one another include articles on democratization of the science granting system, a proposal for world minimum wage, havatars (human, avatar pairs) for a new approach to understanding epidemics such as HIV/AIDS, and a campaign to get up to date books to medical schools in Afghanistan via Books With Wings (http://www.bookswithwings.ca). His varied mentors, to whom he is grateful, include James S. Dwyer, Susan Meschel, E. Peter Geiduschek, Edward Anders, Aaron Novick Terrell L. Hill, Theodore Puck, Stanislaw Ulam, Jack Carmichael, Antone G. Jacobson, Cyrus Levinthal, Robert Rosen, Conrad H. Waddington, James F. Danielli, Lewis Wolpert, Zim Hearon and Lewis Lipkin.

Preface 2
To the Scientist Who Feels Above the Creationist Debate

RICHARD GORDON

Department of Radiology, University of Manitoba, Room GA216, HSC, 820 Sherbrook Street, Winnipeg, Manitoba R3A 1R9 Canada
gordonr@cc.umanitoba.ca

First, let me say that I am not a creationist, but this book demonstrates that creationists vary enormously in their understanding of modern science. For example, the profoundness and vastness of the evidence for evolution evades many of them. Some will give you a good argument, while others just won't understand you. So why am I and other anti-creationist scientists in here with them? Because science has not answered some of the most fundamental questions that people want answers to. This is deeper than the question of missing links in evolution. It extends to basic unsolved problems:

1. origin of our universe
2. origin of life
3. origin of cellular differentiation (origin of progress during embryogenesis)
4. origin of species (origin of progress during evolution)
5. origin of mind
6. origin of purpose

In origin of mind, for example, is tied up the yet unsolved problem of the relationship of the observer to the observed phenomenon, even in our over century old quantum theory. The formation of the observer from matter, i.e., embryogenesis, is just beginning to be explored in the new field of embryo physics. The origin of progress, whether in the self-building of an embryo from egg to organism through cell differentiation or the evolutionary building of the phylogenetic tree from simpler to more complex organisms through speciation (Gordon, 1999), is not yet well enough understood. With up to date conclusions such as "Life was born complex..." (Glansdorff, Xu & Labedan, 2008), we cannot but wonder how it started, or whether the origin of life was not on earth (i.e., arrived as panspermia), and perhaps much more ancient than we have dreamed (Gordon & Hoover, 2007). Step back from the vast study of emergent phenomena of all sorts, and one sees that they are all in the eye of the beholder, i.e. of the observer, who needs somehow to be included in the equation. The above six unsolved problems may be summarized by saying that we do not yet have a firm grasp on the origin of novelty. Physics has arrived at some tentative answers in the realm of the nonliving through the mathematics of nonlinearity and chaos theory (Dixon, Tuszynski & Clarkson, 1997), although the role of the observer has still not been spelled out.

Creationists have assorted agendas, from charlatans having fun making fools of scientists and misleading young minds, to theologians seeking to preserve traditional concepts despite the evidence (which they often selectively ignore), to religious scholars with a deep need to fathom our existence. In all of these endeavors they challenge scientists, sometimes infuriating them to the point of sacrificing the concept of academic freedom, to the embarrassment of the rest of us (Brackman, 2008). Some creationists are scientists themselves, contradicting the presumption that "we are all, of course, atheists" (http://thesciencenetwork.org/BeyondBelief2/). The best response is to listen, to educate, to challenge in return, and to probe, understand and teach where scientific evidence fades off into guessing. Debating creationists sharpens the mind, and, may I say, leads us to new directions for research, where we just might solve some of these deep problems. What they claim is our Achilles heel could prove to be our vein of gold.

In my pass through of editing this volume, I was tempted to argue with many of the authors, scientists and creationists alike. For example, when I hit the three chapters by our Turkish creationist colleagues, I noticed that large blocks of the same text, albeit (ironically) in mutated form, appeared between all three of them, suggesting descent from a common ancestor. I started removing the redundancies but then realized that this is not appropriate. It should be left as a mark of their form of scholarship and mechanism of self-reproduction (with variation). So we shall see if this means of scholarly speciation (akin to that which has given us a multitude of religions) is more fit and continues to evolve into all niches of the world, or goes extinct by being overwhelmed by Darwinian predators. In the latter case, this book will be part of their fossil record. To be fair I must carry this analogy to the other Abrahamic religions. For example, I've pointed out in my own chapter what amounts to evolutionary convergence represented by a Christian creationist speaking before a gathering of Jewish creationists. Undoubtedly, divergence of predator and prey from a common ancestor, convergent evolution, and horizontal meme transfer would link every chapter in this book, including those touching on Eastern religions, save for the one on the parallel universe of the Mexican Nauhas. The latter makes us wonder how quaint or profound our current global mix of science and religion may look a century hence.

In regards to Charles Darwin himself, it is repeated by many of our authors, almost as a catechism, that Darwin was a gradualist. But "it ain't necessarily so", as the song goes. His clear statement of what we now call punctuated equilibrium and stasis, (and also the fractal nature of the phylogenetic tree, an even more "modern" concept), which I discussed in Gordon (1999, p. 1037), occurs in the 1866 4th and later editions of *Origin of Species* (Rhodes, 1983; Penny, 1983, 1985; Clark, 1984, p. 330; Gingerich, 1984; Gingerich & Rhodes, 1984; Scudo, 1985). Stephen J. Gould (2002, p. 1014) defended his punctuated equilibrium priority over Darwin, dismissing "later editions that compromise original statements".

We all keep coming back to Charles Darwin, praising or belittling him, because he did indeed start a revolution in the way we think of ourselves, which has not yet seeped into the core of our being.

Brackman, L. (2008). Anti-Darwinists turned away by Israeli academia. http://www.ynetnews.com/articles/0,7340,L-3568715,00.html.

Clark, R.W. (1984). *The Survival of Charles Darwin, A Biography of a Man and an Idea*. New York: Random House.

Darwin, C. (1859). *On the Origin of Species by Means of Natural Selection, Or The Preservation of Favoured Races In The Struggle For Life*. 1st ed., London: John Murray. 4th ed.: 1866.

Dixon, J.M., J.A. Tuszynski & P.L. Clarkson (1997). *From Nonlinearity to Coherence: Universal Features of Nonlinear Behaviour in Many-Body Physics*. Oxford, Oxford University Press.

Gingerich, P.D. (1984). Punctuated equilibria — where is the evidence? *Systematic Zoology* **33**: 335-338.

Gingerich, P.D. & F.H.T. Rhodes (1984). Darwin's gradualism and empiricism: discussion and reply. *Nature* **309**: 116.

Glansdorff, N., Y. Xu & B. Labedan (2008). The Last Universal Common Ancestor: emergence, constitution and genetic legacy of an elusive forerunner. *Biology Direct* **3**(29), http://www.biology-direct.com/content/pdf/1745-6150-1743-1727.pdf.

Gordon, R. (1999). *The Hierarchical Genome and Differentiation Waves: Novel Unification of Development, Genetics and Evolution*, Singapore & London: World Scientific & Imperial College Press.

Gordon, R. & R.B. Hoover (2007). Could there have been a single origin of life in a big bang universe? *Proc. SPIE* **6694**, doi:10.1117/1112.737041.

Gould, S.J. (2002). *The Structure of Evolutionary Theory*. Cambridge: Harvard Univ Press.

Penny, E.D. (1983). Charles Darwin, gradualism and punctuated equilibria. *Systematic Zoology* **32**: 72-4.

Penny, E.D. (1985) Two hypotheses on Darwin's gradualism. *Systematic Zoology* **34**: 201-205.

Rhodes, F.H.T. (1983). Gradualism, punctuated equilibrium, and the Origin of Species. *Nature* **305**: 269-272.

van Wyhe, J., Rookmaaker, K., Basman, A., Secord, J., & Browne, J. (2008). The Complete Work of Charles Darwin Online, http://darwin-online.org.uk/.

Dialogue

Juan G. Roederer
I have followed with interest the case of the visit of two Turkish creationists to Hebrew University. Why is it that I am reminded so much of our own evangelical creationists in the US? As director and dean here at the University of Alaska in the past, I had several encounters with such individuals. But I never denied them a forum in my institute and college. What I deliberately refused to do was to engage personally in a *public* "debate" on creationism (to which they had challenged me repeatedly), on the grounds that in the official capacity as a university professor and administrator in the environmental sciences, I do not mix science with religion. That might be done within another school or department--for instance, philosophy, psychology, history or political sciences. The challengers of course retorted with the matter of "*scientific* creationism". That, I always did and still do denounce publicly as being both phony science and phony religion, which can flourish only in a country with a dismal public education record like ours.

Richard Gordon
In this book we do indeed debate the Turkish gentlemen, our authors Drs. Oktar Babuna and Cihat Gündoğdu, who were invited to Hebrew University and then told they could not speak. Now I don't agree with their conclusions, but their reasoning is as compelling as many a distinguished rabbi, and in fact we have Jewish scholars writing in this book who pretty much concur with them. The political fact of the matter is that this group in Turkey has single handedly kept the question of Darwinism central to modern Islamic thought around the whole world, and this affront will not go unnoticed. As with problems of innumeracy and de facto scientific illiteracy, Darwinism has not really taken hold anywhere in the West, in Islamic nations, nor in Israel. Even many people who pay it lip service do not understand its deep, and to some, disturbing, implications. The revolution in thought that Darwin started is far from over. I go for academic freedom and the desirability of a clash of ideas. The world is too small for separate solitudes of science and religion.

Preface 3
Monotheism: The Basis for Unifying Abrahamic Religion and Science?

RICHARD GORDON[✡]

Department of Radiology, University of Manitoba, Room GA216, HSC,
820 Sherbrook Street, Winnipeg, Manitoba R3A 1R9 Canada
gordonr@cc.umanitoba.ca

Our new rabbi has set a dangerous precedent, asking a university professor to give a so-called short sermon. Rabbi Larry Lander doesn't know that my typical lecture takes three hours, which I rush through in 1 hour with 100 PowerPoint slides. I thought of setting up a computer projector here on Friday afternoon, to let it run continuously. But this would change our ambience more than a mechitza [partition to separate men and women during religious events]. So I'll struggle with no slides, and condense my thoughts so you fall asleep in 10 minutes instead of the usual hour. It's okay. I've slept through many lectures myself, and even an occasional sermon.

I was asked to give a book review, and of course I chose Judaism versus science as the topic, for honestly, in the chronology of my world path, I am a scientist first and a Jew second. This is common amongst Jewish scientists (Cantor & Swetlitz, 2006). Let me elaborate. I was raised knowing I was Jewish, but

---

[✡] After I drafted what is now Preface 2 for this book, I was asked to give a sermon at my Conservative synagogue, Congregation Etz Chayim in Winnipeg, Canada, so in what follows I have incorporated and expanded upon it. The original title was "Science and Judaism".

without Judaism. I decided in 1953, at age 10, to become a scientist. I did not begin with Judaism until I immigrated to Winnipeg in 1978.

I have annoyed Reconstructionist (Kaplan, 1937) and Lubavitch rabbis (Ginsburg, Branover & Schneerson, 2004), and every kind in between, with my questioning that was clearly framed to seek hints in Judaism for answers to scientific questions, not vice versa. For this sermon, I picked one book, then another, and the more I dug the more I found (Branover & Attia, 1994; Cantor & Swetlitz, 2006; Carmell & Domb, 1978; Efron, 2006; Jammer, 1999; Nelson, 2005; Pollack, 2000; Slifkin, 2006; Wahrman, 2002). Jews have been muddling over the relationship between science and Judaism at least since Moses Maimonides' *Guide of the Perplexed* published in 1190 CE (Maimonides, 1190), and it's clearly an unsolved problem. So I've given you a whole reading list. There will be a quiz next week.

Traditionally, Jews have approached the relationship between science and Judaism four ways (Robinson, 2006):

1. Rejection of science. This is the head in the sand approach.
2. Looking for weaknesses and limitations of science. While reaping the benefits of science, we say it's not so great anyway. One problem this generates is that as science grows, religion shrinks, and God becomes the God of the gaps, which keep getting filled in.
3. Reinterpret *Torah* to make it look consistent with science. Deny there's a conflict. So we don't take Genesis literally, and try to stretch the 5768 years since the Creation to cover the 13.7 billion years since the Big Bang, a ratio of 2 million to one. But major events, like the global Canadianization of the earth 2 billion years ago when it all froze over, are left out (Walker, 2003).
4. Claim Kabbalistic transcendence of science, that there is a deeper mystery that science just can't handle. Science and religion just have different domains. Well, maybe, but that's just another gap.

I think there may be a fifth, and perhaps better way. Let's start at the top of the problem, with the relationship between monotheism, for which Judaism takes credit, and science. The shema, the central prayer of Judaism is:

> "Hear, O Israel: The Lord our God, the Lord is One" (Harlow, 1985).

Science is also based on the faith that the universe is one, subject everywhere and at all times to the same laws (de Duve, 2008). This is not changed one whit by new Star Trek theories of parallel universes or the crazy notion that every quantum event splits the universe into two new universes. These are just bizarre arguments between scientists about what the really universal laws are. In this regard, nothing changed when we made the transition from Newton's Euclidean geometry universe to Einstein's relativity and curved space, or from Newton's deterministic universe to quantum statistics and chaos, or from separate creation of each species of plant and animal to one universal evolutionary tree. The law is the law. Scientists are just trying to figure out exactly what it is.

Where does this scientific belief come from? I would suggest that it stems from monotheism itself, the idea of one God, one set of rules for everyone and everything, for all times and places. If I am correct, then science is historically rooted in monotheistic religion. Therefore I am in profound disagreement with anyone who says that science and religion each has its own domain. The world is one, and if we truly believe that, we will seek a way to combine science and religion and live accordingly.

Historically, monotheism was not, however, sufficient to bring about the growth of science. If it were, Jews would have been millennia ahead of all other people. Both Islam and Christianity, outgrowths of Judaism, are monotheistic. In Islam, Abu Hamid Muhammad al-Ghazali declared 100 years before Maimonides that to restrict God to His laws was blasphemous (al-Ghazali, 1095). Science by Jews, Christians and Muslims had flourished under Islam until then (Spencer, 2007). In contrast to Islam, Judaism retains the concept that God is subject to God's rules, similar to the idea that a king is subject to the same rules as his subjects, as in the story of David and Batsheva.

It was left to Christianity, not Judaism, to spark modern science. In fact, it seems to have been a particular sect of Christianity, the Puritanism of the 1600s CE in England, that got things moving. This did not at all involve a separation of science from religion, but rather, to quote:

> "Perhaps the most directly effective element of the Protestant ethic for the sanction of natural science was that which held that the study of nature enables a fuller appreciation of His works and thus leads us to admire the Power, Wisdom, and Goodness of God manifested in His creation" (Russell, 1973).

Torah Umadda (Lamm, 1990), the slogan "Torah and secular knowledge", is a comparable, but small movement in Modern Orthodox Judaism. The current creationist, so-called Intelligent Design movement, rampant in fundamentalist Christianity but rapidly penetrating both Islam and Judaism, has the opposite effect, for it denies science (Perakh, 2004; Seckbach & Gordon, 2008).

These themes, the parallel between the oneness of God and the lawful nature of the whole universe, God subject to God's rules, science as a way to God, and the knowability of the laws of God through science, suggest a strong and deep bond between science and religion. However, science has achieved something remarkable that religion has barely begun to contemplate. Science has a universality that crosses all cultures, and a means of cooperation and consensus building that transcends all boundaries between nations and peoples. If science and religion are one, because the universe is one, because the Shema is correct, then science may show the way for religion to join it in one universal conception that replaces both.

The Big Bang, the notion that the indeterminacy of quantum mechanics somehow gives us free will, the crude and probably wrong parallels between Genesis and the presumed (Gordon & Hoover, 2007) origin and evolution of life on earth (Schroeder, 1990; Schroeder, 2001), have led to a false sensation that science and religion are now in harmony. On the contrary, I suggest that we have a lot of work ahead. I would like to repeat the oft quoted relationship between modern science and religion, formulated by Robert Jastrow:

> "For the scientist who has lived by the faith in the power of reason, the story ends like a bad dream. He has scaled the mountains of ignorance; he is about to conquer the highest peak; as he pulls himself over the final rock, he is greeted by a band of theologians who have been sitting there for centuries" (Jastrow, 1992).

But I add a new ending:

> who have been sitting there for centuries… killing each other and their descendents in the name of God. But the scientist and his comrades, who although they had also argued for centuries, formed together a single world effort at understanding the universe. They then convinced the younger theologians to descend the mountain and consider whether science itself might provide a way to bring an end to their mutual

slaughter. In exchange, the scientists had to broaden their view, and start working on some of the tougher problems that they had previously fobbed off on religion.

References

al-Ghazali, A.H.M. (1095). *Tahafut Al-Falasifah [Incoherence of the Philosophers]*, 1963 translation ed., Lahore: Pakistan Philosophical Congress.

Branover, H. & J. Attia (eds.) (1994). *Science in the Light of Torah, A B'Or Ha'Torah Reader*, Northvale, New Jersey: Jason Aronson.

Cantor, G. & Swetlitz, M. (eds.) (2006). *Jewish Tradition and the Challenge of Darwinism*, Chicago: University of Chicago Press.

Carmell, A. & Domb, C. (eds.) (1978). *Challenge, Torah Views on Science and Its Problems*, 2nd ed., Jerusalem: Association of Orthodox Jewish Scientists.

de Duve, C. (2008). Scientists and beliefs. In: Seckbach, J. & R. Gordon, *Divine Action and Natural Selection: Science, Faith, and Evolution*, Singapore: World Scientific.

Efron, N.J. (2006). *Judaism and Science: A Historical Introduction*, Greenwood Press.

Ginsburg, J., H. Branover & M.M. Schneerson (2004). *Mind Over Matter: The Lubavitcher Rebbe on Science, Technology and Medicine*, Shamir.

Gordon, R. & R.B. Hoover (2007). Could there have been a single origin of life in a big bang universe? *Proc. SPIE* **6694**, doi:10.1117/12.737041.

Harlow, J. (1985). *A Prayerbook for Shabbat, Festivals, and Weekdays*, New York: Rabbinical Assembly, United Synagogue of America.

Jammer, M. (1999). *Einstein and Religion*, Princeton: Princeton University Press.

Jastrow, R. (1992). *God and the Astronomers*, 2nd ed., Norton.

Kaplan, M.M. (1937). *The Meaning of God in Modern Jewish Religion*, 1994 reprint ed., Detroit: Wayne State University Press.

Lamm, N. (1990). *Torah Umadda, The Encounter of Religious Learning and Worldly Knowledge in the Jewish Tradition*, Northvale, NJ: Jason Aronson Inc.

Maimonides, M. (1190). *The Guide of the Perplexed [Translated with an introduction. and notes by Shlomo Pines]*, 1963 translated ed., Chicago: University of Chicago Press.

Nelson, D.W. (2005). *Judaism, Physics and God: Searching for Sacred Metaphors in a Post-Einstein World*, Woodstock, Vermont: Jewish Lights Publishing.

Perakh, M. (2004). *Unintelligent Design*, Amherst, NY: Prometheus Books.

Pollack, R. (2000). *The Faith of Biology and the Biology of Faith: Order, Meaning, and Free Will in Modern Medical Science*, New York: Columbia University Press.

Robinson, I. (2006). "Practically, I An a Fundamentalist": Twentieth-Century Orthodox Jews contend with evolution and its implications. In: Cantor, G. & M. Swetlitz, *Jewish Tradition and the Challenge of Darwinism*, Chicago: University of Chicago Press, 71-88.

Russell, C.A. (ed.) (1973). *Science and Religious Belief: A Selection of Recent Historical Studies*, London: University of London Press.

Schroeder, G.L. (1990). *Genesis and the Big Bang: The Discovery of Harmony Between Modern Science and the Bible*, Bantam Doubleday Dell Pub.

Schroeder, G.L. (2001). *The Hidden Face of God, Science Reveals the Ultimate Truth*, New York: Simon & Schuster.

Seckbach, J. & Gordon, R. (eds.) (2008). *Divine Action and Natural Selection: Science, Faith, and Evolution*, Singapore: World Scientific.

Slifkin, N. (2006). *The Challenge of Creation: Judaism's Encounter with Science, Cosmology, and Evolution*, Brooklyn: Zoo Torah/Yashar Books.

Spencer, R. (2007). *Religion of Peace? Why Christianity Is and Islam Isn't*, Washington, D.C.: Regnery Publishing.

Wahrman, M.Z. (2002). *Brave New Judaism: When Science and Scripture Collide*, Hanover: Brandeis University Press.

Walker, G. (2003). *Snowball Earth, The Story of the Great Global Catastrophe that Spawned Life as We Know It*, New York City: Crown Publishers.

Dialogue

Jack A. Tuszynski

I'm not sure I'd agree with your giving credit for the launching of science to the Puritans. First of all, science has been at the service of religion for millennia (e.g. the astronomers of Babylon, Egypt) but not only. Greece has given us non-religious mathematicians (Euclid, Pythagoras — he was also a mystic though) and physicists (Democritus). They did not have universities but small schools of disciples. However, the first university with hundreds if not thousands of students was founded in Alexandria some 2000 years ago. Then you had the method of inquiry that has a lot in common with the Midrash and the organization of Talmudic scholarship in yeshivas. In modern times, the first European universities (late 1100s to early 1200s) were created largely (if not exclusively) by the Catholic Church in places like Bologna, Paris, Oxford or Padova. The first real scientists were closely associated with the Church (Copernicus, Galileo, Giordano Bruno) although they did their work largely against the Church's doctrine. Newton may have been a singular scientific genius of the greatest caliber but he was also a theologian, a mystic and an alchemist. I guess, what I'd like to say is that while religion doesn't directly promote science, it does stimulate scientific thinking. I guess the big divide has to be traced back to Bacon who distinctly separated religion and philosophy but maintained that the two can coexist. The main distinction he drew was that religion is based on "blind" faith while science on a method of inquiry. I think, had he been raised in Judaism, he would not have seen such a sharp difference between science and religion.

Richard Gordon

The hypothesis that modern science started with England's Puritans is not mine, but rather Robert K. Merton's. See:

Merton, R.K. (1936). Puritanism, pietism and science. *Sociological Review* **28**(Part 1), 574-606.

Hall, A.R. (1963). Merton revisited, or Science and Society in the Seventeenth Century. *History of Science* **2**, 1-16.

Both are reprinted in:

Russell, C.A. (ed.) (1973). *Science and Religious Belief: A Selection of Recent Historical Studies*, London: University of London Press.

Merton himself was Jewish:

Efron, N.J. (2007). *Judaism and Science: A Historical Introduction.* Westport, CT USA, Greenwood Press.

In retrospect, my idealistic, universalistic view of "Science as a Value" above is placed in its American-Jewish historical context by Efron (his p. 201).

Part 1

Background in Theology, Philosophy and Science

Biography

Dr. Christian de Duve was born in England and educated in Belgium. He is emeritus professor at the Catholic University of Louvain (Belgium) and at the Rockefeller University in New York. He shared the Nobel Prize in Physiology or Medicine 1974 with Albert Claude and George Palade, "for their discoveries concerning the structural and functional organization of living cells." He is known for the discovery of lysosomes and peroxisomes and has, in recent years, become interested in the origin and evolution of life. Dr. de Duve is the author of several books, including "*Life Evolving: Molecules, Mind, and Meaning*" (Oxford University Press, 2002) and "*Singularities: Landmarks on the Pathways of Life*" (Cambridge University Press, 2005).

1

Scientists and Beliefs

CHRISTIAN DE DUVE

Rockefeller University, 1230 York Avenue
New York NY 10065 USA
deduve_private@icp.ucl.ac.be

Science is based on the postulate that the universe obeys natural laws and is intelligible within the framework of those laws. I intentionally call it a postulate, that is, a working hypothesis to be accepted until proven wrong, in order to avoid founding the scientific approach on a belief or *a priori* affirmation. This precaution, which may strike the more sanguine proponents of science as unnecessary, is actually profoundly pertinent to distinguish science from religion.

There can be no doubt that the naturalistic postulate has been spectacularly validated by the achievements it has permitted, not only in physics and chemistry, but also in biology. Research based on the assumption that life is entirely explainable in terms of physics and chemistry has already provided a deep understanding of living processes, leading in turn to powerful technological applications; it has thrown a revealing light on the origin and evolution of life on our planet.

An idea that has evoked considerable interest in recent years is the fact that the universe happens to have just the right physical properties to be capable of giving rise to life and mind. Embodied in the so-called anthropic principle, this observation has moved a number of physicists to state that there is something special about our universe, which some have not hesitated to describe as fine-tuned, or even designed, for the appearance of life and mind.

Others have proposed, instead, that our universe is only one in a multitude of universes, or multiverse, one that happens by chance to be endowed with the appropriate properties and, therefore, is devoid of significance. In my opinion, the two interpretations differ only by a matter of degree. Even if the multiverse theory should happen to be true — a point that may be difficult, if not impossible, ever to establish — the fact remains that a universe capable of generating life and mind exists. Even diluted by any number of inert universes, our universe remains remarkably significant.

The existence of this universe raises the obvious question of its origin. Scientists have very diverse attitudes with respect to this problem. I know of no statistic on the topic and will simply list some of the opinions that have been publicly expressed by scientists on one occasion or another.

A number subscribe to the view, propagated by the major monotheistic religions, that the universe was created *ex nihilo* by God, considered as a "Supreme Being" that exists outside the universe. This belief, which goes under the general heading of theism, may be privately held or go together with affiliation to some organized religion, occasionally even with membership in its clergy.

Others are not satisfied with theism, which either raises the question of "Who created God?" and so on, ad infinitum, or stops by seeing God as an uncreated being, making creation an unnecessary step. Because of this, they dispense with the notion of a creator and adhere to the view that the universe and God form a single entity, making up what I have called "Ultimate Reality" Next to this view, generally defined as pantheism, there is a variant, known as panentheism, according to which the universe lies within ultimate reality, being only part of it.

Yet others take refuge in agnosticism. Content with accepting the universe as it is, they consider its origin as unknowable and, therefore, outside the realm of scientific inquiry. Some go beyond mere agnosticism and actually affirm the nonexistence of a divine creator (atheism). A few scientists are militant defenders of this view, which they propagate with a zeal akin to religious fervor.

It must be emphasized that the vast majority of scientists, whatever their beliefs, agree in accepting the naturalistic postulate as the basis of their investigations. This is even true of many of the most devout theists, who generally take it that God created the universe as it is, once and for all, with all its intrinsic laws and properties, and then left it to operate on its own steam, so to speak. Their God, in other words, respects the naturalistic postulate. Note, however, that belief in supernatural phenomena, such as miracles, divine incarnation, resurrection, and the like, may be a cause of embarrassment for some, who either elude the problem or deal with it by assigning the phenomena involved to a special realm.

A few scientists actually deny the naturalistic postulate or, at least, put limits to it, claiming that certain phenomena, especially in the biological sphere, cannot be explained simply in terms of physics and chemistry and must have involved the intervention of some kind of purposeful agency, often, though not always, identified with God. This theory, known as intelligent design, is really a modern resuscitation of finalism, the view long held by a number of biologists who argued that certain evolutionary phenomena — the development of animal eyes and the conversion of reptiles into birds are favorite examples — could not possibly have taken place naturally and must have been "guided."

Intelligent design is professed by only a very small minority of scientists. It would deserve no more than a passing mention were it not for the enormous political significance this theory has acquired in recent years, as a rallying point for many fundamentalist creationists who, while recognizing that the biblical account of Genesis cannot be accepted literally, remain wedded to the concept of an interventionist creator who goes on directing the course of his work, not hesitating to break the laws of his own making in order to achieve a certain purpose.

This is not the place for a detailed discussion of intelligent design. Only two brief points will be made. First intelligent design is not a scientific theory in the usual sense of the term, in that it can neither be proved nor disproved. It is an *a priori* affirmation that some things cannot be explained naturally, a condemnation before the fact of any attempt at finding such an explanation. Such an attitude flies in the face of the scientific endeavor.

Next, some of the arguments put forward in support of intelligent design have been convincingly refuted, their main weakness being that they fail to take into account the fact that the phenomena mentioned were extremely slow, long, drawn out processes involving a very large number of individual steps, each of which could have a natural explanation, which, in some cases, is already beginning to be recognized. The evolution of animal eyes, the subject of many books, is a good example.

Further Reading

C. de Duve, *Life Evolving* (New York: Oxford University Press, 2002).
C. de Duve, *Singularities* (New York: Cambridge University Press, 2002).

Dialogue

Charles H. Lineweaver

I share your dismissal of intelligent design as non-scientific. I also agree that it might be useful to be explicit about the naturalistic postulate shared by scientists that "the universe obeys natural laws and is intelligible within the framework of those laws". Another way to say the same thing is "Stop invoking gods to explain things away and try to investigate things experimentally." Some of us, inspired by the work of John Wheeler and irreligiously-minded cosmologists like Stephen Hawking, are willing to take the naturalistic postulate further and try to explain the origin of the universe and its laws in a naturalistic, rather than a supernaturallistic way. We work under the extended version of the naturalistic postulate; that the origin of the universe is not beyond the purview of science. Maybe the laws of the universe emerged at the time of the big bang in some self-consistent way that a Theory of Everything could shed some light on. We don't know. Quantum cosmology is hard going and very speculative — contact with observations are hard to come by. The traditional views that populate your list of plausible explanations for the origin of the universe hold that the origin of the universe and its laws are beyond science and are therefore eligible to be explained away by invoking the gods… again. Just as Intelligent Design is not a scientific theory that can be proved or disproved, these traditional views can neither be proved or disproved. They are unscientific. And if one can live with uncertainty, they are unnecessary. I believe that invoking spirits and gods to explain things we don't understand should no longer be even a last resort for a scientist. We should be able to tolerate uncertainty. Our disagreement is closely related to the debate about the limits of science, embodied in the debate about over-lapping or non-over-lapping magesteria discussed in my contribution to this book.

Further Reading

J.A.Wheeler, "Frontiers of time" in *Problems in the Foundations of Physics*, Ed. G. Toraldo di Francia (North-Holland, Amsterdam, 1979)

R. Dawkins, *The God Delusion* (Bantam Press, London 2006)

Biography

Professor George V. Coyne, S.J. (Societas Jesu), obtained his PhD in Astronomy from Georgetown University (Washington, DC) in 1962 and the Licentiate in Theology from Woodstock College (Maryland) in 1966. In addition he has received PhD degrees *honoris causa* from St. Peter's University (New Jersey), Loyola University (Chicago), The University of Padua (Italy), the Jagellonian University (Krakow, Poland), Marquette University (Milwaukee) and Boston College (Chestnut Hill, Massachusetts). Since 1966 he has been associated with astronomy programs at the University of Arizona (Tucson) and from 1976 to 1980 he served in various capacities in the administration of the astronomical observatories at that university. From 1978 to 2006 he was Director of the Vatican Observatory (Specola Vaticana), which has its headquarters at Castel Gandolfo (Rome, Italy) and a research branch at the University of Arizona in Tucson. His research interests have ranged from the study of the lunar surface to the birth of stars; and he pioneered a special technique, polarimetry, as a powerful tool in astronomical research. Currently he is studying cataclysmic variable stars, the interstellar dust in the Magellanic Clouds and the detection of protoplanetary disks. Parallel to his scientific research he has developed an interest in the history and philosophy of science and in the relationship between science and religion. In 2008 he received the Mendel Medal.[†] "The Mendel Medal is awarded to outstanding scientists who have done much by their painstaking work to advance the cause of science, and, by their lives and their standing before the world as scientists, have demonstrated that between true science and true religion there is no intrinsic conflict."

[†]Villanova University (2008). Villanova Co-sponsors "Gregor Mendel: Planting the Seeds of Genetics," at Academy of Natural Sciences, May 24 - Sept. 28, http://www.villanova.edu/events/yearofmendel.

2
Evolution and Intelligent Design. Who Needs God?

GEORGE V. COYNE

Vatican Observatory
V-00120 Vatican City State
gcoyne@specola.va

1. Introduction

One of the principal issues involved in the debate about the supposed insufficiency of neo-Darwinian evolution and the contested validity of intelligent design explanations for biological phenomena is that to do with the meaning of science itself, as it has been understood since the time of Galileo. As a background to defending my claim that the intelligent design explanations are not science, I wish to present a fundamental distinction which is at the basis of understanding what science does and what it cannot, as such, do. This is the distinction between what is meant by "origins" and what is meant by "creation." In so doing I wish to set a cosmological background to the discussion of biological evolution. Then I will give a brief history of scientific methodology in order to substantiate my claim that the intelligent design movement lies outside of that methodology.

2. Origins and Creation

The key to understanding the difference between creation and origins is the notion of change (Carroll 2003). Changes in nature are the object of study for the natural sciences. From pure energy to matter, from hydrogen to hydrocarbons, from giant molecular clouds to star clusters, from single cells to organs, from amino acids to the human brain — these are all objects of investigation for the natural sciences. They all require an existing entity which changes. The natural sciences do not deal with the issue of existing at all; they deal with existing in a specific way and the changes in nature, which bring about specific ways of existing.

Creation, on the other hand, speaks to the very existence of whatever exists. It does not speak to change. Creation does not deal with the chain of events which bring about a specific kind of being. It deals with the source of being of whatever exists. It does not address the evolution of one kind of being from another. To create, therefore, is not to work on or with some already existing material. Creation is not, therefore, a cause in the usual sense of the word. Or, if you wish, creation is the complete cause of all things. Such a complete causing is precisely what the act of creation is. Thus, to create is to give existence to whatever exists in a specific way. To create does not mean to take "nothing" and make "something" out of it, in the sense of changing it from not being to being. To exist means to depend upon a source of existence. So, creation is not exclusively, nor even primarily, some distant event; to create is the continual, complete causing of the existence of whatever is.

So there can, in principle, be no necessary conflict between the doctrine of creation and any scientific explanation of origins. The natural sciences seek to account for change and the origins of change. Whether the changes described are biological or cosmological, have a beginning or not, are unending or temporally finite, they remain processes. Creation accounts for the existence of things, not for changes in things. So, given that something exists, how did life originate from this something is a scientific question. Why is there something rather than nothing is not a scientific question.

Religious perceptions of beginnings speak of creating out of nothing (*creatio ex nihilo*). There is a persistent confusion between cosmological and metaphysical/religious conceptions of "nothing." Quantum cosmological views of the beginning of the universe speak of "vacuum fluctuations" and some are tempted to equate this "vacuum" with the "nothing" of the theologians. This is quite incorrect. The "vacuum" of quantum mechanics is something, if only a mathematical concept. To speak of "creation out of nothing" in philosophy or theology means that one is denying that any matter at all is changed or transformed into something else. The expression "out of nothing" or "from nothing" is, at its root, a denial of any material cause whatsoever in the act of creation.

In our discourse on beginnings we find it necessary to speak in a temporal framework. The creator is considered to be prior to what is created, but the priority is actually not temporal. The relationship is metaphysical not temporal. To be created out of nothing does not mean that the creature is first nothing and then something. It means that the creature has a dependent existence. Ancient philosophers of nature thought that the universe was eternal in the sense that it had no beginning. Some cosmologists propose that the very notion of temporality is a subsidiary concept. Neither view challenges the fundamental metaphysical truth that the universe is created. Some also propose that there is "eternal inflation," an endless series of universes within universes. Still, all such universes would require creation in order to be. Nor is there a contradiction in the notion of an eternal created universe. For, even if the universe had no temporal beginning, it still would depend upon a creator for its very being. The radical dependence on a fundamental source of being as the "cause" of being is what creation means.

To affirm creation or to deny it on the basis of scientific theories represents a fundamental misunderstanding of both origins and creation. The Big Bang described by modern cosmologists is not creation. The evolution of life in the universe is not creation. Creation is not one of the ways in which the universe itself or life in it might have come to be. Creation speaks to the ultimate reason for the existence of all things. The natural sciences seek for the ways in which things came to be, their origins. It is mistaken, therefore, to conclude that there are implications for a creator if the universe is completely self-contained, with no singularities or boundaries, and completely described by a unified theory. One mistake which is made by those who use scientific theories to deny creation is the old error of thinking that to create means to be an agent cause of change.

Some cosmologists speak of "self-reproducing universes"; but these are not self-creating universes. Others speak of the universe's inflating itself "out of nothing in an instant." Such physical-mathematical theories serve to explain much that is observed of the universe as a whole. However, the need to explain the existence of things does not disappear as a result of new scientific explanations which propose to account for various changes which have occurred in the universe.

3. A Brief History of Scientific Methodology

In order to understand the nature of modern science it will be useful to review its history. All ancient civilizations in Egypt, Mesopotamia, Greece and many other parts of the world have left evidence of an early stage of intellectual development at which the discourse about nature was framed in the ordinary language of interpersonal communication between human beings. Consequently, nature was conceived as a kind of all-embracing society or state, the rulers of which were a number of more or less powerful gods, spirits and demons (Jacobsen 1949, Pritchard 1955). The arbitrary will of the gods of nature was behind everything, serving the human as a reasonable, or at least intelligible, explanation of all phenomena. There was no split between nature and culture.

The old mythological conception of nature gradually began to yield to the new idea that the phenomena of nature did not appear as a consequence of the free decisions of its gods, but because they had to appear as a consequence of an inner necessity which forced them to do so. This was a stark denial of the wisdom of all previous ages and as such it provoked an intellectual upheaval compared with which all later "scientific revolutions" appear as mere ripples on the surface of that ocean of thought which was first stirred by that band of innovators who have become known an the Presocratic Philosophers.

Looking back upon this development, which ultimately changed the intellectual outlook of a great part of all mankind, one cannot help but be struck by the sheer linguistic difficulties of the whole undertaking. We have no evidence that the Ionian thinkers stated this problem in so many words; but we can clearly see how they grabbled with it and tried to solve it in two very different ways: by metaphor and by mathematics.

Throughout the centuries the Greek philosophers pursued numerous experiments in the metaphorical use of ordinary language. The result was a new vocabulary of technical terms the metaphorical origin of which went into oblivion in the course of the long process which gradually made the Greek world familiar with the idea of a non-mythological account of the ways of nature. The age-long mythological discourse on nature had been framed in the ordinary language of human affairs in which there simply were no words for the abstract ideas which the new discourse tried to work out. For example when Herodotus tried to say that the sun was the "cause" of the inundation of the Nile river he had no such word at his disposal. What he actually said was that the sun was the *aitia*

(Herodotus, Historia, II, 26) of the swelling of the waters, using a well known Greek word which was in common use as a term denoting the guilt which a criminal brings upon himself by committing his offence. In other words, Herodotus said that the sun was "guilty" of the inundation. Taken in the literal sense this was a rather astonishing and perplexing statement. It made the sun a criminal and even an habitual criminal since he committed his "crime" year after year. Moreover, his crime was no crime at all, but a universal blessing for all the land of the Egyptians.

What has happened is a good example of a general pattern in which a common word is lifted out of its everyday context and used metaphorically in a different field of thought as a means of expressing an idea that is homeless in ordinary language. Examples of similar procedures are legion. For instance, the fundamental idea of the inherent necessity in nature with which the new discourse would stand or fall was expressed by the word *ananke*. This belonged to ordinary language in a sense which appears in Herodotus' story about a criminal cowherd who was apprehended by the guards and forced to confess his offence under stress of "necessity" (*ananke*, Herodotus, Historia, I, 116; cf. the similar use of the word in Homer, Odyss., VI, 136). In general the word *ananke* was used of all the means, from persuasion to torture, by which a criminal could be forced to confess because he was unable to resist them. Now it was adopted by the new philosophers who used it to denote that hidden connection in nature which forces the phenomena to appear in an irresistible way.

A very different solution than metaphor was discovered by the Pythagoreans. This discovery of a mathematical alternative to the metaphorical discourse about nature had far-reaching consequences. Since then science has never forgotten that nature contains necessary, internal connections which only mathematics is able to disclose and express. However, this new insight had to fight for its survival. Aristotle had already fought this special conception of the mathematical discourse on nature on several fronts. In the final chapter of the Metaphysics he raised his voice against numerological speculations in a rhetorical manner in marked contrast with his usual style, as if he were almost emotionally involved in this question (Heath 1949). This stems from Aristotle's particular concept of philosophical knowledge in general and natural knowledge in particular. Here the key word is "cause." The point is that any account of nature must remain incomplete if it ignores one or more of the four causes: material, formal, efficient and final. If a philosopher does not discover them he has not reached his goal. While the mathematician is not concerned with final causes the natural philosopher is obliged to study all the four types of causation (Aristotle, Phys., II, 7, 198a). When all is said and done Aristotle would have refused to admit not only the mathematician but also the mathematical physicist to the kingdom of final causation with the obvious inference that the purely mathematical discourse on nature contributes nothing to the quest for wisdom and is unable to shed any light upon the ultimate questions of human existence.

But it is impossible to realize what happened in both Hellenistic and later science without admitting the existence of another great tradition, which may be properly named after Archimedes. This tradition is characterized by a consistent use of the language of mathematics and by a general disregard of causal and teleological explanations (Drachman 1967). The Archimedean approach was fruitfully adopted by medieval scholars and both Galileo and Kepler used it to lay the foundations of modern mechanics and astronomy. So, even if Archimedes failed to comply with Aristotle's insistence on causal explanations as the hallmark of a scientific description, it is impossible to ignore the fact that over the ages the Archimedean tradition was able to produce an ever increasing body of insights into the connections of the phenomena of nature, insights that were obtained thanks to mathematical discourse and could not have been obtained or expressed in any other way. And it gives food for thought that Archimedes' results in mechanics are valid even today when Aristotle's causal explanations have largely fallen into oblivion.

Into this world torn by conflicting views on the proper discourse on nature and the true relations between God and human beings Christianity emerged from its obscure origin in Palestine. At first sight it would seem that it must stay out of the philosophical battle as a non-combatant who was singularly uninterested in the scientific achievements of the Greeks. There is no treatise on cosmology in the New Testament and extremely few references to particular elements of the Greek account of the universe. All efforts are spent on the proclamation of the belief that the birth, life, death and resurrection of Jesus had radically changed the way in which the relations between God and the world should be envisaged.

From the religion of Israel, Christianity also inherited the belief that the one Lord of the world is also its Creator (Clifford 2000, Anderson 1984). Time and again the Old Testament underlines the fact that the world is created. This is almost always understood in the sense that it has come into being independently of man and without human assistance. "Where were you when I laid the foundations of the earth?" (Book of Job, 38, 4) was God's question to Job.

However, the Biblical doctrine of creation seems to be marked by a paradox. On the one hand there is a chasm between God and His creatures. Nothing in nature is divine. On the other hand, the created world is said to testify to the divinity of its creator. God must be present within it in such a way than man can recognize it as created. The beginning of the gospel of St. John indicates a solution of this dilemma. "In the beginning was the logos, and the logos was with God, and the logos was God. He was in the beginning with God. All things came into being through him, and without him not one thing came into being." (Gospel of John, I, 1) When the fourth Gospel opens by saying that: "In the beginning was the logos," it looks at first sight like the introduction to a Greek philosophical treatise. To use the word in a Christian context was an important step towards assimilating the conception of the world as a rational structure according to the basic tenet of Greek philosophy.

Despite its apparent ignorance of all matters scientific the New Testament presented Christianity in a way that contained a number of seminal ideas out of which the future relationships, between the scientific discourse on the laws nature and the religious belief that these laws revealed a divine plan, would develop. The belief in One God implied a demythologization of the discourse on nature. That nature was created meant that its inner connections were established independently of the human mind which had to respect them when they were discovered. Finally the logos Christology made the idea of an all-permeating rationality at home in a religion which hailed Christ as the Lord of the World. It is difficult not to see a connection between this insight and the emergence of experimental methods in science.

But in later centuries there were to be diverse Christian traditions as to the implications involved in affirming a rational structure to the universe. What is characteristic, for instance, of Thomas Aquinas is his insistence that the natural knowledge of God must be acquired in the same way as all other knowledge (Summa contra Gentiles, III, 47. Cf. I, 3). Bonaventure represented a much more traditional theology which gave natural reason a more limited scope (Itinerarium mentis in Deum). Bonaventure is imbued with the Augustinian notion of the interior light by which God illuminates the soul so that it cannot look at the world except as something which is related to him. This was consciously a polemical stand against Aquinas who upheld the autonomy of human reason within its proper bounds without the special assistance of grace. Another great tradition in Christian thought is that of Duns Scotus (Gilson 1952). He claims that the philosopher is unable to describe creation except in terms of cause and effect with the consequence that the world derives from God by necessity. On the other hand, the theologian knows that the world came into being through a free act of God just as man is saved by a free gift of grace. This meant that the laws of nature are such as they are because of a free decision by God. If God had so willed, they might have been different. The recognition of the laws of nature as contingent upon the Divine will was more than a theological subtlety. It had implications of immediate importance for the scientific approach to nature.

The increasing use of mathematical arguments in the 14th century went hand in hand with a new awareness of how thought experiments based on common sense and everyday experience could contribute to the critical re-examination of the discourse on nature. Johannes Kepler and Galileo Galilei became the heralds of a new era in which mathematical physics would go from strength to strength. They both speak of the Book of Nature, a metaphor which goes back to the age of the Fathers but it took quite a long time before it got off the ground. Its prehistory is as old as theology itself since the fundamental idea was already expressed by St. Paul's assertion that the works of God disclose His divinity, invisible being and eternal power (Romans 1, 18 - 20). With Kepler the Book of Nature reached the summit of its metaphorical life as the vehicle of the self-understanding of a first rate scientist who was deeply committed to the Christian Faith. But with Galileo the Book of Nature was confronted with the Book of Scripture in a dramatic encounter which has ever since been regarded as one of the most decisive interactions between the world of science and the world of belief. Many polemicists have even taken it as the final proof of the alleged incompatibility of these two worlds and evidence of an essential enmity between the Church and the scientific attitude. However, the framework of traditional cosmology, based principally on Aristotle and Ptolemy, had no room for such discoveries as those reported by Galileo in his Sidereus Nuncius (Drake 1957) and it would collapse under their weight.

The results of Kepler and Galileo provided a completely new point of departure for the science of mechanics. The philosophers were duly impressed and already in 1637 Descartes proposed a general theory of the universe in terms of purely mechanical interactions between various types of fundamental particles supposed to fill all space and influencing each other by their mutual collisions. On the other hand more mathematically inclined scientists became increasingly aware that Descartes had built his physics on shaky foundations. In Book I of the Principia (Koyré and Cohen 1972) Newton showed how all problems of motion could be mathematically stated on the basis of a few fundamental axioms, now called Newton's Laws, so that their solutions would depend only on appropriate mathematical techniques.

Newton argued that nature exhibits a number of mechanical phenomena for which no theoretical explanation could be found within a theory that was designed to comprehend all the motions of the bodies in the whole universe. From these premises he had constructed his argument for the existence of a Deity whose direct intervention would explain the gaps in the theoretical discourse. But this manner of reasoning made Newton's natural theology extremely vulnerable. His argument would clearly lose its strength at the moment when this discourse itself became sufficiently advanced to close the gaps by its own force. In the beginning of the 19th century the work of Laplace and his colleagues produced a growing feeling that at long last Newtonian mechanics itself had become able to stop the gaps in which Newton had found room for the Deity. This is the background of the popular anecdote of Laplace replying to Napoleon, when the Emperor asked him why God did not figure in his Mécanique céleste (1799 CE and later): "Sir, I have no need of that hypothesis."

From these first centuries in the development of modern science we move rapidly to today. There appear to be two strains in modern science which are in tension with one another. On the one hand, there is the increasing mathematization of physics. On the other hand, through studies in chaos and complexity there is the recognition that the world of sense experience has an innate unpredictability which prevents it from being subject to ultimate mathematical analysis.

From this historical overview we can garner the following characteristics which contribute to our understanding of the laws of nature and the search for purpose in the universe. In the age of mythology there was no split between nature and culture. For the Pythagoreans nature contains necessary, internal connections which only mathematics is able to disclose and express. While Aristotle insisted that nature could only be understood by searching out the four causes, Archimedes emphasized that knowledge of nature came through sense experience and experimentation with the use of mathematics. Christianity at its very birth asserted that the Lord and Savior was also the Creator of the world and, through the logos theology of John, that there was a rational structure in creation which derived from the very triune nature of the Creator. Thus, the world of the senses was worth investigation through the experimental method.

The question arose, however, as to whether there is a necessary connection between the Creator and the rationality of the universe or whether God freely chose that rational structure. With the birth of modern science the delicate balance between the search for necessity and for spontaneity in the evolution of the universe was threatened and no scientist could afford to be too facile in arguing for intelligent design from our knowledge of the laws of nature.

4. The Life Sciences

In the context of this discussion of origins and creation and of the historical overview of the development of scientific methodology I would like to introduce a discussion of the life sciences. Such a scientist observes changes that take place in living systems and he seeks to understand those changes by looking for natural processes. In seeking such natural explanations a scientist, as such, takes no position on any elements that lie outside nature. So creation, a creator, an intelligent designer are simply outside the confines of scientific investigation. To be more specific it is simply not possible within the confines of the sciences to have recourse to an intelligence at the origin of natural phenomena. Always realizing the limits of their methodology, scientists by profession seek only natural causes for natural phenomena. If they do not succeed today, they seek to do so tomorrow. That methodology places no limits on the total reality of the universe and of life. It simply admits that it cannot as such say anything about what lies outside natural causes. Anyone who does so is not doing science. And this is precisely what a recourse to intelligent design to explain natural phenomena does.

The great achievement of Charles Darwin was precisely to bring the study of life into the ambit of the sciences already well established in physics and chemistry (Ayala 1998). With him the origins of the many life forms about us became truly a scientific study. It attempted to explain all natural living phenomena by natural causes. And the attempt is just that: an attempt. And it has to our day had immense success. To date there is no other scientific explanation that rivals that whereby all living beings, including ourselves, come about by chance mutations in the original being which result in stepwise changes in the products carried out by natural selection in the environment in which the products come to exist. Those products survive which can best adapt to their environment. There is, therefore, an apparent destiny towards more perfect beings, i.e. better able to adapt, in this process; but the apparent destiny can be explained by the natural process itself. Needless to say, we do not yet know the natural processes whereby life first came to be.

5. The Fallacies of Intelligent Design

The principal fallacy of the intelligent design movement is to have recourse to an explanation for the origins of life forms, which is both non-scientific and not necessary. What I have said above should suffice to establish that intelligent design is non-scientific. It has recourse to explanations that are not natural, not within the ken of scientific explanation. A general statement should first be made about why I say such explanations as intelligent design are not necessary. To repeat, a fundamental tenet of the sciences is to seek for natural causes for natural phenomenon. When these natural causes are found, science has succeeded. When they are not yet found, scientists continue to search but they will not allow that it is necessary to seek for a cause outside nature, an intelligent designer, a "God of the gaps." History has shown, as we have seen with Isaac Newton, that the "God of the gaps" eventually surrenders to a natural explanation. For this reason and for methodological consistency science will always find recourse to non-natural causes unnecessary.

In the case of intelligent design every case of a biological system that has been proposed as requiring intelligent design has failed. A natural explanation within evolutionary biology has been found. This holds for the flagellum, the cilium and the blood clotting cascade in vertebrates (Miller 2004), all of which have been claimed to require intelligent design. The fallacy here is the failure to accept what is at the heart of neo-Darwinian evolution, namely, that by a step-by-step process of mutations and adaptation through natural selection an organism which is the result of former mutations and adaptations and which has a certain function before mutating and adapting again, can have another function afterwards and can, in fact, be integrated into a more complex organism of which it now constitutes a part. Evolution is a creative process. The claim of intelligent design that there are complex systems which could not function unless all of their parts were assembled at the same time according to a design is wrong. At least no such system has been yet proposed which passes the test of requiring design.

Intelligent design is seen by most scientists, and has thus far been judged by the judicial system in the United States, to have a hidden religious agenda and a fundamentalist inspiration at that. Whether this is true or not, one might examine the influence of the intelligent design movement upon religious belief. In the next section I would like now to show that here again the intelligent design movement fails.

6. Biological Evolution and Religion

The intelligent design movement claims that certain complex living organisms require an intelligent design and, therefore, a designer. While it is claimed that this designer is not necessarily the God of religious faith, it is difficult to imagine whom it might be. At any rate, at the heart of life in the universe is placed a designer. This belittles the God of religious faith by making him one who plans or assigns his minions to plan every step in the coming to be of life in the universe and in its evolution. This is far from the God who has truly revealed himself in the universe he created. But before elaborating upon that let us review a bit of history.

For historical reasons, and not truly religious ones, biological evolution has been the enigma of religions. Fundamentalist religious thought denies it. Catholic thought, as it has matured, accepts it as scientifically verified, but hesitates in how to deal with it. Why the denial and the hesitancy? Because God must be omnipotent and have everything under his control. The dynamism intrinsic in the universe in evolution seems to escape this omnipotence. I would like to discuss the most recent example of 'Catholic hesitancy" in light of the positive turn that it has taken.

A message of John Paul II on evolution was received by the members of the Pontifical Academy of Sciences on 22 October 1996 during the Plenary Session of the Academy being then held at the seat of the Academy in the shadow of St. Peter's Basilica and was subsequently made public (John Paul II 1996). It stirred a vast interest among both scientists and the public, an interest that went well beyond the usual attention paid to Papal statements. While the encyclical of Pope Pius XII in 1950, Humani Generis, considered the doctrine of evolution a serious hypothesis, worthy of investigation and in-depth study equal to that of the opposing hypothesis, John Paul II states in his message:

> "Today almost half a century after the publication of the encyclical [Humani Generis], new knowledge has led to the recognition that the theory of evolution is no longer a mere hypothesis."

In order to set the stage for dialogue the message distinguishes in traditional terms the various ways of knowing. The correct interpretation of observed, empirical, scientific data accumulated to date leads to a theory of evolution which is no longer a mere hypothesis among other hypotheses. It is an established scientific theory. But since philosophy and theology, in addition to the scientific analysis of the empirical facts, enter into the formulation of a theory, we do better to speak of several theories. And some of those theories are incompatible with revealed, religious truth. It is obvious that some theories are to be rejected outright: materialism, reductionism, spiritualism. But at this point the message embraces a true spirit of dialogue when it struggles with the opposing theories of evolutionism and creationism as to the origins of the human person. And this is obviously the crux of the message.

The dialogue progresses in the following way: The Church holds certain revealed truths concerning the human person. Science has discovered certain facts about the origins of the human person. Any theory based upon those facts which contradicts revealed truths cannot be correct. Note the antecedent and primary role given to revealed truths in this dialogue; and yet note the struggle to remain open to a correct theory based upon the scientific facts. The dialogue proceeds, in anguish as it were, between these two poles. In the traditional manner of Papal statements the main content of the teaching of previous Popes on the matter at hand is reevaluated. And so the teaching of Pius XII in Humani Generis that, if the human body takes its origins from pre-existent living matter, the spiritual soul is immediately created by God. And so, is the dialogue resolved by embracing evolutionism as to the body and creationism as to the soul? Note that the word "soul" does not reappear in the remainder of the dialogue. Rather the message moves to speak of "spirit" and "the spiritual".

If we consider the revealed, religious truth about the human being, then we have an "ontological leap", an "ontological discontinuity" in the evolutionary chain at the emergence of the human being. Is this not irreconcilable, wonders the Pope, with the continuity in the evolutionary chain seen by science? An attempt to resolve this critical issue is given by stating that:

> "The moment of transition to the spiritual cannot be the object of this kind of [scientific] observation, which nevertheless can discover at the experimental level a series of very valuable signs indicating what is specific to the human being."

The suggestion is being made, it appears, that the "ontological discontinuity" may be explained by an epistemological discontinuity. Is this adequate or must the dialogue continue? Is a creationist theory required to explain the origins of the spiritual dimension of the human being. Are we forced by revealed, religious truth to accept a dualistic view of the origins of the human person, evolutionist with respect to the material dimension, creationist with respect to the spiritual dimension. The message, I believe, when it speaks in the last paragraphs about the God of life, gives strong indications that the dialogue is still open with respect to these questions.

I would like to use the inspiration of those closing paragraphs to suggest that reflections upon God's continuous creation, in light of what we have said above in the section on "Origins and Creation," may help to advance the dialogue with respect to the dualistic dilemma mentioned above. We might say that God creates through the process of evolution and that creation is, therefore, continuous. Since there can ultimately be no contradiction between true science and revealed, religious truths, this continuous creation is best understood in terms of the best scientific understanding of the emergence of the human being, which I think is given in the following summary statement by the eminent evolutionary chemist, Christian de Duve, in his paper at the very Plenary Session of the Pontifical Academy of Sciences to which the Papal message on evolution was directed (de Duve 1997):

> "Evolution, though dependent on chance events, proceeds under a number of inner and outer constraints that compel it to move in the direction of greater complexity if circumstances permit. Had these circumstances been different, evolution might have followed a different course in time. It might have produced organisms different from those we know, perhaps even thinking beings different than humans."

Does such contingency in the emergence of the human being contradict religious truth? Not, it appears to me, if theologians can develop a more profound understanding of God's continuous creation. God in his infinite freedom continuously creates a world which reflects that freedom at all levels of the evolutionary process to greater and greater complexity. God lets the world be what it will be in its continuous evolution. He does not intervene, but rather allows, participates, loves. Is such thinking adequate to preserve the special character attributed by religious thought to the emergence of spirit, while avoiding a crude creationism? Only a protracted dialogue will tell. The spirit of the closing paragraphs of the message of John Paul II on evolution is, I believe, an invitation to just such dialogue.

It is obviously much too early in his papacy to discern how, with Pope Benedict XVI, the sequel to the Church's view in modern times on evolution will go. Nevertheless, on several occasions Pope Benedict has given some indications. At his general audience on 9 November 2005 he continued the series of talks in his catechesis of the prayer of the Church as derived from the Psalms (Benedict XVI 2005). On this occasion he addressed the so-called "Pascal Hymn" of God's ancient chosen people (Psalm 135) which expresses the glory of God revealed in his creation as it celebrates God's love and fidelity to his alliance with his chosen people. The Pope uses the opportunity to speak indirectly of evolution.

"The first manifestation of this love and fidelity," says the Pope, "is to be found in God's creation: the heavens, the earth, the waters, the sun, the moon and the stars." "Consequently, there exists," affirms His Holiness, "a divine message, inscribed secretly in creation as a sign of God's love and fidelity ..." The discourse then moves on to more modern concerns with allusions to evolution as the Pope, recalling the thoughts of St. Basil the Great, states: "There are some who, tricked by their deeply imbedded atheistic stance, imagine a universe with no guidance or order, as if floating along by sheer chance." The Pope, at that point departing from his written text, wonders about how many of those "some" among scientists today, drawn by atheism, see only chance in the world's unfolding, when we know from God's love and fidelity that he created the world out of love according to an intelligent design.

The Pope is speaking, of course, from a purely theological point of view in expressing God's love in creating a world which, to respect his fidelity, is orderly and does not evolve by sheer chance. I must recall at this point that neo-Darwinian evolution does not claim that the world evolves by sheer chance. The Pope says nothing about whether the natural sciences, respecting their own methodology, are capable of discovering God's intelligent design — and this is the critical issue. The Pope's position is that God's love and fidelity are at the source of his creation of the universe. If we use our best scientific knowledge of the "fertile" expanding and evolving universe to reflect upon the nature of God the Creator, we will find, as the Pope suggests, that God is not primarily a "designer", an attribute which diminishes his magnificence. He is primarily a lover who in creating shares his love.

In his homily at the Easter Vigil liturgy of 2006 Pope Benedict again alludes to evolution when he suggests that the greatest "mutation" in the history of mankind is found in the Lord's Resurrection (Benedict XVI 2006). Through God's special intervention the human and the divine have been definitively united. These are, of course, religious and theological reflections but it is interesting that the Pope clearly adopts the language of evolution in expressing them.

7. The God of a Believing Scientist

Cosmological and biological evolution reveal a God who made a universe that has within it through evolution a certain dynamism and thus participates in the very creativity of God. If they respect the results of modern science, religious believers must move away from the notion of a dictator God, a Newtonian God who made the universe as a watch that ticks along regularly. Perhaps God should be seen more as a parent. Scripture is very rich in this thought. It presents, indeed anthropomorphically, a God who gets angry, who disciplines, a God who nurtures the universe. Theologians already possess the concept of God's continuous creation. I think to explore modern science with this notion of continuous creation would be a very enriching experience for theologians and religious believers. God is working with the universe. The universe has a certain vitality of its own like a child does. You discipline a child but you try to preserve and enrich the individual character of the child and its own passion for life. A parent must allow the child to grow into adulthood, to come to make its own choices, to go on its own way in life. In such wisdom does God deal with the universe.

These are very weak images, but how else do we talk about God. We can only come to know God by analogy. The universe as we know it today through science is one way to derive analogical knowledge of God. For those who believe modern science does say something to us about God, it provides a challenge, an enriching challenge, to traditional beliefs about God. But there is always the temptation in this reasoning to make God into our own image and likeness. This would be idolatry. And I am afraid that the intelligent design movement has unwittingly fallen into this idolatry by making God or his minions designers.

References

Anderson, B.A. (ed.) (1984), Creation in the Old Testament, Philadelphia.

Ayala, F.J. (1998) Darwin's Devolution: Design without Designer, In: R.J. Russell, W.R. Stoeger and F.J. Ayala (eds.) Evolution and Molecular Biology, Vatican Observatory Publications, Vatican City, pp. 101-116.

Benedict XVI 2005, L'Osservatore Romano, 6 November 2005, pp. 1 ff.

Benedict XVI 2006, L'Osservatore Romano, 17 April 2006, p. 1 ff.

Carroll, W.E. (2003) Thomas Aquinas, Creation, and Big Bang Cosmology, In: C. Impey (ed.) Science and Theology: Ruminations on the Cosmos, Vatican Observatory Publications, Vatican City, pp. 1-18. I am indebted to Carroll for much of the discussion in this section.

Clifford, R.J. (2000) Creation in the Hebrew Bible, In: R. Russell, W. Stoeger and G. Coyne (eds.), Physics, Philosophy, and Theology, Vatican Observatory Publications, Vatican City, Second Edition, pp. 151-170.

De Duve, C. (1997) Life as a Cosmic Imperative, In: The Origin and Early Evolution of Life, Pontifical Academy of Science, Vatican City, Part I, 311-331.

Drachman, A.G. (1967) Archimedes and the Science of Physics, Centaurus, 12, 1-11.

Drake, S. (1957) Discoveries and Opinions of Galileo, New York; English translation of Sidereus Nuncius, Venice, pp, 55-96.

Gilson, E. (1952) Jean Duns Scot: Introduction à ses positions fondamentales, Etudes de Philosophie Médiévale, 42.

Heath, T.L. (1949) Mathematics in Aristotle, Oxford University Press, Oxford 1949. Almost all mathematical passages in the works of Aristotle have been collected and translated into English.

Jacobsèn, Th. (1949) The Cosmos as a State, In: H. Frankfort, H. A. Frankfort, J. A. Wilson, Th. Jacobsen (eds.) Before Philosophy, London, pp. 137-199.

John Paul II 1996, The original message in French was published in L'Osservatore Romano for 23 October 1996 and an English translation in the Weekly English Edition of L'Osservatore Romano for 30 October 1996.

Koyré, A. and Cohen, I.B. (eds.) (1972) Philosophiae Naturalis Principia Mathematica, Harvard University Press, Cambridge, Massacusetts.

Miller, K.L. (2004) The Flagellum Unspun: The Collapse of "Irreducible Complexity," in Debating Design, ed. W.A. Dembski and M. Ruse, Cambridge University Press, Cambridge, UK, pp. 81-97.

Pritchard, J.B. (1955) Ancient Near Eastern Texts Relating to the Old Testament, Second Edition, Princeton University Press, Princeton.

Dialogue

Victor J. Stenger

Introducing myself, I am a retired professor of physics and astronomy from the University of Hawaii where I spent 37 years in research on elementary particle physics and astrophysics. In 2000 I moved to Colorado where I am an adjunct professor of philosophy at the University of Colorado in Boulder. I am the author of seven published books and countless articles and essays on the interface between physics, cosmology, theology, pseudoscience, and alternative medicine. My last book, *God: The Failed Hypothesis--How Science Shows That God Does Not Exist* made the *New York Times* bestseller list in March 2007.

You can access most of my writing at:
http://www.colorado.edu/philosophy/vstenger/.

I enjoyed reading George's essay (I am assuming we will go by first names), which is well written and deeply thought out.

I found myself agreeing with much of section 2 (except for the most important assertion, which I will discuss later). This agreement comes, I think, from the fact that George has come a lot closer to the modern cosmological view than any other theologian I have read. He seems to accept the fact that science can, in principle, offer an explanation for the transition of nothing to something, at least where nothing is as defined by physicists as the state of no particles and zero energy. Such a state can be defined perfectly well mathematically as the bottom step in the ladder of energy levels one gets when one quantizes free boson and fermion fields under the assumption of supersymmetry. One simply keeps applying the annihilation operator to states with particles until no particles remain.

Another way to look at nothing is the state of complete lack of structure, the state of maximum entropy. If a system has structure then it is something. Nothing would be expected to have no structure. And this is what we get when we extrapolate back to the earliest definable moment — a state of maximum entropy. This does not violate the second law of thermodynamics since the expansion of the universe that follows allows increasing room for order to form.

Thus the universe began in total chaos and retains no memory of a "creator" in the traditional temporal sense. Furthermore, one can write down a complete mathematical description of (at least) two possible scenarios: (1) our universe appearing from that chaos and (2) our universe appearing by quantum tunneling through the chaos from an earlier universe.

Now, George says that the physicist's nothing is not the "nothing" of theologians since it is still "something." I hate this playing with words. Certainly the nothing of the theologians is something, too! Making a concept vague and undefined does not give one the right to claim it is meaningful.

Similarly George redefines the traditional meaning of "creator" as not temporal but the "source of all being." I am actually OK with that since, in my view time is just a human invention anyway. Of course, it has something to do with reality. But we have no reason to believe that time is anything more than what we read on a clock and the arrow of time of common experience is anything more than a definition, too, namely the direction in which the total entropy of the universe increases. All of the basic equations and interactions of both classical and modern physics are time reversible (though not necessarily exactly time symmetrical, as in the case of macroscopic observations). Thus, notions of cause and effect must not distinguish between the two. It follows that, the traditional notion of a creator is meaningless, so it is nice that George replaces it with something timeless or tenseless.

But, then, what does his definition mean? I don't see the need for any creation at all.

Parenthetically, it is to be noted that Pope Pius XII was happy to associate God with the originator of the big bang. At least George admits that what most people believe about the creation of the universe, including most theologians, is ruled out by science.

I found section 3 very informative until George took off his philosopher's hat and put on his apologist's one and moved to Christianity. Then it ceased to be very convincing to me. It seems really stretching it to say that the simple use of the word "logos" in John's gospel implies any kind of connection with the "emergence of experimental methods in science" and a "rational structure in creation which derived from the very triune nature of the Creator." Is there any concept in human thought more irrational than the triune God?

I do not agree with George's main theme, which is that intelligent design is outside the confines of scientific investigation. This is an application of the principle he asserts in section 2 that the natural sciences just deal with changes and can say nothing about origins. I have written a paper published in a philosophical journal that works out in mathematical detail the Hartle-Hawking model of the origin of the universe (Stenger VJ: A scenario for a natural origin of our universe using a mathematical model based on established physics and cosmology. *Philo* 2006, 9(2):93-102). Other physicists and cosmologists have published their own scenarios.

I am of the opinion that the intelligent design theorists Dembski and Behe fully comprehended that they needed to do more than just exploit gaps in scientific knowledge to prove the existence of God. That had to demonstrate that there was no way that a natural explanation ever could be found for the phenomena they discussed. Of course they did not do this, and that's why intelligent design is a failure, not because it wasn't science.

Biography

Dr. Edward T. Oakes, S.J. teaches theology at the University of St. Mary of the Lake in Mundelein, Illinois. He earned his Ph.D. in Systematic Theology at Union Theological Seminary in New York City in 1988. He is the author of *Pattern of Redemption: The Theology of Hans Urs von Balthasar* (Continuum, 1997), editor of *German Essays on Religion* (Continuum, 1995), co-editor of *The Cambridge Companion to Hans Urs von Balthasar* (Cambridge, 2004), and the author of numerous articles on religion and science, especially on questions of teleology, evolution and original sin.

3

The Enigma of Final Causality

Biological Causality in Aristotle and Neo-Darwinism

EDWARD T. OAKES

University of St. Mary of the Lake. Mundelein Seminary,
1000 East Maple Avenue, Mundelein, Illinois 60060 USA
eoakes@usml.edu

1. Introduction

St. Augustine famously said that he knew perfectly well what the nature of time was—until he started to think about it. Something similar can be said about causality: we think we have a pretty good idea of what causes are, how they ineluctably lead to certain effects (like a nighttime thunderstorm causing wet sidewalks the next morning). But then, like Augustine with time, we start thinking about the matter; and we suddenly realize David Hume (Hume 2000) was right: we never actually see a cause at work. The only thing we can nail down with certainty is that one event follows upon another. And if that is the case, then, as Hume says, "All our reasonings concerning causes and effects are derived from nothing but custom."

The rejoinder of the Aristotelian to this perfectly obvious psychological fact is a response with an insight equally obvious: assertions about causality are not grounded in acts of perception but in acts of judgment, a quite different faculty of the human mind. But such a retort, while perfectly true on its own terms, does not automatically resolve the fundamental conundrum that arises when we make assertions that some event has caused another event. For to locate the validity of assertions of causality in the faculty of human judgment raises another problem:

what does the human mind contribute to the operations of nature? Don't causes bring about effects whether the human mind recognizes them or not? Should not assertions of causality in nature be true independent of the human mind? Indeed is this not the real lesson that Immanuel Kant drew when Hume's alarm clock woke him from his dogmatic slumbers? (Kant 1965, 32)

2. The Enigma in Kant, Hume and Copernicus

The problem as Kant saw it was that, if Hume was right, events follow one another arbitrarily, in a kaleidoscope of ever-changing sense impressions. But we all instinctively know that cannot be true: surely something more than a Skinnerian psychological reinforcement is involved in the expectation that a rainstorm leads to wet sidewalks (which is why Hume intriguingly called causality "the cement of the universe"). But if we never see causes at work, then it must be the judgment that contributes to the human mind's realization of causes at work. And once Kant had made that realization, his turn toward the transcendental ego was locked into place.

Commentators like Walter Kaufmann have often been puzzled that Kant dubbed this philosophical turn toward the transcendental ego his "Copernican revolution," since the whole point of Copernicus' heliocentrism was to refute our subjective ("egotistical") impression of geocentrism, whereas Kant tried to ground all valid assertions of objectivity (including most especially assertions about the operation of causality) in the transcendental ego:

> "Kant claimed to have accomplished a Copernican revolution. Actually, his appeal is inseparable from the fact that in his *Critique of Pure Reason* he brought off an anti-Copernican revolution. He reversed Copernicus' stunning blow to human self-esteem. Before Copernicus the Western world had believed that man was at the center of the universe and that the sun revolved our earth. Copernicus' doctrine involved what Freud liked to call a 'cosmological mortification' of man's self-love. ... [But Kant] restored man to the center of the world and actually accorded even greater importance to man than the Book of Genesis had done. He tried to prove that it is the human mind that gives nature its laws" (Kaufmann 1980, 87-88).

These commentators have a point, although the objection would have surprised Kant, who would have alerted these critics to his distinction between the transcendental and the subjective, empirical ego. At least in relation to the latter, Kant was as Copernican as Johannes Kepler and Isaac Newton, whose lesson for Kant was that theory (true theory, that is) always trumps the senses. In other words (and here we follow the formulation of the Hungarian-American chemist and philosopher of science Michael Polanyi), theory is inherently Copernican inasmuch as all theory shares three features:

a) A theory is something other than myself.
b) A theory, moreover, cannot be led astray by my personal illusions.
c) Since the formal affirmations of a theory are unaffected by the state of the person accepting it, theories may be constructed without regard to one's normal approach to experience (Polanyi 1962, 4).

3. Objective, Inexorable Causes vs. the "Accidents" of Biology

This is the single greatest theoretical attraction in the scientist's search for causes: they are objective; and it is also why Hume had to be answered. Effects must inexorably follow from their causes, otherwise the assertion of causality at work would be pointless. Admittedly, some events follow other events inexorably without the one set causing the other, as day follows night. But that is only because both day and night are the effects of other causes at work, which do follow inexorably (position of the sun, rotation of the earth, and so forth). Furthermore, the cause/effect nexus operates temporally, along "time's arrow": first cause, then effect. This temporal relation is what gives causes both their predictive power and their overpowering connection to reality, as Polanyi saw:

> "One may say, indeed, quite generally, that a theory which we acclaim as rational in itself is thereby accredited with prophetic powers. We accept it in the hope of making contact with reality; so that, being really true, our theory may yet show forth its truth through future centuries in ways undreamed of by its author. Some of the greatest scientific discoveries of our age have been rightly described as the amazing confirmations of accepted scientific theories. In this wholly indeterminate scope of its true implications lies the deepest sense in which objectivity is attributed to a scientific theory" (Polanyi 1962, 5).

This (very sketchy) account of the human mind's encounter with the reality of the cause-effect nexus highlights the conundrum of teleology, that is, the study of the operation of final causes. What has made final causality so problematic for science is the apparent way teleology reverses time's arrow. For if the giraffe's neck develops its extraordinary length with the end in view of reaching the vegetation of tall trees, does that not place the cause in the future, in the final end (in the giraffe's case of reaching vegetation)? But if the later effect (the final form of the organism) causally determines the initial moments of the organism's construction, does that not mean that a later event is the cause of an earlier effect? But how can that be, if causes must always precede effects?

4. Idealism and Final Causality

That problem can only be resolved by positing the presence of a preceding mind that imaginatively conceives the final form and then constructs the initially available elements into the final form that it "has in mind," as we say. No wonder, then, that the question of teleology is so controversial, for if final causes are required to explain biological complexity, this will imply that the mental, in some way, precedes the material.

The philosophical position that gives precedence to the mental over the physical is classically called "Idealism," a highly generic terms that can cover any number of otherwise differing philosophies. It would of course far exceed the bounds of this paper to provide a taxonomy of all the schools of Idealism in the history of philosophy, from Plato through Hegel to today. Suffice it to say that at least in this regard even Aristotle can be regarded as an Idealist (of the Platonic variety), a point recently made by Lloyd P. Gerson in his *Aristotle and Other Platonists* (Gerson 2005), not least because of Aristotle's heavy reliance on teleological explanations. It obviously cannot be the burden of this chapter to outline all these differences, nor am I much interested here in the fact that, generally speaking, Idealism has struck believers as more compatible with religion than materialism, as became clear when the secular atheist philosopher Thomas Nagel found, to his discomfort, that he had just argued himself into Platonism:

> "If we can reason, it is because our thoughts can obey the order of the logical relations among propositions—so here again we depend on a Platonic harmony. The reason I call this view alarming is that it is hard

> to know what world picture to associate it with, and difficult to avoid the suspicion that the picture will be religious or quasi-religious. Rationalism has always had a more religious flavor than empiricism. Even without God, the idea of a natural sympathy between the deepest truths of nature and the deepest layers of the human mind, which can be exploited to allow gradual development of a truer and truer conception of reality, makes us more at home in the universe than is secularly comfortable. The thought that the relation between mind and the world is something fundamental makes many people in this day and age nervous. I believe this is one manifestation of a fear of religion which has large and often pernicious consequences for modern intellectual life" (Nagel 1997, 129-130).

Just why feeling "at home" should make someone "uncomfortable" is, I assume, part of the secular dilemma that Nagel is exploring, a sure sign of the "perniciousness" of the secular phobia of religion. Even more distracting is the example of the early nineteenth-century Anglican Archdeacon William Paley shows, who famously cited biological complexity as his major proof for the existence of God (Paley 1828).

No wonder, then, that Charles Darwin created such a stir with his theory of the evolution of species by means of natural (meaning, non-mental) selection. For finally biologists had a plausible way of explaining heterogeneous complexity without having to invoke a prior mind that can provide—in its marvelous beneficence—a long neck on the giraffe to enable that otherwise hapless creature the wherewithal to reach high-borne vegetation.

To speak personally here, I have no dispute with natural selection as a factor, even the most crucial factor, in the explanation of evolution, meaning here, "descent with modification." (The word "evolution" carries so many connotations and means so many different things to different people that it is important to confine its meaning in biology exclusively to that which biology can verify: the genetically verifiable connection between offspring and parent, which can be traced back through the genetic record over several billions of years [see Cavalli-Sforza 1994].) But the matter is otherwise, as we shall see, when natural selection becomes the all-purpose explanation for every event in nature. For then the intent is to expel all teleological explanations in the universe. At that point "natural selection" becomes a stand-in term for materialism *tout court*. This sleight-of-hand, taking natural selection out of biology and into all the events of nature, can only work under the assumption that Darwin has expelled Aristotle, which assumption the rest of this paper will seek to show is unfounded.

5. Aristotle, Our Contemporary

Let us begin with what might seem like an odd remark in Aristotle's Physics: "If purpose is present in art, it is present also in nature" (Aristotle, Physics II, 8, 199b1, in: Aristotle 1984, I: 341). At first this sounds like Aristotle is begging the question: after all, no one disputes that artists and artisans design things with an ulterior end in view (literally "in mind"), but that same finality is precisely what is in dispute in nature. So what does Aristotle mean by asserting that because we can detect purpose in art, we can locate it in nature as well? Here is his explanation:

> "It is absurd to suppose that purpose is not present just because we do not observe the agent deliberating. Art does not deliberate. If the shipbuilding art were in wood, it would produce the same results by nature. If, therefore, purpose is present in art, it is present also in nature. The best illustration is a doctor attending to and healing himself: nature is like that. It is plain then that nature is a cause, a cause that operates for a purpose" (ibid.).

But what is his evidence that nature is a doctor doctoring herself? Now Aristotle admits that the contrary can be conceived, for nature also knows events that occur not for a purpose but by the necessity of physical laws: rain doesn't fall in order to make corn grow or to let harvested corn rot on the threshing floor. So why not say the same of teeth: that they grow not for the sake of mastication but because it just happens that way, the way rain happens to fall in temperate climates and not in arid ones?

A difficulty presents itself: why should not nature work, not for the sake of something, nor because it is better so, but just as the sky rains, not in order to make the corn grow, but of necessity? What is drawn up must cool, and what has been cooled must become water and descend, the result of this being that the corn grows. Similarly if a man's crop is spoiled on the threshing-floor, the rain did not fall for the sake of this—in order that the crop might be spoiled—but that result 'just happened.' Why then should it not be the same with the parts in nature, for example, that our teeth should come up of necessity? (ibid., 339; emphases added)√

Aristotle's answer is simple: because if that were the case, then the same would also hold true of man as well, that quintessentially tool-using, design-driven creature—*Homo faber*. But to assert that man "just happened" to appear in the world in the same way that rain "just happens" to be beneficial to farmers would be to make man not a product, but a freak, of nature. Yet man's own designs in fact piggyback on those of nature: natural forms testify to purpose, and the inventor merely extrapolates those natural forms by analogy for his own purposes. (Aristotle did not have available to him the obvious examples of airplanes and submarines, which imitate the natural forms of birds and fish, respectively; but he would not have been surprised at that kind of inventiveness. In fact, would it ever have occurred to anyone—Leonardo da Vinci, the Wright brothers, whoever—to invent airplanes if birds had never evolved?) And that ability to extrapolate by analogy has to say something about nature itself:

> "Where there is an end, all the preceding steps are for the sake of that end. Now surely as in action, so in nature; and as in nature, so it is in each action, if nothing interferes. Now action is for the sake of an end; therefore the nature of things also is so. Thus, if a house, for example, had been a thing made by nature, it would have been made in the same way as it is now by art; and if things made by nature were made not only by nature but also by art, they would come to be in the same way as by nature [think here again of birds and airplanes]. The one, then, is for the sake of the other; and generally art in some cases completes what nature cannot bring to a finish [think prosthetic limbs], and in others imitates nature. If, therefore, artificial products are for the sake of an end, so clearly also are natural products. The relation of the later to the earlier items is the same in both. This is most obvious in the animals other than man: they make things neither by art nor after inquiry or deliberation. That is why people wonder whether it is by intelligence or by some other faculty that these creatures work—spiders, ants, and the like. By gradual advance in this direction we come to see clearly that in plants too that is produced which is conducive to the end—leaves, for example, grow to produce shade for the fruit. If then it is both by nature and for an end that the swallow makes its nest and the spider its web, and plants grow leaves for the sake of the fruit and send their roots down (not up) for the sake of

> nourishment, it is plain that this kind of cause is operative in things which come to be and are by nature. And since nature is twofold, consisting of both matter and form, of which the latter is the end, and since all the rest is for the sake of the end, the form must be the cause in the sense of 'that for the sake of which'" (ibid., 340).

The Darwinian does not go along with this argument, although he will often concede that such reasoning might have seemed eminently plausible at the time (and indeed all the way up to Paley). The reason the Darwinian balks is because of the element of a late-developing consciousness. Yes, man can design because he "inquires and deliberates." But Darwin showed that nature develops her complex forms even though she does not, cannot, so "inquire and deliberate." It was precisely that dilemma—not being able to imagine purpose-driven organs and organisms without a designing consciousness—that gave Paley the plausibility of his argument; and only when Darwin was able to show that natural forms could arise by non-deliberate means did Paley lose the argument. But Aristotle is not Paley. For one thing, he did not use his teleological concept of organic forms to establish his philosophical notion of G_d, a point it would behoove theologians to bear in mind more than they do. The whole point of Aristotle's philosophy of nature is simply to be true to nature. As Etienne Gilson, with admirable lucidity, puts it:

> "The analogy with art, then, assists us to recognize the presence in nature of a cause analogous to that which is intelligence in the operations of man, but we do not know what this cause is. The notion of a teleology without consciousness and immanent in nature remains mysterious to us. Aristotle does not think that this should be a reason to deny its existence. Mysterious or not, the fact is there. It is not incomprehensible because of its complexity, which we can only hope science will one day clarify, but because of its very nature, which does not allow it to be expressed in a formula" (Gilson 1984, 10).

The real block to accepting Aristotle's reasoning here does not stem so much from Darwin as from the whole ethos of modern science, which balks at teleological explanations because of their unverifiable appeal to mental presences, a point trenchantly adumbrated by the early seventeenth-century apologist for modern methodologies in science, Francis Bacon, who compared the invocation of final causality to a lamprey clinging to a ship that only impedes the vessel on its way to port: "For the handling of final causes, mixed with the rest in physical inquiries, hath intercepted the severe and diligent inquiry of all real and physical causes, and given men the occasion to stay upon these satisfactory and specious causes, to the great arrest and prejudice of farther discovery" (Bacon 1871, I: 37). But this is where the real begging of the question is located. Again, in Gilson's lucidly Gallic formulation:

> "If the scientist refuses to include final causality in his interpretation of nature, all is in order; his interpretation of nature will be incomplete, not false. On the contrary, if he denies that there is final causality in nature, he is being arbitrary. To hold final causality to be beyond science is one thing; to put it beyond nature is something completely different. ... Explanations which rely on final causality have often been ridiculed, but mechanist explanations have often been ridiculed also, and this does not disqualify the legitimacy of either point of view" (Gilson 1984, 26).

Again we must stress that the point of Aristotle's philosophy of biology will be lost if we read him through Paley's lens. Facts are facts, and biological facts are biological facts, and they require explanation according to the testimony of those facts: "Aristotle inevitably came, in each science or each class of sciences, to the positing of principles which are indemonstrable because primary; but they were also evidently true, because an entire order of nature becomes intelligible by their light. The notion of 'end' is for him among this number. It signifies for him the limit, the achievement of growing of all living beings, animal or plant. Since their development always leads to a limit which is at least provisionally felicitous, and since the reason for this success is not met with in any of their parts as parts, it is necessary that this future limit preside from the beginning over the ordering of the parts" (Gilson 1094, 15).

The great irony in this alleged rout of Aristotle by Darwin is that Aristotle, unlike Paley, was perfectly willing to concede the late arrival of man, with his conscious construction of things with an end in view. But Aristotle did not see consciousness as an inherent requirement for espying the presence of final causes (which is why Paley was no Aristotelian!). Rather, he saw that, in the case of art, the principle of finality is exterior to the work, whereas in nature finality is an interior component of the work. But not having Bacon's hang-ups about the lamprey dragging down a ship's speed, he was able to draw the right conclusions about man's seamless connection with nature, a point that Darwinians are always hammering away at but whose metaphysical implications elude them because of their phobia about invoking final causes. Perhaps nowhere is Gilson more brilliant in his analysis of Aristotle's contribution than here:

> "Aristotle conceives the [designing] artist as a particular case of nature [the realm of finality]. This is why, in his natural philosophy, art imitates nature, rather than nature imitating art. The contrary is imagined because—every man being more or less an artist, an artisan, and a technician—we know, more or less confusedly, yet with certitude, the manner in which art operates. But insofar as we are natural beings, we are the products of innumerable biological activities of which we know practically nothing, or very little. The manner in which nature operates escapes us. Her finality is spontaneous, not learned. ... In nature the end, the telos, works as every artist would wish to be able to work; in fact, as the greatest among them do work, or even as the others work in moments of grace when, suddenly masters of their media, they work with the rapidity and infallible sureness of nature. Such is Mozart, composing a quartet in his head while writing down its predecessor. Such is Delacroix, painting in twenty minutes the headpiece and mantle of Jacob on the wall of Saint-Sulpice. A technician, an artist, who worked with the sureness of a spider weaving its web or a bird making its nest would be a more perfect artist than any of those that anyone has ever seen. Such is not the case. The most powerful and the most productive artists only summon from afar the ever-ready forces of nature which fashion the tree and, through the tree, the fruit. That is why Aristotle says that there is more purposefulness [to *hou heneka*], more good, and more beauty, in the works of nature than in those of art" (Gilson 1984, 9-10).

6. Mental Air

What is missing in all versions of naturalism, at least those "scientific" versions that have swept the field since the days of Darwin is any sense of what it means for naturalism to make man such a freak exception to the outcomes of the laws of nature. This has always been the great snag in scientific naturalism. On the one hand, it wants to incorporate man into the cosmos fully, understanding him only as the outcome of blind laws operating blindly, and yet have him emerge as the great non-blind exception. This is what the German philosopher Paul Natorp labeled the essential superficiality of naturalism, "which," he says, naïvely "transforms the world into a single great calculation which, as a result of an immense intertwining of factors, finally produced the calculator as well as the calculation" (Campo 1967. 5: 446).

It really requires an extraordinary amount of mental gymnastics to hold to this view consistently. Witness Stephen Jay Gould's (Gould 1990) endlessly reiterated assertion that if one rewound the tape of evolution at any juncture, it would be highly unlikely that *Homo sapiens* would re-emerge. Of course, as Daniel Dennett pointed out in his book on Darwinian theory, *Darwin's Dangerous Idea*, this view would make nonsense of any attempt to search for extra-terrestrial intelligence, another research project dear to the naturalist's heart.

But suppose this search succeeds, against all the odds of the vast distances in space and of the time it takes to traverse that space, even at the speed of light. Success would certainly confirm the naturalist's thesis that consciousness and intelligence are somehow natural to the universe. But then one could hardly turn around and say how freakish we are and how utterly exceptional is our natural endowment of intelligence. Again, Dennett captures the point exactly when he says:

> "Suppose SETI [the Search for Extra-Terrestrial Intelligence] struck it rich, and established communication with intelligent beings on another planet. We would not be surprised to find that they understood and used the same arithmetic that we do. Why not? Because arithmetic is right... The point is clearly not restricted to arithmetic, but to all "necessary truths"—what philosophers since Plato have called *a priori* knowledge. ... It has often been pointed out that Plato's curious theory of reincarnation and reminiscence, which he offers as an explanation of the source of our *a priori* knowledge, bears a striking resemblance to Darwin's theory, and this resemblance is particularly striking from our current vantage point. Darwin himself famously noted the resemblance in a remark in one of his notebooks. Commenting on the claim that Plato thought our 'necessary ideas' arise from the pre-existence of the soul, Darwin wrote: 'read monkeys for pre-existence'" (Dennett 1995, 129-130).

Now I wonder if Darwin himself was aware of how radical that concession was. When he said rather cryptically "read monkeys for pre-existence," he is clearly groping toward a recognition that the course of evolution takes place within a nature that already is structured with ideal realities.

This is because in Darwinian theory any functional feature of an organism almost invariably reveals something about the environment of that organism. That is, just as the evolution of the eye on fourteen separate occasions (according to current theory in evolutionary biology) testifies to the ubiquity of light and just as the existence of the fin and wing testifies to the density of water and air, respectively, so too we are permitted, nay, required, to come to certain conclusions about nature's environment from the existence of the human brain. Evolutionary theory makes no sense unless it posits a fit between organism and environment, a point that Dennett drives home with an interesting thought experiment:

> "Any functioning structure carries implicit information about the environment in which its function 'works.' The wings of a seagull magnificently embody principles of aerodynamic design, and thereby also imply that the creature whose wings they are is excellently adapted for flight in a medium having the specific density and viscosity of the atmosphere within a thousand meters or so of the surface of the Earth. Suppose we carefully preserved the body of a seagull and set it off into space (without any accompanying explanation), to be discovered by Martians. If they made the fundamental assumption that the wings were functional, and that their function was flight (which might not be as obvious to them as we, who have seen them do it, think), they could use this assumption to 'read off' the implicit information about an environment for which these wings would be well designed. Suppose they then asked themselves how all this aerodynamic theory came to be implicit in the structure, or, in other words: How did all this information get into these wings? The answer must be: By an interaction between the environment and the seagull's ancestors" (Dennett 1995, 198).

In other words, we must radicalize the concept of man's place in nature and see anything that man does as inherently an expression of nature's capabilities; and that therefore our minds, our experiences, our theory-building activities say something about the essence of nature. The emergence of mind must, on

Darwinian terms, be environmentally advantageous; and that can only be the case if the universe already displays that intelligibility for brains to evolve as minds into. Wing's evolve into air, eyes evolve into light, and brains into what I call—being admittedly metaphorical—mental air.

I am not of course claiming, as Paley would have done, that this "mental air" indicates a divine consciousness at work. Such an argument would have to introduce other factors. One such factor would be the recognition that the "Fifth Way" of proving God's existence in St. Thomas Aquinas' *Summa Theologiae* has been misnamed as the "Teleological Argument." Actually, it's an argument from the order of the universe, not its directionality. Darwinian complexity cannot arise except in a prior environment of order, which is why Darwinism cannot refute Thomas' Fifth Argument, although many people think so because they labor under two false assumptions: that Thomas' Fifth Way is teleological; and that Darwin expelled teleology from biological explanations. Neither assumption can withstand scrutiny. My only point is that Darwinism not only does not refute Idealism; it depends on it.

7. Abstract

It is widely assumed that Darwin expelled teleological explanations from evolutionary biology with his theory of natural (that is, non-mental) selection, thereby rendering Aristotle otiose. A close examination of the logic behind both Aristotle's philosophy of causation as well as the logic of neo-Darwinism itself shows that such an expulsion of teleological explanations is impossible if Darwinism is to follow through on the logical implications of its own theory.

References

Aristotle (1984) Physics, in: J. Barnes (ed.) The Complete Works of Aristotle, Princeton University Press, Princeton, New Jersey.

Bacon, F. (1871) On the Proficience and Advancement of Learning, Divine and Moral, in: The Works of Lord Bacon, Bohn, London.

Campo, M. (1957) "Paul Natorp" in: P. Edwards (ed.) The Encyclopedia of Philosophy, Macmillan & The Free Press, New York. Vol. 5:445-448.

Cavalli-Sforza, L.L. (1994) The History and Geography of Human Genes, Princeton University Press, Princeton, New Jersey.

Dennett, D.C. (1995) Darwin's Dangerous Idea: Evolution and the Meanings of Life, Simon & Schuster, New York.

Gerson, L.P. (2005). Aristotle and Other Platonists, Cornell University Press, Ithaca, New York.

Gilson, E. (1984) From Aristotle to Darwin and Back Again: A Journey in Final Causality, Species, and Evolution, trans. John Lyon, University of Notre Dame Press, Notre Dame, Indiana.

Gould, S.J. (1990) Wonderful Life: The Burgess Shale and the Nature of History, W. W. Norton, New York.

Hume, D. (2000) A Treatise of Human Nature, Oxford University Press, New York.

Kant, I. (1965) Critique of Pure Reason, trans. Norman Kemp Smith, St. Martin's Press, New York.

Kaufmann, W. (1980) Discovering the Mind, Vol. I, Goethe, Kant, Hegel, McGraw-Hill, New York.

Nagel, T. (1997) The Last Word, Oxford University Press, New York.

Paley, W. (1828) Natural Theology, 2nd edition, J. Vincent, Oxford.

Polanyi, M. (1962). Personal Knowledge: Toward a Post-Critical Philosophy, University of Chicago Press, Chicago.

Dialogue

Lev V. Beloussov

I have read Dr. Oakes essay with pleasure and benefit. So far as I could understand, one of his main points is that Darwinism cannot work without teleology. Yes, certainly. But I would go further and am brave enough for claiming that Darwinism does not work under any circumstances. What is this theory about? Does it "explain" how new organisms arise? For doing this, it must first of all formulate the laws of morphogenesis which coin these organisms during each subsequent set of embryonic development. Unfortunately, Darwinism (in a form, to a great extent created more by enthusiastic Darwin's followers than by Darwin himself) has done its best to divert contemporary embryology from approaching this task (already started to be resolved by so called "idealistic morphologists" of the XIX century) and involving this science in vicious circles of phylogenetic-ontogenetic relations.

Certainly, in no way I am a creationist. Nobody in his right mind will deny evolution. I only claim that Darwin did not explain the latter, because a scientific theory of evolution can be based only upon still unknown laws of morphogenesis which Darwin, most probably, did not see at all (his theory might be true if the living world consisted only of adults and embryonic development did not take place at all).

Now about the final causes proper. As Dr. Oakes, I admire Aristotle and, even stepping outside the teaching curricula, talk each year in my Developmental Biology course about his 4 causes. However, to bring back the final cause in a strict Aristotelean form to modern science (500 years after its expulsion by Francis Bacon) would be, by my view, a kind of anachronism. Meanwhile, a modern theory of self-organization gives us another way for doing practically the same. What I mean, is the notion of an *attractor*. Many models of self-organization are constructed in such a way that even if started from quite different points of a phase space they will reach quite a definite final state. The latter can be regarded as a final cause, which can be known from the very beginning of the process, if we are informed about the *parameters* of a system, that is about the variables that are an order of magnitude or more slower than the dynamical ones (which are immediately observable). Some attractors look rather simple (energy wells), while other are more or less sophisticated (like Lorentz' strange attractor). I would not like to state that introduction of attractors gives a final solution of an endless problem of teleology, but it is at least a heuristically useful step for combining the elements of teleology with more ubiquitous modes of causality.

Biography

Professor Julian Chela-Flores was born in Caracas, República Bolivariana de Venezuela and studied in the University of London, England, where he obtained his Ph.D. in quantum mechanics (1969). He was a researcher at the Venezuelan Institute for Scientific Research (IVIC) and Professor at Simon Bolivar University (USB), Caracas until his retirement in 1990. During his USB tenure he was Dean of Research for six years. He is a Fellow of The Latin American Academy of Sciences, The Academy of Sciences of the Developing World, the Academy of Creative Endeavors (Moscow) and a Corresponding Member of the Venezuelan "Academia de Fisica, Matematicas y Ciencias Naturales". His current positions are Staff Associate of the Abdus Salam International Center for Theoretical Physics (ICTP), Trieste, Research Associate, Dublin Institute for Advanced Studies (DIAS) and Profesor-Titular, Institute of Advanced Studies (IDEA), Caracas. His particular area of expertise is astrobiology, in which he is the author of numerous papers. He organized a series of Conferences on Chemical Evolution and the Origin of Life from 1992 till 2003. In 2001 he published the book: *The New Science of Astrobiology From Genesis of the Living Cell to Evolution of Intelligent Behavior in the Universe.*

4

Astrobiological Reflections on Faith and Reason

The Issues of Agnosticism, Relativism and Natural Selection[1]

JULIAN CHELA-FLORES

The Abdus Salam International Centre for Theoretical Physics
Strada Costiera 11; 34136 Trieste, Italy
and Instituto de Estudios Avanzados, Caracas 1015A,
Republica Bolivariana de Venezuela
chelaf@ictp.trieste.it

1. Is There a Crisis in the Dialogue Between Faith and Reason?

At a time when agnosticism is widespread in large sectors of society, it is relevant and timely to evaluate those insights into our origins that the science of astrobiology has given us, keeping in mind other sectors of culture. There are both philosophical and theological issues to discuss. First of all, is it conceivable to reject with scientific methods concepts that are of a theological nature? It should be emphasized that the scientific method is itself self-correcting. In spite of this remark, currently it seems to be a widely spread attitude to attempt going beyond the rational boundaries of science, a discipline that should restrict its statements only to hypothesis that can be verified by experiment, in a tradition that goes back to Galileo Galilei. The fact that science is based on a research

[1]The text has been expanded in a forthcoming book: *A Second Genesis: Stepping Stones Towards the Intelligibility of Nature.* World Scientific Publishers, Singapore.

process that makes it continually and permanently susceptible to improvements and corrections is indeed the main strength of the scientific method. Extending such a boundary into the domain of theology has the difficulty that theology is based on tradition and revelation, both concepts lying strictly beyond the scope of science. We argue that such debates are sterile even in the reverse direction. Our faith in divine action is independent of experiments and rational hypothesis.

At present we are faced with two new aspects of postmodernism that have not always characterized Western thought, namely relativism and agnosticism (St. Augustine, 417-427). We may distinguish two forms of relativism: cultural relativism means that a given culture should be understood on its own merits. Morality for one culture could be considered immoral in a different culture. Ethical relativism implies that what is ethical for a person is not an absolute concept but depends on circumstances, or the given society to which the individual belongs. On the other hand, agnosticism proclaims that humans cannot know the existence of anything beyond the phenomena of their experience. The roots of this doctrine can be traced back as far as the Enlightenment (cf., Sec. 4). The term is sometimes used loosely implying skepticism about religious matters, especially rejecting traditional religious beliefs. The growth of science is assumed to have influenced this doctrine.

The dialogue between faith and reason has been prominent since classical times in ancient Greece. An early example is expressed in the defense of the Christian Church that had to survive in a secular Roman world. "The City of God" should be seen in the broad perspective of all the Abrahamic traditions, including Judaism, Christianity and Islam (St. Augustine, 417-427 CE).

Indeed, Augustine advocated the worship of one true God and the rejection of all false gods. He completed Chapter XVII by the year 427 CE. Section 3 of that Chapter has the title: "The threefold meanings of the prophets, referring sometimes to the earthly Jerusalem, sometimes to the Heavenly City, sometimes to both at once". At the end of the section he dwelt on a problem that has reached our own times: *Is it erroneous to suppose that the Holy Scriptures have any significance beyond the purely historical record?* But he also denied that every statement is a complex of allegorical meanings. The creation narrative of the book of Genesis is a recurring theme. His *De genesi ad litteram* was the result of many years of work, from the late 390s to the early 410s (St. Augustine, 401–414/415 CE). In this work he assumed that when a view has been expressed about the Earth, size and distance of the stars, the nature of animals derived from reason and experience, it is damaging to confront this view by purporting to speak in accordance to holy writings (O'Meara, 1984).

Modern science evolved a thousand years later as a separate aspect of our civilization, mainly due to emphasis on the experimental foundations of science. In a modern context the dialogue between science and faith has not disappeared. In fact, it became even more prominent. The root of agnosticism, a second aspect that affects the dialogue between faith and reason, reappeared in the meetings of The Metaphysical Society. This learned association of intellectuals was a monthly discussion group that met in London starting in 1869. The Society included clergymen, Positivists, Deists and Unitarians. Thomas Henry Huxley was a Victorian biologist and a keen supporter of Charles Darwin. He attended these meetings (Desmond and Moore, 1991). Rather than 'atheism' he preferred the term 'agnosticism'. In his view an agnostic avoided religious issues. He rather advocated a positivist position that is shared by contemporary science, in which biology and other sciences deal only with the knowable world.

At the beginning of the 21st century globalization has had the net effect of allowing ideas to disseminate more efficiently. A consequence has been the widespread discussion of questions such as the following: *To what extent is knowledge dependent on context?* These topics define the philosophical topic of relativism (O'Grady, 2002; Ambrosi, 2005). According to this doctrine we no longer have absolute values. It would even be argued that modern thinking should be adjusted to incorporate relativism. Inevitably a crisis of reason in confrontation with faith can be perceived in Western thought (Ratzinger, 2004).

In the context of these two aspects of postmodernism (agnosticism and relativism), we discuss the phenomenon of life, with natural selection as its mechanism of evolution. Our main objective is to discuss how agnosticism, relativism and natural selection affect our society. Science, through research in astrobiology, is close to answering basic questions on the nature of life in the universe (its origin and evolution). The new perspectives that astrobiology might make available would contribute to the dialogue between faith and reason. One of them is the possibility of identifying a 'second Genesis'. Up to the present we are only certain that life has originated on Earth, but space missions for the exploration of the Solar System are attempting to investigate whether elsewhere in the Solar System life may have originated a second time.

2. Has there been a Second Genesis in the Solar System?

The satellite Europa is a candidate for life in the Solar System. This follows from the indication that a large proportion of the spectroscopically detectable material on its surface is water (Soffen, 1976; Horneck, 1995). The Voyagers are two US spacecrafts that explored the outer Solar System during the period 1977-1989. They still continue their travel beyond the orbit of Pluto. The two Voyagers were launched the same year with different objectives. Voyager 2 completed a set of images of every planet in the Solar System.

According to the results obtained on the Jupiter system by the Voyager 2 mission, we know that there may exist an ocean of water under Europa's icy surface. The lure of this Jovian moon is due to the possibility of finding traces of life on that satellite. The Galileo Mission gathered much information during its activity (1995-2003). Galileo has changed the way we look at the Solar System and especially Europa. The Galileo spacecraft was the first to conduct long-term observations of the Jovian system from orbit. It found evidence of subsurface water not only on Europa, but also in Ganymede and Callisto. It also revealed the intensity of volcanic activity on Io.

Up to the present time we do not fully understand the divergence into the three 'domains' of life (previously called 'families') that arose from the evolution of the earliest ancestor of all life on Earth. Indeed, plate tectonics has obliterated fossils of early organisms from the crust of the Earth, which is the only record available to us on the evolution of early life. Europa is a candidate for the search of extra-terrestrial microbes either extant or extinct. There is at present a possibility for returning to Europa with LAPLACE (Blanc et al. 2008), a mission to Europa and the Jupiter System for the European Space Agency's Cosmic Vision Programme. The exploration of Europa raises significant questions: *Does it represent the "habitable zone" of the Jupiter system? Does Europa actually harbor life?* These questions on habitability can be answered in the future by the identification of reliable biosignatures, namely the identification of characteristic biochemical substances, or minerals that can be taken to be a reliable indicators of the biological origin of a given sample. This is a major priority in the current exploration of the Solar System. The options for selecting the right instrumentation for the purpose of searching for biosignatures have been discussed elsewhere (Chela-Flores and Kumar, 2008).

Previous work has suggested that there are at least two sources of heating of Europa independent of the Sun. They are tidal and radiogenic heating (Reynolds et al. 1987). Of these two forms of heating, on Earth we are familiar with radiogenic heating, which is a consequence of the heat produced by the accumulation within the Earth crust of radioactive compounds. The new factor in the Jupiter system is that unlike the Earth, Europa is influenced by its two neighboring satellites: the very volcanic satellite Io, which is closer to Jupiter, and the giant satellite Ganymede, which is even larger than planet Mercury. The dynamics of this three-body system keeps the satellites from perfect circular orbits. The consequence is that the eccentricity of the Europan orbit varies significantly in the points of closest and farthest approach to the giant planet, thereby creating thermal gradients that we have called earlier 'tidal heating'.

Fortunately, we have some clear observational evidence of this form of heating, since Io, its nearest Moon-sized neighbor, is not covered with ice like Europa itself, neither does it have a thick atmosphere such as that of Titan, the satellite of Saturn. It is possible then that the Europan internal ocean has consequently been formed (underneath a relatively thin ice cover) through dehydration of silicates, the heating source being due to tidal heating with an addition due to radiogenic heating. From the similarity of the processes that gave rise to the solid bodies of the Solar System, we may expect that hot springs may lie at the bottom of the Europan ocean. On Earth there is an analogous environment in Antarctica: Lake Vostok lies underneath the Vostok station of the Russian base. It appears to be harboring hydrothermal vents beneath the water surface. This is suggestive of what may be occurring on Europa. This lake is about 1,000 km from the South Pole and beneath 4 km of ice. In the southern region of Antarctica many bacterial species have been found in zones of accreted ice, about 120 m above the water-ice surface (cf. Dudeja et al. 2008 for references). The main thesis of the proponents of the existence of an Europan biota is that, as Jupiter's primordial nebula must have contained many organic compounds, then possibly, organisms similar to heat-loving microbes can evolve at the bottom of Europa's ocean (Oro et al. 1992).

3. On the Implications of Darwinism

The search for life on Europa represents one of the major efforts that will have to be faced in the foreseeable future to give us insights into life's origin that would go beyond what we have been able to learn from research in organic chemistry. The general outline of life's emergence on Earth is nevertheless understood, and its implications for the humanities have been discussed elsewhere (Chela-Flores, 2005). For example, life is known to have emerged early in the history of the Earth, some 3.5 billion years before the present. The evidence comes form micropaleontology (Schopf, 1993). The microorganisms involved are fossils of stromatolites. These are geological features consisting of a stratified rock formation, which are essentially the fossil remains of bacterial mats. The bacteria that gave rise to these formations were mainly cyanobacteria. Similar mat-building communities can develop analogous structures of various shapes and sizes in the world today.

Besides, it is safe to assume that over three billion years ago there was a flora of cyanobacteria, although the exact date for the earliest ancestor is a hotly debated issue (Brasier et al. 2002). The early forms of life are also known to have been a major factor in the evolution of the hydrosphere and the atmosphere. The time sequence of these events has been inferred following standard procedures and hypothesis. Like other branches of science, these dates are subject to improvements by new experimental techniques and observation procedures. This state of affairs is in sharp contrast with the questions of faith that are based on tradition and revelation. The statements of science are closer to philosophy in the sense that both systems attempt to make their statements on a rational basis. Charles Darwin, the English naturalist, was the author of the theory of evolution through the mechanism of natural selection (Darwinism). His theory was published in 1859 in his book "*The Origin of Species by Means of Natural Selection or the Preservation of Favored Races in the Struggle for Life*". This work led to the definitive theory of evolution that had been anticipated in earlier incomplete forms by Charles' grandfather Erasmus Darwin and independently by Jean-Baptiste Lamarck.

The case of Lamark is especially relevant in the present context. He published in 1778 a book on French plants, *Flore Française*. At the beginning of the 19th century Lamarck in his late fifties began the revolutionary steps that led him to develop an evolutionary theory, rather than accepting that the living world was fixed and harmoniously organized. Lamarck believed that a change in the environment causes changes in the needs of organisms living in that environment, which in turn causes changes in their behavior, and once again in turn this leads to differential use of internal organs. (Eventually organs either continuously improve or gradually disappear.) Such a mechanism for evolution of life on Earth — called Lamarckism — was different from Darwin's (particularly since Lamarck stated that these changes in the organisms would be inheritable). But, in the end, Lamarckism leads to adaptive change in lineages driven by environmental change, over geologic time. In spite of its limitations, the originality of evolution of life on Earth being driven by the environmental changes disrupted a view of life that persevered in the life sciences since the time of Aristotle. According to Lamarck, therefore, living species are interrelated through reproduction, slowly evolving through the course of generations.

Throughout his life Darwin avoided the problem of the origin of life, except from making a few speculative remarks about the possible environments where life could have originated, the so-often-quoted 'warm little pond'. This was a very reasonable attitude for the late 19th century, before experimental science began to address the question of the origin of life with Fox, Haldane, Oparin, Miller, Oro, Ponnamperuma and others.

4. Darwinism, Philosophy and Theology

Darwin, wisely for his time, avoided the question of the origin of life. He rather focused on the origin of the species by introducing the term 'natural selection' for reproductive success, allowing adaptation to changing environmental conditions. In other words, natural selection is the non-random element in evolution that gives evolution its direction. The magnitude of this revolutionary contribution to science is evidenced by the perennial difficulty to insert this fundamental aspect of science into the mainstream of cultural knowledge. This may be illustrated with the dialogue between his Eminence Cardinal Christoph Schönborn and George Coyne SJ. The main point made by Schönborn is that evolution in the sense of common ancestry might be true, but evolution in the neo-Darwinian sense — an unguided, unplanned process of random variation and natural selection — is not (Schönborn, 2005). Any system of thought that denies or seeks to explain away the overwhelming evidence for design in biology is ideology, not science.

However, the point made by Coyne, and shared by most scientists, is that science is completely neutral with respect to philosophical or theological implications that may be drawn from its conclusions. Those conclusions are always subject to improvement. As we have emphasized above this is the nature of science. But to deny today's science on religious grounds is to go beyond the natural boundaries of theological thoughts, which are interpreted by Coyne in very clear terms as (Coyne, 2005):

> "Theologians already possess the concept of God's continuous creation with which to explore the implications of modern science for religious belief."

Likewise, to attempt to make changes in theological thought on the strength of science is to go beyond the natural boundaries of science. Western civilization has faced this dichotomy before during the Enlightenment. This was an intellectual movement of the 17th and 18th centuries in which ideas concerning God, reason, nature, and man went into a synthesis that had many supporters. Amongst the most distinguished thinkers of this period we have: Descartes, Diderot, Montesquieu, Pascal, Rousseau and Voltaire. This movement was influential in the development of art, philosophy, and politics. Reason was the main theme underlying most innovations of this period. The thinkers behind this movement searched for a deeper understanding of the cosmos. Rationalists strived towards more freedom, knowledge and happiness. Under the influence of the Enlightenment the French philosopher Auguste Comte (1798-1857) founded a movement advocating that intellectual activities should be confined to observable facts.

The reason why this movement was called "positivism" is that Comte called observable facts 'positive'. Indeed, positivism can be considered as a philosophical system of thought maintaining that the goal of knowledge is simply to describe the phenomena experienced, not to question whether it exists or not. This point of view was developed much later by a group of philosophers working in Vienna in the 1920s and 1930s. They were known as the "Vienna Circle".

The nucleation of the group began with Moritz Schlick, when he settled in Vienna in 1922 (Feigel, 1963). Some of the Vienna Circle's members were Rudolf Carnap, Hans Reichenbach, founder of the Berlin Circle, Herbert Feigl, Philipp Frank, Kurt Grelling, Hans Hahn, Carl Gustav Hempel, Victor Kraft, Otto Neurath, Friedrich Waismann. Also K. R. Popper and H. Kelsen were related to the Vienna Circle, although they did not strictly belong to it.

This group of philosophers maintained that scientific knowledge is the only kind of factual knowledge. The distinctive aspect of this version of extreme positivism was an attempt, referred to as 'logical positivism', to develop knowledge based on experience (empiricism), with the help of mathematics and logic. They insisted on the soundness of logical analysis of scientific knowledge. Indeed, logical positivists appealed to the earlier contributions of Russell and Wittgenstein. The Vienna Circle maintained that all traditional doctrines are to be rejected as meaningless. They were generally hostile to metaphysics and theology. The Vienna Circle went beyond positivism in maintaining that the ultimate basis of all knowledge rests on experiment. Although some scientists have adopted this philosophy, either consciously or unconsciously, the fact remains that modern science begins with Galileo, who initiated the tradition of formulating theories based on observation and experiments. No underlying philosophy was adopted then beyond the dialogue between theory and experiment. On the other hand, there are a large number of issues that science cannot handle, or even formulate. In his *History of Western Philosophy* Bertrand Russell makes this point (Russell, 1991): "Almost all the questions of most interest to speculative minds are such as science cannot answer".

Positivism avoided all considerations of ultimate issues, including those of metaphysics and religion. However, as anticipated by Russell, the reduction of all knowledge to science is a matter that debate has not yet settled. Natural selection was expanded by the gradual growth of the science of genetics, especially molecular genetics. The origin of life on Earth is on similar grounds; the phenomenon is understood in its broad outline. The details of the specific organic pathway from biochemistry to a microorganism are still being sought. But our theories imply that once the biomolecules of life were self-assembled into a functional living cell, natural selection led to evolutionary pathways that left clear records in the fossils.

5. Discussion

So far the above discussion of the emergence of life on Earth is subject to experimental refutation. The example mentioned earlier in relation to dating of the oldest microfossil of 3.5 billion years before the present is a clear example of the way science progresses. Refutation of arguments previously assumed to be on secure basis constitutes a normal procedure in science. The philosophical discourse, on the other hand, is in a 'no-man's land' between faith and reason. One clear illustration is provided by the question of relativism. One universal truth in a theological sense is the omnipresence of divine action, which is an act of faith that requires no scientific support, neither is it subject to the relativist constraints. As a question in the philosophy of religion, relativism may be marginally more interesting. But relativism is irrelevant to faith that is based on tradition and revelation. These premises of faith are common to the whole Abrahamic tradition that is professed by Jews, Christians and Muslims alike.

With Augustine we believe that it is erroneous to attempt to confront the scientific implications of natural selection with religious statements from the Holy Books. This would take us beyond the natural frontiers of science. It would be equally erroneous to confront the theological statements related to divine action with scientific concepts: theological thinking should only concern revealed truth and related rational thinking. No controversy should arise in relation to teaching science and religion. If natural selection were eventually faced with scientific facts and observations that suggest it not to be the best hypothesis for the evolution of life on Earth, then a better mechanism would be suggested, due to newer facts that would not allow a traditional Darwinian interpretation. Likewise if new prophecies were to be lead to new revealed truth in the realm of theological thought, then this would lead to a new theologies, not to new scientific approaches. This has occurred repeatedly in Western civilization when, for instance, the revealed truth of Judaism was supplemented with the revealed truth of Christianity, and later by that of Islam. Nevertheless, these parallel avenues of thought often address the same questions, such as what is the relation between humans and the universe.

Darwinism and theological thought have been shown to be compatible within the framework of kenotic process theology. Indeed, kenosis as a concept is taken to mean self-emptying and voluntary sacrifice on behalf of others, based on genuine and freely given love for others, and resulting in the generosity and respect that flow from it (Ellis, 1998). God is the sole ground for the world's being (Haught, 1998, 2005). This approach to natural theology leads us to explain the world in terms of evolution, as understood within the Darwinian tradition (Russell et al. 1996). Haught focuses on process thought. This philosophical system is considered to be particularly helpful in the task of constructing an evolutionary theology that may throw some further insights on Darwinism.

Our main conclusion on the question of faith and reason is that space exploration, especially the exploration of Europa, can encourage a constructive dialogue between faith and reason for the benefit of culture in general.

6. Glossary

Agnosticism is the doctrine that humans cannot know of the existence of anything beyond the phenomena of their experience. The term is sometimes used loosely implying skepticism about religious matters especially rejecting traditional Christian beliefs. The growth of science is assumed to have influenced this doctrine.

Astrobiology is the research into the origin, evolution, distribution and destiny of life, not only in the Solar System, but also in the whole universe. Astrobiology is currently in a period of fast development due to the many space missions that are in their planning stages, or indeed already in operation.

Divine action The Mosaic traditions (Judaism, Christianity and Islam) assume a process of deliberate self-revelation of God to humanity. Accompanying such forms of piety is the attribution to God of both intentions and the capacity for action as a means of expressing those intentions. Making sense of the concept of divine action is a current challenge to both philosophy and theology.

Darwinism A theory of the mechanism of evolution due to Darwin. It is assumed to explain organic change. It is brought about by three principles: (1) variation, (2) heredity and (3) the struggle for existence.

Faith is a subjective response to divine truth as well as a supernatural act of the will.

Natural selection is a term suggested by Darwin for the struggle for existence, differences in survival, in fertility, in rate of development, in reproductive success. It is a process resulting in the adaptation of an organism to its environment by means of selectively reproducing changes in its genetic constitution. The magnitude of this revolutionary contribution to science is evidenced by the perennial difficulty to insert this fundamental aspect of science into the mainstream of cultural knowledge.

Natural theology is the body of knowledge about religion that can be obtained by human reason alone, without appealing to revelation.

Neo-Darwinism. The evolutionary synthesis of Charles Darwin's theory of the evolution of species by natural selection, Gregor Mendel's theory of genetics as the basis for biological inheritance, and the mathematical formulation of population genetics. This aspect of evolution has been due to the work of several scientists, including Theodosius Dobzhansky, Thomas H. Morgan, Ronald Fisher, J.B.S. Haldane, William D. Hamilton, Julian Huxley, Ernst Mayr, George Gaylord Simpson, G. Ledyard Stebbins and several others. These evolutionary biologists introduced the connection between the units of evolution (genes) with the mechanism of evolution (selection). The term also applies to the unification of several branches of biology especially genetics, cytology, systematics, botany, and paleontology.

Positivism is a philosophical system of thought maintaining that the goal of knowledge is simply to describe the phenomena experienced, not to question whether it exists or not. In other words, in philosophy positivism is any system that restricts itself to the data of experience to the exclusion of arguments based on metaphysics.

Reason A faculty contrasted with experience, passion or faith.

Relativism (cultural) A given culture should be understood on its own merits. Morality for one culture could be considered immoral in a different culture.

Relativism (ethical) What is ethical for a person is not an absolute concept but depends on circumstances, or the given society to which the individual belongs.

Revelation A process by which communication of truth by God takes place. Christian philosophers have distinguished between 'truths of reason' and 'truths of revelation'. (cf., natural theology.) The monotheistic religions state that God has chosen to manifest himself through the prophets, but He can manifest Himself through his main creations: the universe and the life that has evolved in it. In the Judeo-Christian tradition the prophets are witnesses and interpreters of God's divine action, both in their transmission of God's messages, as well as in the way to interpret His divine action.

Theology (Kenotic Process Theology) An approach to natural theology emphasizing God as the sole ground for the world's being. It attempts to explain the world in terms of evolution as understood within the Darwinian tradition.

References

Ambrosi, E. (2005). *Il bello del relativismo*. Marsilio Editore, Venice, Italy.

Blanc, M. and the LAPLACE consortium (2008). *LAPLACE: a mission to Europa and the Jupiter System*, Astrophysical Instruments and Methods, in press, a digital version can be consulted in: http://www.ictp.it/~chelaf/ss186.html (Team members full list is available at http://www.ictp.it/~chelaf/ss164.html).

Brasier, M. D., Green, O. W., Jephcoat, A. P., Kleppe, A. K., Van Kranendonk, M. J., Lindsay, J. F., Steele, A. and Grassineau, N. V. (2002). Questioning the evidence for Earth's oldest fossils, *Nature* **416**, 76-81.

Chela-Flores, J. (2005). Fitness of the Universe for a Second Genesis: Is it Compatible with Science and Christianity? *Science and Christian Belief*, **17** (2), 187-197.

Chela-Flores J. and Kumar N. (2008). Returning to Europa: Can traces of surficial life be detected? *International Journal of Astrobiology*, in press.

Coyne, G., S. J. (2005). God's chance creation, *The Tablet*, **6** August. http://www.thetablet.co.uk/cgi-bin/register.cgi/tablet-01063

Desmond, A. and Moore, J. (1991). *Darwin*. M. Joseph, London, p. 568.

Dudeja, S., Bhattacherjee, A. B. and Chela-Flores, J. (2008). *Microbial mats in Antarctica as models for the search of life on the Jovian moon Europa*, ICTP preprint IC/2008/34, to be submitted by invitation to Microbial Mats (2009). Eds. J. Seckbach and A. Oren in the Series Cellular Origin, Life in Extreme Habitats and Astrobiology, Springer, Dordrecht, The Netherlands.

Ellis, G. F. R. (1998). The thinking underlying the new 'scientific' world-views, in R. J. Russell, W. R. Stoeger, S. J. and F. J. Ayala, (eds.) *Evolutionary and Molecular Biology: Scientific Perspectives on Divine Action*, Vatican City State/Berkeley, California: Vatican Observatory and CTNS, p. 276.

Feigel, H. (1963). Physicalism, Unity of Science and the Foundations of Psychology, in *The Philosophy of Rudolf Carnap*, (The Library of Living Philosophers, Vol. 11), P. A. Schilpp (ed.), London, Cambridge University Press, pp. 227-267.

Haught, J. F. (1998). Darwin's gift to theology, in R.J. Russell, W. R. Stoeger, S. J. and F. J. Ayala, (eds.) *Evolutionary and Molecular Biology: Scientific Perspectives on Divine Action*, Vatican City State/Berkeley, California: Vatican Observatory and CTNS, pp. 393-418.

Haught, J. F. (2005). The Boyle Lecture 2003: Darwin, Design and the Promise of Nature, *Science & Christian Belief*, **17** (1), 5-20.

Horneck, G. (1995). Exobiology, the study of the origin, evolution and distribution of life within the context of cosmic evolution: a review, *Planet. Space Sci.* **43**, 189-217.

O'Grady, P. (2002). *Relativism*. Acumen, Buks, UK.

O'Meara, J. J. (1984). *Introduction to St. Augustine's Concerning the City of God against the Pagans*, Penguin Classics, London, pp. 1-15.

Oró, J. Squyres, S. W., Reynolds, R. T. and Mills, T. M. (1992). Europa: Prospects for an ocean and exobiological implications, in G. C. Carle, D. E. Schwartz and J. L. Huntington (eds.), *Exobiology in Solar System Exploration*, NASA SP **512**, pp. 103-125.

Ratzinger, J. (2004). *Europa I suoi fondamenti oggi e domani*. San Paolo, Milano.

Reynolds, R. T., McKay, C. P. and Kasting, J. F. (1987). Europa, tidally heated oceans, and habitable zones around giant planets, *Adv. Space Res.* **7**, 125-132.

Russell, B. (1991). *History of Western Philosophy and its Connection with Political and Social Circumstances from the Earliest Times to the Present Day*, Routledge, London, p. 13.

Russell, R.J., N. Murphy and C.J. Isham, (eds.) (1996). *Quantum Cosmology and the Laws of Nature. Scientific Perspectives on Divine Action*, 2nd ed., Vatican Observatory, Vatican City State, cf., Russell's Introduction, pp. 1-31.

Saint Augustine (417-427 CE). *Concerning the City of God against the Pagans*, Penguin Classics, London.

Saint Augustine *De genesi ad litteram* (*Literal Commentary on Genesis*) (401–414/415 CE). http://www.holycross.edu/departments/religiousstudies/alaffey/Augustine-Genesis.htm

Schonborn, C. (2005). Finding design in nature, *New York Times*, July 7. http://www.nytimes.com/2005/07/07/opinion/07schonborn.html?ex=1278388800&en=95804823e49fb832&ei=5088&partner=rssnyt&emc=rss

Schopf, J. W. (1993). Microfossils of the Earth Archaean Apex Chert: New evidence of the antiquity of life, *Science* **260**, 640-646.

Shapiro, R. (1994). L'origine de la vie, Flammarion, Paris, pp. 402-404. For a more detailed account we refer to: Feinberg, G. and Shapiro, R. (1980). *Life beyond Earth: The intelligent Earthling's Guide to Life in the Universe*, William Morrow and Co., New York, pp. 328-332.

Soffen, G. A. (1976). Scientific results from the Viking Mission, *Science* **194**, 1274-1276.

Dialogue

Tom Barbalet

From Section 1:

There are a number of points made as truisms I find concerning.

Cultural relativism is used haphazardly in contemporary discussions. No contemporary society allows isolated individuals to commit acts of murder or wantonly damage property. Whilst the state may be able to do these things in a variety of contemporary societies, no society allows the individual at an undetermined time to do these acts. From that grounding cultural relativism has limits and can not be used to justify all acts as relative across cultures. Somewhat non-trivially this also undermines the assertion of ethical relativism, although this is more controversial.

> "... it is erroneous to suppose that Holy Scriptures have any significance beyond the purely historical record."

It would appear to be erroneous to suppose that Holy Scriptures have any significance with regards to the historical record either.

> "Modern science evolved a thousand years later as a separate aspect of our civilization, mainly due to emphasis on experimental foundations of science."

Bar perhaps the sciences with a mathematical component that can be traced back to Classical mathematics, if not earlier. Not to forget sciences that require engineered/applied experimentation that also can be traced to ancient Egypt. In fact, to say modern science comes from the modern age is a temporal truism. To say modern science evolved from somewhere between 1000 and 1800 CE (depending on which prophet starts the thousand year counter) neglects the foundations of science that predate this time.

From Section 2:

There is a popular misuse with regards to astrobiology, similar to quantum mechanics in new age spirituality and genetics in memetics, that lays credit to a wide variety of farcical possibilities. The language used by scientists with regards to astrobiology needs to be metered with a substantial degree of caution otherwise it is likely to be used as evidence rather than possibility.

Even with primary contact to a crime scene, for example, the ability to find genetic evidence is still non-trivial. In terms of planetary exploration, in order for astrobiological evidence to be credible to the definitions of modern science alluded to in section 1, there will need to be substantial detailed exploration in the future. The contemporary remote and piecemeal sensing of planets and their satellites give a quality of information analogous to grainy photographic plates as an attempt to prove a person's identity.

From Section 4:

> "... science is completely neutral with respect to philosophical or theological implications that may be drawn from its conclusions. Those conclusions are always subject to improvement."

Early quantum mechanics and Kantian philosophy could be a counter-example to this hypothesis. Whilst the dialogue of the early practitioners of quantum mechanics seems to indicate that they where interested in basing their mechanics in Kantian perception, they metered their conclusions based on this philosophical view as well. This has been challenged thoroughly by practitioners following them, without question.

To a certain extent, I think biological phyla are a counter example to this as well. This is a structural methodology with philosophical and theological roots which predate contemporary Darwinian biology and continue to circumvent conclusions.

Perhaps I am predisposed to the idea that particular dominant philosophies perturb all aspects of the scientific process.

Julian Chela-Flores
I welcome these comments hoping that our readers will not only take both points of view into account, but that they will also participate in a constructive dialog.

Biography

Bruce Damer has been a pioneer of the medium of virtual worlds and avatars over the last dozen years, creating the first conferences, large scale experiments and writing about the medium including *Damer, B. (1997). Avatars: Exploring and Building Virtual Worlds on the Internet, Berkeley: Peachpit Press.* He is currently leading a team to build virtual worlds for the design of NASA missions. In 1996 he founded Biota.org (Barbalet 2007) which hosts the Biota conference series and podcast, now a leading community resource for artificial life developers. In 2008 he initiated the EvoGrid project to create a large scale evolutionary simulator in the Internet. Bruce also serves as curator and historian of one of the world's largest collections of artifacts of personal computers housed in his Digibarn Computer Museum. Here he is shown in real life and as his avatar in Second Life.

5

The God Detector

A Thought Experiment

BRUCE DAMER

bruce@damer.com

Disclaimer

I am a technologist, and in this piece I shall approach the key questions of this book as a programmer and historian of technology. In my treatment I will not consider influences of the Divine in our lives in terms of matters of the heart, faith, hope, or the rest of the human cultural milieu. I will simply take on the claim made by some that God plays an active ongoing role in the mechanics of the universe and in the evolution of life. To me this seems like a question best approached from an engineer's frame of reference. A good starting point is to consider the lessons learned and the questions raised by those of us engaged in the new field of "artificial life".

1. The Artificial Life Programmer, the New Alchemist?

Like medieval alchemists before them, programmers developing artificial life software (often shortened to "A-life") are drawn to the elusive yet seductive proposition that they have the power to animate inanimate matter (Farmer & d'a Belin 1991). In this modern reincarnation of alchemy the inanimate medium is a microscopic substrate of billions of transistors. Popular media from science fiction to Hollywood often depicts A-life as computer viruses and self-reproducing robotics running amok. This means that A-life practitioners (in academia or the hobbyist community) attract quite a bit of press, much of it sensational. As a result, in these reports we are rarely treated to the subtle issues and challenges faced by coders of biologically-inspired virtual worlds.

Another key point is that there is often confusion between the fields of artificial life and artificial intelligence (AI). A-life developers agree that theirs is a "bottom up" approach wherein they simulate a large number of interacting components employing relatively simple rules from which complex behaviors of a whole system emerge (Langton 1991). AI on the other hand tackles the ever receding goal of creating a "conscious" entity with which we would one day be able to communicate. The apocryphal moment of the coming of walking, talking machine intelligence is sometimes referred to by pop-culture practitioners as "the singularity" (Kurzweil 2005). To complicate matters further, developers of A-life software cannot even agree on what defines an "authentic" A-life implementation.

Still, out of all of this confusion emerge some insights we could apply to the Intelligent Design/Creationism vs. Evolution/Science discussion. But before we can draw a hasty conclusion as to whether an artificial life programmer is acting as an "artificial god" (Adams 1998) and "intelligent designer" of his or her own authentic little virtual universe we have to understand the two diametric poles of the A-life continuum.

Two Kinds of God in the A-life Universe

Figure 1: Karl Sims' Evolving Virtual Creatures (1994).

Figure 2: Will Wright's game Spore (2007).

Perhaps the best way to classify A-Life software is to look at two ends of a continuum represented on the one hand by Karl Sims' Evolving Virtual Creatures (Figure 1) and on the other by Will Wright's game Spore (Figure 2). Karl Sims' creatures started life as a simple pair of hinged blocks in a virtual universe that simulated basic physical properties such as fluid, collisions, gravity, and surface friction (Sims 1994). From that point on the simulation was allowed to continue on its own without human intervention. The creatures would perform simple tasks such as swimming or walking, or competing with other creatures for control of a block of "food". The best performers were allowed (by the system software) to reproduce. Random mutations were introduced automatically into the "genome" of the creatures between generations, affecting the external body shapes or internal control functions. In this completely "hands off" A-life system the virtual creatures "evolved" many of the same mobility strategies found in nature (swimming with four paddles like a turtle, slithering like a snake, or perambulating like a gorilla). All of these behaviors emerged without human programmer intervention.

In contrast, the computer game Spore, which is being developed by Will Wright of the Maxis-Electronic Arts company, bears only a passing resemblance to an A-life environment. The release of Spore in 2008, will, however, be heralded as an "evolution" or "biological" game and yet most activities are largely directed by the human player and built-in procedures. Players use editor tools to design creatures, landscapes, dwellings and vehicles, guiding virtual creatures who inhabit toy planets to live out virtual lives from primordial soup to the space age. The populations "evolve" through procedural algorithms until the player (or game code itself) again intervenes to keep the action moving forward.

Given this continuum, we posit that there two kinds of God in the A-life universe: the Karl Sims' *God the Mechanic* building the machine that is the whole simulation, setting its initial conditions and then returning only occasionally to view the current state of the simulation; and Will Wright's *God the Tinkerer*, constantly poking and prodding to tweak the mechanisms of virtual creation. Clearly these definitions might also apply to different extremes of *god traditions* found in human cultures.

There are two key kernels of truth that we can winnow from these early decades of A-life alchemy:

> Kernel 1: That all attempts to render life down into its basic elements and then represent it abstractly come down to: a) creating an algorithm for making copies of the blueprints to make yet more algorithms and b) that imperfect copies of these blueprints are sometimes passed on, creating variations and, possibly, advantageous adaptations.
>
> Kernel 2: That after these algorithms run for a while, passing on a great number of blueprints and interacting within some kind of a simulated virtual environment, the whole system reaches a tipping point where, to our perception, it becomes opaque to complete understanding. Thereafter even the A-life developers themselves must assume the role of a biologist, dissecting the genomes of their virtual creatures or examining their "fossil record" looking for clues to what the process of artificial evolution hath wrought.

2. Lost in the Noise of the Data Explosion

Thus, the observer of the biologically inspired software simulation soon becomes "lost in the noise" (Negroponte 1995), much as a biologist might spend a lifetime to grasp one small aspect of the stupefyingly complex machinery of a single cell.

I propose that this property of *onset opacity* also holds for the world's religious traditions. For each there was an original prophet, and an original set of core stories and concepts (some new, some drawn from prior traditions). Once the copying of these stories got underway, a mutation and adaptation process began. The resulting data explosion of writings, stories, laws, debates, schools, conflicts, extinct lines, and new branches soon obscured many of the original statements attributed to the founding prophets. Religious seekers (and even many serious researchers) are unable or unwilling to apply reductionist methods to prune out later inserted, contradictory or inconsistent yet closely held beliefs or writings. In addition, modern monotheistic religions stand upon foundations of earlier belief systems, most of which carry no written record. Therefore, fundamental questions about God and the universe that might emerge from any religious tradition are likely to remain lost in the largely opaque "tree of noise" of religious histories and discourse. In other words, if at any time God ever made Himself unequivocally visible to a human being and uttered or physically manifested anything about life or the universe, that original direct experience of

God's existence has become irretrievably lost. In modern times, no verifiable experience of God's presence in the physical universe that is not explainable by other means has been observed. Therefore, if we cannot validate the original claims, or detect any direct physical influence today, we have to look for evidence of God's Hand at another level.

3. The God Detector

For some of the other authors of this book, prior writings about God, or personal (but unverifiable) experiences of God is evidence enough of His existence. However, when making a strong claim about God the Intelligent Designer, such empirical evidence is not good enough to make the case. If God is a programmer tweaking the code of the physical observable universe (not just affecting our own independent thoughts) his influence has to be detectable and independently verifiable. To sense the hitherto unseen Hand of God, we hypothesize that is might be possible to employ a *God Detector* which could either be *found* or *built*. We will first take on the challenge of identifying an existing natural God Detector and later on in this chapter, consider building a God Detector using human technology. If you will indulge me, dear reader, I invite you to join me in the following thought experiment surrounding the *quest for the God Detector*.

4. Finding the God Detector

How to look for signs of God's influence comes down to *where* to look for them, and that comes down to *what* you look at and what you *exclude* looking at within the universe.

For a time, I set down my pen and declared to myself that this was an unsolvable problem. A few days later I was reading a history of the Institute for Advanced Study in Princeton in the USA where I am a currently a visitor. A brilliant account of John Von Neumann's digital computer designed and built at IAS in the late 1940s contained an account of an impassioned researcher named N. Boracelli who was developing "numerical symbioorganisms" for this pioneering digital computer (Dyson 1997). I was stunned to realize that on a machine of such tiny capabilities, Boracelli was able to run basic artificial life code thirty five years before the term was coined.

This led me to the following insight: what if the universe could be reduced down at the lowest levels to a programmable machine running algorithms? Several theories of how the universe works at the quantum level propose that this is in fact how things work (Lloyd 2006). I realized that if you can render the universe's operation down to simple algorithms, basic questions could then be asked, and a natural God Detector could be found at a key code location found within one of the universe's algorithms.

5. God and the Copying Rule

A living organism differs from bare rock, gases or a pool of liquid in one very specific way: the living organism contains instructions that are copied, for the most part unaltered, from one version to the next. In fact the organism *must* copy these instructions or face extinction. Thus, there would be no copying mechanism if previous copying mechanisms ceased to work, so copying mechanisms can and must continue to copy. This is the Copying Rule, and, as we have seen previously, it can also be found at work in human culture, where language permits the telling and retelling of a story, and also within the new medium of digital computers and networks, where programs are copied between computers.

The universe contains a large number of seemingly copied objects, from rocks to stars to galaxies, but the process by which these objects were made did not involve construction from a blueprint, instead their existence is owed to the laws of physics applied to starting conditions. Therefore, as far as we know, all matter and energy in the universe inhabits one of two organizational regimes:

Regime 1 which is governed by…	**Regime 2 which is governed by…**
• Formulaic Laws of Nature • An element of uncertainty, or randomness we might call "R"	• Formulaic Laws of Nature • An element of uncertainty, or randomness we might call "R" • The Copying Rule

As we infer from the above table, along with the Laws of Nature and the Copying Rule, another of the distinct organizing operators of the universe is the element of uncertainty. This could be thought of in terms of unpredictable (i.e. random) effects either from some source in the very small (quantum fluctuations for example) or the very large (the mass overlapping effect of gravitational forces from atoms, stars and galaxies for example). We will take up this operator "R" later as it is the pivot on which this simple thought experiment turns.

The Copying Rule is well understood in molecular biology. For each genotype (information blueprint encoded in a cell's nucleus) a phenotype (a living body or other resulting output) is produced. The Copying Rule as seen operating in human culture is less well understood but clearly occurs. Copy-able cultural objects such as ideas, stories, music or instructions are sometimes referred to as "memes" within a new field called "memetics" (Dawkins, 1976). Clearly, technological objects (where copies of objects are made with reference to instructions) also execute the Copying Rule. As we addressed previously, a sub-specialty of computer software called artificial life attempts to emulate the biological implementation of the Copying Rule by creating software analogues to genotypes and phenotypes. More radical thinkers consider all software, such as ordinary applications like word processors, to also execute the Copying Rule with humans acting as the phenotype (the host) that is the mechanism to enable the copying of these programs (Dyson 1997).

A Simple Model of the Copying Rule

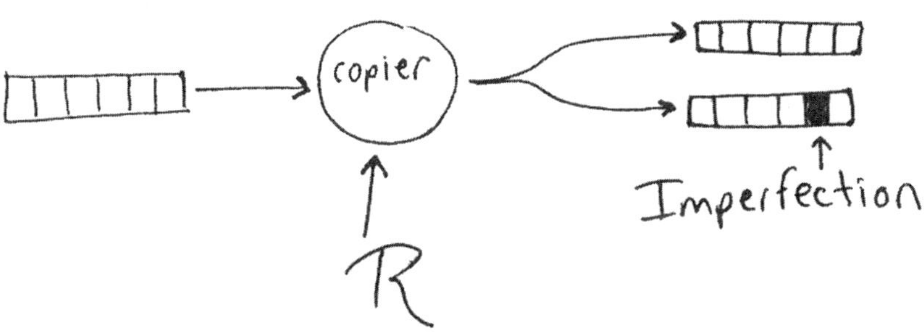

Figure 3: The Copying Rule.

A simple model of the Copying Rule is depicted in Figure 3. An input sequence of information, which could be encoded in molecular material, language or computer code, enters a copying mechanism upon which some random input R may or may not act, and two or more resultant output sequences are produced, some of which may contain random changes. There are variations of this mechanism, one that would take two input sequences and combine them into an output sequence. A Copying Rule could be said to have been "successfully executed" if the output information sequence is not so altered that it could not be used to produce a future copy. A "failed" application of the rule produces a sequence that can never again be copied.

6. Scope and Time Scales of the Copying Rule

The Copying Rule is the central driving mechanism within biological evolution, cultural evolution and technological evolution and operates across a range of time scales and scopes: from billions of years to kilo-years for molecular evolution to years or days for cultural evolution, and days to milliseconds for evolution in information systems (see table below).

Molecular copying **4 billion to1 kilo-years**	**Cultural copying** **1 kilo-year to 1 day**	**Digital copying** **1 day to 1 millisecond**
• Development of multi-cellular life • Divergence of Galapagos finch populations	• Rise and fall of a great empire (or religion) • Spread of hoax on the Internet	• Spread of virus on the Internet • 1 millisecond of computation in SETI@Home grid

7. How God the Intelligent Designer Engages the Copying Rule

A "designer" is someone who makes a plan for the future and instructs other people or mechanisms to bring that plan into reality. If God is acting in the universe as an "intelligent designer" and desires to operate in places where there are living things, then He has no choice but to engage the Copying Rule.

God has two obvious ways to interact with Copying Rule:

1) God would engage the Natural Laws that make the copying happen or
2) God would influence the operation of the Copying Rule by engaging the nondeterministic forces we call R, which create the imperfections or mutations in the copying process.

Common sense dictates that God cannot use both of these mechanisms at the same time as they work in opposition. For example, while the laws of gravity cause a feather to fall predictably, the random motions of the air through which the feather travels produce an unpredictable landing place.

By calling God a "designer" it is implied that He an actor upon the forces that shape the universe and is not those forces themselves. A God who is operating solely through deterministic laws is a God with no free-will. These laws affect the universe in pre-ordained ways with predictable outcomes. After the creation of the universe (and these laws) this *God the Mechanic* would simply leave the universe to run on autopilot and thereafter be undetectable.

If God cannot suspend or change the natural laws, then He might operate by introducing imperfections as a *tinkerer* in the mechanics of the Copying Rule shifting the application of the randomizer R to cause accumulated errors in the copying process that would give rise to our world (Figure 4).

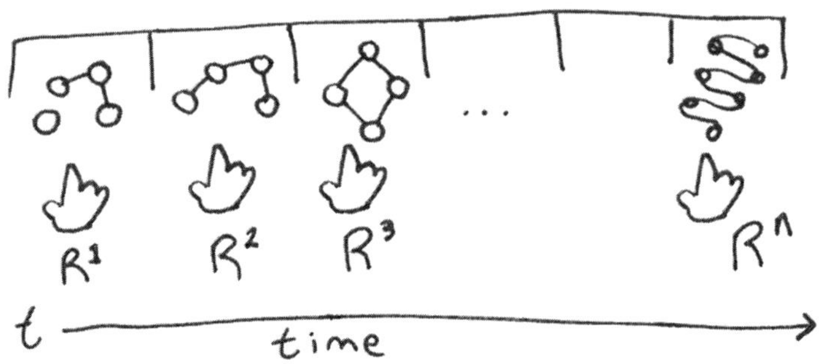

Figure 4: The accumulating effects of R through time.

Perhaps God could decide to permit R to affect the copying mechanism or not, or He could choose to proactively add to or subtract from the influence of R by a large enough amount to "tip the balance" in favor or one copying outcome or the other. In this case the Hand of God should be detectable as localized statistically anomalous behavior in an otherwise uniformly distributed random landscape of R. The monkey wrench in these works is that R itself is by definition unpredictable. If R is governed by some Natural Law or mathematical formula then it would not be R. If God could predict the future value of R and act accordingly then we would have our God the Mechanic back. So God, just like the rest of us, has to live with the unpredictability of R (Figure 5) and would seem to us to operate not by absolute *Will* but by *Whim*. This kind of God would hardly be able to exercise much design upon the universe.

Figure 5: What is affected by R?

8. The Monk and the Copying Rule

Here is where our thought experiment could use a bit of help from a hypothetical real-world example. Picture a literate monk, working at his table some time in the early years of Christianity. He is given a book written in Hebrew to translate into Greek. In it is a section of a passage:

> "...and Jesus was born to the young girl Mary"

Reaching for his Hebrew-to-Greek dictionary scroll and not finding it, he sighs and makes a guess, translating the phrase to:

> "...and Jesus was born to the virgin Mary"

Perhaps random fluctuations in air molecules contributed to a puff of air that nudged the dictionary scroll from the table, and hence caused the translation of "young girl" to "virgin". Many scholars believe that this translation error actually occurred (Brown 1977, Pagels 2003) and led to the concept to the "virgin birth" or "immaculate conception" in the Catholic tradition. The resulting impact of this was substantial for the future of Christianity, leading to its wider adoption throughout the Mediterranean, where there were existing religious movements that also believed in spiritual power emanating from a virgin birth. The virgin birth idea also led to the suppression of women (whom evidence suggests were treated more equally in the early church) by enforcing male celibacy and sequestering devout and intelligent females away in convents. Male domination of the early church was therefore assured, which eased the integration of the religion into Roman power structures. The supernatural aura of the virgin birth also propelled the character of Jesus along a road that led to his elevation to Godhood following the council of Nicea in the fourth century.

Would God have had a hand in this fateful application of R to the translation of early Christian texts? Certainly if it was God's intention as an Intelligent Designer to promote Christianity as a new religious force (at the cost of existing belief systems) then we might say "yes", God influenced the movement of air molecules, at the quantum level, at that critical juncture.

However, God would have to have done more than just arrange for the translation error. God would also have to ensure that the proof-reading senior monk, upon seeing this one term, would not find it erroneous and send it back for correction. In addition, the natural error correcting mechanisms of the followers of the Hebrew version would have to be overcome. In practice, each small change affected through an influence of R (which is by no way guaranteed to work given the unpredictable nature of R) is followed by a virtually uncountable large number of subsequent required adjustments that require almost total foreknowledge of every action. It seems that God's task in influencing history in this way would require a brain that would be large enough to store all possible outcomes while executing perfect adjustments of random effects to guide each step. The size of the required decision-tree for even relatively small scale design changes might exceed the size of the countable particles in the universe. Amazingly, each Monk's brain contains a number of unique pathways through its neurons that already exceed this number. At the finest level of detail, God's brain would have to account for each of these neural pathways and be able to affect the journey of each electron. We are fast approaching an event horizon of total implausibility.

Cultures all over the world attribute "god-like" powers to those who seem to be able to repeatedly "beat the odds" in dice-tossing, in war, in procreation, or in longevity. However, no documented case of the conquest of tremendous odds has ever been produced. Methuselah' 969 year lifespan, and other miracles live strictly in the domain of mythology. It would seem that God is as powerless to affect next toss of the dice as the rest of us.

Many believers might state here that God is a separate, all-knowing, omnipotent actor for whom the universe is a mere toy. In this case then He could choose to be detected or not and we would be powerless to make inquiries about His existence or nature (in which case there is no reason for this book to exist). So let us return to reason and consider God as an actor within the universe subject in some way to its laws, rather than an incalculably large and immeasurable actor separate from everything.

9. God the Intelligent Adapter

But wait, there is another way to affect the application of R in the Copying Rule, and that is *through adaptation, after the copying is completed.* Every single celled organism in Earth's early seas that suffered an injustice due to physical or chemical fluctuations, heat, cold or an attack had a chance to adapt to the situation and survive to reproduce another day. The machinery of adaptation adjusts for the ravages of R and therefore diminushes and redirects its impact into the future.

So could God in fact be living at "the output end" of the Copying Rule, in the land of adaptation? If so, God's Hand would be busy helping adapt everything from cellular machinery on up to guiding the entire biosphere through the slings and arrows of large scale misfortunes such as meteor impacts.

In human culture, intelligence emerged as a critical adaptation. Might intelligence therefore be a place where the mark of God is the strongest? Would God then not be an Intelligent Designer but instead be a Designer of Intelligence? Would any act of intelligence be an act of God, regardless of the outcome? If God is trying to effect some kind of perfect design upon the universe then influencing outcomes of adaptation might be just as numerically challenging as trying to control individual applications of R. Just as in our monk's brain example, God is again relegated to being an imperfect player, making do with a limited ability to influence adaptations to direct the future of life.

10. God, Life, the Universe and Everything

So we return to our original question: if God is an actor in the universe and we render the universe down to its simplest organizing principles, then God must have some kind of fundamental relationship with the Copying Rule. We have decided that, for our purposes, we are not considering a God the Mechanic, who simply sets up the initial Laws of Nature and then departs the scene. If our God is actively tinkering then He could only affect the progress of life and culture in two ways: by affecting the unpredictable R value that randomly affects copying machinery, or by working His miracles on the output side of the Copying Rule that adjusts for the influences of R through adaptation.

We concluded that God could not affect any kind of predictive design on the universe by trying to influence the unpredictable effects of R as copying occurs. God's information processing capability would probably have to be many times the size of the universe for even minor adjustments to the future and therefore He could not be an actor in the universe.

This left God only one place to act, as a player in assisting the power of adaptation at the output end of the Copying Rule. Thus, God would not be an Intelligent Designer but instead could be thought of as an *Intelligent Adapter*. If God is indeed operating at the adaptation end of the spectrum, then there is no difference between God's work and the work of evolution through Natural Selection or engineering through human intelligence.

For example, a human technologist using his or her own intelligent genetic engineering skills or the processes of Natural Selection over eons could both create a fish that can live in near-boiling water. To those who did not witness the processes of the engineer or Natural Selection, this fish would be indistinguishable from a miracle from God. Would then believers be forced to conclude that Natural Selection or human genetic engineering must be equivalent to the Hand of God or that God's Hand need not be present at all?

In conclusion, given all the above uncertainties the Copying Rule, when pressed into service as a natural God Detector, is unable to permit us to unambiguously detect any unique sign of the Hand of God.

Where does this leave the believer and the non-believer? Those who still wish to include the presence of a *God the Tinkerer* in the universe could still invoke a vision of God the *Intelligent Adapter*, playing an ongoing (but by no means exclusive or unique) hand in the survival and glorious diversification of life as well as the blossoming richness of human culture and technology. Those who find no need to place an actor like God in the picture can celebrate and seek to better understand the process of billions of years of evolution by cumulative copying and adaptation, made even more astonishing by the very fact that *no hand guided it.* Stuart Kaufmann may show us another way, in which he redefines God "...to mean the vast ceaseless creativity of the... universe" (Kaufmann 2008). If God is embodied in the artful adaptation on the output end of the Copying Rule then He is the agency of the seemingly miraculous processes of Natural Selection and Emergent phenomena.

11. Afterthought Experiment: Building a God Detector

What if our cumulative technology including computers, networks, robotics, sensors, and Cyberspace, is creating a set of tools which we can use to determine, once and for all, whether God exists? And if so, might we also be able to use these tools to determine God's nature and the exact means by which He manifests in the world? If we as a species could answer the question of the presence of deity in the world it would save untold future strife and focus our intellectual and artistic pursuits like never before.

What if we could "set a trap for God", a place where God could not resist manifesting His Will? What I am proposing is to engage all of the best programmers, artists and philosophers of our generation to create a gigantic network of software and computers, working to create a sort of "Evolution Grid" or "EvoGrid" (Damer 2008). This EvoGrid would start out as *God the Mechanic* (like Karl Sims' creatures) in which we build the simulation, set the initial conditions and then let the artificial ecosystem go from there.

Indeed, such a simulation might satisfy Richard Gordon's challenge in the chapter *Hoyle's Tornado Origin of Artificial Life, A Computer Programming Challenge* found in this volume. The EvoGrid would therefore seek to show that in amongst the vast machinery of the natural laws, and despite the chaos of R, the universe (or God acting within the universe) possesses the innate property to instantiate the Copying Rule and generate us.

However, the EvoGrid, could be set up to also embody some of Will Wright's *God the Tinkerer*, with people in the loop. The way this might work is that the creatures of this simulated ecosystem would systematically consume all of human language and culture available to them in the semantic flow of the Internet. Every piece of text, image, music or video, blog, or other cultural artifact would be both the landscape and foodstuffs for the EvoGrid. The creatures of the EvoGrid would continuously adapt to the myriad streams traveling along the growing cyberspace synapses of the collective human mind. The EvoGrid would communicate in its own language which we might be able to understand. If there was ever any medium through which God could speak to us, this would be it. Gerald de Jong claims that artificial life and the EvoGrid might be our way to finely polish a mirror we could then hold up to ourselves (de Jong, 2008). Would we then see the face of God?

12. Giving Birth to God

In our age old quest to detect and define God, there might be another ultimate outcome in store for us. Over the coming eons, would our own divine creations, such as the EvoGrid, allow us to merge with all living things, and transform and connect all of technological and biological reality? Would we then survive long enough to contact and combine with the EvoGrids of other sentient civilizations? If we never detected God in our own EvoGrid it would no longer matter because in some far distant time all sentient minds, biological bodies, and technological creations would ultimately merge into one total universal life form. If the universe succeeds to birth itself as one conscious entity, everything, including us and all of our past selves, will unify into a single being which we will then call… *God the Universe.*

So perhaps God is nothing more and nothing less than an expression of our hopes and dreams for that distant possibility.

"*God who created all things in the beginning is himself created by all things in the end*" *(Stapledon 1937).*

References

Adams, R. 1998, Is there an Artificial God, in the online proceedings of Digital Biota 2 available on the web at: http://www.biota.org/people/douglasadams/

Barbalet, T. 2007, Biota.org, available on the web at: http://www.biota.org

Brown, R. E. 1977, *The Birth of the Messiah*, Appendix V: The Charge of Illegitimacy, Doubleday & Company.

Damer, B. F. 1997, *Avatars: Exploring and Building Virtual Worlds on the Internet*, PeachPit Press, Berkeley CA.

Damer, B. F. 2008, The EvoGrid, an Evolution Technology Grid, available on the web at: http://www.evogrid.org

Dawkins, R. 1976, *The Selfish Gene*, Oxford University Press.

Dawkins, R. 1986, *The Blind Watchmaker: Why the Evidence of Evolution Reveals a Universe Without Design*, W.W. Norton & Company.

De Jong, G. 2008, from a conversation on the Biota Live Podcast #22, June 20, 2008, *What is the Philosophy of Artificial Life?*, available on the web at: http://www.biota.org/podcast

Dyson, G. 1997, *Darwin Among the Machines: The Evolution of Global Intelligence*, Perseus Books.

Farmer D. J., d'a Belin A. 1991. Artificial Life: The Coming Evolution. In: Langton, C., C. Taylor, J. D. Farmer, & S. Rasmussen [eds], *Artificial Life II, Santa Fe Institute Studies in the Sciences of Complexity*, vol. XI, pp. 815-840, Addison-Wesley, Redwood City, CA.

Gordon, R. 2000, The emergence of emergence: a critique of "Design, observation, surprise!", *Rivista di Biologia /Biology Forum*, 93(2), 349-356.

Wright, W. 2008, From a public conversation with Will Wright, designer of Spore, NASA Ames Research Center, Virtual Worlds Workshop, January 26, 2008.

Kauffman, S. 2008, *Reinventing the Sacred: A New View of Science, Reason, and Religion*, Basic Books.

Kurzweil, R. 2005, *The Singularity Is Near: When Humans Transcend Biology*, Penguin Books.

Langton, C., C. Taylor, J. D. Farmer, & S. Rasmussen [eds], 1991, *Artificial Life II, Santa Fe Institute Studies in the Sciences of Complexity*, vol. XI, pp. 815-840, Addison-Wesley, Redwood City, CA.

Levy, S. 1993, *Artificial Life: The Quest for a New Creation*, Vintage Books.

Lloyd, S. 2006, *Programming the Universe: A Quantum Computer Scientist Takes on the Cosmos*, Alfred Knopf.

Negroponte, N. 1995, *Being Digital*, Alfred Knopf.

Pagels, E. 2003, *Beyond Belief: The Secret Gospel of Thomas*, Random House.

Sims, K. 1994, Evolving Virtual Creatures, *Computer Graphics SIGGRAPH 1994 Proceedings*, pp. 15-22.

Stapledon, O. 1937, *Star Maker*, Methuen, UK.

Dialogue

Tom Barbalet

Re: Two Kinds of God in the A-life Universe

As is frequently discussed through our recorded conversations[1], I think the one dimensional continuum between Karl Sims' and Will Wright's creations is problematic.

In an applied critique, Will Wright's Spore is not an artificial life simulation. It is a series of procedural games with a shared theme. Whilst it was originally marketed as a surreal biological evolution game, Spore's marketing has changed focus. From your [Bruce Damer's] and Rudy Rucker's recorded prompting of Will Wright[2], Wright failed to use the term artificial life and in fact showed little interest in contemporary artificial life in his analysis.

Let's take an idealistic view of Karl Sims' work and Will Wright's game. Karl Sims' work is still based in a created environment that he set up to demonstrate a series of ideas. In contrast Will Wright's game allows tens if not hundreds of thousands of users to design creatures to interact. There is one intelligent designer in Karl Sims' environment and sufficiently many in Will Wright's game to be considered noise rather than intelligence.

In addition to the problems of the idealistic view's deconstruction there are many examples which fall outside this one dimensional "A-Life Universe".

Even if the designer is intelligent, the designer does not have to manufacture intelligent creations. In this regard Karl Sims is more sympathetic than Will Wright. The early Grey Thumb experiments[3] offered some interesting results too. Here a small collective of hobbyists spent a Saturday or two trying to produce a more refined version of Tom Ray's Tierra[4]. My understanding is that these experiments either led to the creation of, or heavily utilized, Adam Ierymenko's Nanopond[5].

This idea of the most potent artificial life soup, where digital micro-organisms can develop, seems to be a much lower level attempt to model artificial life than the engineered example of Karl Sims or the multi-million dollar Will Wright creation.

Re: The Copying Rule and Memetics

Through the discussion and the table, I'm not clear if rocks are inherited from a single Platonic rock or whether rocks exist through a geological naming convention, independent of the Copying Rule. Underneath the Copying Rule there can be many different kinds of chemistry and as attributed to human culture I would assume psychology.

In terms of the "new field called memetics", memetics is as old as I am. In fact, truth be told, memetics is slightly older than I am. I digress.

As genetics is an established field, memetics is not. There are a number of problems using memetics as cultural genetics. I'd like to offer a few of them here.

Genetics as science offers predictable and usable metrics. Memetics offers no predictive analysis. If memetics were cultural genetics, it should be extraordinarily useful to predict cultural trends. Yet it seems to be used to explain things after the fact, not into the future.

Death of an individual is central to genetics. The fact that there are life cycles is critical to genetics. There is nothing analogous to the death of an individual in memetics.

The definition of a meme as a cultural gene means that it should be part of a cultural idea, not all of a cultural idea. In fact what is commonly referred to as a meme should be a number of memes. There is nothing analogous to a cultural atom or gene through memetic analysis.

Re: The Monk

The Monk is about entropy (human inertia) which offers an improvement to memetics.

The Monk exists today in the archetype of the popular historian discarding unimportant facts, the journalist leaving information they don't understand from a story and fundamentally the retelling of a story with particular emphasis to the point of falsifying the story.

The "virgin" error was a remarkable marketing tool that the editorial monks could have encouraged rather than missed reviewing. In one part, entropy (human inertia) forces a reduction and, in another part, the need to produce more excitement in the reduction. This isn't genetics, this is entropic physics. You can derive genetics from entropic physics. I would put to you that entropic physics are more primary than genetics and can be used to construct a truer cultural genetics.

References

[1] Barbalet, T. et al. (2006-2008), Interviews and Conversations from Biota.org, http://www.biota.org/podcast/, Specifically: "Can Artificial Life Explain the Cambrian Explosion?" with Roy Plotnick.
[2] http://www.biota.org/podcast/biota_wright_rucker.mp3
[3] http://www.greythumb.org/blog/index.php?/archives/9-Getting-started-in-computer-based-artificial-life.html
[4] http://www.his.atr.jp/~ray/tierra/
[5] http://www.greythumb.org/wiki/Nanopond

Biography

Tom Barbalet created the Noble Ape Simulation in 1996 and continues its development to this day. Noble Ape is used by Apple and INTEL as well as a number of universities to teach biodiversity, multi-media education, vector processing, real-time graphical interfaces and a number of other technologies. He has been the editor of Biota.org (since 2005), a leading community resource for artificial life developers. He is the host of the weekly Biota Live internet radio show where he discusses a variety of topics relating to artificial life, artificial intelligence and simulation philosophy with a variety of guests. He is also the co-chair of the International Game Developers' Association's Intellectual Property Rights Special Interest Group.

6
Welcome to the Simulation

TOM BARBALET

Las Vegas NV USA
tom@nobleape.com

1. Introduction

I was awakened from my dogmatic slumber by the Nick Bostrom paper, "Are We Living in a Computer Simulation?" (2003). Simulators may want to be pigeon fanciers waiting for Darwin (Secord, 1981). I believe simulators can be both pigeon fanciers and Darwin.

2. Field of Reference

When I lived in England I would sit in my study and look out on a field. It was a large field that was sometimes used as a fair ground, sometimes as a sporting field and at least once a year as the town's location for fireworks. Aside from a large grassy center that could easily accommodate a football game and a group of spectators, there was a good combination of wild plants and some cultivated trees on the fringes. In the far corner of the field was the town's garden lots, where townsfolk would grow their vegetables in small plots. Much to the discomfort of people who strayed from the surrounding pathways there was poison ivy growing close off the path.

In all seasons, but winter, the field would be a sea of rich green vegetation. There was a good quantity of insects and a vast slew of birds that would fly around the field showing their aerobatic prowess. Following frequent rain, these birds would spend hours pulling worms from the field. It was clearly a rich source of organic production to support such a healthy worm population.

In the winter, the field was covered with snow. I peered onto it through the ashy silhouettes and complex geometries of the exposed tree limbs guarding the field. Footprints and paw prints would be left in the snow as a memory of humans, dogs, foxes, and prints that could only be attached to wolves. The prints in the fresh snow served as a memory of the life that moved through the field.

Fifty years before I sat looking over the field, Alan Turing lived in the town (Hodges, 1983; Copeland, 2004). Although it was the latter part of his life and he had already broken the Nazi's Enigma codes (in the process inventing modern computing), I imagined him walking through the field. Perhaps he had worked in the garden lots or perhaps he had watched a local football game being played.

About five minutes walk down the street from the field was a chain of ponds and an English swamp, affectionately known as a bog by the locals. There a prehistoric man had been found and named after the bog - Lindow man (Robins & Ross, 1989). Strikingly, the man had been exceptionally well groomed for an Iron Age man. Lindow man had no-doubt walked through the Iron Age version of the field on the way to his near-eternal preservation. Perhaps he had been part of a large group, perhaps he walked alone. He could have died naturally and been buried there. Maybe the field had been the site of Lindow man's home. What had this man seen? What kind of conflict had he experienced? Were his genes flickering in the local population, even now, crossing the field?

The field and its surrounds represented a theater of life. Every aspect of the field provided some connection with life. So while life may have an ethereal descriptive quality, every aspect of the field and its surrounds could be modeled with a simulation derived from observation; the field's operation in relation to the town as a social gathering place, the rationale of the human placed tree cultivation and the vegetable plots, the growth and variety of the competing ecosystem of plants and insects through summer, the paced routes of humans and animals through winter, also the chemical and bacterial interchange that preserved Lindow man, not to mention the possibilities of the Lindow man Iron Age society, and, even Lindow man's genetic descendants through to Turing and the snow walking humans.

Surely replicating the observable qualities of the field would yield no additional information?

Such a process is analogous to an animated film or perhaps a historical re-enactment where you can move through the observations but get no more information about history except through observation. You can't say to the re-enactor, "If your side had lost the battle rather than won, what would the world look like today?"

Simulated life solves this problem in the movement from creating environments that have elements of realism into what-if experiments — exploring limiting and expanding parts of the simulated environment. Rather than merely recreating observable life, simulation provides an environment where what-if experiments can be enacted.

There may be a bias to presuppose simulation refers to something deterministic, designed by intelligence or in some way constructed to be based on a small set of rules. This isn't the way the word simulation is used here. In fact, the simulations described in the field can (and should) in most cases be chaotic.

The purpose of this text is to argue that simulation developers (simulators) aspire to perfect their simulations to the point that simulation is reality. This change of reference where reality is simulation enables a better understanding of contemporary and future simulation development and also provides additional analytical tools which are not available if simulation is considered as a created model of a small fraction of reality.

3. Simulation Metaphysics

The word simulation comes from the Latin *simili* which means like. A simulation is a system constrained to reproduce another system.

Suppose you wanted to build a time tracking simulation of the Mississippi River to look at shipping along the river in the late nineteenth century (Twain, 1883). You could take a piece of paper and draw the traversing Mississippi River along the paper identifying points where the boats would stop. You could then move counters along the Mississippi River route to show the shipping commerce. You created something like the Mississippi River but rather than using the waterway itself or collecting information directly from the waterway, you created a simulation on paper of the waterway and the routes taken. It may seem rather strange to say by adding the route of the Mississippi River to the piece of paper you were constraining the paper in some way, but you were adding constraints to the system that the paper created. You could, however, add a number of fictitious tributaries to the Mississippi River and speed up some sections of water commerce. This identifies how the constraints on the simulation can be expanded to allow for greater freedom.

A computer simulation is another good example. If you wanted to simulate the interaction of two marbles striking, you would write a computer program that provided mathematical constraints. The idea of constraint is critical to understanding the purpose of a simulation (Winsberg, 1999). The computer is just as programable to make the marbles sprout arms and legs and have a multi-round boxing match. The constraints are designed to make the simulation close to what is being simulated. You want the simulation "like" the simulated.

Freedom (in a simulation) relates to possibility rather than actuality. For example a computer has near limitless possibilities which makes it an ideal simulation tool. We shouldn't be predisposed to thinking of simulation solely as part of some mechanistic construction. Simulations don't have to just exist within a computer or through the imaginations of intelligent agents as the paper Mississippi River example shows. The most powerful simulations can exist in organic environments by placing constraints on the participants. These constraints can be physical, they can exist in social mores, written laws, best practices and perceived constraints.

With this broad definition of simulation, it is possible to look at road systems as simulations, legal systems as simulations, the financial systems — from buying a pint of milk to the stock market — as simulations. In fact we are the willing and unwilling participants in a number of simulations that shape every aspect of our lives.

Aside from the appeal to the Platonic form of simulation as reality, an appeal for our collective participation in simulations could come through a contemporary Cartesian Cogito, ergo sum;

I am constrained, therefore I am simulated.

Consider that our environment is far richer an environment than we could possibly utilize. In fact our environment is constantly being constrained not by itself but by our interactions with the environment. The simulation dichotomy of two environments — the primary environment and the simulation environment — requires the simulation environment to be far freer than the primary environment. The simulation environment has the addition of constraints to bring it to the primary environment.

If the real world is the primary environment, the simulation of the real world will also allow for possible worlds. But through the constraints of the real world, the simulated world is the primary environment. Simulation allows the exploration of possible worlds as well as the real world.

4. Alive without Intelligence

The word "alive" has a number of different meanings through contextual use. If we ask the question "Is simulated life alive?" the first problem that is posed is which meaning of alive to use in the question. It would be trivial to define alive as being something biological — requiring respiration — or something that required a physical presence. These kinds of definitions of alive would result in an immediate negative in terms of simulated life being alive.

Aside from our intimate connection with "alive", we observe others around us that are "alive". The observation is easiest to see in those around us in physical proximity, both those we know and have conversed with, and also the people we may see driving or walking down the street. This observation is important because we don't need to interrogate other entities in order to establish that they are "alive". We don't need to talk or interact with people on the street in order to establish they too are "alive". It is something we are very familiar with on an implicit level. This idea of "alive" is also applicable to people we see in films and on television. In fact, when we see human actors — whilst we understand they may be performing a play — they are "alive".

This characteristic of "alive" becomes particularly interesting with animated cartoons. The actors that give the voices to the cartoons are "alive". But are the cartoons "alive"? This is a question of suspension of disbelief. This is particularly prevalent with films or television programs that use computer generated actors that interact with real world actors. The better the computer generation, the easier it is to suspend the disbelief and assume for an instant (or longer) that all the participants — human and computer generated — are "alive".

Through this analysis, "alive" can be independent of scripting. Even if we are watching a play, we know the actors are "alive". We can also acknowledge that there can be computer generated entities that also have this "alive" characteristic. This is particularly important for simulated life because "alive" here can have one of two components. Either the entity needs to look good — look "alive" — or they need to act well — act "alive". Now whilst all simulated life developers strive to create stunningly beautiful representations of their simulated agents, the characteristic simulation developers strive to create is something that acts "alive". That is the aim of simulated life.

It would be fair to presume that making something that acted "alive" required intelligence, intelligent design or an intelligent designer.

There are two simple means of showing this isn't the case and that intelligent design is not part of the simulated life developer's process. The first comes through paleobiology and the second comes through a simple thought experiment.

The fossil record shows that there was life before there was intelligence. Intelligence is a cunning adaption that aids life immeasurably, but life clearly existed before there was any intelligence in the life. The paleobiologist Roy Plotnick (Barbalet, Daigle, Kerr, DeJong & Plotnick, 2006) narrates this beautifully by describing the pre-Cambrian floating fauna that slowly started adapting to floating between feeding grounds. It was this understanding of feeding grounds that allowed for the amazing variety of fauna that came through the Cambrian period. The most primitive marker for intelligence comes through changing movement to optimize for feeding grounds. This was the first glimmer of intelligence but life came first. If you were to simulate this period, you couldn't do it if you designed with intelligence. Intelligence is the outcome you would desire, but if you wrote it in explicitly you would defeat the experiment.

The thought experiment for this is relatively easy too. Consider what life is fundamentally, not over generations, but in an instant. Life is an ability to survive through a changing environment. Consider that the changing environment is chaotic — there are things that may be predictive in some regard — but fundamentally there is just chaos. So life is an ability to survive chaos.

This definition of life is removed from intelligence. Intelligence is very good at understanding the predictive part, but intelligence can't quantify and work through chaos. There needs to be something that is sub-intelligence that deals with chaos.

There are a number of solutions to this problem. The most trivial example comes through tuned dynamic equations — which are clearly not intelligent but sufficiently reactive to act "alive". From this, there are higher orders of response which act "alive" too. This emerges into patterns which look intelligent but the components that create this perceived intelligence are dynamic and not something that could begin to be quantified as intelligence.

In order to understand this with real world examples, consider a moving object on a fluid — a surf board with a surfer — or a moving object that needs adjustment for balance — a bicycle with a rider — these are fundamentally dynamic and the responses to maintain them upright have to be equally dynamic and not intelligent. It is this reflex which the simulation developer needs to master if their creations are to act "alive".

Consider a surfer or a cyclist that thought about their response to the dynamic environment. This is a starting surfer or beginner cyclist fundamentally. It is also a response that would not yield balance. If you start from intelligence without reactive dynamics, the agent may be able to act clumsily, but they won't act "alive".

5. Intelligence and the Game Hunter

I have offered the briefest definition of intelligence as being a continuation of the survival element of life, life being the ability to survive chaos. With this definition a number of interesting properties of intelligence can be found.

Whilst this is an implicit property of this definition, I think it is important to slay the dragon early in this exploration. The human has no primacy in terms of intelligence. As someone who has developed artificial intelligence (in a simulated environment) for a number of years, I've always found the popular obsession with human intelligence rather curious. In fact it is a good litmus test for hucksters in the artificial intelligence community when they start talking about how we are looking to create a digital version of human intelligence.

When I have been interviewed about this issue, I have returned to the following parable (Barbalet, Trumbule, VanNuys, 2007):

Two horses are standing in a field overlooking a freeway. One horse looks at the other horse and then looks out at the cars driving along the freeway and says,

"Sure they are fast, those cars, but they'll never be horses."

This also illustrates a secondary point that people developing artificial intelligence rarely look to make human intelligence when they have something faster, and through utility, superior.

Once you lose the pinnacle of human intelligence as the height of intelligence you start to realize something very powerful. There are a lot more intelligent systems out there.

Looking at the survival property of intelligence, you start to find systems which hold this survival property very strongly. A personal favorite of mine is the road system. If you have ever driven through an area of natural disaster, you'll realize that it is amazing how felled trees are moved or new disaster roads are created. The need for movement produces a survival in the road system which is quite fantastic. Taxes, motor vehicles and roadworks crews go into making the road system, that's true. There can be no bias against systems that require additional components. The human brain requires a huge conditional upkeep in order to keep it functioning and until a few paragraphs ago, it was the pinnacle of intelligence.

There are three additional systems I like to discuss as simulations of intelligence. The legal system, the financial system and the information system containing the Internet. The latter is a favorite point of discussion particularly considering how one would begin to rate the intelligence of the Internet.

How do you begin to rate the intelligence of these systems?

I have a slight bias in thinking that all these systems including the road system are vastly more intelligent than humans. They have certain elements of fragility without question but they exist and survive for far longer and achieve far more through their survival than any human.

A pragmatic quantity of intelligence may relate to the number of humans it would take to stop a particular system. This is the big game hunter calibre test (Roosevelt, 1909) for intelligence.

It took but ten humans to slay this intelligent system.

Because this is such a rough measure, I'd like to empower the logarithm base ten as the way to fairly weight this measure. So in the case of ten humans to slay the intelligent system, it would have the intelligence value of 1, a hundred humans then the value of 2, and so on.

Whilst these ideas may seem a little radical, I am not appealing to conspiracies. I understand the road system is probably best measured with regards to a functioning city. Living in Las Vegas, I have observed the road system can survive with 10 human caused obstructions (Las Vegas Metropolitan Police Department, 2008) but I doubt it could survive with 100. So the intelligence value of Las Vegas' road system is somewhere between 1 and 2.

Similarly, I wouldn't suppose that there is an international legal system. But within a small county legal system, I have recently seen hundreds of immediate cases can effectively shut the legal system down (Dougherty & Holusha, 2008). This would indicate an intelligence value of a small county legal system as being somewhere between 2 and 3.

Take a large city's legal system – the effective division of labor – and this number would no-doubt be a lot larger.

When you consider a national or the global financial system, you begin to realize that it takes millions of humans to have but the tiniest effect. Now it could be argued that times of interpreted catastrophe involving a relatively small number of people can cause fluctuations in the global financial system but the metric is based on slaying the system, not leaving it fluttering.

How did these systems become so powerful?

There appears to be a combination of factors — a lot of time, divergent and competing engineering principles. Fundamentally a "survival first" pragmatism. This pragmatism isn't intelligent design, it's life.

6. Computational Power

When I started developing Noble Ape, I used the computers that were readily available to me as a nineteen year old university student of limited means. Not the university's computer resources, but the kinds of machines I would keep running near me for regular updates when I had a moment to spare. Machines that I could find for next to nothing (Barbalet, 2005b). I started developing Noble Ape on technology that was already seven to ten years older than the state-of-the-art of the time. For this reason, I have seen the same fundamental simulation run on more than two decades of advancing hardware. In addition to this, I have also had the privilege of working with some of the most brilliant hardware and software engineers at Apple Inc. (since 2003) and INTEL (since 2005) with Noble Ape (Barbalet, 2005a). Looking at how INTEL uses Noble Ape in particular I have a deep respect for the advances in contemporary computer hardware.

It is difficult to quantify the advances in computational power over the past twenty years because all applicable popular cliches fail to grasp the increase in power but also the kinds of computation that can currently be achieved which would never have been considered possible even a decade ago. In this context, the greatest advances of recent years have also provided the revolution that simulation (as discussed in this text) must embrace. This is the idea of atomic computation – not of quantum computing – but of taking computer algorithms and translating them into something that can be run in a vast parallel way over a number of different processing cores and potentially over a distributed network. In addition to this, the advances in vector processing means that single processor cores can now be thought of as multi-stream processing pipelines.

This kind of processing power is inconceivable in a popular context. Contemporary computing hardware has the power to simulate anything we would wish to throw at it. The simulation problems merely need to be optimized for the current processor architectures. This is a non-trivial exercise.

The problem with translating contemporary computing power to something which is meaningful for simulation is a software problem. Software has always been the great lag in terms of translating the power available in computing to a popular audience. Ironically the simulation systems discussed – the financial system and the information system containing the Internet – utilize this power considerably greater than simulated environments that would translate well to human-scale understanding.

Consider the Newtonian model of the physical world. Mathematics was never the shortfall in the adoption the Newtonian model. No one ever said to Newton, "Your physics is great but unfortunately the mathematics isn't powerful enough yet to run your physics." The same is true with contemporary computing. The power is there to run vast simulations. The issue currently is there is a need for new mathematics to translate these simulation environments into a context that contemporary computing can embrace.

This is a challenge.

7. New Science

Simulation has existed in science prior to computation. The broad definition of simulation offered here includes most (if not all) of experimental science to-date. A confined hypothesis-testing experiment is a beautiful subset of the definition of simulation I've offered.

Computer simulation for testing specific scientific hypotheses isn't new either. To frame the problems discussed here, the simulation discussed in this text – with regards to the language, philosophy and mathematics of simulation – doesn't refer to serial computer simulation to find particular variables or test a simple hypothesis. Simulation in the context of this text relates to vast parallel simulation that has only been computational reality in recent years.

This text is arguing that the ground rules have fundamentally changed and contemporary computing can offer insights back to science which could never have been considered with the "simulation to test scientific hypotheses" interface (Barbalet, Damer, Gordon & Schafer, 2008).

The value of implementing existing science in simulations (as a starting framework) relates to a few basic problems. It is important to note the scope of science in this context because the use of mathematics in simulation is critical. The applied mathematical analysis that physics offers is also critical but one of the main problems with science in the context of simulation can be described through writing simulation that solely replicates physics. To-date this has been shown to be problematic outside simple physics simulation (Barbalet & Damer, 2007). There is nothing wrong with simple physics simulation but it isn't the same context of simulation discussed here.

The kinds of problems in physics relating to complex systems are the bread and butter of contemporary simulation. There is also a boundary condition problem with simulation of purely physics (or chemistry, or molecular biology, or another existing non-simulation derived scientific method). The equations and ideas from traditional science are created in a tight frame of reference and generally don't translate well to simulation.

In the face of these boundary conditions, simulators have used a wide variety of modeling methods that either under-approximate the effects of physics (for example) or even better allow the simulation to push the boundaries of contemporary physics. It is the ability for simulation to prompt and develop new science that is particularly exciting for many simulators and scientists alike but requires a new kind of thinking (Barbalet, Damer, Gordon & Schafer, 2008).

There are two major obstacles to this new science.

The first relates to the mathematical language of simulation. Whilst there are a number of traditional computational methods, contemporary computing creates new shortcuts which traditional computational methods underutilize. As much as this text is a call to change perspective on the philosophy of simulation, it is also a plea for improvement in the mathematics of simulation. As physics was based on mathematics, simulation is based on computation. In short, mathematics plus atomic computation – vast parallelism, intercommunicating mathematical processes with shared and local memory – is the new simulation paradigm that requires a new kind of mathematics.

The second obstacle relates to the corrosive nature of simulation on traditional methods. In short, science will need to have the capacity to change in order to embrace the potentially radical insight that contemporary simulation can offer.

The first and the second obstacles are fundamentally intertwined because there needs to be an education process that goes hand-in-hand with the new simulation methodology. The outline and methods of this education are outside the scope of this text. Except to say that the philosophical problems contemporary simulation offers to metaphysics are dwarfed by the education problems contemporary simulation poses.

There is a lead-by-example solution to these obstacles. If the new mathematics were in place, if the first obstacle has been overcome, the second obstacle melts away as the insight begins to flow. In this regard, contemporary simulators should focus on communicating and developing the new mathematics. The people who hold the best initial capacity for this are the super skilled employees of contemporary semiconductor manufacturers. My work with engineers at INTEL has identified the capacity for the new mathematics through these kinds of developers. Hopefully the new mathematics can move outside a proprietary context.

8. Moving the Discussion Forward

It may seem rather curious that a text of this nature would appear here in the context of science and belief. Within the simulation community there has been a lot of discussion about whether simulators are intelligent designers. My own view is that physicists and mathematicians don't have such a burden about their endeavors (Barbalet & Daigle, 2008). They just get on with doing what they are doing. In this regard simulators should follow suit and explore the amazing new scientific and philosophical landscape contemporary computer simulation offers.

This new landscape is about computation exploring aspects of the real world that may be discussed at a very high level but are still fundamentally unknown. Independent and inquiring minds should begin to push boundaries and explore new possibilities — simulation is the way forward.

References

Barbalet, T. S. (2005a). Apple's CHUD Tools, INTEL and Noble Ape. Noble Ape Website. Retrieved June 20, 2008, from http://www.nobleape.com/docs/on_apple.html

Barbalet, T. S. (2005b). *Original Manuals, Noble Ape 1996-1997*. San Mateo, CA: Cafe Press.

Barbalet, T. S., Daigle, J. P., Kerr, D., DeJong, G., & Plotnick, R. E. (2006). Can Artificial Life Explain the Cambrian Explosion?. *Biota Podcast*. Retrieved July 21, 2008, from http://www.biota.org/podcast/biota_020307.mp3

Barbalet, T. S., & Damer, B. (2007). Bruce Damer Chat. *Biota Podcast*. Retrieved July 21, 2008, from http://www.biota.org/podcast/biota_102307.mp3

Barbalet, T. S., Trumbule, J., & VanNuys, D. (2007). Shrink Rap Radio Live #3: Artificial Life and Artificial Intelligence. *Shrink Rap Radio*. Retrieved July 21, 2008, from http://www.nobleape.com/transcript/srr126.html

Barbalet, T. S., & Daigle, J. P. (2008). More Outstanding Questions. *Biota Podcast*. Retrieved July 21, 2008, from http://www.biota.org/podcast/biota_052308.mp3

Barbalet, T. S., Damer, B., Gordon, R. & Schafer, S. (2008). the EvoGrid (May Update). *Biota Podcast*. Retrieved July 21, 2008, from http://www.biota.org/podcast/biota_053008.mp3

Bostrom, N. (2003) Are We Living in a Computer Simulation? *The Philosophical Quarterly*, Vol. 53, No. 211, pp. 243-255.

Copland, B. J. (2004). *The Essential Turing*. Oxford University Press.

Dougherty, J., & Holusha, J. (2008) Chaotic Hearing in Polygamist Case. *New York Times*. Retrieved July 21, 2008, from http://www.nytimes.com/2008/04/18/us/17cnd-polygamy.html

Hodges, A. (1983). *Alan Turing: the Enigma of Intelligence*. London: Burnett Books.

Las Vegas Metropolitan Police Department (2008). *Current Traffic*. Retrieved July 21, 2008, from http://www.lvmpd.com/cgi-bin/traffic/traffic.pl

Robins, D., & Ross, A. (1989). *Life and Death of a Druid Prince*. New York: Simon & Schuster.

Roosevelt, T. (1909). *African Game Trails: An Account of the African Wanderings of an American Hunter-Naturalist*. New York: Syndicate Publishing.

Secord, J. A. (1981) Nature's Fancy: Charles Darwin and the Breeding of Pigeons, in *Isis*, Vol. 72, No. 2 (June, 1981), pp. 163-186.

Twain, M. (1883). *Life on the Mississippi*. London: Chatto & Windus.

Winsberg, E. (1999). Sanctioning Models: The epistemology of simulation, in Sismondo, S. & Gissis, S. (eds.) (1999) *Modeling and Simulation*. Special Issue of *Science in Context* 12.

Biography

Stephen P. McGrew received a Master of Science degree in Physics from the University of Washington. Steve is an inventor/entrepreneur who developed many of the technologies used for making holograms; and he is co-founder and a board member of GenPrime Inc., a biotech firm specializing in high-speed, high-sensitivity microbial detection technologies. He developed Generator™, a genetic algorithm software package widely used in diverse fields; and initiated a successful project at Eastern Washington University to evolve bacteria to accumulate and concentrate zinc. His interests are evolutionary theory, geology, paleontology, philosophy of science, and science education.

7

Darwinian Evolution: A Practical Tool in Industry

STEPHEN P. McGREW

President, New Light Industries, Ltd.
9713 W. Sunset Hwy.
Spokane WA 99204 USA
stevem@nli-ltd.com

1. Introduction

I'll start this chapter with a sad story.

> Nearly 20 years ago I placed an ad to hire an administrative assistant for a research and development company I was launching. After winnowing down several dozen resumes to ten, I began interviewing applicants. After asking the usual questions about experience and skills, I told each applicant, "At our company we use a computer program based on Darwinian evolution as a crucial tool for things like designing electronic circuits and for scheduling. Would you be able to work in that environment?"
>
> About a fourth of the applicants said they would have to ask their minister or spouse for permission.

Although the very idea of evolution frightens or offends some religious fundamentalists, evolution is useful. In the "real world", the important measure of a process or principle is to see if it is useful. Do machines based on the principle work well? Do systems using the process do what they are intended to do? Can we use the principle to make money?

In fact, Darwinian evolution does make money. It is a valuable tool for the engineer who wants to design a better machine or the financial consultant who wants to maximize clients' investment income. Computer programs based on evolutionary principles can solve difficult problems, often coming up with solutions no human being would be likely to discover. A "Genetic Algorithm" or "GA" is one kind of evolutionary computer program deliberately structured to closely mimic Darwinian evolution and is designed as a general-purpose problem solver.

In the following pages I will describe briefly how GAs are structured, how they are used, and how they perform.

2. Background: Structure of a Genetic Algorithm

Though evolutionary computing methods can take on a wide range of different forms, the term "Genetic Algorithm" or "GA" refers to a specific class of algorithms having the following properties:

a. An initial population (a first generation) of trial solutions is created at random, each trial solution (regarded as an "individual") represented as a string of numbers or symbols.
b. A method is provided for gauging how well each individual performs in comparison with others, to yield a "fitness" value.
c. New trial solutions (new individuals) are formed by "recombining" portions of number or symbol strings representing two "parent" trial solutions. (This is something like a simulation of the progeny resulting from a sex between the "parents".)
d. Each new set of "children" are subjected to mutation, which amounts to randomly changing a small fraction of the symbols in the string.
e. Higher-fitness individuals are selected at random from the resulting generation to survive for the next round of reproduction.
f. Generations are formed iteratively by repeating steps c, d and e over and over until a "good enough" solution emerges.

GA users refer to the strings as "chromosomes" and to the symbols as "genes"; and they refer to the typical recombination mechanism as "crossover".

A GA exploits a simplified, but useful and reasonably accurate, version of the process Charles Darwin proposed in his evolutionary theory: selection acting on random variation. Random mutations and random recombinations produce new genes and genomes and therefore new individuals; and allocating more opportunities to higher-fitness individuals to produce offspring is analogous to natural selection.

Of course, in the nearly 150 years since 1859 when Darwin published *Origin of Species*, we have learned a lot about the mechanisms of genetics and about complex system dynamics. Taken as a model of Darwinian evolution, a Genetic Algorithm does not take into account the myriad of factors that influence mutation rates. A GA normally does not use the relatively complex process of meiosis, recombining genetic material from two maternal grandparents and two paternal grandparents to construct chromosomes inherited from mother and father (i.e., haploidy rather than diploidy is usually assumed). A GA normally does not model the interactions between individuals in a group, nor does it model mate selection, ecological interactions, geography or other complex factors that contribute to natural selection. However, a GA does exhibit the overall dynamics of recombination, mutation and selection, the foundations of the modern theory of evolution.

3. A Toy Example: Use a GA to Kill Wasps

Suppose you have a plot of ground 100 meters long and 100 meters wide; and on that plot are scattered twelve wasp nests whose locations and sizes are known. One day you set out to eliminate as large a percentage of the wasp nests as you can, by placing cans of insecticide at three specific locations on your plot. Suppose further that you have three cans of insecticide, each with a particular kill radius. You haven't studied mathematics, so you don't know how to calculate exactly where to put the cans for optimum effect. How can you solve the problem? Use a Genetic Algorithm!

This is actually a rather simple problem to give to a GA. A trial solution can be a set of six numbers: two X,Y coordinates specifying the location of each of three cans. Those six numbers are the symbols in a string representing the trial solution. The "fitness" of the trial solution simply can be the percentage of wasps killed if the cans are placed in those locations. Mutation can be a random change of the X,Y coordinates. Recombination can be accomplished by randomly grabbing some of the 6 numbers in the string representing one parent, and exchanging them for corresponding numbers in the string representing the other parent.

So, one trial solution might be "A": (20, 17, 86, 3, 75, 25) which means the first can is set at a location 20 meters East and 17 meters North of the southwest corner of the plot. The second can is set at 86 meters East and 3 meters North; and the third can is set at 75 meters East and 25 meters North. If trial solution "B" is (80, 20, 27, 33, 81, 81), then the parents A and B could produce a child by recombination, for example "child" C: (80, 20, 86, 3, 75, 25). A small mutation rate might change one of the numbers, making the new child have the coordinate values of (80, 23, 86, 3, 75, 25).

Your objective is to kill as many wasps as possible, so "fitness" corresponds to the number of wasps the cans of insecticide can reach. Fitness is calculated by counting how many wasps are within range of the three spray cans in your plot of land.

Rank the members of the population according to their fitness. Give each member a chance to pair up with another member to produce a child, but let the number of chances depend on fitness ranking. Start the GA out by giving it a population of 20 members (individual trial solutions), each consisting of a random string of six numbers between 0 and 100. Start with 20 members in the population, produce 20 children from those members, and use those children to form the population of the next generation.

Note that the GA does not have any information about how to solve this particular problem. It is not designed for this problem; it merely has a fitness measure plus Darwin's mechanisms of recombination, mutation and selection.

Most importantly, the GA works beautifully. I ran this problem in Generator a moment ago; and it found an optimal solution in a mere 100 generations (about 75 seconds).

4. Real Applications

Genetic Algorithms are used in a staggering range of different engineering applications. One of the most interesting to me is in controlling quantum resonant systems: adjusting the structure of a laser pulse so that when the pulse strikes a particular type of molecule, the molecule is placed into a specific excited state.

Most users of Generator are financial planners – investment portfolio managers, for example. If their task is to maximize the expected profit from a set of investments while keeping risk within an acceptable range, they have an optimization problem roughly analogous to the Wasps example. "Genes" correspond to the amount of money allocated to each investment, while "Fitness" corresponds to the expected profit (perhaps divided by how much the risk exceeds the allowed value). Recombination works approximately the same way as in Wasps, except that the sum of all the investment amounts must add up to the fixed dollar value of the portfolio. That can all be set up in an Excel spreadsheet rather easily; and Generator finds an optimal solution within a few minutes merely by executing the same steps Nature uses in evolution: recombination, mutation and selection.

The first user of Generator was a graduate student working on her Ph.D. in mechanical engineering. Her dissertation compared different types of optimization algorithms for designing a mechanical linkage for a fork lift tractor. The Genetic Algorithm outperformed all of the other algorithms she tried.

I have used Generator to design electronic circuits, lens systems, and optical coatings. Occasionally I use it to solve mathematical puzzles posed in popular science publications. Other people have used it to discover the chemical processes operating in a gas mask, to determine the best routes and locations for roads and culverts in a state park, and to optimize a model for interactions between various aquatic species in a river system. Aircraft designers use GAs to design seats for their airplanes; NASA has used GAs to optimize scheduling and logistics.

5. Other Uses of Evolution in Industry

A GA differs from natural evolution in two very important ways. First, in a GA reproductive success is allocated to individuals according to how well they meet externally imposed fitness criteria, while in natural evolution fitness is simply a measure of reproductive success without any particular predetermined goal or direction of progress. Second, in a GA all the recombination, mutation and selection occurs in a computer while in nature the same processes occur in "real life".

Plant and animal breeders artificially impose criteria for reproductive success on populations of organisms. Dogs are bred for size, color, personality, shape or intelligence, by selecting individuals with desirable characteristics to produce the next generation of dogs. Cattle, sheep, llamas, and chickens are bred similarly to maximize meat and milk production, wool quality, or egg production. Flowers and garden vegetables are selected and artificially pollinated to improve size, color, shape or flavor. Darwin recognized some of this in Chapter 1 of *Origin of Species* on "Variation Under Domestication".

In recent decades, microbes have been subjected to directed evolution including artificially induced mutations and recombinations in addition to targeted selection, to produce "super" microbes that can pump out large quantities of desired proteins or specific metabolic products. In about 1996 under the direction of Don Lightfoot at Eastern Washington University I started a project to develop a "super" bacterium capable of accumulating large amounts of toxic metal (zinc) from its growth medium and converting it to a harmless form. The project continued for more than 12 years, subjecting selected kinds of bacteria to gradually increasing concentrations of zinc so that only the ones best able to tolerate the zinc survived to produce offspring. Now, after many thousands of generations, the bacteria can survive zinc concentrations a thousand times higher than the original bacteria. By analyzing the genes and metabolisms of these bacteria we hope to learn how to design bacteria to help clean up sites contaminated with toxic metals.

A quick Google search using the terms "directed evolution bacteria" yields 245,000 links, to articles describing the development of microbes for such things as:

- Bacterial fuel cells

- Cleanup of contaminated soil and water
- Making improved detergents
- Improving pharmaceutical drugs

6. Evolution Works

Pleasantly isolated from the trench warfare of creation-vs.-evolution argumentation, an engineer or researcher has no doubts that evolution works. Evolution occurs every day in computers and laboratories, fields and breeding pens all over the world. Not only does evolution work, but it works fast and it works well. Evolution can be used productively by anyone with access to a computer and the Internet, and the ability to set up an Excel spreadsheet. Evolution can be used productively by back yard flower breeders and by road engineers. Though it is fun to know the intricacies of the evolutionary process, it is not really necessary to know why evolution works in order to use evolution to make money or to incrementally improve the human condition, just as it is not necessary to know why a computer works in order to use it to send an email.

Creationists may debate the extent to which evolution has contributed to emergence of the diverse organisms inhabiting the Earth, or even just to emergence of human beings; but they really cannot debate the efficacy of evolution. Evolution works.

References

More detailed information about Genetic Algorithms is given at http://www.nli-ltd.com/products/genetic_algorithms/ga_description.php. A free demo version of my genetic algorithm, Generator™, is downloadable at http://www.nli-ltd.com/products/genetic_algorithms/demos.php, complete with sample problems and a simple user manual.

Dialogue

Richard Gordon

We recently bought a robotic vacuum cleaner that wanders around picking up dust and dog hair, obviously driven by some clever adaptive algorithms, perhaps genetic algorithms. Now this robot doesn't have the capacity to self-reproduce, at least in the ordinary biological sense, though in a round about way it does. Its "fitness" is determined by its utility and success in cleaning. More are made back at the factory if people keep buying them, so its "reproduction" is by remote control. This reminds me of the case of symbiotic organisms in our cells, such as mitochondria or (in plants) chloroplasts, whose genes, over evolutionary time scales, slowly migrate to the bigger genome in the cell nucleus. I've discussed the cases of "self-reproducing" hydrogenosomes and centrosomes whose genomes seem to have entirely migrated to the cell nucleus, so they indeed appear to "reproduce" by remote control (Gordon, 1999). In most industrial products of genetic algorithms, as you mention, the device produced is separate from the computer that ran the computer codes for it, so that we have in general (but not always), a physical separation of genotype from phenotype. If something both acts and reproduces by use of genetic algorithms, who are we to say that it is not, in some sense, alive? Genetic algorithms thus may be at the core of the question of what is life, and whether artificial life not simply imitates life, but is alive.

To round out your presentation of the usefulness of genetic algorithms, I've appended a list of papers from the biomedical literature (PubMed) and books on the subject. It is clear that from vacuum cleaners to breast cancer detection, we are already surrounded by the robotic products of Darwinian evolution.

Further Reading

Burke, D.S., K.A. De Jong, J.J. Grefenstette, C.L. Ramsey & A.S. Wu (1998). Putting more genetics into genetic algorithms. *Evol Comput* **6**(4), 387-410.

Cagnoni, S., A.B. Dobrzeniecki, R. Poli & J.C. Yanch (1999). Genetic algorithm-based interactive segmentation of 3D medical images. *Image and Vision Computing* **17**(12), 881-895.

Goldberg, D.E. (1990). *Genetic Algorithms in Search, Optimization and Machine Learning*. Reading, Addson-Wesley.

Gordon, R. (1999). *The Hierarchical Genome and Differentiation Waves: Novel Unification of Development, Genetics and Evolution*. Singapore & London, World Scientific & Imperial College Press http://www.wspc.com.sg/books/lifesci/2755.html.

Handels, H., T. Ross, J. Kreusch, H.H. Wolff & S.J. Poppl (1999). Feature selection for optimized skin tumor recognition using genetic algorithms. *Artif Intell Med* **16**(3), 283-297.

Judson, R. (1997). *Genetic Algorithms and Their Use in Chemistry (Reviews in Computational Chemistry, Vol. 10)*. New York, VCH Publishers.

Kermani, B.G., S.S. Schiffman & H.T. Nagle (1999). Using neural networks and genetic algorithms to enhance performance in an electronic nose. *IEEE Trans Biomed Eng* **46**(4), 429-439.

Levin, M. (1995). Use of genetic algorithms to solve biomedical problems. *MD. Comput.* **12**(3), 193-199.

Mitchell, M. (1996). *An Introduction to Genetic Algorithms*. Cambridge, MIT Press.

Olmi, R., M. Bini & S. Priori (2000). A genetic algorithm approach to image reconstruction in electrical impedance tomography. *IEEE Transactions on Evolutionary Computation* **4**(1), 83-88.

Pedersen, J.T. & J. Moult (1996). Genetic algorithms for protein structure prediction. *Curr Opin Struct Biol* **6**(2), 227-231.

Siepmann, P., C.P. Martin, I. Vancea, P.J. Moriarty & N. Krasnogor (2007). A genetic algorithm approach to probing the evolution of self-organized nanostructured systems. *Nano Lett* **7**(7), 1985-1990.

Willett, P. (1995). Genetic algorithms in molecular recognition and design. *Trends Biotechnol* **13**(12), 516-521.

Zheng, B., Y.H. Chang, X.H. Wang, W.F. Good & D. Gur (1999). Feature selection for computerized mass detection in digitized mammograms by using a genetic algorithm. *Acad Radiol* **6**(6), 327-332.

Part 2

Towards Harmony Between Science and Religion

Biography

Professor Ernan McMullin is Professor Emeritus of Philosophy and Director Emeritus of the Program in History and Philosophy of Science at the University of Notre Dame, U.S.A. He obtained his Ph.D. in philosophy from the University of Louvain, Belgium in 1954 and taught at the University of Notre Dame from 1954 until his retirement in 1994. His interests are in the philosophy of science, the history of the philosophy of science, and the relations between the natural sciences and theology.

8
Could Natural Selection be Purposive?

ERNAN McMULLIN

Department of Philosophy, University of Notre Dame,
Notre Dame IN 46556 USA
mcmullin.1@nd.edu

> "Evolution in a pure Darwinian world has no goal or purpose: the exclusive driving force is random mutations sorted out by natural selection from one generation to the next. Many who accept the fact of evolution cannot, however, on religious grounds, accept the operation of blind chance and the absence of divine purpose implicit in natural selection." Edward O. Wilson (2005, Introduction, 12.)

1. Introduction

Why do so many Christians in the United States resist the teaching of Darwinian biology in the public-school classroom? Europeans tend to marvel at the debate that has dragged on there, in one form or another, for more than a century. One answer that has taken on a real importance in late years may be found in the declarations of some eminent spokesmen for evolutionary biology like Edward O. Wilson and Richard Dawkins. They argue, with a degree of passion, that Darwinian biology has fatally undermined all forms of religion. Wilson goes on from the quotation above, in his Introduction to a new edition of Darwin's major works, to proclaim as a logical consequence of those works:

> "Thus was born scientific humanism, the only world-view compatible with science's growing knowledge of the real world and the laws of

> nature.... humanism based on science...is the light and the way at last put before us." (Wilson, 2005, Afterword, 1480).

Hardly surprising, then, that some should take Wilson and Dawkins at their word and conclude that if one really has to choose between evolution and one's Christian faith, they would prefer their faith as guide. And when they urge, with equal passion, that a limit be set on the explanatory ambitions of Darwinian science by postulating an active role for the Creator in directing the process of evolution, it is Wilson's version of scientific humanism that is, implicitly, the enemy. If Darwinism can be taught in the public-school classroom, they say, given the negative consequences for religion that its most vocal advocates see as necessarily flowing from it, why should not an alternative that would supplement natural selection by invoking the shaping powers of the Creator be presented also? Does not its exclusion itself amount to a religious statement, surely not a scientific claim?

There is overstatement on both sides of this by now familiar debate, to be sure. But that will not be the topic of this essay. Rather, the focus will be on a misunderstanding shared by both parties to the debate: that evolution by natural selection is a "chance", a "blind" process and thus that any attempt to portray it as purposive must necessarily fail. Were this indeed to be true, it would be easy to conclude that the God whose involvement in his creation lies at the heart of the Judaeo-Christian tradition has finally receded into the shadows. But is it true?

2. Getting the Terminology Straight

The terminology involved in this discussion is dangerously ambiguous. And it is an ambiguity that carries all the way back to the beginning, to the notion of *telos* in Plato and Aristotle (McMullin, 1968). For Plato, *telos* in the sense of goal or purpose means a consciously entertained goal, an intentionally pursued purpose. As he surveys the universe in his great cosmological work, the *Timaeus*, he sees in Form the unmistakable work of Reason, imperfectly realized in the world of sense though it is. Though he does not use the term, "teleology" would therefore amount to the recognition of that fact. A teleological explanation would point to mind as the source of order, an order itself conceived as in some sense a good. Plato deliberately leaves open whether mind in the cosmological narrative of the

Timaeus is to be understood as an agency extrinsic to the sense-world (the Demiurge) or an Intelligence intrinsic to that world and working through it.

As Aristotle looks at that same world, on the other hand, he sees a specific form as intrinsic to each kind, manifesting itself by the characteristic activities of that nature (Gotthelf, 1976). There is no need to introduce intelligence extrinsic to the nature itself; the ensemble of natural kinds just is intelligible by nature. With the living world as the primary focus of his studies, he recognizes everywhere the presence of goal and purpose. But these terms are now understood in a very different way to Plato's. The growth of every living thing to maturity is surely a goal-directed process, the goal or purpose being the full expression of the appropriate form (Lennox, 1992). Likewise, the other activities that define the particular nature are oriented to the good for that nature. Teleological explanation in this understanding of *telos* makes no reference to mind; rather, it identifies the finality of the nature in question and shows how the characteristic structures and behaviors of that nature contribute to that finality.

The selfsame ambiguity reappears in the contemporary discussions of goal and purpose. In the remainder of this essay, it will be helpful to use suffixes to distinguish between the two senses. When the action of intelligence is implied as it is in Plato, reference will be made to goal_1 and purpose_1, and the associated mode of explanation will be described as teleological_1. When the intention is to refer to the regularities of nature perceived as furthering the finalities of nature, "goal_2" or "purpose_2" will be used, and the mode of explanation will be teleological_2.

3. Teleology, Purpose and Natural Selection

A number of eminent biologists have sought of late to rehabilitate the long-maligned term, "teleology", arguing that a dominant mode of explanation in biology can best be described as "teleological", teleological_2 in our terms. And they turn to Aristotle as the pioneer in developing this mode of explanation to good effect in his magisterial survey of the living world. They hasten to add that they do not extend this mode of explanation to the non-living world as he did. Aristotle defines the nature of a thing by means of a cluster of goal-related activities whose *telos* or end is the good of the individual or of the species to which it belongs. What teleological_2 explanation amounts to, therefore, is explanation of those activities and hence the nature they define in terms of the function or purpose they fulfil in the organism. This is, of course, a standard mode of explanation in contemporary biology. Ought this mode of explanation be extended from the characteristic activities to the organs of the living body? It would seem that it should since these can be equally goal-directed: "The goal/purpose/function of the liver is... ." But Ernst Mayr demurs, maintaining that the term "goal", and hence the term "teleological", ought to be restricted to behaviors only (Mayr, 1988, 52). Be that as it may, these writers leave the reader in no doubt that purpose$_2$ and hence teleological$_2$ explanations have a key role to play in biology.

But what about natural selection? Can it be characterized as purposive, its mode of explanation "teleological", in the second sense of those terms, of course? Biologists seem to be averse to any such extension, with some notable exceptions (Ayala, 1970). The reasons are not far to find. The term "teleology" inevitably implies for many the actively shaping action of mind, that is, Platonic teleology$_1$. And this they reject. Natural selection clearly does not lend itself to a teleological$_1$ form of explanation. It proceeds in an automatic way where heritable variations offer differential advantage in terms of average number of progeny. The adequacy of natural selection to give a complete account of the mechanisms impelling evolutionary process is, of course, debated and a number of other complementary mechanisms are still actively under discussion. But though this is the case, biologists are adamant in disallowing the possibility of an intelligence actively steering the process or intervening to help it over awkward spots. Their grounds are that such an hypothesis could not be explored and tested in the way that scientific hypotheses normally are.

One further reason why the term "teleology" tends to inspire a negative reaction among evolutionary biologists is a widespread misunderstanding of the notion of final causality that has traditionally accompanied it. Confusing two of the four causes or types of explanation listed by Aristotle, the critics suppose that postulating a "final" cause implies a sort of backward efficient causality, somehow causing, in an agency sense of that term, a process already in the past. It was this misunderstanding that led C.S. Pittendrigh in 1958 to introduce the term "teleonomy" to cover the sort of goal-directed (but not actively mind-guided) explanation described here as teleological$_2$. He proposed this alternative "in order to emphasize that the recognition and description of end-directedness does not carry a commitment to Aristotelian teleology as an efficient causal principle" (Pittendrigh, 1958, 394). Responding to a query from Ernst Mayr, he later wrote: "I wanted a word that would allow me (all of us biologists) to describe … this end-directedness of a perfectly mechanistic system. Teleology would not do, carrying with it the implication that the end is causally effective in the current operation of the machine. Teleonomic, it is hoped, escapes that plain falsity…" (Letter of February 26, 1970 to Mayr, in Mayr, 1988, 63). Falsity indeed! But certainly not attributable to Aristotle who was at pains to distinguish final from efficient causality. Citing the goal of a living activity is one way (Aristotle would add: the best way) to understand that activity but in no way ought this be taken to endorse a counter-intuitive sort of backward-in-time efficient causal action.

So we shall retain the term "teleology" in its broadly Aristotelian sense in order to describe an important feature of contemporary biology. But can it be extended to natural selection? After all, natural selection is goal-directed, that $goal_2$ being the improvement of the inclusive fitness of a specific population. The material that the selection works on includes "chance" events in several senses of that elastic term: random relative to the needs of the population or perhaps unpredictable if involving a quantum process. But the selection process is not itself a chance affair, strictly speaking, though popular writers can be cavalier in so describing it. Natural selection serves a definite goal in a systematic way, sifting through a series of uncorrelated events for those that serve its purpose. It is important to bring this out, so frequent is the misunderstanding generated by the careless use of the term "chance" in this context.

There is, however, a very important reservation that marks off the sort of teleology appropriate to natural selection from that advocated by Aristotle. For him, the goal is long-term, it envisages the goal toward which the process as a whole is tending. Whereas the goal of the process of natural selection is strictly short-term and one cannot infer from it that the larger process of evolution has any similar goal or indeed, strictly speaking, any definite $goal_2$. The goal of natural selection, to repeat, is the good of the population immediately at hand, not some future population of far-distant descendants. Natural selection lacks any long-term reach; the direction taken in this generation may indeed lock descendants into a course that proves deleterious in the long run. As critics of Intelligent Design like to emphasize, many features of the animal body are poorly designed for the functions they serve; they are not at all what they would have been were those functions anticipated and designed around (Miller, 1999). The end instead is a summation of a myriad of small steps, each of which served on balance the inclusive fitness of the particular generation. But in no way is that end itself a target of selection, anticipating long-term needs or long-term consequences.

This is another reason why terms like "purpose" and "goal", and most especially "teleology", tend to evoke nervousness among many biologists. Making them common currency in the context of natural selection, they fear, could lead some to misapply them to the larger-scale evolutionary process itself where teleological modes of explanation are inappropriate. It is true that there have been long-lasting trends in the course of evolution like the steady increase in size of animal bodies during the Jurassic period or of brain size in the more recent hominid line. And striking instances of parallel evolution in widely-separated niches might suggest something like long-term goal-directed development of certain clusters of traits (Conway Morris, 2003). But even in these cases, the goal-directedness holds, strictly speaking, only at the ground level of natural selection.

Why, then, seek to validate these controversial terms, particularly "teleology", in the context of natural selection? It seems worthwhile to do so simply to counter the all-too-common tendency among both proponents and critics of evolution to describe it as a "chance" process and to draw far-reaching inferences from that careless usage. The sort of explanation afforded by natural selection is teleological in a sense that Aristotle would have understood but that Plato would have regarded as inadequate.

Now for a much more controversial point. Is there any sense in which the evolutionary process could itself be described in teleological$_1$ terms, that is, as the product of intelligence? We have seen that from the scientific standpoint this is inadmissible. But physical science is not the only source of wisdom... theology and metaphysics may have something to say too. To see what that might be, a detour is necessary, this time through territory as unfamiliar to many scientists as evolution is to many theologians.

4. Creator of the Heavens and the Earth

The relevant theological doctrine here is that of the Creation, a doctrine common in one shape or another to all three of the major Abrahamic faiths, Judaism, Christianity, and Islam. A brief outline of a doctrine with a very complex history will have to suffice (McMullin, 1985, 8-16). The various segments of the

Hebrew Scriptures were composed in a somewhat different order than the one in which they are now arranged in the *Torah.* If one follows them in the order in which, to the best of our investigative ability they were composed, one sees the realization gradually dawning that the Lord who had led the children of Israel out of captivity was also the Mighty One who had shaped the heavens and the earth. The powerful poetry of Isaiah, of Job, of the Psalms, celebrates the Maker in lines that have rung out over the ages:

> You stretch out the heavens like a tent.
> You build your palace on the waters above.
> Using the clouds as your chariot,
> You advance on the wings of the wind.... (Psalm 104, 2-3.)

The opening chapter of Genesis recalls the Creation itself: "God said: Let there be light! And there was light." This chapter was of relatively late composition compared to the chapter that follows; it reflects the growing conviction that the Creation is a response to a command: Let it be! And the universe springs into being. But what exactly did that amount to? The authors of the Hebrew texts did not speculate; they were not given to philosophical speculation. Enough that the world and everything in it were subject to the Being who had first called it into existence.

With the coming of Christianity in a world dominated by Greek culture, a new genre began to take shape: a theology drawn from the Old and New Testaments but more and more strongly influenced as time went on by the categories and ways of thought of Greek philosophy. From the second century onward, questions began to be asked about how the notion of creation was to be understood. And the conviction began to grow that it had to be expressed in categories quite alien to Greek thought: it was a beginning-to-be in an absolute sense from nothing preceding, a creation *ex nihilo* as it was described. No Greek philosopher had ever ventured so daring a conception, bringing with it as it did new and troubling philosophical questions about the reality of human freewill and the Creator's apparent responsibility for the manifest evils of the world.

At the turn of the fifth century, the greatest of the earliest Christian theologians, Augustine of Hippo, finally gave the doctrine of creation *ex nihilo* the form it would maintain over later ages. Greek philosophy had taken the existence of the world itself as a given, as something that needed (or perhaps could be given) no further explanation. Perhaps it had always existed (Aristotle), or it took shape from a pre-existent matter (the Presocratics), or it was the work of a Demiurge imposing form on a recalcitrant matter (Plato). Augustine on the other hand argued that what first had to be explained was the existence of the world itself, matter and all. Only a Being altogether different in kind from the changing and corruptible world around us, a necessary Being of a kind whose existence needed no further explanation, could provide an adequate answer. The explanatory quest could not be indefinitely extended. Augustine simply posed a choice: which is the better place to end that quest, the universe as we know it or a Being whose very nature it is to be. And he left no doubt about his own answer.

He went further, taking more or less for granted that the story of the creation in Genesis should be interpreted in some respects as metaphor carrying a serious theological message. Instead of a seven-day series of separate makings of living kinds, he proposed that the Creator implanted a "seed-principle" for each kind in the universe at the first moment of its coming-to-be. The potentialities of the immense diversity of the living world were thus already present from the beginning. Each would lie dormant until the conditions of earth and water were right and then the ancestor of each kind would make its appearance. Augustine recognized that the processes whereby this sort of generation could come about were unknown, but, he added in his defence, so were those involved in the everyday phenomena of growth and aging. It would, he suggested, take a hitherto unknown sort of knowledge to discover how all this might come about. But he seems quite confident that this would have been the way in which the Creator he envisaged would have acted rather than having to intervene at a later time to bring each kind to be in a special way.

Augustine was particularly interested in clarifying the Creator's relationship with time. Time is a feature of the created world just as is space and is brought to be, therefore, with that world. Temporality involves limitation: the past resides with us only in fading memory; the future exists for us only in our expectations. The Creator is not eternal in the sense of an unending succession — which would be a form of time. Rather, the Creator is eternal in the sense of atemporal, a Being to whom time-predicates of any sort simply do not apply. Thus the Creator's knowledge of the future is not a form of foreknowledge based on knowledge of the present. It is the timeless knowledge of the Being who brings all of time to be in a single act. It is not limited by what we would regard as chance or by processes we would call chaotic. Augustine was well aware of the troublesome consequences of this way of representing God's relationship with the created world but evidently saw no other way in which the Creator could be related to a property which is specific to the world the Creator brings to be.

One particularly prominent feature of the Genesis account is the central role assigned to the human race. Humans were made in the Creator's own image, their combination of intellect and freewill setting them apart from the rest of the animal creation. The chronicle of Yahweh's relationship with the people of Israel over the centuries conveys in a powerful way the Creator's special concern for this part of the creation, a concern that is gradually revealed in the Christian dispensation to be as broad as the human race itself. There could scarcely be any clearer testimony to the special place accorded to the human race and its destiny than the Christian doctrine of the Incarnation. From the Jewish as from the Christian standpoint, it was made clear that the created universe would be one in which the human race would find a special place.

5. Evolution and Purpose

We have seen that natural selection can be described as purposive$_2$, as goal-directed, but that the larger evolutionary process cannot be. But now in the light of what has just been said about the doctrine of creation, might there be a sense in which both could be said to be purposive$_1$, involving, that is, the active engagement of intelligence? From the theological standpoint, the answer fairly obviously is: yes. The act of creation was itself the work of intelligence, according to the traditional theological view. More specifically, given the special role attributed to the human race, whatever was required in order that that race would naturally make its appearance within the creation would be part of the original dispensation, a seed-principle for beings that would one day consciously offer (or deny) homage to the Creator. Evolution by natural selection was evidently critical to that purpose's being carried through. Thus, in that light, it can properly be said to be purposive in the first, and stronger, sense.

Doing so is likely to raise all the old objections deriving from the confusion between purposive$_1$ and purposive$_2$. As we have seen, evolution is not purposive$_2$; there is no goal towards which it naturally tends. And a further likely objection to calling it purposive in the second sense might be that it would imply some sort of causal action outside the normal between the Creator and the universe. The issue of how exactly the Creator's purposes are to be achieved as the universe develops is, however, a much controverted one. Attempts to invoke quantum action or chaos theory in order to provide room for "special" action on God's part without disturbing the "natural" order have not been especially successful (Russell, 1992-2000; Saunders 2002). Fortunately this difficulty does not arise in the case of evolution. There are two ways in which one might proceed.

One could begin from the premise that in a universe as vast as ours, the number of locations favorable to the development of life is likely to be quite large. There could be a (qualified) assurance that in one or more of these the evolutionary line leading to a form of human-like life would be very high. What would be purposive$_2$ here would be the entire evolutionary process that would somewhere issue in a form of humanity; the Creator would have instituted this process as a means of bringing about that goal. The drawback here, of course, is the contingent nature of that outcome and the representation of the Creator as depending on everything coming out right.

But in the Augustinian view of the creation, this is not the correct representation. The Creator does not depend on knowledge of the initial state of the universe for knowledge of the future. The temporal terms simply do not apply. God creates a universe in a single act in which past, present, and future are contained, and a universe is chosen in which a single evolutionary line, at least, will bring about the appearance of human beings and the fulfillment of the Divine purpose (McMullin, 1998). The outcome is a contingent one from our viewpoint but not from God's. Asserting the reality of purpose in this perspective takes for granted that one already believes in the existence of a Creator whose action is sufficiently like ours to allow us to call it purposive, in an admittedly analogical sense. It does not mean that we are privy to that purpose, though the traditions of the *Torah*, the *Bible*, the *Koran*, would imply a recognition of one small part of it, at least.

Returning to the quotation from E.O. Wilson from which we began, we have seen reason, then, to challenge both of the claims he makes: that the operation of natural selection can be characterized as "blind chance" (on the contrary, it is purposive$_2$) and that implicit in that operation and thus, more generally, in evolution there is an "absence of divine purpose" (on the contrary, both are purposive$_1$). Evolution does not imply an absence of Divine purpose, nor from the starting-point of evolution can one go on legitimately in this way to a denial of the Creator's existence. Without the presupposition that the universe cannot be the work of a Creator — hardly a scientific presupposition — one cannot validly conclude that the success of evolutionary theory implies that the universe is lacking in purpose.

References

Ayala, F. (1970) Teleological explanation in evolutionary biology. *Philosophy of Science* **37**, 1-15.

Conway Morris, S. (2003) *Life's Solution: Inevitable Humans in a Lonely Universe*. Cambridge University Press, Cambridge.

Gotthelf, A. (1976) Aristotle's conception of final causality. *Review of Metaphysics* **30,** 226-254.

Lennox, J.G. (1992) Teleology. In: E. Fox-Keller and E.A. Lloyd (eds.) *Keywords in Evolutionary Biology*. Harvard University Press, Cambridge Mass. pp. 324-333.

Mayr, E. (1988) The multiple meanings of teleological, chapter 3 of his *Toward a New Philosophy of Biology*. Harvard University Press, Cambridge Mass., 38-64.

Miller, K. (1999) *Finding Darwin's God: A Scientist's Search for Common Ground between God and Evolution*. Cliff Street Books, New York.

McMullin, E. (1968) Cosmic order in Plato and Aristotle. In: P.G. Kuntz (ed.) *The Concept of Order*. University of Washington Press, Seattle, pp. 63-76.

McMullin, E. (ed.) (1985) *Evolution and Creation*. University of Notre Dame Press, Notre Dame, Indiana.

McMullin, E. (1996) Cosmic purpose and the contingency of human evolution. *Theology Today* **55**, 389-413.

Pittendrigh, C.S. (1958) Adaptation, natural selection and behavior. In: A. Roe and G.G. Simpson (eds.) *Behavior and Evolution*. Yale University Press, New Haven, pp. 390-416.

Russell, R.J. (series editor) (1992-2000) *Scientific Perspectives on Divine Action*. Five volumes. Vatican Observatory Publications, Rome; distributed by the University of Notre Dame Press, Notre Dame, Indiana.

Saunders, N. (2002) *Divine Action and Modern Science*. Cambridge University Press, Cambridge.

Wilson, E.O. (ed.) (2006) *From So Simple a Beginning: The Four Great Books of Charles Darwin*. Norton, New York.

Dialogue

Richard Gordon
Having heard many a molecular biologist claim that this or that protein "wants to" do something or other, I'm a bit more dubious than you about claims for purpose in biology, or that biologists eschew purpose for nonmental events. Biology is indeed steeped in teleological language, which I regard as an impediment to progress. See §1.18 A Word on Language and §1.19 The Embryology/Psychology Merry-go-round in:

> Gordon, R. (1999). *The Hierarchical Genome and Differentiation Waves: Novel Unification of Development, Genetics and Evolution*, Singapore & London: World Scientific & Imperial College Press. http://www.wspc.com.sg/books/lifesci/2755.html, 2 vols., 1836p.

I could back up your statement:

> "One could begin from the premise that in a universe as vast as ours, the number of locations favorable to the development of life is likely to be quite large."

with:

> Gordon, R. & R.B. Hoover (2007). Could there have been a single origin of life in a big bang universe? *Proc. SPIE* **6694**, doi:10.1117/12.737041.

where I estimate a minimum of 50,000 independent origins of life.

Ernan McMullin
Your [Richard Gordon's] own uneasiness about the use of terms like 'goal' in biology I can understand. I noted that there are two senses of terms like 'goal' in the extensive early discussions of Plato and Aristotle, one involving mind or intelligence and the other not. It is the second sense that I have in mind: a law-like activity serving the good of some kind of organism was said by Aristotle to have a "*telos*" (end, goal), namely the good of the organism (in the case of

natural selection, this would be the good of the population). Perhaps the English terms, goal and purpose, are irremediably tied to the suggestion of mind at work. But might one not ask: what is the goal of natural selection? responding: increasing the inclusive fitness of the population? In the end, perhaps only the original Greek term will do: *telos* (Ayala's point). It may be too difficult to rescue the English terms from the Platonic overtone of mind involvement.

I am skeptical of efforts to calculate even the approximate number of independent origins of life: I wrote a critique of Sagan's efforts in that direction in *Icarus* many years ago. Too many unknowns in the Drake "equation"... although Drake (and Sagan) had a more ambitious aim than simply focusing on the origins of life! But that the number may be very large indeed I am perfectly willing to concede... just so long as numbers aren't attached!

Richard Gordon
I hunted for an *Icarus* paper by Carl Sagan matching your description, and the closest I came up with is:

> Newman, W.I. & C. Sagan (1981). Galactic civilizations — population-dynamics and inter-stellar diffusion. *Icarus* **46**(3), 293-327.

Is that the one you had in mind? Now, you allude to my paper:

> Gordon, R. & R.B. Hoover (2007). Could there have been a single origin of life in a big bang universe? *Proc. SPIE* **6694**, doi:10.1117/12.737041.

whose goal is a bit more cosmic than Newman & Sagan (1981). I started with the idea that the number of origins of life could be one, but was dissuaded by the calculation, which set a lower bound of 50,000. Of course, it could be orders of magnitude larger. Maybe my computation is no better than counting on two fingers: "one, many". But the distinction between the two values seemed worth talking about.

Stephen P. McGrew
In Ernan McMullin's chapter entitled "Could Natural Selection be Purposive?", Professor McMullin asks the question and offers an answer. Unfortunately, the answer seems to be simply "maybe". I guess we need to accept "maybe" as the only honest answer that can be given at this point on our learning curve.

Some of the points he makes en route to the final "maybe" are particularly interesting. The idea of evolution being contingent from our time-bound perspective while being entirely purposeful from a Creator's timeless perspective is philosophically interesting but is of course not scientifically testable. It is a grand thought, along the lines of those variations of the Anthropic Principle that suggest an infinity of alternative, inaccessible worlds exist covering all possibilities — and of course we find ourselves only in the one universe that we inhabit. The two ideas are as unsatisfactory for the same reasons: they have no predictive power.

It took me a while to grasp McMullin's expansion of the meaning of "goal". In particular, his statement, "The goal of natural selection, to repeat, is the good of the population immediately at hand, not some future population of far-distant descendants", required some pondering. By analogy, it seems that the goal of gravity would be to pull water to the ocean or to concentrate salt in low-lying areas. "Goal" seems like an odd word to choose for the purpose. "Function" might suit it better.

Oddly enough, I agree with McMullin that there are "goals" in nature, though I would suggest that the term "goal" be assigned a different meaning. In particular, I would say that *goals are future states or conditions toward which an intelligent system attempts to move by selecting from among its behavioral options*. Starting from that point and addressing the same issues that you have addressed, one should ask the meaning of "intelligent system", and then decide whether natural selection per se constitutes an intelligent system. Or, one might consider Nature itself as an intelligent system.

I also agree with McMullin that evolution does not imply an absence of Divine purpose, nor deny the existence of a Creator. But the fact of evolution certainly should put some very solid constraints on what role, if any, a Creator might play in the world. We can say with confidence no Divine intervention occurs to the extent that simple physical principles operating throughout the universe and since the beginning of time can accurately model the past and present events we deduce from physical evidence. We cannot explain (yet) where the simple physical principles came from, already pregnant with the potential of producing living creatures like ourselves; so it is okay to postulate that a Creator designed those principles with all their consequences, and maybe with ourselves, in mind. Unfortunately, that postulate has no predictive power.

The task of science is to seek consistent, simple universal principles that explain the observed present and accurately predict at least near-future phenomena. If a scientist says that he has disproved the existence of a Creator, I agree with McMullin that the scientist has made a statement of faith rather than a scientific deduction. The real disconnect between religion and science seems to be that the faithful are obligated to reserve as large a role as possible for a deity in their world models, while scientists are obligated to see how far they can go *without* writing a deity into their equations. But, as McMullin says, this does not mean a scientist needs to be an atheist, or that a religious person cannot be a scientist.

Ernan McMullin

My answer to the question posed in the title of my essay is “yes”, not “maybe”. Someone (like myself) who believes that the universe depends for its existence on a provident Creator will quite naturally infer that natural selection, the engine of evolution, is purposive, since it involves the mind of the Creator. The cosmic conditions that made natural selection possible were the work of the Creator, the means by which the Creator ensured in a natural way the vast diversity of the living world. Someone who is agnostic in regard to a Creator would still have to answer “yes” to my question, since natural selection *could* be purposive provided that the Creator *did* exist. Only someone who *excludes* the possibility that the universe might be created could consistently argue that the answer to the question is “no”. But such a person would have to provide a (necessarily) philosophical argument in support of his position. It obviously will not to do to claim (as E.O. Wilson does) that it follows from the fact of natural selection itself. What *would* follow from the sufficiency of natural selection as a scientific explanation of diversity is that the active interposition of intelligence (and hence of purpose) in guiding the direction of evolutionary process would be excluded.

I agree with my interlocutor that terms like ‘goal’ and ‘purpose’ are dangerously ambiguous in this context: that is why I traced that ambiguity back to its origins in Greek thought. I distinguished in my paper between purposive$_1$ and purposive$_2$. A process is purposive$_1$ when the agency of mind or intelligence is invoked; a process is purposive$_2$ when reference is made to the regularities of nature as serving the finalities of nature, though with no reference to the activity of mind. The reason why some leading biologists have called for the recognition of a teleological element in the operation of natural selection is that selection serves the *good* of the relevant population: the improvement of its inclusive fitness in the short term (and thus purposive). But this sort of teleological explanation is restricted to the living world where natural selection operates.

Aristotle supposed that stones, disturbed, fall back to their natural place, in that way restoring order as a goal and thus serving the overall cosmic "good". We would not say today that gravity works in that way! But perhaps it would be best in the end to avoid terms like 'goal' and 'purpose' in this context because of their almost ineliminable mind-relatedness. The Greek term '*telos*' might be the safest choice.

Stephen McGrew and I agree on most of the issues that I raised. Our main difference is in regard to what counts as knowledge. In his view, science appears to provide the only acceptable form of knowledge of the world. Furthermore, "science" is understood by him in a quite restrictive way: predictive power seems to be the sole criterion. It is a very common view nowadays, an appealing one evidently, since it proposes a neat demarcation between "real" knowledge and everything else. This was the appeal that carried logical positivism to its brief period of dominance in the mid-twentieth century, the same appeal that made Popper's falsificationism so attractive to scientists in the later part of the century.

But it simply won't do. Tracing the demise of positivism, as of the Popperian alternative, would be instructive but quite impractical here. Some simple reflections will have to suffice. How about the work of the historian, for example? The biographer of Dante or the scholar who sets about tracing the causes of the French Revolution will not rely much on predictive power. The moralist who asserts the unrelieved evil of the Holocaust does not appeal to it either. The philosopher who defends a realist account of the physical structures postulated by the geneticist or the astrophysicist against those who would see science as being warranted merely by its practical ends does not call on prediction in support of his position. The legal scholar who makes the case for a particular legal precedent would not call on predictive power in his brief. We speak of historical knowledge, moral knowledge, philosophical knowledge, legal knowledge, without apologizing for that use of the term. In the Greek tradition, "knowledge" tended to have the idealized sense of something known beyond all possibility of doubt. That is no longer true: even the most firmly-held scientific theory may be open to revision, though for scientists it would count as "knowledge".

Even within the sciences themselves, Stephen McGrew's succinct definition of the task of science: "to seek consistent, simple universal principles that explain the present and accurately predict at least near-future observable phenomena" has rather limited applicability. Like so many of the generalizations about science, it is based, fatally, on perhaps the least typical, if most familiar, science, physics. Molecular biologists are not seeking universal principles; evolutionary biologists are not (for the most part) trying to predict the near future; biochemists would not get very far with simple principles; explaining the present is not, on the whole, what palaeontologists are interested in. Drawing a line around what counts as "science", let alone what counts as knowledge, has not had a notably good record!

My point in all this is to underline the risk of setting aside the idea of a cosmic Creator on the grounds of its not being "scientifically testable" or "having no predictive power". Were these indeed to be adequate grounds for its dismissal, too much else would go in its train. There are distinctively philosophical, as well distinctively theological, arguments for the existence of a Creator. But it was never my purpose to put arguments of this time-honored sort forward. To answer the question posed in my title with an affirmative, it is not necessary to do so.

Stephen P. McGrew
Thank you for your response!

Given your definition of "purposive", the question posed in the title of your essay, "Could natural selection be purposive?", can only be answered honestly in the affirmative.

That could easily enough be the end of the discussion; but to me it would be an unsatisfying end. A purpose that lies outside of time and space, that cannot reveal itself through physical evidence, and doesn't offer scientifically useful insights may be satisfying to contemplate, but can never be "known" to be true.

Of course there are ways that unprovable articles of faith can be useful. Certainly a person who invests his entire being in a belief and lets the tenets of that belief guide all his choices may well accomplish things that a person guided by more mundane principles would not accomplish. However, the usefulness of such beliefs, and even the provability of the real consequences of beliefs, does not make the subjects of the beliefs real. "Belief" has nothing to do with facts and truth. It is a person's wholesale investment in a concept *as if* the concept were factual knowledge.

As you pointed out, you and I use different meanings for the word "knowledge". To you, "knowledge" does not need to be factual; it can be a working hypothesis or a current understanding. While I am glad to accept that meaning, to do so creates a gap in available language. What word should we use to mean the sort of knowledge that is absolutely true — that is factual? Maybe that gap *should* exist. Maybe nothing is factual. But scientists, whether they be physicists, biologists or paleontologists, nonetheless *seek* facts about the world. Knowledge as you use the word is expressed in theories and in compilations of observations and their interpretations. What I'll call "factual knowledge" is never attained, but a belief that it might someday be attained motivates scientists to keep working. And, the belief is useful because scientists keep coming up with improved theories that more completely and accurately represent (and predict) the world as we observe it.

So, we agree that natural selection *could* be purposive, but that its purposiveness cannot be proven. A belief that natural selection *is* purposive can perhaps be comforting to a person who needs that sort of comfort. I am comfortable enough merely looking for the next small contribution I can make to a scientific understanding of the world: a dynamic, growing, working model whose completeness and accuracy always improves.

Ernan McMullin
I must admit that I did not realize that Richard Gordon would interpret my remarks in my first E-mail as part of a printable dialogue. They were, rather, a casual response to his original E-mail where he was encouraging me to take part in the DINA enterprise itself [this book].

I said that I wrote a critique of Sagan's efforts to quantify the number of technological civilizations in our galaxy. But the reference was to a paper of my own, not one by Sagan. My paper was called "Estimating the probabilities of extraterrestrial life" and it appeared in *Icarus*, **14** (1971): 2911-4. My reference to Sagan in that article was to his big book with Shklovskii back in the 'sixties. [Shklovskii, I.S. & C. Sagan (1966). *Intelligent Life in the Universe*, Delta.]

I took your figure of 50,000 to be an actual estimate, following perhaps from the Drake equation as Sagan's calculation did. Even a lower bound seems to me, at this stage, still to be premature. We still only have an empirical frequency of just one to go on! But in any event, this issue has little to do with DINA [this book] or with my paper. I mentioned it in my first E-mail, not because of its relevance to the DINA discussion, but only because this numerical estimate caught my eye (and my interest) somewhere in your bibliography, as I recall. I doubt whether this part of our exchange ought to go into DINA. It simply isn't sufficiently relevant.

Richard Gordon

One could argue that we don't even have a frequency of known life in the universe. After all, given a case of 1, we still need a denominator for frequency >= 1/N.

Dialogue is permitted to go off on tangents. The question of the number of origins of life is certainly important in the greater discourse on the three origins that are bones of contention: humans, life and the universe.

Ernan McMullin

To Steve McGrew from Ernan McMullin

Before we break off our discussion, I would like to make a couple of responses to the points you just raised.

First, a clarification of two of the terms you use. The term, 'fact', comes from the Latin term, 'factum', meaning 'something made' (as in 'manufacture'). Facts are made by us in the sense that they are expressed in our language and are conveyed by propositions we construct. Facts are not part of the physical world: you can't sit on one or cut it in two! A fact is something that is conveyed by a true proposition; it is necessarily mind-related. So that insofar as one is trying to tell the truth about something, one is trying to convey a fact. Success in doing so depends in part, then, on the adequacy of one's language for the task in hand. That two plus two equals four is a fact; that Napoleon was defeated at Waterloo is a fact. The closer that an experimental claim or a theory, like the theory of natural selection as an engine of evolution, comes to expressing the truth, the closer it comes to being regarded as a fact; though, as you say, "factual knowledge" (if one takes that to mean knowledge of whose truth we are totally assured), is more like a goal than something readily achieved. When we include the best-supported theories in science as part of our "knowledge", we are admittedly using this latter term more broadly: what we know is that, being well-supported, they are likely to come close to the truth.

The term, 'faith', is even more slippery. The commonest sense might be this: faith is the acceptance of something on the word of another; it rests on trust. Faith in this sense is essential to the work of the scientist: without faith in the experimental reports of others, for example, physical science could not go on. When that faith is violated, as occasionally (but fortunately only occasionally) happens, good science suffers. In large-scale team-science of the sort that is increasingly the norm in many parts of physics and biology today, trust is an indispensable part of everyday research work. Calling the experimental results reported by such science "facts" ought not be allowed to obscure the "fact" that in an important sense they also rest on faith. Faith in this sense is not blind; it rests on reasoning of one sort or another. It can be tested. But in the end, an act of faith is just that, rational or irrational though it be.

Religious faith is also of that kind, although the reasoning that undergirds it is of a quite different sort. It varies a great deal from one variety of religious belief to another. It may derive from trust in a person, based on what is known of that person's life or message. Or it might be in a book, regarded as privileged, as worthy of trust in its author.

Thus, it is problematic simply to characterize "articles of faith" as "unprovable". It depends, of course, on what you mean by 'provable'. But if you mean not rationally justifiable, that seems open to question, at the very least.

Finally, then, to the main point you are making: that belief in a Creator whose purposes infuse the universe, including the evolutionary processes that have brought human beings like ourselves about, is a matter of "faith" and thus can never be "known" to be true. And you conclude that the purposiveness of the universe, in the sense in which I defined it, cannot thus be proven. (You say that we both agree on that; I would want to register caution, if not dissent, depending on what you mean by 'proven'.)

Belief in the existence of a Creator can have many sources, just as belief in the existence of viruses or electric fields can. One source is a very ancient argument that the existence of the universe requires an explanation, that is, a cause. There has been an immense amount of commentary on this argument down through the ages. Some would argue that if the universe began to be some fourteen billion years ago, this argument is even more coercive. Be that as it may, the important point is that this is a matter of argument, of reasoning, not simply a matter of faith (in the sense defined above). Belief in a Creator can, of course, also be a matter of faith, but faith founded on trust for which some reason can be given.

In short, then, the purposiveness of natural selection and of the evolution which rests on it (in the sense, mind, in which I defined purposiveness in this context) can be argued for both on philosophical grounds and as a matter of a (reasonably-based) faith. It cannot be "proven" as a "scientific" claim. But why should that be regarded as a norm for our knowledge-claims generally?

Stephen P. McGrew

It would be very difficult to convince me that there is no objective truth, only mental constructs residing in the minds of human beings. It brings to mind a question I once heard a four year old child ask: "When I close my eyes, does the world disappear?" I think the only truthful answer he could have been given was a definite "No". Before human beings emerged in the course of evolution, there was an earth, and there were plants and animals. That is as objectively true as anything can be, and would still be objectively true if all human beings cast themselves into the ocean and drowned.

You do put me in a bit of a spot by saying in effect that any belief in objective truth is based on faith; and that good science depends on trust (on faith) that experiments or reasoning done by others is reliable. You are right, of course, about the crucial role that trust plays in science. However, I think there is a very solid basis for believing in the existence of an objective truth that is quite independent of human opinions: the observation that some explanations are "reliable", while others are not.

Science gives us a solid basis for placing our bets, even if science cannot directly reveal objective truth. Science tells us it is a very good bet that the sun will rise tomorrow morning and that the phase of the moon will continue to go through its 28-day cycle. It a very good bet that if I pray for an anvil to be levitated, it will *not* happen regardless of how much anyone believes in the efficacy of prayer. I think there is a reasonable basis for estimating the *probability* that there is an objective reality; and the estimate approaches 100% probability.

It should be possible to develop experimental tests to provide a reasonable basis for estimating the probability that purposiveness exists in natural selection. If it is possible to do so, then the question should be moved into the scientific arena.

The argument that the existence of the universe must have been *caused* (because everything must have a cause) should not be persuasive. Experimental evidence demonstrates that the outcome of any single quantum mechanical measurement cannot have a cause, though the *statistics* of many such measurements are clearly caused by the setup of the experiment and the underlying physics. However, evolution has shaped our genes to bias us toward believing that everything has a cause, because the belief is selectively advantageous in those cases where causes do exist and is selectively neutral in most cases where causes do not exist. So, our instinct that everything has a cause is usually right in ordinary circumstances but unreliable when we move far beyond ordinary circumstances, such as in trying to understand how the universe came to be.

Biography

Dr. Oren S. Harman is an Assistant Professor at the Interdisciplinary Program on Science, Technology and Society at Bar Ilan University. He received his doctorate from Oxford University in 2001, and was a Visiting Scholar and lecturer at Harvard University during 2000-2002. Harman is the author of *The Man Who Invented the Chromosome: A Life of Cyril Darlington* (Harvard University Press, 2004), and the co-editor of *Rebels, Mavericks and Heretics in Biology* (Yale University Press, 2008). He is a contributing science editor for The New Republic and is working on a book on altruism to be published by W.W. Norton.

9

Neo-Darwinism and Random Variation, or Why It's Important to Read Darwin Carefully

OREN S. HARMAN

Program in Science, Technology and Society
Bar Ilan University, Israel
harman@vms.huji.ac.il

1. Introduction

Any schoolchild who has had a year of biology will tell you that the logic of Darwinism is simple: as Darwin conceived it, it can be summed up by five facts and three inferences. To begin with, every population in nature has such high fertility rates that its size would increase exponentially if there were no constraints on its fertility. This was clear to the early nineteenth century reverend and writer William Paley as it was clear to his contemporary, the reverend and economist Thomas Malthus. A second fact, universally observed, is that, excepting temporary annual fluctuations, population size remains stable over time. Thirdly, also widely observed and re-enforced by Malthus again, is the fact that the resources available to any species are limited. From these three opening truths, the inference that there is intense competition, or, as Herbert Spencer called it, a "struggle for existence", within species in nature, was self-evident — first to Malthus when he applied this logic from nature to man, then to Darwin when he re-applied Malthus' logic from man back to nature. The fourth fact, strengthened by centuries of home-spun knowledge garnered by animal and plant breeders (and also taxonomists) is that no two individuals in a given population are ever exactly the same. From here, the jump to conceiving that individuals of a population differ from each other in their probability of survival was, historically speaking, a great leap, though logically not one of great difficulty — and it was achieved, first by Charles Darwin, and then later, though still with no publication by either man, by the young naturalist Alfred Russel Wallace. This principle was given the name Natural Selection, and, though it is often spoken of in shorthand as a force, it is in actual fact rather a result. Finally, the fifth and last fact of Darwin's explanatory model of evolution — that many of the differences between individuals in a population are, at least partly, inherited — led evenly and logically to the inference that natural selection over time would result in evolution.

This is a remarkably simple logic for the greatest, most robust theory about the natural world ever produced by science. And while schoolchildren who are taught about Darwin will tell you that the great man himself new little of the actual mechanism of heredity that produced the variation upon which natural selection was supposed to work, they will also be able to point out that the logic of Darwin's general explanatory model is flawless, and serves as the backbone of our understanding of evolution till this very day (Darwin, 1859).

Yet still, what was the source of variation upon which selection must work? The answer to that question was made clear by the application of the results of the work of the Mendelian biologists of the early twentieth century to the problem of evolution by a number of talented population genetics mathematicians during the 1920s and 1930s, principally, and famously, R.A. Fisher and J.B.S. Haldane in England, and Sewall Wright in the United States. By defining evolution as the differential inheritance and propagation of gene frequencies in a population over time, these men and their colleagues were able to demonstrate formally that the selection of heritable genetic variation in a population is strong enough a mechanism to account for the evolution of species over time. Lamarckian notions of the inheritance of acquired traits, either through direct effects of the environment on the organism or through the forces of use and disuse, rather than unequivocally refuted, were simply considered superfluous, and therefore unceremoniously dumped from the great twentieth century achievement in biology already referred to at the time as the Evolutionary Synthesis (Mayr and Provine, 1998).

Like Michelangelo's *Creation of Eve* in the Sistine Chapel, the synthesis rendered mutation at the center of the universe, only this time conceived in evolutionary terms. And, again, as every school child will tell you, mutation is the true source of new variation in nature, and is, besides, always random. Genetic randomness and natural selection became complementary and mutually dependent pillars of the edifice of the new biological worldview, suffusing the natural sciences with a respectability and rigor once afforded only to those physical or chemical. The champions of this new worldview called themselves Neo-Darwinists, and Darwin, who had come perilously close to being expunged altogether from biological textbooks in the first decades of the twentieth century, was now fashioned the very prophet of the new biology — the marriage of Mendelism and gradual evolution by natural selection.

But was Darwin really a Mendelian in the twentieth century sense, meaning that he believed that all variation was random, the consequence of what the Dutch biologist Hugo de Vries in 1904 called 'mutation' to the genetic factors responsible for heredity? Was he, like his twentieth century followers, a believer in the principle that genetic, and only genetic, hereditary variation counts when it comes to evolutionary phenomena — all other variation, while perhaps interesting, being nonetheless meaningless in generational, biological terms? This article will argue that a close reading of Darwin would suggest that, in fact, Darwin was much less of a strict Darwinian than would be imagined were we to infer that the Neo-Darwinists were being scrupulously accurate in calling themselves by his name. Reading Darwin carefully, it will be shown, is not only an exercise in revisionist history, but also, in light of recent developments in evolutionary theory and current political battles being waged over evolution alike, both a testament to the great English sage's perspicuity and farsightedness, and a useful tool with which to advance the hopelessly stagnant debates.

2. Darwinian Variation

Arguably Darwin's most difficult task was to explain where all the variation upon which natural selection acts comes from, for without variation natural selection would sadly remain unemployed and little in Nature would ever evolve. It was clear to Darwin that the origin of variation was closely related to the mechanism of heredity, but unfortunately little was known about how traits are passed on from one generation to the next. "Our ignorance of the laws of variation is profound," Darwin wrote in the Origin. "Not in one case out of a hundred can we pretend to assign any reason why this or that part is varied" (Darwin, 1859). Variation could be the result of some dimly perceived law of development, something Darwin called "correlation of growth" whereby any change in the embryo or larva would entail a change in the mature animal, sometimes with extremely distinct results. Thus did Darwin observe cats with blue eyes to be invariably deaf, or hairless dogs habitually the proud owners of imperfect teeth (Darwin, 1859). But variation could arise alternatively from the direct action of the conditions of life, irrespective of the organism's actions, such that coarse ground might directly produce protective padding on the camel's knee, or grainy wind slanted eyes in the inhabitant of Oriental deserts. Active use or disuse of an organ or function, such as the blacksmith's forearm, or the mole's eyes, might also be a source of variation, for the inherited change, as the Frenchman Jean-Baptiste Lamarck and Darwin's own English grandfather Erasmus had suggested at the turn of the nineteenth century, would pass from one generation to the next and thereby augment the pool of new things. Darwin was not quite sure why, but he knew that there existed much more variation in nature than most naturalists were happy to admit, and that this variation had to come from somewhere.

"Variability is governed by many unknown laws," was Darwin's reply to those who would have God responsible for every hue of feather and every angle of tooth (Darwin, 1859). Once it existed, natural selection would take care of the rest. But this simply would not fly for most of Darwin's contemporaries. Relinquishing variation to natural forces was for them tantamount to claiming that variation was arbitrary, with no ordained or foreseeable end. The idea of evolution could be stomached by most (including many liberal Christians), but could anyone really believe that the origin of the variation in Nature was random? Drawing a triumphant analogy with Swift's sardonic account in Gulliver's Travels of the Laputan practice of writing books by combining words randomly, the astronomer and philosopher John Herschel answered for most. "We can no more accept the principle of arbitrary and causal variation and natural selection as a sufficient account, per se, of the past and present organic world, than we can receive the Laputan method of composing books (pushed á l'outrance) as a sufficient one of Shakespeare and the *Principia*" (Herschel, 1861).

So Darwin's contemporaries wrongly interpreted Darwin to have meant that variation was random, because, by attempting and failing a naturalistic explanation of its origin, Darwin had ostensibly stripped the variability in nature of any well defined direction of which to speak. Paying little heed to the mechanisms he suggested that might be responsible for the creation of variation — after all, Darwin himself conceded that these were little understood — those who came after him, whether they were his followers or not — wrongly interpreted Darwinism to mean the selection of random variation.

Those who were convinced by evolution but could not bring themselves to believe that it was based on a cold, thoughtless mechanism of selection of random variation, sought out other explanations. Asa Grey of Harvard, for instance, an evolutionist and devout Christian, offered what the philosopher John Dewey mockingly called 'design on the installment plan' as an alternative: according to this theory, God provides a pool of preeminently suitable variations for selection to work upon (Gray, 1876; Dewey, 1951).

But Darwin himself, of course, had in fact taken a different tack. Lamarck had died poor and blind and was buried in a pauper's grave in Montparnasse, yet Darwin had resurrected the Frenchman's old theory with his own model of an hereditary mechanism to allow for the inheritance of acquired traits. The 'provisional hypothesis of pangenesis,' introduced in Darwin's *The Variations of Animals and Plants Under Domestication* in 1868, argued that invisible gemmules, spread throughout the body, act as memory cells recording the experiences of the organism, and, as they migrate from the arms and legs and lungs and toes to the gonads to create the sex cells, bring with them the knowledge gleaned by experience, and pass it down to the next generation. The blacksmith's son's forearms would bulge with pride on account of the blacksmith's toil (Darwin, 1896).

It has been noted, not without truth, that Darwin's provisional hypothesis of pangenesis was just that — provisional. In fact, it was something of a desperate attempt to counter what seemed to be a death blow to Darwin's theory of evolution by natural selection delivered by the Scottish engineer Fleeming Jenkins (Hull, 1973). On the supposition of blending inheritance, Jenkin argued, natural selection would be rendered forceless in the face of the law of regression to the median character of the population. In other words, since parents blend their hereditary material when passing it on to the next generation and since variation has been observed to revert to the mean over time, slight changes will be swamped in the great cauldron of life, and the evolution of whole populations over time will therefore become impossible. "Suppose a white man," the nineteenth century Scotsman wrote:

> "...to have been wrecked on an island inhabited by negroes... Suppose him to possess the physical strength, energy, and ability of a dominant

> white race...grant him every advantage which we can conceive a white to possess over the native... yet from all these admissions, there does not follow the conclusion that, after a limited or unlimited number of generations, the inhabitants of the island will be white. Our shipwrecked hero would probably become king; he would kill a great many blacks in the struggle for existence; he would have a great many wives and children, while many of his subjects would live and die as bachelors... In the first generation there will be some dozens of intelligent young mulattoes, much superior in average intelligence to the Negroes. We might expect the throne of some generations to be occupied by a more or less yellow king; but can any one believe that the whole island will gradually acquire a white, or even a yellow population... for if a very highly favored white cannot blanch a nation of Negroes, it will hardly be contended that a comparatively dull mulatto has a good chance of producing a tawny tribe" (Jenkin, 1867).

Modern readers will not appreciate Jenkin's racial sensibilities, but they should note nonetheless the ubiquity of the paradigm of "blending heredity" reflected in his argument. Jenkin offered that large sports might provide, in principle, the material for the production of evolution by natural selection, if only there existed a mechanism that would ensure that such sports would always breed among themselves — a mechanism patently unavailable in nature. Darwin, it seems, though he advocated small, almost imperceptible variation rather than large sports or leaps, floated up pangenesis nevertheless in order to provide just such a mechanism: if a population of organisms shared the same environmental challenges, and if those challenges were met physiologically and then handed down to the next generation (having been assimilated forthright through the inheritance of acquired characters), then it stands to reason that large proportions of the population will gain similar variations, and a variational directionality strengthened through selection will thereby ensue.

Darwin, of course, had no evidence whatsoever for his pangenetical claims, and he even admitted in a latter to a friend that if he were wrong, Fleemin Jenkin's critique would mean a true blow to his theory. Yet even if one accepts that Darwin was treading on thin empirical if not theoretical ice, it nonetheless remains the case that he was also making an exceedingly interesting point: The biology of organisms and their conditions of life are intimately entwined. Variation is not wholly random, but rather often directed, by the very forces of necessity encountered by the organism as it goes about the business of life and development. "If we suppose any habitual action to become inherited — and I think it can be shown that this does sometimes happen —" Darwin wrote in *The Origin*, "then the resemblance between what originally was a habit and an instinct becomes so close as not to be distinguished" (Darwin, 1859). Nature did not stand opposed to nurture, then, as Darwin's first cousin Francis Galton intoned, resurrecting what he mistakenly took to be a novel Shakespearean alliteration — (which actually originated in the pen of the English educator Richard Mulcaster in 1582) — but rather biologies react to environments producing new forms as they march along, hand in hand.

3. Historical Perversions

Darwin died in 1882, and was buried beneath Isaac Newton's tomb in Westminster Abbey. But there was a deep irony concealed in this august proceeding, for Darwin's originality was to propose the mechanism for evolution, natural selection, rather than the idea of transmutation itself. By the time of his internment, natural selection was already in its premature death throws. Darwin had won the battle for evolution, but failed miserably to convince that natural selection was its device (Bowler, 1983).

A greater irony still, and one more important for our purposes here, was that those men who began to call themselves 'Darwinists' championed natural selection to the exclusion of Darwin's own Lamarckian bent. August Weismann, for example, countered Darwin's 'provisional hypothesis of pangenesis' with his own equally literary 'immortal river of the germ-plasm' (Weismann, 1893). Weismann's notion was that the body (soma) and the sex cells (germ-plasm) are totally and wholly distinct, meaning that life experiences could never be passed down to the next age, Lamarck's giraffes and Darwin's pangenes notwithstanding. The germ-cells were like an immortal river, unsullied by the dirty banks of life, flowing from egg to egg and from sperm to sperm from one generation to the next. Heredity and environment were two opposing entities, never entwined, never penetrating each other. The only place where nature and nurture met face to face was when Nature blindly selected those variations (produced randomly in the germ-plasm) that rendered their lucky bearers fitter for this world. If there was still some telos (though not Godly) encrypted in Darwin's Lamarckian idea that through its actions the organism could direct the biological fate of its offspring, Weismann now stripped nature of any such ability. Natural selection, like Weismann himself towards the end of his days, was truly and hollowly blind.

Darwin's provisional hypothesis of pangenesis, of course, was wrong, not only because pangenes do not exist, and therefore do not acquire memory from the environment which they transmit to the sex cells, but because sex cells do not acquire DNA changes that have occurred to the body, full stop. Weismann's immortal germ line remains unsullied by life, and in this sense — the sense of direct adaptive feedback from the soma to the germ line — Lamarckism is indeed, for all we know today, very much dead.

But there is of course, another meaning of Lamarckism which, however wrong his attempt at a hereditary mechanism may have been, Darwin seems to have intuited, and his followers seem to have forgotten. This is the idea that the fit between an organism and its environment is much tighter and more intimate than would be expected if natural selection always acted on totally random variation. Some variation, at least, has to be instructed or directed.

Today we are smarter about such things than Darwin could have been in his day. Take, for example, the impressive ability of certain bacteria to increase the mutation rate of a specific gene involved in the metabolism of a given amino acid (a component of its diet) precisely when that amino acid becomes scarce in the bacteria's environment (Foster, 2000; Rosenberg, 2001). How on earth can this simple single-celled creature accomplish what is obviously a complex reactive computational trick? Until recently it has been assumed that mutations are non-adaptive and not developmentally controlled, but such examples beg to differ. If the apparatus responsible for increasing the mutation rate of the specific gene (which increases the chance of bringing about a variant that will be able to produce the necessary amino acid) is itself heritable, then the conclusion must be that adaptive changes can arise not only from natural selection of random, though 'fit' DNA, but also from the action of evolved internal systems that generate non-random 'guesses' in response to environmental challenges. Chance will, indeed, favor the prepared genome. "Rather than being restricted to contemplating a slow process depending on random (i.e. blind) genetic variation and gradual phenotypic change," writes the geneticist James Shapiro, "we are now free to think in realistic molecular ways about rapid genome restructuring guided by biological feedback networks" (Shapiro, 1999).

So mutations can sometimes be 'instructed' or non-random. But notice that Darwin actually said nothing of changes to the DNA because DNA was totally unknown to Darwin and his generation. It has been argued persuasively that precisely because Darwin was innocent of any true knowledge of the hereditary substance, he held a much broader and pluralistic notion of what constitutes heredity in the first place. In fact, it is perfectly possible to be a loyal Darwinian without being a Mendelian, for it turns out that natural selection is not as genocentric as twentieth century biologists have been. Unconstrained by intellectual fashions or shortcomings, it does not insist on moving all valuable information through Weismann's germ line (Jablonka and Lamb, 2005).

That variation is not always random is intimately connected to the fact that there is much more to heredity than simply the passing of genes (DNA) from one generation to the next. Jablonka and Lamb, in their recent book, provide a beautiful analogy to make the point: Think of a piece of music represented by a system of notes — the score. The relationship between the score and the music is analogous to the genotype/phenotype distinction: only the genotype (the score) is transmitted from generation to generation; the phenotype (the particular performance, or interpretation, of the score) is not. As Weismann instructed, changes in the genotype (mutations) are passed on; changes in the phenotype (acquired characters) are not. But consider what happened to music with the invention of the tape recorder: now interpretations of the score, not just the score itself, could be passed on from one generation to the next. And while in the past only changes to the score could alter changes to performances, the recorder made it possible for a particularly popular performance to lead to notational changes being incorporated in the score that make it easier for the interpretation to be reproduced. The very mechanisms that produce variation, as Jablonka and Lamb and others show, have undergone evolution, and the variation they produce is not all genetic. Here is one example: every cell in your body has the exact same DNA, but the particular pattern of genes that is expressed during development helps to turn one cell into heart tissue, and another into brain, and yet another into liver. If DNA is a recipe for a cake, then development is the process of baking. The molecular machinery responsible for the differential expression of genes (baking) is itself not made of DNA but rather mostly of protein, and is referred to as the epigenetic system (from the Greek epi — meaning beside). We've known for a while that the way a cake is baked can impact which ingredients are expressed in the final product, even if they all exist to begin with in the original paste. But if the epigenetic baking apparatus is itself heritable and influenced by the organisms' interaction with the environment (notice that this doesn't violate Weismann's axiom), then the neo-Darwinian divide between nature and nurture duly collapses, and the incorporation of the organisms' needs into its biology once again emerges as an important part of evolution. (The influence of "nurture" on "nature" can also have adverse affects: the way DNA is packaged in the nucleus is an inherited epigenetic trait, and influences which genes are turned on and which turned off. A major new direction in cancer research is exploring how particular packagings due to particular noxious environments can lead to cancer being passed down from one generation to the next). The important point is that while the environment has traditionally been viewed as the agent of selection — determining which variants survive and reproduce — it actually helps to determine which variants are there to be selected in the first place, too, because of the role it plays in development, or the baking of the cake.

I will not here consider the role of other hereditary systems, such as the behavioral and symbolic ones, that play a role in evolution, and that complicate the facile Weismannian divide between the evolutionarily relevant, randomly generated gametic variation and all other kinds of variation accrued by the soma, and therefore of no evolutionary significance. What is already clear is that genetic, epigenetic, behavioral, and cultural-symbolic evolution are more channeled than the singularly genetic evolution imagined by the architects and followers of neo-Darwinism. The targeted or constructed variation they produce may smell to some like a Lamarckian form of evolution smacking of teleology because it makes it sound as if variation arose on purpose. But, of course, as Jablonka and Lamb write, "it is simply a consequence of the various inheritance systems and the way they respond to internal and external influences" (Jablonka and Lamb, 2005). Nature is in fact blind, but it is nevertheless more than the mere sum of endless tosses of the genetic dice. Since information may be passed from generation to generation through other channels than the genes alone, the game of life is less a game of chance and necessity, as the great French biologist Jacque Monod called it, and more a game of chance, development, and necessity (Monod, 1974).

The general and banal conclusion must be that we are smarter today than we were in the past. When Mendel's results were rediscovered at the turn of the twentieth century, genes were identified with those entities that are transmitted from parents to offspring at conception and that are responsible for the development of the inherited traits. And as Darwinism and Mendelism became wedded in the great evolutionary synthesis of the 1930s and 1940s, the conceptualization of all variation as random was now perceived to be the shibboleth that rendered biology unburdened by teleology, allowing it to take its rightful place alongside physics and chemistry (Mayr and Provine, 1998). For most Neo-Darwinians "inherited" now became synonymous with "genetically determined" and "instinct" became synonymous with "genetically determined behavior". A version of this can be found in Konrad Lorenz's writings in the 1930s (Lorenz, 1937; 1939). Lorenz thought that all basic units of behavior are either instinctive (genetically determined) or acquired (determined by experience). He also thought that innate/inherited/genetically-determined traits can be explained evolutionarily but not developmentally, and that acquired traits can only be explained developmentally and not evolutionarily (Burkhardt, 2005).

Darwin may have gotten heredity wrong, but as the aforementioned and other recent examples of studies in development and evolution show, it is still perfectly possible to be a loyal Darwinian without being a Mendelian (Rollo, 1994; Schlichting and Pigliucci, 1998; West-Eberhard, 2003). This new understanding renders the almost automatic association between genetic randomness and natural selection championed for so much of the twentieth century somewhat naïve and simple, and at the same time trains an illuminating light on Darwin's own perspicacity from the perspective of one hundred and fifty years.

4. Conclusion: Intelligent Design and Current Debates

Why is any of this important? I would like to suggest that precisely in this day and age, when a fierce political battle is being waged, primarily in the United States, between proponents of evolution and those now calling themselves the champions of "Intelligent Design", the return to Darwin's broad, pluralistic approach to evolutionary phenomena might serve as a useful exercise not only, and obviously, in scientific terms, but also in those political. That is because, scientifically speaking, a broader approach to heredity might help to solve age-old conundrums, such as the origin of novel structures like wings of birds or flippers of once terrestrial whales. Even more dramatically, it provides a much firmer basis for the appearance during evolution of what the proponents of Intelligent Design refer to, non-scientifically, as examples of 'irreducible complexity' such as the eye, the brain, or the intricate workings of the kidney. Where Darwin (and Lamarck) failed in trying to describe a system of heredity, he nevertheless intuited a very basic truth, or the second, 'softer' meaning of Lamarckism: The fit between an organism and its environment is much tighter and more intimate than would be expected if natural selection always acted on totally random variation. Explaining nature's seeming perfection would be very difficult indeed, if there existed no mechanism or mechanisms to somehow allow for the transmission and assimilation of acquired information, even if this information is not strictly genetic. Darwin had no way of discovering such mechanisms; the irony is that it is the partial failure of the immensely successful enterprise of the men who championed an ultra-reductionist, DNA-centered approach to evolution, that called itself by Darwin's name, that is now providing the evidence for their mentor's prior claims.

These truths gain further meaning when they are considered in light of the political battles now being waged amongst evolutionists and Intelligent Designers. In facing the political challenge of the latest re-incarnation of creationism, this scientific path is a better one to follow than to try to argue for either an outdated neo-Darwinian theory (the one usually portrayed popularly by evolutionists simplifying for the laymen in the attempt to ridicule the creationist position) (Coyne, 2005) or that Darwin was a religious pluralist (he was — so what?) and really a teleological theist (he most assuredly was not!), as some good-intentioned interlocutors in the recent Intelligent Design-Evolution debates have done (Himmelfarb, 2005). Intelligent Design is an important political issue and a worthy cause for battle, but it is a hollow scientific debate. And while wars are usually characterized by a loss in self-reflection, the kind of self-questioning of the Neo-Darwinian dogma suggested by a closer reading of both Darwin himself, and of the historical reasons for the (mis)use of his name — coupled to the openness to recent developments in biology itself, may actually end up providing more dividends than either waging a battle the other side isn't listening to, or trying to assuage the other side with meaningless compromises. The really interesting stuff, it turns out, is happening not on the frontiers of a metaphoric or actual Bible Belt, or at the backwaters of a needlessly orthodox neo-Darwinian dogma, but immediately under our noses, in the fascinating new insights springing from the scientific realm of evolutionary theory itself. If we are to fight those forces who wield ignorance as a sword with which to dampen evolutionary claims, let us do so with the help of a hard-headed glance, unrestrained and unconditioned by dogma, at nature and her ways. And as for the glance, let us once again use Darwin as our guide: If we take the time to read him more closely than we usually do, we are bound to find that he was even more of a prophet than we consider him to be.

References

Bowler, P., (1983). The Eclipse of Darwinism: Anti-Darwinian Evolution Theories in the Decades Around 1900 (Baltimore: John Hopkins University Press.

Burkhardt, R.W. Jr., (2005). Patterns of Behavior: Konrad Lorenz, Niko Tinbergen, and the Founding of Ethology. Chicago, The University of Chicago Press.

Coyne, J., (2005). "The Faith That Dare Not Speak It's Name," The New Republic, August 22.

Darwin, C., (1859). On the Origin of Species by Means of Natural Selection. London, J. Murray.

Darwin, C., (1896). The Variation of Animals and Plants Under Domestication, Vol II. D. Appleton and Company: New York, Chapter XXVII.

Dewey, J., (1951). "The Influence of Darwinism on Philosophy," Popular Science Monthly, reprinted in Dewey, J. The Influence of Darwinism on Philosophy and Other Essays in Contemporary Thought. New York, P. Smith.

Foster, P.L., (2000). "Adaptive Mutation: Implications for Evolution," BioEssays 22, pp. 1067-1074.

Gray, A., (1876). Darwiniana: Essays and Reviews Pertaining to Darwinism. Appleton: New York, 1876.

Herschel, J., (1861). Physical Geography. Adam and Charles Black, Edinburgh, p. 12.

Himmelfarb, G., (2005). "How the Debate Over Darwin Hasn't Evolved," The New Republic, December 12.

Hull, D., (1973). Darwin and His Critics. Chicago and London, The University of Chicago Press.

Jablonka, E. and Lamb, M., (2005). Evolution in Four Dimensions: Genetic, Epigenetic, Behavioral and Symbolic Variation in the History of Life. Cambridge, Mass, MIT Press.

Jenkins, F. (1867). "Review of the Origin of Species" in The North British Review, 46, reproduced partially in Darwin: A Norton Critical Edition, Philip Appleman ed., (2001). Third Edition. New York and London: W.W. Norton and Company, p. 271-275.

Lorenz, K. (1939). "Vergleichende Verhaltensforschung," Zoologische Anzeitung 12, pp. 69-101.

Lorenz, K., (1937). "Uber den Begriff der Instincthandlung, Folia Biotheoretica 2, pp. 17-50.

Mayr, E., and Provine, W.B., (1998). The Evolutionary Synthesis: Perspectives on the Unification of Biology. Cambridge, Mass, Harvard University Press.

Monod, J., (1974). Chance and Necessity. Glasgow: Collins Fontana Books.

Rollo, C.D., (1994). Phenotypes: Their epigenetics, Ecology, and Evolution. London, Chapman and Hall.

Rosenberg, S.M., (2001). "Evolving Responsibility: Adaptive Mutation," Nature Reviews Genetics 2, pp. 1382-1383.

Schlichting, C. and Pigliucci M., (1998). Phenotypic Evolution: A Reaction Norm Perspective. Sunderland, Mass, Sinauer.

Shapiro, J., (1999). "Genome system architecture and natural genetic engineering in evolution," Annals of the New York Academy of Sciences 870, p. 32.

Weismann, A., (1893). The Germ Plasm: A Theory of Heredity, trans. W. Newton Parker and Harriet Ronfeldt. London.

West-Eberhard, M.J., (2003). Developmental Plasticity and Evolution. Oxford, Oxford University Press.

Biography

Dr. Charles H. Lineweaver is currently a Associate Professor at the Planetary Science Institute in the Research School of Astronomy and Astrophysics and the Research School of Earth Sciences at the Australian National University, Canberra, Australia. He obtained his Ph.D. from the University of California at Berkeley in 1994 and continued his studies and research at Strasbourg Observatory and the University of New South Wales. Dr. Lineweaver's scientific interests are in the areas of: planetology, cosmology and astrobiology.

10

Increasingly Overlapping Magisteria of Science and Religion

CHARLES H. LINEWEAVER

Planetary Science Institute, Research School of Astronomy and Astrophysics and Research School of Earth Sciences, The Australian National University, Canberra, ACT, 0200 Australia

charley@mso.anu.edu.au

1. Science and Religion: Separate Magisteria?

As explorers, missionaries and anthropologists of literate tribes came in contact with pre-literate tribes, we found out that every human tribe has religious ideas about gods, the supernatural and ethics. We can conclude from this that having some kind of religion is a universal part of human nature (Brown 1991). However, to be a scientist means depriving ourselves of the traditional inclination to invoke supernatural explanations for natural events. Scientists cannot invoke gods to explain mountains, lightning, stars, volcanoes, rain, birth, death or any other of the natural events that are part of our lives. But what about the origin of the universe and its laws, the origin of ethics and the purpose of life? Does science have a role to play in explaining these? Or are these phenomena within the exclusive domain of religion?

Occasionally scientists reflect on the relationship between science and religion (Townes 1966, Feynman 1999, Weinberg 2001) and debate whether our understanding of nature can help us make moral decisions or tell us how we ought to behave (Huxley 1894, Williams 1988ab). In *Rock of Ages* (1999) Stephen Jay Gould defends the idea that science and religion reign over separate conceptual kingdoms — separate "magisteria". The magisteria of science covers empirical facts: What is the universe made of? How does it behave? While the magisteria of religion extends over "the search for proper ethical values and the spiritual meaning of our lives." (Gould 1997). Gould calls this conceptual segregation "NOMA" (= Non Overlapping MAgisteria):

> "Science can work only with naturalistic explanations; it can neither affirm nor deny other types of actors (like God) in other spheres (the moral realm, for example)" (Gould 1992).

According to Gould science can tell us how the heavens go, not how to go to heaven. Science can tell us how we behave not how we should behave. Science can tell us what is, not what ought to be. Science is descriptive, not prescriptive. Apparently, we cannot rely on science for all the answers. We must turn to a religion and a holy book or we must look inside ourselves and ask our consciences how we should behave.

Gould is not alone in this opinion:

> "Facts do not constitute a criterion for the judgment of the moral content of human acts." (Cardinal Joseph Ratzinger, 1970, head of The Vatican's Congregation for the Propagation of the Faith).

From science, one "will never obtain a proposition which says: do this, or do not do that; that is to say a proposition which confirms or contradicts ethics" (Gillispie 1973).

> "The evaluation of moral codes or human actions must take into account biological knowledge. But for deciding which moral codes should be accepted, biology alone is palpably insufficient" (Ayala 1987).

Theologian John Haught's testimony against Intelligent Design being taught in Pennsylvania schools, was an expression of NOMA:

> "In deliberately omitting theological or 'ultimate' explanations for the existence or characteristics of the natural world, science does not consider issues of 'meaning' and 'purpose' in the world." (Kitzmiller v. Dover Area School District 2005).

Critics of and commentators on Gould's NOMA include Hall (1999), Dennett (2006) and me (this article):

> "Gould is right, of course, that there should be no conflict, since, from the religious point of view, anything that science finds to be the probable truth about how things actually are, can be incorporated into religion by just saying 'that's how God did it;' and, from the scientific point of view, all of the religious talk about 'souls' and 'afterlife' is as meaningful as the story of Frodo's voyage to the Gray Havens" (Hall, 1999).

> "Although Gould's desire for peace between these often warring perspectives was laudable, his proposal found little favor on either side, since in the minds of the religious it proposed abandoning all religious claims to factual truth and understanding of the natural world (including claims that God created the universe, or performs miracles or listens to prayers) whereas in the minds of the secularists it granted too much authority to religion in matters of ethics and meaning" (Dennett 2006, p. 30).

As science begins to address more fundamental and previously taboo issues (the origin of the universe, the origin of life, the origin and evolution of humanity, the evolutionary biology of human nature, sexuality, consciousness and morality), the magisteria of science expands and increasingly overlaps with the traditional monopoly that religion has had on these areas. Big bang cosmology and Darwinian evolution are revolutionary scientific ideas that have replaced our traditional religious conceptions of the origin of the world and the origin of humanity.

Science also makes large contributions to our search for "proper ethical values". For example, anthropologists studying other cultures often open our minds to the morals of other cultures. Margaret Mead's book "*Coming of Age in Samoa*" (1928) showed us that there are other acceptable ways to raise children and this knowledge changed our behavior. Science has an important voice in many areas that have been taboo to scientific inquiry. The scientific study of human sexuality removed much of the guilt associated with supposedly (but not in actuality) deviant sexual behavior (Kinsey 1953, Masters and Johnson 1966). As scientific studies explore more issues and find out how we behave, our view of ourselves changes and this often changes how we think we should behave. The is and the ought cannot be confined to separate realms.

2. Divine Action and Gods of the Gaps

Religions of literate societies have tried hard to distance themselves from the animistic traditions from which they emerged because these traditional superstitions were in obvious conflict with emerging scientific understanding of lightning, thunder, earthquakes, volcanoes, what goes on at the top of mount Olympus and what pulls the Sun across the sky. Many expressions of religious philosophy accept scientific knowledge of the empirical world, but try to carve out areas in which science has no validity. The areas in which religion is held to have superior knowledge include knowledge of the spirit, the soul, the origin and purpose of human life and ethics. As scientific explanatory power increases, science begins to have something to say about these areas (more in some than others). Religious explanations become marginalized and confined to gaps. The phrase 'God of the gaps' was coined by Coulson (1958) to describe this process.

It is difficult, however, to sustain religious descriptions of the physical world, in which gods act as a cause complementing physical causes (divine action) — filling in the gaps left by scientific explanations (Saunders 2002). Acceptance that force could act at a distance eliminated the need for gods to mediate the force of gravity. An understanding of Darwinian evolution eliminated the need for gods to create people and more generally eliminated the need for gods to design each creature individually.

On the other hand, although biochemistry and genetics indicate that the molecules of life are made of normal atoms and behave according to the laws of chemistry, we still don't understand the origin of life very well; so there is a gap for gods there to breathe the breath of life or "elan vital" into the living. We also do not understand the origin of the laws of physics at all so there is a really big gap there.

At the Vatican cosmology conference in 1981, Pope John Paul II addressed the attendees. White and Gribbin (2002) described the proceedings:

> "[T]he Pope warned the physicists against delving too deeply into the question of how or why the Universe began, reminding them that this

> was solely a matter for theologians. He went on: Any scientific hypothesis on the origin of the world, such as that of the primeval atom from which the whole of the physical world derived, leaves open the problem concerning the beginning of the Universe. Science cannot by itself resolve such a question; what is needed is that human knowledge that rises above physics and astrophysics which is called metaphysics; it needs above all the knowledge that comes from the revelation of God. Hawking sat impassively in his wheelchair listening."

It was at this 1981 Vatican conference that Hawking announced his controversial "no-boundary" theorem, a scientific attempt to explain the origin of the universe; precisely what the Pope was warning against. The magisteria of science seems to be expanding and overlapping increasingly with the magisteria of religion.

3. Biology-based Ethics?

Nature is "neither moral nor immoral, but non-moral" (Huxley 1894):

> "Although during his career he [Thomas H. Huxley] contributed energetically to the advance of Darwinism, he came at last to perceive an impossible cleavage between the Darwinian view of nature 'red in tooth and claw' and human values. Finding it impossible to reconcile the two, he abandoned the quest for a comprehensive evolutionary world view." (Oates 1988).

Huxley's focus on "red in tooth and claw" competition meant that Darwinian evolution could not explain the development of human values. The dilemma Huxley faced was not, however, irresolvable. Current research seems to be uncovering the large, but less explicit contribution that cooperation makes to adaptation and survival (Ridley 1997, Sober and Wilson 1998; see, however, Williams 1988a,b).

The evolutionary origins of human behavior and the evolutionary approach to ethics are the subject of much research and debate: (Darwin 1871; Williams 1966, 1988a,b, Wilson 1976, 1978, Hamilton 1975; 1998, Dawkins 1976; Ayala 1987, Barkow et al. 1992, Ridley 1996, Sober and Wilson 1998, Lieberman et al. 2003 and references therein). Leading evolutionary biologists have argued that selection has "shaped" human morality. The extent of that "shaping" is a scientific question that can be addressed with difficulty, but has important consequences for our view of ourselves and how we think we should behave (Pinker 2002).

> "The biologist, who is concerned with questions of physiology and evolutionary history, realizes that self-knowledge is constrained and shaped by the emotional control centers in the hypothalamus and limbic system of the brain. These centers flood our consciousness with all the emotions, hate, love, guilt, fear, and others-that are consulted by ethical philosophers who wish to intuit the standards of good and evil. What, we are then compelled to ask, made the hypothalamus and limbic system? They evolved by natural selection" (Wilson 1975).

Sociobiology is the systematic study of the biological basis of all social behavior (Wilson 1975). This study can and should be used to help us modify our behavior. If we can understand how our "oughts" came to be inside our heads, then we can look at ourselves and reevaluate what our "oughts" ought to be. There are scientific facts about being human (Wilson 1978) which are important and have important moral implications that help me (and many others) make moral decisions about how to behave. See Bleier (1984) for a criticism of this sociobiological program.

Incest taboos (Jones 1997, Lieberman et al. 2003) produce ethical decisions about who we ought, and ought not, to marry. The origins of these taboos can be understood as the result of an evolutionary selection to eliminate inbreeding and troublesome double recessive genes. Thus, the scientific understanding of the reasons for taboos tells us how we should behave and why. We can modify our behavior to avoid the real problems. We are not left blind with the procrustean fears induced by our traditional supernatural narratives.

Food taboos can also sometimes be explained scientifically. The scientific understanding of trichinosis and the undoing of the taboo against eating pork is an example. Roadside police making scientific measurements of alcohol blood levels is a graphic example of science determining how much you should drink. But this is not a moral issue is it? Are we dealing with a moving target? I.e. once science has understood a phenomenon and determined how one should act in a given situation to minimize harm (who to marry, how to prepare food, how much to drink), are our decisions to act rationally no longer classified as "moral decisions"? If so, then by definition, this excludes science from influencing "moral" behavior and is equivalent to a moral-decisions-of-the-gaps approach.

Ethical values and a sense of meaning are the products of natural selection. Our ideas of "good" and "evil", and our consciences that we rely on to help us make moral decisions (like Pinnochio's Jiminy Cricket), are features of consciousness that have evolved under selection pressure, just like skin color, intestinal pH and fingernail growth rates. The fact that moral decision-making is complex and contextual does not mean that the process — like our other brain functions — did not evolve (Barkow et al. 1992, Pinker 2002). This being so, we can try to understand their origin, why they are there, what purpose they serve and based on this information decide how we should behave. Even speculative explorations of evolutionary psychology give us a better (or at least a complementary) understanding of ourselves than Moses-went-up-to-the-mountain-and-came-down-with-some-rules-that-we-are-supposed-to-follow-because-God-said-so explanations of human ethics.

4. The Origin of Life, High Expectations and Looking for a Purpose

"It is mere rubbish, thinking at present of the origin of life; one might as well think of the origin of matter." (Darwin 1863). The magisteria of science has expanded since Darwin's time. Studies of the origin of matter (baryogenesis), is one of the most important topics in modern cosmology (e.g. Sahkarov 1967, Kolb and Turner 1990). And studies of the origin of life (biogenesis) is one of the most important fields of modern biology (e.g. Schroedinger 1944, Fox 1986, de Duve 1995, Lahav 1999), and has spawned the new science of astrobiology (e.g. Seckbach et al. 2004, Lunine 2005, Hazen 2006).

What about purpose? The first thing we need to know about purpose is: Why do we want to find one? Here are three quotes about purpose from a fundamentalist Christian, a movie star and a Nobel prize winner in physics that point to an answer.

> "Darwinian theory of evolution contradicts not just the Book of Genesis, but every word in the *Bible* from beginning to end. It contradicts the idea that we are here because a creator brought about our existence for a purpose." (quote from Johnson 1991, cited in Kitzmiller v. Dover Area School District 2005).

> "Nature, Mr. Alnutt, is what we were put in this world to rise above." (Katharine Hepburn to Humphrey Bogart in the 1951 film "The African Queen")

> "Below the Earth looks very soft and comfortable — fluffy clouds here and there, snow turning pink as the Sun sets, roads stretching straight across the country from one town to another. It is very hard to realize this all is just a tiny part of an overwhelmingly hostile Universe. It is even harder to realize that this present Universe has evolved from an unspeakably unfamiliar early condition, and faces a future extinction of endless cold or intolerable heat. The more the Universe seems comprehensible, the more it also seems pointless." (Weinberg 1977)

Thirteen years later Weinberg reflected further on his reflections:

> "If you say things are pointless, you have to ask, 'Well, what point were you looking for?'" (Lightman and Brawer 1990 p. 466)

The universe is only pointless to the degree that we insist it have a point. Our expectations, not the universe, are to blame for the apparent pointlessness. We have high hopes for top billing — hopes to see our names writ large on the universe, above nature and at the express orders of the gods. Real observations of nature cannot compete with such expectations. We are left disappointed. Expectations high enough can always preclude one from accepting objective facts not specifically constructed to boost our egos. In many ways, scientific observations engender in us a sense of humility about our position in the universe. This is another way in which science shapes our self-opinion and shapes our moral stance with respect to the Earth and universe.

Believing you have a purpose is probably adaptive. It makes you feel worth something and increases self-esteem. However, it also produces vanity and eventually disappointment when science finds out (as it inevitably keeps finding out) that the universe does not rotate around us. The resultant disappointment is seen in Steven Weinberg's casual comments about the pointlessness of the universe, in the fundamentalist Christian's irate anti-Darwinism and in Hepburn's pre-Darwinian airs about humanity belonging above nature. Human beings apparently need a sense of purpose.

> "[E]ven a "scientific" world view must answer the needs of the human mind for a sense of order and meaning. It is not enough to offer the intellect a vision of how the whole is put together. It must also offer the emotions a sense of belonging, the imagination a sense of delight, the heart a sense of goodness. These are much of what distinguish a world view from mere theory or speculation. And the skill with which these affective elements are integrated with the intellectual and empirical is certainly one of the crucial measures of a world view's power-in Darwin's day, in Huxley's and in our own." (Oates 1988)

The expectation that humans have a more noble purpose that matters in the grand scheme of things, is not easily jettisoned. Expectations exist for a reason. We are life forms that have been specifically selected to harbor the adaptive belief that we are important. Thinking that you are worth more than the thing you need to kill to stay alive has adaptive value. Evolutionary psychologists might explain that encephalated heterotrophs who did not believe this have gone extinct for lack of psychological reinforcements. Natural selection of encephalated beings has as much to do with selecting adaptive ideas as it has to do with physical adaptiveness. In large-brained beings, the survival of genes and memes are correlated (Dennett 2006). I look to science to help me understand why we think that humans are worth more than other species, and why my in-group, my nation, my culture is better than other groups, nations and cultures. The current scientific debate over group selection may help answer some of these questions (e.g. Sober and Wilson 1998). For a perceptive discussion of the status of such adaptive explanations, see Gray et al. (2003).

5. Is the Purpose of Life The Production of Entropy?

Another topic within the traditionally religious magisteria is the purpose of life. In contrast to most scientists, I think that science has a lot to say about it. If one gives up the idea that human beings are somehow special and have a purpose that is different from other life, one is free to explore the question without fear of disappointment. Dawkins articulated the case for the purpose of life as the promotion of the survival of selfish genes:

> "We are survival machines — robot vehicles blindly programmed to preserve the selfish molecules known as genes. This is a truth which

> still fills me with astonishment." (Richard Dawkins in the preface of *The Selfish Gene* 1976)

But if our purpose is to preserve genes, what is the purpose of the genes? If self-preservation is their purpose, where did this purpose come from? Where did the information in the genes come from that promotes this purposeful activity?
All of the information content of genes came from the environment, i.e. from the selection pressures, on molecules, of the architectural complexity of a low entropy environment (e.g. Fox 1986). A low entropy state seems to have been the initial state of the universe (Frautschi 1982, Lineweaver 2005). In an environment in equilibrium, without structure — without specific chemical free energy gradients, no information could be inserted into genes — no information could be extracted by genes — because there would be no information. There would be no selection pressure — no advantage to one type of behavior over another. No purposefulness could evolve in the form of self-preserving genes. Life requires non-equilibrium. You can't get into the self-preservation business in the first place without a structured, information-rich environmental architecture to push you around in specific ways. Without free energy gradients, life in particular, and far from equilibrium dissipative structures ("FEDS") in general, are not possible. With such gradients, FEDS (in general) are inevitable. We see them everywhere in the universe. Whether free energy gradients inevitably produce the type of FEDS called life is an open question (Prigogine 1980, Prigogine and Stengers 1984, Schneider and Kay 1994, Schneider 2004, Schneider and Sagan 2005).

Given a low entropy environment, there will be lots of structure, lots of free energy and the second law of thermodynamics will spontaneously create FEDS which seem to have as their "purpose" (at least that is what they do) the production of entropy over and above the amount of entropy that would be produced in their absence by diffusion and other relaxation processes. These far from equilibrium dissipative structures (galaxies, stars, convection cells, hurricanes, whirlpools, life forms) promote the flow of material, and heat and reduce gradients (Lineweaver 2006). To take seriously the idea that the purpose of life is to increase the entropy of the universe we need to show that the second law holds in far from equilibrium situations. This is not straight-forward (Dewar 2003, 2005, Kleidon and Lorenz 2005), however, it seems to be the most reasonable working assumption.

Dawkins tells us that it's not people who are in control of their destiny, it is the selfish genes who control us. But one can also postulate that genes are not selfish. Genes are the unwitting pawns of the second law, which has conjured them up to increase the entropy of the universe, much as the genes, in their turn, have conjured us up to do their bidding. Thus we can take Dawkins' logic one step further and propose that:

Genes are entropy machines — robot vehicles blindly programmed to increase a quantity known as entropy. This is a tentative truth which is beginning to fill me with astonishment.

Purposes are linked to origins, and finding out the origins of life; what prerequisites drove life to become what it is; can go far — possibly as far as one can go — towards revealing purpose. The "purpose" of a hurricane is to reduce the pressure and temperature gradients and increase entropy (at least that is what it does). The "purpose" of a convection cell is to reduce the temperature gradient faster than diffusion would allow and increase entropy. In the same sense, I suspect that the "purpose" of life is the same as these other far from equilibrium dissipative structures: to reduce gradients and increase entropy. The fact that this "purpose" does not conform to the type of answer we want and expect increases my confidence in its validity because it has not been constructed to flatter us. See Sagan and Whitesides (2004), Schneider and Sagan (2005), Schneider and Kay (1994).

A convection cell is nature's way of reducing a thermal gradient, or nature's way of increasing the entropy of the universe faster than it would without the existence of a convection cell. Thermal gradients produce convection cells and these convection cells reduce thermal gradients. There is a chronological order here. The same chronological order was followed when chemical gradients produced proto-life, proto-life reduced chemical gradients and along the way, proto-life centralized some information and became life, free to search out chemical gradients to survive off of.

6. Ancestors Become Food

Complex structures form whenever there is an energy gradient to drive them. It is appealing and possibly useful that the "purpose" of these structures is to get rid of the gradient that produced it. That is, the purpose of these structures is to increase the entropy of their environments.

The second law of thermodynamics brings about spontaneously the existence of far from equilibrium dissipative systems "to reduce" (or "that reduce") free energy gradients. One of those systems (life) learned how to centralize some information (genes) and learned how to reproduce itself. With this autonomy it could actively seek out the chemical free energy gradients. Gradients which were needed for its existence now could be actively sought. Today life sees free energy gradients as food, rather than ancestors. Self-preservation became a more important goal than the original goal of gradient reduction. Gradient reduction became a secondary goal or by-product of its self-preservative activity. The origin of life and the original purpose of life are connected by the second law. Life's new purpose is preservation. But life still shares its primary purpose with convection cells, stars, hurricanes and other far from equilibrium dissipative systems — and that purpose is the reduction of gradients and the production of entropy.

Consider the following two hypotheses:

1) Entropy production is a by-product of life (usual assumption)
2) Life is a by-product of entropy production

The first places life at the center while the second gives primacy to entropy production. I am arguing qualitatively for 2), that the entropy production of the second law is the driving force responsible for the origin of FEDS and life. But, I'm not sure these ideas are precise enough to test quantitatively. Could both be true? Consider two more hypotheses:

a) An egg is a chicken's way of making another chicken (usual assumption)
b) A chicken is an egg's way of making another egg

There is an attractive symmetry here that leads one to say that both could be correct. The real difficulty for those who seek "purpose" lies in the unanswered question: Who is "in control"? Who is the "active agent"? In the absence of a way to determine the "active agent" we could base our answer on which is more fundamental — Which came first, the chicken or the egg? The easy answer to this question is that the egg came first since reptiles laid eggs and birds evolved from reptiles. The egg is more fundamental and thus the chicken is the egg's way of making another egg. Similarly, entropy production and non-living FEDS preceded, and are fundamental to, life. Life is a subset of FEDS, not the other way around (Lineweaver 2006). Thus, life is a by-product of entropy production and statement "2)" is probably closer to the truth than "1)". Whether these ideas are precise enough to test more rigorously is an open question.

Quantum cosmologists are comparing their models of the origin of the universe with observations of the cosmic microwave background. Organic chemists are exploring alternatives to DNA as they try to create life in the laboratory. Molecular biologists are retracing phylogenetic relationships to the last universal common ancestor of all life to sniff out what the earliest life was like. Primatologist and evolutionary psychologists are trying to uncover the sources of human morality. Science has dispelled many myths and solved practical problems. The scientific method of observation, testing, hypothesis building, testing, rethinking and re-testing is, I think, the most powerful tool we have for producing knowledge about the origin of the universe, the origin of life, who we are, how we behave and the sources of our ethical behaviour. This knowledge has changed the way we behave, modified our self-opinions and changed the way we think we ought to behave. The conceptual realms of science and religion are increasingly overlapping magisteria.

References

Ayala, F.J. 1987 "The Biological Roots of Morality" *Biology and Philosophy*, **2**, 235-252.

Barkow, J.H., Cosmides, L., Tooby, J. 1992, *The Adapted Mind: Evolutionary Psychology and the Generation of Culture*, Oxford Univ. Press: Oxford.

Bleier, R. 1984, *Science and Gender: A Critique of Biology and Its Theories on Women*, Pergamon Press, Athene Series.

Brown, D.E. 1991. *Human Universals*. McGraw-Hill: NY.

Coulson C.A., 1958 *Science and Christian Belief*, Fontana: Great Britain p. 32.

Darwin, C.R. 1863 in letter to J.D. Hooker, March 29, 1863, cited in Darwin F., ed., *The Life and Letters of Charles Darwin*, [1898], Basic Books: New York NY, Vol. II., 1959, reprint, pp 202-203.

Darwin, C.R. 1871 *The Descent of Man, and Selection in Relation to Sex*. Murray: London.

Dawkins, R. 1976 *The Selfish Gene*. Oxford Univ. Press: Oxford.

de Duve, C. 1995, *Vital Dust: The Origin and Evolution of Life on Earth*, Basic Books NY.

Dennett, D. 2006, *Breaking the Spell: Religion as a Natural Phenomenon*, Viking: NY.

Dewar R.C. 2003 "Information theory explanation of the fluctuation theorem, maximum entropy production and self-organized criticality in non-equilibrium stationary states". *Phys. A Math, Gen* **36**:631.

Dewar R.C. 2005 Maximum Entropy Production and Non-Equilibrium Statistical Mechanics, in: A. Kleidon and R. Lorenz. *Non-equilibrium Thermodynamics and the Production of Entropy: Life, Earth and Beyond.* Springer: Berlin pp 41-53.

Feynman, R. 1999 *The Meaning of It All*, Perseus.

Fox, S. 1986, "Molecular Selection and Natural Selection" *Quarterly Review of Biology*, **Vol 61**, No. 3 (Sept 1986) pp 375-386.

Frautschi S. 1982 "Entropy in an Expanding Universe". *Science* **217**:593-599.

Gillispie, G.C. 1973, *The Edge of Objectivity*. Princeton Univ. Press: Princeton, N.J. pp 154-155.

Gould, S.J. 1992, "Impeach a Self-appointed Judge", review of P.F. Johnson's "*Darwin on Trial*" in *Scientific American.*

Gould, S.J. 1997, "Nonoverlapping magisteria", Natural History 106, March, 1997, pp 16-22 Reprinted in *Leonardo's Mountain of Clams and the Diet of Worms*, 1998, Harmony Books: NY pp 269-283.

Gould, S.J. 1999, *Rocks of Ages*, Ballantine Books: NY.

Gray, R.D., Heaney, M. and Fairhall, S.2003, "Evolutionary Psychology and the challenge of adaptive explanation" in: K. Sterelny and J. Fitness (eds) *From Mating to Mentality: Evaluating Evolutionary Psychology.* Psychology Press London, pp. 247-268, Chapter 11.

Hall, N. 1999, "Science and Religion: Conflict or Conciliation" *Skeptical Inquirer*, Special Issue on Science and Religion, July-August 1999.

Hamilton, W. D. 1975 "Innate social aptitudes of Man: an approach from evolutionary genetics". in R. Fox *Biosocial Anthropology*, pp. 133–155. Halsted Press: NY.

Hamilton, W.D. 1998 *The Narrow Roads of Gene Land, Vol 1, Evolution of Social Behavior*, Oxford Univ. Press Oxford.

Hazen, R.M. 2006, *Genesis: The Scientific Quest for Life's Origin*, Joseph Henry Press: Wash. DC.

Huxley, Thomas H. 1894. *Evolution and Ethics and Other Essays*, D. Appleton: NY.

Jones, S. 1997 *In the Blood: God, Genes and Destiny*, Flamingo: London.

Johnson, P. 1991, *Darwin on Trial*, Regnery Gateway Publishing: Wash. D.C.

Kinsey, A.C. 1953, *Sexual Behaviour in the Human Female*, W.B. Saunders: Philadelphia.

Kitzmiller v Dover Area School District, 2005, 400 F. Supp. 2d 707, 735 (M.D. Pa 2005) (quoting testimony of John Haught) (also quoting Johnson's book *Darwin on Trial*).

Kleidon, A. and Lorenz, R. 2005 *Non-equilibrium Thermodynamics and the Production of Entropy: Life, Earth and Beyond*, Springer: Berlin.

Kolb, E.W. and Turner, M.S. 1990, *The Early Universe*, Addison Wesley: Redwood City CA.

Lahav, N. 1999 *Biogenesis: Theories of Life's Origin*, Oxford Univ. Press: Oxford.

Lieberman, D. Tooby, J. and Cosmides, L. 2003, "Does morality have a biological basis? An empirical test of the factors governing moral sentiments relating to incest" *Proc. R. Soc. Lond. B* (2003) **270**, 819–826.

Lineweaver, C.H. 2005, "Cosmological and Biological Reproducibility: Limits on the Maximum Entropy Production Principle" in: A. Kleidon and R. Lorenz, *Non-equilibrium Thermodynamics and the Production of Entropy: Life, Earth and Beyond*, Springer: Berlin pp 67-76.

Lineweaver, C.H. 2006, "We Have Not Detected Extraterrestrial Life, or Have We?" in J. Seckbach *Life as We Know It* Springer: Dordrecht.

Lightman, A. Brawer, R. 1990, *Origins: The Lives and Worlds of Modern Cosmologists*, Harvard Univ. Press: Cambridge, MA.

Lunine, J.I. 2005, *Astrobiology: A Multi-disciplinary Approach* Pearson: San Francisco.

Masters, W.H. and Johnson, V. 1966, *Human Sexual Response*, Lippincott/Williams and Wilkins: NY

Mead, M. 1928, *Coming of Age in Samoa*, William Morrow: NY.

Oates, D. 1988, "Social Darwinism and Natural Theodicy" *Zygon*, **24**, No 4 Dec.

Pinker, S. 2002, *The Blank Slate*, Viking: NY.

Prigogine, I. 1980 *From Being to Becoming*, WH Freeman: NY.

Prigogine, I. and Stengers, I. 1984 *Order Out of Chaos*, Bantam NY.

Ridley, M. 1997 *The Origins of Virtue, Human Instincts and the Evolution of Cooperation*, Viking: NY.

Sagan, D. and Whiteside, J.H. 2004 "Gradient Reduction Theory: Thermodynamics and the Purpose of Life" in: S.H. Schneider, J.R. Miller, E. Crist and P.J. Boston *Scientists Debate Gaia: the Next Century*, MIT Press, Cambridge, MA p. 173.

Sakharov, A.D. 1967, "Violations of CP Invariance, C Asymmetry and Baryon Asymmetry of the Universe" *Pis'ma Zh. Eksp. Teor. Fiz* **5**, 32 [JETP Letters **5**, 24, 1967].

Saunders, N. 2002 *Divine Action and Modern Science*, Cambridge Univ. Press UK.

Schneider, E.D. 2004 "Gaia: Towards a Thermodynamics of Life" in S.H. Schneider, J.R.Miller, E. Crist and P.J. Boston *Scientists Debate Gaia: the Next Century*, MIT Press: Cambridge, MA p 45.

Schneider, E.D. and Kay, J.J. 1994, "Life as a Manifestation of the Second law of Thermodynamics" *Mathematical and Computer Modeling*, **Vol 19**, No. 6-8 pp 25-48.

Schneider, E.D. and Sagan, D. 2005, *Into the Cool: Energy Flow, Thermodynamics and Life*, University of Chicago Press, Chicago Chapter 20 "Purpose in Life".

Schneider, S.H., Miller, J.R. Crist, E. and Boston, P.J. 2004, *Scientists Debate Gaia: The Next Century*, MIT Press, Cambridge, MA.

Schroedinger, E. 1944 *What is Life?* Cambridge Univ. Press: UK.

Seckbach, J. Chela-Flores, J., Owen, T and Raulin, F. 2004. *Life In The Universe: From the Miller Experiment to the Search for Life on Other Worlds*, Kluwer: Dordrecht.

Sober, E. and Wilson, D.S. 1998, *Unto Others: The Evolution and Psychology of Unselfish Behavior*, Harvard Univ. Press: Cambridge, MA.

Townes, C. 1966 "The Convergence of Science and Religion" *Think*, published by IBM March-April 1966. Vol 32, No 2.p 2-7.

Weinberg, S. 1977 The First three minutes, Basic Books: NY.

Weinberg, S. 2001, "A Designer Universe?" *Skeptical Inquirer* Sept/Oct 2001, **25**:5 p 64.

White, M. and Gribbin, J. 2002, *Stephen Hawking: A Life in Science*, Second Edition, Joseph Henry Press (p 201).

Williams, G. C. 1966 *Adaptation and Natural Selection*, Princeton Univ. Press:Princeton, NJ.

Williams, G.C. 1988a, "Huxley's Evolution and Ethics in Sociobiological Perspective", *Zygon*, **23**, 4, 383-407.

Williams, G.C. 1988b, Reply to Comments on "Huxley's Evolution and Ethics in Sociobiological Perspective", *Zygon*, **23**, 4 383-407.

Wilson, E.O. 1976, *Sociobiology*, Belknap Press of Harvard Univ. Press, Cambridge, MA.

Wilson, E.O. 1978 *On Human Nature*, Harvard Univ. Press, Cambridge: MA.

Dialogue

Oren Harman

You challenge the notion that science is descriptive rather than prescriptive. That is to say, the idea that science and religion represent different and distinct magesteria that do not intersect seems to you at best naïve, and, perhaps more menacingly — deeply misguided. You present a number of examples, such as big bang cosmology, Darwinian evolutionary theory, and the work of Margaret Mead and of Alfred Kinsey, in which you claim that when science deals with problems such as the origins of the universe and of life, the origins and evolution of humanity, sexuality, morality and consciousness, it cannot but affect the way we think about these issues outside the realm of science, namely — our moral attitudes towards such topics.

The argument you present is based on the underlying assumption that facts and values can never really be separated one from the other. The reason for this is that facts (not all, but some) compel us (in the sense that they leave us no other option) to view things in the moral sphere differently.

I hold the view that, to the contrary, facts about the world and values are distinct from one another. Science really can never tell us how to behave, or "how to get to heaven", or, for that matter, whether or not there is a god waiting for us up there. Of course, what we learn about the world impacts upon how we see our proper place in it, but the important thing to remember is that that relationship is not determinative. Facts about the world can never compel us to feel a certain way versus a certain other way. There are disagreements about the implications of facts. When a certain fact emerges from scientific inquiry — let's say, for the sake of example, that we discover that animals can feel and understand pain – this alone does not compel anything. What it does do is to lead to a kind of moral and political negotiation about the implication of such knowledge for our attitudes and behavior towards animals. The fact that different peoples in different cultures will handle this negotiation in different ways, reaching different conclusions and feelings, is enough to show that the magesteria are indeed separate.

Charles H. Lineweaver
In the first paragraph you describe my position as:

"...when science deals with problems such as the origins of the universe and of life, the origins and evolution of humanity, sexuality, morality and consciousness, it cannot but affect the way we think about these issues outside the realm of science, namely — our moral attitudes towards such topics."

I do not agree that "our moral attitudes towards such topics" are outside the realm of science. Brains are organs of our bodies, that, like all other organs, have evolved to help us survive. Despite the claims of traditional religion and some secular humanist scholars, I do not believe that our moral attitudes are immune from, or outside the influences of natural selection. Rather, our moral attitudes, like the color of our hair, the balance of bacteria in our guts, our excellent primate vision and our terribly shrunken olfactory lobes have evolved to help us, social beings that we are, survive. In the language of Dennett (1996), moral attitudes have evolved by cranes, not by skyhooks.

I see two main ways in which moral attitudes are not outside the realm of science. One is that science (specifically evolutionary psychology) can study the origins and evolution of morality in humans and other species and can give us a much more meaningful descriptive understanding of who we are as animals (e.g. Barkow, Cosmidies and Tooby 1995, Dennett 2003). In addition to this descriptive self-knowledge there is the almost inevitable feedback into how we think we should behave. This feedback between is and ought is not without subtleties. For example some schools of psychoanalysis work under the assumption that by identifying the sources of our own anomalous behaviour, this self-knowledge will allow us to change our behavior — that is, change the way we think we ought to behave. This may be the case but it is not obvious...more obvious is that our "oughts" have not fallen from the sky or tumbled down from some mountains.

> "...that relationship [between facts and feelings] is not determinative. Facts about the world can never compel us to feel a certain way versus a certain other way."

I agree with that. Moral attitudes change and specific facts we learn about other cultures do not change our feelings in determinative ways, but there are limits and constraints that have evolved. In many ways we are on an evolutionary leash, not a pillory. Take our brain's ability to use language as an example. There are universals of human language (Chomsky 2002) that indicate a strong genetic component to our linguistic abilities, but there is much linguistic variety within the constraints of those universals. The exact venation of the leaves on the tree outside my window are all different in detail, but each leaf has the same DNA, the same leash, the same constraining valleys but not the same paths through those valleys.

> "There are disagreements about the implications of facts. When a certain fact emerges from scientific inquiry — let's say, for the sake of example, that we discover that animals can feel and understand pain – this alone does not compel anything."

Ouch. We are animals. We discovered a long time ago that we feel pain.

"What it does do is to lead to a kind of moral and political negotiation about the implication of such knowledge for our attitudes and behavior towards animals."

Yes, but I believe the valleys and landscape in which this moral and political negotiation takes place have been set up by millions of years of social living. Our very ability, to have such negotiations inside ourselves and with others, evolved to help our ancestors survive in the social contexts that have come up.

> "The fact that different peoples in different cultures will handle this negotiation in different ways, reaching different conclusions and feelings, is enough to show that the magesteria are indeed separate."

Not at all. Does the fact that there are different detailed venation patterns on genetically identical leaves show that there is a supernatural magisteria of venation that is responsible for those details and that all this is beyond the purview of science? Of course not. The existence of different languages — variations on the evolved universals of human language — do not imply that there is a supernatural linguistic magisteria responsible for the differences between English and French (Pinker 2002). Similarly, the existence of different cultural variations on the universal themes of human morality (themes that organizations like the United Nations and Amnesty International are valiantly trying to codify) in no way implies that there is some supernatural magisteria of religion beyond the scope of science.

Dennett, D.C. 1996 *Darwin's Dangerous Idea: Evolution and the Meanings of Life*, Simon & Schuster

Dennett, D.C. 2006 *Breaking the Spell*, Viking

Chomsky, N. 2002 *Syntactic Structures*, Walter de Gruyter; 2nd Edition

Barkow, J.H. Cosmides, L. & Tooby, J 1995, *The Adapted Mind: Evolutionary Psychology and the Generation of Culture*, Oxford University Press

Pinker, S. 1994, *The Language Instinct: How the Mind Creates Language*, William Morrow

Oren Harman

Of course brains are "organs of our bodies" and as such have undergone selection. We are biological beings, and our brains are no exception. Yet the fact that the human brain has undergone a process of selection does not mean that all of its products, or, more precisely, all that is made possible by it, exists because it was selected in order to enhance the fitness of its bearers. The Lewontin anti-optimality and Lewontin and Gould anti-adaptationism arguments seem especially relevant to the complex organ called the human brain. In an earlier age, even the hyper-selectionist Alfred Wallace, co-discoverer of the principle of natural selection, saw that this must be true: he took a look at Hottentot brains, took another look at the brains of Fellows of the Royal Society, and observed that they looked pretty much the same. How could this be, for heavens sake? After all, the Royal Society Fellow could accomplish the greatest feats of mathematical abstraction, and yet the Hottentot had no concept of a number above 3! On the other hand, a Hottentot could be *taught* mathematics and philosophy of the highest level, making clear that he has latent abilities which he never seems to use. But this would be impossible on a selectionist view: after all, nature can only select what is immediately useful for survival, and never anything that remains, sleepily, of no use!

Incidentally, this train of thought led Wallace to the conclusion that a supernatural force or forces had to be responsible for the evolution of man, having breathed a kind of "spirit" into him after his biological evolution had reached a certain stage. But we need not fall into the trap of belief or unbelief (however you want to look at it!) into which poor Wallace stumbled. Today we understand that many features in organisms exist not necessarily because they were selected to increase fitness. There are developmental constraints, there is genetic linkage, there are exaptations, there is the phenomenon of epiphenomena, and then, of course, there is chance. I need not go into each and every of the many examples for the above phenomena — others have already done so convincingly.

None of this it to say, of course, that I do not believe that the *capacity* for moral reasoning is not a feature of *Homo sapiens* that has been selected for over evolutionary time. And yes, we have doubtless developed proclivities (some of them contradictory, by the way) for general moral attitudes that made sense in the context of the balance between cooperation and competition which defines life in small (and growing) social groups. Only a blind or foolish man (or someone adamantly opposed to the idea that we are evolved apes) would disagree. But from this "originalist" argument (which exists in evolutionary time) to the much more extreme "constraint" argument (which applies to historical time) and which you seem to be espousing, there is quite a bit of ground that your position does not account for. On an innatist view of morality, it becomes relevant to ask: Does the fact that the brain is an evolved organ stave off the threat of moral skepticism? Can it serve as the basis for some version of moral realism? These are interesting philosophical questions (see Joyce, 2006). But they are a different species from the ones being asked from the *applied* sociobiological perspective. What precisely is meant by "the valleys and landscape in which this moral and political negotiation takes place has been set up by millions of years of social living" — in *moral* terms? When you are faced with the challenge of actually cashing out that statement, I think it becomes quite clear that it is ultimately rather general and therefore empty. *Of course* we cannot *but* think with the apparatus that nature has provided us, that is, we can only think — and feel — as humans. But in what sense that is relevant to moral *negotiation* does this obvious fact restrict us? And in what sense does the fact that our evolved brain serves as the organ of such negotiation does it follow that the negotiation itself should be thought of as "part of science" and reducible to the dictates of an adaptive evolutionary past?

Venation patterns on genetically identical leaves is an example of the extent to which biology is sensitive to environment. Of course one need not — should not — invent a supernatural magesterium to explain such phenomena, since we know they are the result of interactions between developmental programs and environmental contingents. In the same vein (sorry!), one need not invent a supernatural magesterium to explain moral behavior in humans; that's not what I am arguing. The point is that to subsume the rather complex interaction between psychology, sociality, custom and biology under "science" seems, at best, not terribly informative. And while I don't take the following to be an argument for "letting sleeping dogs lie", it remains the case that history has been quite a powerful guide that, at worst, the reductive approach can be shockingly destructive.

The extreme form of the evolutionary approach to thinking about culture results in a rather shallow appreciation of what it means to be human — one that discounts what is not biologically "functional". The approach suffers from the bias of originalism, a kind of cousin to, and no lesser a sin than, the naturalistic fallacy. Just because we have evolved doesn't mean that we can be explained entirely by evolution. That's like arguing that because chocolate is made with cocoa, "deliciousness" can be reduced to farming practices in Brazil. Even if we knew precisely why our ancestors acquired certain taboos and dispositions, such facts would remain irrelevant to the kinds of moral negotiations in which we partake in our lives. After all, how does knowing that we value secrecy because discretion was adaptive help in any way to navigate, or to contribute to, the current debates about medical information and insurance? In what sense does our proclivity to policing help to explain the dilemmas resulting from the clash between the allocation of limited state resources for law enforcement and the psychological demand for personal safety? The two kinds of explanations are of an entirely different order of resolution. One need only further consider that certain taboos, such as incest, square well with evolutionary logic while others, such as promiscuity, certainly do not. Culture cannot be reduced to biology, and talk of a "leash" betrays a fundamental confusion between historical and evolutionary scales.

The principled argument is what counts, but for the sake of record, let's consider empirics, too. In truth, we are far from establishing the evolutionarily ultimate reasons for even one general moral inclination. Where the "just-so-stories" told by evolutionary psychologists ring true they are banal; in any case, they are unfounded. In fact, if anything can be said to be squarely "outside of science" I would submit that much of what is produced these days in evolutionary psychology research qualifies quite nicely. Just because an explanation is plausible, doesn't mean it is true. Unlike in the humanities, in science, truth must be established via falsifiable experiment; if that's impossible then on the very terms used to delimit science from other kinds of intellectual pursuits, we're merely dealing with more or less interesting conjecture. The "*science* of morality"?! Let's be real. We're just beginning to establish the existence, let alone the evolution, of a simple genetic motif that allows for stress-reactive behavior in yeast! (Alon, 2007).

Lewontin, R.C, "Sociobiology as an adaptationist program", *Behavioral Science* 24, 1979, pp. 5-14

Gould, S.J. and Lewontin, R. C. (1979) "The Spandrels of San Marco and the Panglossian Paradigm: A Critique of the Adaptationist Programme", *Proceedings of the Royal Society of London*, Series B, Vol. 205, No. 1161, pp. 581-598.

Joyce, R., *The Evolution of Morality* (Oxford: Oxford University Press, 2006)

Alon, U., "Network motifs: theory and experimental approached", *Nature Reviews Genetics* 8, June 2007, pp. 450-461.

Charles H. Lineweaver

You make some important points about the limitations of sociobiology and evolutionary pyschology, many of which I agree with. I hadn't read Lewontin's criticism of sociobiology before, but I find his suggestions for improving the science of sociobiology very useful. I imagine I will also find Joyce's book of interest. I also sympathize strongly with the fear that reductive science can be "shockingly destructive" — Social Darwinism, eugenics and Nazi science are all good examples of arrogantly simplistic misinterpretations of evolution.

You make the useful distinction between evolutionary time and historical time but you mistakenly describe the constraints that I have evoked (which have effects in historical times on our morals) as being constraints that have evolved over historical time. Let's use the venation of a particular maple leaf as an example. There is a general pattern to all maple leaves that distinguish them from other species. Let's call this pattern strongly genetic. Then there are the fine details of the venation of this particular leaf that are different in subtle ways from other leaves. These fine details are probably best explained by the vagaries of the environment of the developing leaf: the amout and timing of the light, water and nutrients that the leaf gets. However, the level of sensitivity of a developing leaf, to an environmental variable, is a characteristic that has come under strong selection pressure. This level of sensitivity is the result of evolution over evolutionary times. In this sense, the exact details of the venation are the result of the dance between evolution and environment. This is not to say that evolution determines, or compels the leaf into the exact configuration in which we find it...rather, the interaction between evolution, genes and environment is complicated. It is obvious to me (and possibly to you too) that science can and should investigate and understand this complexity.

This leaf analogy can help us understand the influences that shape our behavior, mental states and moral attitudes Replace the general patterns of venation with universal human moral attitudes (attitudes like “my body is more important than that rock”, “my species is better than that other species”, “take good care of your children”). And replace the fine venation that varies from leaf to leaf, with the less universal moral attitudes — attitudes that vary from culture to culture and often lead to cultural conflict. These fine venation aspects of culture are also the ones that are most variable over historical time within a given culture and the ones we seem to pay the most attention to. The evolutionary constraints on these finer aspects of our moral attitudes are similar to the evolutionary constraints on the details of the venation of a particular leaf — that is, the sensitivity of our morals to a changing environment has evolved over evolutionary time. The degree of flexibility of our morals to a changing social environment, i.e. the level of sensitivity to the environment, has been selected for over evolutionary time. For example, the general contours of our abilities to empathize with other humans and other animals is probably a universal but it is modulated and modified (in historical times) by our cultures and experiences. Consider our moral attitudes towards non-human animals. There is a general fundamental pattern of a preoccupation with our fellow mammals, a comparative apathy towards our more distant reptilian cousins and the absence of any moral attitude with respect to insects. Superimposed on this general pattern is a very recent heightened awareness among animal rights groups of the suffering of mammals. Such finer changing aspects of our moral attitudes are the ones we are the most conscious of because they vary from person to person and from culture to culture and can vary noticeably over a human life time: human rights, slavery, abortion, gender equality as well as the issues you cite, the debate over medical information and insurance and allocations for law enforcement. The incest taboo would be a general pattern, while levels of promiscuity would be a finer detail. I suspect that our inability to understand the origin of our moral attitudes and place them on some kind of spectrum (from universal to quirky) as I’ve tried to describe here, is the source of much diplomatic friction and avoidable cultural conflict.

Does any of this affect my own moral attitudes and choices? Yes, in very specific ways. When I have a better idea of which of my moral attitudes are universal and which are quirky products of my specific culture, then when push comes to shove, I rely more on the universal ones and less on the quirky ones...particularly in cases where the source of the quirkiness is identified and found to no longer exist in my particular culture. The impact of the work of

Margaret Mead and Alfred Kinsey are examples of how morals can shift when human behaviours are subjected to scientific scrutiny.

Evidence affects moral choices. It does not dictate or determine a moral choice, but it can push us in one direction or another For example, a scientific understanding of the reasons behind a moral injunction against eating pork, can inspire a less strict adherence to this moral injunction in the presence of new cooking or refrigeration capabilities.

The effects that an understanding of Darwinism and modern cosmology has on our beliefs in gods and traditional animistic/theological explanations of the origin of morals, is another example of how science is providing explanations for, and changing our moral attitudes. This is not to say that evolutionary psychologists and sociobiologists are not severely limited in what they have been able to tell us. They are investigating some of our most deeply held beliefs and feelings. Thus, the implications of any results, whether they are well-done analyses with proper controls or otherwise, are distorted, misread, hotly contested, debated and ridiculed. It is our job as scientists to help improve the controls, remove the biases (to the extent possible), correct for systematic errors, recognize the limitations of the results and arrive at an evidence-based understanding of ourselves. Of course this is difficult and complex and plagued with ideologies of vested interest and national identities, as you and Lewontin and others have pointed out, but good evidenced-based science (combined with much good will) is how I think we should proceed. Cordoning off values and religion and moral attitudes from scientific investigation and pretending that they belong to some separate magesteria is not the way forward; helping to improve the science of who we are and how we got here and why we believe the things we do, is.

Finally, regarding the stress-reactive behavior in our fellow eukaryote, yeast. I'll bet you a Guinness that the same or similar proteins reponsible for the yeast behavior are present and performing similar functions in vertebrates like us. A detailed phylogeny of the protein family should be able to tell us this.

Oren Harman

I am in complete agreement with you about the necessity of improving the science of who we are and how we got here. I think self knowledge is the greatest riches humanity possesses, a veritable evolutionary "gift", though clearly it comes with its complications, too. Yes, we are evolved apes, and our evolutionary history has much to teach us about who we are.

But I find the conclusions you draw from this somewhat odd. You write: "When I have a better idea of which of my moral attitudes are universal and which are quirky products of my specific culture, then when push comes to shove, I rely more on the universal ones and less on the quirky ones…". Do you mean by this that when the chips are down, you'll revert to xenophobia since it (let's just say) was selected for eons ago when we were still small-brained scavengers who needed to be wary of other groups we encountered in the forest? In other words, when the going gets tough, you chuck culture out of the door? That's a scary thought. It also betrays, I think, a mistaken view of the relation between evolutionary expediency and questions of right and wrong.

A final point: I think that for the purposes of our discussion, the philosopher Hillary Putnam got it dead right when he illustrated that explanations need to be of the correct resolution for the problems they set out to solve. The metaphor Putnam used was of pegs and holes: Why doesn't a square wooden peg fit into a round hole? Well, if you try to explain this by referring to the lattice properties of the lignin of the wood you'll be right in a strict sense, but your explanation won't be as relevant as one referring to the geometry of the pegs and holes. I think we agree that our culture is in some sense related to our biology, and, also, that it won't ever be explained by it entirely. Our argument is really about the *extent* to which biology is informative of culture and psychology, and I'm of the view that much that is really interesting about our great human adventure gains little clarity through recourse to our evolutionary past.

Are we really any better off than the ancient Greeks were when it comes to wrapping our minds and our hearts around our existential realities? Has Motherhood really been fathomed any more by invoking the importance for the mother-child bond of oxytocin levels in the blood? Have the depths of Jealousy or Love or Happiness really been illuminated by the understanding that in some general sense emotions must have conferred an advantage in our evolutionary past, and were therefore selected? Even if science will one day be able to explain consciousness as the complex outcome of all interactions between the nerve cells in our brain, will we be able to say that we have gained a deeper understanding of those things that really matter, but which we already know that we can never wholly comprehend? Just like the ancient Greeks and the medieval Moguls, just like the Bushman and the Chinaman and the Turk, we continue our odyssey in the world of unrestrained imagination and undeniable death. All cultures have always known that there are places so deep, not even knowledge can penetrate. "Even if all possible scientific questions be answered,"

Wittgenstein wrote in his *Tractatus*, "the problems of life have still not been touched at all. Of course there is then no question left, and just this is the answer." Amen!

Hillary Putnam, "Philosophy and Our Mental Life", in Putnam, *Mind, Language and Reality. Philosophical Papers*, vol. 2. (Cambridge, Mass.: Cambridge University Press, 1975), pp. 291-303.

Ludwig Wittgenstein, *Tractatus Logico-Philosophicus* (London: Routledge, 2001)

Charles H. Lineweaver

Amen! Shaman! Earth to Wittgenstein! For human beings and other life forms, staying alive is one of the most important problems of life. Scientific and medical understanding, Louis Pasteur and public sewers have touched everyone. For example, I have two children who are still alive and haven't died from some evil curse associated with swamp water or masturbation. We, the life forms on earth continue our odyssey in the world of restrained imagination, tinkered metabolisms and undeniable death...well, death may be deniable: genotypes may be eternal and phenotypic death seems to have been a eukaryotic innovation that evolved with the innovation of eukaryotic sex (Clark 1998). The science of evolution sure helps me deal with existential realities — if you're going to have sex, you're going to die.

Regarding xenophobia, if we can identify the factors in our evolutionary past that probably shaped our xenophobia, and further, we make an estimate of how common those factors are today and find that they are much reduced, then we can more reasonably try to reduce the xenophobia that seems to live in us all. Also, as the world gets smaller and we get more familiar with each other, this familiarity seems to be reducing xenophobia — maybe we have an evolved sensitivity — we don't have to be as afraid of things when we become more familiar with them.

You wrote "I think we agree that our culture is in some sense related to our biology, and also, that it won't ever be explained by it entirely." Yes, I agree...nothing is ever explained by anything entirely.

Clark, W. R. *Sex and the Origins of Death* (Oxford: Oxford University Press, 1998)

Biography

Dov Berger is currently lecturer in the Lifshitz College of Jewish Education, Jerusalem, Israel. He also serves as a Rabbi in the Hameiry Yeshiva, Jerusalem. Mr. Berger is a graduate student working towards his Ph.D in Bar Ilan University, Ramat Gan, Israel. His scientific interests are in the areas of Theory of Evolution; the Philosophy of Rabbi Kook, and the relationship between sciences and faith.

11

A Cabbalistic Interpretation of the Theory of Evolution based on the Teachings of Rabbi Kook

Natural Selection and Its Moral Implication

DOV BERGER

Committee on Interdisciplinary Studies, Bar Ilan University
Ramat Gan 52900, Israel
berger1@macam.ac.il

1. Introduction

This article examines Rabbi Abraham Isaac HaCohen Kook's view of the Theory of Evolution and, in particular, the struggle for existence. Rabbi Kook (1865-1935) who served as Chief Rabbi of Israel, was an unconventional figure in the rabbinical circles of his generation. He was a learned *Torah* Scholar and an authority on Halachic (Jewish law) issues who wrote many books on Jewish law and Jewish philosophy. However, his typically ultra-orthodox appearance notwithstanding, he was also fully conversant with contemporary philosophical and scientific culture and was a prominent public figure. He differentiates between the theory of biological evolution, which concerns the evolutionary mechanisms that work in the animal world, and the philosophical interpretation of Darwin and his successors. He does not question the data and the scientific theory. On the other hand, however, he absolutely opposes the philosophical materialistic interpretation.

There is an unbridgeable gap between Rabbi Kook and Darwin's evolutionary philosophy. The latter is, on the whole, materialistic, random, based on blind mechanisms, and devoid of value meaning. Rabbi Kook's understanding includes a contrary approach. All of reality is a single unity, all the warring entities are limbs and permutations of a sort of general organism, the true reality is spiritual, and it is only the way it is reflected in our own senses that is materialistic. Reality follows a process of development, and this trend has value significance. Through these conceptual spectacles, the Theory of Evolution and the struggle for survival undergo a revolutionary transformation.

Natural selection occurs in nature as a result of a cruel struggle for existence that determines which individual or species continue to survive, and which will become extinct. The Alfred Lord Tennyson expression that is often used to describe the struggle for existence — "Tho' Nature, red in tooth and claw" (Gould, 1996, pp. 63-75) — leaves no room for misunderstanding, the Theory of Evolution portrays this cruel struggle for existence as the driving force behind the development of the animal world. Disturbing moral issues arise from the struggle for existence and its centrality in the evolutionary process. Does the behavioral pattern of nature reflect the desired mode of conduct of humanity? However, such behavior is not morality, but natural behavior of the animal world, which is based on instinct. But why should humanity act differently from animal life? Humanity itself is part of the animal kingdom, so what possible reason could there be for it to behave differently? Moreover, humanity also lives within a struggle for existence. As such, this gives rise to the question regarding what moral consideration could dictate to it to behave in a moral manner which contradicts the struggle for existence in which humanity exists?

Evolutionary philosophy followed a number of approaches to confronting the moral significance of this struggle.* Rabbi Kook, on the other hand, accepted the

---

* Some thinkers, such as Spencer in discussing social Darwinism and Galton who commented on eugenics, argued that natural selection does not evoke a moral issue as it reflects the true morality. See, for example: Hofstadter, 2001; Gould, 1996, pp. 285-319.

Others argued that the struggle for existence and natural selection were misunderstood. The driving force behind the development of life is not the battle between members of the same species, over sources of food and partners, but precisely the struggle of a community with its environmental conditions. For this reason, those who developed cooperative relations, such as mutual help in rearing offspring and shared hunting, etc., stood a better chance of surviving. Among human beings, this is reflected in social and moral behavior. The whole of Kropotkin's book *Mutual Aid* is devoted to this subject (Kropotkin, 1902).

Theory of Evolution while ignoring the struggle for existence and — even more surprising — ignoring the ensuing problematic moral significance (Kook, 1964 A, p. 537). This is not the result of non-recognition of the concept — Rabbi Kook was perfectly aware of the struggle for existence and he cites the concept several times (Kook, 1999 A; Kook, 1999 B). It is unlikely that he did not appreciate the centrality of the struggle for existence in the Theory of Evolution, as Rabbi Kook was familiar with the popular philosophical systems of the day (Ish-Shalom, 1993, pp. 308-309). It is unclear, therefore, why Rabbi Kook does not address this moral issue.

The object of this paper is to examine how Rabbi Kook deals with this problem, despite not addressing it directly, and to understand how radical a transformation the Theory of Evolution undergoes through the notions of the Cabbalistic world. Rabbi Kook did not write an orderly treatise on the Theory of Evolution. In order to achieve an understanding of his view one must reconstruct his approach to the subject from a number of different angles. This reconstruction, however, is not speculative. Rabbi Kook referred to those topics as pertaining to development.

Rabbi Kook addressed these subjects extensively. It is not possible to cover these topics fully and comprehensively and, thus, they will be explained only as required as part of the discussion of the paper's central theme. A brief discussion may appear to be superficial. In any case, the problem stems from the limited scope of the article, and not from Rabbi Kook's writings. Rabbi Kook did not see himself, in these central parts of his philosophy, as an innovator, rather as elucidating the approach of Judaism based on its sources (Kook, 1962 B, p. 132). Each subject is discussed, it seems, divorced from natural selection. The last section incorporates the various topics, presents Rabbi Kook's interpretation of natural selection, and explains how it addresses the moral problem. These issues in Rabbi Kook's philosophy have been thoroughly researched in literature, and this paper does not seek to augment this. The fresh insight offered by this paper appears only in the last section, in which his approach is reconstructed in an effort to understand how the struggle for existence and natural selection are understood in Rabbi Kook's philosophy.

There is a third approach to reconciling the moral problem that believes that one cannot learn anything about the human species from events in the cosmic world. Thus, it follows that one cannot derive the desired moral behavior of the human species from the struggle for existence that takes place in nature (Huxley, 1898).

2. Unity

The unity of reality is an important principle in the teachings of Rabbi Kook. The whole of existence is a single organic entity — a sort of body of infinite separate limbs of which the movement of each impacts on all the other limbs and, naturally, on the overall whole entity. It is only through this form of observation that it is possible to understand reality and its implications. In reality we see an infinite number of separate, contradictory items that combat each other. However, this view is inaccurate:

> "...although the axiom of reality is but a great unity, and the many and various individuals are but special manifestations, different parts, colors and shades of the same unified entity" (Kook, 1964 A, p. 456).

This unity implies that the entire Creation is a single entity; like one body (Kook, 1964 A, p. 417), just as an individual is a single entity, despite the fact that a person has different limbs, and they comprise a large number of different cells, and the cells die and are replaced throughout the person's lifetime. But a person undergoing a gradual process of change is the same person. Similarly, the entire universe is a sort of single organism — from the smallest innate object through animals and vegetation to human beings. The parts of the universe are countless and they change and are replaced, but they still pertain to the same single entity. This fundamental motif in Rabbi Kook's teachings is termed "the total unity". This does not imply mere technical unity that links different parts that do not share any substantial common denominators and which, at the most, offer a new aspect on reality. Rather, this refers to organic unity of the entire world — "the whole of existence is contained in a single point" (Kook, 1964 A, p. 391). S.H. Bergman emphasized that Rabbi Kook refers less to the overt existence in which the individual parts fight against each other, and more to the initial stratum from which reality derives. In this stratum reality stems from the single divine source that is common to us all (Kook, 1964 A, pp. 354-355; Bergman, 1959), similar to a tree with a single trunk (the divinity) which produces numerous leaves, flowers and fruit (our reality). While we may only perceive the leaves they are all, in fact, connected to the same trunk. The warring individuals are unable to negate the common source and the one common objective shared by all beings and, thus, the correct approach is one of unity. The entire world is considered to be a single living organism whereby each and every part contributes to the general organism. This does not refer exclusively to a philosophical approach but to an understanding that creates a new experience of reality. The sense of overall unity — this does not imply a lack of awareness of contradictions and differences, rather an understanding that all the discrepancies and contradictions, between the various items of the Creation, actually express the limitations of the human consciousness and its inability to discern the overall picture of unity (Kook, 1983 A; Ish-Shalom, 1993, pp. 208-209).

In the scientific world Lovelock offered a suggestion that the entire planet Earth — the crust, atmosphere, oceans and all forms of animal and vegetable life — is a single organism. This theory is known as Gaia and argues that life ensures the stability of the chemical, physical and climatic conditions essential to its existence by regulating the carbon dioxide and oxygen in the air and by breaking up the minerals in the rocks and oceans (Lovelock, 1979). One should not confuse the two methods. Rabbi Kook did not specifically talk about the Earth, rather about the entire Universe, and the entire Creation. Rabbi Kook does not offer a biological view of reality, instead he expresses a Cabbalistic philosophy. Rabbi Kook also stresses that the influence is also acting in spiritual terms, and not necessarily on a physical level:

> "We see the organic attribute of the material and spiritual reality. We can confidently observe the overall ratio between everything and, in any case, the action of everything being impacted on by everything, and everything affects everything. We do not have a material atom, or a spiritual thought that does not act upon or is activated by everything" (Kook, 1964 A, p. 355).

Unity is explicitly linked to the subject of development (Kook, 1984, p. 16). This single entity, the unified Creation, is the one that progresses as a single unit. "The entire existence develops and emerges, as is evident in its parts" (Kook, 1964 A, p. 537). This development influences each of the individual items that make the whole. On the other hand, the development of each part affects the other parts, and thus, in the development of only a single item, in practice, everything develops.

3. Death and Immortality

Every individual and all substances have both a material and a spiritual foundation. The spiritual basis is the most important component, and is not perishable, and is a part of the evolving overall unity. It is for this reason that every individual that has lived, at any time, has eternal existence in a certain dimension. In the divine reality, which is the absolute reality of the world, the individual exists despite the fact that, in physical reality, he is dead (Kook, 1964 A, p. 340). Terrestrial death does not signify the end of life.[†]

Following on this perception one can arrive at a revolutionary approach to the concept of death. Death is an error of consciousness, an illusion from which one must awaken (Rosenberg, 1991). The more spiritual a person is, the closer he is to understanding that death does not exist at all (Kook, 1963 A, p. 8). We judge reality with the tools of human perception but these tools do not describe true reality in full. Consequently, there may be situations in which the dead may be more alive than the living, as these dead people are active in practical life more than living people (Kook, 1993, p. 95). The moment when the soul is freed from the body, and moves on to a more spiritual phase, is understood by those who remain in the material world as death, as cessation. However, the reason for this is our perceptual limitations with regard to non-material concepts. For the soul the transition may be a very happy one. Human death is a change in the physical state of life. Just as water takes on different states, and is still the same water whether it is a liquid, or a gas or a solid; its transition from liquid to gas does not signify cessation, instead it indicates a change in the physical state. Death is also a change in the physical state of life, and is not nullification.

This can be compared to the situation of twins in their mother's womb, which is their entire world. When the first twin emerges, from the point of view of the other twin this is perceived as a disaster; his sibling has vanished from reality. In fact, he was born. But, from the perspective of the mother's belly, it looks like cessation. Death is a similar transition. From the point of view of our reality, death also looks like an end, although, from the wider perspective, it can look like a birth. The experience of life is an incomplete experience. We do not perceive reality on all its various levels. We do not perceive the reality we call death and, therefore, it is understood as a state of non-existence.

[†] See Ross, 1996. The paper discusses Rabbi Kook's view of how the power of repentance will eradicate death from the world to come, but the beginning of the paper also relates to our topic.

4. "Everything Transcends"

The overwhelming majority of the species that existed in the world have become extinct. The number of species alive today is negligible compared with the total number of species that lived on the Earth throughout all the eras of its existence.‡ The number of individual creatures that lived during this period and died is inestimable. At the most basic level, there is no significance to all the creatures and species that died out. A creature or species that died is lost forever; its death is absolute cessation. Rabbi Kook does not accept this line of thought. According to his evolutionary developmental perception, "everything transcends" (Kook, 1964 A, p. 548); "and this is the general rise that no individual can remain behind" (Kook, 1964 A, p. 537), including the dead ant and the extinct mammoth.

As we noted above, according to Rabbi Kook's viewpoint, all individual items exist in a dimension that is beyond time. Thus, the ongoing existence of each individual is simultaneous and common to all, including those whose terrestrial lives took place in different periods. The concept of unity now takes on an additional meaning. It now refers not only to unity between individuals living at the same time but also the complete unity of everyone who has ever lived in our world. The entire life chain, the entire living world throughout its generations — all belong to the complete unity. Each individual has a value by virtue of being a part of the general single organism. Based on this understanding, there is no room for the thought that a species or individuals that have died have lost their significance. Death is not cessation. The bond with the general organism is perpetually maintained and, consequently, dead individuals and species have perpetual significance and ability to develop. Thus, in evolutionary development according to Rabbi Kook, the single individual also continues evolving. The expression "everything transcends" includes everything — the species that have become extinct as well as the individual creatures that have died. Even the species that the popular perception referred to as "inferior" — are part of the whole, that same "whole" which Rabbi Kook perceives as transcending. Everything, the entire reality, including every individual item in it, is undergoing

‡ According to the theory the animal world underwent periods of mass extinction during which many species died out completely. In one such period about 95% of all species on Earth disappeared, and in another mass extinction, around 75% became extinct as the result of a single event. Mass extinctions occurred several times on Earth (Ridley, 1993).

a process of development, and it is not possible to detach individuals from the unity (Ben Shlomo, 1984, p. 298).

The issue of "progress" is incongruous with modern thinking, partly due to the anthropocentric approach. Development relates to the evolution of humanity that stands at the top of the pyramid of progress, and it is humanity that is developing. Rabbi Kook refutes the anthropocentric interpretation of progress. It is not only humanity that is developing. Everything is developing. Every organism develops towards a state of perfection. Every unit participates in this process, with all its individuality (Ben Shlomo, 1984).

5. Unity and Humanity's Place in Nature

Humanity has a special role within the general organism. Only humanity can develop the entire Universe, both spiritually and physically[§] (Kook, 1964 A, pp. 427-429). Thus, humanity is the individual capable of imparting a spiritual meaning to every item in Creation. The degree to which the world progresses and develops is dependant on humanity. Humanity, through its actions, impacts on the transcendence of the world (Kook, 1962 A, pp. 339-340). When a person develops, every individual in the Universe is imbued with new meaning. The implication is that when a person transcends to a higher plane he adds new meaning to each of the interim stages in his development. Some have already elapsed although they have a part in all the deeds that humanity performs — they live on within it. This understanding imposes great responsibility on a person. A person's importance does not decrease with his understanding of the infinite expanses of the Universe. On the contrary, humanity's responsibility increases all the more (Kook, 1964 A, p. 433). The extent of the value of the Universe, with all its individual parts, is determined according to the deeds of humanity. Through his deeds the amoeba becomes a different amoeba. This ability allows humanity to ascend above the plane of the created and to become a partner to the Creator (Kook, 1964 A, p. 527). Humanity is asked to forsake its individual privacy and to understand and sense its bond with reality in its entirety, and to act on its behalf:

[§] Several decades after Rabbi Kook, Julian Huxley put similar thoughts to paper (Huxley, 1943).

> "The most elevated form of sanctity is the sanctity of silence, the sanctity of existence, whereby a person knows himself devoid of his personal inner world and lives a common life, the life of all. He feels animal, mineral and vegetable life, life in its entirety, the life of every speaker, of each and every person, the life of all intelligence and all recognition, every intellectually attaining and every feeling being, and the whole of existence transcends with him to its origin..." (Kook, 1964 A, p. 297).

Rabbi Kook's words appear to be wandering. If there is a developmental trend in reality, which serves to advance reality, what is the point in the actions of humanity? Humankind will act as it sees fit and the developmental forces in reality will do theirs. Why shouldn't humanity become passive and wait for the world to develop on its own? The answer to this appears in the Talmud. The development process can be rapid and easy, and can also be slow and painful. Humanity, with its deeds, determines the manner in which events take place. Humanity can change a long and painful process into a rapid and easy one.[**] Consequently, humanity is responsible for the state of the world.

6. Evil, Unity and Reality

The unity Rabbi Kook describes appears heavenly, utopic and unreal. History and reality indicate the workings of evil, and that the world is full of good and evil forces battling with each other. So, where is the unity? Where is the world working as a single living organism?

Rabbi Kook lived reality and was fully aware of the existence of evil (Kook, 1983 B). He refers to concepts like war (Kook, 1999 A) and politics (Kook, 1964 A, p. 457) with deep pain. He is sensitive to evil but explains it in an unconventional manner. According to his approach, the evil that manifests itself to us as an independent force is "imaginary reality" which takes on the appearance of reality (Kook, 1964 A, p. 475; Kook, 1964 B, p. 168). Evil manifests itself as an orderly and organized phenomenon, and cannot be treated as an incidental phenomenon. One must, therefore, recognize that evil was

---

[**] The tractate of *Sanhedrin*, page 98. The Talmud explains that the process of Redemption can happen rapidly and easily. The process may also be slow and painful. This is contingent on deeds.

created[††] with a clear structure and objective: "which was created only to serve to enhance the reality of good… that without the evil the reality of good would not be so perfect" (Kook, 1964 A, p. 479).

Rabbi Kook highlights the unity concealed beneath the visible layer of the world. Our world is one of contrasts and opposites. Deeper recognition reveals "the unity of opposites", that even the most contradictory ideas and forces are joined at the root. On the surface the forces appear opposite to each other but, from the side of truth, they are parts of a single organic unit (Kook, 1962 A, p. 133; Yaron, 1991; Rosenak, 2003). Good and evil, from a comprehensive viewpoint, are two sides of the same coin. In fact, evil is part of good. Evil helps processes that are oriented towards good, only it does so in a concealed manner. In certain situations, it is evil that fuels and activates good (Kook, 1963 B). Evil as "evil", without any good purpose, is imaginary reality. It is an error of our perception. It is a conscious illusion, as if perceived through a dream. This continuing dream may have a strong presence and generates senses, but it is only a dream from which we must awaken (Rosenberg, 1991). In other words the life which we know is a partial perception of reality as in a dream. In this sort of perception "evil" indeed seems evil but is not the true. The basis and role of evil are good. Paradoxically, when an evil activity reaches its zenith — when it completes its role — it disintegrates, and the truth transpires (Kook, 1962 A, pp. 303-304).

According to Rabbi Kook all manifestation of evil, all pain and suffering, are like labor pains (Kook, 1984, pp. 360-362). Just as the birth of new life goes through stages of pain and suffering, birth that leads to development often passes through stages of pain in which evil appears to be the dominant force. However, just as a newborn baby gives meaning and understanding to the long and painful process of birth, all pain and suffering gives fruit, all sorrow produces something in reality, and it is this result that confers it with meaning. Nevertheless, this does not imply that a person can always fathom the meaning of evil and suffering. This lack of understanding gives rise to the experience of the tangible presence of evil and suffering which appear to be devoid of meaning.

[††] *Isaiah*, 45, 7: "I make peace and create evil."

The view of evil as an independent element, rather than an integral part of a whole system, is analogous to a situation in which we see a knife cutting into a person's body. Blood is pouring out and bones are breaking. The conclusion drawn from this is one of violence and evil. At a later stage, the camera pans out and we see the full picture: the knife is held by a surgeon in an operating room, he is surrounded by doctors, nurses and medical equipment. Only the full view, in which we see more than just the knife, but also the surgeon's clothes and the full scene, allows us to arrive at a full understanding of the events. The usual view of evil shows us only the knife, without the hand holding it, so the evil appears so misunderstood.

Rabbi Kook explicitly links the decline in the appearance of good and its manifestation as evil to the concept of development. It is the limitations of the perception that give rise to the perception of evil as evil that generates the process of development. If everything was revealed in its entirety there would be no need for development. In fact, it is the problematic phenomena of evil that advance reality (Kook, 1964 A, p. 462), just as a decayed seed can provide fertile ground for a new tree. A seed that does not undergo the process of decay will not grow into a tree. It is decay that makes the tree's "birth" possible.

However, a person who has experienced the horrors of evil will find it difficult to accommodate this approach. The perception of evil offered by Rabbi Kook looks like philosophical deliberation detached from daily suffering and pain. The reason for the difficulty in accepting this explanation of evil stems from the difficulty in accepting philosophical responses that have no effect in the world of human experience, while the experience of evil is so powerful. When we are faced with a question of evil we intuitively expect to be given an answer with emotional force commensurate with the sense of evil. Rabbi Kook only offers an intellectual response, and it is not easy to accept it. In reality, it is this very philosophical perception that provides humanity with the means to face up to evil. The awareness that evil is a means that works for good, and is not an independent force, forges a different approach to the experiences of evil and, thereby, also individual ability to withstand evil (Kook, 2000, p. 113).

Despite the aforesaid about the role of evil, humanity is capable of strengthening the forces of good and to channel them so that they can also fill the role of evil. In such a situation, the world no longer has a need of evil, and evil becomes extinct (Kook, 1964 A, p. 480).

7. Materialism

According to popular perception, the secular is reality that contradicts the holy. Rabbi Kook views the secular not as separate from the holy but as deriving from it. It is the human perception that deceives humanity and creates the erroneous dichotomy between the holy and the secular, by concealing the divine root of manifested reality (Kook, 1963 A, p. 98). Everything that manifests itself in the material world is only a shadow of the spiritual entity (Kook, 1963 A, p. 2). The material world is realized as spirituality that takes on a material guise that conceals the true spiritual content. The entire world exists in a divine existence that gives it life. The divine life is the driving force behind the whole of reality and channels it towards development (Kook, 1964 A, p. 515).

Darwin, however, believed in materialism. In the "*Origin of Species*" the materialistic interpretation is not explicitly mentioned, but it does appear explicitly in Darwin's earlier notebooks. In notebooks M and N, written in 1838 and 1839 ("*The Origin of Species*" was published only in 1859), Darwin explains numerous behavioral types in materialistic terms — senses, excitement, memory, instincts, free will, the ability to think, and even morality and conscience. In so doing, he explains all life phenomena in a materialistic way (Gruber, 1980, pp. 51-57). In one place in notebook Darwin says:

> "Plato ... says in Phaedo that our 'necessary ideas' arise from the preexistence of the soul, are not derivable from experience — read monkeys for preexistence" (Gruber, 1980, p. 30).

Darwin attached great importance to the principle of natural selection he discovered and considered it the foremost mechanism, the primary driving force behind the evolution of species in the animal world. (Darwin didn't think it was the exclusive force behind the evolution. He did accept that other forces, such as randomness and sexual selection play an important role.[‡‡]). Darwin believed that, as this mechanism is so powerful, there is no need for any other element or force, thus there is no role for the Creator or other metaphysical force to play in

---

[‡‡]Contemporary thinking emphasizes the importance of those random forces. With regard to phenomena relating to mass extinction, see Freeman, 2001. With regard to phenomena of genetic drift processes see ibid., pp. 168-179.

this picture.[§§] Based on this approach, Darwin created a biological base for a materialistic view of reality. Stephen J. Gould notes that uncompromising materialism differentiates Darwinism as a philosophical approach from other theories of Evolution. Marx and Engels, the leading materialists of the nineteenth century, understood that Darwin formed the scientific infrastructure for their philosophical standpoint and, therefore, attached great importance to the Theory of Evolution and to Darwin himself (Gould, 1977, pp. 21-27).

It is at this juncture that the sharp contrast between the materialistic Darwinist philosophy and Rabbi Kook's beliefs comes to the fore. Materialism views matter as the primary element, whereas Rabbi Kook considers divinity as the primary factor. Darwin's great achievement was in finding a natural mechanism that explains the formation of the range of species in nature without the need for a spiritual-metaphysical explanation. Darwin himself lost his religious faith as a result of his evolutionary understanding (Darwin, 1974). Rabbi Kook argued that evolution, with all its mechanisms, is one of the ways by means of which the divine forces function in reality (Kook, 1964 A, p. 537). The belief that the entire Universe came about by chance is not reasonable (Kook, 1998). With regard to the conundrum of life — the question of the meaning of human existence — the materialist response simply replies that human existence has no significance. The entire spiritual world has no real existence. It follows that neither the Universe nor humanity has any spiritual or valuable significance (Dennett, 1995). Rabbi Kook, on the other hand, offers the Cabbalistic response: Every action of humanity has a spiritual importance. Humanity, through its deeds, can maintain the development of the world and, becomes a partner to the Creator in the work of Creation. According to the materialistic approach, humanity's natural ambition to attain meaning, for personal fulfillment, the striving for perfection, the attempt to decode the secrets of the Universe and to achieve spiritual development are no more than the invention or distortion of the human perception. According to Rabbi Kook these points specially express the uniqueness of humanity — uniqueness that implies the realization of spiritual potential. Evolution cannot be limited to the material dimension. The peak of evolution lies specifically in humanity's achievement of the spiritual world, which is also the peak of life as a whole (Kook, 1964 A, pp. 547-548).

[§§]This approach is not simple, as natural selection does not explain the formation of physical reality — the universe, substance, the chemical principles and the physical constants. This, the discovery of natural selection does not explain the question of the formation of life.

According to the materialist approach, the moral deed is devoid of value as, in reality, there is no moral significance whatsoever. All laws only have a social value, and no objective moral — super-social — value. The concept of moral meaning is a human invention. This does not imply that Darwin encouraged immoral behavior. On the contrary, for example, Darwin objected strongly to slavery. But he left out the moral significance of deeds. He argued that morals are a human instinct, just like sexual instincts or eating instincts (Darwin, 1871; Gruber, 1980, pp. 56-57) and, thus, a good deed has no moral meaning. Rabbi Kook argued that every deed has a moral value. He said that we express the world's spiritual-moral infrastructure, through the moral behavior of humanity and through any event that expresses social justice (Kook, 1962 A, p. 45).

The many differences between Rabbi Kook and Darwin's philosophical understanding place the very validity of comparisons between their approaches in question. Is there is a connection between these two different approaches? Here Rabbi Kook's innovative approach comes into play. He incorporates biological evolution in the general philosophy based on the Cabbala, thereby creating a new evolutionary interpretation. In so doing he frees the Theory of Evolution from its dependency on a — one and only — materialistic explanation. Rabbi Kook offers a wide evolutionary-developmental approach, which applies to both materialist and spiritual dimensions. This approach restores the moral significance to humanity and its actions, and provides humanity and mankind with an idealistic, superhuman plane of aspirations to which humanity intuitively strives (Kook, 1964 A, pp. 368-369; Ben Shlomo, 1984, p. 297). This is an objective that the materialist approach to the Theory of Evolution cannot offer, as materialism negates its very existence.

8. Humanity, Morals and the Struggle for Existence

Humanity lives within a complex reality, which is both material and spiritual. The life activities within the material dimension represent the low level of humanity, which is also common to animals. Life entrenched in the material dimension is described by Rabbi Kook as a disease (Kook, 1978). Humanity's life is an ongoing struggle between corporeal and spiritual inclinations.

> "Darkness must be continually battled. Humanity must strive to free itself from the force of lust..." (Kook, 1964 B, p. 132).

Research indicates the existence of moral behavior, on certain levels, in the animal world too (De Waal, 1996). This provided support for the approach that morals are an instinct that developed during the struggle for existence, and does not indicate a spiritual dimension. This approach was broached by Darwin himself (Darwin, 1871; Gruber, 1980, pp. 51-57). It is not reasonable that Rabbi Kook's approach has a problem with the argument that some basic moral behaviors are an instinct. He believes that it is not morals that are required from humanity at all. Humanity has to make moral decisions that are not a submission to instinct but actually control it. Moral behavior that is intrinsic to instinct is only a low level created in order to safeguard the minimal ability of life. The purpose of Creation is specifically, spiritual moral development (Kook, 2000, p. 24). Humanity, in its spiritual essence, belongs to the world of values and is able, and must, choose spiritual existence (Kook, 2004).

> "Life itself, the whole of life, the essence of life, is nothing but... the striving to achieve closeness to God" (Kook, 1984, p. 34).

Humanity has the ability of choice and can decide whether to adhere to the laws of the struggle for existence or opt for a system of higher values, whether to succumb to instinct or control it. It is for this reason that humanity is required to live its life within spiritual moral parameters, and not within the moral parameters of the material world, and its laws of the struggle for existence (Kook, 1999 B). The realization of human freedom lies in controlling the laws of the struggle for existence, and not succumbing to them. The fact that humanity has developed evolutionarily, and has achieved moral and ethical abilities, means that this is his "theater of operations", that he must not fill the role of an animal or live according to an animal's moral rules. Humanity is expected to behave in a spiritual-moral manner. It must not accept the instinct, it should control it.

9. Rabbi Kook's Interpretation of Natural Selection and Its Moral Implication

This paper began with the question of how Rabbi Kook accepts the Theory of Evolution without critically addressing the problematic moral significance ensuing from the struggle for existence, which has a central role in the Theory of Evolution. We will try to understand from his philosophical teachings, which we examined above, why the moral implication does not arise in his understanding of the Theory of Evolution.

According to Rabbi Kook's perception, the entire universe is a single developing unity. This unity includes interchanging individual items, such as a person's cells that interchange during the person's lifetime. There is no item that is not part of the developmental process. Every organism has an eternal existence in the spiritual dimension. Every development in this unity impacts on each of its other parts. It thus follows that each individual item continues to progress even after it undergoes physical death. Each of the items and species that are no longer alive remains a part of the existence of the general organism. The development of each individual item of the general organism influences and imparts meaning to each of the other parts, including those that have died. Humanity has a central role in this unity. Through his deeds, humanity impacts on the sublimation of the world. He can help the world advance or inflict damage on it. When humanity transcends and the Universe transcends with it, each and every individual in the world also transcends, and gains a new value and a new meaning.

The understanding of the Theory of Evolution from the viewpoint of unity of every being confers a positive role on the struggle for existence and natural selection. It is only when each force or individual stands apart from each other and fights for survival that individual progress reaches its peak. It is only in so doing that it expresses itself in the most complete manner. Consequently this unity is the unity of forces at their very peak rather than a unity of the weak and unfocused. Based on this understanding all individual items belong to one large entity. There is no place for hatred or wars with others as they, too, belong to the same unity. In that sense "our heroism is delicate" does not refer to the heroism of killing and destruction, as the other does not hate but is part of the same unity (Kook, 1993, p. 110). For example, the cells of the immune system — the lymphocytes — identify cancerous cells in the body. These are cells that have undergone a carcinogenic process and pose a threat to the entire organism. Certain lymphocytes are dispatched to fight the cancerous cells. This is a battle of life or death. Other cells in the immune system swallow the red blood cells, which, after functioning for 120 days, have completed their role. Is it possible that this is a bloody and violent war? The actual understanding that this is a single living, breathing and functioning organism, with activity control between the various parts, gives these examples a significance of healthy living and negates the militant concept. This understanding of Rabbi Kook for the Theory of Evolution is the principal response to the question of his ignoring the moral issue. The moral issue does not arise with this approach — the war for survival is an internal battle between the cells of a general developing organism.

Rabbi Kook incorporates the Theory of Evolution in his Cabbalistic view of development, and gives it a new moral perspective. There is no item or species whose life is without value. Every individual item and every species that has died or become extinct is a part of a single great unity progressing along a transcending path. Every individual or species that has died out exists in a spiritual dimension after its death. It is part of the chain of development that led to the present situation, and it lives on inside us. It has a part in every substantial action of humanity, and thus its life is full of value. Humanity grants them life through its actions. This understanding imposes heavy moral responsibility on humanity, along with great hope for a better future. The Universe's life of value entirely depends on the deeds of humanity, as it is now clear to us that the physical condition of the world depends on the actions of humanity. The individual person's membership in the great collective, and his ability to impact on it, grants humanity its full importance, importance that no single individual — as important as he may be — is not capable of achieving.

10. Epilogue

Rabbi Kook views evolutionary theory as an incomplete work, a scientific theory that appears to champion heresy but actually contains a deep spiritual meaning. It should, he believes, be freed from the immature interpretations it got (Kook, 1964 A, p. 467). Rabbi Kook did not write an orderly treatise expounding all the main points in his approach towards the Theory of Evolution. In this paper, we have reviewed some of the principles of his approach with a view to understanding the way he understood the spiritual-moral significances of the struggle for existence and natural selection.

Despite the fact that Rabbi Kook accepts the Theory of Evolution he rejects the philosophical and materialistic significances that have been given to it. He incorporates the Theory of Evolution in his own developmental thinking and changes the spiritual perception of evolution, in accordance with *Torah*, and particularly Cabbalistic concepts. Thus, his teachings restore the moral significance eradicated by materialism, both to life and deeds, and grants humanity and mankind an idealistic spiritual objective the existence of which has been negated outright by materialism.

11. Acknowledgements

I wish to express my thanks to Avshalom Elitzur, Oren Harman, Noah Efron and Shalom Rosenberg, for reviewing earlier versions of this manuscript, and for their thoughtful comments and generous help. I am also grateful to Menachem Fisch, Raz Chen-Morris, Uriel Barak, Hana Yekel, David Patchi, Boaz Tamir, Samuel Chayen, Rafi Chiko, and Barry Davis. I would like also to thank the Lifshitz College of Jewish Education and its directors, for their support of this research.

References

Ben Shlomo, Y. (1984) "Shelemut VeHishtalmut BeTorat HaElohut shel HaRav Kook", in *Iyyun*, Vol. **33**, pp. 289-309.

Bergman, S. H. (1959) "Mavet VeAlmavet Bemishnato shel HaRav Kook", in *Hogim VeMaaminim*, Tel Aviv, pp. 104-105.

Darwin, C. & Huxley, T. H. (1974) *Autobiographies*, Oxford University Press, London, p. 50.

Darwin, C. (1871) *The Descent of Man*. The Modern Library, New York 1962. p. 912-913.

De Waal, F. (1996) *Good Natured: The Origin of Right and Wrong in Humans and Other Animals*, Harvard University Press, Cambridge, Massachusetts.

Dennett, D. C. (1995) *Darwin's Dangerous Idea: Evolution and the Meanings of Life*, Simon & Schuster. New York, p. 153.

Freeman, S. & Herron, J. C. (2001) *Evolution Analysis*, Prentice Hall, New Jersey, pp. 527-536.

Gould, S. J. (1977) *Ever Since Darwin*, W. W. Norton & Company, New York.

Gould, S. J. (1996) *Dinosaur in a Haystack*, Harmony Books, New York.

Gruber, H. E. (1980) "M Notebook Commentary", in P. H. Barrett, *Metaphysics, Materialism & the Evolution of Mind, Early Writings of Charles Darwin*. The University of Chicago Press, Chicago.

Hofstadter, R. (2001) The Vogue of Spencer. In P. Appleman, ed. *Darwin, a Norton Critical Edition*, Third Edition, W. W. Norton, New York, pp. 389-395.

Huxley, J. (1943) "Evolutionary Ethics", in *Evolution and Ethics 1893-1943, T. H. Huxley & J. Huxley*, Kraus Reprint co., New York 1961, pp. 105-142.

Huxley, T. H. (1898) Evolution and Ethics, In *Evolution and Ethics and Other Essays*, D. Appleton and Company, New York, pp. 46-116.

Ish-Shalom, B. (1993) *Rav Avraham Itzhak HaCohen Kook — Between Rationalism and Mysticism*, State University of New York Press, Albany.

Kook, Rabbi A. I. (1962) *Igrot HaReAYaH*, Mosad HaRav Kook, Jerusalem, Vol. I.

Kook, Rabbi A. I. (1962) *Igrot HaReAYaH*, Mosad HaRav Kook, Jerusalem, Vol. II.

Kook, Rabbi A. I. (1963) *Olat ReAYaH*, Mosad HaRav Kook, Jerusalem, Vol. II, p. 262-264.

Kook, Rabbi A. I. (1963) *Orot HaKodesh*, Mosad HaRav Kook, Jerusalem, vol. I.

Kook, Rabbi A. I. (1964) *Orot HaKodesh*, Mosad HaRav Kook, Jerusalem, vol. II.

Kook, Rabbi A. I. (1964) *Orot HaKodesh*, Mosad HaRav Kook, Jerusalem, vol. III.

Kook, Rabbi A. I. (1978) *Lights of Return,* translated by A. B. Z. Metzger, Yeshiva University Press, New York, p. 39.

Kook, Rabbi A. I. (1983) *Arpeley Tohar*, Jerusalem, p. 39.

Kook, Rabbi A. I. (1983) *Hazon HaTzimchonut VeHaShalom*, Nezer David, Jerusalem, pp. 13-14.

Kook, Rabbi A. I. (1984) *Maamarey HaReAYaH*, Jerusalem.

Kook, Rabbi A. I. (1993) *Orot*, translated by B. Naor, Jason Aronson Inc, New Jersey and London.

Kook, Rabbi A. I. (1998) *Orot HaEmuna*, Jerusalem, p. 41.

Kook, Rabbi A. I. (1999) *Ginzei ReAyaH,* HaMachon Al Shem HaRav Zvi Yehuda Kook, Jerusalem, vol. I-VII, p. 157.

Kook, Rabbi A. I. (1999) *Shmona Kvazim*, Jerusalem, vol. I, p. 12.

Kook, Rabbi A. I. (2000) *Ein AyaH,* HaMachon Al Shem HaRav Zvi Yehuda Kook, Jerusalem, The tractate of Brachot Vol. I.

Kook, Rabbi A. I. (2004) *Pinkas YudGimele,* HaMachon Al Shem HaRav Zvi Yehuda Kook, Jerusalem, p. 5.

Kropotkin, P. (1902) *Mutual Aid – A Factor of Evolution*, Penguin Books Ltd, 1972, pp. 17-49.

Lovelock, J. (1979) *Gaia- A New Look at Life on Earth*, Oxford University Press, Oxford, pp. 1-11.

Ridley, M. (1993) *Evolution*. Blackwell Scientific Publications, London, pp. 591-622.

Rosenak, A. (2003) "Halacha, Agada Unevua BeTorat Eretz Israel LeOr Achdut Hahafachim BeMishnat HaReAYaH Kook" in A. Sagi, D. Schwarz (ed.), *Meaa Shnot Zionut Datit*, Ramat Gan, Vol. **I**, pp. 261-285.

Rosenberg, S. (1991) "Introduction to the Thought of Rav Kook" in *The World of Rav Kook's Thought*, Avi Chai, United States of America, pp. 33-36.

Ross, T. (1996) "Al-Mavet, Hukiut HaTeva, VeTafkid Hahasaga HaEnoshit BeKitvei HaRav Kook", *Daat*, vol. **36**, pp. 35-46.

Yaron, Z. (1991) *The Philosophy of Rabbi Kook*, translated by A. Tomaschoff, The World Zionist Organization, Jerusalem, pp. 83-88.

Dialogue

Jack A. Tuszynski

In my dialogue with Dov Berger, I would like to focus on Rabbi Kook's Kabbalistic world view that emphasizes the interconnectedness of all beings and the ancient principle stating that the development of the world is taking place through the reapplication of basically similar principles at all hierarchies of its organization. This has some overtones reminiscent of the Hermetic dictum (Hall, 2003): "That which is below is above, that above is also below." Also the belief that humanity is the bearer of a divine, immortal spark was developed in both Kabbalistic and Hermetic sources. Hermetic philosophers suggested the existence of two realms of reality: heaven and earth, spirit and matter, God and man. This idea can be found in Kabbalah, too, one example being the divine image of the archetypal Man, the *Adam Kadmon:* man below reflected his Creator form above. The terms macro-cosmos and micro-cosmos also reflected this duality, a duality that is still seen in modern physics as I argue below. The macro-cosmos was an outer creation of the universe which reflected the micro-cosmos, the inner creation and seed of God in man. I will try to provide a scientific perspective on the relationship between Kabbalistic thought and aspects of modern science that goes beyond Darwinian law of natural selection and concentrate on the building blocks of physical and biological systems and their functional relationships that may have been presaged by Kabbalists such as Rabbi Kook in their deep insights into the inner workings of the universe.

Democritus is credited with the idea of atoms that were endowed with mass, size and motion. According to classical physics the world is made up of atoms, irreducible and indestructible behaving according to Newton's laws of motion. Modern physics asserts that the building blocks of matter are still smaller entities called elementary particles: quarks and electrons which have mass, charge, spin, strangeness and charm, among other properties, but the basic view of their autonomy and interactions remains the same. Recently, however, a new viewpoint has emerged with not a clockwork world of independent entities as a blueprint but an entangled interactive world whose constituents derive their identities and properties from one another through unceasing interactions. This brings physics closer to being an example of "it from bit," a phrase coined by John Wheeler to describe the grand unified theory of the universe based entirely on logic without any physical parameters. Curiously this model of the world is not about particles anymore, but about their physical and mathematical relationships, especially their symmetries, fundamental aspects of nature that remain invariant and universal. Perhaps the first such example can be traced back to elementary particle physics where the "Eightfold Way" is a term coined by Murray Gell-Mann for a theory organizing subatomic baryons and mesons into octets alluding to the Noble Eightfold Path of Buddhism. This elegant theory led to the development of the quark model later on proved to be correct by experiment, both theorists and experimentalists being awarded Nobel Prizes for their work. Moreover, quantum mechanics also enforces its own version of relatedness between particles. According to it, microscopic particles can become "entangled" meaning they remain connected even if they are separated by macroscopic distances, so that measuring a property of one particle instantaneously affects the outcome of measuring the other. Thus, in the quantum world there is an underlying deep connection among all entities in the universe, as eloquently argued by David Bohm in his theory of the "implicate order" of the world (Bohm, 1980). One can therefore assert that a sort of Kabbalistic curtain separating the 1% outwardly manifest from the 99% of the hidden part of the universe does indeed exist between the two worlds. The curtain, the edge between the classical macroscopic world and the microscopic quantum world can be viewed as the phenomenon called the "collapse of the wave function". This reduction from the world of multiple possibilities of the quantum superposition to the classical world of definite values is also referred to as a reduction of the quantum state vector. It still remains one of the mysteries of modern science. Amazingly, early experiments in quantum mechanics seemed to show that quantum superpositions if hidden, or isolated from the environment would remain in superposition until measured by a conscious observer. This direct clash between the micro-cosmos and the macro-cosmos of modern physics is still a hotly debated issue.

Moving to a cosmic scale, according to Mach's principle (the broad notion is that mass there influences inertia here), it makes no sense to consider a single particle alone in the universe (Mach, 1960). As Lee Smolin in his book (Smolin, 1997) wrote: *"It can no longer be maintained that the properties of any one thing in the universe are independent of the existence or nonexistence of everything else"*. Hence, no particle is an island. To the extent that the science of the day reflects the way our society views itself, these new emerging paradigm shifts may be a manifestation of our alienation in a society driven by the laws of economics (Overbye, 2008). Lynn K. Nyhart stated that *"science is surrounded by society"*, pointing out that the phrase "natural selection" was first used in economic circles before Darwin applied it to his theory of biological evolution (Nyhart and Broman, 2002). Nyhart said that the utopian language of entanglement in quantum physics sounded like a reaction against the atomization of our society. *"We are so atomized by the markets and people are trying to find ways to reassert their connections."* Kabbalah and one of its main 20th century figures, Rabbi Kook perceived the world as being much more interconnected and hence closer to the entangled picture. In fact, this entanglement existed not only on the physical plane but transcended into time and beyond as explained by Dov Berger in the preceding chapter.

In addition to the quantum reality of matter, there is another way of seeing physical particles forming a coherent whole through the use of nonlinear methods of modern physics. The prevailing "linear" scientific methodology conforms with the Cartesian framework of deterministic philosophy which underlies the edifice of science inherited from the founders of the rationalistic movement of Western civilization some 300 years ago. This mindset is clearly orthogonal to the mystical thought of Kabbalistic judaism (Epstein, 1988). Until recently, within physics, all the major conceptual building blocks seemed to neatly fit into the scientific framework where cause and effect are in a proportional relationship to one another and all physical objects are autonomous albeit interacting. Therefore, it is commonly expected that the larger the perturbing influence, the more pronounced the deviation from its unperturbed state. More and more phenomena in the world around us strike us as completely contradictory to this viewpoint. In the past few decades, several inherently nonlinear phenomena have been observed giving rise to compelling new ideas that force a transformation of the scientific viewpoint towards one based on nonlinearity which may give rise to structural integrity and inter-relatedness of systems of particles. These ideas include solitons, chaos theory, self-similarity and fractals. In the latter, for example, a vast set of natural patterns and structures has been uncovered, all of which exhibit a new type of symmetry which was not recognized before. This symmetry links the form of a system on all length scales and hence it has been called the property of self-similarity. Examples of self-similar structures include many shapes seen in everyday life, such as the branching of trees, the delicate structure of ferns, the arteries and veins of the circulatory system to name but a few. This philosophically brings us back to the Hermetic doctrine of similarities between categories above and below and to the connections between everything as taught by Rabbi Kook.

One way of seeing the interconnectedness of all physical matter is to view an unfolding universe from the smallest scale of matter to the largest (or vice versa) including the emergence of consciousness out of biological substrates (Scott, 1995). Starting at the physical scale of elementary particles and their constituent quarks, the formation of the smallest grains of matter is described as an "interactive dance" in which mysterious matter fields participate. This is where nonlinearity first enters the physical world. At the next level of size scales, the elementary particles (i.e. protons and neutrons) can be seen as the building bricks in the construction of stable atomic nuclei. Curiously enough, we can now afford to forget that there is an intricate structure inside each elementary particle and simply focus on the nature of the interactions between neighboring particles. This is how organization of nuclear matter may be understood. The next step involves the atom. The basic principles are quite similar. The constituent members, i.e. electrons, are treated on an equal footing and their mutual interactions, together with the attractive coupling to the nucleus, are essential to the development of the atomic structure. This emergent coherent physical entity is spatially localized and is stable against external influences reaching its extreme resilience in the case of noble metals. However, unlike standard atomic theories put forward in twentieth century physics, where the onus was placed on individual particles, we should stress the predominant role of interactions in the determination of what an atom is. Linear theories were all based on the picture of quasi-independent electrons orbiting the nucleus, much like the planets orbit the Sun, and their mutual coupling was typically incorporated at a later stage through a sequence of corrective measures. The nonlinear approach to the problem of atomic structure is based on a mysterious envelope field for the atom as a whole, i.e. its overall charge density map (see Dixon et al., 1997). This field describes the electronic charge density and hence it self-interacts due to electrostatic repulsion while it is being pinned down in space by the attractive force of the nucleus. Once again, the lower level entity, in this case the nucleus, can be adequately regarded as structure-less (a geometrical point) at this higher level of description. Amazingly, endless fascinating possibilities spring to life when we arrange identical atoms in various regular spatial configurations. The question of how these possibilities can be realized in practice and what properties they are expected to reveal is a task for many a scientist. What at first sight appears to be a mundane orderly sequence of atoms leads to an array of materials such as semiconductors, crystals, metals, superconductors and magnets.

Moving to the next "plane" of organization, the infinite set of chemical compounds brings new possibilities due to the diversity of atoms that are their building blocks. However, chemical components are not just a "sum of their constituent parts". For instance, the properties of water have virtually nothing in common with those of oxygen and hydrogen which form it. With this in mind, we can safely state that chemistry's potential for complexity is virtually limitless. Chemical compounds may be viewed as nonlinear coherent structures whose stability and identity is mainly derived from the interactions (chemical bonds) between the "coupled components" or ions. The internal structure of the reactants, however, is of utmost importance to the chemical properties of the component created. In particular, depending on how many valence electrons the interacting atoms have makes a dramatic difference to their chemical reactivity and the type of bonding that results. Again, the interconnectedness here is self-evident. It is technically referred to as the ability to make chemical bonds. A metaphor to represent human interactions springs to mind readily.

The next level of complexity must necessarily involve biological matter and living systems in general. The incredibly complicated world of biology has not completely revealed its organizing principles to us yet except for the simplest of laws such as the genetic code. We can point to a picture of biological organization arising from the interlocking hierarchies, each of which is brought about by the onset of coherence with a multitude of similar functional sub-units. For example, a given type of cell when multiplied and arranged purposefully gives rise to a tissue or an organ. On a higher level of organization, autonomous organisms can be combined to create a social unit such as a termite colony, a slime mold or a human tribe. Over the course of history, human societies evolved so that, in spite of their individual differences and diversity, a level of cohesion and stability has been achieved which is necessary for their survival. Internal organization (e.g. language, culture, religion or moral values, economy, etc.) facilitated through human interactions, prevents the disintegration of a social structure, be it an ancient city-state, a medieval principality or a modern nation-state. Interconnectedness and mutual interactions leading to structure and function, consistent with the Kabbalistic world view are evident.

We can also look at the largest sized structures in the universe, i.e. our planetary system, the Milky Way galaxy and distant clusters of thousands of galaxies. We can draw parallels between the astronomical objects and the micro-world of the quantum physicist. In spite of the enormous differences in the magnitudes involved, we see a compelling unity of the universe in which form and structure emerge from a shapeless primeval soup as a consequence of the binary interactions between the building blocks of matter submerged in it. For a more detailed description of the role of nonlinearity as a new paradigm of modern science the reader is referred to the book by Dixon, Tuszynski and Clarkson (1997).

There is in Jewish mysticism a long-held belief that we are all interconnected, as expounded by Rabbi Kook, and that in creating this universe God implanted everywhere within it divine sparks, sacred elements within the fabric of energy that makes up all creation. In Kabbalah, the universe is composed of sefirot, spheres of existence, that are made up of energy and possess circular motion, spin-like characteristics, to put it in physical terms (Cooper, 1997). These sefirot, these spheres are connected, yet distinct, located in a variety of places both above us and within us and yet simultaneously linked to the *Ein Sof*, the Divine Source of all energy, and thus linked to all creation (Sholem, 1974). The Tree of Life is a relatively simple but profound diagrammatic representation of the grand design of creation as graphically represented by Kabbalistic mystics (see Fig. 1 below).

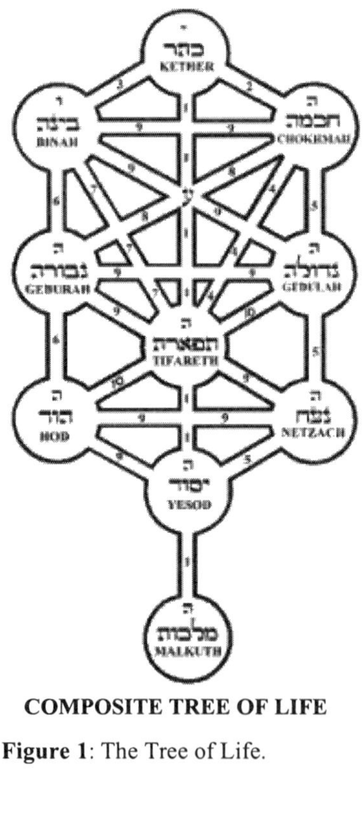

COMPOSITE TREE OF LIFE

Figure 1: The Tree of Life.

As briefly discussed above, there are organizing principles of physical and biological matter that can be used to understand their functional roles. As such it should have relevance also to the organization of matter. The four worlds of creation (*Asiyyah*-Action, *Yezirah*-Spirit, *Beriah*- Creation, *Azilut*-Emanation) are interconnected and arranged hierarchically, one giving rise to the next, similarly to the hierarchical organization of matter from physical to chemical to biological to societal (Whitehouse, 2008).

It is tantalizing to think that the simple diagram given by the Tree of Life can be used to connect the various elementary entities and provide meaningful relationships between them. For example, if we start by considering the 6 main elements indispensable for the formation of bio-molecules, namely: C, H, N, O, P and S, they lead to the 6 main functional groups of bio-molecules which are: hydroxyl: H-O, carbonyl: C=O, carboxyl: O=C-O-H, amino: $N\text{-}H_3$, sulfhydryl: S-H and phosphate: $P\text{-}O_4$. It is not too difficult to arrange them in the tree-of-life configuration starting with NH_3 at the bottom with N taking the position of *Yesod*, then moving on to COOH, CO and HO right above it with C taking the position of *Tiferet*, then PO_4, with P in the position of *Daat*, etc. This is sketched in Figure 2 below.

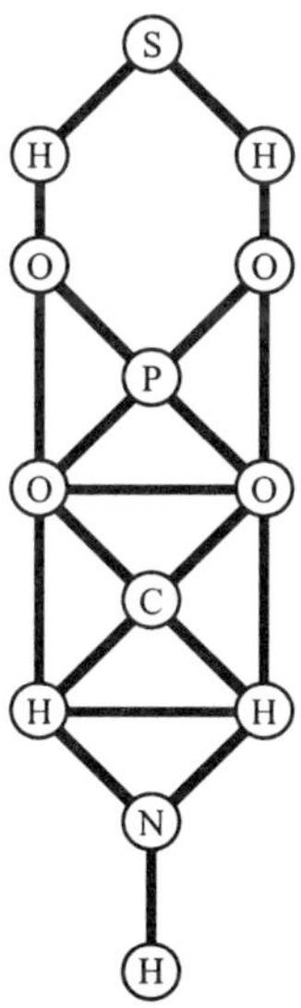

Figure 2: Schematic representation of the emergence of key biomolecular functional groups using the Tree of Life. In addition to H-O, C=O, O=C-O-H, $N\text{-}H_3$, S-H and $P\text{-}O_4$ groups, molecules such as H_20 and CO_2 can be seen.

	Seconed Position								
	U		C		A		G		
First Position	code	Amino Acid	code	Amino Acid	code	Amino Acid	code	Amino Acid	Third Position
U	UUU	phe	UCU	ser	UAU	tyr	UGU	cys	U
	UUC		UCC		UAC		UGC		C
	UUA	leu	UCA		UAA	STOP	UGA	STOP	A
	UUG		UCG		UAG	STOP	UGG	trp	G
C	CUU	leu	CCU	pro	CAU	his	CGU	arg	U
	CUC		CCC		CAC		CGC		C
	CUA		CCA		CAA	gln	CGA		A
	CUG		CCG		CAG		CGG		G
A	AUU	ile	ACU	thr	AAU	asn	AGU	ser	U
	AUC		ACC		AAC		AGC		C
	AUA		ACA		AAA	lys	AGA	arg	A
	AUG	met	ACG		AAG		AGG		G
G	GUU	val	GCU	ala	GAU	asp	GGU	gly	U
	GUC		GCC		GAC		GGC		C
	GUA		GCA		GAA	glu	GGA		A
	GUG		GCG		GAG		GGG		G

Figure 3: The Genetic Code.

It has been noted that the double helix structure of DNA resembles the Jacob's ladder extension of the Tree of Life obtained by repetition of the motifs (Whitehouse, 2008). Similarly, the 4 nucleic acids (U, C, A, G or T, C, A, G, respectively) can be ordered in triples in RNA or DNA to form a genetic code for transcription and translation into amino acids (see Fig. 3). While there are 64 possibilities of forming triple sets of the bases, there are numerous redundancies such that there is correspondence to the 20 naturally occurring standard amino acids, a stop codon and a start codon giving rise to 22 elementary building blocks for the construction of proteins. A group theoretic analysis was done (Hornos and Hornos, 1993) to represent the genetic code as a branching tree in the graph theory sense (see Fig. 4 below).

One of the Kabbalistic practices is referred to as the Path of Letters. It may be of major relevance in our effort to make a connection with molecular biology. There are 22 letters in the Hebrew alphabet and each of them can be placed over a horizontal, vertical or diagonal connector line in the Tree of Life. Thus, one can make a connection between a Hebrew letter and an amino acid symbol (plus the start and stop symbols required for proper translation of DNA and RNA sequences into polypeptides). Interestingly, there is a widely-used in molecular biology single-letter designation for amino acids using the Latin alphabet. On the other hand, the 22 Hebrew letters used in the Tree of Life represent consonants in Hebrew and the 10 sefirots refer to vowels adding up to what is called the 32 mystical paths to wisdom. While there are only four nucleic acid bases for building either a DNA or RNA structure leading to exactly 64 of their triplets, the group theoretic bifurcation diagram in Figure 3 shows 10 bifurcation points that lead to the separation of triples into the amino acid code. There may be a rigorous mathematical link between the connected graph of the Tree of Life and the branching tree of the supersymmetric model of the genetic code but this is still an open problem to be solved in the future. It would be very interesting to find out if, for example, a transformation such as a graph inversion (replacing nodes with links and vice versa) may be used to connect these two seemingly related graphs. Likewise, a one-to-one correspondence between the letters of the Hebrew alphabet and the amino acid designations should be achievable through thorough mathematical analysis.

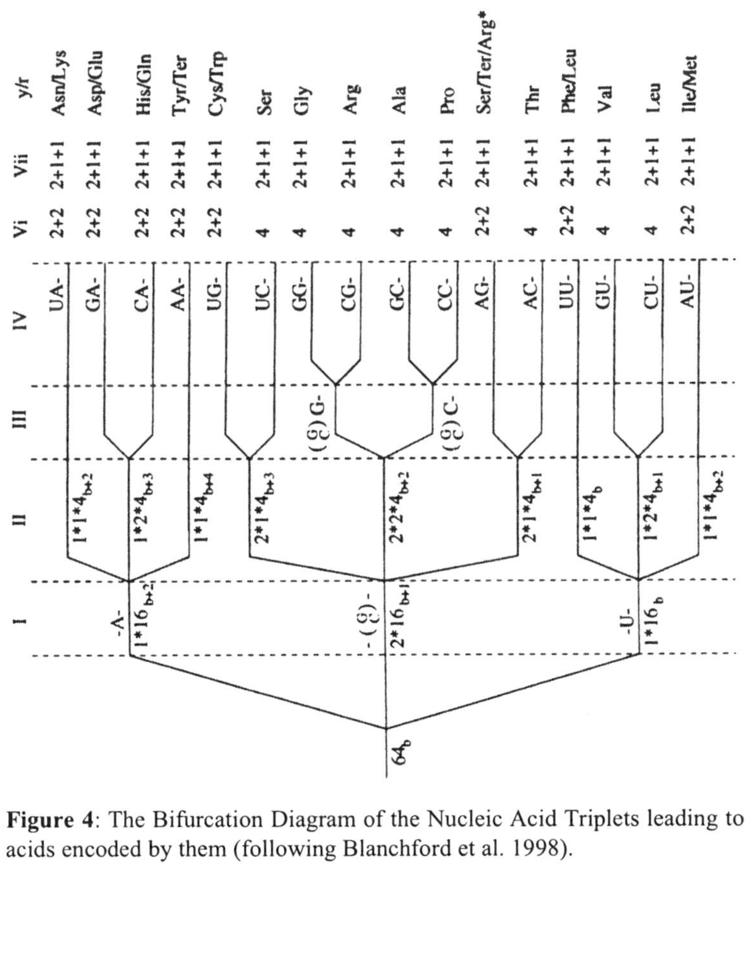

Figure 4: The Bifurcation Diagram of the Nucleic Acid Triplets leading to the 20 amino acids encoded by them (following Blanchford et al. 1998).

In summary, the main message is that, from the beginning of time, the universe has not only been expanding, but has continually unfolded and in the process, myriads of diverse coherent structures have been exposed at every level. While within each plane of existence the principal elements are inter-related, there is also a more intricate interconnectedness between planes of existence that modern science is uncovering. In each coherent structure similar general principles are at work and to uncover the hidden mysteries behind them one must employ nonlinear concepts of inter-relatedness and coherence (both spatial and temporal co-localization and integrity) which are just now becoming available to us through science although ancient mystical experience may have glimpsed millenia ago as can be appreciated by studying the work of Rabbi Kook. Rabbi Kook used the metaphor of the human body and its parts to describe the unity of all things and people. I see a direct parallel between this description and the coherent structure term in nonlinear physics.

Finally, in addition to the knowledge aspect of the process at work, Kabbalah implores the followers to restore the beautiful symmetry of the universal creation by adding a moral aspect according to the principle of *"tikkun olam"* (Matt, 1997). The shattered vessels and their shards became sparks of light trapped within the material of creation. This vivid image resembles the symmetry breaking of the pre-inflation Universe following the Big Bang. Prayer, contemplation of the various aspects of the divinity encapsulated in the *Sephirot* and engaging in *mitzvot* according to Kabbalists help to release these sparks and allow them to reunite with God's essence thereby repairing this world and the world to come (Waite, 1996).

L. Smolin, (1997) *The Life of the Cosmos*. Oxford University Press, Oxford.

D. Overbye, (2008) *The New York Times*, February 17, 2008: In the new physics no quark is an island"

A.C. Scott, (1995) *Stairway to the Mind, The Controversial New Science of Consciousness*. Springer, Berlin.

J.E.M. Hornos, Y.M.M. Hornos, Algebraic model for the evolution of the genetic code, *Physical Review Letters* 71 (1993) 4401.

M. Whitehouse, (2008) *Total Kabbalah*, Chronicle Books, San Francisco.

D.A. Cooper, (1997) *God is a Verb; Kabbalah and the Practice of Mystical Judaism*, Riverhead Books, New York.

J.D. Blanchford, I. Tsohantjis and P.D. Jarvis, A supersymmetric model of the evolution of the genetic code, *PNAS USA* 95 (1998) 987.

M.P. Hall, (2003) *The Secret Teachings of All Ages*, Penguin, New York.

P. Epstein, (1988) *Kabbalah: The Way of the Jewish Mystic*, Shambhala, Boston.

D.C. Matt, (1997) *The Essential Kabbalah: The Heart of Jewish Mysticism*, Castle Books, Edison, NJ.

G. Sholem, (1974) *Kabbalah*, Meridian Books, New York, 1974.

A.E. Waite, (1996) *The Holy Kabbalah, A Study of the Secret Tradition in Israel*, Oracle Publishing, Royston.

J.M. Dixon, J.A. Tuszynski and P.L. Clarkson (1997) *From Nonlinearity to Coherence; A Study of Universal Features of Many-Body Systems*, Oxford University Press, Oxford.

D. Bohm, (1980) *Wholeness and the Implicate Order*, London: Routledge.

L.K. Nyhart and T. Broman (eds) (2002) *Science and the Civil Society*, University of Chicago Press, Chicago.

E. Mach (1960). *The Science of Mechanics; a Critical and Historical Account of its Development*. LaSalle, IL: Open Court Pub. Co.

Richard Gordon

In this round, I'll just pick up on some of Jack's points:

1) While I acknowledge the importance of nonlinearity, I don't think it is sufficient. Just because WE recognize new phenomena arising from nonlinear interactions does not mean that they matter one whit to the operation of the universe, except insofar as we observe them.

2) There is a subclass of nonlinear phenomena in which the whole alters the behavior of the parts. Our joint work on cortical rotation in the one-cell axolotl embryo is a manageable example*. But I'm not yet convinced that there is anything unique about this subclass that alters comment #1.

3) Since we can alter what physics happens (or can we? — the question of free will), through our observation of these emergent nonlinear phenomena, the role of the observer has to be examined well beyond that considered in quantum mechanics.

---

*Nouri, C., R. Luppes, A.E.P. Veldman, J.A. Tuszynski & R. Gordon (2008). Rayleigh instability of the inverted one-cell amphibian embryo [Invited: "Physical Aspects of Developmental Biology" special issue, Guest Editor Shane Hutson, Vanderbilt University]. Physical Biology 5, 015006.

4) The emergence of novelty in the universe has yet to be addressed properly by physicists. This is at the core of all the debates about origins and creationist cop outs.

Dov Berger
I've read your dialogue. This is a very interesting paper on the cabbalistic interpretation of the building blocks of the universe.

Your dialogue concentrates on the fundamental basic of the cabbalistic notion of Unity. I can only say that I agree with you. I didn't want to concentrate in my paper on this subject, because my main goal was the interpretation of the Natural Selection. From the point of view of the great unity, Natural Selection became an internal battle between the cells of a general developing organism.

Jack A. Tuszynski
Dear Dov and Dick,
Many thanks for your kind emails. I realize the focus on natural selection was Dov's main intention in his chapter while I went off on a different tangent (unity and self-similarity in the architectural design of the basic elements of the universe). However, I think there is a way of incorporating Natural selection within the unity argument. There is never a single force at work for a balanced and sustainable process, be it physical, chemical, biological selection emphasizes competition but one also requires collaboration (symbiosis, nurture, education, marriage, etc.). Thus, a cabbalistic world view is more complete than a purely biological approach exemplified by Darwin. On the other hand, a fully "nonlinear" scientific methodology that I tried to give a glimpse of will include both tendencies in a way similar to the cabbalistic analysis of positive and negative aspects (e.g. Hesed and Gevurah) yielding an equilibrium. That's why Tipheret and Daat are located centrally. In nonlinear physics, in order to create a stable coherent structure such as a soliton (localized propagating wave), both dispersion and nonlinear superposition of individual waves must be present for this unusual objects to emerge. Dispersion favors spatial diffusion (scattering) of the wave while nonlinear superposition piles the individual waves on top of each other increasing their overall amplitude. When proper balance between these two opposite tendencies is achieved, a coherent structure emerges. Many examples of coherent structures exist in various fields of science, ranging from laser action to galaxy formation, and hence it may be an important organizing principle for the emergence of order in general. In this vein, Kabbalah shows very profound knowledge of the universe by emphasizing a balance of opposite

forces as an organizing principle of God's creation including our psyche. This is much more enlightened and in tune with modern science than any philosophical/religious construct that I'm familiar with. I hope I'm making sense. I'll be happy to add some thoughts along these lines to the dialogue. I just didn't want to write too long a piece. I find the subject fascinating as you may have guessed.

Dov Berger
The dialogue of Jack A. Tuszynski is a very interesting paper on the cabbalistic interpretation of the building blocks of the universe, concentrating on the fundamental basics of the cabbalistic notion of Unity. I didn't want to concentrate in my paper on this subject, because my main goal is the interpretation of Natural Selection. From the point of view of the great unity, Natural Selection becomes an internal battle between the cells of a general developing organism.

In my paper I tried to stay within the framework of Rabbi Kook's philosophy. Rabbi Kook deals with the general principles of the material world according to the principle of the Kabbalah. Even though Rabbi Kook doesn't deal with specific analogies from the Kabbalah to the "building blocks" of physical and biological systems, and I followed the path of Rabbi Kook.

Biography

Dr. Richard Gordon is a theoretical biologist whose endeavors range from AIDS prevention to breast cancer imaging on the medical side and from the effects of microgravity on amphibian embryos to the delights of diatom motility and morphogenesis on the basic science side. He inadvertently wrote the first paper on diatom nanotechnology. He has a B.Sc. in Mathematics from the University of Chicago (1963) and a Ph.D. in Chemical Physics from the University of Oregon (1967), and is now a Professor of Radiology at the University of Manitoba, where he has also held appointments in Botany, Computer Science, Electrical & Computer Engineering, Pathology, Physics and Zoology, perhaps foretold by his 1981 Rh Institute Grant for Outstanding Contributions to Scholarship and Research in the Interdisciplinary Category. His forays into how people treat one another include articles on democratization of the science granting system, a proposal for world minimum wage, havatars (human, avatar pairs) for a new approach to understanding epidemics such as HIV/AIDS, and a campaign to get up to date books to medical schools in Afghanistan via Books With Wings (http://www.bookswithwings.ca). His varied mentors, to whom he is grateful, include James S. Dwyer, Susan Meschel, E. Peter Geiduschek, Edward Anders, Aaron Novick Terrell L. Hill, Theodore Puck, Stanislaw Ulam, Jack Carmichael, Antone G. Jacobson, Cyrus Levinthal, Robert Rosen, Conrad H. Waddington, James F. Danielli, Lewis Wolpert, Zim Hearon and Lewis Lipkin.

12
Over-Confident Anti-Creationists versus Over-Confident Creationists

RICHARD GORDON

Departments of Radiology, Computer Science and Electrical & Computer Engineering, University of Manitoba, Room GA216, HSC, 820 Sherbrook Street, Winnipeg, Manitoba R3A 1R9 Canada
gordonr@cc.umanitoba.ca

The question of religion versus science raises some profound issues, alongside the political ones and the tug of war for turf. Any dogmatic reaction of scientists to creationists is not in the spirit of science itself, but unfortunately there is a lot of it. "If we divest ourselves of our humility under pressure from creationists to appear certain of ourselves, we end up producing dogma, not science" (§10.05: Gordon, 1999). Here I survey and critique a few books to show how such dogmatism would better be replaced with humility in the face of the unknown.

1. Book #1: My Unwelcome Introduction to "Intelligent Design" (ID)

Behe, M.J. (1996). *Darwin's Black Box, The Biochemical Challenge to Evolution*, New York: The Free Press.

When I came across a copy of this intriguingly titled book (Behe, 1996), I settled in for something I thought would be of the gender of *The Great Evolution Mystery* (Taylor, 1983). Naively plowing ahead, it took me 100 pages before I realized that Behe's book was a creationist tract. I closed it in revulsion, feeling taken in and betrayed.

I have generally been allergic to creationism, after earlier unsatisfactory encounters with parapsychology (Rhine, 1934) and disembodied morphogenetic fields (Casti, 1990; Sheldrake, 1981) that are a fantasy beyond the embryological evidence (Beloussov, 2001; Gordon, 1999). But today's creationism is not just a matter of sloppy or wishful thinking. There is a political agenda of intelligent design creationists to penetrate into elementary school curricula and beyond (Ratliff, 2004). My political comments here concentrate on their efforts to infiltrate universities, which are succeeding: in 2001 I attended a seminar "Dare a Scientist Believe in Design?" (Gingerich, 1994). Gingerich has the right credentials, including fitting our North American infatuation with Harvard as the presumed center of our intellectual universe (Ronald, 2001).

Out of curiosity and dismay, and perhaps to explore the dimensions of my own spirituality, I plowed through some of the plethora of recent books heralding the harmony between religion and science (Branover & Attia, 1994; Kaplan, 1937; Matt, 1996; Raymo, 1998; Schroeder, 1990), but generally found the science amateurish and the conclusions unconvincing. For example, somehow in the presumed one-to-one correspondence between the stories of Genesis and science, albeit with a distorted time scale, G-d's text left out what would seem in retrospect a major event, namely the freezing over of the whole earth for 3 to 15 million years (Bodiselitsch et al., 2005).

While I sought out rabbis, Reconstructionist, Conservative, Orthodox and Lubavitch, and read the *Guide of the Perplexed* (Maimonides, 1190) and Torah Umadda (the study of the world outside of the *Torah*: Lamm, 1990), I found myself looking to them and religion for hints of answers to scientific questions: how to solve the problems of evolution, the origin of perception and consciousness, and the origin of life. If this was soul searching or Adult Jewish Education, it was not what the rabbis had in mind. How were these transformations of matter accomplished *without* a guiding hand was my question? Science stops when we assume "G-d did it". Failure to explain is not proof of G-d, just proof of giving up.

I read works on emergence (Emmeche, 1997; Holland, 1998; Johnson, 2001; Morowitz, 2002) and wrote, myself, about the "emergence of emergence" (Gordon, 2000). My work on embryos had shown me that the sense of awe need not vanish when one analyzes this "miracle of life" in terms of math, chemistry and physics. I was picking up where the pioneers of *Entwichlungsmechanik* (developmental mechanics) had started with real physics:

> "To think that heredity will build organic beings without mechanical means is a piece of unscientific mysticism" (His, 1888),

and then lost their way.

For example, I realized the loss to science of Hans Driesch, whose brilliant work on sea urchin embryos (Driesch, 1891) had unnecessarily driven him to vitalism (Driesch, 1929; Gordon, 1999): his nonphysical "entelechy" may have a basis in active (excitable) waves in tissue (matter) (Gordon, 2006; Markin, Pastushenko & Chizmadzhev, 1987). Driesch seems to have failed to recognize that a process in matter (like waves) need not imply a separate substance. von Uexküll (1926) got it (Gordon, 1999). If we can explain how active, nested differentiation waves in embryos launch, propagate and terminate, and trigger cell differentiation (Björklund & Gordon, 1994, 2006; Gordon, 1999; Gordon & Björklund, 1996; Gordon, Björklund & Nieuwkoop, 1994), then we may indeed be able to achieve a physics of embryos (Beloussov & Gordon, 2006; Forgacs & Newman, 2005), one that bridges the immense size scale between molecules and organisms (cf. the previous attempt of *Topobiology*: Edelman, 1988). That would indeed be awesome.

2. Book #2: A Breath of Fresh Air

> Perakh, M. (2004). *Unintelligent Design*, Amherst, NY: Prometheus Books.

Perakh has done what I avoid here: tackled the whole modern creationist enterprise book by book, shredding their arguments (Perakh, 2004). He has done a needed service, clearing the cobwebs for those of us for whom this barrage of pseudoscience masquerading as science has created some niggling doubts (and after all, doubt and questioning is the essence of the scientist (Russell, 1935), so our very strength is also our Achilles heel).

Perakh finds he has to argue on many levels: authoritarianism in science, probability theory, logic, etc. He tries to assiduously avoid those areas in which he is not expert, especially biology. Thus you won't, unfortunately, find a vigorous defence of Darwin's theory of evolution here, nor even a clear exposition. Behe's "irreducible complexity" in biochemistry (Behe, 1996) is not debated in biochemical terms. Too bad, because most biologists like to see the creationists trounced on our own turf, with concise arguments about redundancy in genetic networks, robustness of regulation in embryonic development, and the radiation of species into new niches. But that's available elsewhere (Dembski & Ruse, 2004a; Pennock, 2001). While Perakh may mention gene knockouts, for example, he does not bring to bear the weight of evidence about their importance in demonstrating gene redundancy (Jablonka & Lamb, 2005; Shastry, 1998). His lack of medical experience leaves out the vast array of (unintelligently designed?) inherited metabolic deficiencies with which humans nevertheless suboptimally muddle on through their lives (Habener & Williams, 2004). Of the more difficult problem of origin of life (where the original conditions are long ago, perhaps far away: Gordon & Hoover, 2007, and have to be inferred), he only gives us hints of the possibility of chains of probable events, without any details. This contrasts with the despondent view of Conway Morris that 70 years of laboratory chemistry has not produced anything about to crawl out of our test tubes (Conway Morris, 2003). But nor has it produced the magic bullet to destroy cancer (Raso, 1990). We plod on. Science is a multigenerational trip.

Perakh's strengths are in chemistry, statistical mechanics, and the clear thinking of a working scientist. For those of us who are Jewish, we finally have an overview of Jewish creationism (Spetner, 1994) treated on equal footing, and squashed just as well as Christian creationism (which it closely parallels, a form of convergent evolution?). Indeed, the first Jewish/Christian interfaith symposium of creationists has taken place (Dembski, 2005).

I can illustrate Perakh's approach by having some fun at his expense. In his last chapter "The rise and fall of the *Bible* code", he discusses the statistics of ELSs (Equidistant Letter Sequences), in which the *Torah* is "shown" to predict future events. Hebrew letters selected *n* letters apart are strung together to reveal startling prophecies. Perakh concludes: "Despite a number of attempts, so far [no one]… succeeded in proving that sets of ELSs in the *Bible* form meaningful phrases, let alone whole sentences" (p. 412).

However, for contrast, let us define the VLS (Variable-distant Letter Sequence) as an ordered set of letters, with variable distance between them, representing indeed a whole, meaningful sentence. For mathematicians, the algorithm for a VLS is the following. Given a sentence of letters, spaces, and punctuation A_i, i=1,,,L, and a string B_i, i=1,,,P of characters on a book page holding P characters, where L<P, we find the first s such that $B_s = A_1$. Then we continue along the page until we find the first t>s, such that $B_t=A_2$, etc. The calculation to be made is the probability that all L matches will occur within the finite limitation of P characters on a page. Empirically I find it is high. If we take the sentence "Perakh is wrong", it is found as a VLS five times on the subsequent page 413 of Perakh's book! (For balance note that "Behe is wrong" occurs 6 times on the same page.) Of course, this sentence would even be found in similar multiplicity in the King James version of the *Bible*. But then every text resounds with the message: "Perakh is wrong"! (For the mathematician, the argument I have just given is RAA, i.e., Reductio Ad Absurdum.)

Perakh's book, however, leaves us cold, perhaps because: "The heart of the problem, I believe, is to explain how it might be that we, a product of evolution, possess an overwhelming sense of purpose and moral identity yet arose by processes that were seemingly without meaning" (Conway Morris, 2003). One would think that Conway Morris was making a clarion call for the science of biosemiotics, which seems well under way (Barbieri, 2003; Farina, 2005), despite suggestions that "questions of meaning… cannot be solved by the methodologies developed by modern science" (Klostermaier, 2004). The facile classification of morality as a nonscientific problem flies in the face of many attempts to address the issue (Shermer, 2004; Stent, 1980). One more gap being filled?

3. Book #3: Over-Confident Anti-Creationists

Young, M. & Edis, T. (eds.) (2004). *Why Intelligent Design Fails: A Scientific Critique of the New Creationism*, Piscataway, NJ: Rutgers University Press.

Creationism is, perhaps smugly by those who live elsewhere, thought to be mostly a problem of the "Bible belt" in the USA. This is far from the truth. Most non-Americans can find it in their own communities in forms that are, or have the potential to be, just as pernicious. So, although *Why Intelligent Design Fails* (Young & Edis, 2004a) seems designed to gird American school teachers against the onslaught of the new vanguard of intelligent design creationists, it refreshingly comes out of work on Muslim creationism (Edis, 2002, 2003, 2004a). Muslim creationism may date back to the rejection (al-Ghazali, 1095; Hoodbhoy, 1991; Malik, 1980; Nasr, 1968) of the very philosophy that accompanied the introduction of Arab (and via them, ancient Greek) ideas into Europe. Not only are Islamic fundamentalists creationists, but creationism permeates much of Islamic culture (Edis, 2003). Of course, this is true of Christian and Jewish cultures too, which, as Dennett points out (Dennett, 1995), pay lip service to evolution but have not come to see the world from its perspective: "most religions, including Judaism, continue to have great difficulty in absorbing the detailed facts and implications of the record of natural selection written in every corner of the world, from the DNA of our own cells to the fossils of our ancestors deep within the earth" (Pollack, 2000). As a result, we act as if G-d created the world, as if G-d created us, and as if G-d can intervene in our lives on our behalf. To act alternatively as if our "pocket universe" (Susskind, 2006) started 13.7×10^9 years ago (Hawley & Holcomb, 2005), as if we are descended from bacteria (Schopf, 1999), some of whom took 10^9 years to form the symbioses that led to eukaryotes (Margulis & Dorion Sagan, 2002) and then through the Cambrian "explosion" (McMenamin, 1989) and rapid subsequent evolution to us (rapid on the geological time scale), and that we are (machines?) chock full of crawling, rotating and sliding nanomotors (Schliwa, 2003), is a bit humbling. We have yet to think all this new information through for its consequences for our "moral compass", let alone for consistency with Genesis.

The tenor of *Why Intelligent Design Fails* is openly belligerent: "The contributors to this book are legitimate academic researchers who… refute the neocreationists' pretensions" (Young & Edis, 2004b). So I see my job as taking them down a peg, because science somehow has to oppose dogmatism without entering into its own.

3.1. *Anti-Creationists Against Progressive Evolution*

> "Evolution in the minimal sense of descent with modification could be accommodated if it could be seen as a progression toward higher orders of being; indeed, such was the initial response of even evangelical theologians to Darwin (Livingstone, 1987). Interpreting evolution as an explicitly guided development would retain a sense of intelligent design; and this approach is still alive among more liberal thinkers,

both Christian and Muslim. Darwinian biology, however, strains this view since it relies on nothing but blind mechanisms with no intrinsic directionality" (Edis, 2004a).

> "Progress is a noxious, culturally embedded, untestable, nonoperational, intractable idea that must be replaced if we wish to understand the patterns of history" (Gould, 1988).

The notion that we are part of a great scheme for improving the world is deeply rooted in Judaism (tikkun olam). This creates a problem for scientists studying evolution, who either see a directionality to it (reviewed in: Gordon, 1999), or, like the late Stephen J. Gould, deny progress exists. Gould pushed the concept that evolution has no direction and is therefore not progressive (Gould, 1988), despite the evidence of the fossil record. This is a strange position for a paleontologist to take.

Gould came to this conclusion by an abstract mathematical thought process, weighting all living organisms equally, placing *Homo sapiens* at the (insignificant?) statistical tail of a probability distribution of complexity. Indeed, we are outnumbered: there are over 10,000 times as many bacteria on and in each one of us than there are people on earth (Todar, 2006), and despite our predilection to eat everything in sight, we still are but 0.33% of the world's biomass (Wikipedia, 2006).

Unfortunately for his argument, Gould's concept of complexity is one dimensional, which ironically means that he reinstated the long defunct, one dimensional "ladder of life" or "great chain of being" (Lovejoy, 1936), 150 years ago replaced by the tree of life. He considered the single celled organism to represent a "wall" of minimal complexity, and evolution to merely do an unbiased, one-dimensional random walk with what is called a reflecting boundary (Feller, 1968), bouncing off this "wall" (i.e., an organism can't have less than 1 cell), which mathematically consists of nothing more than the point at the origin. But if complexity is multidimensional, then a random walk would rarely bounce off just one point and would usually go around it (§10.03: Gordon, 1999). A single point is a significant barrier along a line (one dimensional), but not at all in two or more dimensions. The probability of even just ever hitting this "wall" (the mathematical origin of a coordinate system) plunges rapidly from 100% for 1D and 2D to 34% for 3D reaching only 7% by

8D (Weisstein, 1999). With 20,000 to 30,000 genes (Venter et al., 2001), our complexity is $10^{20,000}$ D or more, and such a "wall" loses all significance.

There are at least two general ways that evolution could proceed through random mutation from less complex to more complex organisms. One is an unbiased random walk in a higher dimensional space of the genes, just discussed. Another is (on average) directional progress, in the sense of a biased random walk (Feller, 1968). Biased random walks are generally considered anathema by anti-creationists. However, in a situation of thermodynamic *non*equilibrium, as occurs in all living organisms, biased random walks are possible at the intramolecular (Brandsburg-Zabary et al., 2000), molecular (Thomas, Imafuku & Tawada, 2001), cellular (Alt, 1980; Hill & Häder, 1997), and speciation (Gavrilets, 2000; Katz, 1987) levels. This is not evolution towards a goal. It is just as "blind" (Dawkins, 1986) as random mutation. Our language makes it difficult to distinguish "directional" from "directed". However, if a loose stone, in a nonequilibrium position on the side of a hill, ends up downhill, we no longer attribute this to intention on its part or direction imparted by a deity. If evolution is, objectively, progressive in this directional (not directed) sense, then we need to find something that acts like the slope of a hill. Gene (Gordon, 1994; Ohno, 1970), and gene cascade (Proposition 179 and §10.02: Gordon, 1999) duplication, which are ubiquitous, provide potential mechanisms: it is generally acknowledged that the two copies of the gene present after a duplication can drift in function from the original, single gene in the course of subsequent evolution (though Behe has challenged this: Behe & Snoke, 2004). If a species achieves higher fitness as a result, it is reasonable to assume that subsequent loss of one of the genes in an individual could put it at a disadvantage relative to its compatriots. There is the bias: it's often better to keep the genes you've got than lose some (albeit not always, since parasites lose genes: Andersson & Andersson, 1999). Another objective source of direction is the arms race or coevolution between predator and prey (Van Valen, 1973).

If progress in evolution is in part a consequence of DNA mutations (such as duplications), then we cannot so lightly dismiss the creationists' urge to call such a process G-d given, though it may just be an example of a multitude of irreversible processes (Campbell, 1988; Keizer, 1987; Tyson & Novak, 2001), so irreversibility, rather than reversible thermodynamic equilibrium, may be fundamentally what is "G-d given". Of course, this kind of progress can be as "mindless" as the progression of frost across a window, beautiful to behold, but hardly a result of cosmic intention (cf. von der Malsburg, 2004), or even Jack

Frost (Melling, 2003). My point is that some retreat from Gould's dogmatic position may be warranted, and that without allowing the possibility of actual progress in evolution through a nonsupernatural mechanism, anti-creationists are being overconfident.

3.2. *Anti-Creationist Cookbook Recipes for Making Organisms*

> "The genome... is a recipe, not a blueprint. The genome tells the mouse to have hair, for example, but it does not specify where each hair is located, precisely as a recipe tells a cake to have bubbles but does not specify the location of each bubble" (Young, 2004).

We could excuse this language as shorthand for the real process of embryogenesis, and its still unknown relationship to the genome (Beloussov & Gordon, 2006). The problems here are manifold. A genome is not a thinking entity that "tells" cells what to do by "reading" a recipe (§1.16: Gordon, 1999). While an embryo is a life cycle stage that *contains* its genome as a "subset", it actually contains multiple copies of the genome after the one cell stage. Which cell's genome is "telling" which cell what, and why are the messages different? Who reads the recipe? Where is their kitchen? The silliness belies some real problems. Many embryos do just fine in distilled water, with no environmental cues except gravity (and even that is not necessary at early stages: Dournon, 2003; Ubbels et al., 1989), so that the vague idea that variations in gene expression between cells are somehow caused by the external environment has been narrowed to the "environmental" effects of cells *within* the embryo on one another (Gilbert, 2004). How an embryo or its genome starts or spatially organizes cell differentiation is left up in the air. For the creationist, the hand of G-d cuts through all this nonsense, and makes more sense. We have our work cut out for ourselves if we are to achieve a plausible explanation for embryogenesis, and need to express humility that we haven't solved this one yet.

3.3. *Anti-Creationists Against Reductionism*

> "Emergent properties are the result of self-organization (Shanks & Karsai, 2004) and force reality into a series of levels: biochemical, organelle (an "organ" within a cell), cell, organ, organism, ... for example. No one level is more fundamental than any other. Liquid

> water is no more fundamental than isolated water molecules are. Rather, each level is an alternate way of looking at reality" (Young, 2004).

Reductionism has had some bad press (Koestler & Smythies, 1969; Rothman, 2002). But there are problems (Gordon, 1996) with the alternative view that new scientific laws come into play at higher "hierarchical" levels (Gould, 1982, 1998). I've argued that we need to put effort into "linking disciplines", of which statistical mechanics is the best example, to relate one level to the next, which would seem to preserve reductionism (Gordon, 1996, 1999). But while I now think there might be much more to these levels, the question of whether higher levels have any real independence is far from settled. This is the major, ancient "mixtio" problem of the origin of new qualities in substances from their "elements" (Dijksterhuis, 1950), which may involve the poverty of physics and the evolution of perception (Martin & Gordon, 2001), and has led to the concept of holism (Smuts, 1926). The reductionist view is that physics, like mathematics, permits deduction from primitives and axioms. If such a physics encompasses life and mind, then all science could be deductive and reductionist. Humility is again needed, as has been well expressed:

"My own scientific career was a descent from higher to lower dimension, led by the desire to understand life. I went from animals to cells, from cells to bacteria, from bacteria to molecules, from molecules to electrons. The story had its irony, for molecules and electrons have no life at all. On my way life ran out between my fingers" (Szent-Györgyi, 1972).

3.4. *Anti-Creationists Create "Self-Organizing" Systems*

Creationist Michael Behe's main argument is that many biochemical systems are "irreducibly complex", which seems to mean that they cease to function if even one component is removed or defective (Behe, 1996). The *reducibly* complex nature of the evolution of vision (Gehring & Ikeo, 1999; Osorio, 1994), bird flight (Gishlick, 2004), or the bacterial flagellum (Musgrave, 2004; Ussery, 2004) have been demonstrated and contradict Behe. Such analyses are based on the fossil record, careful reading of DNA sequences and their inferred alteration over time, and computer simulations demonstrating plausibility of paths of evolution. Moreover, convergent evolution indicates more than one way to solve what would seem to be the same evolutionary problem (Conway Morris, 2003; Fernald, 1997).

But let's consider another approach: the study of so-called self-organizing systems. Two such systems are given as counterexamples to intelligent design creationism (Shanks & Karsai, 2004): Bénard cell convection and the Belousov-Zhabotinski reaction (Tyson, 1976). Honeycomb patterned Bénard flow cells may be observed by gently heating a pan of shallow water with pepper added. Belousov-Zhabotinski reactions require chemicals that react in certain nonlinear ways.

The Bénard cell convection as a model for morphogenesis is not self-contained (what makes the pan and stove? §1.16: Gordon, 1999). But rather than dwell on this, perhaps what is more important in the current context is what seems to me to be a contradiction:

> "The version of the [Belousov-Zhabotinski] reaction that one of us (Niall Shanks) has used in classroom demonstrations has the following ingredients: potassium bromate, malonic acid, potassium bromide, cerium ammonium nitrate, and sulfuric acid. When the ingredients are placed in a beaker, the system self-organizes to form a repeating cycle of reactions.... You can use it to tell the time: it is a watch that forms in a beaker without help of a watchmaker.... The reaction is important because advocates of intelligent-design theory claim that irreducible complexity can appear only as the result of the actions of an intelligent designer who takes the components of the system and assembles them into a functioning whole.... Apparently, the unguided laws of chemistry will generate irreducibly complex systems" (Shanks & Karsai, 2004).

It seems quite clear that Niall Shanks, as an experimentalist, *is* the "intelligent designer" of the Belousov-Zhabotinski "watch" in his lab, which is analogous to any assembled modern watch that contains a feedback circuit driving a crystal into resonant oscillations by the "unguided laws" of mechanics and electricity. Fundamentally, what the anti-creationists are ignoring is that they are part of these systems, by constructing them. Chemical "watches" do not arise spontaneously (unless we trace back the origin of Niall Shanks himself, generation by generation, to the spontaneous origin of life).

This ignoring of human designer context extends to arguments against the probability arguments of William Dembski (Dembski, 1998; Dembski, 1999):

> "If... we were to find a haphazard jumble of letters on the paper, something like IHJFL/BLACV?GYUFHRFWWVHBMD..., we would think differently; apparently, a mindless process had produced meaningless nonsense.... we would not find intelligent design" (Edis, 2004b).

If I came across such a clearly laid down line of letters on man-made paper, I would indeed assume that they were placed there directly or indirectly (via a computer) by an intelligent person. I might guess they formed a code or come from a language with which I'm not familiar, but the last thing I would suspect is a "mindless process". On the other hand, we have to look darn hard to find even single letters "written" in and by nature, such as on butterfly wings (Amato, 1990). Analogies to manmade objects always miss the point, in that we are their designers. We have to confront the actual self-construction of embryos, the real process of novelty generation in evolution, including consciousness, and the difficult question of the origin of life from non-life, all without our assistance, or, dismissing deities, anyone else's.

3.5. *Anti-Creationists are on the Verge of Creating Life?*

> "Given the known laws of physics and chemistry, we can easily imagine life based on silicon (computers, the Internet?) or other elements chemically similar to carbon" (Stenger, 2004).

This supposes either a much greater flexibility of silicon chemistry than we have discovered, or success of the artificial life program in software or hardware (Sipper, 2002), neither of which I can "easily imagine", without a lot more work ahead (Gordon, 2001). Such glib assurances do not persuade. In general, by analogy to the "G-d of the gaps" approach to theology, scientists often promise a "science of the gaps", i.e., more than they can presently deliver. It is a matter of faith that future science will fill in these gaps, or that they won't. The retreat of theologians has a long history (Russell, 1935; White, 1896), but that doesn't automatically mean that we can extrapolate a "victory" for science on all questions we ask.

4. Book #4: Christian Anti-Creationists

Dembski, W.A. & Ruse, M. (eds.) (2004a). *Debating Design: From Darwin to DNA*, Cambridge: Cambridge University Press.

So far I have dealt with Jewish and Muslim anti-creationists, who take a firm stand, if they sometimes overreach: "...we risk legitimizing intelligent design simply by engaging it. Let us make clear, then, that we do not consider intelligent design to be a legitimate scientific endeavor" (Perakh & Young, 2004). The authors of *Debating Design* (Dembski & Ruse, 2004a), where they discuss a specific religion at all, almost always refer to Christianity. For this reason I'll refer to them as "Christian" anti-creationists or "Christian" creationists, as *Debating Design* is a collaboration of both, with three strange characteristics: A) It does not refer to any literature written by the Muslim or Jewish anti-creationist authors in (Young & Edis, 2004a), although their non-Christian authors refer to them. B) It legitimizes intelligent design by co-editorship ("even doing something like this can be seen as giving one's opponents some kind of status and legitimacy" (Dembski & Ruse, 2004b); cf. a similar legitimizing volume in artificial intelligence: Richards et al., 2002). C) Despite the title, there is *no* debate, just sets of papers that talk past one another.

For example, one exposition of the plausible evolution of the bacterial flagellum in *Debating Design* (Miller, 2004) is not discussed in another article in the book that overconfidently asserts "Darwinists have... provided few if any causally specific reconstructions of the pathways that lead to the formation of irreducibly complex structures" (Menuge, 2004). But then the bacterial flagellum article (Miller, 2004) also makes claims about creationists unsupported by references. This is neither debate nor dialogue. Other "dialogues" have been just as unproductive (Milner & Maestro, 2002).

Debating Design is published by a major university press (cf. Campbell & Meyer, 2004; Pennock, 2001), and is thus a big step towards what creationists say they want: "Acceptance by the Educational Mainstream.... One of ID's long-term goals is to place at major universities more scientists whose work is explicitly shaped by an ID research program. For that reason, it is essential for the ID movement to build bridges with its opponents and to find sympathetic ears in the academy" (Menuge, 2004).

Mistakes of overconfidence are also made by the anti-creationist authors in *Debating Design*. For example, one would again think that the solution of embryogenesis is at hand: "It is possible, at least in principle, to give a causal account of the various physical and chemical processes in the development of an egg into a chicken..." (Ayala, 2004). We are asked to swallow contradictions, like *Alice in Wonderland* (Carroll, 1866), such as seeing science as both unlimited and limited at the same time: "...nothing in the world of nature escapes the scientific mode of knowledge.... There are matters of value and meaning that are... forever beyond science's scope" (Ayala, 2004).

The protein folding problem, how a linear polymer of amino acids forms a 3D functioning protein, which is still outstanding (§1.17: Gordon, 1999; Daggett & Fersht, 2003), and the assembly of protein complexes (Kuthan, 2001), are stated as solved problems: "...if one gets the sequences of all the proteins right, localization and assembly will take care of themselves" (Miller, 2004).

5. Conclusion

There is the historical case of "the breath of life" that provides some perspective: "Genesis 2,7 Then the LORD God formed man of the dust of the ground, and breathed into his nostrils the breath of life; and man became a living soul" (*Torah*, 1917).

Respiration is no longer an issue of science versus religion, no longer a gap, because no one questions our science and understanding of it. Respiration thus no longer inspires religious awe. If and when we come to understand morphogenesis ("the miracle of life": Nixon, 1983), the mechanistic working of the mind, and perhaps consciousness and the soul, the issues firing creationism will likewise be absorbed into our culture. The awe may remain, for those of us who can see it, but if these gaps are filled our awe won't be of the "G-d done it" variety.

Present day religion would appear to have left the definition and filling of gaps to science, which is why it is always in retreat, although historically the intellectual effort came from both sides, often in the same person (Dijksterhuis, 1950; Klostermaier, 2004). If religion and science ultimately merge or reunite into one discipline, one common understanding of the universe and ourselves, then en route both may need to set the problems to be solved, and work at them.

Thus for now I see a need to achieve a balance between religious G-d of the Gaps dogmatism and Science of the Gaps dogmatism. The answer is an open mind that advocates neither: "I doubt that the proponents of [intelligent design]... would support placing a sticker in the bibles, prayer-books, and hymnbooks used by churchgoers, asking them to approach their deism with skepticism" (Johns, 2005). But maybe they should. Dogmatic certainty is a hindrance to good science and the teaching of it, and perhaps to good religion too (Watkins & Winston, 2005). None of us have all the answers yet.

6. Acknowledgements

Supported in part by the Manitoba Institute for Child Health and the Canadian Space Agency. Thanks to Natalie K. Björklund, Fred L. Bookstein, Robert Craigen, Frederick W. Cummings, Bruce Damer, Brian Fristensky, Stephen P. McGrew, Khaleel Mohammed, Joseph Seckbach, Frithjof A.S. Sterrenburg, Anne P. Ussing, Christoph von der Malsburg and the reviewers for constructive comments.

References

al-Ghazali, A.H.M. (1095). *Tahafut Al-Falasifah [Incoherence of the Philosophers]*, 1963 translation ed., Pakistan Philosophical Congress, Lahore.

Alt, W. (1980). Biased random walk models for chemotaxis and related diffusion approximations. *J Math Biol* **9**(2), 147-177.

Amato, I. (1990). Insect inscriptions, hunting for human symbols in gently fluttering wings. *Sci. News* **137**, 369, 376-379.

Andersson, J.O. and Andersson, S.G. (1999). Insights into the evolutionary process of genome degradation. *Curr Opin Genet Dev* **9**(6), 664-671.

Ayala, F. (2004). Design without designer: Darwin's greatest discovery. **In**: W.A. Dembski and M. Ruse (eds.), *Debating Design: From Darwin to DNA*, Cambridge University Press , Cambridge, pp. 55-80.

Barbieri, M. (2003). *The Organic Codes. An Introduction to Semantic Biology*, Cambridge University Press, Cambridge.

Behe, M.J. (1996). *Darwin's Black Box, The Biochemical Challenge to Evolution*, The Free Press, New York.

Behe, M.J. and Snoke, D.W. (2004). Simulating evolution by gene duplication of protein features that require multiple amino acid residues. *Protein Sci* **13**(10), 2651-2664.

Beloussov, L.V. and Gordon, R. (2006). Preface. Morphodynamics: Bridging the Gap between the Genome and Embryo Physics. *International Journal of Developmental Biology* **50** (2/3), 79-80.

Beloussov, L.V. (2001). Morphogenetic fields: outlining the alternatives and enlarging the context. *Rivista di Biologia / Biology Forum* **94**(2), 219-235.

Björklund, N.K. and Gordon, R. (1994). Surface contraction and expansion waves correlated with differentiation in axolotl embryos. I. Prolegomenon and differentiation during the plunge through the blastopore, as shown by the fate map. *Computers & Chemistry* **18**(3), 333-345.

Björklund, N.K. and Gordon, R. (2006). A hypothesis linking low folate intake to neural tube defects due to failure of post-translation methylations of the cytoskeleton. *Int. J. Dev. Biol.* **50**, 135-141.

Bodiselitsch, B., Koeberl, C., Master, S. and Reimold, W.U. (2005). Estimating duration and intensity of Neoproterozoic snowball glaciations from Ir anomalies. *Science* **308**(5719), 239-242.

Brandsburg-Zabary, S., Fried, O., Marantz, Y., Nachliel, E. and Gutman, M. (2000). Biophysical aspects of intra-protein proton transfer. *Biochim Biophys Acta* **1458**(1), 120-134.

Branover, H. and Attia, J. (eds.) (1994). *Science in the Light of Torah, A B'Or Ha'Torah Reader*, Jason Aronson, Northvale, New Jersey.

Campbell, J.H. (1988). Evolution as nonequilibrium thermodynamics: halfway there? **In**: B.H. Weber, D.J. Depew and J.D. Smith (eds.), *Entropy, Information, and Evolution*, MIT Press, Bradford Books, Cambridge, MA, pp. 275-284.

Campbell, J.A. and Meyer, S.C. (eds.) (2004). *Darwinism, Design, and Public Education*, Michigan State University Press, East Lansing.

Carroll, L. (1866). *Alice in Wonderland*, D. Appleton & Co., New York.

Casti, J.L. (1990). *Searching for Certainty, What Scientists Can Know About the Future*, William Morrow and Co., New York.

Conway Morris, S. (2003). *Life's Solution, Inevitable Humans in a Lonely Universe*, Cambridge University Press, Cambridge.

Daggett, V. and Fersht, A. (2003). The present view of the mechanism of protein folding. *Nat Rev Mol Cell Biol* **4**(6), 497-502.

Dawkins, R. (1986). *The Blind Watchmaker, Why the Evidence of Evolution Reveals a Universe Without Design*, W.W. Norton & Co., New York.

Dembski, W.A. (1998). *The Design Inference: Eliminating Chance Through Small Probabilities*, Cambridge University Press, Cambridge.

Dembski, W.A. (1999). *Intelligent Design: the Bridge Between Science & Theology*, InterVarsity Press, Downers Grove, Ill.

Dembski, W.A. (2005). The place of Intelligent Design in the natural sciences. **In**: *The Sixth Miami International Conference on Torah & Science, Absolute Standards in an Age of Relativity, 13-15 December 2005, Miami, Florida*, http://www.borhatorah.org/home/conference/absbioconf6.html.

Dembski, W.A. and Ruse, M. (eds.) (2004a). *Debating Design: From Darwin to DNA*, Cambridge University Press, Cambridge.

Dembski, W.A. and Ruse, M. (2004b). General introduction. **In**: W.A. Dembski and M. Ruse (eds.), *Debating Design: From Darwin to DNA*, Cambridge University Press, Cambridge, pp. 3-12.

Dennett, D.C. (1995). *Darwin's Dangerous Idea: Evolution and the Meanings of Life*, Simon & Schuster, New York.

Dijksterhuis, E.J. (1950). *The Mechanization of the World Picture*, 1961 translated ed., Oxford University Press, Oxford.

Dournon, C. (2003). Developmental biology of urodele amphibians in microgravity conditions. *Adv Space Biol Med* **9**, 101-131.

Driesch, H.A.E. (1891). The potency of the first two cleavage cells in the development of echinoderms. **In**: M. Gabriel and S. Fogel (eds.), *Great Experiments in Biology*, Prentice-Hall, Englewoods Cliffs, pp. 210-214.

Driesch, H.A.E. (1929). *The Science and Philosophy of the Organism*, 2nd ed., Black, London.

Edelman, G.M. (1988). *Topobiology, An Introduction to Molecular Embryology*, Basic Books, New York.

Edis, T. (2002). *The Ghost in the Universe: God in Light of Modern Science*, Prometheus Books, Amherst, N.Y.

Edis, T. (2003). Harun Yahya and Islamic creationism. **In**: A. Chesworth, S. Hill, K. Lipovsky, E. Snyder and W. Chesworth (eds.), *Darwin Day Collection One*, Tangled Bank, Albuquerque, pp. 419-422.

Edis, T. (2004a). Grand themes, narrow constituency. **In**: M. Young and T. Edis (eds.), *Why Intelligent Design Fails: A Scientific Critique of the New Creationism*, Rutgers University Press, Piscataway, NJ, pp. 9-19.

Edis, T. (2004b). Chance and necessity - and intelligent design. **In**: M. Young and T. Edis (eds.), *Why Intelligent Design Fails: A Scientific Critique of the New Creationism*, Rutgers University Press, Piscataway, NJ, pp. 139-152.

Emmeche, C. (1997). Defining Life, Explaining Emergence, http://www.nbi.dk/~emmeche/cePubl/97e.defLife.v3f.html.

Farina, A. (ed.) (2005). *Gatherings in Biosemiotics 5, Urbino, Italy - July 20-24, 2005, Abstracts*, Urbino.

Feller, W. (1968). *An Introduction to Probability Theory and Its Applications*, 3rd revised ed., John Wiley & Sons, New York.

Fernald, R.D. (1997). The evolution of eyes. *Brain Behav Evol* **50**(4), 253-259.

Forgacs, G. and Newman, S.A. (2005). *Biological Physics of the Developing Embryo*, Cambridge University Press, Cambridge.

Gavrilets, S. (2000). Waiting time to parapatric speciation. *Proc Biol Sci* **267**(1461), 2483-2492.

Gehring, W.J. and Ikeo, K. (1999). Pax 6: mastering eye morphogenesis and eye evolution. *Trends Genet* **15**(9), 371-377.

Gilbert, S.F. (2004). Mechanisms for the environmental regulation of gene expression. *Birth Defects Res C Embryo Today* **72**(4), 291-299.

Gingerich, O. (1994). Dare a scientist believe in design? **In**: J. Templeton (eds.), *Evidences of Purpose*, Continuum, New York, pp. 21-32, 184-185.

Gishlick, A.D. (2004). Evolutionary paths to irreducible systems: the avian flight apparatus. **In**: M. Young and T. Edis (eds.), *Why Intelligent Design Fails: A Scientific Critique of the New Creationism*, Rutgers University Press, Piscataway, NJ, pp. 58-71.

Gordon, R. (1994). Evolution escapes rugged fitness landscapes by gene or genome doubling: the blessing of higher dimensionality. *Computers & Chemistry* **18**(3), 325-332.

Gordon, R. (1996). Book review: Hierarchy versus Reductionism, a Review of Eldredge, N. (1995). *Reinventing Darwin: The Great Debate at the High Table of Evolutionary Theory.* New York: John Wiley & Sons. J. Evol. Biol. **9**(3), 383-386.

Gordon, R. (1999). *The Hierarchical Genome and Differentiation Waves: Novel Unification of Development, Genetics and Evolution*, World Scientific & Imperial College Press, Singapore & London.

Gordon, R. (2000). The emergence of emergence: a critique of "Design, observation, surprise!". *Rivista di Biologia /Biology Forum* **93**(2), 349-356.

Gordon, R. (2001). Making waves: the paradigms of developmental biology and their impact on artificial life and embryonics. *Cybernetics & Systems* **32**(3-4), 443-458.

Gordon, R. (2006). Mechanics in embryogenesis and embryonics: prime mover or epiphenomenon? *Int. J. Dev. Biol.* **50**(2/3), 245-253.

Gordon, R., Björklund, N.K. and Nieuwkoop, P.D. (1994). Dialogue on embryonic induction and differentiation waves. *Int. Rev. Cytol.* **150**, 373-420.

Gordon, R. and Björklund, N.K. (1996). How to observe surface contraction waves on axolotl embryos. *Int J Dev Biol* **40**(4), 913-914.

Gordon, R. and Hoover, R.B. (2007). Could there have been a single origin of life in a big bang universe? *Proc. SPIE* **6694**, doi:10.1117/12.737041.

Gould, S.J. (1982). The meaning of punctuated equilibrium and its role in indicating a hierarchical role to macroevolution. **In**: R. Milkman (eds.), *Perspectives in Evolution*, Sinauer Association, Sunderland, MA, pp. 83-104.

Gould, S.J. (1988). On replacing the idea of progress with an operational notion of directionality. **In**: M.H. Nitecki (eds.), *Evolutionary Progress?* University of Chicago Press, Chicago, pp. 319-338.

Gould, S.J. (1998). Gulliver's further travels: the necessity and difficulty of a hierarchical theory of selection. *Philos Trans R Soc Lond B Biol Sci* **353**(1366), 307-314.

Habener, J.F. and Williams, G.H. (2004) *Metabolic Basis of Common Inherited Disease*, W.B. Saunders Co., Philadelphia.

Hawley, J.F. & K.A. Holcomb (2005). *Foundations of Modern Cosmology*. Oxford, Oxford University Press.

Hill, N.A. and Häder, D.-P. (1997). A biased random walk model for the trajectories of swimming micro-organisms. *J. Theor Biol.* **186**, 503-526.

His, W. (1888). On the principles of animal morphology. Roy. Soc. Edinburgh Proc. **15**, 287-298.

Holland, J.H. (1998). *Emergence: From Chaos to Order*, Perseus Books, Reading, Massachusetts.

Hoodbhoy, P. (1991). *Islam and Science: Religious Orthodoxy and the Battle for Rationality*, Zed Books,

Jablonka, E. and Lamb, M.J. (2005). *Evolution in Four Dimensions: Genetic, Epigenetic, Behavioral, and Symbolic Variation in the History of Life*, MIT Press, Cambridge.

Johns, D. (2005). Sound advice or just plain absurd? *Science* **309**(5735), 698.

Johnson, S. (2001). *Emergence, The Connected Lives of Ants, Brains, Cities, and Software*, Scribner, New York.

Kaplan, M.M. (1937). *The Meaning of God in Modern Jewish Religion*, 1994 reprint ed., Wayne State University Press, Detroit.

Katz, M.J. (1987). Is evolution random? **In**: R.A. Raff and E.C. Raff (eds.), *Development as an Evolutionary Process*, Alan R. Liss, New York, pp. 285-315.

Keizer, J. (1987). *Statistical Thermodynamics of Nonequilibrium Processes*, Springer-Verlag, New York.

Klostermaier, K.K. (2004). *The Nature of Nature: Explorations in Science, Philosophy and Religion*, Theosophical Publishing House, Adyar, Chennai, India.

Koestler, A. and Smythies, J.R. (eds.) (1969). *Beyond Reductionism, New Perspectives in the Life Sciences*, Hutchinson, London.

Kuthan, H. (2001). Self-organisation and orderly processes by individual protein complexes in the bacterial cell. *Prog Biophys Mol Biol* **75**(1-2), 1-17.

Lamm, N. (1990). *Torah Umadda, The Encounter of Religious Learning and Worldly Knowledge in the Jewish Tradition*, Jason Aronson Inc., Northvale, NJ.

Livingstone, D.N. (1987). *Darwin's Forgotten Defenders: The Encounter between Evangelical Theology and Evolutionary Thought*, Erdman, Grand Rapids, Michigan.

Lovejoy, A.O. (1936). *The Great Chain of Being: A Study of the History of an Idea*, Harvard University Press, Cambridge.

Maimonides, M. (1190). *The Guide of the Perplexed. Translated with an introd. and notes by Shlomo Pines*, 1963 translated ed., University of Chicago Press, Chicago.

Malik, H. (1980). *Sir Sayyid Ahmad Khan and Muslim Modernism in India and Pakistan*, Columbia University Press, New York.

Margulis, L. and Dorion Sagan (2002). *Acquiring Genomes: A Theory of the Origins of Species*, BasicBooks, New York.

Markin, V.S., Pastushenko, V.F. and Chizmadzhev, Y.A. (1987). *Theory of Excitable Media*, John Wiley & Sons, New York.

Martin, C.C. and Gordon, R. (2001). The evolution of perception. *Cybernetics & Systems* **32**(3-4), 393-409.

Matt, D.C. (1996) *God & the Big Bang: Discovering Harmony Between Science & Spirituality*, Jewish Lights Publishing, Woodstock, VT.

McMenamin, M.A.S. (1989) *The Emergence of Animals: The Cambrian Breakthrough*, Columbia University Press, New York.

Melling, D. (2003). *The Tale of Jack Frost*, Barron's Educational Series, Hauppauge, NY.

Menuge, A. (2004). Who's afraid of ID? A survey of the intelligent design movement. **In**: W.A. Dembski and M. Ruse (eds.), *Debating Design: From Darwin to DNA*, Cambridge University Press, Cambridge, pp. 32-51.

Miller, K.R. (2004). The flagellum unspun: the collapse of "irreducible complexity". **In**: W.A. Dembski and M. Ruse (eds.), *Debating Design: From Darwin to DNA*, Cambridge University Press, Cambridge, pp. 81-97.

Milner, R. and Maestro, V. (2002). Special report: Intelligent Design? *Natural History* (April), reprinted online October, 2005.

Morowitz, H.J. (2002). *The Emergence of Everything, How the World Became Complex*, Oxford University Press, Oxford.

Musgrave, I. (2004). Evolution of the bacterial flagellum. **In**: M. Young and T. Edis (eds.), *Why Intelligent Design Fails: A Scientific Critique of the New Creationism*, Rutgers University Press, Piscataway, NJ, pp. 72-84.

Nasr, S.H. (1968). *Science and Civilization in Islam*, Harvard University Press, Cambridge.

Nixon, B. (1983). *The Miracle of Life [videorecording]*, WGBH Educational Foundation, Boston.

Ohno, S. (1970). *Evolution by Gene Duplication*, Springer-Verlag, New York.

Osorio, D. (1994). Eye evolution: Darwin's shudder stilled. *Trends in Ecology and Evolution* **9**(7), 241-242.

Pennock, R.T. (ed.) (2001). *Intelligent Design Creationism and Its Critics: Philosophical, Theological, and Scientific Perspectives*, MIT Press, Cambridge.

Perakh, M. (2004). *Unintelligent Design*, Prometheus Books, Amherst, NY.

Perakh, M. and Young, M. (2004). Is intelligent design science? **In**: M. Young and T. Edis (eds.), *Why Intelligent Design Fails: A Scientific Critique of the New Creationism*, Rutgers University Press, Piscataway, NJ, pp. 185-196.

Pollack, R. (2000). *The Faith of Biology and the Biology of Faith: Order, Meaning, and Free Will in Modern Medical Science*, Columbia University Press, New York.

Raso, V. (1990). Antibodies in diagnosis and therapy. The magic bullet--nearing the century mark. *Semin Cancer Biol* **1**(3), 227-242.

Ratliff, E. (2004). The crusade against evolution. *Wired* **12**(10), 154-159, 161, 202-203.

Raymo, C. (1998). *Skeptics and True Believers: The Exhilarating Connection Between Science and Religion*, Walker and Company.

Rhine, J.B. (1934). *Extra-Sensory Perception*, Boston Society for Psychic Research, Boston.

Richards, J.W., Gilder, G.F., Kurzweil, R., Ray, T., Searle, J., Dembski, W. and Denton, M. (2002). *Are We Spiritual Machines?: Ray Kurzweil vs. the Critics of Strong A.I.*, Discovery Institute Press, Seattle.

Ronald, S. (2001). Theological truth versus scientific truth: Dare a scientist believe in 'design? The Manitoban (November 28), http://themanitoban.com/2001-2002/1128/features_11.shtml.

Rothman, S. (2002). *Lessons from the Living Cell, The Limits of Reductionism*, McGraw-Hill, New York.

Russell, B. (1935). *Religion & Science*, 1968 reprint ed., Oxford University Press, London.

Schliwa, M. (ed.) (2003). *Molecular Motors*, Wiley-VCH, Weinheim.

Schopf, J.W. (1999). *Cradle of Life: The Discovery of the Earth's Earliest Fossils*, Princeton University Press, Princeton.

Schroeder, G.L. (1990). *Genesis and the Big Bang: The Discovery of Harmony Between Modern Science and the Bible*, Bantam Doubleday Dell Pub.

Shanks, N. and Karsai, I. (2004). Self-organization and the origin of complexity. **In**: M. Young and T. Edis (eds.), *Why Intelligent Design Fails: A Scientific Critique of the New Creationism*, Rutgers University Press, Piscataway, NJ, pp. 85-106.

Shastry, B.S. (1998). Gene disruption in mice: models of development and disease. *Mol Cell Biochem* **181**(1-2), 163-179.

Sheldrake, R. (1981). *A New Science of Life: the Hypothesis of Formative Causation*, Jeremy P. Tarcher, Los Angeles.

Shermer, M. (2004). *The Science of Good & Evil: Why People Cheat, Gossip, Care, Share, and Follow the Golden Rule*, Times Books, New York.

Sipper, M. (2002). *Machine Nature: The Coming Age of Bio-Inspired Computing*, McGraw-Hill Trade,

Smuts, J.C. (1926). *Holism and Evolution*, 1961 reprint ed., Viking Press, New York.

Spetner, L.M. (1994). The evolutionary doctrine. **In**: H. Branover and J. Attia (eds.), *Science in the Light of Torah, A B'Or Ha'Torah Reader*, Jason Aronson, Northvale, New Jersey, pp. 169-178.

Stenger, V.J. (2004). Is the universe fine-tuned for us? **In**: M. Young and T. Edis (eds.), *Why Intelligent Design Fails: A Scientific Critique of the New Creationism*, Rutgers University Press, Piscataway, NJ, pp. 172-184.

Stent, G.S. (ed.) (1980). *Morality as a Biological Phenomenon: The Presuppositions of Sociobiological Research*, revised ed., University of California Press, Berkeley.

Susskind, L. (2006). *The Cosmic Landscape: String Theory and the Illusion of Intelligent Design*, Little, Brown and Co., New York.

Szent-Györgyi, A. (1972). *The Living State, With Observations on Cancer*, Academic Press, New York.

Taylor, G.R. (1983). *The Great Evolution Mystery*, Harper & Row, New York.

Thomas, N., Imafuku, Y. and Tawada, K. (2001). Molecular motors: thermodynamics and the random walk. Proc Biol Sci **268**(1481), 2113-2122.

Todar, K. (2006). Todar's Online Textbook of Bacteriology, http://www.textbookofbacteriology.net/.

Torah (1917). *Hebrew Bible in English* [http://www.mechon-mamre.org/e/et/et0.htm], Jewish Publication Society, Philadelphia.

Tyson, J.J. (1976). *The Belousov-Zhabotinskii Reaction*, Springer-Verlag, Berlin.

Tyson, J.J. and Novak, B. (2001). Regulation of the eukaryotic cell cycle: molecular antagonism, hysteresis, and irreversible transitions. *J Theor Biol* **210**(2), 249-263.

Ubbels, G.A., Berendsen, W., Kerkvliet, S. and Narraway, J. (1989). The first seven minutes of a *Xenopus* egg fertilised on a sounding rocket in space. **In**: G.A. Ubbels, H. Oser and T.D. Guyenne (eds.), *Microgravity as a Tool in Developmental Biology, Selected papers from a special ESA Symposium held during the 11th International Congress of the International Society of Developmental Biologists, August 20-25, 1989, Utrecht*, European Space Agency, Paris Cedex, pp. 49-58.

Ussery, D. (2004). Darwin's transparent box: the biochemical evidence for evolution. **In**: M. Young and T. Edis (eds.), *Why Intelligent Design Fails: A Scientific Critique of the New Creationism*, Rutgers University Press, Piscataway, NJ, pp. 48-57.

Van Valen, L.M. (1973). A new evolutionary law. *Evol. Theory* **1**, 1-30.

Venter, J.C., et al. (2001). The sequence of the human genome. *Science* **291**(5507), 1304-1351.

von der Malsburg, C. (2004). Self-organization: the unfinished revolution. **In**: W. Greiner and J. Reinhardt (eds.), *Idea-Finding Symposium, Frankfurt Institute for Advanced Studies, Frankfurt, Germany, April 15-17, 2003*, EP Systema Bt., Debrecen, Hungary, pp. 127-138.

von Uexküll, J. (1926). *Theoretical Biology*, Kegan Paul, Trench, Trubner & Co., London.

Watkins, T. and Winston, R.M. (2005). *The Story of God. 3. God of the Gaps* [TV Production], Dangerous Films, London.

Weisstein, E.W. (1999). Pólya's Random Walk Constants [MathWorld—A Wolfram Web Resource], http://mathworld.wolfram.com/PolyasRandomWalkConstants.html.

White, A.D. (1896). *A History of the Warfare of Science with Theology in Christendom*, D. Appleton and Co., New York.

Wikipedia (2006). Biomass (ecology), http://en.wikipedia.org/wiki/Biomass_(ecology).

Young, M. (2004). Grand designs and facile analogies: exposing Behe's mousetrap and Dembski's arrow. **In**: M. Young and T. Edis (eds.), *Why Intelligent Design Fails: A Scientific Critique of the New Creationism*, Rutgers University Press, Piscataway, NJ, pp. 20-31.

Young, M. and Edis, T. (eds.) (2004a). *Why Intelligent Design Fails: A Scientific Critique of the New Creationism*, Rutgers University Press, Piscataway, NJ.

Young, M. and Edis, T. (2004b). Preface. **In**: M. Young and T. Edis (eds.), *Why Intelligent Design Fails: A Scientific Critique of the New Creationism*, Rutgers University Press, Piscataway, NJ, pp. ix-xi.

Dialogue

Tom Barbalet

You present me with a problem. I fundamentally agree with you. I do feel you are caught in a folly based in your specific academic interests and your noble quest to find intelligible meaning through it all.

I don't think Creationism and Intelligent Design are the same movement. It appears to be common to refer to Intelligent Design as a neo-Creationist movement (Wikipedia, 2007) because ID doesn't represent anything close to a majority of Creationist beliefs. Creationism is a grass roots movement and, because of this, it is very hard to survey the issues associated with Creationism through finding academic Creationist or explicitly Intelligent Design texts. In short, I think we shouldn't try to compare ID books with science, we should provide summaries of Intelligent Design books to grass-roots Creationists. Intelligent Design may be too much of a compromise for Creationists.

Creationism is part of a broader "distrust of science" movement. I have found Over-Confident Anti-Creationists can be very welcoming of other Anti-Creationists even if they actively promote a distrust of (other areas of) science through their particular political dogma (Dawkins award 2005). I would seek a similar leafletting of Anti-Creationists showing the divergence of their views from science in other areas.

While the goal, as you have noted, for Intelligent Design is "to find sympathetic ears in the academy" (Menuge, 2004), my assessment of the contemporary discourse is that both sides have taken their arguments to the streets in terms of winning a popular audience and selling a good many books in the process. Popular science books have long been tarnished by "too much certainty" and "too little science". Do you feel some of the Over-Confident Anti-Creationist books you note are just part of this popular science genre flaw, destined for the shelves of the mall rather than the university bookshop?

Thankfully the mall bookstores allow you to read from cover to cover without purchase. In my reading of the popular books, specifically Dawkins' *God Delusion* (2006) closely followed by the chapter in Coulter's *Godless* (2007) dealing with evolution, they both commit the same kinds of fallacies. It is almost a chorus and counter-chorus that is divorced from science. Do you feel the two Over-Confident sides, divided only by the word "Anti-", are becoming meta-critiques of each other rather than any discussion of science? Certainly this is my view although I wouldn't want to lump all popular science books that refute some Creationist arguments as Over-Confident Anti-Creationist books. The latter has a particular quality that merits better definition.

Your thoughts?

Taner Edis
Richard Gordon tries to be evenhanded, criticizing "overconfident evolutionists" as well as creationists. There may be something to this. Even if Darwinian evolution is a brilliant idea, one of the roles of scientists is to stretch ideas as far as possible, to find out where they break down. So some overreaching is inevitable, even a good thing.

But I would have expected Gordon's examples of overconfidence to derive from long-standing disputes over the extent that Darwinian thinking can be applied to the social and cultural domain. After all, there is plenty of controversy there. Gordon focuses instead on scientists who defend evolution against intelligent design. I do not think he succeeds.

Let me start with Stephen Jay Gould. I think Gordon reads too much into Gould's semipopular work. Gould addressed the question of progress in evolution throughout his career, often with considerably more rigor. His central claim, as I see it, is one that would be familiar to a statistical physicist: that trends in evolution take place not due to any inherent directionality in the mechanisms of evolution, but as artifacts of the asymmetry due to the starting point. This seems to be a reasonably secure conclusion; it certainly fits well with current thinking in physics about obtaining macroscopic irreversibility from reversible microscopic physics. In any case, whatever peripheral claims Gould may have made, his overall position is squarely within the mainstream of natural science. It is hard to describe it as overconfident. At present, mainstream science offers no support for inherent progressivity in evolution, or any shaping of life's history by intelligent design.

Gordon also intimates that there is something mysterious about irreversibility. This is difficult to understand. Again, from the point of view of a physicist, the absence of macroscopic irreversibility would be more surprising. Much of thermodynamics, for example, is not very sensitive to the details of the underlying microscopic physics, even to whether it is quantum or classical.

Later, Gordon echoes intelligent design proponents such as William Dembski, pointing out that experiments such as those demonstrating the B-Z reaction have to be carefully designed by scientists. ID advocates like to say that therefore the resulting structure is due to information originating in the fine tuning by scientists. This is demonstrably incorrect. Furthermore, I suspect that such an argument incorporates an important misunderstanding of the role of experiments and models in scientific work. It is like saying that since Darwin used artificial selection to explain natural selection, and artificial selection implies intelligent design by human breeders, selection in nature must also incorporate information injected by an intelligent agent.

In some cases, Gordon appears not to pay attention to the context of an argument, as with the illustration of a haphazard jumble of letters on paper. The context here is the question about the origin of information raised by intelligent design proponents. The actual symbol alphabet in the illustration I used is irrelevant, as is wonder about alphabetical symbols appearing at all.

Gordon also misreads Victor Stenger's example about silicon-based life. Indeed, taking illustrative examples out of context seems to be a mainstay of his critique. Stenger does not make the elementary mistake of supposing that silicon can substitute for carbon in organic chemistry. His argument instead points out that when we consider varying physical parameters to obtain different universes, we should not assume that we are restricted to familiar ways of realizing those processes we call life. Besides some basic physical requirements, such as a long-lived universe, we do not know enough to rule out complex self-reproducing processes in universes with significantly different fundamental physics.

Now, I appreciate the need to be evenhanded when editing a volume such as this one. And certainly the claim that we do not have all the answers yet is unobjectionable—indeed, I disagree with any suggestion that those us who defend evolution against creationism and ID are so confident that we think we have all the answers. But from a scientific perspective, the standing of naturalistic evolution is incomparably better than any creationist or ID alternative. Being evenhanded here can easily cross over into being misleading.

Richard Gordon

Dear Taner: I won't pull the authority thing on you, but I did train under Terrell Hill in statistical mechanics. However, it is embryogenesis that drives my thinking: I don't have much research interest in the "long-standing disputes over the extent that Darwinian thinking can be applied to the social and cultural domain", perhaps because it is such a long leap from stat mech. The New Synthesis of evolution and genetics, of which we might count Stephen Gould as the tail end, ignored embryology. The movement of "evolution and development" that started in the 1970s ("EvoDevo") tried to make up for this deficiency, but has not yet succeeded in producing a three-way synthesis, although I am appreciative of Gould's contribution:

> Gould, S.J. (1977). *Ontogeny and Phylogeny*, Cambridge: Harvard University Press.

In regards to progress, you misinterpreted what I said: progress in evolution may be no more mysterious than the inevitability of change in irreversible processes. Now, the degree of mystery in the latter has to do with the long debates of the nature of time, popularly referred to as "time's arrow". If those debates are settled now to everyone's satisfaction, fine. But I can model progress in evolution as a pretty ordinary irreversible process, and that toy model just might contain the essence of that problem. This was done in:

> Gordon, R. (1999). *The Hierarchical Genome and Differentiation Waves: Novel Unification of Development, Genetics and Evolution*, Singapore & London: World Scientific & Imperial College Press. http://www.wspc.com.sg/books/lifesci/2755.html, 2 vols., 1836p.

Nothing is evenhanded in my essay, except the title. I almost started it by saying: "I have generally been allergic to creationism...". It is not helpful to say that we must not agree with anything creationists say, because they will latch on to it as support. In fact, much of the emptiness of their arguments is exposed by the way they hang on to our every word and discovery. The authority of science has been used (or misused, as you wish) to support the authority of religion for centuries. That's a kind of unadmitted defeat.

The point of my essay is that scientists throw the baby out with the bath water when they claim to have solved problems that are still unsolved. That's called dogma, or more in our context: "Science of the gaps", as I put it. Of course, the source of this difficulty is that most scientists blithely accept the current paradigms in their fields, which is the reason we occasionally have scientific revolutions, a la Thomas Kuhn:

Kuhn, T.S. (1996). *The Structure of Scientific Revolutions*, 3rd ed., Chicago: University of Chicago Press.

Dear Tom: In the USA, much of the ID movement is funded by the Discovery Institute, which in turn heavily involves fundamentalist Christian sources of money, to the best of my understanding. So in this sense I find them inseparable, despite the spectrum of beliefs. But then Christianity itself is also so fragmented, that a unified view in not to be expected. My general view is "scratch just about anyone raised in an Abrahamic-based culture and you will find at root a creationist", part of the observation:

White, M. (2008). "We have met the enemy... and he is us", http://www.igopogo.com/we_have_met.htm.

It's the basic theme of:

Dennett, D.C. (1995). *Darwin's Dangerous Idea: Evolution and the Meanings of Life*, New York: Simon & Schuster.

The relationship of creationism to science is schizophrenic, both distrusting science, as you say, and appealing to its authority. As far as what sells books, I tried to give this book a "street" title precisely to extend its reach beyond the academy, "*God Did It, Eh?*", but was outvoted. Going to the street is an old tradition, and is what kept some of our heroes, despised by the academy in their time, afloat:

Duclaux, É. (1920). *Pasteur, The History of a Mind*, Philadelphia: W.B. Saunders Co.

So I wouldn't knock it.

Taner Edis
I agree that we do not know nearly as much as we would like about the connections between evolution and the developmental process. *If*, in this essay, I had seen examples of scientists claiming too much about evo-devo, I would

have joined in the criticism of such overconfidence. Looking at the exciting new work being done in evo-devo right now, it would also be legitimate to emphasize how little we once knew and how much more we have to find out. I see no such example in the essay.

I also see nothing objectionable in statements that "scientists throw the baby out with the bath water when they claim to have solved problems that are still unsolved." Nothing, that is, provided that we are discussing genuine examples of claiming too much. I don't think the examples put forth in the essay fit this description.

Richard Gordon
Dear Taner: Well, I guess you haven't confronted embryos. They are tough to get a handle on the relationship between embryo physics and the genome. See:

> Beloussov, L.V. & R. Gordon (2006). Preface. Morphodynamics: Bridging the Gap between the Genome and Embryo Physics. International Journal of Developmental Biology 50(2/3), 79-80. (All papers in this special issue are free at: http://www.ijdb.ehu.es/web/contents.php?vol=50&issue=2-3).

The problem is that we have not bridged the gap between the genome and the physics of embryos, nor between the physics of morphogenesis and the physics of consciousness. See:

> Tuszynski, J.A. (ed.) (2006). *The Emerging Physics of Consciousness*, Heidelberg: Springer Verlag.

These gaps are wide open for creationists to run into, while we scientists plod on trying to see if we can fill them. You can say that the cup is half full, by praising our accomplishments to date. Perhaps I'm less impressed by these than you are. Be that either way, I'm more interested in the half empty part.

Creationists have a curious role here, though not one they intend. By causing us to focus intently on the gaps they find, our thinking about what constitutes an explanation that might fill those gaps is sharpened.

Tom Barbalet
There is no mystery where the money is coming from for the Discovery Institute. My concern is that hard-working, grass-roots Creationists aren't getting value-for-money through the Discovery Institute. The deity as petty downtrodden engineer should be offensive both to believers and engineers.

I am not knocking the streets. All I have is street credibility and only on some streets.

Part 3

Is a Purely Scientific Approach to the Origin of Life in the Universe Sufficient?

Biography

Dr. Florence Raulin Cerceau is a senior Lecturer at the National Museum of Natural History of Paris (France). She obtained her Ph.D. in Astronomy and Space Techniques from the University of Paris (Pierre et Marie Curie – Paris VI) in 1983. From 1988 to 1994, she was involved in the preparation of the Grande Galerie de l'Evolution (Museum, Paris), devoted to biological evolution. She was also co-designer of two temporary exhibitions ("Météorites!" and "L'aventure Polaire"), which were held in Paris respectively in 1996 and 1997. Since 1997, her researches take place at the Centre Alexandre Koyré (Center of History of Sciences) and are entirely consecrated to Epistemology and History of Astrobiology. More specifically, her scientific interests deal with the spontaneous generation debate during the second part of the nineteenth century, the forerunners of interplanetary communication during this same period, and the beginnings of interstellar communication with the birth of SETI (Search for Extraterrestrial Intelligence).

13

French Views on the Origin of Life Combined with Theological Thoughts

From Louis Pasteur (1860) to Edmond Perrier (1920)

FLORENCE RAULIN CERCEAU

Centre Alexandre Koyré, (CNRS-EHESS-MNHN- UMR 8560)
Case postale 25, Muséum national d'Histoire naturelle
57 rue Cuvier – 75005 Paris – France
raulin@mnhn.fr

1. Introduction

The question "How did life originate?," framed in terms of the occurrence of a primordial, even perhaps unique, event in time, could have been asked only after naturalists had accepted the concept of organic evolution. As Jean-Baptiste Lamarck (1744-1829), for example, assumed in his *Philosophie zoologique* (1809), the notion of organic evolution implied that the simplest living organisms had, through a series of events, given birth successively to all the others. From this fundamental assumption it would follow that the diversification of life was a continuous differentiation of species in time and space. Many of Lamarck's opponents, however, including the influential paleontologist Georges Cuvier (1769-1832), supported the opposite notion of "fixity" of species. The controversy between biologists who believed in the fixity of species and others who held that species had changed over time was resolved by Charles Darwin (1809-1882), who proposed in his epoch-making book, *The Origin of Species by Means of Natural Selection* (1859), the mechanism that explained species differentiation: natural selection, in other words, a process by which characters of organisms were selected without any pre-established choice.

In 1809 Lamarck believed that microorganisms ("infusoires" in Lamarck's original text) and parasitic worms arose by continuous spontaneous generation and that these two classes were at the origin of the entire range of animal diversity. Exactly half a century later, Darwin hypothesized that the origin of life took place in a very remote past, and that ever since that period, life had come from life. Darwin, however, was not very explicit on the topic of how life came about. He did not develop a clear concept of the origin of life in his *Origin of Species*. His doubts about the origin of life were expressed in his letters, for instance in a letter to the British botanist Joseph Dalton Hooker (1817-1910) dated March 29st, 1863: "It is mere rubbish, thinking at present of the origin of life; one might as well think of the origin of matter" (Darwin in Darwin, F., 1905). The only allusion to Darwin's views on the conditions in which primitive life would have first appeared is found in a letter he sent in 1871 to Hooker, in which he evoked his famous "warm little pond" (Darwin in Darwin F., 1887).

Even though, as is well known, there are major differences between Lamarck's and Darwin's views about evolution, the fact that these two naturalists had proposed theories that attempted to explain the reality of evolution opened the ground for a discussion about the origin of life on Earth. This discussion, inevitably given the times, involved not only scientific ideas but also theological beliefs. In Darwin's and Lamarck's time, and indeed for many years thereafter, scientists had difficulties separating the results of scientific experiments from the beliefs in a Creator. This was as true in France as it was in other countries, even though the background of the science vs. theology debate in France was somewhat different from that in other countries, for example Germany.

In this paper I present a panorama of French views[*] on the origin of life, mainly during the second part of the nineteenth century, by means of an analysis of the thought of some of the most influential scientists concerned with this topic. This overview will highlight first the diversity of opinions current at that time — for

---

[*] I have provided, throughout the text, free English translation of passages from the original texts in French.

example during the Pasteur-Pouchet controversy about spontaneous generation (1860-1866) and, second, will describe the debate about theological aspects of the question of the emergence of life. Reviewing the general concept of spontaneous generation will require me to delve into the dichotomy between the primordial origin of life and a (supposed) continuous generation, still existing nowadays. The diversity of opinions on this controversy will be seen through the works of several key figures who were all more or less influenced by theological thought. Finally, I will present a later opinion (published about 1920) that shows the definitive abandonment of the doctrine of spontaneous generation and illustrates the first steps taken towards the future concept of chemical evolution.

2. The General Context in France Related to the Origin of Life, at the Time of the Debate over Spontaneous Generation

The hypothesis that life had begun by spontaneous generation was generally favored by naturalists who accepted the idea that basic life originated with very simple matter, such as protoplasm, which was viewed as an simple essential substance that could be produced directly from inorganic matter. If life was supposed to be as simple as a single substance, why could it not occur relatively easily (still today) given appropriate environmental conditions? The answer to this question depends on how life is defined. For most of the materialist scientists of the 1860s (especially in Germany, where the materialist trend was very strong), the distinction between organic compounds and simple living organisms seemed to be meaningless. Materialists such as the philosopher and naturalist Ludwig Büchner (1824-1899) asserted that there were no other forces in nature beside physical, chemical and mechanical ones and that organisms must have been produced by these forces, a view strengthened by the publication of Darwin's *Origin of Species*.

For materialists, therefore, if life was as simple as albuminoid or plasmatic compounds, it could consequently arise spontaneously from inorganic matter through evolutionary processes. The possibility of an ongoing life origination process was not excluded, a position held notably by the German naturalist Ernst Haeckel (1834-1919), one of the most ardent defenders of Darwin's theory. Haeckel, as well as other German materialists of that time, believed that there could be continuous evolution from inorganic matter to life (abiogenesis), and that this process still took place if conditions were suitable. At the time of Haeckel, as the environmental conditions of the primitive earth were not known, no clear distinction was made between primeval spontaneous generation from inorganic matter (archeogenesis) and current spontaneous generation (present day abiogenesis). Haeckel's position about the continuity from inorganic to organic matter (and his conception of an extreme simplicity for living matter) led him to propose (in his *Natural History of Creation*; Haeckel, 1868) that archeogenesis occurred through a series of steps. These steps would have taken place during the first ages of the earth: this view of Haeckel's turned him into a forerunner of later scenarios about the origin of life (Raulin Cerceau, 2004), even if he didn't push aside the question of a present day abiogenesis.

Abiogenesis was not widely accepted in France and the controversy about spontaneous generation focused on a different point. In that country, the debate about spontaneous generation clearly started with the publication, in 1858, of a paper written by the biologist Félix-Archimède Pouchet (1800-1872), director of the Rouen Museum, entitled "Note sur des Proto-organismes végétaux et animaux, nés spontanément dans de l'air artificiel et dans le gaz oxygène" (Note on plant and animal proto-organisms, born spontaneously in artificial air and oxygen gas) (Pouchet, 1858). In his major work, "*Hétérogénie ou traité de la génération spontanée*" (*Heterogenesis or Treatise on Spontaneous Generation*) (Pouchet, 1859) Pouchet strongly favored a spontaneous generation occurring by heterogenesis[†]. In that work he indicated that he was convinced by the facts he had observed in his own experiments, namely, the spontaneous birth and persistence of microorganisms in infusions already containing organic[‡] substances (such as albuminous matter) subjected to various treatments. He emphasized that "the appearance of the first organisms in infusions was always preceded by phenomena of fermentation or putrefaction" (Pouchet, 1859). Heterogenesis was the mechanism through which life could arise, either directly from organic matter considered to be endowed with living properties ("living molecules" — an idea initially proposed by Buffon in his *Histoire naturelle de l'Homme*, 1777), or indirectly from living organisms in a

[†] Nicolas Joly, one of Pouchet's colleagues, gave the following definition: "Heterogenesis or spontaneous generation, is not a creation from nothing, but rather the production of a new organized being, without any parent, whose primordial elements are coming from ambient organic matter." Heterogenesis could be considered as a process that was " recycling organic matter".

[‡] Until the end of the nineteenth century the term organic meant, in the mind of heterogenesists, produced and used by living beings. Of course, the primordial origin of these organic compounds escaped such a definition.

state of decay. The process of heterogenesis differed completely from that of abiogenesis, in which life was supposed to arise from inorganic matter.

Heterogenesis was the cornerstone of the debate on the origin of life in France during the second part of the nineteenth century. The famous controversy between the chemist and microbiologist Louis Pasteur (1822-1895) and the biologist Félix Pouchet finds its roots there.

It is generally assumed by naturalists that in France the controversy concerning spontaneous generation was on the decline as a result of Pasteur's statements. If we refer to Farley however (Farley, 1977), spontaneous generation was quite unpopular in France long before Pasteur looked into the questions of whether germs existing in the atmosphere and what their role was in the phenomenon of fermentation. According to Farley, the majority of French scientists — except perhaps Pouchet himself and a few others — were increasingly convinced that the atmosphere was filled with a large number of spherical organic corpuscles. In addition, it did not seem clear to many scientists of the 1850s that high temperatures could destroy all the germs supposed to be present in the infusions they used in their experiments on spontaneous generation. For instance the French physiologist Henri Milne-Edwards (1800-1885) pointed out in 1859 that germs in infusions could resist high temperatures. He also underlined the ambiguity of the term "spontaneous" which could mean that matter would have the capacity of self-organization (Cantor, 1992). Indeed, Milne-Edwards thought that living organisms could not be generated without the assistance of a "vital force" and that "brute matter cannot organize itself in such a way as to form an animal or a plant" (quoted in Farley, 1977). In the mid-1850s, then, to accept the concept of spontaneous generation was the equivalent of accepting materialism (especially the strong, and in Milne-Edwards's view perhaps "dangerous", German materialism evoked earlier) and of rejecting geological evidence showing the non-interrupted "chain" of living beings.

The problem of spontaneous generation was associated in the minds of French scientists with the concepts of materialism and "transformism." Darwin's publication of *The Origin of Species* had of course its portion of influence on the French debate about the origin of life. Farley (1977) pointed out that the theory of biological evolution proposed by Darwin was regarded in France as a politico-theological doctrine allied with the forces that were attempting to overthrow Church and State. French scientists realized that Darwin's theory logically demanded that the beginnings of life be abiogenetic. Darwin's explanation of evolutionary mechanisms led one to accept a belief in abiogenesis, when at the same time experimental findings posed the problem of heterogenesis. Moreover, the first translation into French of Darwin's *Origin of Species*[§] was made by Clémence Royer (1830-1902) who was a materialist, a Republican and an atheist, and prefaced her translation by a lengthy diatribe against the Catholic Church (Farley, 1977). She explained that "it is in vain that Mr. Darwin affirms that his system is in no way opposed to the divine idea". As a matter of fact, the theory of spontaneous generation was viewed by many French scientists quite negatively, as this doctrine attacked the very core of Catholicism. From that viewpoint, Pasteur's disproof of spontaneous generation could have been used as a strong argument against Darwin's theory and its consequences. In 1864, in his attack on Darwinian evolution, the physiologist Pierre Flourens (1794-1867) brought up the fact that Darwin's theory postulated the occurrence of spontaneous generation and thus could not be maintained since (according to Pasteur's demonstration), "spontaneous generation is no more. Mr. Pasteur has not only illuminated the question, he has resolved it" (Farley, 1977).

The question of the origin of life in terms of a primordial switch from inorganic to organic matter seems to have been really discussed within the French scientific community only toward the end of the 1870s, in other words a few years after the Pasteur/Pouchet controversy, and perhaps under the influence of the new Republican context that had begun to prevail in France. Given the political influence of republicanism, and also given the growth of positivism and materialism, the emergence of living matter could be mentioned more openly in France without any reference to a Creator, as Catholicism demanded. In addition, scientific progress during the last decades of the nineteenth century, involving substantial experimental and technical advances, appeared powerful enough to solve many questions, such as the one of the origins of life. This world view, or "scientism" suggested that no other form of knowledge could exist except through the application of the scientific method and, should science enable one to clear up any unsolved question, there would no longer be room for God.

[§] Clémence Royer translated it from the third English edition. This first French translation was published in 1862, in the middle of the Pasteur/Pouchet controversy.

3. The Refutation of the doctrine of Heterogenesis: Towards the idea of a Primeval and Unique Genesis?

Louis Pasteur first focused his research, as a chemist, on molecular asymmetry and fermentation. In *Fermentations et générations spontanées* (*Fermentations and Spontaneous Generations*), one of the series of books "Oeuvres de Pasteur," Louis Pasteur Vallery-Radot (1886-1970) explained how his grandfather was led by his early work to be concerned with spontaneous generation (Pasteur Vallery-Radot, 1922). As early as February 1859, Pasteur wondered about the origin of the organized ferments present in his experiments. Were albuminous materials the origin of the observed fermentations, as claimed by his compatriot, the biologist Félix Pouchet? Experimental data seemed indeed to show that these ferments could result from the contact between albuminous matter and oxygen from the atmosphere (Pouchet, 1858). But this solution didn't satisfy the rigorous researcher that Pasteur was:

> "Either these ferments are organized and oxygen alone gives birth to them by its contact with nitrogenous matter and therefore they result from spontaneous generation; or else these ferments are not spontaneous, the oxygen acting in their formation not alone but as the stimulant of a germ that was carried along with it, or that pre-existed in the nitrogenous matter. This is how, at this point in my studies of fermentation, I decided to make up my mind about the question of spontaneous generation." (Pasteur, 1861)

Pasteur had the intuition that ferments were not dead substances but rather living beings whose germs originally came from the atmosphere. Pasteur's key idea was that ferments were in reality fed with albuminous material and represented living beings. This view had nothing to do with the one defended by Pouchet, who claimed that if germs came from the atmosphere, the latter should be saturated with them; in that case the atmosphere would turn into smog, and finally would become unfit for life. This assertion puzzled Pasteur and the main question in his mind became: Are there germs in the air? And its corollary was: Are these germs present in such high numbers that they could explain the production of organized ferments in infusions heated up beforehand?

Pasteur conducted experiments in different places. These experiments revealed to him that the atmosphere was indeed more or less filled with germs, depending on climatic and meteorological circumstances — such as whether it was raining or not-, or on environmental parameters — germs seemed to become more rare at high altitudes-. He highlighted the crucial fact that germs were not present in constant amounts in the atmosphere: the density of germs differed from one place to another. So, finally, Pasteur came to the following conclusion:

> "Dust suspended in the air represents the exclusive origin, the essential requirement for the presence of life in infusions, in decaying matter and in every liquor likely to ferment." (Pasteur, in Pasteur Vallery-Radot, 1922).

In spite of Pasteur's conclusion the controversy was still going on. Pasteur quickly understood that the main difficulty to counter the opinion of heterogenists consisted in finding an experimental method that was as rigorous as possible, in view of the high probability of contamination. In the historical part of his famous paper "Mémoire sur les corpuscules organisés qui existent dans l'atmosphère" (Essay on organized corpuscles existing in the atmosphere) (Pasteur, 1861), Pasteur reminded readers of earlier works (1837) carried out by the German physiologist Theodor Schwann (1820-1882) about fermentation. Pasteur quoted Schwann's conclusion: "Concerning alcoholic fermentation as decomposition, it is not oxygen, at least oxygen alone existing in the atmosphere, which is involved in these reactions, but rather a principle contained in the ordinary air that could be destroyed by heat." Pasteur also noted that the German biologist Heinrich Schröder (1810-1885) had wondered in 1859 what could be at the origin of alcoholic fermentation and decomposition: "Should we regard this active substance (supposed to be in the air) as organized microscopic germs scattered in the air? Or is it a still unknown chemical substance? I don't know." Both Schwann and Schröder had underlined the existence in the air of a still unknown principle (or an active substance) that constituted the condition for life in infusions. What was it that, present in the air, could produce biological organization? Was it a question of germs, solid bodies, gases, or fluids? Only experiments could answer these questions and Pasteur was going to "force the minds of those who were open to ideas to forcefully reject the whole notion that the air contained a more or less mysterious principle, whether gas, fluid, ozone, etc., endowed of the property of causing whatever kind of organization in infusions." (Pasteur, in Pasteur Vallery-Radot, 1922).

Pasteur stressed the fact that all the experiments carried out on spontaneous generations during past years — such as those supporting heterogenesis — were based on the use of infusions including matter from substances coming from plants or animals. In his work on the mode of nutrition of ferments, Pasteur proved that primitive organisms could appear without the systematic presence of organic material such as albuminous matter. His view was that no "living organic molecules" coming from organic material are capable of gathering and creating other living beings (Pasteur, in Pasteur Vallery-Radot, 1922). In the hypothesis of heterogenesis an essential role was attributed to the fermentation and decomposition processes played by an organic matter endowed with special properties acquired during some earlier formation. But, according to Pasteur, heterogenesis didn't exist for very concrete reasons, particularly since albuminous matter, that was actually only food for the germs already present in the infusions, could be substituted by mineral materials such as ammonium salts and phosphates.

Confronted with such a debate, the French Academy of Sciences sponsored a contest for the best experiment that would either prove or disprove spontaneous generation. The challenge was worded as follows: "Trying, by means of well-conducted experiments, to shed new light on the question of so-called spontaneous generation." Pasteur's famous demonstration, based on his "Mémoire sur les corpuscules organisés qui existent dans l'atmosphère" (1861), took place before a committee composed of the physiologists Claude Bernard (1813-1878), Henri Milne-Edwards (1800-1885) and Pierre Flourens (1794-1867), the botanist Adolphe Brongniart (1801-1876), and the naturalist Victor Coste (1807-1873). Pouchet withdrew from the competition** because he was convinced that the committee members were not in favor of the hypothesis of spontaneous generation. Remaining alone in the scientific arena, and after a demonstration as brilliant as it was rigorous, Pasteur was awarded the Alhumbert Prize on December 29th, 1862. He designed his experiment in order to be very illustrative and to definitely persuade the audience, as specified by Bruno Latour. According to Latour, Pasteur's genius was in what might be called the "théâtre de la preuve" (theatre of the proof) (Latour, 2001). With a simple but a pertinent experiment, Pasteur had both refuted the theory of spontaneous generation and convincingly demonstrated that the microorganisms in the air were the source of contamination (Pasteur, 1862).

**On December 20th, 1858, the French Academy of Sciences received the first work by Pouchet on heterogenesis, in which he announced the production of organisms in an environment deprived of atmospheric air and composed of fragments of hay, heated in order to be sterilized, and of oxygen gas or artificial air. According to Pennetier (Pennetier, 1868), the Academy immediately rose up against Pouchet's assertions.

Pasteur's aim was clearly to put an end to the controversy about heterogenesis, a theory in which he didn't believe from the beginning. Pasteur triumphantly concluded: "The doctrine of spontaneous generation will never recover from the death-blow stricken by this simple experiment." One can state today that Pasteur was entirely correct when he insisted that the controversy was mainly a question of experimental protocol and a misinterpretation of experimental results.

It has often been said that the theory of spontaneous generation was destroyed in France by Pasteur's demonstration before the French Academy of Sciences. In fact, according to Pasteur Vallery-Radot, Pasteur's publications about spontaneous generation, including the discussions with Pouchet, were spread over a six-year period, from 1860 to 1866 (Pasteur Vallery-Radot, 1922). In Pouchet's mind the controversy was not terminated by the decision of the Academy because Pasteur's demonstration did not disprove the doctrine of spontaneous generation. Pouchet believed instead that in such experiments oxygen was required for the generation of life. After the decision of the Academy, the biologist Nicolas Joly (1812-1885), one of Pouchet's closest colleagues, who was a Professor in Toulouse, published a critical review (Joly, 1863) of Pasteur's famous "Mémoire couronné" (the essay that was "crowned" by the Academy in 1862). We see in Joly's text that he still believed Pasteur's paper to be "marred by mistakes and full of contradictions." Raising Pasteur's hypothesis of the non-continuity of germs in the air, Pouchet reproached him to have had the "bright idea" of a limited panspermia. It was a sort of semi-panspermia confined in certain atmospheric areas where germs, more numerous than in other areas, were likely to seed the experiments. Ironically, Joly went as far as asking Pasteur to draw up a map of the parts of the atmosphere concerned by this semi-panspermia. Joly's own point of view was that the theory of atmospheric germs did not provide the right answer to the question of spontaneous generation: semi-panspermia was a deadlock. If germs saturated the air in several places, Joly wondered, why had chemists failed to detect them in their atmospheric analyses?

Several years after Pasteur presented his experiments before the Academy, Pouchet still claimed that heterogenesis was an accepted fact among the most famous physiologists of Germany, Italy, England and North America, but not in France, a country that was in his view too strongly influenced by Pasteur's hypothesis on atmospheric germs. Pouchet called Pasteur's supporters "the reformists of atmospheric panspermia" (Pouchet, in Pennetier, 1868). Definitely resistant to Pasteur's conclusion, Pouchet tried to incite the Academy, in January

1864, to grant another demonstration to check whether or not the experiments conducted by Pasteur were favorable to the doctrine of spontaneous generation. Unfortunately, according to Pennetier (Pennetier, 1868), the Commission (made up this time of Dumas, Milne-Edwards, Brongniart, Balard, and Flourens) did not accept the experimental program proposed by Pouchet and his colleagues (Joly, and Musset). Once again, they withdrew from the competition. Officially (and historically), one can consider that the debate ended for good on April 7th, 1864, after the lecture given by Pasteur at La Sorbonne[††].

Pouchet's heterogenesis sustained the idea that life mechanisms could have started under the influence of inexplicable forces (a " supreme force") capable of having an effect on some groups of molecules. Pouchet could have been influenced in this view by contemporaries studying protoplasm, such as the French biologist Félix Dujardin (1801-1860), who supported the idea of spontaneous generation on the basis of the simplicity of primitive organisms (Farley, 1977). Pouchet also believed in Cuvier's theory of successive creations and therefore dissociated himself from the evolutionary thought of Lamarck and Darwin. According to Pouchet, denying spontaneous generation was equivalent to adopting an atheist position and adhering to Darwinism (Latour, 1989).

Pouchet explained his view on the origin of life in the preface to Georges Pennetier's book (Pennetier, 1868) entitled *L'origine de la vie* (Pennetier's view is presented in the next section). In Pouchet's preface, it seemed obvious to him that life had had "multiples origins." He considered the succession of organic creations to be a fundamental axiom of geology and stated that life would have appeared as many times as there had been important stages of geological genesis. These successive spontaneous generations would have initially emanated from a unique creation, according to an unceasing mutability of organization (Pouchet, in Pennetier, 1868). As the reorganization of matter depended on geological changes, heterogenesis would have been possible repeatedly during the various ages of the Earth. Pouchet's opinion demonstrates that one of the strongest supporters of heterogenesis distinguished without any ambiguity the problem of <u>continuous generation</u> from that of the <u>first emergence</u>

[††]According to Farley (Farley, 1977), in 1871, just one year before Pouchet's death, Pasteur once again was forced to debate the issue (this time with the chemist Frémy and the botanist Trécul – Frémy and Trécul both considered yeast and bacteria to be engendered from albuminous matter) as part of his defense of his own theory of fermentation.

of life. Each of these events was supposed to be dependent on different mechanisms. Heterogenesis would explain a continuous generation through the reorganization of matter already provided with organic characteristics. But in that case what would be the mechanism for an initial genesis? Pouchet did not provide a clear scientific answer to that question. Heterogenists like Pennetier agreed about the correctness of the idea of change from a mineral to an organic state, referring to the results of chemists like the German Friedrich Wöhler (1800-1882) and the Frenchman Pierre-Eugène Marcellin Berthelot (1827-1907), who were able to create artificially some of the substances coming initially from living beings; however, Pouchet probably disagreed with this last opinion.

Pouchet tried to dissociate primeval genesis from continuous spontaneous generation. Perhaps this view permitted him to keep theological considerations as part of a complete theory of the origin of life. According to many authors, Pouchet's theory was perfectly compatible with religion. These theological considerations were apparent when he asserted the two following points: (1) the (first) origin of life was "a true spontaneous generation operating under divine inspiration." (2) God had imposed laws, for matter and for life, fixing the organizational forces able to give rise to new beings (by means of heterogenesis). He distinguished between the (first) origin of life viewed as a "true" spontaneous generation operating under divine inspiration, and further acts of spontaneous generation (by heterogenesis) viewed as the result of the action of Nature, just as God had wished to do in his (first) design. In that latter hypothesis, the laws of heterogenesis were natural ones but had been previously specified by the Creator, when the laws of matter and life were fixed. Keeping faith in the existence of God and of Creation, he manifestly saw a perpetual and dominating force ruling over the evolution of living forms. Pouchet, who was of Protestant origin, was probably influenced by the German Romantic movement and the writings of the philosophers of Naturphilosophie (Cantor, 1992).

Therefore, religious aspects were not ruled out in Pouchet's conception of heterogenesis. Farley emphasized that Pouchet tackled the problem of spontaneous generation in his text of 1859 (Pouchet, 1859) with the utmost precaution: the first 137 pages of his book were devoted to an historical and metaphysical justification of his belief in spontaneous generation (Farley, 1977). According to Raynaud (Raynaud, 1999), during the entire debate he engaged with Pasteur, Pouchet not only spoke about pursuing "a holy cause," but was in contact with many press reporters who turned the issue into a public debate in which religious, ideological and political views were intermixed. However, and unfortunately for Pouchet, his doctrine gave the advocates of spontaneous generation the opportunity of getting rid of the necessity of a Creator and then of keeping their distance from religion and dogma (Debré, 1994). Pouchet never argued against God; he even reproached Pasteur for having put on the "mask of religion" so as to triumph over his opponents. Pouchet was convinced that the "theological illusion" created by Pasteur had contributed to the rejection of spontaneous generation by the Academy of Sciences.

Nicolas Joly's opinion, which also supported heterogenesis, shows how ambiguous the problem raised by the "confrontation" between science and religion can be when one is concerned by spontaneous generation. On the one hand Joly seemed to be not very conciliatory when faced with the connection between science and religion: "Religion is not expected to intervene in a debate in which it is not at all involved" (Joly, 1863). On the other hand, and in the same paper, Joly didn't hesitate to quote Pouchet as saying that "the thesis we have elaborated (heterogenesis) had been supported by the greatest Christian philosophers, Saint Augustine, Saint John and Saint Basil."

It is worth noting here, however, that Pasteur's own position on religious considerations was not devoid of ambiguity. According to Latour (Latour, 1989), Pasteur tried to associate heterogenesis with materialism — and then the negation of a Creator — in order to better persuade the members of the Academy's Commission. And yet he pushed away religious considerations when he used his experiments to dismantle heterogenesis. In a lecture delivered to the Chemical Society of Paris (Pasteur, 1862), Pasteur raised one of the main problems that exist if we consider spontaneous generation to be a spontaneous organization: "Could matter coming from living beings in a state of decay keep a last trace of life and then could organize itself?" The only way to answer this question was, according to Pasteur, to undertake experiments and observations. Pasteur refused to venture onto non-scientific ground and thought that it was as hard to introduce religion into science as it was to introduce science into religion: "It is not here a question of religion, or philosophy, or some other system. Assertions or a priori views matter little. It is a matter of facts. You will also notice that I don't pretend to establish that spontaneous generations never exist. In such topics one cannot prove by the negative." (In the French text: "Dans les sujets de cet ordre on ne peut pas prouver la negative.")

In these last two sentences, Pasteur asserted that his type of experiments couldn't prove the impossibility of spontaneous generations (Dubos, 1955). But here again Pasteur was not unambiguous: did he cautiously approach the problem of the first genesis? The idea of a primeval origin of life considered as a process emerging by abiogenesis didn't seem to have been officially debated by Pasteur. One explanation could have been that Pasteur's works on molecular asymmetry led him to preserve a notable difference between the living and the mineral worlds. However, Pasteur took an active interest in the possibility of abiogenesis, according precisely to his conception of an "asymmetric force," the intervention of which he considered to be essential to the production of asymmetric molecules and, hence, of life (Farley and Geison, 1974). Living matter was not reducible to mineral matter. However, according to Geison, in manuscript notes written in 1870 — no published results emerged from them-, Pasteur jotted down some thoughts about the origin of life and projected to carry out a new series of experiments. They were designed to create asymmetric products with the aid of the elements carbon, hydrogen, sulphur, and phosphorus, and to make them react under the influence of magnets, solenoids, or anyything holding asymmetric actions (Geison, 1995). Geison's opinion is that Pasteur ultimately saw the Creator-God as the source of the original cosmic asymmetric force since, in his view, life was dominated by asymmetric forces and all living species were primordially functions of cosmic asymmetry (Geison, 1995).

Another point of view, avoiding abiogenesis this time, could be held to explain the origin of life in terms of a first genesis. Pasteur's own words suggested the concept of "eternal life," which finally would have allowed him to dismiss the idea of the origin — itself — of life:

> "Spontaneous generation, (...), I have been looking for it for twenty years without discovering it. No, I do not consider spontaneous generation to be impossible. But who indeed entitles you to want it to have been at the origin of life?... What tells you that the continuous advances in science would not oblige the scientists in the next century, or in a thousand years, or ten thousand years... to assert that life has existed from the beginning of times instead of matter? Who will convince me that in ten thousand years one won't consider that life wouldn't lead to matter?" (Pasteur in Debré, 1994).

Pasteur seemed to be in an uncomfortable situation when confronted with religion. Pasteur, who was born into a Christian family, chose to be very cautious about the consequences of his experiments. His suggestion of an eternal life was perhaps, somehow, the sign of a form of spirituality but without clearly embarking on religious ground. However, it seems obvious that Pasteur didn't want his religious convictions to influence his scientific conclusions. Different authors have emphasized the fact that Pasteur wished to keep an objective interpretation of his experimental results at all costs. For instance, Pasteur Vallery-Radot mentioned that Pasteur had such faith in the experimental method that he was convinced it could never mislead him (Pasteur Vallery-Radot, 1922).

4. Other Views in the 1870s: Georges Pennetier (1836-1923) and Hyacinthe de Valroger (1814-1876)

The opinion of less well-known scientists than Pasteur or Pouchet can also be taken into account in order to give examples of the diversity of views on spontaneous generation that existed in France in the nineteenth century. The biologist Georges Pennetier for instance, a student of Pouchet who became director of the Rouen Museum, viewed the question differently from Pouchet.

Pennetier's book *L'origine de la vie* (*The Origin of Life*) (1868) offered a different view on heterogenesis. Pennetier accepted Darwin's evolutionary ideas and seemed to adopt a materialistic position. He thought that modern chemistry had proved that mineral matter could transform itself into a semi-organized state in which every organic element belonging to organisms was present. He specified that this state was not yet alive but was only "waiting" for new combinations of circumstances. In other words, it was a spontaneous organization of matter (which possessed an "organizing force") under particular environmental conditions. According to this hypothesis, matter could be found in three successive states of complexity: mineral, organic, and organized. And in order to give an answer to the problem of the primordial origin of life, it would be necessary to understand the change from the first state to the second one. The third state (spontaneous generation) would have been able to give birth to a large number of animalcules or primitive plants (such as monads, bacteria, and vibrions) or primitive plants by means of heterogenesis, and so by recombination of organic matter coming from the previous state or directly from matter in decay or fermentation. Pennetier's hypothesis solved both the problem of the primeval origin of life and that of a (supposed) continuous genesis.

Pennetier reported that heterogenists like the French botanist Auguste Trécul (1818-1896) — and himself in experiments described in "*L'origine de la vie*" — stated in 1865 to have observed through the microscope organic substances contained in some plant organs being transformed into living beings by the action of decay during their maceration in water. Pennetier didn't seem to question whether the observed facts could be misleading. The question was then: if this spontaneous generation occurring by heterogenesis has been observed experimentally, why does it have to be so disputed within the scientific community? He reported Pasteur's demonstrations performed before the Academy of Sciences. When he presented the confrontation between Pasteur and Pouchet, he clearly emphasized the comparison of Pasteur's doctrine with panspermia. Pennetier's opinion was that this doctrine was generally regarded as an abstract idea, a metaphysical concept with no scientific reality.

The debate about heterogenesis couldn't leave theologians insensitive. The French oratorian Hyacinthe de Valroger (who first studied medicine and was a man of great erudition) published in 1873 "*La genèse des espèces, études philosophiques et religieuses sur l'histoire naturelle et les naturalistes contemporains*" (*Genesis of Species, Philosophical and Religious Studies on Natural History and Contemporary Naturalists*). De Valroger was one of the thinkers who attempted to reconcile science and religion around the theme of biological evolution. Insofar as heterogenesis was concerned, de Valroger defended a very different point of view from Pennetier, as can be expected given his religious position: "not only was inorganic matter unable to create organized beings, but also organic matter was crippled by this natural incapacity," even in the most favorable conditions (de Valroger, 1873). He pointed out that the opinion of the physiologist Pierre Flourens (1794-1867), expressed in his *Ontologie naturelle ou étude philosophique des êtres* (1861), rose up against the possibility that an organized being was the result of the blind assembly of physical elements. De Valroger's harsh words show how he completely rejected a physical action on the components of life: "Whatever the way we take the theory which attributes the origin of organized beings to the influence of physical agents, this theory can tolerate neither examination nor criticism" (de Valroger, 1873). According to him, the only way to explain these phenomena would be to accept the intervention of an "Intelligence" always acting in accordance with the same design. The important point in de Valroger's theory was that there was no relationship of causality between two categories (physical and biological) of phenomena which appeared to him to be so dissimilar: the one always obeys the same laws, whereas the other one obeys new combinations at every new period of the earth. De Valroger totally rejected theories implying that laws of matter had an action upon the organization of living matter.

Eventually, de Valroger advocated a spiritualist theory of spontaneous generation, in which spontaneous generation would not be the result of an inert and "unintelligent" matter. Instead it would be the work of God, in accordance with pre-established laws (acting thanks to predetermined circumstances), or else it would be a special and immediate action of God leading as often as necessary to life's organization. However, in spite of his wish to create links between science and religion, de Valroger appeared too much dogmatic when he tackled the problem of the first genesis of life.

5. Camille Flammarion (1842-1925)

Even though French biologists were at the heart of the problem of the origin of life, astronomers also had their point of view on this matter. The opinion of French astronomers was sometimes clearly influenced by their religious beliefs, showing that science and religion tried to live together. It was the case of the French astronomer Camille Flammarion, founder of the Juvisy-sur-Orge Observatory and of the Société Astronomique de France (SAF), who attempted to build a religious philosophy based on the principles of positivist science. According to Flammarion, experimental sciences should help us to obtain an answer to the real question concerning theology, namely whether God exists or not. Flammarion's thoughts about the place of God in the sciences were developed in one of his numerous publications, *Dieu dans la Nature* (*God in Nature*) (1869). In this book, Flammarion criticized the scientists among his contemporaries who "go through God without seeing him" as well as theologians who "look through a prism that distorts images." His view was that "scientific atheism is a mistake, and religious illusion is another one."

The aim of Flammarion's book was to propose a non-theological refutation of contemporary materialism. Flammarion was indeed convinced that precise knowledge about the construction of the universe, about life and about thought, would be the only effective way to clarify the fundamental question of the existence of God. This knowledge could tell us whether matter alone prevailed in the universe or if we have to admit the existence in Nature of an "organizing Intelligence, a design and a destiny for living beings".

In the chapter "L'origine des êtres" (The origin of beings), Flammarion took up the topic of the debate on heterogenesis. In his critique of the debate, he pointed out that the experiments that had been carried out had not solved the problem of the origin of life. If germs came from the atmosphere and were at the origin of ferments, what then was the origin of the germs themselves? He even appeared to be ironic when he evoked Pasteur's demonstration of the rarefaction of germs in the "pure air of the mountaintops." Flammarion mentioned the reproach of atheism directed at Pouchet, whereas Pouchet himself had never interpreted his experiments in a theological framework. However, without openly supporting heterogenesis, Flammarion considered that it was a wrong solution to transpose the question of spontaneous generation towards a theological interpretation.

Flammarion wondered whether the emergence of life was likely to still occur from inorganic matter (present day abiogenesis), but he used Darwin's arguments (species filiation leading back to the existence of a common ancestor) to reject this hypothesis. Flammarion however, as well as many other scientists concerned by the problem of the origin of life, firmly regretted that Darwin, in *The Origin of Species*, had not clearly given his opinion on that topic:

> "He (Darwin) contents himself with explaining the potential variability of a certain number of primitive types, and it is remarkable to comment that in a large and rich work on the origin of beings, there wasn't any mention of this origin!"

Sensitive to an explanation of the power of God by scientific methods, Flammarion took great interest in Darwin's opinion on a potential action of a Creator at the very beginning of life. Flammarion quoted, in *Dieu dans la Nature*, Darwin's last remarks in *The Origin of Species* (Clemence Royer translation), in which it appeared to him that Darwin didn't dismiss the possibility of "the breath of God" at the origin of the forces animating the primordial forms (or a unique one) (Flammarion, 1869). Flammarion remarked that, just as Pouchet's doctrine on spontaneous generation, Darwin's theory had often been considered to overthrow the idea of God, whereas in fact neither Pouchet nor Darwin accepted such an accusation, which they described as an "illusion" that existed only in the mind of their opponents.

Flammarion concluded that spontaneous generation had shown that the problem of the origin of life was, and perhaps would remain, an inexplicable mystery. In view of the questions raised, but left unsolved by his contemporaries, the problem of the origin of life seemed to him to hold extreme complexity. Many distinguished scientists had been confronted with huge obstacles that can be perceived through some of the opinions expressed by Flammarion.

The French chemist Edmond Frémy (1814-1894) had pointed out that an insurmountable barrier existed between biological organization and the chemical syntheses carried out in the laboratory. This obstacle was based on the assumption that to be generated, life would necessarily have required materials provided with a vital force corresponding to semi-organized matter (as evoked by Pennetier). Despite their acceptance of the basic similarity of organic and inorganic matter, most experimental chemists were aware of the differences between the organic substances obtained in the laboratory and biological organization. The opinion of German materialists (like Büchner, Vogt, and Virchow), as Flammarion pointed out, was quite different and also rather complex, not to say unclear (from Flammarion's viewpoint). The naturalists Ludwig Büchner (1824-1899) and Karl Vogt (1817-1895) generally considered that abiogenesis had played a greater role in primordial times than in recent times. This "organic creation" should have been possible without the intervention of any external force (an opinion also adopted by Haeckel). Very few German scientists at that time were opposed to abiogenesis but slight differences in opinions highlighted the complexity of the problem. The physician and anthropologist Rudolf Virchow (1821-1902) stated that, as far as he was concerned, new combinations of elements should have received a "vital movement" that "must be considered as the result of a definite joint action of physical and chemical forces" (Virchow in Farley, 1977).

Flammarion criticized these German authors for not clearly expressing the way in which the first organic cells could have been produced at the expense of inorganic chemical elements. Nevertheless, we can say today that German materialists were perhaps among the first to approach the problem of the origin of life only from an abiogenetic perspective, even if the conditions of this abiogenesis were not very clear and did not exclude a present day abiogenesis.

Flammarion's opinion about the role of God in Nature and about the genesis of life makes it clear that he felt that there was a "creative force," in other words a "hidden God," behind these still inexplicable processes, such as those described above. Life's laws were directed by causes that were not blind but that must at least "know what they have to do." Life was universally spread in Nature as organic as well as inorganic elements were filled with invisible life, with species (such as insects) affected by a sort of metamorphosis under the influence of environmental conditions. Primitive substance took up space boundlessly.

Seen in that light, as Flammarion himself stated "spontaneous generation should only represent one of the aspects of the universal life that is stirring in every atom of matter." Flammarion expressed in lyrical terms that "there is a design given by God in which primitive substance would one day be condensed in worlds where life and intelligence would display their magnificence." More precisely, he saw an "organic force" which is inherited from generation to generation. This organic force would be the natural and inescapable outcome of the fertile conditions of the earth when the era of life came. To summarize his thoughts, one can say that Flammarion believed a divine Creator to be necessary to organize these fertile conditions and to transmit to Nature a never-ending tendency to complexity. God was the Creator of the forces imparted to matter as a whole, from mineral substances to organic molecules. In this view, he was not very far from Pouchet's own view, as I mentioned earlier.

6. Archeogenesis and Chemical Evolution: Edmond Perrier (1844-1921)

I will end this survey of French views on the origin of life by discussing the ideas of the zoologist Edmond Perrier, who was Director of the National Museum of Natural History in Paris from 1900 to 1919 and a fervent advocate of Darwin's and Lamarck's theories.

During the first decades of the twentieth century, after the doctrine of heterogenesis had been finally abandoned, basic living materials were perceived more and more as the result of a global evolution of matter, which had occurred during the first ages of the Earth. The question of the origin of life could henceforth be limited to an initial origin of the first primitive living beings by archeogenesis. To be specific, the crucial point was then to identify the first stages of the transition from mineral elements to organic substances, in other words to understand how abiogenesis could have occurred in a primitive environment whereas it was no longer occurring in the present day.

In one of his early works, entitled *La Philosophie zoologique avant Darwin* (Zoological philosophy before Darwin) (1884), Perrier examined critically the naturalists of the first half of the nineteenth century who, in his opinion, did not try to explain the laws dealing with nature because they generally believed in an "undefined being dignified with the name of Nature," so that, everywhere Nature intervened, it was not necessary to look for explanations of the observed phenomena. Perrier disagreed with this metaphysical position and claimed that just as in the physical sciences, in the natural sciences problems such as that of the origin of life can be explained by a scientific method that focused especially on connecting effects to causes.

Almost twenty years later, Perrier's view clearly reflected the progress of the scientific method applied to the problem of origins of life. By the early 1900s, scientists were aware of the impossibility for life to emerge nowadays from inorganic matter. In his book, *La Terre avant l'histoire: les origines de la vie et de l'homme* (*The Earth Before History; The Origins of Life and Mankind*), published in 1920, Edmond Perrier devoted an entire chapter to the problem of the origin of life in which he described the advances in that topic, including also the hypothesis of panspermia. According to Perrier, the cornerstone of the problem was how to consider the fundamental nature of living substance. Of course more than thirty years had elapsed since the debate on spontaneous generation and by then it was easier to reject life's simplicity. Perrier was convinced that it was impossible for life to be the work of a single substance (as had been assumed by Haeckel for instance), a special substance (a "physical basis of life," like the protoplasma, as proposed by the British naturalist Thomas Henry Huxley (1825-1895)), or else organic substances endowed with specific forces of organization (Pouchet's heterogenesis). Perrier then went on to give several methods of explanation based on organic chemistry to strengthen his arguments.

Perrier believed that life's emergence would be the result of physicochemical forces. The first thing to do would be to search for processes of synthesis leading to the building and gathering of common and inert substances, such as carbohydrates, fats and albuminoids. He brought up one of the main difficulties one would have when dealing with these syntheses: "mixing in correct proportions the substances which are expected to be associated in order to give birth to life and to preserve it." Of course, the time factor should also be taken into account in order to approach the idea of chemical evolution. Eventually, he highlighted the insolvable problem with which we are still confronted today, namely the impossibility of going back to the initial conditions, which makes the science of the origin of life an historical science.

The last word (so topical) must be given to Perrier:

> "It (the origin of life) would have occurred during a situation that we can reconstitute today by a mental image, a situation that persists perhaps in some stellar systems, but that has disappeared with no return in our solar system" (Perrier, 1920).

7. Conclusion

Although Pasteur, like many other French scientists of his time, was convinced as early as the first years of the 1860s that spontaneous generation by heterogenesis was an impossible process, the possibility of spontaneous generation was not completely dismissed until the end of that decade, in spite of the debate incited by the French Academy of Sciences. The controversy over spontaneous generation in France showed that the problem of "origins" was not correctly identified during the 1860s and 1870s as the debate concerned above all the eventuality of a continuous genesis of life and not a single (and primordial) origin. In his own mind, Pasteur did not solve the problem of the primeval origin of life (Pennetier, 1907). During the years of the controversy, the notion of a primeval origin was not at the heart of the scientific discourse and often ended up being attributed to divine action or relegated to (the supposed) eternal existence of life.

It is worthwhile to speculate here that, during the controversy the use of the term "origin" led to a semantic, rather than a methodological hurdle. In other words, is it possible that the scientists involved in the problem of the origin of life did not ask the right questions? If that were so, it is possible that these scientists reached pseudo-scientific "explanations."

The influence of theological thought was probably also a significant factor in the slow abandonment of the idea of spontaneous generation. Both Pasteur and Pouchet (either positively or negatively) commented on the role of God in the origin of life, although Pasteur did so only in a muted way. Most scientists considering the problem of the origins of life were certainly more or less influenced by their religious convictions, but their attitudes varied depending on how their observations could be interpreted. As for Flammarion, he completely integrated God's power into his conception of the beginnings of life. Perhaps Flammarion, who was an astronomer and not a laboratory scientist like Pasteur or Pouchet, cannot be viewed as a real specialist of the question. He was nevertheless an important thinker whose views mattered.

It seems quite obvious today that the time of the spontaneous generation controversy, which was also a period of spectacular advances in biology, should have led scientists to reconsider (or at least to wonder about) the place of a potential Creator. This should have been so especially when the problem of the primeval origin of life was raised. On the one hand, many scientists were probably tempted by the possibility of a connection between the processes involved in a primordial genesis and a potential place for God. On the other hand, fundamental concepts did start to become clearer during the last decades of the nineteenth century, albeit in a timorous way. In order to decipher the enigma of a first genesis (without the help of the power of God), one had to master concepts such as those of chemical evolution (and the huge time-scale required for this process), and of specific environmental conditions of the primitive Earth. It was only the association of these two concepts that could have led to the new paradigm of a unique (in space and time) starting point for biological processes on Earth.

One had to wait until the end of the nineteenth century to see the dawn of a deeper and more mature reflection, based more strictly on scientific reasoning, about how the transition from inorganic to organic matter could have occurred on the primitive Earth. Only then could early theories of chemical evolution be developed.

8. Acknowledgements

I would like to thank Dr. François Vuilleumier, Department of Ornithology, The American Museum of Natural History, New York, for his extremely helpful comments and suggestions during the preparation of the manuscript.

References

Cantor, M. (1992). *Félix-Archimède Pouchet, savant et vulgarisateur*, Thèse de Doctorat, Université Paris 11.

Darwin, F. (ed.) (1887). *The Life and Letters of Charles Darwin*, Murray, London (3 vols.).

Darwin, F. (ed.) (1905). *The Life and Letters of Charles Darwin*, Appleton, New York (2 vols.).

Debré, P. (1994). *Louis Pasteur*, Flammarion Champs, Paris.

De Valroger, H. (1873). *La Genèse des Espèces, études philosophiques et religieuses sur l'histoire naturelle et les naturalists contemporains*, Didier, Paris.

Dubos, R. (1955). *Louis Pasteur franc-tireur de la science*, Presses Universitaires de France, Paris.

Farley, J. (1977). *The Spontaneous Generation Controversy from Descartes to Oparin*, Baltimore and London, The Johns Hopkins University Press, Baltimore and London.

Farley, F. and Geison, G. (1974). Science, politics and spontaneous generation in 19th century France: The Pasteur-Pouchet debate, *Bull. Hist.Med.* **48**: 161-98.

Flammarion, C. (1869). *Dieu dans la nature*, Didier, Paris.

Flourens, P. (1861). *Ontologie naturelle ou étude philosophique des êtres*, Garnier, Paris.

Geison, G. (1995). *The Private Science of Louis Pasteur*, Princeton University Press, Princeton.

Haeckel, E. (1868). *Natürliche Schöpfungsgeschichte* (*The Natural History of Creation*), Reimer, Berlin.

Joly, N. (1862). Examen critique du Mémoire de M. Pasteur, relatif aux generations spontanées, et couronné par l'Académie des Sciences de Paris, dans sa séance du 29 décembre 1862, In: *Mémoires de l'Académie impériale des Sciences de Toulouse*, VIème série, Tome 1, p. 215.

Latour, B. (1989). Pasteur et Pouchet : hétérogenèse de l'histoire des sciences, In: *Eléments d'Histoire des Sciences*, M. Serres (ed.), Bordas, Paris, pp. 423-445.

Latour, B. (2001). *Pasteur: guerre et paix des microbes suivi de Irréductions*, Editions La Découverte & Syros, Paris.

Pasteur, L. (1861). Mémoire sur les corpuscules organisés qui existent dans l'atmosphère. Examen de la doctrine des générations spontanées, Annales des sciences naturelles, 4ème série, **XVI**, pp. 5-98 et *Annales de chimie et de physique*, 3ème série, LXIV, janvier 1862, pp. 5-110.

Pasteur, L. (1862). Sur les corpuscules organisés qui existent dans l'atmosphère. Examen de la doctrine des generations spontanées, In: *Leçons de chimie et de physique professées en 1861 à la Société chimique de Paris*, Hachette, Paris, pp. 219-254.

Pasteur Vallery-Radot, L. (1922). *Oeuvres de Pasteur, Tome II Fermentations et Générations dites spontanées*, Masson, Paris.

Pennetier, G. (1868). *L'origine de la vie*, Paris.

Pennetier, G. (1907). *Un débat scientifique – Pouchet & Pasteur, 1858-1868*, Actes du Muséum d'Histoire Naturelle de Rouen. – XI, Rouen.

Perrier, E. (1884). *La Philosophie zoologique avant Darwin*, F. Alcan, Paris.

Perrier, E. (1920). *La Terre avant l'histoire. Les origins de la vie et de l'homme*, La Renaissance du Livre, Paris.

Pouchet, F. (1858). Note sur des proto-organismes végétaux et animaux, nés spontanément dans de l'air artificial et dans le gaz oxygène, *Comptes rendus de l'Académie des Sciences*, **XLVII**: 979-982.

Pouchet, F. (1859). *Hétérogénie ou Traité de la generation spontanée*, Paris.

Raulin Cerceau, F. (2004). Analysis of the Works of the German Naturalist Ernst Haeckel (1834-1919) on the Origin of Life, In: J. Seckbach, J. Chela-Flores, T. Owen and F. Raulin (eds.), *Life in the Universe*, Kluwer Academic Publishers, the Netherlands, pp. 349-352.

Raynaud, D. (1999). La correspondance de F.A. Pouchet avec les membres de l'Académie des Sciences: une réévaluation du débat sur la generation spontanée, paper revised for the *European Journal of Sociobiology*, **40** (2): 257-276.

Biography

Dr. Peter R. Bahn is currently Chief Scientist, Patent Attorney, and President of Bahn Biotechnology Company, located in Mount Vernon, Illinois, USA. He received a B.S. in chemistry from Haverford College in 1972, a Ph.D. in molecular biophysics and biochemistry from Yale University in 1978, and a J.D. from the University of Miami School of Law in 1984. Dr. Bahn's main scientific interests are molecular evolution, the origin of life, and astrobiology.

14
The Nature of the Origin-of-Life Problem

PETER R. BAHN

Bahn Biotechnology Company
10415 E. Boyd Rd., Mt. Vernon, Illinois 62864 USA
pbahn@mnv.net

1. Introduction

The origin-of-life problem (Calvin, 1969; Davies, 1999; Fox and Dose, 1977; Hazen, 2005; Matsuno, 1989; Oparin, 1938; Seckbach, Chela-Flores, Owen and Raulin, 2004) is a problem which can most usefully be solved by faith in science. The origin-of life problem is essentially this: How in the world could all of these clever little quantum mechanical machines that we call organic chemicals get together to form this beehive of molecular activity that we call life?

The origin-of-life problem is primarily a colossal problem in the field of organic chemistry. Another way of putting it might be to ask the question: How does organic chemistry, with the proper astrophysical and geological influences, turn itself into biological chemistry?

Faith in traditional human religions and/or deities cannot help use solve the origin-of-life problem in a mechanistic sense because such religions and/or deities do not speak to us in the language of molecules, and to solve the origin-of-life problem in a mechanistic sense we absolutely must speak in the language of molecules since life is composed of molecules.

The above statements and the statements to follow are not an attack upon religion, which this author believes speak mostly to us about the purpose of life. However, because religions were formulated before we humans knew that all living matter was composed of molecules, and that nothing distinguishes living molecules per se from non-living molecules, religions are of little help in describing the molecular interactions that first led to life.

2. What is Science?

Most traditional religions, which have been practiced by mankind for the past several thousand years, have been based on a single founder with various miraculous powers accessible to him alone, whereas science, in a way, is a religion that has been practiced by mankind for the past few hundred years, and which has many founders, with miraculous powers (like lasers and antibiotics) accessible to everyone.

Science is a human activity in which the only deity allowed is Nature, that is, the physical world itself. For this reason science used to be called natural philosophy before the word science itself was invented. For example, the greatest scientific text book ever written (at least in the opinion of this author), by Isaac Newton, when translated from Latin to English, is entitled *Mathematical Principles of Natural Philosophy* (1687).

Science, unlike traditional religions or traditional philosophy is a matter of proof. 'If you can't prove it, it isn't science'. To give just two examples of what is meant by proof, in 1905, it was convincingly shown by Einstein that the electrodynamics of moving bodies with a constant velocity are correctly described by the special theory of relativity (Einstein, 1920), and in 1953, it was convincingly shown by Watson and Crick that the genetic information carried by biological cells resides in DNA in the form of a double helix (Watson and Crick, 1953).

3. Faith in Science

Working toward and waiting for proof in science requires a good deal of faith in the processes by which science is carried on. This faith in science means believing in the formulation of theories that make predictions and have physically testable results, and sometimes having to wait for years to obtain such results. For example, in the 1940s, the 1950s, and the 1960s, when two different theories of the Universe, the Big Bang Theory (most adequately summarized by Weinberg, 1994) and the Steady State Theory (Hoyle, 1975), were in competition, the first theory stating that the Universe had a beginning in time billions of years ago and the second theory stating that the Universe was eternal and unchanging in time, it was not clear at all that either theory could be proved. Scientists worried about how one would prove that the Universe started with a certain event billions of years ago. However, with the discovery of the cosmic background radiation by Penzias and Wilson in 1965, this problem was solved, and it was convincingly demonstrated that the Universe did indeed begin in a huge explosion of matter, energy, space, and time approximately 14 billion years ago. What this example points out is that the laws of science, at least the laws of physics and chemistry, do not change even over billions of years of time and over billions of parsecs of space. The faith that this is so is the kind of faith that is required to solve the origin-of-life problem.

4. What is Life?

There is no single definition of life (Rizzoti, 1996; Ward, 2005) that satisfies everyone, including even scientists. For example, the official NASA definition of life is: *Life is a self-sustained chemical system capable of undergoing Darwinian evolution.* One difficulty with this definition is that Darwinian evolution involves competition between entities, so it would appear that there must be system*s* that are capable of undergoing Darwinian evolution. If there are self-sustained chemical systems capable of undergoing such evolution, then there must be boundaries, presumably, to such systems. In today's biological world, the self-sustained chemical systems are called *cells* (Morowitz, 1992) and the boundaries of such systems are called cell *membranes.* Cells appear to self-assemble from their component molecules based largely on hydrophobic and hydrophilic reactions, the membrane components being based on a spherical shell in which the shell-interior components are hydrophobic, and the cell interior component molecules are associated with hydrophilic interactions.

Here is another definition of life: *Life is an intricate mass coordinated wiggling of different types of carbon based molecules spreading itself indefinitely through time and space.* What is meant by the word *wiggling* is that life appears in the phenomenological sense to be like a wave of wind blowing through a field of wheat. We cannot see the wind directly, but we imply from the coordinated movements of the wheat stalks that a wind is there. Life is like that wind, and all that we can see of it directly at the molecular level is the dance of the molecules. It is for this reason that life is often referred to as an emergent property of groups of molecules. It is also for this reason that it is unlikely that there will ever be found such a thing as a "living" molecule. This does not stop RNA World (Gesteland, Cech and Atkins, 1999) afficionados from searching for an RNA ribozyme that can catalyze the assembly of itself from monomers (which they have not yet found and which they may never find). However, the smallest such self-assembling unit that Nature has been able to conjure up is the system that we call a biological cell.

Is it possible to define life without words? Yes, for example with mathematical equations: *Life is something which follows this set of equations: (equation), (equation), (equation, etc.) that describe such and such mathematical parameters of such and such molecules and their interactions.* Of course, the difficult part of this "mathematical definition is, just what the equations in the set should be. Does anyone really know? It seems that in trying to define life, one encounters something that can be called the "horizon problem" which is this: The vast number of processes that are taking place simultaneously so as to constitute life is so enormous and so complex that it is completely impossible for a human mind, *any* human mind, to comprehend all of them in precise molecular detail. That is why coming up with the right set of definitional equations would be a monster of a problem. I call this the horizon problem because human minds have a considerably limited horizon in being able to simultaneously see and comprehend all of the physics and chemistry that simultaneously goes on in the life process even within a single living cell.

Can there be a definition of life which is neither verbal nor mathematical? Yes, a picture definition: *Life is something which looks like this picture or pictures: (picture or pictures)*. The problem here is, just exactly what picture or pictures does one use? Again, we face the horizon problem. The horizon problem is also the reason that a really accurate and good definition of life in mere words will likely *always* evade the capacities of human beings. We humans would have to evolve to something far more intelligent than human beings to have a wisp of a chance of succeeding in solving the definition-of-life problem in a definitive manner, no matter what mode or modes (verbal, mathematical, or pictorial) we try to use. We thus could say, in a manner of speaking, that trying to define life with words is like try to pick up a dime with big pillows and also is like trying to pick up a boulder with tweezers.

5. A Runaway Chemical Reaction

There can be no doubt that life on Planet Earth is in some respects a runaway chemical reaction that started out almost four billion years ago and which such reaction continues to this very day, having taken over the surface of this Earth, and which has spread itself to visit our Moon and which has sent its artifacts to other planets and other moons in our Solar System.

If a runaway chemical reaction can be said to have a purpose, it is to convert more of the surrounding chemicals into being a part of the runaway chemical reaction. Thus, a combustion reaction of, say, trinitrotoluene (TNT) explosively expands to encompass surrounding or nearby trinitrotoluene molecules into more of the combustion reaction system, along with any flammable materials nearby.

What the final purposes of the long lasting chemical reaction systems called life are, we do not yet know. But, as with any runaway chemical reaction, at least one purpose of life appears to be to make and spread more life around as much as possible.

In other words, life is a runaway chemical reaction that increasingly converts inorganic carbon into biochemical carbon. How this particular process first got started is in this sense really a problem of organic chemistry. Consequently, in the origin-of-life field, a scientist can never really know enough organic chemistry.

There are many different ways of looking at life. The tendency of life to expand at an explosive rate is one of life's more curious attributes.

6. Organic Chemistry

It was the chemist Friedrich Wohler who first showed that there is nothing magical or special about the chemicals found in living systems when compared to the chemicals found in bottles on the shelves of a chemical laboratory. Wohler

(1828) found that when he heated an inorganic compound ammonium cyanate, the reaction produced urea which was *exactly* the same as the urea produced in biological urine. This was one of the first indications that the chemicals of which living entities are composed are just the same chemicals of which non-living entities are composed. What living things are mostly composed of, besides water, is indicated by what remains when living things or once living things are burned up. What remains is a pile of carbon. Living things are made up largely of carbon, but there are other elements involved, namely those found in the formula CHNOPS, that is carbon, hydrogen, nitrogen, oxygen, phosphorous, and sulfur. It is because carbon forms strong chemical bonds with itself and with other elements that carbon chemistry is complicated enough to be a whole field of chemistry in itself called organic chemistry (Wintner, 2004).

7. Biomonomers

The monomers from which life on Earth is composed are sugars, amino acids, mononucleotides (which in turn are comprised of a phosphate, a sugar, and a heterocyclic base), and lipids. Much progress has been made by scientists in figuring out how such monomers were probably made on the Primitive Earth from primordial inorganic compounds.

For example, Butlerov discovered in 1861 that formaldehyde in aqueous solution will self-condense into a wide array of sugars if a small amount of base is added to the formaldehyde solution. This chemical reaction is called the Formose Reaction.

Amino acids were shown by Miller (1953) to result if a mixture of methane, ammonia, hydrogen, and water vapor were subjected to electrical sparking. Harada and Fox (1964) made amino acids by heating methane, ammonia, and water together, and Fox and Windsor (1970) also made amino acids by heating formaldehyde and ammonia together.

Oro (1960) made the purine adenine by heating ammonium cyanide, and Fox and Harada (1961) made the pyrimidine uracil by heating malic acid and urea in the presence of polyphosphoric acid.

Hargreaves, Mulvhil and Deamer (1977) made phospholipids by heating fatty acids, glycerol, and phosphate together.

The above is not to say that all problems relating to monomer biogenesis on the Primitive Earth have been solved as yet. For example, we still don't understand why a relatively obscure sugar, ribose, from the Formose Reaction was chosen by Nature to become the one sugar found in nucleic acid backbones. We also still don't understand how nucleotides were formed on the Primitive Earth from ribose, phosphate, and heterocyclic bases.

8. Biopolymers

The polymers from which life on Earth is composed are polysaccharides, proteins, and nucleic acids. Some progress has been made by scientists in figuring out how at least the beginnings of such polymers were possibly made on the Primitive Earth from prebiotic monomers.

For example, Fox and Bahn (1990) heated glucose and fructose in the presence of glutamic acid to make polyglucofructose. Schwartz (1962) heated ribose to make polyribose.

Schiff (1897) heated aspartic acid to make thermal polyaspartic acid. Fox and Harada (1958) heated mixtures of amino acids to make thermal proteins, which are branched proteins (Pappelis and Bahn, 2004).

Matsuno and his colleagues (Ogasawara et al., 2000) heated mononucleotides under simulated hydrothermal vent conditions to get oligonucleotides. Schwartz (1965) heated the mononucleotide CMP in the presence of polyphosphoric acid to get oligo-C.

The problem of how the first *informational* biopolymers on the Primitive Earth were formed is a massive unsolved scientific problem because we do not yet know how it was that information for conducting molecular operations necessary for life got programmed into the sequences of such biopolymers. One can imagine that short oligomers produced by thermal dehydration condensation (Bahn and Pappelis, 2004) preceded the biopolymers and that some oligomers became more predominant than other oligomers. Such "selected" oligomers were lengthened, and the sequences of the biopolymers that now inhabit the biosphere were the final result. The origin of life on Earth started out as a gigantic combinatorial experiment in the organic chemistry of monomers and became, in time, a gigantic combinatorial experiment in the interactions of ever lengthening sequences of such monomers contained within the resulting biopolymers.

9. The RNA World

The RNA World is a model of early biological evolution based on the fact that RNA, like all nucleic acids, can act as templates for complementary duplication of themselves and also the fact that some RNAs act as chemical catalysts (Shapiro, 1984). In the modern biological world, proteins (Petsko and Ringe, 2004; Richards and Wyckoff, 1973; Tanford and Reynolds, 2001) are the molecules which are the major chemical catalysts and nucleic acids in the form of DNA code for the assembly of proteins. However, the assembly of proteins from the DNA code takes place in ribosomes where the assembly is catalyzed by RNA at the active site of the ribosomes. The structure of the ribosome (see *Cold Spring Harbor Symposia on Quantitative Biology*, Vol. LXVI, 2001) is monstrously complex and therefore could not have been present at the origin of life. What RNA World researchers are trying to find or to synthesize is an RNA molecule that is catalytic for its own self assembly from mononucleotides while being simultaneously catalytic for peptide bond assembly. So far, such a RNA molecule has not been found. One of the most interesting capabilities of catalytic RNA is that RNA sequences can undergo “test tube evolution” which means that various chemical changes in the aqueous environments of RNA molecules in a test tube can induce inheritable changes in such RNA sequences if successive sequences are repeatedly made using RNA polymerase, which is a protein.

The RNA World cannot be a model of the very *origin* of life because nucleic acids are too complicated to have arisen spontaneously from prebiotic biomonomers. Such a process has never been observed to occur in the laboratory under prebiotically plausible conditions. Indeed, even the assembly of single monomer nucleotides has still eluded a prebiotically plausible modeling in the laboratory. Most scientists agree that at the beginning of life, nucleic acids were preceded by some other system of chemical evolution. The question is what? That is the enticing question that draws origin-of life scientists onward. Presumably, at some point of time, primordial catalytic RNA molecules, when they did form, were enclosed within protocell structures where a kind of “chemical alliance” took place between RNAs and prebiotic proteins thus leading to the genetic code, ribosomes, and coded protein synthesis (Shapiro, 1984).

10. Protocells

Life on the planet Earth comes in the form of *cells*. There are three main types of models of protocells that have been developed by origin-of-life scientists: coacervate droplets, lipid vesicles, and thermal protein microspheres. Coacervate droplets are various combinations of biopolymers that separate out of aqueous solutions and which are about the size of biological cells. For example, coacervate droplets made from the protein gelatin and the polysaccharide gum arabic were extensively studied by Oparin (1938) and his colleagues. Lipid vesicles are usually double bilayer hollow spheres, about the size of biological cells, which are made of phospholipids (Segré, Ben-Eli, Deamer and Lancet, 2001). Thermal protein microspheres are partially hollow spheres, about the size of biological cells, that are made from branched proteins or polypeptides that were made by heat polymerizing a suitable amino acid (Bahn and Pappelis, 2005; Bahn, Pappelis and Bozzola, 2005) or mixtures of amino acids (Fox, Harada and Kendrick, 1959). Presumably, such droplets, vesicles, or spheres when forming on the Primitive Earth would have encapsulated other ingredients necessary for life, or would have subsequently been filled with such ingredients as just the right combinations of primordial proteins, nucleic acids, and metabolites to get processes of chemical evolution going, which subsequently led to processes of biological evolution. The job of origin-of-life scientists is to convincingly reconstruct the chemical and geological scenarios which led to the biosphere that we observe today all about us. To say that it is impossible to reconstruct such scenarios of biogenesis because biological structures are "irreducibly complex" (Behe, 1998) is to give up on science.

11. A Very Brief History of Life on Earth

The Earth was born approximately 4.5 billion years ago as a molten sphere of accreting planetesimals. For the first 500 million years of its history, the earth was too hot for informational biopolymers and cells containing them to form. However, after cooling down for half a billion years, the Earth was cool enough to allow the formation of life. Scientists find evidence in the form of suggestive fossil biosignatures that indicate that life probably formed quite quickly geologically speaking, possibly within 100 million years or less (Lazcano and Miller, 1994) once the Primitive Earth had cooled down enough. In any case, during the period when life did form, the Earth was considerably hotter than it is now, a fact which is backed up by accumulated evidence that the Last Common Universal Ancestor (LUCA) of all life on Earth was a hyperthermophilic Archaean microbe. It has been firmly established by detailed study of ancient microbe fossils that cyanobacteria-like organisms were living on the Primitive Earth at least 3.5 billion years ago (Schopf, 1999).

The strange thing about life on Earth (Ward and Brownlee, 2004) is that while simple single cell microbes without nuclei originated in a relatively short period of time after 0.5 billion years of Earth cooling, it took another 1.5 billion years of evolution on the Earth for more complex cells with nuclei, such as eukaryotic cells, to form (Margulis and Dolan, 2002), and it took a further 2 billion years after that for such eukaryotic cells to evolve to form multicellular lifeforms. Before the available micropaleontology evidence made these facts clear to us, one might have supposed that it would have taken much time for life to form out of organic chemicals but little time for life to evolve to more complex cells and then to multicellularity.

12. Panspermia

Panspermia is the idea popularized by Arrhenius (one hesitates to call it a theory) that life did not originate here on Earth, but rather that life originated elsewhere in outer space, perhaps on another planet in this Solar System or perhaps on another planet in another solar system other than that of our Sun, and that such life was subsequently transferred to the Primitive Earth where it evolved further. Is this possible? Yes. But is it probable? No. The only life that we know about so far is the life that we find on this Planet Earth. No other planet in our Solar System and that we are likely to learn about in any extrastellar solar system is more conducive for the formation of life than Earth. Instead of bringing scientific clarity to solving the origin-of-life problem, the idea of panspermia simply removes the problem to another location where it still remains to be solved but in new circumstances about which we know little and about which we can only speculate. The idea of panspermia may be good science fiction, but it is not good science, which should always focus its attention, according to Occam's Razor, on the simplest and most probable explanations of physical phenomena. What the idea of panspermia does is to tend to cut off the possibility of our being able to *do* science leading toward a convincing solution of the origin-of-life problem here on Earth. This does not mean that we should not be looking for life elsewhere in outer space. We most certainly should. Only, it should be kept in mind that finding life elsewhere may help understand, but will not solve, the origin-of-life problem here on Earth.

13. Astrobiology

When we look out at the Universe that we inhabit with our most powerful telescopes and particularly with the fabulous Hubble Space Telescope, we see billions of other galaxies, of which our own Milky Way is just a typical galaxy. Each of these galaxies, like our own Milky Way, contains billions of stars. We have now detected hundreds of extrasolar planets, and so it is becoming increasingly clear to us that most stars probably have planets and indeed solar systems orbiting around them, much as our own Sun has a number of planets orbiting around it to form our own Solar System. As far as we know, there is nothing special about the location of our Solar System within the Milky Way. Our Solar system is located approximately one quarter of the way in from the outer rim of the Milky Way to the center of the Milky Way. Thus, in a sense, we inhabit a typical rocky planet orbiting the "habitable zone" around a typical star, among billions of other stars and planets in a typical galaxy among billions of other galaxies. It would be monumentally presumptuous and naïve to suppose that we are the only life form and even the only supposedly "intelligent" life form to inhabit the Universe.

Astrobiology is the scientific search that attempts to find and characterize other lifeforms that may exist in our Solar System and other solar systems. So far, we have not found any other extraterrestrial lifeforms. However, we really have only just begun to look. It appears, that as far as we are looking for life as we know it (Ward, 2006), Mercury and Venus are too hot for such life. Mars, which looks disturbingly like Arizona without the cacti and the sagebrush, appears to have no life on its surface, but might harbor life in subsurface environments. The outer planet gas giants appear to have chemical properties too bizarre to harbor life as we know it, but it is possible that some of the gas giant moons such as Europa and Titan might have microbial life forms on their surface or subsurface environments.

If and when we do find convincing scientific evidence of life in other parts of our Solar System, or our galaxy, or other galaxies, it will one of the most momentous discoveries ever made by the human species. We would learn much about the wonders and mysteries of chemistry and biology by comparing other life with Earth life. If we do find life on other planets, then we will want to think long and deeply about how such other life originated in comparison with how we may surmise that life originated on the planet Earth. Thus, the origin-of-life problem on Earth is really just a part of a larger origin-of-life problem in the Universe as a whole. Few scientific vistas lead us onward as much as those of astrobiology which beckons us to eventually travel to the stars that we see in outer space.

14. Nature and Evolution

Evolution by natural selection is *the* central fact of biology. Biology and the plethora of species that inhabit the biosphere hardly make any sense at all without the organizing principle of evolution. We and all the plants and animals that live with us on this beautiful Planet Earth are descended from a single progenitor that originated on the Primitive Earth some four billion years ago. We know this because every lifeform on the Earth shares the same genetic code that life uses to make more of itself. Evolution, which is a central truth about Nature, was summed up in the poetry of science by its great discoverer Charles Darwin (2004) in the last paragraph of his immortal book *The Origin of Species* (first published in 1859) as follows:

> "It is interesting to contemplate an entangled bank, clothed with many plants of many kinds, with birds singing on the bushes, with various insects flitting about, and with worms crawling through the damp earth, and to reflect that these elaborately constructed forms, so different from each other, and dependent on each other in so complex a manner, have all been produced by laws acting around us. These laws, taken in the largest sense, being Growth with Reproduction; Inheritance which is implied by reproduction; Variability from the indirect and direct action of the external conditions of life, and from use and disuse; a Ratio of Increase so high as to lead to a Struggle for Life, and as a consequence to Natural Selection, entailing Divergence of Character and the Extinction of less-improved forms. Thus, from the war of nature, from famine and death, the most exalted object which we are capable of

> conceiving, namely, the production of higher animals, directly follows. There is a grandeur in this view of life, with its several powers, having been originally breathed into a few forms or into one; and that, whilst this planet has gone cycling on according to the fixed law of gravity, from so simple a beginning endless forms most beautiful and most wonderful have been, and are being, evolved."

If there is something in this Universe which is truly divine, perhaps it is Nature itself, and the matter, atoms, and compounds of which the world is made, and which, of their own accord, combined and reacted together so long ago, to form things, that would go on to form other things, that would eventually come to form ourselves, who now wonder in amazement and delight that such developments are even possible, and who now strive to understand the mechanics of this miracle.

15. Summary

The origin-of-life problem can only be solved by faith in science. At the present time, it is still largely an unsolved problem. Science is a matter of proof. The laws of physics and chemistry do not change over space and time. There is no single definition of life that satisfies everyone. However, life appears to be an emergent property of groups of molecules that comes in packages called cells. In another sense, life is a runaway reaction of organic chemicals. We have made substantial progress on the origin of biomonomers and some progress on the formation of biopolymers. The origin of informational biopolymers is still a big problem. The RNA World is an attempt to model the origins of ribosomes, but the RNA World could not have arisen spontaneously and must have been preceded by something else at the very origin of life. Coacervate, lipid, and thermal protein models of protocells have been developed. Simple life developed relatively early and complex life developed relatively late on the Earth. Panspermia does not solve the origin-of-life problem but merely displaces it. The origin-of-life problem is part of astrobiology which is the search for life in the Universe in general. If anything in this Universe is divine, perhaps it is Nature itself.

16. Acknowledgments

I wish to thank Hunt Bonan, Arnie Brown, Thad Heckman, Aristotel Pappelis, and Steve Pravdo for helpful discussion.

References

Bahn, P.R. and Pappelis, A. (2005) Thermal polyaspartic acid microspheres, P-50, ISSOL 2005 Conference, Beijing, China.

Bahn, P.R., Pappelis, A. and Bozzola, J. (2005) Protocell-like microspheres from thermal polyaspartic acid, *National Workshop on Astrobiology, Capri, Italy, Oct. 26-28, 2005.*

Behe, M.J. (1998) *Darwin's Black Box: The Biochemical Challenge to Evolution*, Simon & Schuster, New York.

Butlerov, A. (1861) *Justus Liebig's Annalen der Chemie* **120**, 295.

Calvin, M. (1969) *Chemical Evolution: Molecular Evolution of Living Systems on the Earth and Elsewhere*, Oxford University Press, Oxford, England.

Cold Spring Harbor Symposia on Quantitative Biology Vol. LXVI (2001) *The Ribosome*, Cold Spring Harbor Laboratory Press, Cold Spring Harbor, New York.

Darwin, C. (1859). *On the Origin of Species by Means of Natural Selection, Or The Preservation of Favoured Races In The Struggle For Life*. London, John Murray.

Davies, P. (1999) *The Fifth Miracle: The Search for the Origin and Meaning of Life*, Simon & Schuster, New York.

Einstein, A. (1920) *Relativity: The Special and General Theory*, Henry Holt and Company, New York.

Fox, S.W. and Bahn, P.R. (1990) Process for preparing thermal carbohydrates, U.S. Patent No. 4,975,534.

Fox, S.W. and Dose, K. (1977) *Molecular Evolution and the Origin of Life*, Marcel Dekker, New York.

Fox, S.W. and Harada, K. (1958) Thermal copolymerization of amino acids to a product resembling protein, *Science* **128**, 1214.

Fox, S.W. and Harada, K. (1961) Synthesis of uracil under conditions of a thermal model of prebiological chemistry, *Science* **133**, 1923-1924.

Fox, S.W., Harada, K. and Kendrick, J. (1959) Production of spherules from synthetic proteinoid and hot water, *Science* **129**, 1221-1223.

Fox, S.W. and Windsor, C.R. (1970) Synthesis of amino acids by the heating of formaldehyde and ammonia, *Science* **170**, 984-986.

Gesteland, R.F., Cech, T.R. and Atkins, J.F. (eds.) (1999) *The RNA World, Second Edition*, Cold Spring Harbor Laboratory Press, Cold Spring Harbor, New York.

Harada, K. and Fox, S.W. (1964) Thermal synthesis of natural amino acids from a postulated primitive terrestrial atmosphere, *Nature* **201**, 335-336.

Hargreaves, W.R., Mulvihill, S.J. and Deamer, D.W. (1977) Synthesis of phospholipids and membranes in Prebiotic conditions, *Nature* **266**, 78-80.

Hazen, R.M. (2005) *Genesis: The Scientific Quest for Life's Origin*, Joseph Henry Press, Washington, D.C.

Hoyle, F. (1975) *Astronomy and Cosmology: A Modern Course*, W.H. Freeman and Company, San Francisco.

Lazcano, A. and Miller, S.L. (1994) How long did it take for life to begin and evolve to cyanobacteria? *J. Molec. Evol.* **39**, 549-554.

Margulis, L. and Dolan, M.F. (2002) *Early Life: Evolution on the PreCambrian Earth,* Jones and Barlett Publishers, Boston.

Matsuno, K. (1989) *Protobiology: Physical Basis of Biology*, CRC Press, Boca Raton, Florida.

Miller, S.L. (1953) A production of amino acids under possible primitive earth conditions, *Science* **117**, 528-529.

Morowitz, H.J. (1992) *Beginnings of Cellular Life: Metabolism Recapitulates Biogenesis*, Yale University Press, New Haven, Connecticut.

Newton, I. (1687) *Philosophiae Naturalis Principia Matematica.*

Ogasawara, H., Yoshida, A., Imai, E., Honda, H., Hatori, K. and Matsuno, K. (2000) Synthesizing oligomers from monomeric nucleotides in simulated hydrothermal environments, *Orig. Life & Evol. Biosphere* **30**, 519-526.

Oparin, A.I. (1938) *Origin of Life*, The Macmillan Company, London.

Oro, J. (1960) Synthesis of adenine from ammonium cyanide, *Bioch. Biophys. Res. Com.* **2**, 407-412.

Pappelis, A. and Bahn, P.R. (2004) Determinism and the proteinoid theory, **In**: J. Seckbach, J. Chela-Flores. T. Qwen and F. Raulin (eds.) *Life in the Universe: From the Miller Experiment to the Search for Life on Other Worlds*, Kluwer Academic Publishers, Dordrect, The Netherlands.

Petsko, G.A. and Ringe, D. (2004) *Protein Structure and Function*, Sinauer Associates, Inc., Publishers, Sunderland, Massachussetts.

Richards, F.M. and Wyckoff, H.W. (1973) Ribonuclease-S, **In**: D.C. Phillips and F.M. Richards (eds.) *Atlas of Molecular Structures in Biology*, Clarrendon Press, Oxford, England.

Rizzoti, M. (ed.) (1996) *Defining Life: The Central Problem of Theoretical Biology*, University of Padova Press, Padova, Italy.

Schiff, H. (1897) Ueber polyaspartsaeuren, *Berichte der Deutschen Chemiscthen Gesellschraft* **XXX** 16 (S 107) A.

Schopf, J.W. (1999) *Cradle of Life: The Discovery of Earth's Earliest Fossils*, Princeton University Press, Princeton, New Jersey.

Schulze-Makuch, D. and Irwin, L.N. (2006) *Life in the Universe: Expectations and Constraints*, Springer, Berlin.

Schwartz, A.W. (1962) *The Thermal Polymerization of Ribose*, M.S. Dissertation, Florida State Univ.

Schwartz, A.W. (1965) *Condensation of Cytidine-2'(3')-Phosphate in the Presence of Polyphosphoric Acid*, Ph.D. Dissertation, Florida State Univ.

Seckbach, J., Chela-Flores, J., Owen, T. and Raulin, F. (eds.) (2004) *Life in the Universe: From the Miller Experiment to the Search for Life on other Worlds*, Kluwer Academic Publishers, Dordrect, The Netherlands.

Segré, D., Ben-Eli, D., Deamer, D. and Lancet, D. (2001) The lipid world, *Orig. Life & Evol. Biosph.* **31**, 19-145.

Shapiro, R. (1984) The improbability of prebiotic nucleic-acid synthesis, *Orig. Life & Evol. Biosph.* **14**, 565-570.

Singh, S. (2005) *Big Bang: The Origin of the Universe*, HarperCollins Publishers,

Tanford, C. and Reynolds, J. (2001) *Nature's Robots: A History of Proteins*, Oxford University Press.

Ward, P. (2005) *Life as We Do Not Know It: The NASA Search for (and Synthesis of) Alien Life*, Viking, New York.

Ward, P.D. and Brownlee, D. (2000) *Rare Earth: Why Complex Life is Uncommon in the Universe*, Springer, New York.

Watson, J.D., Baker, T.A., Bell, S.P., Gann, A., Levine, M., Losick, R. (2004) *Molecular Biology of the Gene Fifth Edition*, Cold Spring Harbor Laboratory Press, Cold Spring Harbor, New York.

Weinberg, S. (1994) *The First Three Minutes: A Modern View of the Origin of the Universe*, Perseus Publishing.

Wintner, C.E. (2002) *Organic Chemistry Reference Lectures*, http://www2.haverford.edu/wintnerorganic chem/.

Wohler, F. (1828) Ueber kunstliche Bildung des Hamstoffs, *Annalen der Physik und Chemie* **12**, 253.

Dialogue

Taner Edis

I disagree with practically nothing of substance in Dr. Bahn's chapter. I am a physicist, not a chemist, and certainly have no expertise on any chemistry relevant to research on the origins of life. Therefore my agreement possibly means as little as any potential disagreement. Nevertheless it is worth pointing out that, especially for physicists such as myself who have a prejudice to think about chemistry as a rather narrow and unambititious (even boring) area of natural science, the chemistry of the origin of life is a clear counterexample. This is chemistry to get excited about, even while being sobered by the magnitude of the problem. Biologists have fossils and other direct historical evidence. Chemists have to operate very much in the dark about the earliest stages of life, which makes their task even more heroic.

Nevertheless, I think we know enough, and have enough prospects for further progress, to be confident that the origins of life is an unsolved problem for science rather than an unsolvable problem that calls for exploring faith-based alternatives. That, perhaps, is my only objection to Bahn — he is careless in referring to "faith in science." Faith can mean trust, and in that sense, yes, we have good reason to trust our scientific efforts. But I would insist that this is different from faith in the sense of a blind commitment.

To further that trust, let me suggest that Bahn's perspective could be augmented by taking a more functionalist approach to life as well. In other words, while chemical research remains indispensible, I think that we can learn a lot from approaches that downplay details of the chemistry and focus on abstract functional characteristics of life. Some of this is directly related to origin of life research, such as replicator theory and work on the far-from-equilibrium thermodynamics of complex systems. Other relevant work takes the high-level abstraction even further, such as in "artifical life" research. No single approach will crack what is a very tough problem. But what we learn through different approaches can help all aspects of our overall effort. And all of this bolsters our confidence that the prospects for naturalistic explanations of the origins of life are very good.

Let me make an analogy with a similarly tough problem the modern sciences of the mind face. Neither an exclusively neuroscientific approach nor the purely functionalist approach of artificial intelligence should be counted upon to drive scientific progress on its own. In the origin of life as well, chemists legitimately lead the way, but I think it is worth emphasizing that physicists, biologists, and even computer scientists have useful insights to contribute.

Peter Bahn

Dr. Edis, I agree very much with you that the origin-of-life problem is largely a chemical problem, but I agree most emphatically that physicists, biologists, geologists, and computer scientists have very much to contribute to the origin-of-life problem. The origin-of-life problem is very, very much an interdisciplinary problem involving all of the sciences.

Florence Raulin Cerceau

Dear Dr. Bahn, in section 2 entitled *What is Science*, you wrote in the first paragraph that "science, in a way, is a religion (…) with miraculous powers (…) accessible to everyone." But in the third one, you added that "Science (…) is a matter of proof: 'If you can't prove it, it isn't science'."

In my opinion, religion is something in which we believe without the help of any sort of proof, whereas science is, as you mentioned, a matter of proof. Science describes and tries to explain the world around us — that could be called "nature" — with the help of hypotheses and theories based on facts. Hypotheses are tested *via* experiments and/or observations, and can be either accepted or refuted. It seems to me that religion is of another nature. Faith doesn't challenge God's power and believing in God doesn't need experiments or even observations.

In that case, could science be somehow like religion?

So I didn't quite grasp your position, in this paragraph, about the nature of science: did you mean that science, because it has sometimes powers that could be considered as "miraculous", could be assimilated to a sort of religion, at least by some people? (For instance, some can believe in the power of antibiotics whereas, in fact, the efficiency of antibiotics is scientifically proved).

Besides this, you seem to attribute to Nature (with capital N) the status of deity. Do you mean that God's power is transferred to Nature?

You seem indeed to adopt this opinion in section 14 *Nature and Evolution*, in which you suggested "If there is something in this Universe which is truly divine, perhaps it is Nature itself (...)", and you emphasize this point in your last sentence "If anything in this Universe is divine, perhaps it is Nature itself". It is perhaps dangerous to switch this sort of "supernatural" power from God to Nature.

Peter Bahn
I think that many people believe in religions because the founders of the religions essentially convinced such people that they, the founders, had communicated with God. Many people who live in the times after the founders have faith in such religions but such faith is largely based on miraculous occurrences such as the delivery of the Jews out of Egypt by Moses, the resurrection of Jesus from the dead, and the recitation of the Koran to Mohammed by the angel Gabriel.

I think that science, which is the source of almost miraculous power, is slowly, bit by bit, replacing traditional religions as that aspect of the world in which people have most of their faith, as people become more and more secularly minded. Thousands of years from now, I believe that few people will believe in traditional religions and that most people will believe in science. I think that this will especially be the case if we discover intelligent life elsewhere in the Universe. I don't think that traditional Earth-based religions mesh well with the nature of our scientific quest of astrobiology.

As a conservative Jew, I believe in Y-HW-H, the God of Abraham, Isaac, and Jacob. However, as a scientist, I have trouble believing in a personal God, and I am more inclined to believe in the impersonal God of Spinoza. Spinoza defined God in his book *Ethics* as follows: "God is a substance with an infinite number of attributes". I believe that the word "substance" means the material nature immanent in the Universe.

This kind of switch of the power of God to Nature did get Spinoza in trouble since he was ex-communicated from the Jewish congregation as a result. I am trying to stay out of danger, but as a Jewish scientist, I have a bit of a split personality. I believe in my religious heritage, but I wrote my paper as a scientist because this book is being published by a scientific press.

Florence Raulin Cerceau
In section 3 entitled *Faith in Science*, I agree with your remarks stating that, "science requires a good deal of faith in the processes by which science is carried on". Reliability in instrumentation, in my opinion, is necessary to make experimental measures credible. Instruments are intermediaries between "reality" and experimental facts. Then I think we firstly need to have faith in instrumentation, secondly in the universality of the laws of nature.

However, perhaps we have to distinguish what could be called "faith" ("foi" in French) from what could be called "confidence" ("confiance" in French). In science, wouldn't it be above all a matter of confidence rather than faith?

You speak also about the "laws of science" which do not change over very large scales of time and space. I agree with you if you speak about the "laws of nature". The laws of science can only be founded by theories that are not unchanging but on the contrary in constant evolution: if the laws of nature are immutable, the theories used in science are changing. Science is dynamic.

Paradoxically, we are in practice lead to accept a certain durability concerning scientific theories. What I mean is if we want to make progress in a specific research field — such as the Origins of Life — we have to conduct experiments within a framework of research expected by theory. So we have also, in a way, to have confidence in the longevity of scientific theories.

Anyway, as you said, the laws of physics and chemistry are the same all over the universe and we must have confidence in that point if we want to solve (one day) the problem of the origin of life and if we want to be in a position to identify other forms of life, elsewhere in the universe.

Peter Bahn
By the "laws of Science" I mean the "laws of physics and chemistry". I think that we must have not only confidence but also faith as scientists that these laws are the same throughout the Universe and do not change. I agree with you that theories of science do change, and that in this sense, science is a dynamic enterprise. However, science is an enterprise for approaching closer and closer to unchanging objective truths of nature.

Florence Raulin Cerceau
In section 5 entitled *A Runaway Chemical Reaction*, as you speak about life as a runaway chemical reaction, you wrote that "at least one purpose of life appears to be to make and spread more life around as much as possible".

If we parallel the problem of chemical evolution with the one of biological evolution, in that case and in my opinion, no purpose is to be taken into account. Biological evolution, which leads to the diversity of life by mainly two "blind" mechanisms — Darwinian natural selection and mutations — has no purpose: selection first (of organisms), and then diversity and spread of life. The same thing might have occurred with chemical evolution (?): selection first (of molecules), and then diversity and spread of life precursors (?).

Chemical evolution, as well as biological evolution, seems to me to be closely connected with contingency (probably through mechanisms of Darwinian selection). And contingency doesn't match purpose. Of course, it is my own opinion and I would be glad to get yours.

Peter Bahn
I believe that biological organisms, by their very nature, manifest in themselves an immediate purpose of reproducing themselves in geometric proportions, over time. This is what I meant by the sentence of mine that you quoted. I realize that mutation and natural selection are blind processes, but they are blind processes that act upon a runaway proliferation of biological organisms in various populations.

Florence Raulin Cerceau
Many thanks for your very rich paper and your detailed answers which clarify my questioning about the delicate parallel often made between science and religion when we talk about the origin and nature of life and the possible place of faith among these problems. As you mentioned in this discussion, it is probable that the discovery of another "intelligent" life in the universe would change completely the balance of power between science and religion. I think there is here an important point to be pursued.

Biography

Harun Yahya is a pen name used by Mr. **Adnan Oktar**. He is a prominent Turkish intellectual, who studied at Mimar Sinan University's Academy of Fine Arts and at the Philosophy Department of Istanbul University. He has written more than 200 books which are published in 57 languages all over the world. Mr. Yahya is completely devoted to moral values and dedicated to communicating the sacred values he cherishes to other people. He started his intellectual struggle in 1979 during his education at Mimar Sinan University. During his university years, he carried out detailed research into the prevalent materialistic philosophies and ideologies around him, to the extent of becoming even more knowledgeable about them than their advocates. He has written various books against the theory of evolution and dedicated intellectual effort opposing the lines of evolution and materialism that have grown out to be a worldwide phenomenon.

15
Did Life on Earth Begin Suddenly and in Complex Forms?

HARUN YAHYA

Academy of Fine Arts at Mimar Sinan University of Istanbul and,
Department of Philosophy at Istanbul University
seda_aral@yahoo.com

1. Introduction

The theory of evolution claims that all the living species on Earth descended, by means of a series of minute changes, from a common ancestor. To state the theory another way, living species are not separated from one another by absolute differences but exhibit a continuum. However, actual observations in nature have pointed to the lack of such a continuum. What we see in the living world are different categories of organisms, separated by vast and distinct differences. Robert Carroll, an authority on vertebrate paleontology and an evolutionist, admits this in his book *Patterns and Processes of Vertebrate Evolution*:

> "Although an almost incomprehensible number of species inhabit Earth today, they do not form a continuous spectrum of barely distinguishable intermediates. Instead, nearly all species can be recognized as belonging to a relatively limited number of clearly distinct major groups…" (Carroll, 1997, p. 9).

Evolution is merely a process alleged to have taken place in the past, and fossil discoveries are the only scientific source about the history of life. In order for the fossil records to be the judge on this subject, we need to compare what the theory of evolution predicts and what the actual fossil record displays.

According to the theory, all living things have descended from various "ancestral" forms. A species that existed before gradually turned into another, and every present species emerged in this way. According to the theory, this transition took place slowly and gradually over hundreds of millions of years. The theory therefore demands the existence of innumerable quantities of "intermediate forms," emerged and lived during long periods of the suggested transitional processes, e.g. half-fish half-amphibian creatures maintaining some fish-like characteristics but also with certain acquired amphibious features or reptile-birds with both reptilian and avian features. Since these creatures were in a supposed process of transition, they must have been deformed, deficient and flawed.

Logic also demands that abundant traces of any such species, in evolutionist terminology, "intermediate forms," are be found in the fossil record, because the number of intermediate forms have to be much greater than the number of animal species known today. The geologic strata should be full of the remains of fossilized intermediate forms. This logic is articulated by no other than Charles Darwin himself. He wrote in his book, *The Origin of Species*:

> "If my theory be true, numberless intermediate varieties, linking most closely all the species of the same group together, must assuredly have existed... Consequently evidence of their former existence could be found only amongst fossil remains" (Darwin, 1998, p. 146).

Yet Darwin, the author of these lines, was aware that no intermediate forms had yet been found, and regarded this as a major dilemma facing his theory. In the chapter "Difficulties on Theory" of his book he wrote:

> "... Why, if species have descended from other species by insensibly fine gradations, do we not everywhere see innumerable transitional forms? Why is not all nature in confusion instead of the species being, as we see them, well defined? ... But, as by this theory innumerable transitional forms must have existed, why do we not find them embedded in countless numbers in the crust of the earth? ... Why then is not every geological formation and every stratum full of such intermediate links? Geology assuredly does not reveal any such finely

> graduated organic chain; and this, perhaps, is the most obvious and gravest objection which can be urged against my theory" (Darwin, 1998, pp. 140, 141, 227).

In the face of this fundamental dilemma, the explanation Darwin could offer was that the fossil record of his time was insufficient. He claimed that later, when the fossil record had been examined in detail, the missing intermediate forms would definitely be found. Yet despite 150 years of intensive efforts, Darwin's dream never materialized. Edmund R. Leach, author of the book *Rethinking Anthropology*, wrote this in his article in *Nature*:

> Missing links in the sequence of fossil evidence were a worry to Darwin. He felt sure they would eventually turn up, but they are still missing and seem likely to remain so (Leach, 1981, p. 19).

Alfred S. Romer, one of the most eminent paleontologists of his time, said this on the subject:

> "'Links' are missing just where we most fervently desire them [to point to a transition between species] and it is all too probable that many 'links' will continue to be missing" (Romer, 1963, p. 114).

David B. Kitts, professor of geology and the history of science at the University of Oklahoma admits the absence of the intermediate forms required by the theory of evolution:

> "Evolution requires intermediate forms between species and paleontology does not provide them" (Kitts, 1974, p. 467).

Some 250,000 fossil species have been collected to date, and there is absolutely no trace of intermediate forms in any of them. To the contrary, all fossils unearthed belong to flawless creatures. The picture that emerges from the fossil record is completely compatible with creation. The record reveals that living things appeared suddenly and lived for long periods of time without undergoing any change at all.

2. The Fact Revealed by the Fossil Record

What is the origin of the "evolution-paleontology" relationship that has been installed in society's subconscious? Why is it that when the fossil record is mentioned, most people wrongly assume that the record is on the side of Darwin's theory? The answer is set out in an article in the *Science* magazine (emphasis added):

> "A large number of well-trained scientists outside of evolutionary biology and paleontology **have unfortunately gotten the idea that the fossil record is far more Darwinian than it is**. This probably comes from the oversimplification inevitable in secondary sources: low-level textbooks, semipopular articles, and so on. Also, there is probably some wishful thinking involved. In the years after Darwin, his advocates hoped to find predictable progressions. **In general these have not been found yet the optimism has died hard, and some pure fantasy has crept into textbooks**" (Raup, 1981, p. 289).

Following comments of paleontologist Niles Eldredge and anthropologist Ian Tattershall are noteworthy in this context (emphasis added):

> "That individual kinds of fossils remain recognizably the same throughout the length of their occurrence in the fossil record had been known to paleontologists long before Darwin published his Origin. Darwin himself, ... prophesied that future generations of paleontologists would fill in these gaps by diligent search ... One hundred and twenty years of paleontological research later, **it has become abundantly clear that the fossil record will not confirm this part of Darwin's predictions. Nor is the problem a miserably poor record. The fossil record simply shows that this prediction is wrong**. The observation that species are amazingly conservative and static entities throughout long periods of time has all the qualities of the emperor's new clothes: everyone knew it but preferred to ignore it. Paleontologists, faced with a recalcitrant record obstinately refusing to yield Darwin's predicted pattern, simply looked the other way" (Eldredge and Tattersall, 1982, pp. 45-46).

The American paleontologist S. M. Stanley describes how this fact, revealed by the fossil record, is ignored by the Darwinist dogma that dominates the scientific world, and how others are also encouraged to ignore it:

> "The known fossil record is not, and never has been, in accord with gradualism. What is remarkable is that, through a variety of historical circumstances, even the history of opposition has been obscured.... 'The majority of paleontologists felt their evidence simply contradicted Darwin's stress on minute, slow, and cumulative changes leading to species transformation.'... their story has been suppressed" (Stanley, 1981, p. 71)

3. "Stasis": Stability in the Fossil Record

When we investigate natural history, we find not living things "evolving into different anatomical structures," but ones that have remained unchanged, even over the course of hundreds of millions of years. This lack of change is referred to by scientists as "stasis." Living fossils and organisms that have not survived down to the present day, but which have left their fossils behind in various strata of the Earth's history are concrete proof of stasis in the fossil record. And this stasis shows that no gradual process of evolution ever occurred. In an article in the magazine *Natural History*, Stephen Jay Gould describes this inconsistency between the fossil record and the theory of evolution:

> "The history of most fossil species includes two features particularly inconsistent with gradualism: (1) Stasis. Most species exhibit no directional change during their tenure on earth. They appear in the fossil record looking much the same as when they disappear; morphological change is usually limited and directionless. (2) Sudden appearance. In any local area, a species does not arise gradually by the steady transformation of its ancestors; it appears all at once and 'fully formed'" (Gould, 1977, p. 14).

If a living thing survives in a flawless form down to the present day with all the features it displayed millions of years ago and having undergone no change whatsoever, then this evidence is powerful enough to entirely dismiss the gradual evolution model anticipated by Darwin. Moreover, far from there being just one example to demonstrate this, there are in fact millions. Countless organisms exhibit no differences from their original states, which first appeared millions or even hundreds of millions of years ago. As openly stated by Niles Eldredge, this state of affairs is causing paleontologists to avoid the idea of evolution, which is still supported today:

> "No wonder paleontologists shied away from evolution for so long. It seems never to happen. Assiduous collecting up cliff faces yields zigzags, minor oscillations, and the very occasional slight accumulation of change over millions of years, at a rate too slow to really account for all the prodigious change that has occurred in evolutionary history" (Eldredge, 1996, p. 95).

The stasis in the fossil record really does represent the greatest difficulty facing the proponents of evolution. Faced with this problem, evolutionists came up with the idea of "punctuated equilibrium." The authors of the "punctuated equilibrium" model — Niles Eldredge and Stephen J. Gould — admitted that the stasis in the fossil record presented a "problem." But since they considered it impossible to abandon the idea of evolution, they suggested that living things came into being not through small changes, but by sudden and very large ones.

It is widely known today that the punctuated equilibrium model of evolution has now been comprehensively refuted by microbiology and genetics. In addition to all these scientific impossibilities, the adherents of punctuated equilibrium can't explain why even the smallest traces of changes are never found in the fossil record.

This clearly demonstrates that both the gradual model of evolution that Darwin proposed, and the punctuated equilibrium model put forward to cover up its deficiencies, are not able to account for the stasis in the fossil record, the sudden appearance of living forms, and the lack of transitional ones. Whatever theory may be proposed, all claims that living organisms underwent evolution will end in failure and are scientifically condemned to collapse, because living things did not evolve. God has created all living things in their perfect states, from nothing. Stephen J. Gould admitted this in all clarity at a conference he gave at Hobart and William Smith College:

> "Every paleontologist knows that most species don't change. That's bothersome… brings terrible distress… . They may get a little bigger or bumpier. But they remain the same species and that's not due to

imperfection and gaps but stasis. And yet this remarkable stasis has generally been ignored as no data. If they don't change, it's not evolution so you don't talk about it" (Gould, 1980).

4. The Cambrian Explosion: A Dazzling Variety of Life

Life on Earth exhibits the most amazing variety. From the Poles to the Amazon, from mountain peaks to the ocean deeps, our planet is overflowing with an endless variety of life forms. Many organisms, from bacteria to dolphins, from ants to trees, from sea gulls to horses, have been equipped with extraordinarily sensitive systems and gloriously complex structures. Thanks to specialized systems they are able to survive in such close harmony with their environment. These systems, the details of which have been discovered by biologists, contain features that astonish human beings.

The Cambrian Period is the name given to that geological age when all contemporary categories of multi-cellular species appeared suddenly. So sudden and comprehensive was this emergence that scientists have named it the "Cambrian Explosion." Stephen J. Gould described the phenomenon as "the most remarkable and puzzling event in the history of life" while the evolutionary zoologist Thomas S. Ray writes that the origin of multi-cellular life is an event of comparable significance to the origin of life itself.

The last 25 years have seen a major increase in our knowledge of the Cambrian Period, and the extraordinary nature of the Cambrian explosion has attracted enormous scientific interest. Scientists analyzing the discoveries made by various disciplines have realized that this phenomenon is one that took place even more suddenly, and in an even more unique manner than they had ever imagined. The consolidation of our understanding of the characteristics unique to the Cambrian explosion has resulted in concrete facts that reliably explain the origins of multi-cellular organisms and of life in general.

The first rock strata belonging to this period were discovered in North Wales by the British geologist Adam Sedgwick in 1835. Inspired by the original Latin name for Wales, Cambria, Sedgwick named this period the Cambrian. According to dates proposed by the International Subcommission on Global Stratigraphy in 2002, the Cambrian Period is regarded as beginning 545 million years ago and ending 490 million years ago. It is also sub-divided into three epochs, the Early

(542 to 513 million years ago), the Middle (513 to 501 million years ago) and the Late Cambrian (501 to 490 million years ago).

What makes the period important in terms of natural history is the "Cambrian explosion." This is an explosion which took place in the transition from the Precambrian to the Early Cambrian, and refers to the extremely sudden and worldwide appearance of complex organisms with no forerunners behind them. So great is the difference, in terms of biological variety and complexity, between the Precambrian eon and the Cambrian period that this phenomenon is described as an "explosion," a reference to the sudden coming into existence of life forms. Stephen J. Gould describes the Cambrian explosion as follows:

> "The most famous such burst, the Cambrian explosion, marks the inception of modern multicellular life. Within just a few million years, nearly every major kind of animal anatomy appears in the fossil record for the first time" (Gould, 1989, p. 65).

This period was when some 50 separate phyla, including the 35 phyla existing today, suddenly emerged. This is a most important piece of information, because it means that all the features of living things today, and of even more extinct ones, appeared quite suddenly around 530 million years ago. It was claimed that 14 more small phyla emerged after the Cambrian, but in the light of the features these possessed they were included in the 35 phyla existing today. This means that, contrary to Darwinist expectations, there has been no increase in the number of higher taxa such as phyla between the Cambrian and the present day, and that with some phyla becoming extinct, there has even been a reduction. Therefore, the Cambrian Period is more complex than the present in terms of the fundamental structures that determine phyla.

Snails, trilobites, sponges, worms, jellyfish, starfish, shellfish and sea lilies all with very different characteristics, have left behind traces of large parts of their bodies in the fossil record. So sharp and perfect are these that remains of their internal organs and even circulatory systems are crystal clear, and soft tissues are sufficiently clear as to reveal these creatures' life vital systems. What is remarkable is the presence in the majority of the animals in these strata of virtually unchanged structures and advanced physiological features such as the eyes, gills and circulatory and excretory systems. These fossils in the Cambrian rocks are those of creatures with powerful skeletons, muscles, chambers for storing food and hard organs to pierce shells and foodstuffs. (Gould, 2001, p. 51)

Richard Monastersky, an evolutionist writer at *Science News* magazine reflects on the subject in *Discover* magazine as below:

> "A half-billion years ago the remarkably complex forms of animals we see today suddenly appeared... This moment, right at the start of Earth's Cambrian Period, some 550 million years ago, marks the... explosion that filled the seas with the world's first complex creatures... [T]he large animal phyla of today were present already in the early Cambrian and... they were as distinct from each other then as they are today" (Monastersky, 1993, p. 40).

The British biologist Richard Dawkins, one of the most determined proponents of Darwinism, admits facts about the Cambrian with an unintentionally truthful description (emphasis added):

> "... [T]he Cambrian strata of rocks ... are the oldest in which we find most of the major invertebrate groups. And we find many of them already in an advanced state of evolution, the very first time they appear. **It is as though they were just planted there, without any evolutionary history"** (Dawkins, 1987, p. 229).

The Cambrian Period clearly and definitively refutes all evolutionary approaches to the worldwide variety and complexity of living things, their origins, and the way they emerged "suddenly" and with no forerunners behind them. Evolutionists have invalidated their own claims through the condition they have imposed of "gradual evolutionary development over time." There is no gradual development in the Cambrian, nor enough time for there to have been any. All there is are fully developed living organisms that emerged within a relatively short period in their complex forms. In other words, life forms were created.

5. What the Fossil Record Reveals: Evolution Never Happened

> "Animal life today is phenomenally diverse, more so than any other of life's six kingdoms. Over the past three centuries, scientists have described an estimated 1.5 million species of living animals, but so many more species have not yet been studied — particularly small ones in the tropics — that true totals of 5 or even 50 million have been

> guessed at. Most of these species (mostly arthropods and parasites, 75 percent of all species) live on the land. Far fewer species live in the oceans (about 295,000 have been recognized). Yet it is the ocean that contains more of the main divisions of the animal kingdom, the phyla — almost every one of them…" (Gould, 2001, pp. 51-52).

These words, which appear in Stephen J. Gould's *The Book of Life*, provide a summary of the magnificent variety of life on Earth. As it is their theory promising to explain life, the evolutionists should be giving account for the extraordinary variety as well as the disappeared species. They have to show how a bacterial cell can turn into a whale and give rise to millions of animal species. They have to develop an evolutionary process scenario for each one of these species and point to a proof in the fossil record that this process actually took place.

Yet there is no trace of this process in the fossil record. Millions of different species have left behind not a single trace of an intermediate form. According to evolutionists, a bacterium must have turned into a whale by way of various transitional stages, and this imaginary transition must have taken millions of years. Yet throughout this long period of time there is not a single intermediate form to point to such a transition having taken place. Even bacteria have left behind traces of themselves in rocks, and there are countless complete and fully formed fish fossils, yet there is not the slightest trace of any peculiar intermediate forms between the two.

The answer to this contradiction is clear: Evolution never happened. Living things did not develop through evolution, and species did not evolve by turning into one another. The "evolutionary process," which its proponents have spent the last 150 years propagating, is a fantasy. Not one single claim of the theory has been scientifically proven. The fossil record, that should have been the pillars of the Theory of Evolution, has produced not a single piece of evidence. Not a single "evolutionary mechanism" has been observed in support of the theory. No branch of science supports this theory in any way whatsoever; on the contrary, they keep producing evidence demolishing it. To sum it up, living things did not evolve.

Cambrian life forms, which evolutionists encounter in a state of shock, are by themselves striking proofs of this fact. The 530-million-year-old picture, where some 50 phyla containing the fundamental system of the animal world are displayed, radically undermines the theory of evolution. Evolutionists are still trying to recover from this shock and to gloss over this extraordinary phenomenon. Yet the Cambrian explosion is a fact, one in the face of which evolution has melted away. Evolutionary biologist Douglas Futuyma describes this truth as follows:

> "Organisms either appeared on the earth fully developed or they did not. If they did not, they must have developed from pre-existing species by some process of modification. If they did appear in a fully developed state, they must indeed have been created by some omnipotent intelligence" (Futuyma, 1983, p. 197).

The Cambrian explosion clearly shows that living things appeared on Earth in a perfect and fully developed state, as Futuyma says. Jeffrey S. Levinton, professor of ecology and evolution at the State University of New York, admits in the *Scientific American* article "The Big Bang of Animal Evolution" that "something special and very mysterious — some highly creative 'force' — existed then" in the Cambrian period (Levinton, 1992, p. 84).

There is little point for evolutionists to debate evolution after all this and seek evidence for it as it is meaningless in the presence of Cambrian life forms to form creative narrations on how perfect the evolutionary mechanisms are or set out detailed scenarios of the fictitious transition from water to land and from land to the air. Paleontological proof dating back 530 million years has to be accounted for while the theory of evolution is unable to explain how this perfection came into being.

What the Cambrian explosion unambiguously tells us is nothing other than a miracle of creation taking place 530 million years ago, as one did when the Earth was first created. Thousands of individual members of thousands of different species comprising 50 very different phyla were created out of nothing together with hundreds of characteristics such as eyes, nervous systems, gills, extensions for hunting, feet for walking and magnificent shells. All amazing forms of life, including those which emerged in the Cambrian, are the work of Omniscient and Almighty God. Whether or not they choose to accept it, this truth is right before the eyes of the evolutionists.

6. Darwin's Chopped Down Tree of Life

Darwinism famously maintains that life emerged from a single common ancestor and varied by way of small changes. Had that been the case life must first have emerged in very similar and simple forms. According to this same hypothesis the differentiation among life forms and the increased complexity must have emerged over a long period of time. Therefore, according to Darwinism, life must be like a tree, growing from a single root and then spreading into various branches. Indeed, this hypothesis is stressed in Darwinist sources and the term "tree of life" is frequently employed. According to the tree of life, there must initially have been just one phylum, because the first imaginary cell also means the first basic physical blueprint of life, or phylum. This hypothetical first species, over a very long period, must have turned into others. The further the new forms that emerged grew from their illusory evolutionary ancestors, the greater the differences in their appearances must have been. There must also have been a gradual increase in the number of phyla related to this.

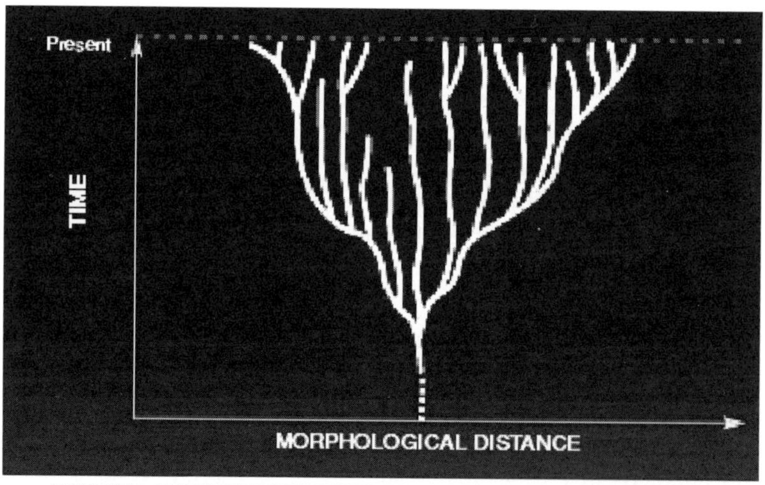

NATURAL HISTORY ACCORDING TO THE THEORY OF EVOLUTION

Figure 1

The theory of evolution maintains that different groups of living things (phyla) developed from a common ancestor and grew apart with the passing of time. Figure 1 states this claim: According to Darwinism, living things grew apart from one another like the branches on a tree.

But the fossil record shows just the opposite. As can be seen from the figure 2, different groups of living things emerged suddenly with their different structures. Some 100 phyla suddenly emerged in the Cambrian Age. Subsequently, the number of these fell rather than rose (because some phyla became extinct).

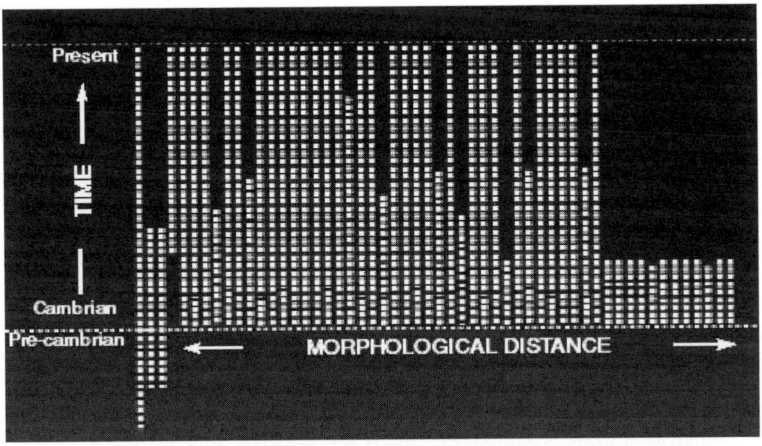

TRUE NATURAL HISTORY AS REVEALED BY THE FOSSIL RECORD

Figure 2

Darwin claimed that species he depicted in the diagram in *The Origin of Species* branched, like a tree, over a long period he divided into 14 time frames, and that differences between varieties would increase over the course of time. An amateur biologist, Charles Darwin expressed his flights of fantasy on this subject in the words, "*I see no reason to limit the process of modification, as now explained, to the formation of genera alone... These two groups of genera will thus form two distinct families, or orders*" (Darwin, 1958, p. 115).

This unrealistic expectation of Darwin's imposes certain conditions:

- According to Darwin, "first of all species" must have diverged, followed by the emergence of higher taxa and eventually phyla. Therefore, the number of phyla should increase with time. Therefore, taxa should follow a "bottom-up" course in their appearance in the fossil record.
- If we think of the tree of life as a cone, then the course of biological variation over time should be one of "increasing diversity." Therefore, the cone of diversity should assume a "V" shape.

However, the fossil record shows that these expectations of Darwinism are fundamentally incorrect.

Taxa follow a "top-down" course in the fossil record, not a "bottom-up" one, and phyla emerged first.

The fact revealed by the fossil record is along the lines of "phyla first." First of all, phyla came into being, followed by the emergence of lower taxa, such as species. The conclusions drawn by the researchers Douglas H. Erwin, James W. Valentine and J.J. Sepkowski from their comparison of the variation in life confirm this:

> "The fossil record suggests that the major pulse of diversification of phyla occurs before that of classes, classes before that of orders, orders before that of families... the higher taxa do not seem to have diverged through an accumulation of lower taxa (Erwin, Valentine, Sepkowski, 1987, p. 1183).

In the presuppositions that lower taxa, such as classes and families, would diverge and vary over time and that species varied from a single phylum would in turn give rise to other phyla, Darwin assumed a "bottom-up" development. Yet the Cambrian explosion reveals the exact opposite, in the words of the science writer Roger Lewin:

> "Several possible patterns exist for the establishment of higher taxa, the two most obvious of which are the bottom-up and the top-down approaches. In the first, evolutionary novelties emerge, bit by bit... The Cambrian explosion appears to conform to the second pattern, the top-down effect" (Lewin, 1988, p. 292).

The course of appearance of taxa in the fossil record is top-down. What is more dramatic is that the number of phyla, which should increase by stages over long periods of time, actually decreases. Although 50 different phyla emerged in the Cambrian, there are today only around 35. Darwin's assumptions have been literally "overturned" in the fact of the fossil record, and paleontology has definitely and clearly invalidated his theory.

The Cone of Diversity Is the Exact Opposite of What Darwin Claimed.

While branching the tree of life Darwin hypothesized that life would diversify in the form of a "cone of increasing diversity." Yet life does not increasingly expand and diversify; on the contrary, it begins with great variety and then narrows. The University of California, Berkeley professor Phillip Johnson describes the contradiction between Darwinism and this fact revealed by paleontology:

> "Darwinian theory predicts a 'cone of increasing diversity,' as the first living organism, or first animal species, gradually and continually diversified to create the higher levels of the taxonomic order. The animal fossil record more resembles such a cone turned upside down, with the phyla present at the start and thereafter decreasing" (Johnson, 1994, p. 12).

The variety assumed by Darwin is not cone-shaped ("V"), but rather in the form of a line ("Λ") in which all varieties are present together. Darwin's famous "tree of life" standing upside down on its head is a source of major disappointment for the theory itself and for its proponents. In the book *Icons of Evolution* American biologist Jonathan Wells describes this in precise terms:

> "Since higher levels of the biological hierarchy appear first, one could even say that the Cambrian explosion stands Darwin's tree of life on its head. If any botanical analogy were appropriate, it would be a *lawn* rather than a tree. Nevertheless, evolutionary biologists have been reluctant to abandon Darwin's theory. Many of them discount the Cambrian fossil evidence instead" (Wells, 2002, p. 42).

Life emerged suddenly and with a perfect variety with the Cambrian explosion. It is clear that there was no process of evolution beginning with a single bacterium and eventually extending as far as human beings, of the sort Darwinists believe in.

7. No Ascent from the Simple to the Complex

Darwin's theory claimed that all the complexity in present-day forms of life emerged as the result of an imaginary evolutionary process alleged to have continued over millions of years. Complex structures such as the sonar system of a dolphin, the tongue of a chameleon or the tentacles of an octopus must, according to this hypothesis, have emerged gradually from inferior systems. The theory placed an imaginary first cell, with none of these complex systems, at the start of the fictitious evolutionary process. Therefore, according to Darwinism, the evolution of life forms claimed to have taken place must have followed a developmental course from the simple to the complex throughout natural history. However, the Cambrian explosion irrefutably demolished that claim.

First and foremost, the living things that emerged in the Cambrian already had very complex structures. University of London biochemist D. B. Gower states this fact in clear terms:

> "In the oldest rocks we did not find a series of fossils covering the gradual changes from the most primitive creatures to developed forms, but rather in the oldest rocks developed species suddenly appeared" (Gower, 1975, p. 4).

Second, the fossil record and the species that lived in the period after the Cambrian are the exact opposite of Darwin's gradual development model. They indicate no gradual development at all. George Gaylord Simpson, one of the 20th century's foremost paleontologists, expresses this as follows:

> "It is a feature of the known fossil record that most taxa appear abruptly. They are not, as a rule, led up to by a sequence of almost imperceptibly changing forerunners such as Darwin believed should be usual in evolution" (Simpson, 1960, p. 117).

Third, despite all search by evolutionists, there is not the slightest evidence that any biological complexity emerged by way of evolution. Ernst Mayr, a Harvard University biologist and one of the 20th century's most influential proponents of Darwinism, displays an example of evolutionist despair on this subject:

> "[Research reveals that there is] no clear evidence ... for the gradual origin of an evolutionary novelty" (Mayr, 1988, p. 530).

Kevin Kelly, a researcher of complexity, makes a confession on the same subject:

> "No one has yet witnessed, in the fossil record, in real life, or in computer life, the exact transitional moments when natural selection pumps its complexity up to the next level" (Kelly, 1995, p. 475).

Life began with already complex entities. There is no evidence that complexity increased through evolution. The idea of an increasing scale of complexity throughout natural history was a deception. It emerged that the course followed by complexity in natural history was totally at odds with the Darwinist scenario. A pamphlet of the American Geological Institute, an authority on fossil strata, makes the following admission:

> "The old Darwinian view of evolution as a ladder of more and more efficient forms leading up to the present is not borne out by the evidence" (Newell, 1984, p. 10).

Scientists seeking to gather evidence that complexity could have developed out of simpler structures encountered the exact opposite. The vertebrate jaw, for example, is a complex structure, every component of which functions in a very sensitive manner. According to Darwinism, this complex structure must have been simpler in fish, at the lower branches of the imaginary tree of life, and then have evolved further in later vertebrates. Yet the facts reveal the exact opposite; the jaw is more advancedly developed in fish, in the supposedly lower section of the vertebrate tree of life. John G. Maisey from the Department of Vertebrate Paleontology at the American Museum of Natural History expresses this fact in these terms:

> "As we move back down our evolutionary ladder, jaw structure becomes more instead of less complex, and in fishes the jaws are very elaborate indeed" (Maisey, 2000, p. 61).

Another important example in this regard is the eye of the trilobite, an extinct arthropod. The complexity of this organ in the body of one of the most ancient animals did not find its way into any subsequent arthropod. The counterexamples are not confined to trilobites. No existing lineage possesses any fossil record of such a kind as to confirm the kind of development hypothesized by Darwinism. Stephen J. Gould writes:

> "The eyes of early trilobites, for example, have never been exceeded for complexity or acuity by later arthropods... I regard the failure to

> find a clear 'vector of progress' in life's history as the most puzzling fact of the fossil record" (Gould, 1984, pp. 22-23).

The fossil record has been a reminder to paleontologists of the extent of the incompatibility between Darwin's theory and scientific facts. As Ernst Mayr admits:

> "Paleontologists had long been aware of a seeming contradiction between Darwin's postulate of gradualism … and the actual findings of paleontology" (Mayr, 1991, p. 138).

As can be seen above, the natural history of life cannot be explained in Darwinian terms. The complexity displayed in the Cambrian explosion is already at an exceedingly high level. Living things in later periods did not progress from the simple to the complex, but remained exactly as they were when first created by Almighty God.

8. The Hidden Facts

From time to time, newspapers and magazines report that a 200-million-year-old mosquito fossil has been found, or a 30-million-year-old lizard fossil discovered. Unsuspecting readers may well assume that there is something extraordinary about these fossils or that they are rare findings. The truth cannot be further away from that.

The Earth is filled with millions-of-years-old fossils of present-day living things. A very large part of these have been unearthed, and everywhere that paleontologists excavate and study, still they find fossil specimens of modern living things with all their flawless attributes. Kept in countries' museums are millions-of-years-old spiders, ants, flies, scorpions, crabs, frogs and many other creatures, extinct and otherwise. Even specimens perfectly preserved in amber in all their detail are to be found in museums in their thousands, or even hundreds of thousands. Yet their numbers are seldom mentioned in books, the media, or scientific journals. Forums and discussions do not address them. And why is that?

The reason is that every "living" fossil discovered is another refutation of evolution. Every single example of such a living species is enough to trashbin the theory to which Darwinists dedicate their professional lives. For that reason, evolutionists attempt to keep large numbers of these fossils hidden.

Keeping Cambrian life forms hidden in the famous Smithsonian Institution for 70 years — fossils of the oldest complex life forms in the history of the planet — is a significant instance of this typical method. Charles Doolittle Walcott, a paleontologist and Secretary (1907-1927) of the Smithsonian, began research in the fossil-bearing Burgess Shale region in the Rocky Mountains, Canada, and unearthed one of the greatest finds in the history of paleontology: the first fossils of Cambrian creatures 530 million years old.

These approximately 530-million-year-old fossils entirely eliminated the false reasoning of gradual evolution. Yet they were brought out from where they had been stored and presented to the world only after 70 years had gone by. Walcott, a committed Darwinist, had decided to conceal the fossils he had obtained rather than sharing his findings with his fellow scientists. The Burgess Shale fossils were brought to light only in 1985, when the museum archives were re-examined. The Israeli scientist Gerald Schroeder comments:

> "Had Walcott wanted, he could have hired a phalanx of graduate students to work on the fossils. But he chose not to rock the boat of evolution. Today fossil representatives of the Cambrian era have been found in China, Africa, the British Isles, Sweden, Greenland. The [Cambrian] explosion was worldwide. But before it became proper to discuss the extraordinary nature of the explosion, the data were simply not reported" (Schroeder, 2000).

These fossils, which made the invalidity of the theory of evolution crystal clear, literally left evolutionists speechless, and are all flawless proofs of creation that represent one of the greatest difficulties facing the theory.

9. Conclusion

Historically, proponents of evolution have committed countless examples of fraud: ape jaw amalgamated to human cranium, imaginary ape-man social reconstructions from a single tooth of fossilized pig, feathers introduced creatively on the dinosaur fossils, etc. The adherents of this theory didn't hesitate to produce fake "intermediate" fossils to support their claims, but have felt compelled to conceal hundreds-of-millions-of-years-old, real and conclusive fossils that would consign their theory to the dustbin. Their untenable position is being gradually exposed, and since the Earth's strata were full of similar fossil specimens, some of the hidden facts were slowly, reluctantly displayed. Yet this deception still persists today, and some fossil specimens are still kept quietly concealed in museums. If all these specimens kept out of public view were made available to add to the store of human knowledge, the obvious facts would be realized.

A theory that is based on concealing scientific facts and establishing its scientific credentials through fraud and deception, declares its own bankruptcy. It is evident that the species that both walk the earth and exist in the fossils of hundreds of millions of years are the work of God, the Creator and Lord of all things, Who created these living things millions of years ago and Who has preserved them in their perfect forms right down to the present day.

References

Carroll, R.L. (1997) *Patterns and Processes of Vertebrate Evolution*, Cambridge University Press, p. 9.

Darwin, C. (1998) *The Origin of Species*, Oxford University Press, New York, pp. 140, 141, 146, 227.

Darwin, C. (1958) *The Origin of Species,* London, Everyman's Library, p. 115.

Dawkins, R. (1987) *The Blind Watchmaker*, London, W.W. Norton, p. 229.

Eldredge, N. (1996) *Reinventing Darwin: The Great Evolutionary Debate*, London, p. 95.

Eldredge, N. and Tattersall, I. (1992) *The Myths of Human Evolution*, Columbia University Press, pp. 45-46.

Erwin, D.H., Valentine, J.W. and Sepkowski, J.J. (1987) "A Comparative Study Of Diversification Events: The Early Paleozoic Versus The Mesozoic," *Evolution*, **41**: 1183.

Futuyma, D.J. (1983) *Science on Trial*, Pantheon Books, New York, p. 197.

Gould, S.J. (1977) "Evolution's Erratic Pace," *Natural History*, **86**: 14.

Gould, S.J. (1980) Lecture at Hobart and William Smith College.

Gould, S.J. (1984) "The Ediacaran Experiment," *Natural History*, pp. 22-23.

Gould, S.J. (1989) "An Asteroid to Die For," *Discover*, p. 65.

Gould, S.J. (2001) *The Book of Life: The Burgess Shale and the Nature of History*, W.W. Norton, pp. 51-52.

Gower, D.B. (1975) "Scientist Rejects Evolution," *Kentish Times*, England, p. 4.

Johnson, P.E. (1994) "Darwinism's Rules of Reasoning," *Darwinism: Science or Philosophy, Foundation for Thought and Ethics*, p. 12.

Kelly, K. (1995) "Out of Control: The New Biology of Machines," *Wired Magazine*, Fourth Estate, London, p. 475.

Kitts, D.B. (Sep 1974) "Paleontology and Evolutionary Theory," Evolution, **28:** 467.

Leach, E.R. (03 Sep 1981) "Men, Bishops and Apes," *Nature*, **293**: 19.

Levinton, J.S. (1992) "The Big Bang of Animal Evolution," *Scientific American*, **267**: 84.

Lewin, R. (1988) "A Lopsided Look at Evolution," *Science*, **241**: 292.

Maisey, J.G. (2000) *Fossil Fishes*, Westview Press, Boulder, p. 61.

Mayr, E. (1988) *Toward a New Philosophy of Biology: Observations of an Evolutionist*, Harvard University Press, Cambridge, p. 530.

Mayr, E. (1991) *One Long Argument: Charles Darwin and the Genesis of Modern Evolutionary Thought*, Harvard University Press, Cambridge, Massachusetts, p. 138.

Monastersky, M. (1993) "Mysteries of the Orient," *Discover*, **14**: 40.

Newell, N.D. (1984) *Why Scientists Believe in Evolution*, 1984, p. 10.

Raup, D.M. (1981) "Evolution and the Fossil Record," *Science*, **213**: 289.

Romer, A.S., (1963) In: *Genetics, Paleontology and Evolution*, L.G. Jepsen, E. Mayr, and S.G. Gaylord (eds.), Athenaeum, New York, p. 114.

Schroeder, G. (2000) "Evolution: Rationality vs. Randomness".

Sepkowski J.J. in: Gould, S.J. (2001) *The Book of Life: An Illustrated History of the Evolution of Life on Earth*, W.W. Norton, New York, pp. 51-52.

Simpson G.G. (1960) In: *Evolution after Darwin*, Sol Tax (ed.), vol. I, "The History of Life," University of Chicago Press, Chicago IL, p. 117.

Stanley, S.M. (1981) *The New Evolutionary Timetable: Fossils, Genes, and the Origin of Species*, Basic Books, New York, p. 71.

Wells, J. (2002) *Icons of Evolution: Science or Myth?*, Regnery Publishing, Washington, DC, p. 42.

Dialogue

Taner Edis

Most of what I said in response to C. Gündogdu [see his chapter below] applies to Yahya's article as well. From the perspective of not just mainstream science but any academic discipline, articles such as Yahya's are valuable only as illustrative examples of a popular but intellectually shallow form of Islamic creationism. Because such articles consistently violate basic scholarly norms, they are not persuasive as attempted critiques of evolution.

Nevertheless, since "Harun Yahya" has become a leading brand name for popular Islamic creationism, an opportunity for a few questions is hard to turn down.

On my bookshelves, there an enormous book called *The Atlas of Creation*, volume 1, one of the many Harun Yahya books I have. It came my way as a result of thousands of scientists in Europe and North America being mailed this, for free. This is just one example that shows the impressive financial resources that must be backing the Harun Yahya operation. What is the source of all this money?

Adnan Oktar has responded to such questions by saying that his books sell well, and that he is blessed with a circle of friends including many wealthy businessmen devoted to his cause. This is a vague, evasive, and ultimately unconvincing answer. Would it be possible to get a more detailed, substantive answer? If not, why not?

Another question. Some Turkish academics I know are concerned that in the present climate, publicly defending evolution means taking on a personal risk. In extreme cases, they may even worry about being targets of Islamist violence, especially if advocating evolution is considered apostasy from Islam. It is quite possible that their fears are exaggerated, but is there anything you can say that might reassure them? Would you, for example, be willing to make a public pronouncement that while you consider Islam and evolution to be incompatible, you also think that Muslims who say they accept evolution are making a mistake in good faith?

And a last question. The Yahya operation has become impressively international in scope. How much success do you perceive your version of creationism is having among Muslim immigrant populations in Western countries? How much of a concern do Western Muslims have about evolution in education?

Cem Sedat Altan

1) Taner Edis has issued an evaluation on the scientific level of Adnan Oktar's article. Millions of readers are aware that Mr. Adnan Oktar's books are written in the light of up-to-date scientific developments and that the views expressed in his books and other writings are entirely confirmed by scientific evidence. On the other hand, it is worthy of note that no scientific evidence is submitted in Taner Edis's own writings, be they his books or articles. Edis criticizes, according to his own lights, the adherents of the Divine religions and creation on the basis of his own preconceptions concerning the theory of evolution. The criticisms in his writings about Adnan Oktar, in which no evidence and no scientific objections are raised, are the result of a personal and biased reaction to genuine belief in Allah (God). Taner Edis has adopted evolution literally as a dogma, and he has not given any scientific answer to the demolition of Darwinism.

In addition, we also wish to remind the reader that the fact that someone has an academic background does not necessarily mean they are speaking the truth, in the same way that the fact that someone who does not come from an academic background does not necessarily mean they are not speaking the truth. What matters is whether what the author says is true. The truths set out in Adnan Oktar's books convince all those who read them and encourage them to question the theory evolution and to reflect on Allah's creation.

The way that some academics defend the theory of evolution with Latin words and complicated language is mere sleight of hand. These people underestimate the public's logic and common sense and impose the precondition that "one has to be a scientist in order to understand evolution." The fact is, however, that not only has the theory of evolution collapsed in the face of the scientific evidence, but at the same time anyone possessed of common sense and looking at matters impartially can see the logical deficiency contained within it. In his criticism of the theory of evolution Adnan Oktar uses both scientific methodology and a logic accessible to all to demolish it. That is why the days when the public could be deceived with the lie that "evolution cannot be questioned, and you have to be a scientist to understand it" have come to an end.

2) Taner Edis's question regarding material gains is one that is far from being scientific and is solely intended to distract the reader's attention. However, we believe it will be beneficial to issue the following statement to the public's attention: Adnan Oktar does not aim to make any profit from his books, and he receives no royalties from sales. The sending of the *Atlas of Creation* to academic circles is a promotional technique by the Book Global publishing house, the body officially authorized to sell Adnan Oktar's books. The fact that Adnan Oktar's books enjoy high global sales means that the publisher can make greater profits and thus set aside greater funds for such promotional campaigns. These questions should therefore be addressed to Global Publishing rather than to Adnan Oktar (http://www.bookglobal.net).

3) What some contemporary Turkish evolutionist academics have is not, as Taner Edis maintains, fear for their own safety, but rather fear for their own high standing as a result of the defeat that is being inflicted on them. Humiliated by having espoused the worst scientific fraud in history, evolutionists have fallen silent, and have lost all their enthusiasm and the energy to defend themselves. These people, grievously hurt by having dedicated their lives to a false theory, are unable to defend their theory because of the doubts they also harbor concerning it.

In the 5 May, 2005, edition of *The Pitch* magazine, published in the USA, the well-known Turkish evolutionist Professor Ümit Sayın expressed that defeat as follows:

> *"There is no fight against the creationists now. They have won the war.... In 1998, I was able to motivate six members of the Turkish Academy of Sciences to speak out against the creationist movement. Today, it's impossible to motivate anyone."*

As can be seen, the reason for evolutionists' silence and their inability to openly defend evolution in Turkey is this defeatist mindset.

Furthermore, Turkey is a democratic and enlightened country. Nobody has ever been subjected to violence for espousing evolution. This is a lie put out solely for demagogic purposes. The identities of those contemporary Turkish professors who hold evolutionist views are in any case well known. Since the writings of these proponents of evolution have been persistently discredited for many years now, it is of course impossible for them to support a lost cause and keep alive a demolished theory. Darwinism in Turkey has been dealt a body blow in the intellectual arena. There is no doubt that the convincing evidence in Adnan Oktar's works has contributed hugely to this, and the theory of evolution in Turkey has today been consigned to the shelves of history.

4) As Taner Edis mentions, there has for a long time been a minority view that evolution is actually supported by the Qur'an. However, the book *Why Darwinism is Incompatible with the Qur'an*, published by Adnan Oktar as a supplement to his scientific works, has demolished the error in which such people find themselves with evidence taken from the Qur'an itself. The erroneous ideas on this subject have now largely disappeared.

5) Taner Edis also questions the impact of Harun Yahya's works among Western Muslims. In summary, the answer is "ever increasing support and affection." One observes increasing fascination with the person of Harun Yahya and enormous interest in his books among Muslims in the West. The impact of these works today is also reflected on just about a daily basis in the foreign press. (See http://www.harunyahyaimpact.com).

We also see that European students are today opposing their teachers with responses taken from Harun Yahya's books, and that they are even responding to university professors by giving them books by Harun Yahya to read. What is more, this support is not restricted to Muslim circles. The reports in the press indicate that Christian and Jewish students are also making use of Adnan Oktar's works.

Adnan Oktar's 250 or so books, so far translated into more than 54 languages, and articles, CDs, internet sites, radio and television programs, documentaries, interactive presentations, audio cassettes and conferences based on them, are a means whereby people all over the world are coming to faith, and increases the enthusiasm of those who already believe by binding them together.

6) Taner Edis also asks whether Western Muslims are concerned by the obligatory nature of evolution in education. The imposition of e**volution on** the education system has today ceased to be a matter of concern. That is because with the widespread reading of *Atlas of Creation* and people exposed to biased teaching being able to find alternative ideas free of charge on the Internet has led to a considerable decrease in numbers of believers in evolution. And those numbers are continuing to decline. Darwinism in Europe has now collapsed

thanks to its demolition with convincing evidence by *Atlas of Creation*. We are currently witnessing that process of collapse taking place.

The reason why the Council of Europe referred to Adnan Oktar's *Atlas of Creation* as "dangerous" stems from its convincing and effective nature. There is no doubt that such fear of an idea, and regarding *Atlas of Creation* as a threat because it proves the existence and oneness of Allah, confirm the truth of the message contained within it. Because of the panic they are experiencing, Darwinist and materialist circles are trying to ban any questioning, criticism or even suspicion of the theory of evolution. The circles in question, that impose a biased education on young people by totally discounting such concepts as democracy and freedom of thought and faith, are struggling in vain. Our young people have already come by correct information, free of charge, over the Internet, and having seen the truth are now questioning the theory of evolution. So great is their espousal of the fact of Creation that they are now standing up to their teachers in high schools and universities.

Biography

Dr. Oktar Babuna is a medical doctor specialized in Neurosurgery, graduated from Istanbul University Medical School in 1987. Until 1991 he did research on molecular biology of the nervous system at the University of California San Francisco Medical School, Department of Neurosurgery and Neurology, and at the University of California Berkeley Donner Lawrence Laboratories and Health Sciences Center. Until 1996 he specialized in neurosurgery in the New York University Medical School Department of Neurosurgery. He has many publications in major international medical and scientific journals. He has done research since more than 10 years into the Theory of Evolution and Creation. Dr. Babuna has lectured at more than 500 conferences internationally on "The Collapse of Theory of Evolution and the Fact of Creation".

16
The Origin and Creation of Life

OKTAR BABUNA, MD

Istanbul University Medical School, Istanbul University Cerrahpasa Medical School, Department of Neurosurgery, Istanbul, Turkey
University of California San Francisco Medical School, Department of Neurosurgery, California, USA, and New York University Medical School, Department of Neurosurgery, New York, USA
oktar_babuna@yahoo.com

1. Introduction

When living things on Earth are examined, a manifest creation is to be observed. Every species is furnished with extremely complex features that enable it to play its role in the overall ecosystem to the best of its ability. Since life is so very organized, it certainly must have a Creator. The theory of evolution that was advanced in the 19th century denies this increasingly evident fact of creation. Charles Darwin's theory holds that the species on Earth were not created by God, but came into being in stages as a result of processes governed entirely by happenstance.

Darwin argued that all species are descended from a common ancestor by means of small incremental changes over long eons of time. But actually Darwin was aware of a great many facts that invalidated his theory. He admitted these in his book in a chapter entitled "Difficulties on Theory" and hoped that these stumbling blocks would be overcome by new scientific discoveries. But in fact, advances in science would refute Darwin's claims one by one. Darwin proposed that all species evolved successively from a common ancestor. But how did that first living thing come into being?

In his book, *The Origin of Species*, Darwin did not address this question at all! He was not even aware that this very point was one of the biggest refutations of his theory. The primitive understanding of 19th century science assumed that life had a very simple structure. According to a theory called "spontaneous generation," which had been popular since medieval times, it was believed that living things could arise by themselves from non-living matter, given just the right conditions.

The first evolutionist to take up the issue of the origin of life in the 20th century was the Russian biologist Alexander Oparin. His aim was to explain how the first living cell—allegedly, the common ancestor of all living beings according to Darwin's theory—could have emerged. In the 1930s, Oparin formulated a number of theories to show how the first living cell could have arisen from inanimate matter by sheer chance. However, his efforts ended in failure. Oparin himself had to confess:

> **"Unfortunately ... the problem of the origin of the cell is perhaps the most obscure point in the whole study of the evolution of organisms" (Oparin, 1953, p. 196).**

The question that renders Darwin's theory meaningless from the very outset is of how life first appeared on earth. In addressing this dilemma, evolutionary theory claims that life started with a cell that itself formed by chance. Four billion years ago, according to this scenario, various chemical compounds underwent a reaction in the primordial atmosphere on the Earth, during which the effects of thunderbolts and atmospheric pressure led to the formation of the first living cell.

However, the possibility that nonliving materials can come together to form life is an unscientific conjecture that has never been verified by any experiment or observation. Life can be generated only from life. Each living cell is formed by the replication of another, previous cell. No one in this world has ever succeeded in forming a living cell by bringing together inanimate materials, not even in the most advanced laboratories.

We will examine why this claim is contrary to the most basic principles of science and reason. The following statement by the geochemist Jeffrey Bada, from the San Diego-based Scripps Institute, makes clear the helplessness of evolutionists:

> "Today, as we leave the twentieth century, we still face **the biggest unsolved problem that we had when we entered the twentieth century: How did life originate on Earth?**" (*Emphasis added*) (Bada, 1998, p. 40).

2. Complex Structure and Systems in the Cell

The biggest impasse confronting evolution is the incredibly complex structure of the living cell. Scientific technology of the twentieth century explored this, the tiniest particles of life, and has revealed that the cell is the most complex system mankind has yet confronted. We know today that the cell contains miniaturized power stations producing the energy it uses, high-tech factories manufacturing the enzymes and hormones essential for life, a databank where all the necessary information about all products it can produce is recorded, complex transportation systems and pipelines for carrying raw materials and their products from one place to another, advanced laboratories and refineries for breaking down external raw materials into their useable parts, and a seemingly conscious cell membrane that controls the cell's incoming and outgoing substances. And these constitute only a small part of this incredibly complex system!

W.H. Thorpe, an evolutionist scientist, acknowledges that "[Even the] most elementary type of cell constitutes a 'mechanism' unimaginably more complex than any machine yet thought up, let alone constructed, by man" (Bird, 1991, pp. 298-299). Actually, the probablity of forming a cell by chance is about the same as that of producing a perfect copy of a book following an explosion in a printing house.

Sir Fred Hoyle, the English mathematician and astronomer, made a similar comparison in an interview published in *Nature*. Although an evolutionist himself, Hoyle stated that the chance that higher life forms might have emerged in this way is comparable to the chance that a tornado sweeping through a junk yard might assemble a Boeing 747 from the materials it contained (Hoyle, 1981, p. 105).

A cell is so complex that even today's high level of technology cannot produce one. No effort to create an artificial cell has ever met with success. Indeed, all attempts to do so have been abandoned. This implies that it is not possible for the cell to have come into being by chance, and that therefore, it must definitely have been "created."

One of the basic reasons why the theory of evolution cannot explain how the cell came into existence is its "**irreducible complexity**." A living cell maintains itself with the harmonious co-operation of many organelles. If only one of these organelles fails to function, the cell cannot remain alive. Thus, the first cell on Earth was necessarily a complete one that had to possess all the required organelles and ther attendant functions. This definitely means that this cell must have been created.

3. The Problem of the Origin of Proteins

So much for the cell, but evolution fails even to account for its building blocks of a cell. Under natural conditions, the formation of just one single protein out of the thousands of complex protein molecules composing the cell is impossible.

Twenty different types of amino acids found in nature are used in protein structure. Depending on the number used and their order, an infinite number of proteins can actually be made from only these 20 different amino acids. If you compare any protein to a chain, amino acids are the links in that chain. The number of amino acids in different protein types within living things is anywhere between 100 and 3,000. Removing, adding, or changing the order of just one amino acid in the chain randomly would render the entire protein completely useless, and in fact harmful. Furthermore, as will be discussed shortly, Darwin's theory cannot even substantiate the claim of the accidental formation of amino acids.

Robert Shapiro, a professor of chemistry at New York University and a DNA expert, calculated the probablity of the coincidental formation of the 2,000 types of proteins found in a single bacterium. (There are 200,000 different types of proteins in a human cell.) The number that he determined was expressed by 1 over $10^{40,000}$ (Shapiro, 1986, p. 127)—or in other words, a figured expressed by the number 1 followed by 40,000 zeros. Chandra Wickramasinghe, a professor of applied mathematics and astronomy from University College Cardiff, Wales, comments:

> "The likelihood of the spontaneous formation of life from inanimate matter is one to a number with 40,000 noughts [zeroes] after it ... **It is big enough to burry Darwin and the whole theory of evolution.** There was no primeval soup, neither on this planet nor on any other, and if the beginnings of life were not random, they must therefore have been **the product of purposeful intelligence**" (Hoyle, Wickramasinghe, 1984, p. 148) (*emphasis added*).

Sir Fred Hoyle comments on these implausible numbers:

> "Indeed, such a theory (that life was assembled by an intelligence) is so obvious that one wonders why it is not widely accepted as being self-evident. The reasons are psychological rather than scientific" (Hoyle, Wickramasinghe, 1984, p. 130).

In fact, many other evolutionists acknowledge this situation. For example, Harold F. Blum, a prominent evolutionist scientist, states that "The spontaneous formation of a polypeptide of the size of the smallest known proteins seems beyond all probablity." (Blum, 1968, p. 158).

Evolutionists claim that molecular evolution took place over a very long period of time and that this duration made the impossible possible. Nevertheless, no matter how long the given period may be, it is not possible for amino acids to form proteins by chance. William Stokes, an American geologist, admits as much in his book *Essentials of Earth History*, writing that the probablity is so small "that it would not occur during billions of years on billions of planets, each covered by a blanket of concentrated watery solution of the necessary amino acids." (Stokes, 1942, p. 186) So what does all this mean? Perry Reeves, a professor of chemistry, answers the question:

> "When one examines the vast number of possible structures that could result from a simple random combination of amino acids in an evaporating primordial pond, **it is mind-boggling to believe that life could have originated in this way]**. It is more plausible that a Great Builder with a master plan would be required for such a task" (Thomas, 1988, pp. 81-82) (*emphasis added*).

4. Homochirality: The Problem of Left-Handed Proteins

Let us now examine in detail why the evolutionist scenario regarding the formation of proteins is impossible. Even the correct sequence of the right amino acids is still not enough for the formation of a functioning protein molecule. In addition to these requirements, each of the 20 different types of amino acids present in the composition of proteins must be left-handed. Biomolecules are "chiral" or handed, in other words there are two different types of amino acids—as of all organic molecules—called "left-handed" and "right-handed." The difference between them is the mirror-symmetry between their three-dimensional structures, which is similar to that of a person's right and left hands.

Amino acids of either of these two types can easily bond with one another. But research has revealed the one astonishing fact that all the proteins in plants and animals on this Earth, from the simplest organism to the most complex, are composed of left-handed amino acids, that is to say they are "homochiral."

If even a single right-handed amino acid becomes attached to the structure of a protein, that protein is rendered useless. Surprisingly, in a series of experiments, bacteria that were exposed to right-handed amino acids immediately destroyed them. In some cases, they went on to produce usable left-handed amino acids from the fractured components.

The Britannica Science Encyclopaedia, which is an outspoken defender of evolution, states that the amino acids of all the living organisms on Earth, and the building blocks of complex polymers such as proteins, have the same left-handed asymmetry. It adds that this is tantamount to tossing a coin a million times and always getting heads. The same encyclopedia states that it is impossible to understand why molecules become left-handed or right-handed, and that this choice is fascinatingly related to the origin of life on Earth (*Britannica*, p. 415).

If a coin always turns up heads when tossed a million times, is it more logical to attribute that to chance, or else to accept that there is conscious intervention going on? The answer should be obvious. However, obvious though it may be, evolutionists still take refuge in coincidence, simply because they do not want to accept the existence of "conscious intervention." In conclusion, the probabilities we have examined prove beyond a doubt that the origin of life cannot be explained by chance. If we attempt to calculate the probablity of an average-sized protein consisting of 500 amino acids being selected from left-handed amino acids only, we come up with a probablity of 1 in 2^{500}, or 1 in 10^{150}.

5. The Indispensability of the Peptide Link

The difficulties that the theory of evolution cannot overcome in regard to the development of a single protein are not limited to those examined so far. It's not enough for amino acids to be arranged in the correct numbers, sequence, and required three-dimensional structures. The formation of a protein also requires that amino acid molecules with more than one arm be linked to each other only in certain ways. Such a "way" is called a "**peptide bond**." Amino acids can form different bonds with each other; but proteins are made up of those—and only those—amino acids that are joined by peptide bonds.

Research has shown that amino acids combining at random will combine with a peptide bond only 50 percent of the time. The rest of the time, different bonds that are not found in proteins are formed. To function properly, each amino acid making up a protein must be joined to others with a peptide bond only, in the same way that it must be selected only from among left-handed forms.

The probablity of this happening is the same as the probablity of each protein's being left-handed. That is, when we consider any protein made up of 500 amino acids, the probablity of all amino acids combining among themselves with peptide bonds only is 1 in 2^{499}, or 1 in 10^{150}.

6. Zero Probability

If we add together the three probabilities for a protein molecule of 500 amino acids (all laid out correctly in the right sequence, all being left-handed, and all joined by peptide links), then we come face to face with the astronomical figure of **1 in 10^{950}**. ($1/10^{650}$ times $1/10^{150}$ times $1/10^{150}$ = $1/10^{950}$ = one chance in $1/10^{950}$). This is a probability only on paper. Practically speaking, there is zero chance of its actually happening. A probability smaller than 1 in 10^{50} is, in mathematics, statistically considered to have a "zero" probability of occurring (Yahya, 2004, pp. 127-129). There are 3 basic conditions for the formation of a useful protein:

- First condition: that all the amino acids in the protein chain are of the right type and in the right sequence
- Second condition: that all the amino acids in the chain are left-handed
- Third condition: that all of these amino acids are linked together by forming a chemical bond called "peptide bond".

In order for a protein to be formed by chance, all three basic conditions must exist simultaneously. The probability of the formation of a protein by chance is equal to the multiplication of the probabilities of the realization of each of these conditions. For instance, for an average molecule comprising of 500 amino acids:

<u>1. The probability of the amino acids being in the right sequence:</u>
There are 20 types of amino acids used in the composition of proteins. According to this:

- The probability of each amino acid being chosen correctly among these 20 types = 1/20
- The probability of all of those 500 amino acids being chosen correctly = $1/20^{500} = 1/10^{650}$ = 1 chance in 10^{650}

<u>2. The probability of the amino acids being left-handed:</u>

- The probability of only one amino acid being left-handed = 1/2
- The probability of all of those 500 amino acids being left-handed at the same time = $1/2^{500} = 1/10^{150}$ = 1 chance in 10^{150}

<u>3. The probability of the amino acids being combined with a "peptide bond:"</u>
Amino acids can combine with each other with different kinds of chemical bonds. In order for a useful protein to be formed, all the amino acids in the chain must have been combined with a special chemical bond called a "peptide bond." It is calculated that the probability of the amino acids being combined not with another chemical bond but by a peptide bond is 50%. In relation to this:

- The probability of two amino acids being combined with a "peptide bond" = 1/2
- The probability of 500 amino acids all combining with peptide bonds = $1/2^{499} = 1/10^{150}$ = 1 chance in 10^{150}

The Probability of a Protein Being Formed by Chance is Zero

TOTAL PROBABILITY = $1/10^{650} \times 1/10^{150} \times 1/10^{150} = 1/10^{950}$
= 1 chance in 10^{950}

Just for a comparison, remember that the number of electrons in the universe is estimated at 10^{79}—which although truly vast, is still a much smaller number. Even if we suppose that amino acids have combined and decomposed by some "trial-and-error" method, without losing any time since the formation of the Earth, in order to form one single protein molecule, the time required for something with a probability of 10^{950} to occur would still hugely exceed the estimated age of the Earth (which is only 10^{17} seconds). Professor Richard Dawkins, one of the foremost proponents of the theory of evolution, states in these terms the impossibility the theory has fallen into:

> "So the sort of lucky event we are looking at could be so wildly improbable that the chances of its happening, somewhere in the universe, could be as low as one in a billion billion billion in any one year. If it did happen on only one planet, anywhere in the universe, that planet has to be our planet—because here we are talking about it" (Dawkins, 1996, p. 283).

This admission by one of evolution's foremost authorities clearly reflects the logical muddle that the theory of evolution is built on. The above statements in Dawkins's book *Climbing Mount Improbable* are a striking example of circular reasoning that actually explains nothing: "If we are here, then that means that evolution must have happened."

7. Is There a Trial-and-Error Mechanism in Nature?

The calculations of probability we made above reach astronomical levels, and these astronomical odds have no chance of actually happening. However, here is a much more important, damaging fact that faces evolutionists: Despite the astronomical odds, no period of trial and error can even begin under natural conditions, because in nature, there is no trial-and-error mechanism exists from which proteins could emerge.

The calculations we cited above, to demonstrate the probability of the formation of a protein molecule with 500 amino acids, are valid only for an ideal trial-and-error environment—one that doesn't actually exist in real life. That is, the probability of obtaining a useful protein is one in 10^{950} only if we suppose that there exists an imaginary mechanism in which an invisible hand joins 500 amino acids at random and then, realizing that this is not the correct combination, disentangles them one by one, and arranges them again in a different order, and so on forever. In each trial, the amino acids would have to be broken apart one by one, and then arranged in a new order. The synthesis should be stopped after the 500th amino acid has been added, and not even one extra amino acid must be involved. The trial should then be halted to see whether or not a functional protein has yet been formed. In the event of failure, everything should be split up again and then tested for another sequence.

Additionally, in each trial, not even one extraneous substance should be allowed. It is also imperative that the chain formed during the trial should not be separated and destroyed before reaching the 499th link. These conditions mean that the probabilities mentioned above can operate only in a controlled environment with a conscious mechanism directing the beginning, end, and each intermediate stage of the process, and where only the selection of the amino acids is left to chance. Clearly, it's impossible for such an environment to exist under natural conditions. Therefore the formation of a protein in the natural environment is not just logically, but technically impossible (Yahya, 2002, p. 204).

Above all, there is still another important point to consider: If any one step in the so-called evolutionary process is proven to be impossible, this proves that the whole theory is totally false and invalid. For instance, by proving that the haphazard formation of proteins is impossible, all other claims regarding the subsequent steps of evolution are also refuted.

8. The Origin of the DNA Molecule

The question of how such an extraordinarily created molecule as DNA originated is one of the countless impasses that evolutionists reach. Seeking to explain life by means of "coincidence," the theory of evolution can never explain the source of the extraordinary information so perfectly and meticulously encoded in DNA.

Moreover, the question is not only that of how the DNA chain originated. By itself, even the existence of the DNA chain with the extraordinary information capacity it contains, means nothing. In order to maintain life, it's essential that the enzymes that "read" this DNA chain, copy them and produce proteins, also exist. Simply put, in order to speak of life, both the 1) data bank we call DNA, and 2) the enzymes to read the data in the bank must co-exist.

To our surprise, enzymes—which read DNA and carry out production accordingly—are themselves produced according to the codes in DNA. This means that there is a factory in the cell that both makes many different types of products, and also manufactures the mechanisms that carry out this production. The question of how this system originated—which would be useless with even a minor defect in any of its mechanisms—is an impasse for the evolutionists. German evolutionist Douglas R. Hofstadler, states his despair in the face of this question:

> "How did the Genetic Code, along with the mechanisms for its translation (ribosomes and RNA molecules), originate? For the moment, we will have to content ourselves with a sense of wonder and awe, rather than with an answer" (Hofstadler, 1980, p. 548).

Another evolutionist, Caryl P. Haskins, expresses the impossibility of the DNA code forming by chance and that this fact is powerful evidence for creation:

> "But the most sweeping evolutionary questions at the level of biochemical genetics are still unanswered. How the genetic code first appeared and then evolved and, earlier even than that, how life itself originated on earth remain for the future to resolve... The fact that in all organisms living today the processes both of replication of the DNA and of the effective translation of its code require highly precise enzymes and that, at the same time, the molecular structures of those same enzymes and that, at the same time, the molecular structures of those same enzymes are specified by the DNA itself, poses a remarkable evolution mystery. How can a particular strand of DNA, dependent for its replication on a highly specific enzyme, have possibly evolved the code to direct the information of that same enzyme, given the obvious condition that the evolutionary survival of the organism containing it must have depended on the success of such replication? This question, of course, raises another. Did the code and the means of

> translating it appear simultaneously in evolution? It seems almost incredible that any such coincidence could have occurred, given the extraordinary complexities of both sides and the requirement that they be coordinated accurately for survival. By a pre-Darwinian (or a skeptic of evolution after Darwin) this puzzle would surely have been interpreted as the most powerful sort of evidence for special creation" (Haskins, 1971, p. 305).

At this point, an important detail deserves attention. One error in the sequence of the nucleotides making up a gene would render that gene completely useless. When you considered that there are between 20 and 25 thousand genes in the human body, it becomes clearer how impossible it is for the millions of nucleotides making up those genes to have formed in the right sequence, by chance.

If we tried to write down the information in the DNA, this would take up approximately a million pages. This is equal to an encyclopaedia forty times bigger than *The Encyclopaedia Britannica*, one of mankind's greatest single accumulations of information. But this incredible information is stored in the tiny nucleus of our cells measuring about a thousandth of a millimeter in size. With its ability to store information in DNA, the genetic code functions like a computer—which further increases the meaninglessness of evolutionist explanations about its origin. According to calculations made by Led Adleman of the University of Southern California in Los Angeles, "One gram of DNA can store [as much] information as one trillion compact discs." (Whitfield, 2003).

Furthermore, the genetic code, although functioning like a computer program, goes far beyond technology. The Windows code, one of the world's most complex computer programs, can have two electronic states—on or off. The DNA code, on the other hand, consists of analogue segments that can contain infinite variations in state. That means that the DNA logic is thousands of times more complex than that of Windows, written and tested by thousands of programmers. As Bill Gates, president and administrator of Microsoft, writes in his book *The Road Ahead*:

> "Human DNA is like a computer program, but far, far more advanced than any software we've ever created" (Gates, 1996, p. 228).

There is absolutely no possibility that genetic information and the genetic code could have emerged in a chance-based manner through natural selection. Huberty Yockey, the prominent information theorist and biophysicist, measured the amount of information need to be contained for life to be possible in even the smallest genome and revealed that the odds of this coming into being by chance were **1 in $10^{186,000}$** (Yockey, 2003, p. 28).

Yockey calculated that for the genetic code to have come into being by chance, natural selection would have to "discover" **1.40×10^{70} different genetic codes** before reaching a universal code. As mentioned before events with a less than 1 in 10^{50} chance of happening are regarded as practically impossible anywhere in the universe (Borel, 1965).

That means that the chance of the genetic code emerging by chance is "0." Science definitively rejects evolutionist scenarios regarding the origin of the genetic code. The effect of this on evolutionists can be seen in the words of highly reputed evolutionist molecular biologist Dr. Leslie Orgel:

> "It is extremely improbable that proteins and nucleic acids, both of which are structurally complex, arose spontaneously in the same place at the same time. Yet it also seems impossible to have one without the other. And so, at first glance, **one might have to conclude that life could never, in fact, have originated by chemical means**" (Orgel, 1994, p. 78) (*emphasis added*).

As shown above, **the origin of the genetic code is crystal clear**—that origin is Creation. The genetic code is just one example of God's flawless creation.

9. Miller's Experiment

The most generally respected study on the origin of life is the Miller experiment conducted by the American researcher Stanley Miller in 1953. (This is also known as the "Urey-Miller experiment" because of the contribution of Miller's instructor at the University of Chicago, Harold Urey.) This is the only "evidence" that evolutionists have to prove their alleged "chemical evolution thesis," which they advance as the first stage of the evolutionary process leading to life. Although nearly half a century has passed, and great technological advances have been made, nobody has made any further progress. But in spite of this, Miller's experiment is still taught in textbooks as the evolutionary explanation of the earliest generation of living things. That is because evolutionist researchers realize that such studies do not support, but rather actually refute, their thesis, and so deliberately avoid embarking on such further experiments.

Stanley Miller aimed to demonstrate that amino acids, the building blocks of proteins, could have come into existence "by chance" on the lifeless Earth billions of years ago. In his experiment, he used a mixture of gasses that he assumed would have existed on the primordial Earth (but which later proved unrealistic), composed of ammonia, methane, hydrogen, and water vapor. Since these gases do not react with each other under natural conditions, he added energy to the mixture to trigger a reaction among them. Supposing that this energy could have come from lightning in the primordial atmosphere, he used an electric current for his purpose.

Miller heated this gas mixture at 100°C—the boiling point of water—for a week and added the electrical current. At the end of the week, Miller analyzed the chemicals which had formed at the bottom of the jar, and observed that three out of the 20 amino acids which constitute the basic elements of proteins had been synthesized.

However, Miller's experiment has since proven to be false in many respects:

1) By using a mechanism called a "**cold trap**," Miller isolated the amino acids from the environment as soon as they were formed. Had he not done so, the environmental conditions in which the amino acids were formed would have destroyed these molecules immediately. The chemist Richard Bliss expresses this contradiction by observing that, "Actually, without this trap, the chemical products would have been destroyed by the energy source" (Bliss, Parker, Gish, 1990, pp. 14-15).

2) The primordial atmosphere that Miller attempted to simulate in his experiment was not realistic. In the 1980s, scientists agreed that nitrogen and carbon dioxide should have been used in his artificial environment, instead of methane and ammonia. So why did Miller insist on these gases? The answer is simple: Without ammonia, it was impossible to synthesize any amino acid. Kevin McKean talks about this in an article published in *Discover* magazine:

> "Miller and Urey imitated the ancient atmosphere on the Earth with a mixture of methane and ammonia... However, in the latest studies, it has been understood that the Earth was very hot at those times, and that it was composed of melted nickel and iron. Therefore, the chemical atmosphere of that time should have been formed mostly of nitrogen (N_2), carbon dioxide (CO_2) and water vapour (H_2O). However, these are not as appropriate as methane and ammonia for the production of organic molecules" (McKean, 1983, p. 7).

3) Another important point that invalidates Miller's experiment is that there was enough oxygen to destroy all the amino acids in the atmosphere at the time when they were thought to have been formed. This fact, which Miller overlooked, is revealed by the traces of oxidized iron found in rocks that are estimated to be 3.5 billion years old.

This situation completely negates Miller's experiment, which completely neglected oxygen. Had oxygen been used in the experiment, methane would have decomposed into carbon dioxide and water, and ammonia into nitrogen and water. On the other hand, in an environment where there was no oxygen, there would be no ozone layer either. Therefore, the amino acids would have immediately been destroyed, since without the protection of the ozone layer, they would have been exposed to the most intense ultraviolet rays.

4) By the end of Miller's experiment, many organic acids had also been formed with characteristics detrimental to the structure and function of living things. If the amino acids had not been isolated, and had been left in the same environment with these chemicals, unavoidable chemical reactions would have destroyed them or transformed them into different compounds.

Moreover, Miller's experiment also produced right-handed amino acids. (Bliss, Parker, Gish, 1990, p. 16) These amino acids' existence refuted the theory even within its own terms, because right-handed amino acids cannot function in the composition of living organisms. To conclude, the circumstances in which amino acids were formed in Miller's experiment were not suitable for life.

All these facts point to one firm truth: Miller's experiment cannot claim to have proved that living things formed by chance under primordial Earth–like conditions. The whole experiment is nothing more than a deliberate and controlled laboratory attempt to synthesize amino acids. In the March 1998 issue of *National Geographic*, in an article titled "The Emergence of Life on Earth," the following comments appear:

> "Many scientists now **suspect that the early atmosphere was different to what Miller first supposed**. They think it consisted of carbon dioxide and nitrogen rather than hydrogen, methane, and ammonia. That's bad news for chemists. When they try sparking carbon dioxide and nitrogen, they get a paltry amount of organic molecules—the equivalent of dissolving a drop of food coloring in a swimming pool of water. Scientists find it hard to imagine life emerging from such a diluted soup" (Monastersky, 1998, p. 68) (*emphasis added*).

Interestingly enough, Harold Urey, who organized the Miller experiment with his student Stanley Miller, made the following confession on this subject:

> "All of us who study the origin of life **find that the more we look into it, the more we feel it is too complex to have evolved anywhere**. We all believe as an article of faith that life evolved from dead matter on this planet. It is just that its complexity is so great, it is hard for us to imagine that it did" (Bird, 1991, p. 325) (*emphasis added*).

10. Conclusion: The Fact of Creation

So far, we have examined how impossible the accidental formation of life is. All scientific data have shown organic matter can reproduce itself only if it already exists as a fully developed cell, with all its organelles. This means that the first cell on Earth must have been formed all of a sudden, together with its incredibly complex structure. So, if such a complex structure came into existence suddenly, what does this mean?

Let us phrase this question with an example, by likening the cell to a high-tech car in terms of its complexity. (In fact, the cell's system is much more complex and developed than a car's.) Now, if you went hiking in the depths of a thick forest and ran across a brand-new car among the trees, what would you think? Would you imagine that various elements in the forest had come together by chance over millions of years to produce such a vehicle? All the parts in the car are made of products such as iron, copper, and rubber—whose raw ingredients are all found on the Earth. But would this lead you to assume that these materials had synthesized by chance and then come together to manufacture such a vehicle?

No doubt, anyone of a sound mind would realize that the car was the product of a design—in other words, a factory—and wonder what it was doing there in the middle of the forest. The sudden emergence of any complex structure in a complete form, quite out of the blue, shows that this is the work of Creator.

Believing that pure chance can produce perfect designs goes well beyond the bounds of reason. Yet every "explanation" regarding the origin of life put forward by the theory of evolution is like that. One outspoken authority on this issue is the famous French zoologist Pierre-Paul Grassé, the former president of the French Academy of Sciences. Grassé is an evolutionist, yet he acknowledges that Darwinist theory is unable to explain life and makes a point about the logic of "coincidence," which is the backbone of Darwinism:

> "The opportune appearance of mutations permitting animals and plants to meet their needs seems hard to believe. Yet the Darwinian theory is even more demanding: A single plant, a single animal would require thousands and thousands of lucky, appropriate events. Thus, **miracles would become the rule: events with an infinitesimal probability could not fail to occur... There is no law against daydreaming, but science must not indulge in it**" (Grassé, 1977, p. 103) (*emphasis added*).

All living things in the world are clear examples of planning we have just been discussing. At the same time, they are living evidence that coincidence had no role to play in their existence. This truth may surprise some scientists, who from the 19th century to the present have viewed creationism as unscientific. But science can only progress by overcoming shocks of this kind and accepting the truth. Chandra Wickramasinghe describes the reality he faced as a scientist who had been told throughout his life that life had emerged as a result of chance coincidences:

> "From my earliest training as a scientist, I was very strongly brainwashed to believe that science cannot be consistent with any kind of deliberate creation. That notion has had to be painfully shed. At the moment, I can't find any rational argument to knock down the view which argues for conversion to God. We used to have an open mind;

now we realize that the only logical answer to life is creation—and not accidental random shuffling" (Wickramasinghe, 1981).

We have examined the scientific evidence for the origin of life, and what emerges clearly demonstrates that life was not the result of chance, as claimed by Darwinism and materialist philosophy in general. All living things were independently and flawlessly created. At the beginning of the 21st century, science offers but one answer to the question of the origin of life: Creation.

References

Bada, J. (February 1998) "Life's Crucible," *Earth,* 40.

Bird, W.R., (1991) *The Origin of Species Revisited*, Thomas Nelson Co., Nashville, pp. 298-299, 325.

Bliss, R.B., Parker, G.E and Gish, D.T. (1990) *Origin of Life*, C.L.P. Publications, California, pp. 14-16.

Blum, H. (1968) *Time's Arrow and Evolution*, p. 158, In: W.R. Bird, (1991) *The Origin of Species Revisited*, Thomas Nelson Co., Nashville, p. 304.

Borel, E. (1965) *Elements of the Theory of Probability*, Prentice-Hall, New Jersey.

Fabbri Britannica Bilim Ansiklopedisi (Fabbri Britannica Science Encyclopaedia), no. 22, **2**: 519

Dawkins, R. (1996) *Climbing Mount Improbable*, W.W. Norton, New York, p. 283.

Gates, B. (1996) *The Road Ahead*, Penguin, London, p. 228.

Grassé, P.P. (1977) *Evolution of Living Organisms*, Academic Press, New York. p. 103.

Haskins, C.P. (May/June 1971), "Advances and Challenges in Science in 1970," *American Scientist*, **59**: 305.

Hofstadter, D.R. (1980) *Gödel, Escher, Bach: An Eternal Golden Braid*, Vintage Books, New York, p. 548.

Hoyle, F. (1981) "Hoyle on Evolution," *Nature*, **294**: 105.

Hoyle, F., Wickramasinghe, C. (1984) *Evolution from Space*, Simon and Schuster, New York, pp. 130, 148.

Mc Kean, K. (1983) "How did the Life Began?," *Bilim ve Teknik* (Science and Technology), **189**: 7.

Monastersky, R. (1998) "The Rise of Life on Earth," *National Geographic*, 68.

Oparin, A.I. (1953) *Origin of Life*, Dover Publications, New York, p. 196.

Orgel, L.E. (1994) "The Origin of Life on Earth," *Scientific American*, **271**: 78.
Shapiro, R. (1986) *Origins: A Skeptic's Guide to the Creation of Life on Earth*, Summit Books, New York, p. 127.
Stokes, W. (1942) *Essentials of Earth History*, p. 186, In: W.R. Bird, (1991) *The* Origin of Species *Revisited*, Thomas Nelson Co., Nashville, p. 305.
Thomas, J.D. (1988) *Evolution and Faith*, ACU Press, Abilene, pp. 81-82.
Whitfield, J. (2003) "Physicists plunder life's tool chest," *Nature*.
Wickramasinghe, C. (14 Aug, 1981) Interview in *London Daily Express*.
Yahya, H. (2002) *Darwinism Refuted*, Goodword Books, 2nd ed., New Delhi, p. 204.
Yahya, H. (2004) *The Evolution Deceit*, Ta Ha Publishers, 8th ed., UK, pp. 127-129.
Yockey, H. (2003) "*Calculating Evolution*," *Cosmic Pursuit*, **3** (1): 28.

Dialogue

Taner Edis

Babuna's chapter is a typical creationist document. It argues by incessant quotation (often out of context), and tedious repetition of improbability arguments. Everything depends on the claim that life is incredibly improbable, *assuming that all the required components came together randomly, at once, in an equiprobable manner*. Serious research on chemical evolution does not make such a naive assumption. And most evolutionary biology proceeds independently of issues regarding prebiotic chemistry. These are obvious to anyone familiar with modern science. A legitimate critic would know the fields he criticizes, rather than attacking grossly distorted straw men.

But let me set my distaste with such poor scholarship aside. Indeed, let us imagine that all of Babuna's improbabilities were not the misrepresentations they really are—that mainstream science had no clue about how life originated. What then?

Babuna ends with a triumphant assertion that "At the beginning of the 21st century, science offers but one answer to the question of the origin of life: Creation." Creation, it appears, is supposed to be an alternative explanation, one that is more successful in making sense of living things. But how is this so? A superior explanation should identify new patterns in data, showing us previously hidden commonalities. It should do better than just restate the data, merely adding that some kind of God is responsible for it all. But nothing of this sort is forthcoming from creationists. In the rare cases they do not misrepresent the data, they list the same information, and invite us to admire the miracle.

Those of us interested in genuine explanations should not, I think, be impressed. If we do not know, if something is a perplexing mystery, well, so it is. We do not gain anything by calling our ignorance a sign of the divine. It is, in fact, possible that a gap in our understanding points to a God. But we should not be convinced of this without some serious argument. Creationists, indeed believers of all stripes, need to tell us why exactly we should find just *these* gaps in our knowledge, and tell us exactly how bringing in a supernatural Creator leads us to better understand nature. We need real patterns, risky predictions, a way of going beyond throwing up our hands and announcing miracles. Otherwise, talk of "creation" is only an unnecessarily mystifying acknowledgment of ignorance. This, I imagine, is hardly the sort of thing devout religious people have in mind.

Oktar Babuna

Mr Edis' response is ideological and furious. In the name of defending his ideology Mr Edis crosses out of the borders of logic and science. By having lack of any scientific evidence to support Darwinism, Mr Edis is also breaching etiquette.

First of all let's define what science means. Science is exploring nature and deriving conclusions from one's findings. If these findings lead to the conclusion that nature is created, science has to accept it. That is the duty of a true scientist; not defending impossible scenarios by clinging to the outdated materialist dogmas of the 19th century.

As we know scientifically, materialism has already collapsed, because the idea that matter has existed since beginning of time has been overthrown by the "Big Bang theory" which shows that the universe was created from nothingness. The claim that matter organised itself and created life is what we call "the theory of evolution," which has also been collapsed with all the scientific evidence.

The reason why Darwinism is still defended against science is to keep atheism upright. An evolutionist scientist Richard Dawkins summarizes as thus:

> ***"Before Darwin, it was difficult to be an intellectually fullfilled atheist and Darwin made it easy to become an intellectual atheist."***
> *(Richard Dawkins on Evolution and Religion, from an interview by Ben Wattenberg Aired on November 8, 1996, PBS Show Think Tank.)*

Now let me explain why Mr Edis is wrong.

First of all starting from the very beginning, there are two possible explanations of how life emerged. One of them is the theory of evolution. Evolution claims that life on Earth began by chance, and that therefore all the species we see around us are the result of mere coincidence.

The other explanation is Creation. No other, third explanation or proposal can be put forward. This fact is now accepted even by all the known evolutionists in the world, and they state this quite frankly.

Douglas Futuyma (an evolutionist from the State University at Stony Brook) is one of these. Douglas Futuyma, as an evolutionist biologist admits this fact:

> "Organisms either appeared on the earth fully developed or they did not. If they did not, they must have developed from pre-existing species by some process of modification. If **they did appear in a fully developed state, they must indeed have been created by some omnipotent intelligence**" (***Douglas J. Futuyma, Science on Trial, New York: Pantheon Books, 1983, p. 197).***

In fact, Futuyma's words underline a very important truth. He says that considering the origin of life, if we see that life emerges all of a sudden with its complete and perfect form, then we have to admit that life is created. Here it can clearly be seen that, even Futuyma who is known as an evolutionist himself, accepts this as a fact.

All the scientific evidence refutes Darwinist claims and proves an exalted Creation by God. Let's summarize why Creation is a scientific fact and Darwinism is an outdated dogma of the 19th century. All observations and experiments indicate that life can only originate from life. And today we know scientifically that life only comes from life. That fact is called biogenesis. This scientific fact is stated in the Campbell-Reece *Biology Textbook* with these words:

> ***"As far as we know, all life today arises only by the reproduction of preexisting life. This "life-from-life" principle is called biogenesis"*** *(Biology, Campbell-Reece, Sixth Edition, 2002, p. 516).*

In this respect, the first life on earth must also have originated from other life. This is a reflection of God's epithet of "Hayy" (The Owner of Life). Life can only start, continue, and end by His will. As for evolution, not only is it unable to explain how life began, it is also unable to explain how the materials essential for life have formed and come together.

Another very powerful evidence which refutes Darwinist claims and proves the fact that living beings are created by God is the sudden emergence of living things in their complete and perfect forms.

3.5 bilion years ago, the first cell emerged all of a sudden. There is the fossil of the very first cell from Australia. These bacteria preserved on this fossil are the same as today's "leptolyngbya" bacterium. (**No evolution for 3.5 billion years.)**

Moreover, any cell must be a complete cell otherwise it can not survive. Today we know that the cell contains power stations producing the energy to be used by the cell, factories manufacturing the enzymes and hormones essential for life, a databank where all the necessary information about all products to be produced is stored, complex transportation systems and pipelines for carrying raw materials and products from one place to another, advanced laboratories and refineries for breaking down external raw materials into their useable parts, and specialized cell membrane proteins to control the incoming and outgoing materials. And these constitute only a small part of this incredibly complex system.

Furthermore, it cannot survive if any of these components is missing or defective. This is a scientific fact and of course refutes the Darwinist claim of "random gradual emergence."

Also if we look at the fosil evidence, we observe the same fact that living beings emerged all of a sudden. I am not sure, how familiar is Taner Edis with paleontology, but the Cambrian Period is the oldest stratum of the Earth in which fossils of living creatures have been found. It can be defined as the most ancient stratum in which the first visible fossils of living things are embedded. Cambrian Period, has an estimated age of 500-550 million years. The living things in that period were invertebrate marine animals, like jellyfish, water lilies, starfish, worms and snails. The living creatures found in the strata belonging to the Cambrian Period emerged all of a sudden in the fossil record with no pre-existing ancestors. Richard Dawkins, one of the best-known advocates of evolution in the world, comments on this reality that undermines the very foundation of all the arguments he has been defending:

> **"For example the Cambrian strata of rocks... are the oldest ones in which we find most of the major invertebrate groups. And we find many of them already in an advanced state of evolution, the very**

> **first time they appear. It is as though they were just planted there, without any evolutionary history."**

Darwin himself recognized the possibility of this when he wrote: "If numerous species, belonging to the same genera or families, have really started into life all at once, **the fact would be fatal to the theory of descent with slow modification through natural selection**." The Cambrian Period is nothing more or less than Darwin's "fatal stroke".

The second most important evidence comes from the fosil record. If it is asserted that species evolved into one another by successive minute variations, and that this happened over billions of years, then evolutionists have to show us these slow, minute changes in the fossilized examples of those species. They must show us half-fish half-reptile, half-reptile half-mammal, half-reptile half-bird fossils. These imaginary species are called **"Transitional Forms"**. There is no doubt that if there is no such fossil record concerning changes among the fossils of stable species, then there is nothing left to discuss as regards to the theory of evolution.

If we consider the capacity of the fossil record, we see that we have millions of fossils in hand which belong to stable species. But the number of so-called transitional species must be much higher than that of the stable species we observe today. Trillions of transitional forms should be present in the fossil record linking these species to one another. That is because if there had been slow and fine gradations that had lasted for billions of years, then the number of stable species should have been smaller, and the number of transitional form fossils reflecting those variations should be much higher. Although there are more than 100 million fossils there is not a singleso-called transitional form.

The number of these transitional forms should have been even greater than that of present animal species, and their remains should be found all over the world. In *The Origin of Species*, Darwin accepted this fact and explained as such:

> ***"...Why, if species have descended from other species by fine gradations, do we not everywhere see innumerable transitional forms? Why is not all nature in confusion, instead of the species being, as we see them, well defined?... But, as by this theory innumerable transitional forms must have existed, why do we not find them embedded in countless numbers in the crust of the earth?..."***

This fact is also admitted by the late evolutionist paleontologist of the 20th century Stephen Jay Gould:

> ***"The history of most fossil species include two features particularly inconsistent with gradualism:***
>
> ***1) Stasis – most species exhibit no directional change during their tenure on earth. They appear in the fossil record looking much the same as when they disappear; morphological change is usually limited and directionless;***
>
> ***2) Sudden appearance – in any local area, a species does not arise gradually by the steady transformation of its ancestors; it appears all at once and 'fully formed'."*** *(S.J. Gould, "Evolution's Erratic Pace", Natural History, vol. 86, May 1977).*

Briefly, it is obvious that life and the universe are created by God and it is proven by all the scientific evidence deriving from all the branches of science.

Mr. Edis either missed the powerful evidence given in my chapter or he is ignoring it not to respond out of distress.

I don't want to go all over again but the impossibility of producing even one single protein by chance refutes all other claims of evolution. Remember that we are talking about one protein. In humans there are about 200.000 different proteins which have to exist at the same time in the same organism. Even bacteria have thousands of proteins.

That disproves the claims of Darwinism at the most basic level. It is impossible for life to emerge by chance as evoultionists claim. Above all, there is one important point to take into consideration: **If any step in the evolutionary process is proven to be impossible, this is sufficient to prove that the whole theory is false and invalid.** For instance, by proving that the haphazard formation of proteins is impossible, all other claims regarding the subsequent steps of evolution are also refuted. After this, it becomes meaningless to take some human and ape skulls and engage them in speculation.

The great wisdom that prevails in nature provides solid evidence for the existence of a supreme Creator dominating over the nature, and this Creator is

God. God has furnished all living beings with extraordinary features and showed men the evident signs of His existence and might.

Taner Edis

Some years ago, when I was a postdoc, I stopped to fill my car up. I went inside to pay, and the cashier, a middle-aged man, must have noticed my university ID while I was looking for my credit card. He asked what I did. On hearing that I was in physics, he started asking me more questions. I answered as best as I can, and it turned into a conversation. After two minutes, he started speaking of UFOs, and how he was sure that the government and the scientific establishment was suppressing evidence of aliens.

I was fascinated, and kept listening. After UFOs, he went on to talk about how the moon landings were faked, again by the scientists and the government. His evidence was incontrovertible. He had, after all, personally seen the movie sets on which the moon landing hoax was filmed. Oh, and Kennedy was assassinated through a complicated conspiracy. He knew this because once he had once shared a car ride across Texas with a man deeply involved in the plot, from whom he had learned all sort of critical details.

I love ideas from the fringes of science; the label on my office door says "Taner Edis: Physics and Weirdness." I spent an enjoyable half hour in that gas station. But after that, the cashier began to repeat himself. I apologized, said I had to leave, and drove back to the physics department.

Biography

Dr. Richard Gordon is a theoretical biologist whose endeavors range from AIDS prevention to breast cancer imaging on the medical side and from the effects of microgravity on amphibian embryos to the delights of diatom motility and morphogenesis on the basic science side. He inadvertently wrote the first paper on diatom nanotechnology. He has a B.Sc. in Mathematics from the University of Chicago (1963) and a Ph.D. in Chemical Physics from the University of Oregon (1967), and is now a Professor of Radiology at the University of Manitoba, where he has also held appointments in Botany, Computer Science, Electrical & Computer Engineering, Pathology, Physics and Zoology, perhaps foretold by his 1981 Rh Institute Grant for Outstanding Contributions to Scholarship and Research in the Interdisciplinary Category. His forays into how people treat one another include articles on democratization of the science granting system, a proposal for world minimum wage, havatars (human, avatar pairs) for a new approach to understanding epidemics such as HIV/AIDS, and a campaign to get up to date books to medical schools in Afghanistan via Books With Wings (http://www.bookswithwings.ca). His varied mentors, to whom he is grateful, include James S. Dwyer, Susan Meschel, E. Peter Geiduschek, Edward Anders, Aaron Novick Terrell L. Hill, Theodore Puck, Stanislaw Ulam, Jack Carmichael, Antone G. Jacobson, Cyrus Levinthal, Robert Rosen, Conrad H. Waddington, James F. Danielli, Lewis Wolpert, Zim Hearon and Lewis Lipkin.

17

Hoyle's Tornado Origin of Artificial Life

A Computer Programming Challenge

RICHARD GORDON

Department of Radiology, University of Manitoba
Winnipeg MB R3A 1R9 Canada
gordonr@cc.umanitoba.ca

Our authors have primarily concentrated on the theory of evolution, its pros and cons and nuances, barely agreeing to disagree. Evolution at least has a huge body of evidence that we can discuss, read about, and see and evaluate for ourselves, because every one of us can have opportunities to see and even dig for fossils (Macdonald, 1974; Murray, 1974; LaPlante, 1977; Arduini, 1987; Burns, 1991; Garcia, Miller & Burns, 1998; Rhodes et al., 2001; Morgan, 2005). It is a profound experience to dig for and find your own fossils, no matter what your beliefs.

At this time we have no evidence for the origin of life that is anywhere near as palpable. The origin of life is one of the outstanding conundrums of modern biology, which has been tackled mostly from a chemical point of view. It is also the playground of creationists who use probability arguments to deny that life could ever have begun "at random". The latter (Babuna, 2008; Gotfryd, 2008; Gündogdu, 2008) especially like to quote scientists who express doubt, such as:

> "In a popular lecture I once unflatteringly described the thinking of these scientists as a 'junkyard mentality'. Since this reference became widely and not quite accurately quoted I will repeat it here. A junkyard

> contains all the bits and pieces of a Boeing 747, dismembered and in disarray. A whirlwind happens to blow through the yard. What is the chance that after its passage a fully assembled 747, ready to fly, will be found standing there? So small as to be negligible, even if a tornado were to blow through enough junkyards to fill the whole Universe" (Hoyle, 1984).

I would like to suggest that artificial life (Alife) enthusiasts take up Fred Hoyle's challenge, that in a way they simulate a tornado going through a junkyard of parts, and come up with something we would all agree is alive, in the Alife sense, from components that are not alive in the Alife sense (Pennock, 2001, 2007).

What is needed is the statistical mechanics approach of the ensemble (Hill, 1960), a large number of "identical" systems, each simulating its own tornado and junkyard of parts. Given the power of modern parallel computation, the size of the ensemble could be considerable. All we need is for one member of the ensemble to make the transition from non-Alife to Alife to prove the point. Thus I am suggesting that Alife enthusiasts have an opportunity to solve the "Origin of Artificial Life" problem well before the chemists will solve the "Origin of Life" problem. By the Strong Alife Hypothesis (Levy, 1992; Boden, 1996; Olson, 1997; Anderson & Copeland, 2002), in which Alife is considered a realization of life, Alife enthusiasts will also have solved the latter problem.

Let's now try to get a handle on the magnitude of the problem. Everyone seems to be willing to have the junkyard already full of scraps that resemble parts of organisms, that are strong enough not to be pulverized by tornados. We don't have to simulate the Big Bang or the condensation of atoms or even the nucleosynthesis of the heavier "metals" (as the astronomers call all atoms more massive than helium). We might even be allowed some pretty complex organic molecules, as these are now found in abundance in the universe, and presumed to be abiotic (Gordon & Hoover, 2007). Abiotic capsules (Fox, 1965, 1972; Brooke & Fox, 1977; Fox, 1980a, b; Luisi et al., 2004), which might isolate components into an ensemble of junkyards, may even be permitted by the nay-sayers. So, taking the assembly problem as the goal, let us assume that capsule j in an ensemble of N capsules has n_{ij} copies of $i = 1,\ldots,C_j$ components, each not Alife alive, but that some subset of their arrangements within the capsule meets the criterion of being "Alife alive". In the acceptance of the tornado and junkyard paradigm for the origin of life, creationists have already conceded a materialistic interpretation of the origin of at least primitive life in putting forth their improbability arguments, so the only question is one of combinatorics.[1]

Combinatorics are more constrained the smaller the volume of a capsule, so the concept of the crowded cytoplasm (Fulton, 1982; Berg, 1990; Han & Herzfeld, 1993; Garner & Burg, 1994; Murphy & Zimmerman, 1994; Lindner & Ralston, 1997; Minsky, Ghirlando & Reich, 1997; Jaenicke, 1998; Rohwer et al., 1998; Burg, 2000; Ellis, 2001a, b; Kinjo & Takada, 2002; Bernado, Garcia de la Torre & Pons, 2004; Ovadi & Saks, 2004; Tokuriki et al., 2004; Weiss et al., 2004) may actually enhance the probability of some capsule making the transition from non-Alife to Alife.

Given an origin of the universe 13.7×10^9 years ago, and firm evidence for life on earth at 3.8×10^9 years ago, real life could have originated any time between 10^6 ABB (years After the Big Bang) to $(13.7\text{-}3.8) \times 10^9 \approx 10^{10}$ ABB. That is time enough for many tornados. So we're not talking about one tornado in one junkyard, but quite a plethora of consecutive tornados in each of a rather large ensemble of capsules. At 10^{11} stars in our galaxy and 10^{11} galaxies in the visible universe (Gordon & Hoover, 2007), half of which may have planets, and a froth of capsules say 200 nm in radius, to correspond to the smallest known cell (Huber et al., 2002; Waters et al., 2003), in a monolayer on half the surface of just one planet per solar system of radius 5000 km, we get 10^{27} capsules per solar system. This comes to 10^{49} capsules in the known universe. That's a rather

---

[1] George Bernard Shaw once found himself at a dinner party, seated beside an attractive woman. "Madam," he asked, "would you go to bed with me for a thousand pounds?" The woman blushed and rather indignantly shook her head.

"For ten thousand pounds?" he asked. "No. I would not." "Then how about fifty thousand pounds?" he continued.

The colossal sum gave the woman pause, and after further reflection, she coyly replied: "Perhaps." "And if I were to offer you five pounds?" Shaw asked.

"Mr. Shaw!" the woman exclaimed. "What do you take me for!" "We have already established what you are," Shaw calmly replied. "Now we are merely haggling over the price."

[This tale is told also of Winston Churchill, has many minor variants, and is probably apocryphal: I have not been able to find an authoritative source. –RG]

large ensemble of capsules within which the transition to life could have occurred in a few. The limit on an Alife approach is really on how many capsules and tornados we can simulate with present day computers, and whether that will be enough to solve the Origin of Alife Problem.

In a limited way, the problem has already been solved in the affirmative. In *The Game of Life* (Gardner, 1970; Pennock, 2001; Beer, 2004) a random starting condition will often produce gliders and other "self-reproducing" patterns. Another example is the spontaneous formation of patterns from subsets of random sequences of RNAs, such as palindromes (Guo, 2005). This suggests a way to achieve modularity in the tornado assembly process. In an RNA palindrome, the binding between complementary portions of an RNA strand can serve to hold it together when a tornado comes through, compared to nonpalindromic portions of the sequence. Thus, in a world of irreversible processes, a bias could occur favoring sequences that produce bigger, more complex structures. Given the enzymatic nature of RNA, and the baby steps that have been taken towards its possible self-replication, we can start to see how the Origin of Alife might be programmed. Experiments on the directed selection of RNA already select out quite rare specific RNA sequences:

> "SELEX (systematic evolution of ligands by exponential enrichment) has proven to be an excellent tool for finding nucleotide molecules that have a high affinity for a particular target from a random pool under specific conditions. It involves three processes, namely: selection of ligand sequences that bind to a target; partitioning of aptamers from non-aptamers via affinity methods; and amplification of bound aptamers… . The most successful aptamers selected by SELEX represent 1 in 10^9 to 1 in 10^{13} of the molecules in the starting library" (Gopinath, 2007).

Suppose only one in 10^{13} capsules made the transition from non-life to life. Life would then have started $10^{49}/10^{13} = 10^{46}$ times in the known universe, i.e., much more frequently than the guesstimated 5×10^4 minimum starts necessary to populate the whole universe (Gordon & Hoover, 2007). With each of the 10^{49} capsules experiencing just one tornado per year to rearrange its parts, with 10^{10} years available the universe could have 10^{59} opportunities to start life. However, diffusion within a capsule will rearrange its components roughly every 0.1 sec, based on a diffusion coefficient of about 10 μm^2/sec (Mullineaux et al., 2006) at temperatures at which water is liquid, raising the number of "tornados" by a factor of 3×10^8, or about 10^{67} opportunities to start life. The nays may not have it.

A creationist could insist that only a limited number of configurations of atoms inside a capsule are consistent with life. Taking just the binary choice of carbon or oxygen and an atomic radius of about 70×10^{-12} meters, a capsule would indeed have at least $2^{10,000,000,000}$ possible configurations, which far exceeds a "mere" 10^{67} opportunities to start life. However, cells don't die just because their shape changes and their atoms move around inside, so that there are many equivalence classes amongst the configurations (cf. Gordon, 1970), which could be further increased by "basins of attraction" (Ott, 2006) driven by atom-atom interactions. Nevertheless, to be fair, and understanding that not all large numbers are created equal, I am willing to leave the conclusion to empirical test.

To summarize, the challenge to the Alife enthusiast in trying to solve the Problem of the Origin of Alife seems to be roughly the following:

1) Choose a criterion for distinguishing non-Alife parts from "live" Alife organisms.
2) By hand or simulated evolution, construct at least one Alife organism from non-Alife parts that you deem alive by your criterion, so we know the goal is attainable. You have acted as the Intelligent Designer in this case (Gordon, 2000).
3) Place random mixtures of these parts into a very large ensemble of capsules.
4) Let the parts interact in a partially irreversible fashion, i.e., simulate a system that is not at thermodynamic equilibrium. This should allow some modularity to form.
5) Simulation of an energy source and sink could create a physics with gradients and fluxes that provide some asymmetry within a capsule. For example, compactness of parts could be interpreted as varying density, resulting in gravitational stratification or electrophoretic separation (Jaffe, 1969; Dante et al., 1995; Nouri et al., 2008).
6) Let the modules interact in a higher order fashion, such as steric hindrance (no overlapping), sticking together if they are "complementary", knot formation, etc.
7) Run simulation of the ensemble of capsules for a long time on many computers.

8) See if your hand constructed Alife organism arises out of this ensemble of capsules, or perhaps something else that meets your criterion of being live Alife, that you didn't anticipate.

Note that I have not constrained the parts to be similar to real molecules. Thus discussions of what kind of physics to simulate, of the parts or their environment, is left open to the programmer's imagination, without constraints to make Alife look like real life, such as:

> "The environment may be defined either by natural geochemistry or artificially, but should involve only simple forms of energy and material, so that the goal is to create an encapsulated biochemical system that can derive energy from simple chemicals or light (a simple form of metabolism, e.g. by redox coupling) and use information carried in primitive genes. This proto-organism, which may be held together by a lipid aggregate, should be able to self-replicate, use energy and nutrients from its environment, undergo evolutionary changes over time, and die" (Bedau et al., 2000).

One can adjust the guesstimates above in many ways. The bottom line is that the number of opportunities for a spontaneous origin of life is huge, that creationists have accepted that simple life is materialistically based and thus just improbable in their eyes, but that the number of opportunities might indeed exceed the constraints of the combinatorics, when equivalence classes and basins of attraction are taken into account. Turning the above pseudocode into a convincing simulation of the Origin of Alife will be a lot of work, but it could produce a major advance in our understanding of the Origin of Real Life.

Acknowledgement

I would like to thank Tom Barbalet for circulating a draft of this article for comments on http://www.biota.org/podcast/.

References

Anderson, D. & B.J. Copeland (2002). Artificial life and the Chinese room argument. *Artif Life* **8**(4), 371-378.

Arduini, P. (1987). *Schuster's Guide To Fossils*, Fireside.

Babuna, O. (2008). The origin and creation of life. In: *Divine Action and Natural Selection: Science, Faith and Evolution [in press]*. Seckbach, J. and R. Gordon. Singapore, World Scientific.

Bedau, M.A., J.S. McCaskill, N.H. Packard, S. Rasmussen, C. Adami, D.G. Green, T. Ikegami, K. Kaneko & T.S. Ray (2000). Open problems in artificial life. *Artificial Life* **6**(4), 363-376.

Beer, R.D. (2004). Autopoiesis and cognition in the game of life. *Artif Life* **10**(3), 309-326.

Berg, O.G. (1990). The influence of macromolecular crowding on thermodynamic activity: solubility and dimerization constants for spherical and dumbbell-shaped molecules in a hard-sphere mixture. *Biopolymers* **30**(11-12), 1027-1037.

Bernado, P., J. Garcia de la Torre & M. Pons (2004). Macromolecular crowding in biological systems: hydrodynamics and NMR methods. *J Mol Recognit* **17**(5), 397-407.

Boden, M.A. (1996). *The Philosophy of Artificial Life*. Oxford, Oxford University Press.

Brooke, S. & S.W. Fox (1977). Compartmentalization in proteinoid microspheres. *Biosystems* **9**(1), 1-22.

Burg, M.B. (2000). Macromolecular crowding as a cell volume sensor. *Cell Physiol Biochem* **10**(5-6), 251-256.

Burns, J. (1991). *Fossil Collecting in the Mid-Atlantic States: With Localities, Collecting Tips, and Illustrations of More than 450 Fossil Specimens*, The Johns Hopkins University Press.

Dante, M., S.S. Michalik, Z. Mathew, B. Doray, L.L. Ho, K.J. Wise & T.W. Christianson (1995). Retardation of RNA electrophoretic mobility due to a palindromic region. *Anal Biochem* **225**(2), 348-351.

Ellis, R.J. (2001a). Macromolecular crowding: obvious but underappreciated. *Trends Biochem Sci* **26**(10), 597-604.

Ellis, R.J. (2001b). Macromolecular crowding: an important but neglected aspect of the intracellular environment. *Curr Opin Struct Biol* **11**(1), 114-119.

Fox, S.W. (1965). *The Origins of Prebiological Systems and of Their Molecular Matrices*. New York, Academic Press.

Fox, S.W. (1972). *Molecular Evolution and the Origin of Life*. San Francisco, W.H.Freeman & Co.

Fox, S.W. (1980a). Metabolic microspheres: origins and evolution. *Naturwissenschaften* **67**(8), 378-383.

Fox, S.W. (1980b). Life from an orderly cosmos. *Naturwissenschaften* **67**(12), 576-581.

Fulton, A.B. (1982). How crowded is the cytoplasm? *Cell* **30**, 345-347.

Garcia, F.A., D.S. Miller & J. Burns (1998). *Discovering Fossils: How to Find and Identify Remains of the Prehistoric Past (Fossils & Dinosaurs)*, Stackpole Books.

Gardner, M. (1970). Mathematical games, The fantastic combinations of John Conway's new solitaire game "life". *Sci. Amer.* **223**(4), 120-123.

Garner, M.M. & M.B. Burg (1994). Macromolecular crowding and confinement in cells exposed to hypertonicity. *Am J Physiol* **266**(4 Pt 1), C877-892.

Gopinath, S.C.B. (2007). Methods developed for SELEX. *Analytical and Bioanalytical Chemistry* **387**(1), 171-182.

Gordon, R. (1970). On Monte Carlo algebra. *J. Appl. Prob.* **7**, 373-387.

Gordon, R. (2000). The emergence of emergence: a critique of "Design, observation, surprise!". *Rivista di Biologia /Biology Forum* **93**(2), 349-356.

Gordon, R. & R.B. Hoover (2007). Could there have been a single origin of life in a big bang universe? *Proc. SPIE* **6694**, doi:10.1117/1112.737041.

Gotfryd, A. (2008). Evolution: myths and facts. In: *Divine Action and Natural Selection: Science, Faith and Evolution [in press]*. Seckbach, J. and R. Gordon. Singapore, World Scientific.

Gündogdu, C. (2008). The collapse of the theory of evolution. In: *Divine Action and Natural Selection: Science, Faith and Evolution [in press]*. Seckbach, J. and R. Gordon. Singapore, World Scientific.

Guo, P. (2005). RNA nanotechnology: engineering, assembly and applications in detection, gene delivery and therapy. *J Nanosci Nanotechnol* **5**(12), 1964-1982.

Han, J. & J. Herzfeld (1993). Macromolecular diffusion in crowded solutions. *Biophys J* **65**(3), 1155-1161.

Hill, T.L. (1960). *An Introduction to Statistical Thermodynamics*. Reading, Addison-Wesley Publishing Co.

Hoyle, F. (1984). *The Intelligent Universe*. New York, Holt Rinehart and Winston.

Huber, H., M.J. Hohn, R. Rachel, T. Fuchs, V.C. Wimmer & K.O. Stetter (2002). A new phylum of Archaea represented by a nanosized hyperthermophilic symbiont. *Nature* **417**(6884), 63-67.

Jaenicke, R. (1998). Protein self-organization in vitro and in vivo: partitioning between physical biochemistry and cell biology. *Biol Chem* **379**(3), 237-243.

Jaffe, L.F. (1969). On the centripetal course of development, the Fucus egg, and self-electrophoresis. *Dev. Biol.* **3**(Suppl.), 83-111.

Kinjo, A.R. & S. Takada (2002). Effects of macromolecular crowding on protein folding and aggregation studied by density functional theory: dynamics. *Phys Rev E Stat Nonlin Soft Matter Phys* **66**(5 Pt 1), 051902.

LaPlante, J.C. (1977). *The Weekend Fossil Hunter*, Drake Publishers.

Levy, S. (1992). *Artificial Life, The Quest for a New Creation.* New York, Pantheon Books.

Lindner, R.A. & G.B. Ralston (1997). Macromolecular crowding: effects on actin polymerisation. *Biophys Chem* **66**(1), 57-66.

Luisi, P.L., P. Stano, S. Rasi & F. Mavelli (2004). A possible route to prebiotic vesicle reproduction. *Artificial Life* **10**(3), 297-308.

Macdonald, J.R. (1974). *Sabretooth Cats and Imperial Mammoths: A Guidebook to Fossil Hunting in Southern California*, W. Ritchie Press.

Minsky, A., R. Ghirlando & Z. Reich (1997). Nucleosomes: a solution to a crowded intracellular environment? *J Theor Biol* **188**(3), 379-385.

Morgan, B. (2005). *Rock and Fossil Hunter (Nature Activities)*, Dorling Kindersley Publishers Ltd.

Mullineaux, C.W., A. Nenninger, N. Ray & C. Robinson (2006). Diffusion of green fluorescent protein in three cell environments in *Escherichia coli. J Bacteriol* **188**(10), 3442-3448.

Murphy, L.D. & S.B. Zimmerman (1994). Macromolecular crowding effects on the interaction of DNA with Escherichia coli DNA-binding proteins: a model for bacterial nucleoid stabilization. *Biochim Biophys Acta* **1219**(2), 277-284.

Murray, M. (1974). *Hunting for Fossils: A Guide to Finding & Collecting Fossils in all Fifty States*, MacMillan.

Nouri, C., R. Luppes, A.E.P. Veldman, J.A. Tuszynski & R. Gordon (2008). Rayleigh instability of the inverted one-cell amphibian embryo [Invited: "Physical Aspects of Developmental Biology" special issue, Guest Editor Shane Hutson, Vanderbilt University]. *Physical Biology* **5**, 015006.

Olson, E.T. (1997). The ontological basis of strong artificial life. *Artif Life* **3**(1), 29-39.

Ott, E. (2006). Basin of attraction. *Scholarpedia* **1**(8), 1701. http://www.scholarpedia.org/article/Basin_of_attraction

Ovadi, J. & V. Saks (2004). On the origin of intracellular compartmentation and organized metabolic systems. *Mol Cell Biochem* **256-257**(1-2), 5-12.

Pennock, R.T. (2001). Getting A-Life: negotiating the border between virtual and real. In: *International Society for the History, Philosophy and Social*

Studies of Biology Meeting, 18-22 July 2001 at Quinnipiac University in Hamden, Connecticut.

Pennock, R.T. (2007). Models, simulations, instantiations, and evidence: the case of digital evolution. *Journal of Experimental & Theoretical Artificial Intelligence* **19**(1), 29-42.

Rhodes, F.H.T., P.R. Shaffer, H.S. Zim & R. Perlman (2001). *Fossils*, Golden Guides from St. Martin's Press.

Rohwer, J.M., P.W. Postma, B.N. Kholodenko & H.V. Westerhoff (1998). Implications of macromolecular crowding for signal transduction and metabolite channeling. *Proc Natl Acad Sci USA* **95**(18), 10547-10552.

Tokuriki, N., M. Kinjo, S. Negi, M. Hoshino, Y. Goto, I. Urabe & T. Yomo (2004). Protein folding by the effects of macromolecular crowding. *Protein Sci* **13**(1), 125-133.

Waters, E., M.J. Hohn, I. Ahel, D.E. Graham, M.D. Adams, M. Barnstead, K.Y. Beeson, L. Bibbs, R. Bolanos, M. Keller, K. Kretz, X. Lin, E. Mathur, J. Ni, M. Podar, T. Richardson, G.G. Sutton, M. Simon, D. Soll, K.O. Stetter, J.M. Short & M. Noordewier (2003). The genome of *Nanoarchaeum equitans*: insights into early archaeal evolution and derived parasitism. *Proc Natl Acad Sci U S A* **100**(22), 12984-12988.

Weiss, M., M. Elsner, F. Kartberg & T. Nilsson (2004). Anomalous subdiffusion is a measure for cytoplasmic crowding in living cells. *Biophys J* **87**(5), 3518-3524.

Dialogue

Steve Grand

The analogy is such a terrible one to start with that it's difficult to see what would be served by extending it, but I imagine that's exactly Dick's point. Creationists (and physicists like Fred Hoyle — it's interesting and illuminating to see how often ID proponents are also physicists) mistakenly think that evolution by natural selection is equivalent to tossing a vast number of dice and expecting them all to come up six. But of course it is nothing of the kind. Natural selection is far from being a random process — it acts as a ratchet — a kind of Maxwell's Demon — that makes the emergence of functional systems vastly, hugely, overwhelmingly more likely than mere tossing of dice (or blowing of tornadoes). It's still a mystery to me how people fail to see that. But they do, and it's easy to cite the lack of available evidence on how life actually got started in order to claim that it just wasn't possible. We know it's a failure of the imagination — the "I just can't see how it's possible, therefore it isn't" fallacy of reasoning — but it is beholden on us to show that it is indeed possible to go from non-living to living systems by accumulation. Nobody has shown it yet. Nobody has used simulated natural selection to go from clearly non-living mechanisms to clearly living ones in a non-trivial way. There's a challenge in the tornado analogy, no matter how obviously wrong it is (or isn't, if you're a physicist).

Pavel Bazant

Several years ago, I amused myself with writing cellular automata (CA) simulators. The simulators visualized the time evolution of the automaton. I experimented with various rules. Some were very interesting to watch. Eventually, the following happened: I was experimenting with a certain 3-state CA. After watching the simulator for several minutes, a self-replicator appeared. It replicated until it filled the whole screen. When two replicators collided, they transformed into rubbish that would kill any replicator that collided with it. The build-up of rubbish eventually led to "extinction" of the replicators, leaving an uninteresting screen full of stationary rubbish. So the message is: replicators can appear spontaneously. Sometimes the simulator would run for half an hour without replicators, but sooner or later they would appear and spoil the "show".

I had to alter the rules ("physical laws") to prevent this from happening. It happened then also, but it took several hours on average.

The rules are as follows (like Conway's 23/3 rules, but with the creation of a 2-cell when the neigbourhood-sum equals 5):

sum of neighbours:
0,1,2,3,4,5,6,7,8,9,...
state 0 ->
0,0,0,1,0,2,0,0,0,0,0,0,0,0,0,0
state 1 ->
0,0,1,1,0,0,0,0,0,0,0,0,0,0,0,0
state 2 -> always to 0

Julian Chela-Flores
I was greatly rewarded by reading the papers you kindly sent to me. I learnt several useful ideas, some of which are quite original, for instance biohorizon and the speed of life (Gordon & Hoover, 2007).

My only perplexity was with the early time for the first onset of life in the universe (half a billion years ABB = After the Big Bang). My difficulty is the following. If the 'Dark Ages' ended 1 billion years ABB, it was only then when the first stars began to be significant. So the metallicity factor (Gordon & Hoover, 2007) probably did not have much of a chance to be significant.

In any case the paper is very thought-provoking.

The chapter (above) in which you provide statistical arguments for the probability of 'Alife' is also interesting. My own feeling though is that the idea of evolutionary convergence (well documented on the Earth biota) greatly constrains the possibilities that chemical evolution (which as you say, evidence for which is everywhere, starting from meteorites) provides for the origin of the first cells a limited number of options making the science of biology of universal validity.

Richard Gordon
As I pointed out in Gordon & Hoover (1977), there is a discrepancy between the theory of nucleosynthesis and astronomical observations: early, "metal" poor stars are predicted, but haven't been found.

The basic premise of Alife is that the substrate for life need not be real matter. Thus, while real life in the universe may indeed be confined to variants on a common chemistry, Alife in the universe may not have to be. This lesser constraint on Alife may also make it easier to solve the origin of Alife than the origin of life.

Part 4

Scientific Views on Divine Action

Biography

Dr. Mordechai Halperin is currently the Chief Officer of Medical Ethics at the Ministry of Health, and the director of the Schlesinger Institute for Medical-Halachic Research, Jerusalem, Israel. He obtained his Mathematics & Physics B.Sc. degree at the Hebrew University of Jerusalem in 1974, his LL.B. degree at Uno Academic Campus in 2005 and his MD degree at the Hebrew University and Hadassah Medical School in 1987. Between 1986 and 2000, he had professional, medical experience at various medical centers throughout Israel, and was the director of the Jerusalem Medical Center for Impotence & Infertility. In the field of medical ethics, he is a Founding Member of the Israeli Society for Medical Ethics, and a member of numerous other ethics committees. Dr. Halperin is the chief editor of *Assia*, the Hebrew quarterly review of medical ethics and Jewish law, the *ASSIA Books series*, the *JME Books series* and various professional books on Jewish medical ethics and halacha [Jewish law]. For 18 Years, until 2004, he was the chief editor of the English international journal *Jewish Medical Ethics*. Dr. Halperin has both lectured and chaired at over 250 international academic conferences in Israel and abroad. In 1989, 1991 and 1994, he was awarded prizes for his dissertations on various medical & ethical topics. Dr. Halperin has authored more than 180 original articles.

18

The Laws of Evolution and Judaism

Lack of Communication

MORDECHAI HALPERIN

4 Man Street, Givaat Mordechai
Jerusalem, Israel, Zip code: 93711
medical.ethics@moh.health.gov.il

1. Introduction*

The word "evolution" arouses strong and conflicting feelings among different groups of people. Religious people are reminded by this term of the endless strugles that the nineteenth-century scientists conducted against the Catholic Church. The essential unity of ecclesiastical and secular institutions was lost during the 19th century, to the point of senseless hostility (Einstein, 1950). Scientists who had been at that time freed from the oppression of the Church fought bitterly against any religious faith, as a reaction against the restrictions that the Church had imposed on scientific thought for many centuries (Cherniavsky, 1965). The burning of Giordano Bruno at the stake in 1600 and the bringing of Galileo Galilee before the Inquisition in the sixteen-thirties together with his quiet muttering "but it will nonetheless move", became essential parts of the anti-religious creed of the nineteenth-century scientists. Darwin's theory of 'the origin of species' (Darwin, 1859), making use of the laws of evolution, became a prominent component of the banner of Liberation

* This chapter is based on a preliminary Hebrew work (Halperin, 2001).

from the Yoke of the Church, and to many it represented the antithesis of religious belief.

To secularists the word "evolution" brings to mind similar associations from the other side. To many, "evolution" was understood to be a scientific alternative to a belief in the Creator. Many people, especially those not closely familiar with scientific research and its principles, see in the Theory of Evolution a scientific proof of the absence of divine creation and of the natural development of man from monkey (or more strictly both from a common ancestor), which developed from lower creatures which in turn developed from a primordial biological cell that came into being by accident as a result of physical and chemical occurrences at some very early time.

Unfortunately, many people fail to appreciate and understand scientific methods. Confusion of terms is common, as well as the inability to distinguish between a theory or assumption and conclusions drawn from controlled experiments. It is no wonder that at times one listens to a "dialogue among the deaf" on evolution, a dialogue based on lack of information and of scientific understanding on both sides. It is now time to clarify the scientific terms, and to differentiate between laws of nature that can be scientifically proved and extrapolative theories which serve as a pivot of faith to sworn atheists although they lack a scientific proof.

2. The Emotional Problem

Man is an emotional, observant and thinking creature. Observation of his surroundings arouses questions already in childhood. Every child who sees an impressive building asks his parents "Who built this building?" Likewise, everyone passes through a stage in his life when he asks "Who created this world?" This latter question has no immediate implication, as opposed to questions concerning the characteristics of our world. Information on the characteristics of the world and the laws under which it operates enable us to develop technology and make use of the results as we wish. By contrast, an answer to the question "Who created this world" will make no direct difference to our conditions of life. There is simply a deep emotional need to know the answer, a need that already exists in childhood. Its practical effect focuses only on the personal conclusions regarding the place of a person in his world and on his personal attitude to religious faith.

In the distant past two conflicting answers were already given to the question "Who created this world?" Aristotle, as many other Greek philosophers, believed in the (past) eternity of the world. In his opinion the world has existed with its laws of nature since all eternity. In simple terms, the world has always existed. This theory is based on the assumption of or belief in the absence of a creator, and conversely the absence of a creator demands as a corollary the eternity of the world.

Against the Aristotelian theory, the religious answer to the fundamental question "who created this world?" is based on the opposite belief — that there is a creator. The world is comparatively young; it underwent a creation process by the Creator.

Each theory contains an element that is beyond human conception. The Aristotelian, that the world goes back to all eternity, incorporates the concept of infinity which is far from ordinary human conception, even though it can be defined in mathematical philosophy. The religious approach, on the other hand, contains the faith in a Creator, whose very nature is beyond human conception. In the words of Maimonides (Laws of Repentance, end of Chapter 5): "Man has not the power to discover the true nature of the Creator, as it is written 'For no man will see me and live' (Ex. 33:20)".

3. The Second Law of Thermodynamics

The nineteenth-century secular scientists, who on being freed from oppression by the Church rejected any connection with the Church or with any religion, had an interesting attitude. They rejected out of hand any scientific explanation that might affect their faith in the eternity of the world. Professor Cherniavsky in his fascinating book *Between Science and Religion*, describes two fundamental scientific principles that are today accepted without dispute, but which were rejected for many years by the nineteenth-century scientific community because of the "danger" that accepting them might cast doubts on the holy principle of the eternity of the world and the absence of a creator. These two principles were the Principle of Least Action and the Second Law of Thermodynamics (Cherniavsky, 1965).

The Second Law of Thermodynamics was born in the mind of a French engineer named Sadi Carnot (1796-1832) at the end of the first quarter of the nineteenth century. The principle, as presented by Ludwig Boltzmann in a popular language, determined that in natural processes the amount of disorder in a closed system increases all the time. The physicist Rudolf Clausius (1822-1888) gave the law a clear and precise mathematical expression by introducing a new quantitative term into science which he called "entropy." This quantity has a special peculiarity: in all natural processes in a closed system its numerical value can only increase, or more precisely, it never decreases.

In other words, in nature there is asymmetry with respect to time and events are unidirectional. The world develops in such a way that its entropy, practically, is always increasing.

Asimov describes this graphically (Asimov, 1970):

> "Another way of expressing the Second Law of Thermodynamics is that the Universe is progressing all the time towards chaos. In this way we can see the realization of the Second Law wherever we turn. We have to invest a lot of hard work in tidying the house. When leaving a room alone you can see how quickly neglect increases. Even if no living creature enters it, a layer of dust and filth accumulates. Consider how much work is needed in maintaining the house or in maintaining machinery. How much treatment we have to apply to our bodies in order to keep them clean. In fact everything deteriorates and wears away if we do not apply proper treatment and care".

Two conclusions follow from this Law. The first conclusion is that an imaginary journey backwards in time will eventually bring us to a time when entropy was a minimum. Before this time no processes were possible, a strong blow against those who believed in the eternity of the world.

The second conclusion is that the world, as a closed system, is progressing towards a state of maximum entropy with complete chaos, a state that could be described as "the death of the world." This implication of the entropy principle led many in the nineteenth century to reject it on metaphysical grounds. In the twentieth century, after Western science had freed itself from the struggle against the Church, the Second Law of Thermodynamics became one of the corner-stones of exact science.

4. The Big Bang

The theory of the eternity of the world received many blows in the last century. A vast amount of scientific evidence indicates the rapid recession of the galaxies and the physical spreading of material in the universe.

There is enough evidence that sometime between 10 billion and 20 billion years ago, the Universe was created from a gigantic cosmic explosion that hurled matter in all directions — the Hot Big Bang. Its subsequent evolution from one hundredth of a second up to the present day can be reliably described by the Big Bang model. This includes the expansion of the Universe, the origin of light elements and the relic radiation from the initial fireball, as well as a framework for understanding the formation of galaxies and other large-scale structures. In fact, the Big Bang model is now so well-attested that it is known as the *standard cosmology*
(http://www.damtp.cam.ac.uk/user/gr/public/bb_home.html;
http://liftoff.msfc.nasa.gov/academy/universe/b_bang.html).

To say the least, these theories do not support the assumption of the eternity of the world. However, as in the past so today there are scientists who zealously uphold their belief in the Eternity Theory, and who have thereby been forced to develop "spring theories" in which everything spreads out, and then springs back and so forth. There is no scientific basis for any of these theories, merely a need to reconcile scientific discoveries with their belief in the eternity of the world. In practice, many leading scientists today reject Aristotle's theory from a scientific standpoint.

5. How Was Man Created?

Rejecting belief in the eternity of the world poses a very difficult and sensitive problem to those scientists who "believe" in the absence of a creator: 'how was man created?'

The theory of the Origin of Species was formed to overcome such difficulties by applying the Laws of Evolution. The source of the theory goes as far back in history as ancient Greece, but it was rejuvenated in the eighteenth and nineteenth centuries, when it enabled European scientists freed from oppression by the Church to live with an explanation of the existence of life in the world.

The conceptual basis of the theory was supplied by fossils in early geological layers. The earliest creatures, which are also the simplest in structure, are generally found in the oldest layers. The oldest fossils discovered were single-celled creatures. In later layers there were algae and simple multiple-celled creatures. Still later appeared mollusks and creatures with legs, followed by insects and amphibians, then reptiles, birds and finally mammals. The order of the discoveries of living creatures from simple to complex led to a further assumption: not only did the complex creatures appear later in time than the simpler ones, but they are also assumed to be derived and developed from the simpler ones who are their ancestors. To explain the mechanism that was claimed to have enabled this radical change to occur, the Darwinists and Neo-Darwinists used the Laws of Evolution.

6. The Laws of Evolution

For our discussion we first need to define the Laws of Evolution. These are laws of nature that explain changes in the characteristics of living organisms over a long period. Examples of these changes are found in many areas. In the course of time bacteria that are initially sensitive to a specific antibiotic medicine develop immunity. The expression "changes in nature" is known to all who are familiar with halakhic (Jewish Law) literature, and is an example of the wider variety of changes in organisms over a long period.

Darwin coined the expression "natural selection" as a basic mechanism for evolutionary changes. The theory of Natural Selection states that all varieties reproduce more than is required to maintain the species in a world in which there is a cruel competitive struggle for survival, and most of those born are destroyed. There are small differences in the characteristics of individuals of each species which affect their ability to survive, so that the organisms which have characteristics most suitable to the environment have a greater chance of surviving long enough to pass on their characteristics to the next generation.

This theory explains observed changes in many species, and with the amendments of the geneticist Theodosius Dobzhansky in 1937, who pointed to mutation as an additional mechanism, further evolutionary changes can be explained (Dobzhansky, 1977; Lammerts, 1965).

In a laboratory it is not difficult to demonstrate changes in large populations of bacteria and insects, changes in certain characteristics that affect their ability to survive in specific situations. But it has not been possible to demonstrate in a laboratory changes in which one species of living creature can produce an offspring which is similar to, and appears to belong to a different species! An analogy is the difference between chemistry and alchemy. It is possible by chemical means to alter chemical compounds, but it is not possible either by chemistry or alchemy to change lead into gold.

From a scientific point of view one can relate to the basic Laws of Natural Selection as to conclusions from controlled experiments. But there is no scientific evidence to support any assumption that these laws can be extrapolated to explain the alleged birth of an offspring whose species is different from that of its antecedents.

7. The Attitude of Halakhic Scholars to Controlled Experiments

Scientific principles based on observations and controlled experiments were always accepted in the world of halakhah (Jewish Law). Rabbi Isaac ben Sheshet, known as Ribash (1326-1408), in his Responsum 447 draws a fundamental distinction between the scientific approach of the (Talmudic) Sages and the Aristotelian approach.

As is well known, Aristotle and his followers in many cases performed no experiments, but simply produced intellectual theories on what ought to happen in the world on the basis of the teleology (purpose) of nature. They presumed that what ought to occur according to human thought certainly does occur. Even when the Greek scientists actually performed experiments, they did not take the trouble to perform controls to prevent doubts that might arise as to the validity of those experiments. For that reason, says Ribash:

> "We place no trust in the words of the Greek and Arab scientists, who gave their opinions only on the basis of their own theories or on some experiments without caring about any doubts that might arise regarding the validity of those experiments."

Consequently, Ribash accepted the opinion of our Sages that offspring inherits both maternal and paternal characteristics, in contrast to that of the Greek scientists who claimed that heredity is only maternal. Similarly, whereas the Greek scientists claimed that intercourse is dangerous for the mother in the eighth month of pregnancy, the Ribash accepted the opinion of the Sages that in the third trimester of pregnancy intercourse is beneficial both for the mother and for the fetus — which is fully in accord with medical knowledge of the past decade.

From the words of Ribash we see that the Sages pioneered the approach that demands drawing conclusions only from controlled experiments, an approach that has been accepted in the medical world only in the past few generations.

A similar conclusion may be reached from the words of Hatam Sofer (Responsa Yoreh Deah 45): "experience is a more trustworthy witness than all theories put together." Similarly Nachmanides (Genesis 9:12) states clearly that we ought to accept a Greek theory only when it is based on a convincing scientific proof. It is a well known approach of Maimonides (*Guide for the Perplexed*) as its roots are in the Talmud itself (Hullin 55b): "We do not conclude in ***terefot*** (fatally injured animals, which their terminal condition makes them ritually unfit for eating…) that one case is comparable with the other." It follows that laws of Natural Selection that can be verified by observation and controlled experiments ought to be accepted in Judaism. The Laws of Natural Selection explain satisfactorily phenomena of characteristics that change in the course of generations, but tell us nothing at all about the origin of species.

On the other hand, Darwin's Theory of the Origin of Species makes use of the Laws of Natural Selection in an attempt to explain the "faith" in the absence of a divine creator, and in the development of man from the ape (or, more strictly, both from a common ancestor, man being a more advanced development), in turn a development from an early single biological cell that accidentally came into being by physical and chemical actions in the distant past.

8. The Origin of Species

Many are persuaded that the theory of the Origin of Species originated in the eighteenth and nineteenth centuries from the Frenchmen Georges de Buffon and Jean Lamarck and the Englishman Charles Darwin. However a brief glance at world history shows the existence of a similar theory already in the sixth century B.C.E. (Kurman, 1974). Anaximander of Miletus, the Greek, a pupil of Thales, explains in his *Book on Nature* how life developed according to his belief: "The Earth was initially in liquid form, and as it settled into shape animals and man were created from it. First came fish, then as the water receded some of these were swept onto the land; of the latter some developed the ability to breathe air, and these were the ancestors of land animals at a later epoch."

Anaximander had nothing with which to back up his development theory. 2500 years later the Neo-Darwinists, believers in the Theory of the Origin of Species, tried to use the Laws of Natural Selection to explain the mechanism of development from prebiotic soup to initial cells and eventually to man.

There have been *Torah* philosophers who found no fundamental contradiction between the Theory of Evolution and Jewish belief. Early in the twentieth century Rabbi Abraham Kook, the first Ashkenazi Chief Rabbi of the Land of Israel, wrote about this. On account of shortage of space, we will not deal here with his surprising approach and will confine ourselves to actual scientific problems that prevent the acceptance of the neo-Darwinian theory. (See D. Berger's chapter in this volume regarding Rabbi Kook's attitude to evolution theory).

What are the statistical chances of a living cell arising spontaneously from mineral matter? Two world-famous astronomers, Sir Fred Hoyle and Chandra Wickramasinghe, calculated the probability of such an event, and published their findings in their book Evolution from Space, (Hoyle and Wickramasinghe 1981). They claim that the probability of such occurring is one in ten-to-the-power-of-forty-thousand ($1/10^{40,000}$). In their words, this number "is sufficient to bury Darwin together with his Theory of Evolution."

It is worth noting that this same Fred Hoyle, one of the greatest astronomers of the past generation, is renowned in particular for his atheism! The *London Times* book critic (7.9.81, p. 6) impartially concluded that it seems that faith in Evolution and the Origin of Species has blinded the scientists.

As an attempt to reach a scientific expression, the Theory of the Origin of Species is far from convincing, despite the beautiful and attractive theories and "convincing pictures" in popular science books and secondary-school text-books. From a scientific viewpoint one may say that the Theory of the Origin of Species is not just unproved but almost disproved. From its very beginning it has been beset with serious scientific difficulties.

When Darwin set it up, the terminology regarding heredity on which it was based was already out of date. Meanwhile it has been re-expressed on the basis of modern terms of mutation (Lammerts, 1965), but that has removed only a few of the problems.

This theory requires that there should be thousands of generations to develop a new organ such that it can help in the fight for survival. Until it reaches a certain stage of development, the organ is usually an impediment; and as such, in the natural process that leads to the survival of the fittest, such an underdeveloped organ is a cause of destruction for those who carry it. Further, there are a number of giant leaps that raise serious difficulties with the theory (Hoyle and Wickramasinghe 1986). In view of all this one needs a very strong and almost fanatical faith to continue to maintain belief in the theory (Crick 1981).

Absurdity reaches its peak in the writings of the biologist Lecomte du Nouy. After attacking the lack of reasonableness of the theory, he nonetheless accepts it, because in our times "it is almost impossible not to believe in evolution"! He answers the unacceptable problems by saying that he assumes direct divine(!) influence, on the assumption that the miracle of converting asexual reproduction to sexual reproduction is less of a miracle than the possibility of divine creation of the sexes separately, one after the other (Du Nouy, 1947).

There is an enormous amount of literature scientifically criticizing the theory of the Origin of Species from a variety of aspects. Hebrew readers interested in learning more about the problems that have arisen following new discoveries of fossils are recommended to *Creation — Origin of Life* (Trop, 1982), which also deals with additional scientific glances at neo-Darwinism and an analysis of recent ideas in the field of evolution research. English readers may use its extensive English references to scientific literature.

9. In Conclusion

One needs to distinguish between results of controlled experiments in evolution on the one hand, and extrapolated results of 'believers' in neo-Darwinist theories on the other.

There is no reason for a religious person to be afraid of exact science that deals with mutations and the Laws of Selection. On the other hand, there is no reason for an atheist to treat the theory of the Origin of Species as proven scientific fact, which it is not.

One who wishes to believe in the Theory because it appeals to him aesthetically or because it provides an alternative emotional faith may do this only on his own responsibility, while recognizing that he is dealing with a personal belief and not with objective science.

References

Asimov, I., Smithsonian Institute Journal, June 1970, p. 6 (retranslated from Trop 1982, p. 152).

Cherniavsky, A., Between Science and Religion (Heb.). Joshua Chechik, Tel Aviv, 1965.

Crick F., Life Itself – its Origin and Nature. Simon and Schuster, New York, 1981.

Darwin, Ch., Origin of Species, 1859.

Dobzhansky T. et al, Evolution. Freeman and Co., San Francisco, 1977.

Du Nouy L., Human Destiny. The New American Library, New York, 1947.

Einstein A., Out of My later Years. Philosophical Library, New York, 1950.

Halperin M., The Laws of Evolution and Judaism (Heb.). Assia 67-68, The Dr. Falk Schlesinger Institute for Medical-Halachic Research at the Shaare-Zedek Medical Center, Jerusalem, 2001.

Hoyle F. and Wickramasinghe, C., Evolution from Space. J. M. Dent, London, 1981.

Hoyle F. and Wickramasinghe, C., Archaeopterix, the Primordial Bird: A Case of Fossil Forgery. Christopher Davis, London, 1986.

Kurman, A., Evolution and Judaism, (Heb.). El-Hamasoret, Tel-Aviv, 1974.

Lammerts, W.E., Creation Research Society Quarterly 2, 39, 1965.

Trop, M., Creation – Origin of Life, (Heb.). Shamir, Jerusalem, 1982.

Biography

Dr. Frits Bienfait obtained his Ph.D. in biochemistry from the University of Amsterdam in 1975, and after a short stay in the Johnson Foundation in Philadelphia, he set up a research group on iron metabolism of plants in the Laboratory of Plant Physiology in Amsterdam. In 1986 he joined the Laboratory of Genetics in Ghent, and in 1987 he moved to the Institute for Nuclear Chemistry in Agriculture in Wageningen. Since 1989 he studies the fundamental forces underneath the Left – Right polarity in politics, in collaboration with the political scientist Prof. Meindert Fennema and the anthropologist Dr. Wouter van Beek.

19

Evangelicals Fighting Doubt, Not Darwin

FRITS BIENFAIT

Oudegracht 285 bis, 3511 PA Utrecht
The Netherlands
bienfait@ision.nl

1. Introduction

If we want to understand the forces that drive the resistance against Darwin's theory on the origin of species, a fruitful approach might be to look for the source of the most strongly and actively formulated arguments against it. In the USA, and also in the Netherlands, the loudest voices are raised by fundamental Christians (Pennock 1999), and especially by members of a movement known as the 'evangelicals'. They consider the theory to be counter to the most fundamental elements of their religion, which says that God created the diverse forms of life in a supernatural process ("creation") not susceptible to scientific falsification.

Darwinists cannot provide evidence for all aspects of the theory but report an increasing quantity of data that support their ideas. The "creationists" hold on to their position, also presented as "Intelligent Design" (Jones, 2005), but without being able, so far, to produce any evidence against Darwin's theory or in favour of their own view.

Problems arise when decisions must be taken on how biology should be taught in schools, as was recently illustrated by the case of Kitzmiller vs. Dover Area District School board (Jones, 2005). In this case, the board tried to introduce a creationist biology textbook against the wish of both teachers and parents of pupils, who demanded a text based on science. But the controversy is not restricted to biology. An ideology based on the notion of supernatural creation as a fundamental force in culture and politics has been formulated in the "Wedge Document" (an internal document of the "Discovery Institute" and available on internet) even including a strategy for establishing a kind of theocracy in the USA [http://en.wikipedia.org/wiki/Wedge_strategy].

One may ask why evangelical Christians should be so keen on fighting a scientific theory. If they are really convinced that their religious ideas, including those on creation, represent the inviolable truth, they could just sit and wait. Their struggle might therefore be interpreted as originating from a fear that the theory will eventually prevail, and in doing so destroy their own conviction — with disastrous consequences, as has been most colourfully depicted by e.g. Ken Ham (1987). However, this interpretation, plausible as it seems, may be too simplistic.

In order to get a clearer understanding of the evangelicals' position, I shall try to identify some elements in their motivation. Arguments can be outspoken, but behind them are often unspoken motives or unconscious forces which can be at least as clarifying as the openly ventilated reasoning. A search for such motives and forces is difficult and certainly risky — but the issues at stake are important enough to justify such an effort, since it may provide us with an answer to the question why, in the past century and a half, all rational arguments in favour of the scientific theory should have had so little effect on its strongest adversaries. I shall do this, inevitably, from a Dutch perspective.

2. The Evangelicals

There is, it seems, no written comprehensive theology for all those who identify themselves as evangelicals. In the USA (Yancey, 2005), as in the Netherlands (Veldhuizen, 1988), the lack of an established theology is characteristic for the movement, which draws its adherents from various protestant denominations and from Roman Catholicism. But there are some essential elements these varying beliefs have in common:

1. Absolute acceptance of the *Bible* as providing the historical truth and the word of God to mankind.

2. A personal relationship with Jesus Christ who, by his death on the cross, sets the believers free from the consequences of Adam's fall in paradise and saves them from damnation. The relationship with Jesus can be initiated by a "born-again" experience.
3. An obligation to share one's faith in Jesus with other people, in order to save them too.

According to Zoba (Zoba 2005), "Evangelicals battle Satan's attempts to sabotage God's plans by preying upon the weakest spot in every human's heart, including (especially) theirs". Satan started right away with Adam and Eve, "planting the slightest seed of doubt with the words: Did God say...?" Evangelicals believe that from the moment Adam and Eve doubted God and ate the forbidden fruit, humankind fell under Satans domination ("the Fall"). That is why the relationship with Jesus, essential for the redemption from the forces of Satan, is so important to evangelicals.

With regard to their social position in the US, Zoba remarked that the lower income Christians are more likely to be evangelicals: among those with household incomes under $35,000, 45% are evangelicals, while for those with higher incomes this figure is 31%. She also notes that since the 1970s evangelicalism has been criticized as angry and rigid in tone.

In the Netherlands, internet sites of evangelical groups contain, far more often than sites of other denominations, cases of joining of wronged, depressed, or in any other way unhappy people ("My life was a mess and had no purpose, but now... ."). Evangelical theology is often characterized as superficial (Runia, 1984, Veldhuizen, 1988), probably because of their rather anti-intellectual stance ("Belief is a matter of the heart, not the brain") (cf. Zoba, 2005). Their political and intellectual influence in the Netherlands is practically nil. All this is in sharp contrast to the characteristics of other orthodox Christians. Self-assured, these see themselves as the staunch builders of the Dutch nation from the 16th century on. They founded their own protestant university in Amsterdam in 1880, and both their contribution to intellectual life and their political influence are stronger than one would have expected on the basis of their numbers.

To the outside world the most conspicuous characteristics of the evangelicals are the religious services, with their extensive display of gladness (singing and clapping of hands) and their missionary drive. Their gladness is about the redemption from evil by Jesus, which secures for them a beautiful future after this earthly life. The evangelical song texts indicate a warm relationship between the believer and Jesus, more so than in the standard protestant songs. Indeed, words like "gladness", "love" and their synonyms, and the name of Jesus occur far more frequently in the texts of the Dutch evangelical songbook (Eschbach and de Ruiter, 1999) than in the songbook of the united Dutch protestant churches (Lazonder, 1996). The missionary drive is based upon the command given in Matthew 28: "Go ye therefore, and teach all nations, baptizing them in the name of the Father, and of the Son, and of the Holy Ghost". Of course, the *Bible* contains more commands. Could there be some extra source to supply the extraordinary energy and sacrifices demanded by this special task?

3. "Methinks the gentleman protests too much"

When a message is propagated with an intensity that is not clearly justified by the context, the listener will ask what might be the reason for all that energy invested. A suspicion tends to arise that something quite different is hidden underneath the gentleman's message. Looking at the evangelicals, we too may ask: why this exuberant display of gladness? Why this urgent need to share one's belief with others? In order to be able to answer these questions, it may be helpful to have a look at a simple example of a "too much", that everyone will recognize:

> When I was 12 years old, I went to my first summer camp with the boy scouts. We put up our small tents among the pine trees, far from home and civilization. In the evening, as the light slowly disappeared and unknown noises started coming from the woods, we became silent and huddled together at the open place where the camp leaders had prepared a big fire. We started a program of camp songs, the most popular of which consisted of nothing but the text: "Wij zijn niet bang!" ("We're not afraid"), endlessly repeated on varying melodies, with a loud emphasis on the word "BANG!". This simple ritual gave us courage. What exactly happened?
>
> Every boy was slightly nervous. But of course, it was out of the question to show any sign of it. Thus, nobody knew that he was not the *only* one to be afraid. To join in with the song was no problem, since it explicitly *denied* any fear. The louder one was singing, the clearer the

> signal that he was not afraid *at all.* And with everybody in this mighty song assuring the others of his heroic feelings, their fears could be smothered more effectively. However, just because of the extraordinary enthusiasm and the heavy accent on "Bang!", each of us might also develop a suspicion that perhaps the others had the same drive to sing so loudly, and consequently the same underlying feelings. Relief all round: in our common weakness, together we stood, and no more fear! The atmosphere cleared, singing became a slightly ambiguous joke: a song to deny, and admit, with a wink.

With this example in mind, we may now take a fresh look at the two remarkable properties of evangelicals mentioned before. To start with the simpler case:

4. Missionary Zeal

Evangelicals obey the command "to teach and baptize all nations" with exceptional energy.

In a conversation with the aim to convince someone of his belief, the missionary speaks not only to the other person but, hearing his own voice and argumentation, also to himself. Not one, but two listeners are the objects of his missionary work; and both can be impressed by his powers of persuasion. In addition, when the other is convinced, it works like a boomerang: if the other goes for it, the message must be true. And also, the result gives the satisfaction of success; the ranks of non-believers have shrunk, and those of the believers grown — can so many people be wrong? What more could be wished for by a believer struggling with doubt?

But doubt is a tenacious enemy and satisfaction has a short life. Then the fight, and also the missionary effort, has to start all over again with doubled energy. There again, I think, we may discern a denial by a display of the opposite: doubt covered by the propagation of exactly that which is doubted. There is also an ambiguity there: a sincere wish to do the right thing and to please Jesus, but inherently a fight against the doubt whether it really is the right thing.

5. Gladness

Do evangelicals have good reasons for their joy? Yes, there are real grounds for gladness. To find shelter in a warm community where one is completely accepted and more than that — a guarantee that sins in the old life have been forgiven by Jesus, and the certainty of a new life until and even after death — are good reasons to be glad.

But many people simply find warmth within their families and their circle of friends and relatives. What reason could there be to look for shelter elsewhere, in an organized community?

Total surrender to the Lord or Jesus, submitting one's own will to that of the Lord, is essential for evangelicals (Zoba, 2005). Consequently, for them, all *natural* impulses and emotions are suspect, and in young uneducated children their will is therefore considered to be an easy prey for the irresistible forces of evil. The child's own will must thus, if not be broken, at least be brought under God's control. Greven (1978) describes the blood-curdling practices in the US in the 17th and 18th centuries to destroy the child's will ("That Monster, Self ") and replace it by obedience to a higher command. Admittedly, this was long ago — but traditions in education are naturally transmitted from generation to generation. Thus the evangelical Andrew Murray (1828-1917) in his "Raising your children for Christ" (1984): "One of the marks of a godly parent is a deep sense of the sinfulness and the sins of his children", followed by what the parents should do about it. The book is still for sale in the evangelical bookshop in Utrecht, anno 2006.

Growing up in an evangelical family and giving up your own will can be a hard thing. Edmund Gosse, whose father was member of the Plymouth Brethren, wrote in 1907:

> "Let me speak plainly. After my long experience, after my patience and forbearance, I have surely the right to protest against the untruth (would that I could apply it to any other word!) that evangelical religion, or any religion in a violent form, is a wholesome or valuable or desirable adjunct to human life. It divides heart from heart. It sets up a vain, chimerical ideal, in the barren pursuit of which all the tender, indulgent affections, all the genial play of life, all the exquisite pleasures and soft

> resignations of the body, all that enlarges and calms the soul, are exchanged for what is harsh and void and negative. It encourages a stern and ignorant spirit of condemnation; it throws altogether out of gear the healthy movement of the conscience; it invents virtues which are sterile and cruel; it invents sins which are no sins at all, but which darken the heaven of innocent joy with futile clouds of remorse. There is something horrible, if we will bring ourselves to face it, in the fanaticism that can do nothing with this pathetic and fugitive existence of ours but treat it as if it were the uncomfortable antechamber to a palace which no one has explored and of the plan of which we know absolutely nothing" (Gosse, 2005).

He could take this distance because he managed to resist the attacks aimed at his will: "Through thick and thin I clung to a hard nut of individuality, deep down in my childish nature. To the pressure from without, I resigned everything else, my thoughts, my anticipations, my assurances, but there was something I never resigned, my innate and persistent self."

Not everybody has this inborn strength. The successful submission of one's own will to another will is effectively summarized by Yancey (2005): "You can get evangelicals to do anything."

Of course, not every family in the 20th century adopted the extreme regime of the Gosses. And certainly there must be evangelical families with a real interest in the healthy development of their children. But too often it is the most outspoken members of the group, those who are most stringent in upholding the group's principles, that are most influential in determining what should be thought and done; e.g., by writing books on education.

For those who have enjoyed a more of less normal youth in an average family, religious or not, it is difficult if not impossible to imagine what the consequences can be of an education in the evangelical tradition. To have been subjected, right from birth onward, to a system inimical to the natural impulses of the human being, will become a burden for life.

The treacherous aspect of this system is that its basically aggressive nature is covered by a thick layer of "love" (another example of denial by the opposite, see Table 1), so that any possibility for resistance is effectively cut off. But what is even worse, the influences of such a regime in the first two years, which cannot be traced back by memory, escape the healthy effects of later reconsideration and rebalancing and, by their hidden force, keep their noxious influence intact and active. The traumatic effect of such destructive forces during those early years cannot be overestimated. And — what kind of companions do they evoke?

Table 1: Frequencies of "love", both as a noun and a verb, in some evangelical and neutral books on education. A frequency of 30 per 10^4 words is about once every page.

	Language	"love" (per 10^4 words)	Author
1. Evangelical	Dutch	37	Schouten (2003)
2. Evangelical	English	44	Murray (1984)
3. Neutral	Dutch	2	Feddema (2004)
4. Neutral	Dutch	3	Heintz (2000)
5. Neutral	Dutch	5	Welgraven (2005)
6. Neutral	English	9*	Spock & Parker (1998)**

* Frequency on the pages without medical subjects.

** Dutch translation, revised by van Andel and van der Pol.

Anger. Not only as a conscious sentiment, as was shown by Edmund Gosse, but more often suppressed and appearing in malicious disguise, e.g., as rancour in word and tone. Its slightly acid scent rises from some texts against the Darwinists, e.g., in attacks on scientific and other elites who rule the public square.

Depression, as a sign of anger that cannot be unleashed against the outside world, and therefore can only be directed against oneself.

Jealousy, towards scientists and others who were brought up in an atmosphere of free thinking and of encouragement to ask questions of all imaginable kinds.

So, what gladness? How loud must one sing and clap to drown out such ever present companions? What Herculean strength should one's belief possess in order to withstand these insistent and formidable forces? Here doubt is the most dangerous enemy. Thus gladness, firm and loud, supported by belief and its promises, shared by the whole community is badly needed to chase out relentlessly any doubt that might arise.

And now, on top of all that, you have these Darwinist scientists. Deeply envied, successful, self-confident, free to have fun in their work, even enjoying their doubts about the truth of their own ideas, here they come with an attack directly aimed at the very foundation of the belief that for many is their last foothold in life.

To arms!

6. Concluding Remarks

The two most striking characteristics of the evangelical movement, missionary zeal and gladness, can be seen as signs of a hidden desperate struggle with doubt. Doubt in itself is not a surprise for anyone who has read the *Bible*: doubt and unbelief keep reappearing throughout the Old and the New Testament. Liberal protestants accept doubt as an integral part of their religious life (Halfaer, 1972), and orthodox protestants in the Netherlands have learned to live with it.

The excessive sensitivity to doubt of evangelicals in a vulnerable emotional state can be understood if one realizes the special service their faith and practice have to supply for them by offering something like a last protective resort against interior or exterior worlds that are far from paradise. Anything that produces doubt about the truth and solidity of that resort is therefore a threat that has to be fought with all forces available (Anderson, 2004). Darwinism, to the eyes of these evangelicals, is such a threat. Their problems with Darwin's theory therefore seem to be caused not so much by a conflict between the texts of Genesis and Darwin, and the inherent insult to the authority of the *Bible*, as by an aggravation of a pathetically denied hidden conflict within themselves.

What, then, can scientists do about such a kind of resistance to Darwin's theory? There will be little use in arguing with the really hardcore evangelicals. But with the more flexible among them, and other Christians as well, a regular discussion must be possible. First, then, it should be realized that practically all human beings have a deeply felt need for the presence of someone or something that is underneath everything in the cosmos. It is of no use to deny or ridicule this need. Instead, to start with admitting that no one can pretend to know exactly how this world came into being and whether a designer was involved or not is still, I think, the most fruitful strategy. I see no problem either in assuming that a designer, or the Christian God, might have included within the physical forces and elementary particles an inbuilt guideline for the development of life as we see it now. Thus, not as Behe (1996) wants it, who sees the hand of a Designer busily acting throughout the whole process of evolution, but in a way that we cannot understand. As the Preacher teaches us: "He hath made every thing beautiful in his time; also he hath set the world in their heart, so that no man can find out the work that God maketh from the beginning to the end."

The orthodox protestant zoologist Lever, at the protestant university mentioned before, lectured along those lines throughout the Dutch "Bible Belt" with considerable success in the middle of the 20th century, using the words of the Preacher as a motto (Lever, 1956). Possibly due to this effort there is at present no strong opposition against Darwin's theory in the Netherlands.

References

Anderson, N. (2004) *Overcoming Doubt*, Regal Book, Ventura, USA.

Behe, M. (1996) *Darwin's Black Box: The Biochemical Challenge to Evolution.* The Free Press, New York, USA.

Eschbach, J. and de Ruiter, A.W.W. (1999) *Evangelische Liedbundel*, Boekencentrum Zoetermeer, The Netherlands.

Feddema, G. (2004) *Opvoeden kun je leren*, Unieboek, Houten, The Netherlands.

Gosse, E. (2005*) Father and Son*, Nonsuch, Stroud, UK.

Greven, P. (1978) *The Protestant Temperament*, Knopf, New York, USA.

Halfaer, P.M. (1972) *The Psychology of Religious Doubt*, Beacon Press, Boston, USA.

Ham, K.A. (1987) *The Lie, Evolution.* Master Books, Green Forest, USA.

Heintz, C. (2000) *Handboek positief opvoeden*, Spectrum, Utrecht, The Netherlands.

Jones, J.E. (2005) *Memorandum opinion*, U.S. District Court for the Middle District of Pennsylvania, Document 342.

Lazonder, A.W. (1996) *Liedboek voor de Kerken*, Boekencentrum Zoetermeer, The Netherlands.

Lever, J. (1956) *Creatie en evolutie*, Zomer en Keuning, Wageningen, The Netherlands.

Murray, A. (1984) *Raising your Children for Christ,* Whitaker House, New Kensington, USA.

Pennock, R.T. (1999) *Tower of Babel, The Evidence against the New Creationism*, MIT Press, Cambridge, USA.

Runia, K. (1984) *Evangelisch, reformatorisch, gereformeerd.* Willem de Zwijgerstichting, Apeldoorn, The Netherlands.

Schouten, E. (2003) *Opvoeden*, Kok, Kampen, The Netherlands.

Spock, B. and Parker, S. (1998) *Baby and child care*, Penguin Group USA, New York, USA.

Veldhuizen, H. (1988) *Vrije Groepen, evangelische bewegingen en de kerk*, Kok, Kampen, The Netherlands. Welgraven, E. (2005) *Opvoeden anno nu*, Jonge Gezinnen BV, Alphen aan den Rijn, The Netherlands.

Yancey, P. (2005) in Zoba (2005).

Zoba, W.M. (2005) *Evangelical Christianity*, Three Leaves Press, New York, USA.

Dialogue

Taner Edis
Frits Bienfait's essay on the psychology of evangelical creationism is very interesting. It is a bit speculative — it would have been better to have some more solid evidence — but it is plausible, and the experience of a close observer of Dutch evangelical religion has to count for something.

1) Still, I wonder what happens if we turn things around. I am as liable as anyone to get frustrated with creationists and ask what psychological blocks they must operate under that prevents them from appreciating the evidence. But I am also familiar with many anti-evolution thinkers and their writings, and it is clear that they can get just as frustrated with people like me. They rail against what they perceive as a rigid scientific establishment, and accuse us of dogmatism. They wonder if the mental blocks of evolutionists (that cause us not to see the clear evidence of creation) have to do with the sin of pride, or perhaps our desire to avoid a supernatural Judge who will condemn our immoralities.

2) So, my question: among Dutch evangelicals, do you see much of this sort of psychological speculation about people who side with evolution? The answer could be interesting, since evangelicals are so much more of a cultural minority in the Netherlands than in the United States.

3) And also: since all sides can engage in psychological speculation, what is the significance of such psychological matters for the debate about whether evolution is correct?

Frits Bienfait
My answers to your questions follow:

1) Evangelicals accusing Darwinists of dogmatism do have a point. Some scientists seem not to be really open to the possibility that Darwin might be (partly) wrong after all.

2) I guess that in this respect there is not much of a difference between Dutch and American evangelicals. The main line of arguing is that the Bible simply

provides the truth, and therefore there is something wrong in the minds of people who are unable to accept this, e.g., they cannot accept a God who is almighty (is this not arrogance?) or do not want to serve Him. They fear that, should they accept Genesis 1 and 2, they would also have to accept the Fall of man, and, consequently, a God that will call us to account. Therefore they will seize any opportunity to reason Him away, and to this end a theory on evolution, not creation, of species, comes in quite helpful (see also Hobrink 2007).

3) The use of this kind of considerations for the debate: getting to know your adversary better is always useful.

Reference:

Hobrink, B. (2007) *Evolutie, een ei zonder kip [Evolution, an Egg without Chicken]. Gideon, Hoornaar, The Netherlands.*

Biography

K. Tahir Shah is the Managing Editor of the *African Physical Review* and the Moderator for International Open Access Archive at the Abdus Salam International Centre for Theoretical Physics (ICTP), Trieste, Italy. He divides his time between editorial activities, research and practice of yoga. His current research interest includes evolution and theories of intelligence, cognitive philosophy and foundation of mathematics. He studied physics at Karachi, Pakistan (undergraduate 1960-64) and graduate studies in theoretical physics at Bologna, Italy (1964-67) and at Stevens Institute of Technology, Hoboken, USA (1969-71). He was invited to become reviewer for *Mathematical Reviews* in 1975 and took chair of mathematics at Tripoli, Libya, in 1976. After designing intelligent systems for industry in North America for seven years, he returned in 1988 to pursue academic life in Trieste, at ICTP and the University of Trieste. He is also on international editorial council of "Problem of Theory and Methods of Education" (in Russian) and "Revista Cubana de Informatica Medica."

20

Creationism
An Investigation into Its Psychological and Cultural Origins

K. TAHIR SHAH

The Abdus Salam International Centre for Theoretical Physics
Trieste, Italy
shah@ictp.it

> *To uphold a tenet that contradicts reason is to undermine one's credibility; to contradict empirical evidence is a still greater fallacy.*
> A Buddhist Dictum

> "*But surely you will not affirm, that the universe bears such a resemblance to a house, that we can with the same certainty infer a similar cause, or that the analogy is here entire and perfect. The utmost you can here pretend to is a guess, a conjecture, a presumption concerning a similar cause.*" David Hume

1. Introduction

The belief that the universe was created by a supernatural entity can be analyzed from two different perspectives. First, it is the metaphysical perspective to which classical philosophers devoted much of their time and effort. Virtually, all arguments in support of belief in the supernatural were rigorously refuted by sceptic philosophers such as Hobbes, Spinoza, and Hume. Others, like Pierre Bayle, put great emphasis on separation of philosophy and theology on the ground that any theological doctrine could not possibly be defended by reason and was therefore a matter of faith alone (see, e.g., Wolff 1998). Second, in recent years many evolutionary psychologists, anthropologists, and neuroscientists provided explanation of belief in the supernatural as a cognitive phenomenon from an evolutionary perspective. Despite such past intellectual efforts, belief in one or more supernatural creators remains among people of all cultures and throughout known history. The origin of belief in the supernatural can be traced back to an innate mental faculty, now referred to as 'Theory of Mind', which reached its evolutionary threshold some 50,000 to 60,000 years ago, around the time of the last migratory wave of *Homo sapiens* from Africa.

What I shall argue in the sequel is that such beliefs are purely subjective mental constructs, which do not have any support in the concrete empirical world. The other important point is the following. Often, in this kind of debate, meta-physical arguments on the existence of a supernatural entity are intertwined with the assertion that such an entity does indeed possesses human-like intentional attributes. But, the fact is that the two issues are entirely different. The meta-physical issue, whether or not there exists a creator (or creators) of our physical universe, refers to the creator's physical, non-physical, or whatever other kind of existence one can possibly extrapolate. The question of existence is difficult to formulate due to varying interpretations of this notion. For instance, what would it mean to say a non-physical intentional agent exists in space and time, and how can it be compared to a physical intentional agent? Moreover, intentionality is a human attribute that plays a crucial role in communication with other similar intentional agents. If it is assigned to the universal creator, then a simple implication is that intentionality existed before the creation of this universe. That is, a human-like mind with all its weakness of committing errors somehow 'designed' and created a complex universe. Were there other minds present too before the creation of the universe? Why should a species-specific human attribute have been present before the creation of this universe? Creationist doctrines in all cultures consider the creator as an intentional agent and extrapolate human attributes in a manner so as to violate intuitive expectations. The human capacity to meditate on metaphysical questions and to have innate intentionality are so distinct from each other that it is very likely that even their neural and cognitive correlates are not the same. The meta-physical question appears to be the domain of mental causality (or causation), whereas all intentional attributes are the domain of Theory of Mind (ToM)*.

---

* *Theory of Mind*: To be aware of other minds, or theory of the other mind, is a well known classical concept of the philosophy of mind. The term 'theory of mind' (ToM)

Any rationalization of the meta-physical hypothesis (i.e., there must exist an ultimate entity) using causation in the reductive sense leads ultimately to the well known problem of infinite regression. Furthermore, in order to solve the infinite regression problem if either self-creation or the so called first cause is enacted, then it leads to impredicativity. In logical terms, impredicativity arises when universal quantifier (for all) is used over a presumed "completed" totality.

This chapter is thus divided into two parts. The first part summarizes how belief in creationism leads to serious logical problems such as impredicativity, whereas the second part explores the root of its innate psychological origin. In conclusion, what can be said is this: belief about a supernatural creator is an evolutionary exaptation of an innate mental faculty, namely, the 'theory of mind', the evolution of which has made us intentional agents.

---

was coined by Premack and Woodruff (1978) in their paper titled: Does the chimpanzee have a 'theory of mind'? Other terms used concurrently for ToM are 'mind-reading' and taking the 'intentional stance'. Recently, a more general term 'empathizing' was proposed by Baron-Cohen (2005). This includes ToM (the attribution component), but goes beyond by including some affective reaction.

More than a century ago, the term 'intention' was introduced by Bretano by asserting that mental states are 'intentional', meaning that mental states are "about" something. Belief and desires are traditional examples of prototypical mental states. To have intentionality is referred as the first-order intentionality. When an intentional being is aware of first-order intentionality of another being, it is referred to as the second-order intentionality. We may have a mental state that is about the mental state of another person. It is an accepted psychological view that human beings possess second and higher order intentionality, or what is normally called (full) ToM. The theory of other minds or simply the theory of mind (ToM), relates communication, cognition and social behaviour (see, e.g., Baron-Cohen 1999, Vogeley et al., 2001). ToM is required for an intelligent social interaction among the members of a society because having causal knowledge of others' behavior may be adaptive. It enabled ancestral individuals to control events that had important consequences for one's own survival through implanting false beliefs, manipulating emotions and other components of ToM.

However, only the 'self' enjoys the conscious privileged epistemic access to one's own mental states. In religious systems, a supernatural agent can have privileged epistemic access to the 'self' mental states. By implication of this, for instance, the supernatural entity knows all my thoughts and intentions. On the other hand, these supernatural agents are not seen as posing any threat because they do not have direct means by which to communicate potentially damaging social information about the self to other members of the group.

2. Metaphysical Perspective

One of the many variants of creationism, the so called 'Intelligent Design' (hereafter called ID) explanation of the origin and complexity of universe and life, is now claiming itself to be equal to science in its empirical method and content despite the logical weakness of its arguments and illegitimate use of species-specific cognitive primitives in the universal context. Such a belief rationalization is a phenomenon that surfaced not only in recent time, but there have been consistent efforts in the past to do so by many illustrious philosophers and theologians, notable among them George Berkeley, Joseph Butler, John Locke, Samuel Clark, and William Paley. Despite the fact that creationism and similar beliefs emerged in the remote historical times of mankind, they continued to resurface almost unchanged even after the birth of modern science. In recent years, there have been attempts to introduce 'Intelligent Design' as a legitimate scientific theory into the school curriculum in many states of the United States. But, such ideas are no different in substance than their precursor, which was proposed more than two hundred years ago by William Paley[†]. Paley was also a strong supporter of the doctrine called utilitarianism and of truth by revelation, who gave a detailed account of what has been called "Argument from Design" as the "proof" of the "existence" of a universal designer. Paley's argument was no different than many others since ancient times, who all believed in some kind of a supernatural entity or entities. His idea, however, was to enact a widely used paradigm of his time, the clock-work paradigm, in support of his rationalization argument. Of course, such a creator was given many of the human attributes, all suitably extrapolated so as to fit an image that was truly appropriate for the task of creation and control of all that is within the universe. In other instances, these human-like attributes varied according to local requirements, the epoch of human history, and the knowledge of the external world that the local population possessed. A large variety of deity worships and related rituals all over the world witness to this fact. Today, the ID paradigm is being forced by its proponents as equal to science itself, but it is uncommon for trained scientists to accept such a teleological view of the origin of universe and life.

Most of the rationalization arguments use analogies, which relate, for instance, the human creation of artefacts for specific purposes linked to their functionality

[†] For an excellent historical overview of natural theology of William Paley and ID, see Barrow and Tipler (1986).

and the creation of the universe by a supernatural entity. Such analogies are carried over from species-specific human activities such as the creation of an artefact, to the whole universe in order to deduce the existence of a creator. Use of analogy is not a deductive process and cannot be construed as a proof.

How does analogy works? It allows a transfer across different concepts, situations, or domains. In analogical mapping, a familiar situation, called the source analogue, is used as a model for inferential outputs about an unknown (or partially known) situation, called the target analogue. Such transfer maps help to understand a new concept, situation or domain. It can also be taken as 'similarity' in which the same relations hold between different domains, and the inferential assumption that if two entities agree in certain respects then they most likely agree in others (Genter 1983, Holyoak 1985). It is a probabilistic extension of source analogue attributes to the target analogue based on a likelihood assumption. The main assumption of ID is that a mouse trap (or a space shuttle) and the universe agree in all respects. The acceptability of an analogy can be judged through an evaluation process using at least two criteria (in this case):

> *Structural Soundness*: Whether the representational alignment and projected inferences are structurally consistent.
>
> *Factual Validity of the Candidate Inferences*: This is not guaranteed and must be checked separately given that analogy is not a deductive mechanism.

Let me add further that a clear distinction between empirically testable knowledge and various types of assertions should be kept in mind. For instance, counterfactuals are not empirically testable because such assertions consist of a different set of epistemic states. Putting counterfactual assertions and empirically testable knowledge in the same basket leads to what can be called the 'confusion of epistemic states'. Rationalization or proving beliefs and counterfactuals using empirical knowledge in any possible way causes unintentional mixing of epistemic states of two different sorts.

Of course, much is written on these issues since ancient times. It is not my intention to review all that is relevant in this context. My aim in this chapter is to provide only a few new ideas, which may provide a ground for sorting out science from theology. For example, does the complexity of a material entity, be it an artefact or otherwise, always imply the existence of an intentional agent as its creator? This question shall remain unsettled if theology is continued to be mixed with science. The real problem with ID lies in the implicit assumption that instances of man-made artefacts can be extended, by analogy or by induction, to the universe itself, which includes inanimate matter, human or any other life form in possession of ToM, or a completely unimaginably complex mental faculty possessed by a life form living elsewhere in the universe.

3. A Few Relevant Concepts

Often, a discussion on issues of such general interest as this may lead to serious divergences or even a deadlock simply because of differing interpretations of the same concept. In order to avoid a discussion, where each participant may have his (or her) own interpretation, a few clarifications are outlined on such notions as 'existence', 'truth', 'proof', 'belief' and 'supernatural'. For instance, what does it mean to say that some statement is 'true' and how do we find it to be 'true'? What does it mean, when we say that something 'exists' and that it is supernatural?

3.1. *Mental Constructs*

The term 'mental construct' seems to have different meaning to different users. Let me clarify what I mean by this term in order not to cause confusion. Essentially, what it means is all that is constructed by the neuro-cognitive system; be it the construction of reality utilizing external perceptual inputs, complex concepts originating from higher cognitive processes, and all other concepts referring to actual or possible worlds. For instance, perceiving a chair, when seen, is obviously a mental construct. Imagining a magic chair, which can fly and talk, is also a mental construct. All conceptual extrapolations are also mental constructs. But, to achieve all such feats, an alive and functioning neuro-cognitive system is needed. It is because a dead brain does neither construct external reality, nor any other cognitive concept belonging to, say, the counterfactual world. However, it is important to keep in mind the difference between two kinds of mental constructs. Mental constructions of external objects belonging to the physical world are of different nature than mental constructs of abstract objects such as algebraic numbers or infinity. The inputs belong to different sources, the external world in one case, and the conceptual world derived from one's private memory in the other case.

In this sense, both terms 'mental' and 'physical' are mental constructs. The 'physical' is constructed from perceptual input and remains more or less invariant for all normal members of a given species, whereas the 'mental' is only a subjective mental construct belonging to a specific member of that species.

3.2. *Reality: The Actual and the Possible*

It is prominent among traditional meta-physicists to accept that reality also has a modal structure. But, this idea has not been very popular among most philosophers. David Hume rejected the view that modality is an objective feature of the world. Many notable philosophers following Hume's tradition considered modal notions either as illegitimate or simply reflections of our decision to use words in certain ways. There is no modality out there in the external world.

I must emphasize that epistemological pluralism (Shah 2006) too does not entertain the idea of a modal reality. There are only actual realities of all biological species and complex mental constructs derived from the actual.

3.3. *Truth*

What is a truth? Or, more precisely, what is the meaning of the truthfulness of a linguistic expression referring either to the physical world or to a purely mental construct? An intuitively obvious definition is given by Russell (1940) as follows: linguistic expressions (sentences) describe events, which in their turn are the evidence for the truth of the sentences (describing the same events). In fact, the most common and widespread way of understanding the nature of truth and its negation is through the use of correspondence theory, the main argument being that "truth" is whatever corresponds to reality. Now, the burden of knowing truth is shifted to the knowledge of reality. We are back to an equally difficult question on the nature of reality. What reality normally means is the species-specific perceptual reality, a strictly anthropomorphic definition. The question then is how can intuitive ontology violating statements about the supernatural correspond to reality? What kind of truth, then, can one define?

In a general social context, there can possibly be two separate kinds of truths in the minds of common men. The first kind of truth is what scientists and philosophers try to establish by means of empirical methods of verifications and rational arguments. Such determination of truth remains approximate and is subject to refinement through further observations. No doubt, it is very difficult to establish. The other kind of truth is simply a matter of faith, which is accepted despite its counterintuitive nature. It is when a large number of individuals simply begin to call repetitively some aphorism (or a belief), a truth, and over a long period of time, that it gains the status of a truth, namely, a sacred truth (see, e.g., Bulbulia 2004 and references therein). Such a truth is neither to be discussed and criticized, nor tested and evaluated in any manner. Otherwise the wrath and violence of its believers against those who question its validity will forcefully settle the issue into its acceptance, its sacredness, and its truthfulness. History is full of such instances of violence against anyone who dared to question the established dogma of whatever origin and social importance.

Although, the concepts of the supernatural cannot be assigned truth values in any possible way, nevertheless, they are considered as *a priori* truth by their believers. Such truths are of the second kind, their origin being in revelations, dreams, hearing voices (verbal hallucinations) due to natural or drug-induced causes, but always purely subjective in nature. Briefly, we may say that knowledge deals with the objective truth, whereas, the truthfulness of beliefs is subjective, though made sometimes a ‘collective truth’ through social processes. The objective truth is empirically verifiable and the subjective truth, which is “truth by assertion”, remains true only for those who believe it to be true. Human knowledge is thus a set of statements that are truths of the first kind, that is, they are true by virtue of empirical verification and rational argument, whereas beliefs are statements that are truths of the second kind.

There are many other theories of truth, but they are not relevant to the present discussion. For example, in classical systems of logic and formal theory of languages, Tarskian semantics tackles this question.

3.4. *Existence and Proof of Existence*

What does it mean to say that an entity exists? In every day common conversation, the existence of an entity normally refers to its physical existence in space and time. The intuitive notion of ‘existence’ is really the physical existence — something out there in space and time. It may not be perceptually observable at that precise moment in time when a specific conversation about it is taking place, but it is 'understood' that the referred object exists somewhere in space and during some period of time (even infinite, if there is sufficient ground to believe so).

However, such a 'feeling' about the existence of an entity changes as we move to rational disciplines such as philosophy, mathematics and exact sciences. Almost all working mathematicians adhere to Platonism and take for granted *a priori* existence of all mathematical objects, but with varying definitions of what it really means. For instance, in the formalist approach to mathematics, 'existence' refers to an absence of any internal contradiction within the system. On the other extreme, for constructivists, a mathematical object exists only if it can be constructed mathematically from various other previously established entities already shown to exist.

In some belief systems, one or more entities are believed to exist in the Platonic sense, but with all human-like attributes. To begin with, such entities are only mental constructs. Next, it is neither the usual physical existence in space and time, nor it is the kind of existence that is taken for granted in any of the mathematical formalisms or in any other rational discipline of science. It is an ambiguous existence, an entity is physically present everywhere and at all times, but it is also not present physically in space and time, because one cannot detect or observe it in an objective and empirical manner. For example, well-known deities exist physically in space and time (a human-like attribute), but they are invisible, a violation of intuitive physics! There is a contradiction, both an attribute and its negation are assigned (e.g., both infinitely merciful and inflictor of eternal punishment) to an entity.

A proof can possibly be given for the truthfulness of any linguistic expression, known as proof of its validity, but the proof of existence of the same entity is a different matter. To prove the validity of an expression, one takes a set of syntactically legitimate true expressions and propagates the truth from these to a more complex expression to be proved using the rules of inference, which are also valid expressions. The word 'validity' in this context means truthfulness in all possible interpretations. Some philosophers took the word "existence" to stand for a predicate in the logical sense. For instance, in the Cartesian viewpoint, the way a proposition that a deity can be proven to exist is the same way as a proposition of mathematics is proven. In classical logic systems, empirically verified premises (axioms or primitive truths) are used to infer other truths (theorems).

But, we must not forget that logical theories are more about how truth propagates from an initial set of assumed 'true' expressions to other more complex syntactically legitimate expressions than about the nature of truth. In logical theories of artificial intelligence and theories of common sense reasoning, especially using natural categories, the truth value may change if new evidence is added to the database. This is known as the non-monotonic form of reasoning, which is different from the reasoning in classical logics. Logic of default is one example in this class (Reiter 1980). Despite the rigour of logical theories, in this case too, both material and causal implications are not without problems; conditions of soundness and completeness (in the meta-mathematical sense) are required. On the other hand, it is the empirical verification that determines the truthfulness of a theoretical construct of a physical theory of any grain size, be it a simple formula or a complex sub-theory. Physical theories subscribe to ontological naturalism. Ontological naturalism is the view that nothing exists other than spatio-temporal beings embedded within the space-time framework. As such, it denies the existence of abstract entities such as propositions and numbers, Platonic universals, disembodied minds, deities and the like.

3.5. *The Actual and the Possible Worlds*

Knowledge and counterfactuals belong to two separate kinds of worlds, respectively as: 1) The world constructed by the neuro-cognitive system through "direct" perceptual inputs and legitimate inferential processes; 2) The world of "pure" mental constructs which uses as its inputs objects from the first world and cranks out new mental constructs, but there is no direct perceptual input. Let me divide such pure mental constructs into two types: conscious construction of ideas and concepts such as counterfactuals, and those which are not under direct conscious control such as dreams, revelation, and hallucinations of various kinds (e.g., verbal hallucinations or hearing voices, visual hallucinations). Again, at the next stage there are conscious interpretations of such mental constructs which are not under individual conscious control; such interpretations are normally culturally based.

Such worlds are located in the human memory. There is a clear distinction here between externally derived memories originating from perceptual inputs and internally derived memories that originate from imagination. External memories are based directly on sensory information, and represent events that really occurred in physical space and time, objects that were really perceived, actions that were really performed, and so on. Internal memories are pure mental constructs, not dependent on real-time input of direct perceptual data and represent events that have only been imagined, dreamed, etc. The capability to distinguish between the two is called 'reality monitoring'. The reality monitoring system separates the two worlds. A malfunction of this system can cause mixing of the first and the second world – a well-known pathology. In humans, a break down of this capability leads to schizophrenia, many types of dementia, and other related mental disorders (Eysenck 1990). The use of the phrase 'perceptual reality' in this article refers to externally derived memories originating from the perceptual data. Many kinds of mental constructs of the second world are subjective and subject to errors due to their dependence on their linguistic description by the subject who experienced it. It is not possible to achieve any empirical verification or any kind of direct quantification for purely mental constructs.

Although counterfactual worlds use notions from our perceptual world, their content points to purely mental events. In counterfactual belief systems, a world beyond the physical world of objects located in space and time is constructed where most objects are either perceptual objects or abstract concepts constructed from them. Often, in a religious context, it is created in one's own culture's image, in particular, in terms of one's own views of cognitive primitives, reward, punishment, and explanations of physical phenomena.

4. Historical Perspective

After this clarification, let me now narrow down the issue further by focusing on two main trends in philosophical thought, one with its origin in monotheistic religions of the Middle-East and the other with its roots in a collection of Vedic and Buddhist schools in India. Much has been written on the history and philosophy of religious beliefs world wide. The essence of many western historical arguments is reflected in the writings of the following philosophers of monotheistic religions (see, e.g., Wolff 1998, Churchland 2002). Kierkegaard asserted that truth lies in what is subjective, inward, and immediate, but not objective and universal. This view is known as existentialism. On the other end, the two prominent schools of epistemology in India, the Buddhist school, and the Samkhaya school (from which yogic philosophy originated), focus on objective truth. This is to the extent that these schools try to eliminate subjective perception of the external world by various kinds of mental and physical procedures. Both schools prescribe rigorous methods such as "switching off" the role of conscious processing to acquire objective knowledge. It appears that philosophers of many Indian schools, such as Upanishadic and Shankara, were well aware that they could not reach the *a priori* reality, but simply could view the reality that is neither purely subjective nor purely objective. I have practiced yoga for years and studied Buddhist thought, and it appears to me that both philosophies, excluding the practice of rituals, are close in their bare essence to modern scientific realism[‡].

4.1. *Anthropomorphic Deitism: Transfer of Human Attributes*

Paley's 'Argument from Design' seeks to prove that there exists a supernatural being by pointing to the evidence in nature of 'purpose' and 'design' and arguing that they must be the result of an intelligent purposive designer. Both, 'purpose' and 'design' are goal-directed species-specific human cognitive primitives. In this case, the analogy between a human designer and his creation of a complex artefact, the clock, is carried over to show that there exists a creator of this

---

‡ *Scientific Realism*: Realists hold the view that entities postulated by scientific theories (electrons, genes, black holes) are real entities in the physical world, with approximately the properties attributed to them by the best empirically established theories. Scientific realism maintains that we can reasonably construe scientific theories as providing knowledge about unobservable entities, forces, and processes. The issue raised by many philosophers is this: Do scientific theories refer to real but unobservable entities, forces, and relations?

Social theories too involve concepts that appear to refer to unobservable entities. So the issue arises: If we have an empirically well confirmed theory that invokes some hypothetical concept X, is this a reason to believe that X exists? Or, is there some reason to suppose that the ontological assumptions of scientific realism are justified in the natural sciences, but not in the social sciences? The answer, it seems, depends on what can be accepted as real.

In both categories of theories, observation is a crucial process. A postulated entity may be observable at one level of abstraction, but not necessarily at another abstraction level. The crucial point is that there should be some interaction between a cognitive system, the observer, and the entity being observed. This is achievable only through some intermediate entity.

universe. Most importantly, the supernatural entity is given the mental faculty of 'theory of mind' in order to be able to understand human desire, belief, anxieties, etc. Let me call such a creation of human-like deities and belief in their supernatural powers as anthropomorphic deitism. Many researchers argued to show that anthropomorphic deitism is an evolutionary exaptation (Bering 2006).

One of the many problems, in my opinion, with such an argument is the use of the universal quantifier, "for all". The argument goes like this: every artefact is made by some intelligent entity and the universe is an artefact, therefore, the universe is made by an intelligent entity. That is, when our human experience is extended by induction to include the universe itself in "all" artefacts, it entails easily that the universe must have been created by some entity having human attributes of intelligence, but extended *ad infinitum*. A first objection, somewhat close to the spirit of Hume, is on the use of the universal quantifier in inductive inference applied to physical world. When applied in this manner, it leads to either infinite regression or impredicativity. The other two objections are empirical in their nature. First, what is intelligence? It is known that many terrestrial species create fairly complex artefacts, but such species are neither considered 'intelligent' nor do they have intentionality like humans. Second, is universe really an artefact? It was already mentioned that the use of analogies does not constitute a deductive proof.

4.2. *Begin at the Beginning!*

> "*Where shall I begin, please your majesty?' she asked. 'Begin at the beginning,' the king said gravely, 'and go on till you come to the end: then stop.*" Lewis Carroll

Another type of argument was put forward by St. Thomas Aquinas, known as the Cosmological Argument, which uses causality and the existence of the 'First' cause. The argument is as follows: there is motion in the universe, which implies that there must be a 'First' cause of motion, or a first unalterable cause of existence, which itself has no underlying cause and neither comes nor passes away. In this case, to avoid the infinite regression of the causal chain, a 'First' cause is postulated, a fundamental violation of reductionism and causality. Why should there be a 'First' cause? There is no justification except that it should be there to "prove" the existence of an entity! In essence, in this class of arguments, one hypothesis, the First cause, is deployed to prove another hypothesis, the existence of a creator, related to it – an argument not far from impredicativity, where a statement is proved by using itself or a closely related entity.

4.3. *Mental Equals Physical!*

The Ontological Argument given by St. Anselm begins with the concept of an infinite being and argues that since such a concept can be formulated, we can therefore conclude that some entity corresponding to an infinite being must exist. What this amounts to is that mental construct equals 'physical' existence in all cases. In this argument, a pure mental construct is given the physical existence (or something like it) by transferring a human attribute, the physical existence, to such a mental construct. This virtually negates the reality monitoring system that distinguishes the real from the virtual. Two different kind of epistemic states are mixed. This is certainly nothing other than Platonism, equating pure mental construct and physical existence. Many great mathematicians such as Kurt Gödel (Rodriquez-Consuegra 1995) and George Cantor (Dauban 1979) strongly believed in the *a priori* existence of mathematical object *on par* with physical existence.

4.4. *Indian Philosophical Thoughts*

Ancient Indian thinkers, in many ways, reached far beyond in their efforts to solve infinite regression problem than some of today's philosophers. I will mention here only the non-theistic doctrines in Indian philosophical thoughts. First, the concept of god in Upanishadic thinking was quite different from the more common definition of god, an intentional agent, who acts as creator and dispenser of reward and punishment. Their concept of god was more abstract and philosophical. It was because they were seriously concerned with the infinite regression problem due to an unbounded causation chain. The ultimate reality according to Upanishads and Vedanta philosophy was such that both schools avoided an impredicative solution to the infinite regression problem. Let me summarize very briefly.

It was already stated in the Brahmana writings that the universe had its origin in non-existence (as at). That is, its existence must be a result of manifestation of some potentialities which remain unmanifested. According to the Upanishads, the Ultimate Reality is Brahman (or Brahma), which is the origin of all — inanimate matter and life with its capability to perceive and reflect on the inanimate universe. There is an eternal cycle of manifested state (the physical universe) and unmanifested state (nothingness). So, what runs this infinite set of cycles? The proposed solution is that the cause of the manifestation process is Brahman's desire, again a human-like attribute!

On the other hand, Shankara, I think, is an observer-based view point of all that exists — reality is nothing but appearance or illusion (maja), the plurality of the phenomenal world. This veil of appearance has to be penetrated in order to 'perceive' Brahman. In my approach to reality, the doctrine of Epistemological Pluralism (Shah 2006) with its origin firmly grounded in scientific naturalism, ultimate reality is unknowable to any kind of cognitive system whatsoever. What is available is simply the cognitive system dependent reality, an idea somewhat similar to Shankara. But, in Epistemological Pluralism, (the so called) ultimate reality is unknowable and no empirical method can determine its existence or non-existence in the sense of ontological naturalism, no rationalistic proof procedure of any species-specific logical system can prove or disprove it. It is indeed a species-specific purely mental construct.

In Buddhist philosophy (see, e.g., Williams 2000), Samsara (external world, external reality) exists as an endless series of causal processes thus rendering any sort of personal divine creator irrelevant. Any idea of a supernatural deity (or deities) is replaced with that of causal dependence. A deity has no place in a highly complex structure of natural contingency, where each contingent object could be explained as a causal result of another contingent object *ad infinitum*. Such a strong emphasis on causality is the central feature of Buddhist ontology. All elements of Samsara exist in some sense or another relative to their causes and conditions.

This whole concept of causal chain and the corresponding mental faculty responsible for making causal connections between two events, objects, etc. has a strong evolutionary origin. An understanding of causality helps us to survive, e.g., connecting movement of tall grass with a predator. In the Theraveda school of Buddhism, there is neither a personification of gods, nor spiritual and material substance that exists by itself as ultimate reality. The world as we know it does not have its origin in a primordial being such as Brahman. It exists only as a mental construct shaped by the senses. Almost two thousand years later, Descartes came very close to the same conclusion: "I think, therefore, I am". We may interpret his famous dictum as follows. That is, all ideas about creation and existence are human species-specific mental constructs. I emphasized this point strongly as the dominating feature of Epistemological Pluralism.

4.5. *Other Teleological Views in Physics and Biology*

Human intentionality or the 'Theory of Mind', it seems, has also been the source of a generative mechanism in many other goal-directed, intentional and anthropocentric ideas. The belief in a supernatural entity is only one such idea. Other notably similar ideas are:

> *The Anthropic Cosmological Principle*: This is an example of an anthropocentric teleological view that the universe was 'tailor-made' for human life by some form of supernatural design (Barrow and Tipler 1986). It uses the transfer of intentional attributes to a supernatural entity, such as goal, whereas the goal itself is to create the universe leading to intentional being, the human species. The principle has an element of impredicativity because of self-reference.
>
> *Evolutionary Convergence Hypothesis in Biology*: Intelligence (i.e., human intelligence) is bound to arise, if the tape is played again. This is heard quite often. It is a variation of anthropic principle, emphasizing that even Darwinian evolution is of such nature as to lead necessarily to human life (Conway-Morris 1998). An obvious implication is that evolution is goal-directed, in stark contradiction to experimental results.
>
> *Weak Anthropic Principle*: The universe must give rise to observers at some stage in its history, that is, an intelligent observer must exist.

5. The Innate Psychology Perspective

I shall now briefly touch upon some of the works by anthropologists and psychologists to trace the origin of ID and similar beliefs in human psychology, or more precisely, in some innate mental faculties initially evolved for entirely different functions and societal pressures to maintain a kind of *status quo* regarding such beliefs. In this respect, my viewpoint slightly differs from a few anthropologists and psychologists (e.g., Boyer and Ramble 2001, Atran et al. 2004). Certainly, I would agree to the universal and recurrent nature of this phenomenon in human history, but the causes are not only innate cognitive faculties, such as the great facility to remember intuitive ontology violating statements. There are also societal principles, pressures and constraints put on its individual members, as well as the sense of belonging to a specific group. Another point is that the origin of belief in the existence of the supernatural is a matter significantly distinct from its propagation to the next generation and its maintenance in a society. The former is derived from an innate capacity, whereas the later is a social phenomenon. It is important not to ignore the second aspect as less significant.

Despite the fact that the hypothesized creator is a supernatural entity, nonetheless it retains many of the species-specific human attributes, which are carried over to their respective extremes (e.g., infinite goodness, infinite mercy, omnipresence, omnipotence, and so on). As we shall see later, it was shown (Boyer and Ramble 2001, Boyer 2003) that such a transfer of attributes violates intuitive ontology.

5.1. *Intuitive Ontology and Domain-Specificity in Cognition*

Let me begin with the innate aspects of human cognition. In recent years, research in cognitive science established that intelligence can best be explained as a collection of different domains of competence. Each of these domains corresponds to a solution of some evolutionary problem from the past. Recent studies in evolution and development paradigms suggest that these competence domains are an outcome of specific developmental pathways. This viewpoint of specialized competences is referred to as domain specificity in cognition.

An example of a domain-specific system is 'intuitive ontology'. The term, intuitive ontology, has been used in the sense that any ontological understanding of the nature of the external world attained through this domain-specific system does not involve deliberate higher level cognitive thoughts. Its central assumption is that some objects in the world are driven by internal representational states such as goals, desires, and beliefs. According to Boyer (2005), intuitive ontology is a multiple-domain system in which each domain is characterized by a specific type of input and output and by specific inferential principles. One may divide it into four separate competences: physicomechanical, biological, social, and numerical. An important functional aspect of these domains is their applicability to wider external reality than that for which they had evolved. This means that the different classes of objects in our cognitive environments do not always correspond to real ontological categories, that is, the actual applicable domain could always be larger than the evolutionary domain. A typical example cited in the literature is the face recognition system, which identifies and recognizes what it was "designed" to expect in its environment. But, it can also be trained to recognize other objects, say, cars or buildings too, which were not present in our evolutionary past. This flexibility has an important advantage: we do not need a new system for each previously unknown object.

What are the features of domain-specific inference? Each inferential system has specific learning logic and thus knowledge acquisition is dependent on it. For example, animal and plant species are intuitively construed in a manner different from the man-made objects — artefacts, for instance, in terms of their functions. Why is violation of expectation easily remembered? A possible explanation can be given if we consider it analogous to the problem of face search in a crowd. All faces being similar causes a certain type of symmetry in the search problem and it requires an exhaustive search to find one specific face. But a violation of this symmetry provides an extra cue, which leads to fast recognition, a sort of 'conversion' of an intractable (NP or exponential resources) problem into a tractable problem that is soluble using polynomial (P) resources.

To summarize, it can be stated that intuitive ontology is a species-specific interpretation of the real world run by the evolved domain-specific inferential system. This has important consequences in the present discussion.

6. The Intelligent Design Doctrine

This doctrine asserts that complexity of the universe and life can *only* be explained by an 'intelligent' cause and not by a random evolutionary process which lacks human intellectual attributes. Its proponents consider it as a scientific theory on equal footing with established empirical sciences and therefore, their stated objective is to investigate whether or not empirical evidence supports the above assertion. In order not to appear religious, ID supporters deliberately do not identify or name the specific agent(s), who designed and created the universe. They merely state that one (or more) such designers must exist.

Let us analyze carefully the notion of 'intelligent design' of an artefact, which involves the following three factors:

1) Intentional Factor: ID implies functionality and purpose before something is designed. It's a goal-directed process. Both 'goal' and 'purpose' are human species-specific concepts. In all human endeavours, it is to gain something, a reward or some benefit for survival.
2) Material Factor: material necessary for the design of something must already exist. In other words, to design some entity, not only the intentional attributes (ToM as an essential component of intelligence of the designer), but the pre-existence of material is assumed in the notion of design. In Behe's well publicized argument, not only the designer exists, but also material for the mousetrap as well – both exist *a priori*! If one takes this argument, then all components and basic ingredients of this universe were already present at the 'time' of design. From Paley's watch to Behe's mousetrap, the paradigm of analogy remains the same — species specific human design of some artefact utilizing already present material.
3) Intelligence: The designer must be 'intelligent', which in itself is very vague notion. It is not specified what really one means by the term 'intelligence' in the context of a supernatural. Unfortunately, we have very little understanding of even human intelligence!

We now run into two serious problems with ID. First is the problem of infinite regression. Who designed the intelligent designer, maybe a super designer, and who designed the super designer, and so on. This takes us back to the infinite regression. But, if the designer can exist without being designed by another entity, why can't the universe exist without being designed? The argument seems rather simple, but it does have serious implications. Second, the doctrine implies that human species-specific attributes existed before the creation of the universe — a contradiction in itself.

Whether Darwin's account needs refinement or an entirely new scientific theory will take its place is not the issue. Science progresses in this manner through refinements, falsification, and paradigm change. The problem with the ID argument is that it is a trivial generalization of an analogy that in itself is utterly simple, not even suitable as a toy model. All in all, it is no more than a pseudoscientific claim, a kind of 'theory of everything' that is neither falsifiable nor empirically testable. Empirical sciences and ontological naturalism should neither be mixed nor equated with belief-based doctrines such as creationism. Creationism and ID can be a part of religious studies, but not science curricula.

6.1. *Regularity Out of Randomness: A Counter Example to the Design Argument*

Let me give an example which clearly demonstrates that how random processes lead to regularity. Applying these rigorously established results to biological macromolecules (e.g., RNA and DNA), one can easily come to the conclusion that purely random molecules become biological molecules with some regularity when they have reached some threshold size. No divine intervention or intelligent design is required!

Van der Waerdon's theorem and Ramsey's theory in combinatorics assert that sequences formed by a finite set of symbols will eventually have sub-sequences with regularity properties (see e.g., Lothaire, 1983). This is known as the *principle of unavoidable regularities*. To apply these theorems to biological macromolecules, DNA or RNA, suppose we interpret four nucleotides as four abstract symbols. This means that whether or not there are chemical constraints, nucleotide sequences longer than a certain length with regularity properties are sure to be formed. Random sequences simply cannot exist if finite (only four in this case) symbols are used to form them. In particular, studies in the ergotic theory of combinatorics show the existence of Morse-Thue type sequences as inevitable (Furstenberg 1987). Such sequences contain, as their sub-sequences, either palindrome or inverse palindrome type — the so called hair-pin structure well-known to molecular biologists. The theory of random graphs, which was developed from Ramsey's theory, has some similarity with biological evolution in the sense that new properties emerge as nodes and edges are added randomly — a kind of phase transition (Bollobás 1985).

Thus, using only a finite set of symbols, it is not possible to produce purely random biopolymer sequences longer than a certain length. The role of randomness is limited. Both van der Weardon and Ramsey theories assert that a large enough number of random events will eventually show some precise form of regularity. Obviously, the principle of unavoidable regularities does in no way imply that regularities of either pRNA, RNA world, or any other kind of 1-

biopolymer species had exactly these kinds of mathematical regularities. Other factors, such as chemical constraints and natural selection, played their role to give them their final shape.

Chemical Constraints: The Crick-Watson base pairing and free-energy change impose strong constraints on the stability of macromolecular sequences (Saenger 1984). Such constraints preclude any fully random formation of large sequences. For instance, some tetramers are more stable than the others and hence likely to survive longer than others under thermal variations. For instance, at 37°C, GCCGGC is more stable than CCGGC and ACCGGC is more stable than CCGGC. Helixes without loops GCGGCG, GCGGCGA, GCAACCA, GCGGUCA are the most stable as compared to helixes with a bulge or a helix with hairpin loops (Turner & Bevilacqua 1993, Turner et al. 1987). Hairpins represent the dominant secondary structure elements in RNA. Such sequences are found with exceptionally high frequency in many RNAs and are characterized by high thermodynamical stability. Such RNA hairpins define nucleation sites for folding, determine tertiary structure in ribozymes, protect mRNA from degradation and are recognized by RNA binding proteins.

Thus, both abstract mathematical and chemical constraints seem to favour palindrome and inverse-palindrome structures that are found in the genome of all life forms. These results demonstrate very clearly the fallacy and weakness of the ID argument. There is no need to hypothesize the existence of a designer for biological molecules; empirical laws and evolution can do the job.

7. Discussion and Conclusion

There are many vague implicit assumptions in this debate on the nature of science versus creationism and its variants. First, there often is confusion as to the nature of reality and existence. Second, it is not clear what constitutes a proof of counterfactual statements about the supernatural. Third, the most important point is the assignment of human intentional attributes to a universal creator.

It is clear that there is a separation between the innate and the culturally learned components of belief in the supernatural. The innate component can be attributed to the emergence of ToM, a mental faculty responsible for many of the species-specific behaviours, whereas the nurture, the learned behaviour, is often imposed by a society on its members from one generation to the next:

a) The innate component, ToM, is the mental module, which is responsible for the emergence of counterfactual and counterintuitive beliefs in supernatural agents such as deities, ghosts, spirits, etc. Since, supernatural agents are supposed to "know" all about people's anxieties such as death, disease, catastrophes, and so on, therefore, they must have intentional attributes. The origin of all ideas related to the supernatural have their origin in ToM, which arose among *Homo sapiens* species around 40,000-50,000 years ago and was reflected in a sudden burst of symbolic activities (Pfeiffer 1982). This mental capacity seems to have emerged after the latest migration from Africa some 60,000-50,000 years ago and replaced all earlier types of humans. Extending human attributes to another entity, be it a con-specific or an imaginary natural and supernatural object, ToM is crucial; without it one cannot create a deity with human-like mental attributes.
b) Many new societal practices then emerge that are not directly related to ToM, such as rituals and material commitments towards supernatural deities. These include, for instance, offering of material goods and sacrifice of animal and even human life.
c) There is then a collective imposition of rituals, which often becomes forced imposition of both (a) and (b) and dissenters are severely punished. Such ritualistic commitments to counterintuitive worlds generated by supernatural agents become accepted sacred practices.

Supernatural agents violate universal cognitive principles that govern the ordinary human perception and understanding of the everyday world. Consequently, such beliefs and experience cannot be proved to be valid (or contradictory) through consistent logical inference or empirical procedures. The proof of validity is obtained only by practice of rituals collectively or individually. This is the truth of the second kind discussed earlier.

Many authors consider convergence of above components as exaptation, a term invented by S.J. Gould (1989) to indicate that some traits were not evolutionary adaptations, but emerged from a capability evolved for some other functionality. For instance, Atran and Norenzaya (2004) propose religion as a converging by-product of several cognitive and emotional mechanisms initially evolved for adaptive tasks. Similarly, Boyer (2003) takes religion as an emergent by-product of numerous domain-specific psychological mechanisms. Religions and religion-like beliefs are an emergent by-product of numerous domain-specific psychological mechanisms that evolved for mundane adaptive tasks (see, also, Bering 2006, Bulbulia 2004, Dunbar 2006, Motluk 2006).

Creationism, ID, and religious ideas involve domain-specific violations and transfer of ontological expectations, which makes all such ideas attention grabbing and memorable (Boyer 2003). Societal preservation of such concepts is simply a matter of cultural indoctrination — a simple matter of exposure from an early age.

All counterfactual ideas (ID and other forms of creationism) are phenomenological products of the mind rather than an objective reality and can be called ascriptions. Ascription is the ability or practice of attributing particular states or properties to persons, objects or events. Ascriptions differ from descriptions in their ontological commitment. In a description, the reality of the thing described is assumed and hence its availability to any observer. In contrast to such an objectivity of a description, all ascriptions are describer relative: they may or may not be available to other observers. Ascription of beliefs is seen as an aspect of social behaviour that presupposes the reality of the belief ascribed (Olson and Deepthi 1999). Given all the above mention problems with the concept of an intelligent designer of the universe, it qualifies only as an ascription.

ID can also be considered as a form of modern hagiology (or hagiography). A hagiological description unifies what "is" and what "ought" to be. It is invented as a desired mixture of 'actual' and 'possible', factual and counterfactual, and human intuitive ontology (e.g., attributes) extended and joined together with counterfactuals. A designer of the universe is nothing but a concept (in the mind of those who believe it). So, what it "is" is just a concept, and what "ought" to be is a super intentional being capable of all feats.

References

Atran, S. and Norenzayan, A. (2004) Religion's evolutionary landscape: Counterintuition, commitment, compassion, communion, *Behavioral and Brain Sciences* **27**, 713-770.

Baron-Cohen, S. (1999) The evolution of a theory of mind, in: M.C. Corballis & S.E.G. Lea (eds.) *The Decent of Mind, Psychological Perspectives on Hominid Evolution*, Oxford University Press, pp. 261-277.

Barrow, J.D. and Tipler, F.J. (1988) *The Anthropic Cosmological Principle*, Oxford University Press, Oxford, UK.

Bering, J.M., (2006) The cognitive psychology of belief in the supernatural, *American Scientist*, **94**, 142-149.

Bollobas, B., (1985) *Random Graphs*, Academic Press.

Boyer, P. (2003) Religious thought and behaviour as by-product of brain function, *TRENDS in Cognitive Sciences* **7**, 119-124.

Boyer, P. and Ramble, C., (2001) Cognitive templates for religious concepts: cross-cultural evidence for recall of counter-intuitive representations, *Cognitive Science* **25**, 535-564.

Boyer, P. (2005) Functional Origins of Religious Concepts: Ontological and Strategic Selection in Evolved Minds, Malinowski Lecture 1-23.

Bulbulia, J., (2004) The cognitive and evolutionary psychology of religion, *Biology and Philosophy*, **19**, 655-686.

Churchland, Patricia S., (2002) *Brainwise, Studies in Neurophilosophy*, The MIT Press, Cambridge (MA), USA (see, part II Epistemology and part III Religion pp. 239-402).

Conway-Morris, S. (1998) *The Crucible of Creation*, Oxford University Press, Oxford, UK.

Dauban, J.W. (1979) *George Cantor: His Mathematics and Philosophy of the Infinite*, Princeton University Press, NJ, USA.

Dunbar, R., (2006) We believe, *New Scientist*, 30-33.

Eysenck, M.W., (1990) *The Blackwell Dictionary of Cognitive Psychology*, Blackwell, UK.

Furstenberg, H., (1987) *Recurrence in Ergodic Theory and Combinatorial Number Theory*, Princeton University Press, NJ, USA.

Genter, D. (1983), Structure-mapping: A theoretical framework for analogy, *Cognitive Science* **7**, 155-170.

Gould, S.J., (1989) *This Wonderful Life*, Norton, New York.

Holyoak, K.J. (1985), The pragmatics of analogical transfer, in ed. G.H. Bower, *The Psychology of Learning and Motivation*, 19, pp. 59-87, Academic Press, NY.

Lothaire, M., (1983) *Combinatorics on Words*, Addison-Wesley, Reading, MA.

Motluk, A. (2006) Particles of Faith, *New Scientists*, ibid, 34-36.

Olson, D.R. and Kamawar Deepthi (1999), The theory of ascription, in ed. P.J. Zelazo, J.W. Astington and D.R. Olson, *Developing Theories of Intentions*, Lawrence Erlbaum, Associates. NJ, USA.

Pfeiffer, J.E., (1982) *The Creative Explosion, an Inquiry into the Origins of Art and Religion*, Harper & Row, New York.

Premack, D. and Woodruff, G. (1978), Does the chimpanzee have a 'theory of mind'? *Behavioral and Brain Sciences* **4**, 515-526.

Reiter, R. (1980) A logic for default, *Artificial Intelligence* **13**, 81-132.

Rodriquez-Consuegra, F.A. (1995) *Kurt Gödel: Unpublished Philosophical Essays*, Birkhauser.

Russell, B. (1940), *An Inquiry into Meaning and Truth, The William James Lecture for 1940*, Penguin University Books, 1973 edition.

Saenger, W., (1984) *Principles of Nucleic Acid Structure*, Springer-Verlag.

Shah, K.T. (2006) Epistemological pluralism: realities as species-specific interpretations of the material world, in ed. J. Seckbach, *Life as We Know It*, Springer.

Turner, D.H. et al., (1987) Improved Parameters for Prediction of RNA Structure, *Cold Spring Harbor Symposia on Quantitative Biology*, vol. **LII**, 123-132.

Turner, D.H. and Bevilacqua, P.C., (1993) Thermodynamic considerations for evolution by RNA, in *The RNA World*, eds. R.F. Gasteland and J.F. Atkins, Cold Spring Harbor Laboratory Press, pp. 447-463.

Vogeley, K. et al., 2001: Mind reading: neural mechanisms of theory of mind and self-perspective, *NeuroImage* **14**, 170-181.

Williams, P. (2000) *Buddhist Thought*, Routledge, London.

Wolff, Paul R. (1998), *About Philosophy*, 7th edition, Prentice-Hall, NJ, USA.

Dialogue

Taner Edis
I largely see eye to eye with Tahir Shah. We may differ on matters of emphasis, but that is not a basis for criticism.

I do, however, want to ask a question. Shah offers a naturalistic view that follows the tendency of modern science to cast doubt on the literal reality of supernatural agents. But the literature Shah refers to on the evolution and cognitive science of religion also suggests that belief in supernatural agents is deeply rooted in human nature. Whether religion is an adaptation or a byproduct, it provides many social services, and is cognitively much more natural, much less costly than more scientific ways of thinking. So if the scientific community were to more explicitly agree with views like Shah's (or mine), this would probably lead to a more pronounced rift between science and our wider cultures.

Can science, *as an institution*, afford such a rift? In the United States, with its constant creation-evolution wars, it is not in the interest of science to be further associated with religious doubt. In Muslim countries, the situation is even more awkward. As Shah has much experience with the institutions of science as well as the practice of science, I would very much like to hear what he thinks about this.

K. Tahir Shah
Let me reply to Taner by dividing his comments and questions into smaller sections:

> Taner: "But the literature Shah refers to on the evolution and cognitive science of religion also suggests that belief in supernatural agents is deeply rooted in human nature."

True, belief in supernatural agents is one of the many evolutionary exaptations. In this case, it is an exaptation of Theory of Mind (ToM), the innate mental faculty underlying social intelligence and social organization. Human and many other species organize socially around the dominant male (or female) who acts as the protector and provider within a group. Most human societies extended this form of social organization *ad extremis*. It is only in this sense that it is deeply rooted in human psychology. Take another example: the human inferential capability and causal reductionism as one of its consequences. An application of it easily leads to a *reductio ad infinitum* for such questions as "who created who/what?" One moves from the object level explanation to the meta-level and then meta-meta-level explanations, and so on *ad infinitum*. But, such human mental capacity and what follows from it, should in no way imply the existence of the "last" or the "first" cause, and any specific code of human conduct that is supposed to follow and be attributed to such a 'last' or 'first' cause. Such codes of conduct often are contradictory and vary from one belief system to another.

> Taner: "Whether religion is an adaptation or a byproduct, it provides many social services, and is cognitively much more natural, much less costly than more scientific ways of thinking."

I have no doubt about religion's limited social organizational value, but this should neither be taken as the proof of existence of supernatural agents, nor that it always benefits and only benefits humanity. It does provide some social service under a few special circumstances, but it also leads to wars, torture and severe punishments. Such punishments include the death penalty for criticism or change of religious viewpoint, the so called "blasphemy" and "apostasy", respectively; words that were invented to justify draconian punishments to those who oppose a specific and established viewpoint. History has proved, over and over again, religions are not all and only beneficial; there are terribly dark sides too.

It is difficult to say what is cognitively cost-effective and what is not. I am not aware of empirical studies to support one assertion or the other, but the scientific way of thinking apparently seems more costly. It is because there always is a consistent effort to use the inferential mechanism (i.e., mental effort) as well as physical effort to perform empirical verifications. Slogans are always cheap, so are pseudo explanations — no effort goes into empirical verification either mentally or physically. But, do slogans and beliefs add anything to human knowledge? At best, they have a placebo-like effect, but beyond it they are unable to cure human suffering, and nothing is added to human knowledge.

> Taner: "So if the scientific community were to more explicitly agree with views like Shah's (or mine), this would probably lead to a more

pronounced rift between science and our wider cultures. Can science, as an institution, afford such a rift?"

The rift, I think, is due to differing viewpoints taken in order to answer the following basic philosophical question: Should ontology be validated by empirical judgment or through a specific cultural-political stance? On this issue, I would tend to take sides with William James and John Stuart Mill, in whose opinion the essence of ontology is tied to questions about creating a better world. But, I would favor a stronger view that beliefs should be consistent with reality within the established laws of logic and empirical science, both ontologically and epistemologically.

Social validation of beliefs through the practice of rituals is not an answer that is acceptable to science. Science cannot and should not depend on social validation of its empirical principles, methods and practices, either through the practice of rituals or compromise them all simply in order to avoid rifts with a certain segment of society. Why should science compromise with the proponents of the ontological primacy of the social or cultural-political stance taken by some interest group (or groups) in lieu of empirical judgment that is supreme in science? Moreover, making ontology social entails allowing existence of an entity, supernatural or not, to be ascribed to anything society finds it convenient, often in the name of some interest group. Looking from another perspective, such beliefs also commit an extreme form of mind-body dualism. The existence of souls assumes a life after death, a belief in the survival of psychological states after death. Emotive factors are powerful contributors to this form of cultural-political stance. Science professes rationality, whereas politics, which exploits emotions, is deeply rooted in the ontological primacy of the social.

Should we make ontology social, and as a consequence of it, the truth social (truth of a second kind in my article), for the fear that otherwise a deep rift in the society between believer and nonbeliever of anthropomorphic deitism will occur? This is a worrisome question. Beliefs may lead to placebo-like effect, limited in what such effects can do to solve only a small portion of human misery, but not beyond it. Can a society afford to live without science and technology in today's world? Many of us, who grew up in "have-not" nations, know it all too well what it means to see one's loved ones dying due to lack of modern medical facilities — the other alternative, prayers, most often were not answered! The difference between "have" and "have-not" nations is only science and technology. People in developing nations know all too well what it means to live in such a society — poverty, ignorance, unemployment, and so on. A list that is too long to be enumerated.

Let me now move away from theoretical arguments and go over to some factual aspects. I would like to point out that political regimes of strong believers of one or the other kind of doctrine, one or the other religion, went against the establishments of rationality, sometime closing down institutions of learning, while at other times, simply castigating intellectuals who challenged their beliefs. But, soon after, they realized that they need science and technology for various reasons: from enhancing their grip on power to the chess of international politics. They come to grip soon with reality that their supernatural agents do not provide WMD, ballistic missiles, or high-tech surveillance devices to be used against their enemies. Even suicide bombers use high tech explosive devices. When a high profile TV preacher gets sick, he doesn't rush to Lourdes, he goes to an expensive high tech clinic.

Experience at international institutions of science nevertheless suggests something quite encouraging: There is a great desire to pursue science virtually in all developing nations from one end of the political spectrum to another. Above all, quite surprisingly, it is largely supported by public funds in almost all cases. It is true, I may add, that there is some resistance in a rather strange way, but it is only a small segment of some societies. For example, in some quarters there were efforts to Islamize science, something not so different from the efforts of creationists today to mix belief and science, but it was not supported by the mainstream scientists. What seems to me quite clear is that the general public in both Islamic as well in non-Islamic world tend to separate science from theology.

I am not worried thus that science will cease to exist. Quite contrary to all this, it will continue to flourish. Whether some segment of our society believe "the sky to be green" or "Earth to be flat" and label such beliefs as sacred truth is simply irrelevant. The society will continue to need air transport, modern medicine to cure diseases, and all that is derived from science and technology. Of course, this does not mean that we should stop telling them that they are wrong and should not mix science and theology. Separation of science and theology had been a supreme achievement of renaissance in Europe and we should keep it — for the benefit of all life on this planet.

Biography

Professor Benjamin Fain is currently on the faculty (Professor Emeritus) of School of Chemistry, Tel Aviv University. He obtained his Ph.D. from the Gorky University (USSR) in 1956. Professor Fain's scientific interests are in quantum electronics (he published the first book in this field, together with Ya. Khanin in 1965), solid state physics, chemical physics and foundations of quantum mechanics. Recently his interests shifted to the problems of science and religion.

21
Evolution and Providence

BENJAMIN FAIN

School of Chemistry, Tel Aviv University
69978 Tel Aviv, Israel
fain@post.tau.ac.il

1. Introduction: Two Worldviews — Divine Providence, and a World Without God

During the last few centuries, and especially in the nineteenth century and the first half of the twentieth, philosophers and intellectuals have tried to explain the world without God. One of these was Karl Marx, who "explained" historical developments in economic terms. Marxism had such a strong influence on the world that entire parties and states identified with this theory. Empires arose, sacrificing millions of individuals on the altar of Marxism. For a long time, many of the Western intelligentsia identified with Marxism. Today, all recognize the bankruptcy of both the theory and the attempt at implementation. Sigmund Freud "explained" the soul as a by-product of sexual motives. Freud's theory is still influential in some sectors today.

Both Marx and Freud were openly anti-religious, and their theories contributed to an anti-religious atmosphere in the intellectual world. In contrast, Charles Darwin, the nineteenth-century biologist, defined himself as a believing man. Ironically, it was his theory of evolution, or "natural selection," which strengthened the anti-religious worldview more than any other contemporary scientific theory. At first glance, his theory (or later versions of it) seems to indicate the possibility of explaining the development of the world without God. According to the theory of evolution, we can explain the development of life on earth as a result of natural factors, both necessary and accidental.

In the eyes of many, the theory of evolution seems to contradict the worldview of the believer, especially that of the believing Jew. Therefore, we must emphasize the difference between facts and assumptions underlying the theory of evolution, and the worldviews that various philosophers and scientists have developed from those same facts and assumptions. Karl Marx and Friedrich Engels, the enthusiastic materialists of the nineteenth century, immediately recognized the implications of Darwin's achievements. Today, many still use Darwinism (or neo-Darwinism, in its modern form) for justifying the materialist and atheist approach. This approach sees the development of the world in general, and specifically the development of life on earth, as an expression of the activity of the forces of *materia*, matter, without the intervention of any external cause. Matter and its movements are sufficient to explain all phenomena in the world, including life, self-awareness, the soul, and human emotion.

In contrast, in Judaism the concept of Divine Providence lies at the foundation of its entire perspective. According to the Jewish viewpoint, God rules the world, and He is the one who directs and determines everything that takes place within it. Judaism attributes great importance to the concept of general and personal Providence. The first of the Ten Commandments is, "I am the L-rd your God who brought you out of the Land of Egypt out of the house of bondage." In this commandment, God conveys to us the fundamental truth of His role in history. Maimonides counts faith in Divine Providence as one of the thirteen basic principles, which each Jew is required to believe: "I believe with complete faith, that the Creator, may His Name be blessed, knows all the deeds of human beings and their thoughts, as it is said, 'He fashions their hearts all together, He comprehends all their deeds." Rabbi J.B. Soloveitchik describes this process precisely: "God created the world as a separate and distinct entity, but did not grant it independent reality. *The world exists because it suckles from the infinite Being of God*" (Soloveitchik, 1992, p. 202, my italics, BF).

We have before us two worldviews. One is the theocentric — God is the center of the world; He runs it, and intervenes in what takes place in it. The development of the world, including the animal kingdom, is under divine supervision or Providence. The other worldview is the atheist-materialist or secular one. This viewpoint sees us, our inner world, the abundance of life forms, human society, economics, beliefs, feelings of love, hate, and jealousy — as a product of only one single cause: *materia* and its movement. These two outlooks constitute metaphysical positions, and as such, they are not derived from experience or from theory. Thus we can see the materialistic position as a form of faith.

Many have derived far-reaching conclusions from Darwinism, as if the 'theory of evolution' leads to the materialistic worldview. They conclude that there is a conflict between this theory (along with neo-Darwinism) and the Jewish faith. This alleged contradiction is what deters many in certain religious circles from any contact with Darwin's theory of evolution. (Israel's national-religious school curriculum does not include this theory at all). I will endeavor to show why this restraint is unjustified.

I set two goals for myself. One of them, quite modest, is to show that the theory of evolution, in its standard and accepted form, is compatible with the concept of Divine Providence, and does not contradict it. There is compatibility between the evolution of life, as described in the theory of evolution, and the belief in Divine Providence. My second goal is to provide a general analysis of the foundations and assumptions of the theory of evolution. The findings of this analysis undermine the basis of any theory of evolution. (See, also: Fain, 2004).

2. The Basic Mechanism of Natural Selection

The theory of evolution rests on the following basic premises:

1. The profusion of life forms on earth originates from several organisms, or even from only one; we speak of an evolutionary "tree", an evolutionary history.
2. The theory of evolution that explains this process includes the following assumptions:
   a. Heredity: Offspring reconstruct their parents' hereditary material, and thus they reconstruct their parents' organism with great fidelity. Heredity is, in general, a stabilizing factor that tends to preserve a common pattern for individuals in subsequent generations. The oak tree that grows from the acorn is similar to the tree that produced the acorn. A chick that hatches from an egg is similar to the hen that laid it. However, we know that offspring are not a precise copy of their parents. Evolution (no matter in which direction) is a process of change, in which the differences between the offspring and their parents are of utmost importance.

b. Change: There are three possible reasons for change. First of all, the development of the individual is influenced not only by heredity, but also by the environment. Two oak trees, even if they sprout from two identical acorns, do not have the same form, size, or number of leaves. Differences in the environment, such as soil type and weather, influence the trees' growth. The environment in general influences plants more than animals. Among types of animals, mammals are the least influenced by the environmental factor. Environmental changes are not hereditary; they are not preserved in the subsequent generations. According to the theory of evolution, an additional factor leads to differences between offspring and their parents: the ***mechanism of sexual reproduction***. This mechanism is nearly (but not completely) universal both among plants and among animals. The offspring receive different genes from two parents. The parents transfer a combination of genes to their offspring, a set identical neither to the father's set of genes nor to the mother's. This mechanism brings about an infinite number of hereditary variations within the same family, but does not cause the creation of a new type or species. The main factor capable of making a long-term evolutionary change is ***mutation*:** a random change in the hereditary material. There are different causes for the appearance of mutations, amongst them radiation.

c. Natural Selection: According to the theory of natural selection, evolution, or the development of life, is neither totally random nor totally directed, but constitutes a combination of these two types of processes. Mutations are accidental, but adaptation to the environment is deliberate. The idea behind **natural selection** is rather simple, although its operation is most complex and delicate. In every population, some individuals have more offspring than others. Individuals whose hereditary changes are more positive will survive, while individuals with unsuccessful hereditary characteristics will die before giving birth to a continuing generation. Changes that are passed down result in phenomena of *differential* survival, which accumulate from generation to generation. In this way natural selection acts to constantly improve and preserve the adaptations of life forms and plants to their environment and way of life.

The basic mechanism under the rubric "natural selection" includes three components: heredity, mutation and natural selection. Since Darwin, biology has made great advances, and today we are well-informed about the physics and chemistry of the mechanism of heredity. We are capable of deciphering man's genetic code, but the mechanism of natural selection is still considered the basis for understanding evolutionary change.

The mechanism of natural selection is a vital tool for understanding biological phenomena. For example, we can explain the adaptation of bacteria to certain antibiotics such as penicillin through the mechanism of natural selection. Another example is *protective mimicry* where animals such as birds and insects develop resemblance to the colors and patterns of their natural surroundings, as a means of hiding from their predators. We have no alternative to natural selection for describing, explaining and understanding these and many other biological phenomena.

3. The Complexity of Life

Living organisms, from the simplest, to mammals, to human beings, are highly complex and delicate mechanisms. What sets living things apart from inanimate objects is the hereditary material that is common to each living entity. From the simplest bacteria to plants and human beings, all have the same hereditary material, DNA, and humans are different only in the length of the DNA chain.

The hereditary material of the biological organism is largely preserved in subsequent generations, but we can identify miniscule changes from generation to generation as well as changes in various characteristics of the same species. Scientists estimate that 99.99% of human genes are identical, and the difference in hereditary material among individuals finds expression in just 0.01% of this hereditary material, the DNA. The truly astonishing accomplishment in biology in the last hundred years has been the attainment of an understanding that every living organism contains uniform hereditary material that is the most conservative component of the organism (Darwin did not have this information). The hereditary material determines most of the physical characteristics of the living creature. A complete living organism develops from semen ("From whence do you come? From a stinking drop..." Ethics of the Fathers, 3:1). The hereditary material includes all the necessary instructions for developing the organism from the semen. These instructions are written in "letters" and "words" that make up a certain language, like a computer language. Of course, this is only a simile: there is no real connection between the language of DNA, and human language or computer language. But this comparison highlights the possibility that other languages exist beyond humanly created ones.

The hereditary material located in the DNA of the organism's cells includes an entire series of directions for developing the organism. The hereditary material in the DNA chain determines the structure of bio-molecules vital for the functioning of the living organism, and for many kinds of proteins. A certain combination of different molecules determines a certain specific protein, and these molecules are like different "letters" of the hereditary "text." Now suppose that we know nothing of genetic biology. We ask ourselves, how can the hereditary instructions be written up? The letters of such a text must be preserved over millions of years. The text itself changes gradually, because of mutations, but the letters of the text are preserved. All material that we are familiar with from daily experience can be eroded. Here we arrive at a crucial point. The letters of the hereditary text are made of molecules. These letters are not the products of human beings, but are "the finest masterpiece ever achieved along the lines of the Lord's quantum mechanics" (Schrödinger, 1948).

In all living creatures and plants, hereditary information is stored in a uniform way, in the form of polynucleotides (a chain of molecules called nucleotides). In general, the size of these chains is in proportion to the complexity of the organism. The hereditary information in a human being is stored in around 10,000,000,000 nucleotides (since the nucleotides are arranged in pairs and the information stored in two chains, the length of one is 5,000,000,000 nucleotides). If we write out the hereditary information of a human being in English letters, for example, we will fill more than two hundred telephone books. Together, these books will weigh about one ton, and their combined volume will be about one cubic meter. Yet the hereditary information of a human being is concentrated in the cell's nucleus, whose volume is 0.001 cubic millimeter! (However, we must take into account that not all the DNA is the source of hereditary information.)

This data offers only meager support for a mode of thinking free from prejudices about life in general and human beings in particular. Unfortunately, habit and routine prevent us from grasping the miracle of life. The hereditary information in man's cells adds nothing to our understanding of the phenomenon of the soul. We are used to being surrounded by creatures that are capable of both speaking on the one hand, and hearing and listening on the other; of feeling, and inspiring feelings of love, hate, empathy, jealousy, sympathy and antipathy; of behaving rationally and irrationally. We are habituated to these things, like a child who is accustomed to television and cars and thus feels no sense of wonder at these modern inventions.

We can imagine (but only imagine!) an opposite situation (that cannot be realized). A physicist lives in another world devoid of life. He is an expert in quantum physics, the general and special theory of relativity, physical chemistry, and solid state physics — in short, he knows everything about inanimate material. His worldview is simple. Physics is capable of explaining all properties of matter. After all, everything in the world is matter (we are ignoring the problem of the scientist's very existence). All the scientist's personal experience and observations support his worldview. He can explain everything in the world. Then it so happens that the scientist moves to our planet, to earth. Here he experiences many miracles, and he views with wonder the profusion of animal and plant forms. In his eyes, the greatest marvels are the living creatures, especially human beings. Cars, television, and supercomputers do not amaze our scientist, for he easily explains them as the work of humans. However, the people, who talk, explain, argue, feel different emotions, react rationally and irrationally, and have freedom of choice, do not fit in with the scientist's previous worldview, i.e. that everything in the world is matter or made of matter. And everything obeys laws of physics only. His objective analysis leads him to conclude the existence of a being that creates everything, including the human beings themselves.

Before our scientist had arrived on earth, he did not think in this direction. He had believed that matter had always existed, and knew how to explain and understand all its qualities. However, from his conversations with his colleagues, the earthling scientists, he realized that they held a different opinion. They thought that the living world, including man, was created as a result of *evolution*, and that this evolution had a *creative* force. The scientists explained to him that they have a theory of evolution that aptly describes this process. Our scientist is not really convinced, and he begins to contemplate this matter. Without entering into the details of biology, he wants to understand if it is logically possible to derive the theory of evolution. Is a law of evolution possible at all? Now we shall leave our imaginary scientist alone, and endeavor to answer these questions by ourselves.

4. The Development of Life on Earth: Can We Prove the Law of Evolution?

When we speak of evolution, we must distinguish between two different concepts. On the one hand, we speak of the *theory* of evolution: this is allegedly based on the mechanism of adaptation called natural selection (see above, **2**). On the other hand, we can also point to the *fact* of evolution, meaning the development of the profusion of life on earth, which began from primitive organisms and developed gradually to self-conscious creatures. Confusing these two ideas may lead to serious misunderstanding. We must also differentiate between the mechanism of adaptation, or natural selection, and the "theory of evolution." Sometimes scientists consider a successful explanation of a certain phenomenon of adaptation by the mechanism of natural selection as proof of the "theory of evolution."

The *fact* of evolution is based neither on one theory or another, but on paleontological data, on the fossilized findings of animals from different periods. Evolutionary scientists today dispute the idea of gradual evolution as a factor for creating new species. Niles Eldridge and Stephen J. Gould explained the lack (or at least the rarity) of intermediate fossil sequences by asserting that evolution is not gradual. They hypothesized that when new species develop, they do so over a relatively short period of thousands of years. Before and after the change, the species remains stable for long periods, running into millions of years. Scientists call this type of development *punctuated equilibrium*. The main thesis of the analysis below is that we cannot derive the *theory* of evolution (that describes evolution) from paleontological data, from the *fact* of evolution, assuming that it is, indeed, a fact.

I will briefly summarize the main tenets of Karl Popper's theory of scientific cognition (Popper, 1981, 1992, 2000; Fain, 1999, 2004). He showed that all scientific theories are deductive, and logically derived from certain basic assumptions, which are the products of human creation. We cannot derive them logically from observations and experiments. We can only test a theory with experiments and corroborate it by applying it to additional experiments. We must be able to refute or falsify the theory. But, as we have said, we cannot verify any theory absolutely. If we cannot test a theory by experiment, then this theory is not scientific but metaphysical.

Now we will ask ourselves: Assuming that it exists, what is the status of the theory of evolution? Here is Karl Popper's answer. The evolution of life on earth...is a unique historical process, one of a kind. "Such a process, we may assume, proceeds in accordance with all kinds of causal laws, for example, the laws of mechanics, of chemistry, of heredity and segregation, of natural selection, etc. Its description, however, is not a law, but a singular historical statement" (Popper, 1999, p. 109).

Clearly, any theory, no matter how it is formulated, must be tested by new experiments before it merits the status of a law of nature. We cannot discover any law of nature if we remain attached to a one-and-only experiment. This general philosophical understanding, based on logic, leads us to the unequivocal conclusion that the theory or law of evolution does not exist. Several biologists have also come to this conclusion in light of their research findings. In 1965 Jacques Monod, Andre Lwoff and Francois Jacob won the Nobel Prize for their contribution to the field of genetic biology. Monod writes, "The thesis that I shall present in this book is that a biosphere does not contain a predictable class of objects or of events, but constitutes a particular occurrence, compatible indeed with first principles, but not *deducible* from those principles and therefore essentially unpredictable" (Monod, 1997, p. 43). Moreover, he adds, "I believe that we can assert today that a universal theory, however completely successful in other domains, will never encompass the biosphere, its structure and its evolution as phenomena *deducible* from first principles" (ibid. p. 42).

If we were to tape the whole evolutionary process on film, every re-rerun would be a new film, showing a completely different history. With this vivid metaphor, Gould presents the idea that evolution is unpredictable. The key word is "contingency," meaning dependency, or the possibility of an event. "A historical explanation does not rest on direct deductions from laws of nature, but on an unpredictable sequence of antecedent states, where any step of the sequence would have altered the final result. This final result is therefore dependent, or contingent upon everything that came before" (Gould, 1989, p. 283). George Gaylord Simpson, in his book, *The Meaning of Evolution*, (Simpson, 1951) uses other words like "opportunistic" and "pragmatic" in order to describe the interdependent chain of events that make up the evolutionary process.

Doubtlessly, the mechanism of natural selection fulfills an important role in understanding the evolution of life on earth. But it is a great leap from this point to claiming that natural selection explains the evolution of life on earth. Of course, this depends upon how we understand the word "explain." The prevalent opinion is that the mechanism of natural selection explains all evolution on earth, and that from the moment life was created (I am purposely abstracting myself from the problem of <u>creation</u> of life on earth) we can predict the development of life on earth. However, we have seen that it is impossible to make this prediction. We can sum up the analysis of this part in Popper's words: "There exists no law of evolution, only the historical fact that plants and animals change, or more precisely, that they have changed. The idea of a law which determines the direction and the character of evolution is *a typical nineteenth century mistake, arising out of the general tendency to ascribe to the 'Natural Law' the functions traditionally ascribed to God*" (Popper, 2000, p. 340, italics mine).

5. Compatibility with Divine Providence

In 1995, the National Association of Biology Teachers (NABT) in the U.S. included on their agenda a vote on the public statement that "evolution is an unsupervised, impersonal, unpredictable and natural process." After hours of argument, the members of the association agreed to remove the words "unsupervised and impersonal." The president of the NABT, W.W. Carley, who insisted on the change, said that this change was in the spirit of good and honest science and that the assumption that "evolution is unsupervised" is a theological assumption (Larson and Witham, 1999). In the end, the NABT summed up, "Evolution is an unpredictable and natural process." However, this is not the accepted opinion among all biologists. Many of them insist that evolution is indeed an unsupervised and impersonal process. We do not know which side is in the majority, and this is unimportant, as the majority does not decide in this case. What is important is that we cannot decide which side is right since the decision is in the metaphysical realm, as we made clear in the previous section. The question of whether or not the evolutionary process is controlled by Divine Providence is a question of faith. What we may say is that Divine Providence is compatible with the accepted picture of the evolution of life on earth; faith in Divine Providence does not contradict the accepted picture of evolution. I have purposely used the inexact term "accepted picture" instead of "the theory of evolution," since the theory of evolution does not exist, as we have clarified above.

We can connect the explanation of evolution with predictability. However, we can also widen this explanation by including in it random factors such as mutations, which cannot be predicted. And many biologists believe that evolution of life on the earth can be explained by the mechanism of natural selection. Here we must immediately express a reservation, as this assumption enters the field of metaphysics, and it is impossible to prove this explanation by experiment. There is no proof (see above) that natural selection alone can explain the evolution of life on earth, even though this is the accepted explanation of many biologists. In other words, this explanation or "theory" is a matter of faith.

Let us return to the problem of compatibility with Divine Providence. Clearly, the concept of evolution activated by natural selection is compatible with Divine Providence. What human beings conceive of as a "random" factor (mutation) may be, in fact, a Divine action, just as we may interpret a consistent message in an unfamiliar language as being totally random. We conclude that the standard explanation of evolution does not contradict the possibility of Divine Providence; in fact, it fits well. Many consider that the theory of evolution threatens religious belief, because of the widespread opinion that every development on the face of the earth, from unicellular organisms to *Homo sapiens*, is predictable and predetermined, and we can explain it naturally. However, the random factor, mutation, defies prediction of the entire process of development, and offers the possibility of different routes of evolution. The choice between the various routes of evolution can be either totally random, or in the hands of Divine Providence. The decision as to which possibility is realized lies outside the realm of the experiment, and therefore, it is metaphysical.

Above we demonstrated the absence of proof that natural selection alone is enough to activate evolution, as well as the impossibility of proving such a claim. Evolution is a specific historical record, which cannot serve as a basis for any theory. This is an epistemological claim, belonging to the theory of cognition. We cannot discover or find a theory of evolution. Another question is whether the law of evolution exists at all, even if we are incapable of discovering it. This is an essential, ontological problem. We shall devote the next section to this problem.

6. Does a Law of Evolution Exist?

Even though we are not able to discover, establish or prove the law of evolution based on a specific historical record, such a law might indeed exist. We cannot, in principle, prove that this type of law exists or does not exist without additional assumptions. However, if we make a certain metaphysical assumption (that cannot be verified by experiment), then we will be able to prove that the law of evolution does not exist. Our basic premise is that man has free will. Indeed, this assumption seems self-evident, but it is metaphysical in character, as it cannot be tested either experimentally or theoretically. Rather, man senses his free will immediately and directly, and needs no logical reasoning in order to prove it. Faith in free will is one of the foundational beliefs of man, akin to the belief in the existence of other human beings. Therefore, we can accept freedom of choice as a basic premise or axiom. In fact, one of the foundations of Judaism is that every development, every instance of evolution in life, is subject to Divine Providence, notwithstanding the assumption that the laws of nature (physics, chemistry, biology) are true, including the mechanism of natural selection. This is Judaism's basic outlook. We have seen that this concept is compatible with the standard explanation of evolution of life. Now we will ascertain whether the materialist, naturalist and atheist view of evolution can coexist with the assumption of freedom of choice. According to the materialist outlook, in the early stages of the development of the world, only matter existed. Spiritual entities, such as the soul, spirit, and mind, appeared at a more advanced stage of the development of the world. In a naturalist, atheist, godless world, new objects cannot appear *ex nihilo*. Every new thing stems from a previously existing one, and, ultimately, from matter. Therefore, the spirit is derived from matter, and operates according to natural law. The worldview, which says that every new thing is derived from something previous to it, is called reductionism. We know of two types of natural law. One is determinist in that previous events determine subsequent events in a definite manner. The second type of law is non-determinist; this kind of law determines only the probability of new events. In this context, it does not matter which type of law we are discussing, or, to be more precise, which law the world follows, so long as this law does not allow for new creation *ex nihilo*. Therefore, according to the naturalistic outlook, even when evolution reached the stage where there was such a thing as the human soul, natural law was the only determinant of its development. On the other hand, we believe that our individual wills also play a substantial role in the development of things, at least on earth. Desires, aspirations and human choices can be influenced by logic, or by religious and moral considerations, but not by laws of nature! Reading a book can change a person's behavior. Clearly, human desire is not the only factor causing change or development. Obviously, the laws of physics and chemistry influence change. However, if free will exists, then the laws of nature cannot be the *only* reasons for all change. We arrive at the conclusion that the existence of free will contradicts the naturalist aspect of the law of evolution. Until now, we have discussed the world as conceived by certain philosophers, who begin by placing limitations on it. In the next section,

I will present my view of the real world, which assumes the existence of creativity and new creation. In order to describe the development of the world, I will use the term "creative emerging evolution".

7. Creative Emerging Evolution

Many philosophers commit the philosophical "sin" of dictating their own categories and worldview of reality. Their worldview might fall under the rubric of materialism with its assertion that the entire world is derived from matter, or under the heading of idealism with its own assertions. Aside from these general views, we are habituated to a number of supposedly self-evident axioms, which become philosophical dogmas. One of philosophical dogmas that is relevant to our analysis, is termed in Latin: ***ex nihilo nihil fit***, and meaning ***ex nihilo*** creation does not exist, or "there is nothing new under the sun." Evidently this dogma stems from our limited daily experience. Long range extrapolations may very well lead to erroneous conclusions. In any case, we may not turn a conclusion that is the product of our limited experience in both time and space, into a philosophical dogma. The measure of truth of this dogma is very important, as it touches upon the very heart of understanding evolution. Philosophers such as Herbert Spencer (Spencer, 1966), Pierre Teilhard de Chardin (de Chardin, 1975) and Friedrich Engels (Engels, 1940) saw evolution as the realization of a plan woven into a fabric within the structure of the world. A widespread trend is to view the stages of fetal development as a prototype of the stages of evolution. By contrast, the researcher J. Monod (Monod, 1997) writes:

> "For them [the above-mentioned philosophers], evolution is not really a creation but uniquely the 'revelation' of nature's unexpressed designs. Whence the tendency to see in embryonic development an emergence of the same kind as evolutionary emergence. According to the modern theory, the idea of 'revelation' applies to epigenetic development [development of the fetus], but not of course to evolutionary emergence, which owing precisely to the fact that it arises from the essentially unforeseeable, is the creator of *absolute newness*" (p. 116, italics mine).

Let us develop this idea. According to any modern scientific theory, development in the world, or to be precise, development of the world, takes place in stages, beginning with lifeless matter. The first creation was the creation of the world itself. I do not mean the story of Genesis, but the standard theory of the Big Bang — a theory in physics that describes a new creation, creation *ex nihilo* (of course, we can always question the truth of this theory). We will skip certain stages of the world's development: the creation of stars, planets, galaxies and heavy elements (these stages also include new creations, but not as clearly as in the creation of the universe) and go straight to the creation of life, which doubtless constitutes an act of creation *ex nihilo*. From this point on, many new things are created: souls, the human spirit, works of human creativity, in fields such as science, mathematics, philosophy, and art. Each of these fields includes a countless number of new creations. A new symphony, for example, is an *ex nihilo* creation. This symphony did not exist before its creation, and is not derived from former musical works. If we identify *ex nihilo* as a one-time act (and because of this, it does not fall under the rubric of science), as a miracle, then miracles are multiplying every day. "We shall give thanks to you…for Your wonders and Your favors that take place in each time, in the evening, morning and afternoon" (daily *Amida* prayer, the blessing of thanks). This picture explicitly contradicts accepted philosophical methods and dogmas. In this situation, only a philosopher who is not bound by prejudiced assumptions is capable of analyzing the matter objectively. Karl Popper is neither a materialist nor an idealist (with regard to philosophical theory). He considers himself a realist. He does not deny *ex nihilo* based on prior philosophical assumptions. He recognizes the idea of emerging development, and that although it contradicts all accepted concepts, it reflects reality:

> "Whether or not we look at the universe as a physical machine, we should face the fact that it has produced life and creative men; that it is open to their creative thoughts, and has been physically changed by them. …the universe that harbors life is creative in the best sense: creative in the sense in which the great poets, the great artists, the great musicians have been creative, as well as the great mathematicians, the great scientists, and as the great inventors" (Popper, 1995, p. 174).

In my opinion, Karl Popper had enough courage to look at reality as it is, even if this reality contradicts "common sense" and accepted philosophical dogmas such as the impossibility of an *ex nihilo* creation. It is hard to say that philosophy has to follow in the footsteps of science. Indeed, the idea of creative evolution, or emerging development, has its source in a scientific approach. It does not contradict the latest scientific theories; on the contrary, it agrees with them. Nevertheless, it is not derived from them in a logical deduction, and it cannot be *reduced* to science. Science cannot explain creative emerging development, as every act of creation is one-of-a-kind. Science does not deal with unique events; they do not belong to the category we call science. The relationship between science and creative emerging development has another aspect, connected to determinism or non-determinism. The determinist world (which is also reductionist, meaning each stage in development is derived from the previous one) acts with clockwork precision, with no deviations. It includes all organisms, animals and people. From this view stems the negation of freedom of the will. Otherwise, determinism cannot be true. Quantum theory has opened up a new option. We have learned that randomness and probability play an important role in the physical world, and this renders the world no longer determinist. Nonetheless, the situation is no clearer. To say that the black characters on the paper of a scientific article are random is no better than saying that these characters were derived at the time of the creation of the world. In both scenarios, creativity approaches zero. Is the combination of chance and necessity likely to improve the situation? If evolution is activated by natural selection, such a combination is what results. We showed above that one cannot prove that natural selection is sufficient to initiate and explain evolution. This is a belief, and it has no basis in daily experience. We have never seen nor experienced new creation as a result of a combination of natural, necessary and random factors. Likewise, no experiment has succeeded in creating a new biological species as a result of natural, necessary and random factors. In the same way, we cannot predict a new biological species based on the theory of evolution.

Popper as a realist (but not as a materialist!) understood better and more clearly than his predecessors that the development of the world includes many incidents of true creativity, when something new is born that did not exist previously — *ex nihilo*. However, his description is phenomenological: he describes phenomena as they are, without entering into the essence of things, without asking (or answering) questions such as "why." The reason is that the essence is located outside of the intellectual, scientific approach. It is impossible to predict truly creative phenomena in a scientific manner; otherwise these phenomena will cease to be creative. Until now, we have acted with restraint and refrained from entering the field of religion. However, as we have arrived at a dead end in our intellectual, rationalistic path, we will try to let Jewish metaphysics assist us. After exhausting the arguments of physics, we turn to metaphysics. "God created man in His image, in the image of God He created him, a male and a female He created them" (Genesis 1:27). From this verse, we learn that some of man's characteristics make him similar to God. When we compare man to other animals, we identify creativity as the trait that most noticeably differentiates between them. If this is so, then we may conclude that man's creativity is one attribute that makes him similar to God. Thus, if we are looking for signs of Divine activity in the development of the world, this activity must be creative. To use metaphorical language, the development of the world reveals signs of *personal* activity to us. These signs point to what we might call the *creativity of the universe*.

8. Conclusion. Evolution: Law of Nature or Divine Providence?

The basic mechanism called *natural selection* includes three components: heredity, mutation and adaptation to the environment by natural selection. Mutations cause random changes in the hereditary material, and the hereditary mechanism passes them on to the following generations. The mechanism of natural selection can be explained as follows: in each population, some individuals will have more offspring than others. Individuals with more positive changes will, on average, survive, while those with less successful heredity will have fewer offspring. Therefore, natural selection acts continually to improve the adaptation of animals to their environment and to their way of life.

We must distinguish between the mechanism of adaptation by natural selection and the theory of evolution, just as we must distinguish between the *fact* of evolution and the *theory* of evolution. The theory of evolution presumes to explain the entire process of evolution of life on earth. Its basic assumption is that the mechanism of natural selection is responsible for the whole process of evolution. Ernst Mayr, Darwinism's enthusiastic supporter, wrote:

> "Darwinism rejects all supernatural phenomena and causations. The theory of evolution by natural selection explains the adaptedness and diversity of the world solely materialistically. It no longer requires God as creator or designer (although one is certainly still free to believe in God if one accepts evolution)" (Mayr, 2000).

Marx and Engels hastened to react to Darwin's *Origin of the Species*. In their opinion, this work could have been the basis for a natural history that was compatible with their materialist worldview. Many critics think that Darwinism, or, more precisely, neo-Darwinism, offers a meaningful challenge to the Jewish worldview, especially to faith in Divine Providence. The enormity of this challenge becomes clear if we note, in brief, the complexity of the biosphere, the system of life on earth. Man is the pinnacle of this complex system. If we write the hereditary text of man in ordinary letters, it will occupy about two hundred volumes of the *Hebrew Encyclopedia*. However, the hereditary text is written in molecules, and it occupies a volume of around 0.001 cubic milliliters. The hereditary text mainly determines the material or physical characteristics of man. However, man's complexity is determined first of all by his soul. We cannot imagine that such a sophisticated system can be created by a combination of random and necessary factors. Darwin spoke of "the extreme difficulty or rather impossibility of conceiving this immense and wonderful universe, including man with his capacity of looking far backwards and far into futurity, as a result of a blind chance or necessity" (E. Mayr, 1993, p. 59).

On another occasion, Darwin writes, "The mind refuses to look at this universe, being what it is, without having been designed" (ibid.). We must distinguish between the process of evolution, which is a type of historical record, and the theory of evolution. We have shown (above, **5**) that we cannot establish a theory of evolution based on a unique record, just as we cannot determine a law based on one experiment. However, the impossibility of deriving a theory does not necessarily mean that the theory does not exist. Here we enter the metaphysical area, as the existence or non-existence of this law cannot be tested by experiment (since we have only one "experiment," one historical or evolutionary record). The premise of *freedom of the will* is a metaphysical assumption, since we cannot test it through experimentation. This premise contradicts the existence of any <u>law</u> of evolution. We can imagine that a certain law directs evolution, but when it arrives at the stage of humanity, its development no longer follows this law, thus contradicting the basic truth of the law. The development of humanity is determined by the free choices of man together with the laws of nature. Freedom of choice is, in fact, the most basic faith without which human life, morality, and responsibility are meaningless. In the Jewish view, the *Torah* and the commandments have meaning only when man has freedom of choice. Until now, we have discussed the question of the law of evolution. We cannot derive the theory of evolution from paleontological data. However, does this law still exist in principle? We concluded above that the law of evolution does not exist. Thus we can ask the following question: what causes evolution? We know of two kinds of developments. One is development directed by the laws of nature only — we concluded that evolutionary development is not of this kind. We can explain the second kind of development using human life as an example. An individual life is directed not only by the laws of nature, but also by personal choices and decisions. The development of this individual is described as a biography. The concept of evolution applies to every slow, gradual change of a personality or reality, both material and spiritual. We spoke of two kinds of development. Let us begin with the first. The laws of nature control the first type, and the system is the activating factor for its own development. We may use the Big Bang theory of cosmology as a paradigm for describing this type of development. The laws of physics in general, and Einstein's equations of the general theory of relativity in particular, describe the development of the universe *ex nihilo*. At the moment of creation, the universe is concentrated in a tiny point. Suddenly, an explosion occurs; the universe spreads; and galaxies, stars and planets are created — amongst them, the earth. The laws of physics seem to describe this process. I write "seem to," since this description is only theoretical and not an actual description of nature. Physics as we know it today cannot describe a universe concentrated within a point. The laws of physics known to us cannot describe matter or energy concentrated in a very small volume and comprising enormous energy. In other words, the stage of the creation of the universe, the "first moment," or to be precise the range of time approaching "zero," lies outside the field of modern physics. Nevertheless, one can imagine that we are able to describe a certain stage of the development of a world as run by the laws of physics only. This stage includes neither the "first moment" nor the present reality, when man's free choice influences development, in addition to the laws of physics.

Thus we cannot describe the stage of the evolution of life as a mere derivative of the laws of nature. To aptly describe this stage in the development of the world, we can use *human life* as a paradigm. During his lifetime, man makes many decisions based on his free will, while the laws of physics, chemistry and biology are kept as well. This paradigm assumes the concept of creation *ex nihilo*, or change due to the activity of an external factor. *Ex nihilo* is a phenomenological concept that ignores the essence of things, their metaphysics. A rational explanation assumes the activity of an 'external' factor that supervises overall development — Divine Providence. "When we use the word *nihilo* [*ayin* in Hebrew] we must recall that we do it for convenience of speech only. Because before creation, there was *being* [*yesh* in Hebrew] — the Holy One, blessed be He, in the glory of His being" (Soloveitchik, 1998, p. 226).

The description of evolutionary development based on the paradigm of human life is only an analogy, which can be understood in human terms. When we study all of reality and its *creative emergent* development, we conclude that instead of two paradigms, there is one and only one process of development of the world, and it takes place under Divine Providence. The analogy for evolutionary development that is closest to us as men is the life of man. We might use the first paradigm, of development following only the laws of nature, as an approximation. But this prototype is not suitable for the stage of evolution of life. We cannot use the laws of nature alone to explain change in this stage.

9. Abstract

The fact that a chain of events took place in the development of life on earth does not prove the truth of the theory of evolution. A one-time historical event is not enough to establish a law of nature, nor does a past process of evolution prove evolution in the future. Logical analysis based on Popper's cognition theory (Popper, 1992, 2000) leads to the conclusion that no law of evolution exists. Distinguished biologists have also come to this conclusion in light of their research findings. Description of evolution is not a law, but a singular historical statement. Indeed, such a process proceeds in accordance with all kinds of causal laws including natural selection, but scientific description alone is not enough. Development of life on earth is accompanied by many instances that fall under the rubric of creation: of life itself, of various species, of man with his creativity and freedom of will, of creations of human genius. These — and numerous other examples — are new creations *ex nihilo*, unique events, and as such they cannot be described by science and could be accounted for by Providence.

References

de Chardin, T.P., (1975) *Toward the Future*, Harcourt Brace Jovanovich, San Diego.

Engels, F., (1940) *Dialectic of Nature*, International Publishers, New York.

Fain, B., (1999) Comprehensibility of the world: Jewish outlook, *BDD Journal of Torah and Scholarship*, No. 9, pp. 5-21.

Fain, B., (2004) *Yesh me-ayin*, Rubin Mass Ltd., Jerusalem; English version: *Creation Ex Nihilo*, Gefen publishing house, Jerusalem, New York, 2007.

Gould, S.J., (1989) *Wonderful Life*, Penguin Books.

Larson, E. and Witham, L., (1999) "*Scientists and Religion in America,*" Scientific American, September, 81.

Mayr, E., (2000) *Darwin's Influence on Modern Thought*, Scientific American, July, 69.

Mayr, E., (1993) *One Long Argument: Charles Darwin and the Genesis of Modern Evolutionary Thought*, Harvard University Press.

Monod, J., (1997) *Chance and Necessity*, Penguin Books.

Popper, K.R., (1981) *Objective Knowledge*, Clarendon Press.

Popper, K.R., (1992) *The Logic of Scientific Discovery*, Routledge.

Popper, K.R., (1995) *The Open Universe,* Routledge.

Popper, K.R., (1999) *The Poverty of Historicism*, Routledge.

Popper, K.R., (2000) *Conjectures and Refutations*, Routledge.

Schrödinger, E., (1948) *What Is Life?* Cambridge.

Simpson, G.G., (1951) *The Meaning of Evolution*, New American Library.

Soloveitchik, J.B., (1998) *Ha-Adam Ve-Olamo (Man and His World)*, Jerusalem, 5758.

Soloveitchik, J.B., (1992) *Ish HaHalacha — Galui vNistar* (*Halakhic Man — Revealed and Hidden)*, Jerusalem, 5752.

Spencer, H., (1966) *The principles of biology*, in The works of Herbert Spencer, v. 2-3, Osnabruck: O. Zeller.

Biography

Dr. Yehuda (Jerome) Gellman is professor of philosophy at Ben-Gurion University of the Negev, Beer Sheva, Israel, specializing in philosophy of religion and in Jewish thought. He received his Ph.D. in 1969 from Wayne State University, in Detroit, Michigan, and taught at Yeshiva University before taking up residence in Israel. He is the author of four books and over ninety articles. His latest book is: *Abraham! Abraham! Kierkegaard and the Hasidim on the Binding of Isaac,* Ashgate Publishers, 2004. He is the recipient of the Alvin Plantinga Distinguished Fellowship at the Center for Philosophy of Religion at University of Notre Dame, for 2006-2007.

22
God and Chance

JEROME YEHUDA GELLMAN

The Center for Philosophy of Religion, University of Notre Dame, IN, USA
Ben-Gurion University of the Negev, Beer Sheva, Israel
Gellman@bgumail.bgu.ac.il

1. Introduction

What follows is a defense of current evolutionary science from a Jewish religious point of view. To start with we should distinguish between evolution *per se* and *explanations* of evolution. By "evolution *per se*," I mean that species have evolved from one or a very few simple forms of life over an immense period of time. Evolution (hereafter, I will drop "*per se*") does not include an explanation of how species evolved. Prior to Charles Darwin, some biologists, including Jean-Baptiste Lamarck (1744-1829), already believed in the evolution of species (Lamarck, 1984), giving their own explanation for it. Darwin *explained* evolution largely, though not exclusively, in terms of random, biologically unpredictable variations of organisms and the survival, and reproductive success, of those most suited to their environment. (It is a common misunderstanding that Darwin explained evolution solely in terms of natural selection. For a discussion of this misunderstanding, see Gould, 1983). Darwin had no idea of the biological mechanisms involved in producing variations between individuals. *Neo-Darwinism* provided an underlying genetic mechanism for Darwin's variations in organisms. Current evolutionary science is a still developing trajectory of modifications of Darwinism and neo-Darwinism, incorporating advanced genetic mechanisms and various adjustments and improvements onto earlier explanations of evolution. These adjustments include new views concerning the rate of species evolution as well as of non-genetic inheritable changes in organisms. (On rates of evolution see for example Eldredge and Gould 1972: and Gould and Eldredge 1993. On non-genetic causes of evolution see for example Jablonka and Lamb, 2005. For recent research into the connection between embryology and evolution, see Carroll, 2005.) There is massive evidence for evolution *per se*. Indeed, in our day there is an exciting convergence of paleontology, embryology, geo-biology, genetics, cell-biology, geology, and anatomy into a tight circle of evidence for evolution.

2. Young-Worlders

Some religious Jews wish to deny evolution *per se* in the name of a "young world," one no older than about 6,000 years. Since evolution requires hundreds of millions of years, to affirm a young world is to deny evolution, regardless of its explanation. These religious people wish to protect the biblical story of creation in six days of species of distinct origins and the biblical chronology from the creation on. To do so they must come up with a counter-explanation for geological layers of fossils, fossils dated back hundreds of millions of years, and rocks dated back a few billion years. One way of explanation is to invoke the idea that God created the world so that from the first moment of its existence it looked like an old world, billions of years old. A Rabbinic statement describes Adam and Eve as being created as twenty year olds: "Said Rabbi Yochanan: Adam and Eve were created as twenty year olds." (*Midrash Rabbah*, Section 14,7). If Adam and Eve could be created as if they were twenty year olds, why could the earth not have been created as if it were 3.5 billion years old?

2.1. *Bertrand Russell*

If you are thinking of opting for young-worldism, you might want to think twice before adopting it. Here's why. Bertrand Russell once observed that it was logically possible that the world came into being five minutes ago. Put from a religious perspective, if an all-powerful God could create a world with dinosaur bones looking millions of years old, why couldn't the same God create a world with skyscrapers, cemeteries, computer networks, and wheat fields all ready-made? As for our memories of the past, these too, God could plant in our brains and minds at creation, so that what we seemed to remember never really would have happened. Here is how Russell puts this possibility:

> "There is no logical impossibility in the hypothesis that the world sprang into being five minutes ago, exactly as it then was, with a population that 'remembered' a wholly unreal past" (Russell, 1921).

Such a hypothesis is irrefutable, yet it is basic to our thinking that it be false. It is a skeptical hypothesis to be ignored from the outset, in favor of the principle *that the past was what it looks to have been.*

Now, suppose you are tempted to depart from accepting an old world, billions of years old, to claim that the world came into existence 6,000 years ago. If you countenance the *practical* possibility, not just the mere abstract, theoretical one, that the world was created as an old world, then you will be considering rejection of the above principle about the past. In that case, you create a situation where Russell's hypothesis arises as an equally *relevant* possibility, namely, that the world was created five minutes ago. However, that the world was created five minutes ago entails the falsity of Judaism, for it entails that Abraham, Isaac, and Jacob never existed, that the Israelites never left Egypt, that God never revealed the Ten Commandments, that the Israelites never entered Canaan, and so on.

Indeed, if the five-minute-old universe has to be taken into consideration, then you will have to address the possibility that you might not be able to possess any *evidence* for the truth of any Jewish beliefs about history, including that the world was created 6,000 years ago. Any evidence you might cite (e.g. an historical chain of transmission from generation to generation) could be denied as having been "planted" in the world when created, say, five minutes ago. It just *seems* as though the world has existed long enough for there to be that evidence. You could not claim even to have received a personal, direct communication from God that your religion was true. For such a direct revelation would have to be *remembered* to have really occurred to count as evidence. Yet, God — or a powerful Demon, for that matter — could have created the world with false memories of just such a revelation planted in you.

Of course, what I have said here does not provide a refutation of young-worldism. Yet, the temptation to adopt young-worldism should be carefully weighed against the epistemological costs that one might wish best to avoid.

An alternative way of defending a young world would be to maintain that the evolutionary processes supposedly taking billions of years were really speeded up to take only a few days, or so. Since evolution was so fast, we get a young world looking old. Remember, though, that the point of defending a young world was to defend the biblical account of creation. However, a "speeded-up world" scenario would not reflect the Biblical account of God specially creating *discrete* species directly, so it has little to say in its favor from a literalist religious point of view. It also has the potential of playing havoc with aspects of Biblical chronology in general. For once one acknowledges that natural processes can be speeded up, one must be on guard always against such a possibility before making historical claims dependent upon natural processes.

3. God and Chance

Turning now to current evolutionary science, the convergence of various branches of science strongly supports the general trajectory of this still developing explanation. (For a comprehensive presentation of the evolutionary biology fairly accessible to the layperson see Futuyama, 1998. For some exciting new developments see Jablonka and Lamb, 2005). A common religious objection to current evolutionary science – as well as to Darwin's original explanation of evolution – is that current evolutionary science maintains that the basic mechanisms behind evolution are ruled by "chance." For example, chance genetic mutations and gene recombinations produce changes in life forms, producing variations within species, and ultimately new species. The objection takes two forms: (1) It is impossible for all complex life forms to have evolved in the gradual manner envisioned by evolutionary science and (2) An explanation that recognizes the evolutionary development of species is inconsistent with a religious belief in the order of the world being due to the intelligent working of God. I consider these two objections in turn.

3.1. *Irreducible Complexity*

One form that (1) has taken is the well-publicized contention of Michael Behe that current evolutionary science fails to explain "irreducible complexity." Something is irreducibly complex if, roughly, it were to fail to function if any of its parts were to be removed from it. Behe argues that irreducible complexity exists in nature, and could not have evolved by "numerous, successive, slight" modifications. The simple reason is that before it had all of its pieces an irreducibly complex system would fail to function, and thus lack any adaptive advantage. Others have forcefully responded to these arguments. (For Behe see Behe, 1996. For some of the better replies to Behe see: Pennock, 2001, inter alia, Draper, 2002, and Shanks, 2004, Chapter 5). To give just one example of a counter-argument, there is great difficulty in determining for any given case whether the complexity is indeed irreducible and great difficulty in establishing that the irreducible complexity could not have evolved in a gradual way. For example, Niall Shanks argues that irreducible complexity could have evolved gradually by way of a mechanism that no longer exists, leaving us with what would fail to function were we to remove any part. He gives the following analogy:

You cannot, of course, gradually build a self-supporting, free-standing arch by using only the component stones, piling them up, one at a time. But if you have scaffolding — and a pile of rocks will suffice to support the growing structure — you can build the arch one stone at a time until the keystone is in place, and the structure becomes self-supporting. When this occurs, the (now redundant) scaffolding can be removed to leave the irreducibly complex, free-standing structure. In this way, the redundant complexity of biochemical systems, whose existence Behe concedes, can be employed to explain the origins of irreducibly complex systems. (Shanks, 2004, pp. 184-185). Shanks' point is that something could be generated gradually, with the generating mechanism then falling away, leaving us with irreducible complexity. For further replies to Behe, I refer the reader to a volume edited by Young and Edis (Young and Edis, 2004).

The objection that complex forms of life cannot arise "purely" by chance sometimes suggests a picture of tissue and bone just coming together over time, helter-skelter, by random, to become a living organism. Current evolutionary science endorses no such picture. First, new evidence supports the idea that genetic changes are not all that random, rather that by natural selection some organisms have come to be able to generate quite a restricted range of adaptive genetic changes in response to conditions such as stress. (For an early statement of such genetic changes see McClintock, 1984. See also Moxon, Rainey, *et al.*, 1994.) Secondly, while genetic variations and recombinations and genetic drift are biologically unpredictable events in current evolutionary theory, what happens subsequent to such events is not a matter of chance. The fixed laws of nature will allow advantageous chance genetic (and non-genetic) changes in organisms to yield enhanced reproductive opportunities, given the environmental conditions, the laws of genetics, competition for resources, and physical restraints on organisms (e.g. constraints on mass to surface ratio). There is no "pure chance" here.

3.2. *Divine Providence and Chance*

The second objection — that the evolutionary development of species is inconsistent with a religious belief in the order of the world being due to the intelligent working of God — is more to the point, but not really a problem. Indeed, there are Jewish theologies that maintain that Divine providence governs every member of every species. In Jewish thought, the view of the Baal Shem Tov, founder of Hasidism, provides a good example of this sort of theology. The Bal Shem Tov is reported to have taught, "If a leaf is turned over by a breeze, it is only because this has been specifically ordained by God to serve a particular function within the purpose of creation." Such a theology would have to deny any element of "chance" in mechanisms of evolution, independent of God's guiding providence. Every genetic change, for example, would have to be a divinely ordered turn of events. However, religious people could accept this view of providence and also accept current evolutionary science and proceed on their merry way with no more difficulty than they have in acknowledging that leaves are turned over by the wind. All they need deny is the rash claim by some that God had nothing to do with evolution.

3.2.1. *Medieval Jewish Philosophers*

However, other Jewish theologies severely limit divine providence. Indeed, that all animals and plants and most human beings are excluded from divine providence was a standard position in Medieval Jewish philosophy. This exclusion from God's providence means that God neither protects nor punishes, nor acts in any other way to guide, direct, or influence the fate of what is outside God's providence. Maimonides taught that there was no providence for individual members of plant and animal species. In *The Guide of the Perplexed* Maimonides writes:

> "For I do not by any means believe that this particular leaf has fallen because of a providence watching over it; nor that this spider has devoured this fly because God has now decreed and willed something concerning individuals.... For all of this is in my opinion due to pure

chance" (Maimonides, 1963, 3:17, p. 471). (For a discussion of Maimonides' view of providence see Raffel, 1987, 25-71.)

Divine providence for human beings, on the other hand, is graded according to the degree of human perfection:

> "Divine providence for human beings is graded according to the degree of human perfection: Accordingly divine providence does not watch in an equal manner over all the individuals of the human species, but providence is graded as their human perfection is graded.... As for the ignorant and disobedient, their state is despicable... and they have been relegated to the rank of the individuals of all other species of animals" (*Guide*, 3:18, p. 475).

Being relegated to the rank of individuals of other species subjects almost all of humankind to the vagaries of the world, most persons more subject to them, a few, less. Now, Maimonides' views on providence here would seem to contradict what he writes in his legal work, concerning the wrong response when suffering visits the Jewish people "If they will not cry out and will not sound [the shofar], but will say 'This is the way of the world, and this suffering is by chance,' this is cruelty and will cause them to stick to their evil acts.... This is what is written in the *Torah*...If you say this is chance, I will increase the wrath of the same 'chance'" (*Mishneh Torah*, Laws of Fasts, 1:7). However, this passage can be understood as referring to providence over the people as a whole, not over individuals. (See below on "holistic providence.")

Nachmanides (1194-1270) took an even more restrictive view of divine providence, restricting it to the "saintly" only:

> "God's knowledge, which is His Providence in the lowly world, pertains to the preservation of species. And also human beings are given over in [the world] to chance, until their time of judgment. However, to His saintly ones He gives attention to know him as an individual, to have His protection cling to him always" (Nachmanides, 1994/5, on Genesis 18:19. My translation.)

Save for the saintly, all human beings are given over entirely to chance. Similarly, Bahya ben Asher (13th century) writes that:

> "The providence to save one from chance events does not exist for all humans, even in Israel, except for the saintly among them, whom God saves from chance events, *to which other people are given over*." (Bahya bar Asher, n.d., on Genesis 18:19. My translation and my emphasis.)

Levi ben Gershom (Gersonides, 1288–1344) also held views implying that individual human beings were given over to chance, unless they could extricate themselves by moral and intellectual attainment (see Eisen, 1995).

Given these theologies that explicitly acknowledge a wide scope of events outside of divine providence, there should be no danger of heresy in recognizing the truth of current evolutionary science were it too to recognize events outside of God's providence. Of course, none of the Jewish philosophers I have mentioned believed in the evolution of species. My point is only that these thinkers readily recognized the pervasive existence of such chance events, even with regard to human affairs. If so, there need be no *prima facie* fear of heresy to current evolutionary science on the grounds that it recognizes the existence of chance events.

3.3. *Holistic Providence*

Nonetheless, a religious devotee might object that even granting the latter view of divine providence, still current evolutionary science allows that the very species — including *Homo sapiens* — that result from chance genetic events are themselves chance constituents of the universe. And this would contravene a deeply held religious belief that at the very least humanity was not a chance occurrence in the history of the world. At this point, therefore, I must explain how chance events on the micro level need not entail chance events at a macro level.

Maimonides, who held that individual plants and animals were not under divine providence, also held that species as a whole were maintained in existence by divine providence. Of course Maimonides was an Aristotelian concerning the permanent existence of every species, at least from the moment of creation. So he was not an evolutionist. And of course we now know that Maimonides was wrong to think that any species that once existed would continue to exist forever. However, my point is that Maimonides' denial of individual providence was maintained along with providence for an entire species. And what makes this distinction possible is the notion of "holistic providence," providence that works not by determining every element in a system, but by imposing a holistic grid that creates boundary conditions beyond which the system cannot stray. In this way, for Maimonides species are provided for without providence for species members as individuals. The notion of "holistic providence" I borrow by analogy from Paul Weiss who explained holism in biology as follows:

> "A macroevent can be fully determinate in the sense that, given the premises, we can predict the general outcome, the confidence of our prediction resting on the infallibility of countless earlier experiences; yet at the same time, the component microevents involved might take courses that are in detail absolutely unpredictable and unique.... Our mistake is to hypothetically extend macrodeterminacy downward to the molecular level" (Weiss, 1973, p. 17).

On this view, there exists a hierarchy of biological structures, wherein what is determined at a higher structural level need not be predictable at lower structural levels. Thus, what is determined for a macro-level need not be determined at a molecular level. The idea is that occurrences at a lower level can occur by chance, relative to what is determined at higher levels, but occur within the boundaries allowed for by the higher structures, and then are funneled along a causal path by, as it were, the curvatures of the higher grid. Holistic systems occur outside of biology as well, as, for example, in radioactive decay, where it is precisely determined how much material over-all will decay, whereas which atoms in the material will decay is purely random.

While the extent of the application of this idea is quite controversial in biology, I have no interest in insisting on its extent in biological explanation, only in explaining how chance events can be consistent with Divine providence over-all. The analogy to biology shows how it can be possible for evolutionary micro-events to be chance events while the emergence of *Homo sapiens* is not.

Think of God, then, acting as the highest-level framework in which individual events occur. Then, individual occurrences within the framework can occur by chance, within the boundaries allowed for by the framework's structure, and then be funneled along a causal path by the "curvatures" of the grid. In that way, micro-chance is consistent with macro-Providence, where the Providence will be systemic and natural, rather than intrusive into the natural order. This conception of evolutionary providence would find an ally in the recent work of Simon Conway Morris who has noted with great detail the high degree of evolutionary convergence across species and geographical locations (Morris, 2003). A theological interpretation of Morris's work could well involve my notion of higher-level providence within random variances.

There is an intimation of an especially interesting example of this holistic idea in Maimonides. Maimonides addresses the question whether God deprived the Egyptians of their free will by informing Abraham (Genesis 15:13) that they would enslave the Israelites. Maimonides replies that God did not decree on any particular Egyptian that *he* would enslave the Israelites. God revealed only "the way of the world." Maimonides continues, "Each and every Egyptian was free not to do evil to [the Israelites], for there was no decree on a specific person. God only informed [Abraham] that in the end his descendents would be enslaved in a foreign land" (Maimonides, 1967, Laws of Repentance, 6:11-12, my translation). Maimonides may be interpreted to be advancing here a conception similar to the one I am proposing concerning grid-structures that impose direction upon micro events undetermined by the grid. The "way of the world" refers to the shape of the reality imposing the holistic conditions in which human freedom operates. Macro-regulation of unregulated micro-events is the result.

One might wonder whether the idea works for micro free-willed decisions, since holistic constraints impose restrictions on free will. However, we should keep in mind that in any case human free will is constrained by what the existentialists called "facticity." Hard facts, such as an individual's height, weight, sex, time of birth, etc., as well as general human limitations, such as the inability to jump buildings, are constant constraints on human freedom. In any case, with evolution we are dealing with natural micro-events only. (For a discussion of Maimonides' view on free-will, see, Gellman, 1989.) Chance is consistent with Divine Providence.

Given alternative views on the extent of divine providence in the world together with a holistic conception of God's activity in the world, there should be no insuperable barrier for a devout religious Jew to accept current evolutionary science. Accepting it might require a more sophisticated theology than most people hold. But sophistication is all to the good, especially when falsehood looms as the other alternative.

4. Arguments from Design and Daily Miracles

Some religious people insist on trying to prove that current evolutionary science by itself is not capable of explaining the evolution of species. Appeal must be made to a divine power to explain the facts about fauna and flora. In this, they are continuing the tradition of "Arguments from Design" that for centuries have been attempting to prove the existence of a designer of the order we witness in the universe. Alas, such arguments give little support to religious belief. To mention only a few reasons for this: (1) Even if successful, these arguments do not establish that the designer of the universe also was its creator, (2) Such arguments fail to establish that there was exactly one designer. Indeed, in our experience, the more complex an object is the more designers had to be involved in making it; and, the most telling: (3) Arguments from Design do nothing to establish that the one who once-upon-a-time designed the world still exists. No line leads from Arguments from Design to a living infinite God whose will is revealed to the world. (For a battery of objections to the Argument from Design see David Hume, 1991. For a recent attempt to revive the argument and to argue for there being one designer, see Swinburne, 1983).

Rather than insist that there is a proof of the existence of God in biological facts, religious people would be better engaged in acknowledging that their faith requires no proof, and instead dedicate themselves to enhancing appreciation for the wonders of God's creation. Here I would invoke what Martin Buber has written about "miracles":

> "The concept of miracle... can be defined at its starting point as an abiding astonishment.... Miracle is not something 'supernatural' or 'superhistorical,' but an incident, an event which can be fully included in the objective, scientific nexus of nature and history. Miracle is simply what happens; in so far as it meets people who are capable of receiving it, or prepared to receive it, as miracle. What is vital is only that what happened was experienced, while it happened, as the act of God.... The real miracle means that in the astonishing experience of the event the current system of cause and effect becomes, as it were, transparent and permits a glimpse of the sphere in which a sole power, not restricted by any other, is at work" (Buber, 1958).

After all of astrophysics, chemistry, biology, and current evolutionary science have had their say and have been accepted as the closest we have to the truth, an abiding wonder and even astonishment concerning the nature of the world continues undiminished, indeed should be thereby intensified, in the person of faith. After all of the evolutionary explanations will have been exhausted, and duly accepted, the religious person will look on in amazement and bow in thankfulness at the conception and birth of a child. This religious response to existence is reflected in the *Siddur*, the Jewish prayer book, when the Jew thanks God for "The daily miracles you do for us, and for your wonders and goodness at all times, evening, morning, and afternoon." The world and all its fullness is indeed a miracle.

A crucial pursuit of religion is to nurture and expand this awareness of God, an awareness that does not rely on refuting the best endeavors of science. The religious mind will marvel at the laws of nature whose operation has brought about our existence and made possible our communion with God. For example, the religious mind will marvel at the awe-inspiring multiple repetitions in nature of the Fibonacci series. Discovered by the twelfth century mathematician, Leonardo Pisano (whose nickname was "Fibonacci"), the series begins with: 1, and is generated thereafter by the addition of the last two consecutive numbers in the series. This results in the series:

1, 1, 2, 3, 5, 8, 13, 21, 34, 55, 89, 144, 233…

Biological phenomena, from the location of plant leaves on the stem to the swirls of sunflowers and conch shells, widely illustrate Fibonacci proportions. (For a popular presentation of Fibonacci numbers and nature, see Hulme, 2005.)

The religious person will face with humility and a sense of utter finitude the question why God should choose to design the world in such a strange way as current evolutionary science proposes. In this way, and others, scientific advances should compel a new sophistication of "agnosticism," where warranted, in the religious life.

Jewish religious educators should be preparing the young for a world dominated by scientific advances never before imagined, and in which religious sensitivity to the mystery thrives beyond explanation, a world in which religion is well equipped to progress beyond the sorrowful excesses of scientism and secularism.

5. Acknowledgements

I am grateful to participants of the colloquium at the Center for Philosophy of Religion at Notre Dame for their most helpful comments. I also thank Rabbi Nathan Cardozo and Ira Schnall for their comments on an earlier version of this paper. I am also indebted to Joel Rutman for long hours of fruitful discussion on evolution and Judaism.

References

Bahya bar Asher (n.d.) *Commentary to the Torah* No Publisher. Israel.

Behe, M.J. (1996) *Darwin's Black Box: The Biochemical Challenge to Evolution*, The Free Press, New York. Buber, M. (1946) *Moses*, Phaidon Press, Oxford.

Buber, M. (1958) Moses; the Revelation and the Covenant, Harper, New York.

Carroll, S.B. (2005) *Endless Forms Most Beautiful, The New Science of Evo Devo and the Making of the Animal Kingdom*, W.W. Norton and Company, New York.

Draper, P. (2002) Irreducible Complexity and Darwinian Gradualism: A Reply to Michael J. Behe. *Faith and Philosophy* **19**, 3-21.

Eisen, R. (1995) *Gersonides on Providence, Covenant, and the Jewish People, A study in Medieval Jewish Philosophy and Biblical Commentary,* State University of New York, New York.

Eldredge, N. and Gould, S.J. (1972) Punctuated Equilibria: an Alternative to Phyletic Gradualism. in: T. M. Schopf (ed.) *Models in Palaeobiology*, Freeman, Cooper and Co., San Francisco, pp. 82-115.

Futuyama, D.J. (1998) *Evolutionary Biology, Third Edition*, Sinaeur Associates, Sunderland Ma.

Gellman, J. (1989) Freedom and Determinism in the Philosophy of Maimonides, in E. Ormsby (ed.) *Maimonides and His Time*, Washington: Catholic University of America, pp. 139-150.

Gould, S.J.(1983) *The Panda's Thumb*, Penguin, Harmondsworth, England.

Gould, S.J.and Eldredge, N. (1993) Punctuated Equilibrium Comes of Age. Nature **366**, 223-227.

Hume, D. (1991) *Dialogues Concerning Natural Religion*, Routledge, London.

Hulme, J.N. (2005) *Wild Fibonacci, Nature's Secret Code Revealed*, Tricycle Press, Berkeley.

Jablonka, E. and Lamb, M.J. (2005) *Evolution in Four Dimensions, Genetic, Behavioral, and Symbolic Variation in the History of Life*, MIT Press, Cambridge, MA.

Lamarck, J.B. (1984) *Zoological Philosophy: An Exposition with Regard to the Natural History of Animals*, tr. H. Elliot, University of Chicago Press, Chicago.

Maimonides, M. (1963) *The Guide of the Perplexed*, Translated by Shlomo Pines, University of Chicago Press, Chicago.

Maimonides, M (1967) *Mishneh Torah* (Hebrew) Hebrew Publishing Company, New York.

McClintock, B. (1984) The Significance of Reponses of the Genome to Challenge. *Science* **226**, 792-801.

Morris, Simon Conway (2004) *Life's Solution, Inevitable Humans in a Lonely Universe*, Cambridge University Press, Cambridge, UK.

Moxon, E.R. Rainey, P.B., Nowak, M.A., and Lenski, R.E. (1994) Adaptive Evolution of Highly Mutable Loci in Pathogenic Bacteria. *Current Biology* **4**, 24-33.

Nachmanides, M. (1994/5) *Commentary to the Torah*, Mosad Harav Kook, Jerusalem.

Pennock, R.T. (ed.) (2001) *Intelligent Design Creationism and Its Critics: Philosophical, Theological, and Scientific Perspectives* MIT Press, Cambridge, MA.

Raffel, C.M. (1987) Providence as Consequent upon the Intellect: Maimonides' Theory of Providence. *AJS Review*, **12**, 25-71.

Russell, B. (1921) *The Analysis of Mind*, G. Allen & Unwin, London.

Shanks, N. (2004) *God, The Devil, and Darwin: A Critique of Intelligent Design Theory*, Oxford University Press, Oxford.

Swinburne, R. (1983) *The Existence of God*, Clarendon Press, Oxford.

Weiss, P.A. (1973) *The Science of Life: The Living System — A System for Living*, Futura, Mt. Kisco, NY.

Young, M. and Edis, T. (2004) *Why Intelligent Design Fails, A Scientific Critique of the New Creationism*, Rutgers University Press, New Brunswick NJ and London.

Biography

Dr. Stephen M. Barr is Professor of Physics at the Bartol Research Institute and the Department of Physics and Astronomy of the University of Delaware. He received his Ph.D. from Princeton University in 1978, and has held positions at the University of Pennsylvania, the University of Washington, and Brookhaven National Laboratory. His field of research is theoretical particle physics, and he specializes in theories of grand unification, CP violation [http://en.wikipedia.org/wiki/Cp_Violation], theories of the origin and spectrum of quark and lepton masses, and the cosmology of the early universe. He has written over 120 papers in refereed journals, as well as the articles on Grand Unified Theories for the second and third editions of *The Encyclopedia of Physics* published by the American Institute of Physics. He writes extensively on science and religion, primarily for the journal *First Things*, on whose editorial advisory board he sits.

23
The Concept of Randomness in Science and Divine Providence

STEPHEN M. BARR

Bartol Research Institute, Department of Physics and Astronomy
University of Delaware, Newark, DE, 19716 USA
smbarr@bartol.udel.edu

1. Introduction

In this essay, I will consider whether the concept of "random mutations" in neo-Darwinian theory is compatible with the doctrine of divine Providence found in the major monotheistic religions. This question naturally arises if one considers statements that are commonly made by both proponents and opponents of neo-Darwinism. For example, in the 1995 textbook *Biology*, by K.R. Miller and J. Levine, one finds the statement, "evolution works *without either plan or purpose*", and later on the same page, "Evolution is random and *undirected*." Cardinal Schönborn, in an opinion piece in the New York Times in 2005 defined neo-Darwinism as "an *unguided, unplanned* process of random variation and natural selection." [In all cases, the italics are mine.] On the other hand, as we shall see, the traditional notion of divine Providence involves the idea that God does guide and direct the world and all the events therein according to some plan and purpose.

Not necessarily contained in the notion of Providence, but related to it, is that of God as cosmic Designer. Again, there are both proponents and opponents of Darwinian evolution who see a contrast between the chance character of events in evolutionary theory and the idea that the biological realm exhibits design. For example, Richard Dawkins, who attacks religious belief as incompatible with

Darwinian evolution, argues that the randomness of genetic variation shows that organisms are "designoids", i.e. not truly the products of design, but only apparently so. And William Dembski, a leading figure in the Intelligent Design (ID) movement (which accepts evolution, but attacks neo-Darwinism as an insufficient explanation of it), has developed an "explanatory filter", which categorizes patterns according to whether they are the result of "law", "chance" or "design".

In this essay I will discuss four closely related theological questions arising from the role of chance and randomness in biology. I will contend that some of these are pseudo-problems arising from a confusion of language and the rest are instances of more general theological questions that long predate evolutionary theory.

The four questions I will discuss are these: (1) Does "randomness" in nature necessarily imply that nature is "unguided", "unplanned", and "undirected" in a sense that would contradict the notion of Providence? (2) If certain effects arose as a consequence of randomness, does that necessarily imply that those effects were not "intended" or "purposed" in themselves but were at most accidental by-products of something intended? (3) Does the fact that randomness is involved in certain natural explanations mean that God "left certain things to chance"? (4) Why would God have created a world in which randomness and chance play such a prominent role?

2. The Notion of Providence

First we must be clear about terms, and, in particular, what is meant by the religious concept of divine Providence and the scientific concept of randomness. Fortunately for our purposes, the notions of divine Providence found in Catholicism, Eastern Orthodoxy, classical Protestantism (specifically Lutheranism and Calvinism), traditional Judaism, and Islam are not significantly different with respect to the issues we are discussing here. While there are important differences in belief concerning human free will between Muslim and Christian, Catholic and classical Protestant, Calvinist and Arminian, Thomist and Molinist, these are not relevant to a discussion of events in the natural world that involve only inanimate matter or sub-rational organisms. To be sure, there are modern theological movements, especially within Protestantism, that propose weaker notions of divine Providence — for example, "process theology" and "open theism". However, we need not consider these, for if the traditional notion of Providence is compatible with evolutionary theory, then obviously the weaker notions would be too. Since the major religions in their orthodox versions do not differ significantly on the matters under discussion here, we may take any one of them as representative. It will be convenient to choose Catholicism, as its doctrines are most easily specifiable.

According to Catholic doctrine (cf. Ott), God is the source of being, and thus gives being to all that exists outside himself. (God himself exists necessarily). God is outside of time, and thus creates (and knows) everything, past, present, and future, by a single eternal (i.e. a-temporal) act. He is the creator of all finite beings in every aspect of their being, and hence creates them with all their natural potentialities, powers, and mutual relationships. Consequently, while created things do have natural causal relationships to each other ("secondary causality"), God is nevertheless the direct cause of every thing ("primary causality"): God is the First Cause — "first" causally, not temporally. (A simple analogy is that a dagger and the playwright Shakespeare are both causes of Polonius's death in the play *Hamlet*: the dagger kills Polonius, while Shakespeare caused the whole scene, including Polonius, the dagger, Polonius's death, and the fact that the dagger caused Polonius's death. Shakespeare is analogous to the First Cause, the dagger to a secondary cause. Secondary causes only operate because of the First Cause.) It is Catholic dogma that "God, by his Providence, protects and governs all that he established, 'reaching mightily from end to end and ordering all things sweetly'." ("Universa vero, quae condidet, Deus providentia sua tuetur atque gubernat, 'attingensa fine usque ad finem fortitur et disponens omnia suaviter'." Vatican Council I, citing Wis 8:1, cf. Denzinger).

One sees that the doctrine of Providence is not only that God governs the world, but that his governance is wise and beneficent. Divine Providence is distinguished into "mediate providence" and "immediate providence", the former exercised through secondary causes and the latter directly, without such mediation. Therefore, saying that an event is governed by Providence implies nothing about whether it has natural causes and can be naturally explained. Nevertheless, according to traditional teaching, the ordinary means of divine Providence is through natural secondary causality. As Suarez put it, "God does not intervene directly with the natural order where secondary causes are sufficient to produce the intended effect."

3. The Notion of Randomness

Having explained the traditional notion of divine Providence, we must now ask what the word "random" means in science in general and evolutionary biology in particular. It is a term that it is very widely used in science. According to the ISI Scientific Citation Index there are about 50,000 papers in the scientific literature that have the word "random" in the title, coming from every major branch of science. (Compare this to the words "unplanned" which appears in only about 500 titles, mostly having to do with unplanned medical procedures or pregnancies, and "unguided", which appears in only about 50 titles, mostly having to do with guided missiles. Clearly, the word "random" is used as a technical scientific term, whereas the words "unplanned", "unguided", "undirected", and the like are not.) The concept of randomness is applied to wide range of natural phenomena, including the motions of molecules in a gas, quantum fluctuations, noise in electronic devices, and errors in a data set.

The concept of randomness is closely related, of course, to the concept of "probability"; and indeed the assumption of randomness in science would be of little use were it not the basis of probabilistic calculations. Entire branches of science, such as Statistical Mechanics and quantum theory are based on such calculations.

Though the word "random" is widely used in science, the concept is notoriously slippery for philosophers and mathematicians (cf. Bennett). Indeed, there are various notions of randomness: statistical randomness on the one hand, and information-theoretic (or Chaitin-Kolmogorov) randomness on the other. The concept of probability is equally hard to pin down; for example, one has the "classical", frequentist, and Beyesian approaches to understanding it, among others. Fortunately, for our purposes it is not necessary to plunge too deeply into these foundational and definitional subtleties. I will take the somewhat pragmatic approach, thinking of "randomness" and "probability" as they are used in everyday practice in natural science. Perhaps it is best to consider a familiar example of randomness rather than lay down a definition.

Consider a series of coin tosses. One would expect the outcomes to form a "random sequence". This is because the different tosses are "independent" of each other, and therefore their outcomes have no reason to be correlated. One can apply statistical tests to look for correlations. (For example, if we assign the number +1 to heads and −1 to tails, the expectation value of the product of successive tosses in a random sequence should be zero.) The absence (or sufficient smallness) of correlations is evidence (though not proof) of randomness.

Of course, in a short string of coin tosses one would not be surprised to find a pattern, such as HTHTHT (where H = heads and T = tails). One would nevertheless still say that the sequence was random *if it was produced by a "random process"*. Any such interesting pattern that appeared in a short sequence produced by a random process would be regarded as accidental. However, if interesting patterns appeared far more frequently than the probabilities would lead one to expect, one might begin to doubt that the process that produced the sequence was really random. Nevertheless, statistical tests of randomness can never yield absolute certainty. Not only can a truly random processes yield results that seem non-random (as in the coin tosses that come up HTHTHT), but a non-random process (like computing the digits of pi) can generate a sequence that passes all statistical tests for randomness.

This raises the question of how we can know that a certain process is truly random. As just noted, we cannot tell for certain by applying statistical tests to the outcomes. Nor can we tell by knowing the details of the process itself. For example, suppose we lived in a universe obeying deterministic classical laws. Then every detail of the whole process of repeated coin tosses, including exactly how each coin was tossed and how it landed, could be traced back to some set of "initial conditions". A claim that the coin-tossing process was random would then translate into a claim that the initial conditions were random. The problem would only be pushed back to another level, but not resolved. Ultimately, the assumption of randomness in certain aspects of the physical world is a heuristic principle on the same footing as the assumption that nature obeys "universal laws". One might try to justify such assumptions on the grounds of economy of explanation. However one justifies them, they are part of how we understand the "natural order" of the world. Just as something that violated a universal law would be regarded as a "miracle", so also would be something that violated the natural assumption of randomness — i.e. required correlations among the motions of atoms and fields that under the assumption of randomness would be absurdly improbable (such as the unsmashing of a glass, the unspilling of a liquid, or any other process that involved a huge downward fluctuation of entropy.)

At a practical level, in testing physical theories, one is satisfied to use statistical tests of randomness. For instance, in doing "Monte Carlo" numerical simulations of quantum processes researchers employ "random number generators", even though the numbers generated by such algorithms are not truly random, but (like the digits of pi) "pseudo-random". Nevertheless, as long as the numbers are weakly enough correlated with each other, as measured by various statistical tests, they are deemed sufficient to simulate the (presumably) truly random quantum processes. So, from an empirical point of view, the idea of randomness used in science boils down to a lack of correlation.

Physicists assume that such things as quantum fluctuations and the motions of molecules in a gas are random. Biologists assume that genetic mutations are random. Such assumptions are intrinsic to modern science's way of explaining the world and to our notions of what is "natural".

4. Randomness and Providence

This leads to our first question. (1) Does "randomness" in nature necessarily imply that nature is "unguided", "unplanned", and "undirected" in a sense that would contradict the notion of Providence?

This question is best answered by a couple of analogies. Suppose one is driving down an interstate highway in the United States and observing the license plates of the cars that pass by. They will be from various States of the Union, and the sequence will exhibit some degree of randomness: New Jersey, Delaware, Pennsylvania, New Jersey, New Jersey, Florida, etc. (There are, of course, probabilities involved, as in any random process. On certain highways a license plate from New Jersey is more likely to appear than one from Kansas, just as in rolling a pair of dice a total of 7 is more likely to appear than a total of 3.) The sequence of license plates is random in the sense that knowing where one car is from does not tell you where another is from: the cars are "independent" and thus "uncorrelated". This randomness, however, in no way implies that the cars' movements and locations are "undirected", "unguided", and "unplanned". Quite the reverse is true: the cars are directed by the wills of their drivers, who are guided by maps and pursuing plans. It is just that the plan of one driver has no direct connection to the plans of the other drivers, because the lives of the various drivers are (generally speaking) uncorrelated with each other.

Of course, in this first analogy it is a question of many wills separately directing events, whereas in the case of divine Providence there is only one will. So consider another analogy. Suppose a writer is composing a work of prose on a word processor. The final syllables of the lines will form a sequence that exhibits some degree of randomness, in the sense that there is a lack of correlation (or perhaps a weak correlation) among them. (This would not be the case if the author were composing a sonnet, say, where the final syllables would have to satisfy a well-defined pattern, such as ABAB CDCD EFEF GG.) However, the fact that the final syllables of his prose lines exhibit randomness, does not mean that the author did not plan his work, or that his words were unguided and undirected. On the contrary, he may have drafted the work with great deliberation, choosing every word with care. It is simply that he did not choose to impose on the final syllables of his lines the specific kind of correlation called a rhyme scheme. In the same way, God, though he may have planned the universe with infinite care may not have chosen to impose on the movements of molecules in a gas, say, certain kinds of correlations. Thus, in the sense meant by the physicist, the molecular motions would exhibit statistical randomness.

This answers the first of our questions. These examples also shed some light on why statistical randomness and "chance" events do occur in our world. It is *not* necessarily because events are uncaused. As the license plate example illustrates, every event may be part of a chain of causality, but because there are many *independent* chains of causality that intersect each other and impinge on each other, sequences or juxtapositions arise that exhibit a lack of correlation and thus "statistical randomness". The world does not have one storyline, but many storylines that have little direct relation to each other. If one examines a set of events that form an orderly pattern, then events that are not part of that patterned set can appear as random "external" influences. As an analogy, consider a large dining hall with many separate conversations going on simultaneously. From the viewpoint of any one conversation, all the other conversations seem as "background noise". Events that impinge on a subsystem of the world from other parts of the world, or that result from some adventitious juxtaposition, seem as "chance" events. These chance events can disrupt or substantially change the course of events in the subsystem. Events in this world, therefore, do not follow a predictable pattern, but are caught up in a vast web of contingency.

Not surprisingly, the great theological traditions took cognizance of this aspect of the world. Saint Augustine, whose teaching on divine foreknowledge, predestination, and Providence is uncompromising, nevertheless notes in *The City of God* that no one can in this life "escape being tossed about by chance and accident" (*City of God*, 19,4).[9] St. Thomas Aquinas devotes Book 3, chapter 74 of *Summa Contra Gentiles* to proving that "Divine providence does not exclude fortune and chance". The fourth reason he gives is that "the large number and variety of causes stem from divine providence and control. But granted this variety of causes, one of them must at times run into another cause and be impeded, or assisted, by it in the production of its effect. Now from the concurrence of two causes it is possible for some chance event to occur…" (SCG 3, 74). This is connected to the distinction he makes between "necessary causes", which unfailingly produce their natural effects, and "contingent causes", which can be impeded by the action of *other* causes, such as a "contrary agent" or the "resistance of another thing" (SCG 3, 72).

St. Thomas gives another reason why chance events occur. Every material thing has "accidental" properties. To use one of his examples, a builder may be white. The fact that he is white does not follow from the fact that he is a builder, but is an "accidental concomitant" (SCG 3, 14). Therefore, something that acts as a cause, like a builder, since it has accidental properties itself, can produce effects accidentally, and "things which result accidentally from any cause are said to happen by chance or fortune" (SCG 3, 74). This source of chance is related to the previous one that we discussed. If a being is at once a man, a builder, white, French, and a husband, that means that he is simultaneously a part of many orderly subsystems or intelligible structures or patterns of relations within the world. He is a member of a species within a larger biological system, of a profession within an economic system, and so on. These various systems, biological, economic, physical, political, familial, etc., *intersect in him.* This inevitable intersection of disparate systems leads to a certain lack of correlation in things: knowing a man's nationality does not tell you his profession or marital status, for example.

The *Bible* itself speaks of the role of chance in the world. "I returned and saw under the sun, that the race is not to the swift, nor the battle to the strong, neither yet bread to the wise, nor yet riches to men of understanding, nor yet favor to men of skill, but time and *chance* happeneth to them all," as Ecclesiastes notes. (Eccles 9:11) Or, to put it in the terms we have been using: there is not a perfect correlation between being strong and winning or between having bread and being wise.

Let us now turn to evolutionary biology and see how the word "random" is used there. In their college-level textbook *Modern Genetics*, Francisco Ayala and John Kiger explain three senses in which geneticists say that mutations are random: 1. as "rare exceptions to the regularity of the process of DNA replication"; 2. because "there is no way of knowing whether a given gene will mutate in a particular cell or in a particular generation"; and 3. because "they are unoriented with respect to adaptation." They note that this last meaning "is very important for evolution ... Mutations occur independently of whether or not they are adaptive in the environments where the organisms live." Not surprisingly we find here the same basic elements as in our previous discussion of randomness and chance. One has a chain of causation, pattern of events, or orderly subsystem ("the regularity of the process of DNA replication"), which is interfered with by something external to or at least extrinsic to it. There is a lack of correlation between the factors causing these mutations and the adaptive exigencies of the creatures in which they occur (i.e. the former are "unoriented" with respect to the latter), and this is because the mutations occur "independently" of whether they are adaptive. The concept of randomness used in evolutionary biology is essentially the same as that used in physics and other branches of science. And thus our answer to question (1) applies just as well to biological randomness.

However, so far we have only shown that a process may be directed, guided, and planned and yet exhibit "statistical randomness" in some of its aspects. However, in our examples, the randomness occurred in some feature of the process that was not relevant to the purposes of those directing it. The prose writer did not care whether the final syllables rhymed or made any particular pattern. The drivers on the highway were not trying to arrange their license plates to form an interesting sequence. And, as Ayala and Kiger note, genetic mutations are not "oriented" toward any adaptive result. However, the monotheistic religions we are discussing certainly teach that God's eternal purposes included the emergence of the human race.

So we come to the second question: (2) If certain effects arose as a consequence of randomness, does that necessarily imply that those effects were not "intended" or "purposed" in themselves but were at most accidental by-products of something intended?

The first thing to note is that something may be unintended at one level of causality and intended at a higher level. An example used by St. Thomas is that a master may send two servants to the same location unbeknownst to either of them. To the servants, their meeting would appear a matter of chance, whereas in fact it may have been intended by the master. The "purpose" of one servant may be to get water from a well, while that of the other may be to deliver a message. The servants' own purposes are independent and uncorrelated at their own level of understanding, but there is at work a higher and hidden purpose. Some "blind dates" are undoubtedly arranged in this way today; and one can see how it might be necessary to the accomplishment of the matchmaker's purpose that the meeting of the pair appear to them to be a matter of chance. This is what we might call "purpose employing apparent chance".

There are also cases which one might call "purpose employing genuine chance". A pollster, for instance, will try very hard to obtain a truly "random sample" in order to get a meaningful result. And in a game of chance, the participants will want to use a really random method, such as rolling a pair of dice or shuffling a deck. It would defeat the purpose of the game for there to be a predictable pattern or correlation among the rolls of the dice or cards of the deck.

Finally, there is "purpose despite chance". A man may want to deal a "royal flush" from a deck of cards. He may achieve this purpose in several ways. He might take a single deck and introduce the right correlations into it before dealing, a procedure commonly known as "stacking the deck". Or he may prefer to take a huge number of shuffled (i.e. randomized) decks and deal one hand from each. The probability of obtaining a royal flush from any one of the decks is only 1 in 649,740. However, if he uses a million decks he has a good chance of achieving his purpose. And if he uses a billion decks, he is virtually certain to do so; indeed, his chance of failure then is only 1 in 10^{669}. In this case the randomness of the decks is not helping the man to achieve his purpose. Rather, he must overcome the randomness by using large numbers.

In biology we find many instances of "purpose despite randomness", if we understand purpose to include unconscious purpose. A single human sperm has very little chance of finding the egg. Nature therefore arranges that a huge number will be sent in search of it. Similarly, the larvae of certain insects have a very small individual chance of surviving to adulthood, as they are mostly eaten by predators. Nature compensates by having a vast number of larvae produced. Such "survival strategies" are not consciously adopted by the organism or the species, but they do serve a "purpose", in the same sense that eyes are "for seeing".

So there are several ways in which an effect can be the outcome of random processes and yet still be the result of a conscious or unconscious purpose. This answers our second question, though it does not tell us whether the evolution of human beings *was* actually the result of any purpose. However, it certainly could have been, for anything biologists can say. If nature so operates that many sperm are emitted *in order* that one may "win through" to produce a new life, why should God not arrange the universe so that many genetic mutations take place *in order* that a relatively few shall "win through" to produce new and interesting types of organisms? Recalling what Suarez said about God ordinarily achieving his purposes through the natural operation of secondary causes, there is no reason at all.

In any event, we see that no *new* theological issue is raised by neo-Darwinian evolution in this regard. While it is only in modern times that the existence of human sperm and egg have been known, people have been aware since ancient times of cases where chance operates in nature and nature overcomes low probabilities by using large numbers.

We now come to our third question. (3) Does the fact that randomness is involved in certain natural explanations mean that God "left certain things to chance"?

In the poker example, the man who shuffles many decks still leaves something to chance. Even with a billion decks, his chance of failure is not *exactly* zero. Moreover, *which* decks produced the desired result is completely a matter of chance to him. Similarly, even if the chances of fertilization are high, nature is leaving to chance *which* sperm will fertilize the egg and *which* particular human being will be conceived. And in the process of biological evolution nature would also seem to be leaving to chance exactly which particular kinds of organisms will evolve. And yet, according to the traditional doctrine of divine Providence, God knows and wills from all eternity the existence of all the *particular* things that will come into being. Is there a contradiction after all?

There is no contradiction if we understand the significance of the concept of randomness in scientific explanation. An astrophysicist asserts that randomly moving gas and dust particles in interstellar space will, under the right conditions, condense to form very orderly structures called planetary systems. The point of the assertion is that one needn't assume any special, highly correlated motions of the gas and dust in order to account for the fact that a planetary system arises; a fairly "generic" set of initial gas and dust motions will produce *some* planetary system. Of course, a *particular* planetary system will require *particular* motions of the gas and dust; and God, if he is omniscient, has known and willed those particular motions from all eternity.

One can say the same things about the formation of ice crystals from randomly moving molecules of water. No "fine-tuning" of molecular motions is needed to produce *an* ice crystal. Obviously, a specific choice is required to produce *this* one. The same holds for genetic mutations and forms of life. The evolutionary biologist is simply saying that God could have thrown the genetic dice and with high probability gotten *some* kinds of complex organisms. To get the specific ones that actually exist, there had to be specific mutations.

This is where the analogy of the deck shuffler or dice thrower breaks down. The poker player does not know what is in the shuffled decks. The traditional understanding is that God does know from all eternity what is "in the cards". Does this destroy the claim that there is randomness in nature? Not as the scientist speaks of randomness. If a triangular array of pool balls is "broken" and the balls roll around in an apparently random way, there is a certain trivial sense in which one can say that they are not moving randomly — their movements are actually highly correlated. In particular, they are just exactly the movements that, if exactly reversed, would cause them to reassemble to form the triangular array. In the same way, as entropy increases in a physical system, and "disorder" increases, the particles and fields actually continue to preserve *all* the subtle correlations in their motions that are the consequence of having originally come from a state of lower entropy. This information is never really lost, it is there lying hidden underneath the statistical randomness. However, this does not prevent the physicist from saying that the final state is more "disordered". Nor does the fact that the motions of water molecules will eventually produce a very specific crystal make them any less random as far as the physicist is concerned.

What appears to empirical investigation as statistically random is foreknown and willed by God. If this is a paradox, it is one that long predates neo-Darwinian evolution. The *Bible* itself asserts it: "The lot is cast into the lap; but the decision is wholly from the Lord" (Prov 16:33).

We come now to our final question: (4) Why would God have created a world in which randomness and chance play such a prominent role?

There are several reasons that are consistent with the traditional notion of Providence:

(A) As we saw, chance events may be inevitable in any universe that is multifarious, with many intersecting chains of causality. The alternative is a universe where everything marches in lockstep according to a single causal pattern. However, such a universe would not be consistent with a multiplicity of freely acting creatures, for there would be no scope for responsible action whereby a created agent imposed patterns of its own devising on events or things.

(B) Moreover, it is not likely that a world as complex and as lawful as ours could be without chance events. Consider the weather, the paradigmatic example of chance phenomena in everyday life. It might seem a more purposeful world if rain, say, only fell when and where it was needed by some living thing. (Indeed, in the Book of Job, God taunts Job with the seeming irrationality of the world, asking him, "Who has cut a ... path for the thunderstorm to cause it to rain upon a land where no man is; on the wilderness where there is no man" (Job 38:26).) However, it is not at all clear that it would even be mathematically possible in a universe governed by universal laws to tune the "initial conditions" so that the weather always conformed to what was "useful".

(C) And of course, what is "useful" in a dynamic world such as ours is not the same as what is useful in a static world. In a dynamic economy there is what Joseph Shumpeter called "creative destruction", and the same is true in a dynamic natural order. The death of the prey is the life of the predator. The extinction of the dinosaurs is an opportunity for mammals. Thus "chance" also plays a creative role in nature. As Prof. Michael Tkacz of Gonzaga University notes, "One of the reasons Thomas Aquinas thought it was necessary to acknowledge the chance element in nature is that it is the way that novelty is introduced into nature." He cites the following passage from St. Thomas: "Nothing prevents the generation of something to be in itself designed when referred to one cause and, nevertheless, accidental and chance when referred to another cause.... This is because it belongs to [the divine] intention that all forms which are in the potentiality of matter be drawn into act", i.e. realized. This conception of chance and contingency drawing forth novel forms is certainly consistent with the role of chance in neo-Darwinian evolutionary theory.

(D) A world without "chance" events, where all things fulfilled their purposes without possibility of failure, would have to be miraculously governed, and as a result creatures would be robbed to a large extent of their natural secondary causality. Divine providence would eliminate the need for natural "pro-vision" on the part of creatures. There would be no need to seek food, be alert to danger, or strive toward any goal. There would be no need for creatures to be sentient let alone intelligent or free. Virtues such as courage, prudence, and willingness to sacrifice would be superfluous. Nor would it be necessary to seek God in such a world, as divine Providence would be blatant, and God's activity obvious on the surface of events. However, the *Bible* portrays a *Deus absconditus*, a God who "hides his face" (Ps 44:24), and who for the most part reveals himself only to those seek him (Mt 7:7) out of a desire for truth. The moral universe of the traditional monotheistic religions, a world of striving, seeking, and struggle, is consistent with the biological world described by Darwinian evolution.

References

Aquinas, St. T. (1975) *Summa Contra Gentiles*, University of Notre Dame Press, Notre Dame, IN, USA.

Augustine, St. (1950) *City of God*, Modern Library, New York, N.Y., USA.

Ayala, F.J. and Kiger, J.A., (1980) *Modern Genetics*, Benjamin/Cummins Pub. Co., Menlo Park, CA, USA.

Bennett, D.J. (1998) *Randomness*, Harvard University Press, Cambridge, MA, USA.

Dawkins, R. (1986) *The Blind Watchmaker*, Norton, New York , N.Y., USA.

Dembski, W.R., (1998) *The Design Inference – Eliminating Chance Through Small Probabilities,* Cambridge University Press, Cambridge, U.K.

Denzinger, H. and Schönmetzer, A. (1965) *Enchiridion Symbolorum, Declarationum, et Definitionum*, Herder, Barcelona, Spain.

Miller, K.R. and Levine, J.S. (1995) *Biology*, Prentice-Hall, Englewood Cliffs, N.J., USA.

Ott, L. (1954*) Fundamentals of Catholic Dogma*, B. Herder Book Co., St. Louis, MO, USA.

Schönborn, Christoph Card., *New York Times* (Op-ed page, July 7, 2005).

Suarez, F. *De opera sex dierum*, 2,10, 13 **In** Berton, C. (ed.) (1856), *R.P. Francisci Suarez opera omnia*, Paris, France.

Tkacz, M. (2004) *A Designer Universe: Chance, Design, and Cosmic Order* (Talk at the Physics and the God of Abraham Conference, Gonzaga Univ. 2004).

Dialogue

Taner Edis

Much of what Dr. Barr says is theological. I have no competence to comment on how it fits in with his particular religious tradition. It also strikes me as irrelevant to those not already committed to his tradition; it certainly has little bearing on scientific matters.

That being said, I think there is some confusion about randomness exhibited throughout the article, and that might be worth pointing out. I do not by this mean issues concerning philosophical interpretations of probability; like most physicists, I am also perfectly happy to set those aside for pragmatic reasons. What I suspect is that Barr does not give algorithmic randomness as defined by Kolmogorov/Martin-Löf/Chaitin its due as a rigorous definition that captures most of our intuitions about randomness, including lack of correlations.

For example, Barr states that "a non-random process (like computing the digits of pi) can generate a sequence that passes all statistical tests for randomness." This is false. Indeed, the notion of sequences that pass *all possible* statistical tests has been historically important in developing the concept of algorithmic randomness. The digits of pi, or, for that matter, the output of pseudorandom number generators, only pass a limited set of statistical tests and are therefore useful in certain circumstances. They certainly do not pass all tests.

In this context, speaking of randomness is not purely a matter of choosing a pragmatic label. Since almost all (in the measure-theoretic sense) integer functions (or digit sequences, or any number of equivalent descriptions) are random, it makes sense to allow for randomness in physical models. Moreover, exactly all sequences can be expressed by combining algorithms and random functions. It would be surprising if we did *not* find randomness in fundamental physics.

The main significance of randomness in nature for debates over claims such as intelligent design is in the way we build models by recognizing patterns in finite data. Algorithmic randomness is equivalent to a complete lack of pattern; hence, random data *alone* cannot support inferences such as a causal structure. For example, if random variation and selection is sufficient to explain adaptive change in the history of life, then that history cannot be used to infer an intelligent designer. This does not, however, preclude an inference to a designer that does not depend on data from biology.

I am not offering these observations as a correction to Dr. Barr, only as a clarification. He does not defend intelligent design (at least not explicitly). And his examples of how randomness can result from design underline the point that the inference to design in these cases has to rely on information that goes beyond the patternless sequence under discussion. Nevertheless, all this does raise the question of what extra information is available that is relevant to biology. Merely stating that certain forms of theology are compatible with evolution is not controversial. But it may also be empty — even fundamentalist views can be compatible with any amount of data, depending on how much we are willing to massage interpretations of religion on one hand and proposed scientific models on the other. Barr does not tell us why his views of divine design are plausible explanations rather just being minimally compatible metaphysical glosses.

Stephen M. Barr

As I explained in the beginning of my article, there is relatively little difference among the major monotheistic religions on the issues I discuss. Therefore, I believe that most of what I say would be of interest to people outside my "particular religious tradition." Are the theological issues considered in my article "relevant"? They are certainly relevant to the subject of this book, which is about the theological implications of evolution. Are they relevant to science? No. My article is not about whether certain other things are relevant to science, but about whether science is relevant to certain other things, and if so how.

My statement that the digits of pi satisfy "all" tests of randomness was somewhat carelessly phrased. There certainly is at least one obvious respect in which the digits of pi are testably non-random: namely, in that there are algorithms for generating the digits of pi. I do not disagree with the point Dr. Edis makes about pi, but I don't see how it affects any of the points I was making.

Much of what Dr. Edis says has to do with inferences of "design". My article is not about that.

His last point is essentially that I offer the reader no reasons for believing in God in the first place. That is also not the subject of my article. The point of my article is not that science furnishes reasons for believing in God (though I happen to believe that), but that science furnishes no reasons for disbelieving in traditional religious notions of divine providence.

If I were "massaging the interpretation" of either religious doctrines or scientific theories, then my article would indeed be an "empty" exercise. However, Dr. Edis does not pretend to have shown that I am. I think I am rather clearing away crude misinterpretations, based on ignorance, of what science and certain religions have actually said.

Biography

Dr. Donald R. Prothero is Professor of Geology at Occidental College in Los Angeles, and Lecturer in Geobiology at the California Institute of Technology in Pasadena. He earned his Ph.D. degree in geological sciences from Columbia University in 1982. He is currently the author, co-author, editor, or co-editor of 22 books and over 200 scientific papers, including five leading geology textbooks and three trade books as well as edited symposium volumes and other technical works. Dr. Prothero is on the editorial board of *Skeptic* magazine, and in the past has served as an associate or technical editor for *Geology*, *Paleobiology* and the *Journal of Paleontology*. He is a Fellow of the Geological Society of America, the Paleontological Society, and the Linnaean Society of London, and has also received fellowships from the Guggenheim Foundation and the National Science Foundation. Dr. Prothero has served as the Vice President of the Pacific Section of SEPM (Society of Sedimentary Geology), and five years as the Program Chair for the Society of Vertebrate Paleontology. In 1991, he received the Schuchert Award of the Paleontological Society for the outstanding paleontologist under the age of 40. He has also been featured on several television documentaries, including episodes of *Paleoworld* and *Walking with Prehistoric Beasts*.

24
The Truth About Transitional Fossils

DONALD R. PROTHERO

Department of Geology, Occidental College
Los Angeles, CA 90041
prothero@oxy.edu

1. Of Fossils and Creationists

In the United States, the contentious debate about "intelligent design" (ID) has focused on many issues: complexity and its possible explanations; the explanation for biological structures; the naturalistic approach to science; and many other philosophical and biological topics. The books of the "intelligent design" creationists are filled with such examples and discussions, but they pointedly avoid discussing the fossil record or its implications. Behe (1996, p. 27) mentions paleontology only in a few paragraphs in his book-length treatment, and most of the other ID books are similarly silent about the fossil record. Only the ID creationist textbook *Of Pandas and People* (Davis and Kenyon, 2004) discusses fossils at any length, and only in a single chapter of a 170-page book. The ID creationists have been quoted on numerous occasions as conceding that microevolution occurs, and that the earth may be millions of years old. This distinguishes them from the more extreme fundamentalist "Young-Earth Creationists" (YEC), who believe the earth is only 6000 years old and will not admit that microevolution occurs.

The reasons for this lack of interest in fossils are apparent when one scans the contributors and critical reviewers of ID textbooks such as *Of Pandas and People* (Davis and Kenyon, 2004, p. iii). Although the ID creationists include scientists with backgrounds in biology or chemistry, almost none earned an advanced degree in paleontology or geology from a recognized accredited non-creationist institution (with the possible exception of Kurt Wise, who was a student of Stephen J. Gould at Harvard). To my knowledge, not a single ID creationist has ever published a paper on fossils in the peer-reviewed scientific literature. Reading through the little bit that they write about the fossil record, it is very clear that they have no firsthand experience with collecting or interpreting fossils, because they rehash old myths and misconceptions from the YEC literature. Like the YEC, their "research" on fossils consists of reading popular books about paleontology and pulling quotes out of context. ID creationists may impress the uninformed layperson with their Ph.D.'s in biochemistry, but that background has absolutely no relevance to understanding paleontology and fossils. Without the appropriate background or training, they are no more qualified to make statements about the fossil record than they are to discuss music theory or auto mechanics. Thus, their statements about fossils must always be read with the caveat in mind that they don't actually work on these fossils, and *have probably never even looked at the actual specimens* (nor do they have the training to tell one bone from another if they did).

In a volume such as this, it is useful to examine these myths and misconceptions about the fossil record, and give a short update about the truth about the fossils. For a longer account, the reader is referred to my book *Evolution: What the Fossils Say and Why it Matters* (2007).

2. Cambrian "Explosion"—or "Slow Fuse"?

This is the myth about the fossil record that the ID creationists mention most frequently, including Behe (1996, p. 27), Davis and Kenyon (2004, p. 94-96), and several others. It even popped up when I debated an ID creationist on a radio talk show, but he quickly got in trouble when his ignorance of the actual Cambrian fossil record was exposed.

Like all their other arguments, this one was recycled from many of the old YEC books, which have long tried to use it to support their cause. Creationists (both ID and YEC) point to the early appearance of most of the living phyla of animals in the Cambrian as some kind of "creation event." They hijack the commonly used phrase "Cambrian explosion" as if paleontologists were suggesting that all life arose instantaneously in the Cambrian, and no fossils were known from rocks older than the Cambrian.

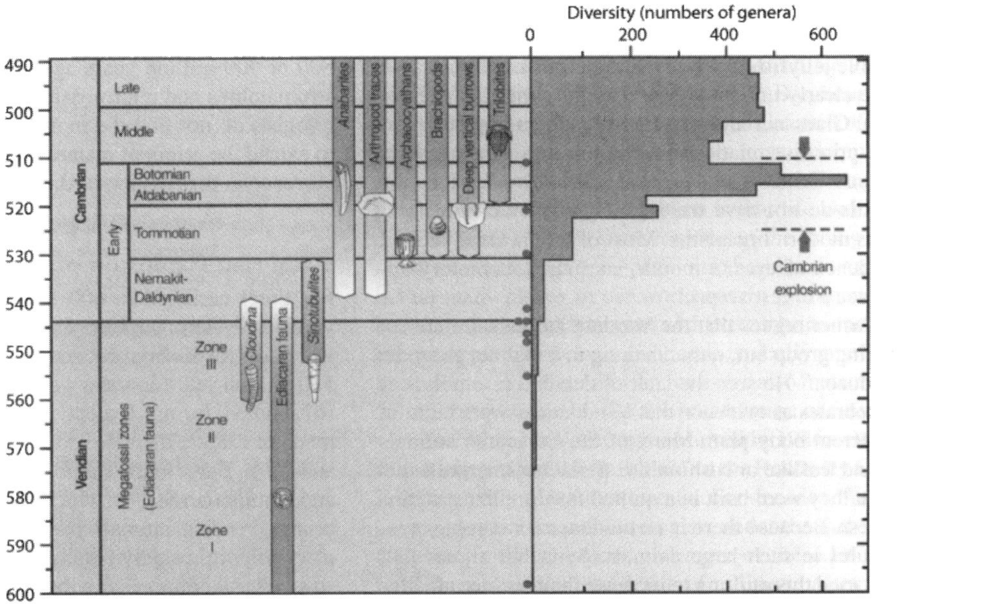

Figure 1: Diagram showing the time ranges of the major fossil groups in the late Precambrian (Vendian) and Cambrian. Generic diversity of these fossils shown on the histogram on the right. (After Prothero and Dott, 2004).

Such an idea about the fossil record might have been credible in Darwin's time (and creationists often quote his puzzlement about the lack of Precambrian fossils known in 1859), but it was rendered completely invalid by the discovery of numerous Precambrian fossils since the 1940s. In the past fifty years, the Precambrian fossil record has become enormous, with thousands of localities and tens of thousands of fossils—none of which the creationist literature ever acknowledges. The key difference is that most of these Precambrian fossils were single-celled microscopic organisms for the first 85% of the 3.5 billion years of life's history, so they can only be found with a microscope and the appropriate techniques in the right kinds of rocks. Nonetheless, there are many extraordinarily well-preserved microscopic fossils (see Schopf, 1999; Knoll, 2004; Prothero and Dott, 2004, for further details). After three billion years of nothing but single-celled fossils, the next stage is the famous soft-bodied Ediacaran fauna, which is found worldwide in rocks as old as 600 million years ago, and vanishes in the early Cambrian (Fig. 1). Following this is the earliest Cambrian, where large shelly invertebrates are rare (and no trilobites occur), but the commonest fossils are tiny forms nicknamed the "small shellies." These show that the factors that permitted the secretion of mineralized skeletons had finally occurred, but (as would be predicted by evolution) the first skeletons and shells were tiny and simple, not complex invertebrates such as trilobites. It is not until the third stage of the Cambrian (the Atdabanian) that larger more complex invertebrates such as trilobites appear, and the rest of the animal phyla appear even later in the Cambrian. Contrary to the outdated creationist literature, there are several major animal phyla with skeletons (such as the Bryozoa) that do not appear until much later (in the Ordovician).

Thus, the fossil record now shows a gradual series of steps from 3 billion years of single-celled organisms, to the multi-cellular but soft-bodied Ediacarans, to the tiny first shelled invertebrates of the early Cambrian, and only by the late early Cambrian do we see the appearance of more complex forms such as trilobites (and much later still the rest of the animal phyla). This is hardly an "explosion" but much more like a "slow fuse," with multiple steps spread out over thousands of feet of strata (well documented in the rocks of Russia, Namibia, and Canada). This sequence has now been subjected to numerous of the highest-quality radiometric dates by the top labs in the world, so it is now extremely well calibrated. From the first Ediacarans to the first trilobites (600-520 million years ago) is a span of 80 million years! Even the sequence from the first "small shellies" to the first trilobites (545-520) is still 25 million years in duration. Hardly an "instantaneous explosion"! The misleading term "Cambrian explosion" was coined by geologists decades ago to describe the fact that the majority of animal phyla appear in the Cambrian, and not later or earlier. But it was never intended to suggest that the event was instantaneous or ever rapid by human standards. Geologists and paleontologists are so accustomed to dealing with hundreds of millions to billions of years that anything that is only tens of millions of years in duration (compared to the 540 million years since the event) seems relatively rapid. But "geologically rapid" is still thousands to millions of years in duration, not a rapid "explosion" in the sense of human time scales of hours, days, years, and decades.

Even if one rejects all of the evidence of radiometric dating (which the YEC do, but the ID creationists apparently don't), the clear sequences of more and more complex fossils found in many places around the world (single cells to Ediacarans to "small shellies" to trilobites) completely rules out the use of the

"Cambrian explosion" as some sort of support for an instantaneous creation event as described in the *Bible*. Any time a creationist mentions the "Cambrian explosion" as some sort of geological conundrum possibly supporting their viewpoint, it is clear that they have only read old outdated YEC accounts of the Cambrian, and have not bothered to learn anything about the past 50 years of discoveries which show it was actually a "slow fuse" that fits perfectly with the predictions of evolution.

3. Transitional Forms, "Missing Links", and the Quality of the Fossil Record

Much of the public (including most creationists) has a mistaken notion of what evolution is. Some will ask, "If humans evolved from apes, why are apes still around?" This misconception goes back to pre-Darwinian seventeenth- and eighteenth-century notions of life as a "great chain of being" or a "ladder of life" (*scala naturae*) up from lowly invertebrates to fish to amphibians to reptiles to mammals to humans to cherubim and seraphim and angels and archangels and ultimately to God at the top. But as Darwin and many other scientists have shown, life is not a "ladder" or "chain" but a branching bush, with many ancestral lineages living alongside their descendants. When humans evolved from apes, they branched out from a lineage that is still around. Apes did not have to become extinct when some of them evolved into the ancestors of humans.

Closely related to this false notion is another metaphor, where each organism on the "chain of being" is like a "link" in the chain. From this kind of thinking comes the notion of "missing links" which tie together two organisms in the chain. Biologists and paleontologists seldom use the term "missing link" because of its erroneous connotations that life is a chain of being, but the public is still confused about this. In reality, there are hundreds of fossils (and a few living forms) that could be called "missing links" or "transitional forms" between major lineages and species. Yet creationists cannot admit the existence of these forms because that would be tantamount to admitting that evolution occurs. So they go through all sorts of rhetorical tricks to deny an obvious reality. In some cases, they blatantly deny the truth that is easily demonstrated. When provided with a picture of a transitional fossil in a debate, they will ask the evolutionist debater to provide even more transitional forms between that fossil and the fossils that came before and after it! No proof of no-longer-missing links is ever enough! No matter what evidence they are given, their denial mechanisms are so strong that they cannot see what is right before them.

Unfortunately, the ID creationists have borrowed one of the worst habits of the YEC authors: quoting scientists out of context to support their arguments. Quoting anyone of out context to indicate the opposite of what they meant is clearly dishonest, a political and rhetorical trick that reflects badly on whomever does it. Usually, when the true context of the quote is revealed, it shows that the person who quoted out of context either couldn't or didn't understand what the quote really means—or that they were intentionally trying to mislead the reader. Davis and Kenyon (2004, p. 96) in their ID creationist book for high school readers provide a particularly egregious example (borrowed directly from YEC books). They quote distinguished paleontologists such as Stephen Jay Gould and David Raup to say that the gradual transitions between fossils groups are rare, and that most fossil species are static and unchanged through millions of years. This is from the "punctuated equilibrium" debate that began with the famous Eldredge and Gould (1972) paper. Anyone who bothers to read this subject carefully, or read the full context of the quotations will realize that what these paleontologists are saying is that transitional forms are indeed rare, *but they are not unknown*. Contrary to the gradualistic evolutionary expectations that were widely held prior to 1972, we now know that there are good biological reasons for most species to stay stable and unchanging for millions of years once they have evolved, *but nevertheless there are good transitions between many of these species within transforming lineages*. More importantly, we can view each step in a transforming lineage as a transitional form, even though each individual species is relatively unchanging during its time on earth.

When ID creationists write about the quality of the fossil record and the supposed lack of transitional forms, they show most clearly that they have borrowed from the old YEC literature without reading anything more up-to-date. For example, Davis and Kenyon (2004, p. 95-98) repeat the old YEC myth that the fossil record must be nearly complete, and that the apparent absence of transitional forms therefore is proof that evolution did not occur. Ironically, what the past 40 years of paleontological research has shown is that the fossil record is much less complete than most people believe. At this moment, biologists have described and named about 1.5 million species on earth (mostly insects), and some estimates say that the earth harbors at least 4 or 5 million species in total. Yet there are at best only about 250,000 known species of fossil animals and plants, or about 5% of the species living today. But today is only one time slice among millions in the past 600 million years that multicellular life has existed. If we total up all those time slices as well, then the total number of species that are represented in the fossil record is a tiny fraction of 1%.

Consequently, the fossil record of some groups that are entirely soft-bodied without hard skeletons or shells (especially insects, worms, jellyfish, and the like) is so poor that most paleontologists do not study them, and do not attempt to say much about their evolution. In certain groups with hard skeletons, however, the potential for preservation is much higher. If we focus just on groups with excellent skeletons and a good chance for preservation (including microfossils, sponges, corals, mollusks, sea stars and sea urchins and their relatives, trilobites, the "lamp shells" or brachiopods, and "moss animals" or bryozoans), the fossil record is not nearly so incomplete. These groups have about 150,000 living species, but over 180,000 fossil species. Depending upon how you do the calculation, between 2% and 13% of all the species that have ever lived in these groups may be fossilized (Prothero, 2004). That's still not great, but much better than the fraction of 1% estimate we just discussed. In some places, the record of fossil shells is very dense and continuous, and these are places where paleontologists focus their attention to study things like evolution. They know that not every species is preserved, of course, but they have enough data to see how evolution occurs in the groups that do fossilize.

In such groups, paleontologists have found abundant evidence of transitional forms between major groups. Here is where the ID creationists are the most in denial of reality. For example, Davis and Kenyon (2004, pp. 95-96) write that "we cannot form a smooth, unambiguous transitional series linking, let's say, the first small horse to today's horse, land-dwelling mammals to today's whales, fishes to amphibians, or reptiles to mammals." This is a terrible deception to put in a high-school textbook. They could not have asked for better-documented cases than the evolution of horses, whales, early mammals or amphibians. Let us look at each of these examples in turn.

4. Of Horses and Whales

Davis and Kenyon (2004, p. 96) deny the existence of the evolutionary sequence of horses, yet make no further mention of it anywhere else in their book. The evolution of horses was one of the first transitional series documented after Darwin's book was published. It began with studies published by Thomas Henry

Huxley and O.C. Marsh in the 1870s and 1880s, and still stands today as one the best transitions we have. As early as 1870, we had fossils of early Eocene (55 million years old) horses such as *Protorohippus* (once called "*Eohippus*" or "*Hyracotherium*") which were the size of small dogs, had four fingers on their hands and three toes on their feet, and primitive low-crowned teeth. As nearly every textbook in evolution and biology shows, from these simple primitive ancestors, horses went through an amazing sequence of changes: their side toes were reduced until modern horses run only on the middle digit; their legs got longer for fast running; their teeth became more and more high crowned for eating gritty grasses; their body and brain size increased, snout became elongated, and overall skull and body proportions changed dramatically until they resembled horses that we know today. All of those transitional horse fossils are real and well documented. I have personally published research on the *Mesohippus-Miohippus* part of the sequence (Prothero and Shubin, 1989; Prothero, 1994), and I have collected, identified, and studied horse fossils from many parts of the sequence.

Some parts of this story have been changed and modified as more fossils are discovered. For example, many of the early renditions of horse evolution were necessarily oversimplifications that showed a simple linear trend in these anatomical changes through time. But we have known for over a century that horse evolution, like that of nearly every other family of organisms on earth, is bushy and branching, with multiple lineages overlapping in time. Not only do we have this well-documented transformation within the horse lineage, but in recent years we have discovered their primitive ancestors, including a fossil from Mongolian rocks 58 million years old known as *Radinskya*, which links horses and their close relatives, the tapirs and rhinos, to all the other lineages of hoofed mammals. I have spent much time working with early Eocene horses such as *Protorohippus* and its close relative, the earliest tapir relative *Homogalax*. In most features, their teeth and skeletons are nearly indistinguishable, yet there are subtle differences in the cusps and crests of teeth that show that one of them was ancestral to horses, and the other to tapirs and rhinos.

The ID creationist websites follow the YEC model when faced with this reality: quote out of context. Usually they cite very outdated references about specific details of horse evolution, and leave out just enough information to give the complete opposite impression about what the quote means—and deliberately deceive their reader. Most of these quotations concern the replacement of the old oversimplified straight-line evolution model with our more modern "bushy" branching model. None of those quotations deny that horse evolution occurs, only that it is more complex than originally thought. None of ID creationists bothers to read the more recent literature or deal with new transitional fossils such as *Radinskya.* Better still, they could stop deceptively mining old papers for quotes out of context and go study the fossils themselves!

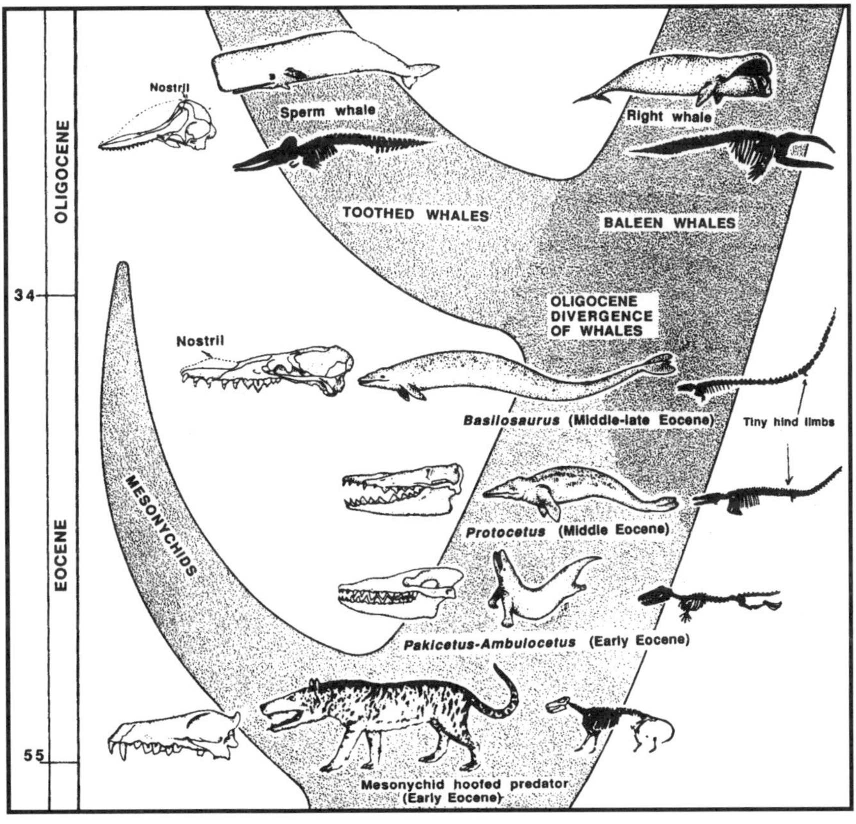

Figure 2: Evolutionary history of whales (from Prothero, 1994).

Even more spectacular is the origin of whales. Davis and Kenyon (2004, p. 101-102) claim "there are no transitional fossils linking land mammals to whales." *They could not be more wrong*. This false statement is carried over from their 1989 edition into their 2004 edition, yet the 1980s and 1990s yielded an amazing array of transitional whale fossils that clearly link terrestrial land mammals to full-fledged aquatic whales (Fig. 2). These fossils have been well documented in many television shows, in popular books such as Carl Zimmer's (1998) *At the Water's Edge*, and published in high-profile scientific journals such as *Science* and *Nature*, so there is no excuse for creationist ignorance or denial of these fossils. Davis and Kenyon (2004, Fig. 4-5, p. 101) illustrate two extremes of the whale evolutionary sequence (the terrestrial mesonychids and the aquatic archaeocetes) but falsely state that there are no transitional forms between them.

If you look at the fully aquatic dolphins, orcas, and blue whales, you would have a hard time imagining them walking on land. Yet scientists have long known that whales are related to hoofed mammals, and even living whales retain tiny vestiges of their hips and thighbones deeply buried in the muscles along their spines. In 1983, specimens of *Pakicetus* were discovered in Pakistan from early Eocene beds about 52 million years old. Although its body was primarily terrestrial, it had the skull and teeth of the archaic archaeocete whales that swam the world oceans in the middle Eocene. Then in 1994, another specimen was found in Pakistan. Dubbed *Ambulocetus natans* (literally, "walking swimming whale"), it was the size of a large sea lion, with broad webbed feet on both front and hind limbs so it could both walk and swim—yet it still had tiny hooves on its toes, and also had the primitive skull and teeth of the archaeocetes. *Ambulocetus* apparently swam much like an otter with an up-and-down motion of the spine, precursor to the vertical motion of the flukes of a whale's tail. Then *Dalanistes* was discovered, with its shorter legs with webbed feet and longer tail, but with a much larger and more whale-like skull. Today, there are over a dozen excellent transitional whale fossils, making this one of the best insensibly graded series that could be expected for such rarely fossilized animals. Their DNA suggests that whales are most closely related to the artiodactyls (even-toed hoofed mammals) and the hippopotamus in particular. This was dramatically confirmed by the 2001 discovery of the characteristic artiodactyl "double-pulley" ankle bone in two types of primitive whales. By the late Eocene, we see the first members of the modern toothed whales (sperm whales and dolphins) and the baleen whales (filter-feeding whales like the blue, right, humpback, and gray whales). There are excellent transitional forms of some of the extremely primitive baleen whales that are not yet toothless (like all modern baleen whale species), but still retain teeth while also developing their baleen filters. As the years go by, more and more transitional whales are being discovered, so that by now the amazing transformation from terrestrial mesonychid to whale is one of the best examples of evolutionary transitions in the fossil record (Fig. 2). This may not make creationists happy, but the fossils cannot be denied.

Whales are not the only aquatic mammals with terrestrial ancestors. Modern sirenians (manatees and dugongs) are huge, docile aquatic plant eaters with no external hind limbs, and front limbs modified into flippers. Their detailed anatomy and molecular biology, however, show that they are closely related to the ancestors of elephants. In 2001, Daryl Domning described a remarkable complete skeleton of *Pezosiren portelli* from Jamaican deposits about 50 million years old. This fossil had the typical skull and teeth of a sirenian, and even the thick sirenian ribs made of dense bone (for ballast)—yet *Pezosiren* had four legs with feet, not flippers. The origin of seals and sea lions from bear-like ancestors is also well documented. *Enaliarctos mealsi* from beds about 20 million years old in California is a perfect transitional form. Although it retained many primitive features of bears, it has some specializations seen in seals and sea lions, such as enlarged eyes, an enlarged nasal cavity for regulating the temperature of the blood as it swims, and larger openings for the muscles that control lips and whiskers. More importantly, its hands and feet are developed into very crude flippers, so superficially it looks like a modern seal.

The origin of mammals is also very well documented. The lineage that led to mammals is known as the Synapsida (once known as "mammal-like reptiles," although they are a separate and parallel branch that had nothing to do with true reptiles). The earliest synapsids include the finback *Dimetrodon* (familiar from kids' toy dinosaur kits, even though it was not a dinosaur), which was the largest predator on earth about 280 million years ago. Even though it was a very primitive form, it already had large stabbing canine teeth and some of the specialized skull features of mammals. Over the next 80 million years, synapsids evolved into a variety of wolf-like and bear-like predators, as well as a variety of bizarre pig-like herbivores. Through their evolution, they show progressively more and more mammalian features: expansion of the area for jaw muscles, with additional jaw muscles for complex chewing motions; a secondary palate covering the old reptilian palate and nasal region, so they could breathe and eat at the same time; specialized multi-cusped molars for chewing, rather than gulping, food; enlarged brains; more upright (rather than sprawling) posture; a muscular diaphragm in the rib cage for efficient breathing; and even evidence that they had that characteristic mammalian feature, hair.

Most remarkable of all is the transformation of the lower jaw. In the reptiles and primitive synapsids, the jaw consists of a number of bones besides the tooth-bearing dentary bone. During synapsid evolution, the dentary bone becomes larger and larger until it takes over the jaw joint and point of jaw muscle attachment. The other reptilian jaw bones shrink until they vanish. Yet two of these extra jaw bones (the quadrate bone of the skull and articular bone of the lower jaw) do not vanish completely. Instead, these bones are crowded out by the new jaw joint between the dentary and the squamosal bone of the skull. Indeed, one fossil (*Diarthrognathus*, "double jaw joint") has the old reptilian jaw joint and the new mammalian jaw joint both operating side-by-side on each side of the mouth. Eventually, the mammalian jaw joint (dentary-squamosal) takes over completely, yet the tiny quadrate and articular bones do not vanish, but instead shift to the middle ear, where they become the "anvil" and "hammer" bones with which you hear. This seems surprising until you realize that reptiles hear with their lower jaw, and sound is transmitted from the jaw through the quadrate and articular to the inner ear. Further proof can be seen in embryology: your middle ear bones were in your jaw when you were an early embryo, but shifted to the ear during later development. The story of the synapsids culminates with the evolution of the earliest true mammals (tiny shrew-sized creatures) from beds about 200 million years old in China, Texas, and South Africa.

How do the ID creationists handle this extraordinary transitional series? Davis and Kenyon (2004, p. 100-101) quote a few evolutionists out of context and even concede, "without a doubt, the Therapsids are highly suggestive of a Darwinian lineage." But then they betray their complete lack of understanding of evolution and try to discredit the entire example by arguing that it is not a single ancestral lineage but many different lineages. That is *exactly* how most evolutionary transitions work in a bushy, branching system—not as "missing links" on a non-existent "chain of being" (the common creationist misunderstanding) but as multiple closely related lineages which each show progressively more mammalian characteristics.

Examples of evolution within the mammals could be enumerated almost endlessly. We have excellent transitional forms that document the evolution of elephants from creatures without tusks or trunks, giraffes from creatures without long necks, rhinoceroses from creatures without horns, camels from creatures without humps, primitive cats and dogs that looked nothing like their living descendants, and hundreds of other beasts familiar from the zoos and circuses. The reader is referred to Prothero (2007) for full details.

5. Of Walking Fishes and Flying Dinosaurs

Another transitional sequence that creationists must deny is the amazing series of fossils that show how some fish evolved into land animals. To some people, it seems amazing that any creature can make the dangerous transition from water to land. However, it turns out to be much easier and more common that you would think. A number of living bony ray-finned fish, including mudskippers, walking catfish, rockfish and sculpins of the intertidal zones, can live out of the water for hours, and many have modified their fin bones into crude devices for crawling across the ground. Thus, when the opportunity presents itself, many different groups of vertebrates have found a way to switch from aquatic to at least a partially terrestrial life.

For decades, the only good transitional fossil between fish and amphibian was *Ichythyostega* from the Late Devonian (about 360 million years ago) of Greenland and Spitzbergen. Although *Ichythyostega* was like many amphibians in having well developed legs, with a complete shoulder girdle, and hips fused to the backbone, it still had the fish-like gill slits, a lateral-line system on its face for detecting underwater currents, and a long fish-like tail fin. More recent discoveries, such as *Acanthostega* from the same beds, show that the picture is much more complicated and interesting. *Acanthostega* had ear bones adapted for underwater hearing, a longer tail fin, and better-developed gills (so it is more primitive and aquatic than *Ichythyostega*), and up to eight fingers and toes on its hands and feet (rather than the standard five of most tetrapods). Apparently, its limbs were primarily adapted for swimming and walking along the bottom than for crawling out on land. Contrary to the popular story that four legs evolved for crawling on land (to escape drying ponds or predators, or chase new food sources, or whatever), it appears that legs evolved for walking underwater (as most salamanders still do today), and only secondarily became useful on land.

The clinching piece of evidence was announced as this chapter was being written. Nicknamed the "Fishapod" but formally named *Tiktaalik*, this Late Devonian fossil from Ellesmere Island in the Canadian Arctic was even more fish-like than *Ichthyostega* or *Acanthostega*, yet its limbs show the perfect transition between fins and feet (Fig. 3). Thanks to this discovery, we now have a beautiful transitional sequence from fully aquatic lobe-finned fish like *Eusthenopteron* to more amphibian-like forms such as *Panderichthys* and *Tiktaalik*, to fully four-legged forms like *Acanthostega* and *Ichthyostega* (which still retain fish-like gills, tail fins, and lateral line systems on the face). This sequence is now so smoothly gradational that it's hard to tell where the fishes end and the amphibians begin—yet is it clear even to a creationist that *Eusthenopteron* is a fish and *Ichthyostega* is an amphibian.

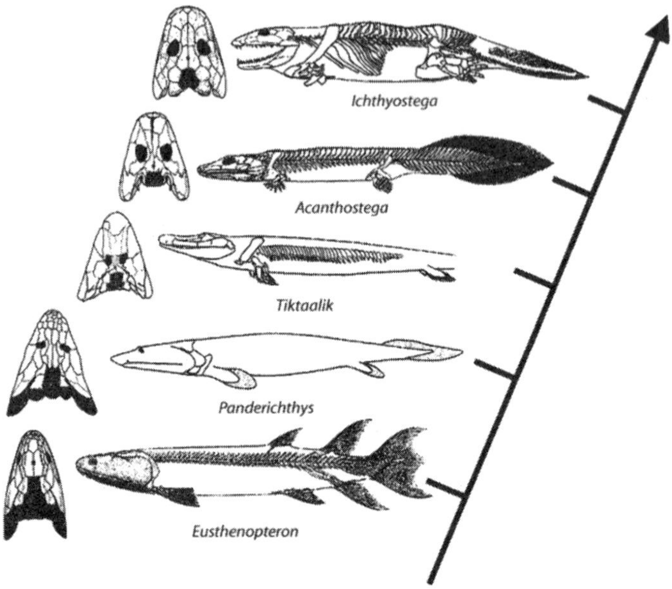

Figure 3: Evolutionary transition from lobe-finned fish such as *Eusthenopteron* through intermediates like *Panderichthys* and *Tiktaalik* to primitive four-legged aquatic amphibians such as *Acanthostega* and *Ichthyostega* (from Ahlberg and Clack, 2006).

How do the ID creationists deal with this extraordinary evidence? They cloud the issue by denying these fossils exist, or by distortion and misstatements. Davis and Kenyon (2004, Fig. 4-8, p. 103) show a fifty-five-year old sketch of *Ichthyostega* and *Eusthenopteron* but make no mention of all the other transitional fossils that were well documented before their book was published in 2004. They show (Fig. 4-9, p. 104) the fin and limb bones of each of these creatures, but ignore all the beautiful transitional fossils that have been documented in the past 50 years. They make a big deal about how dramatic this transition was, yet falsely claim that "we do know that no such transitional species have been recovered." Thanks to *Panderichthys* and *Acanthostega* and now *Tiktaalik*, that falsehood can be safely laid to rest—but I have no expectation that creationist books will ever acknowledge the existence of these fossils, but simply replay their discredited and outdated arguments.

The creationist blinders about transitional fossils are nowhere better demonstrated than their denial of the classic transitional fossil from reptiles to birds, *Archaeopteryx*. Discovered just two years after Darwin's book was published, paleontologists and biologists could not have wished for a better example of a fossil which bridges the gap between two major classes of vertebrates. Now known from eleven specimens (I have studied most of them), they show the classic mixture of bird-like and dinosaurian features. Most of the skeleton is in fact almost indistinguishable from the tiny dinosaur *Compsognathus* (the "compys" of *Jurassic Park* fame), with a long bony tail, long heavy hind limbs, primitive dinosaurian hip bones, well developed clawed fingers, and a dinosaurian skull with teeth—all features found in no modern bird. Yet even the first specimens showed clear bird-like characteristics, including a wishbone and feathers.

Naturally, this fossil has been a thorn in the side of creationists for decades. The usual strategy of the YEC authors (since they can't ignore it or deny its existence, as they have with so many other transitional fossils) is to say it's "just a bird" since it has feathers and a wishbone. Never mind the fact that the entire skeleton is dinosaurian, and no living bird has the long clawed fingers, long bony tail or teeth found in *Archaeopteryx*. The ID creationist authors are even more subtle and misleading. They use a few out-of-context quotations that do not apply, and fall back on the old misconception that evolution must be a smooth gradual "chain of being" within a single lineage. Davis and Kenyon (2004, p. 106) write "it is transitional only if it is part of lineage—one of series of generations in which in-between stages led gradually from one group to another" (illustrated clearly in their Fig. 4-11, p. 106). In one sentence, they have shown their complete misunderstanding of the fundamental concepts of evolution. *Archaeopteryx* does NOT have to be part of single gradually evolving lineage to be a transitional form—those are all misunderstandings about evolution discredited decades ago. It only needs to be one of many species that show transitional features on the bushy, branching tree of life. And in this respect, *Archaeopteryx* could not be a better intermediate transitional form.

But all this creationist focus on discrediting *Archaeopteryx* has been rendered moot by the amazing new Jurassic and Cretaceous bird fossils that have been discovered in the past 15 years, especially from the extraordinary Liaoning beds of China. There are now hundreds of bird fossils that show each step in the evolutionary transition from predatory theropod dinosaurs to full-fledged modern birds. A Late Jurassic bird from China, *Confuciusornis*, had developed the first toothless bird beak, but still had long fingers with claws. Slightly younger fossils (from the Chinese Lower Cretaceous Liaoning beds about 140 million years in age) have even more birdlike features. *Sinornis* could fold its wings against its body, and had grasping feet with an opposable big toe, and its tailbones were fused into a single element (the pygostyle, or "parson's nose" of your Thanksgiving turkey). *Iberomesornis* from the Lower Cretaceous rocks of Spain had a large keeled breastbone for powerful flight muscles, yet still had the primitive long backbone of dinosaurs. *Concornis* not only had these advanced features, but also had the upper row of ankle bones fused to the shin bone (which you see in the cap of cartilage on the turkey or chicken drumstick), a characteristic of all modern birds. Thus, *Archaeopteryx* is now one of hundreds of new Mesozoic bird specimens that show the remarkable transformation from dinosaur to bird. Most surprising of all, however, was the recent discovery of numerous non-flying dinosaurs (closely related to *Velociraptor* of *Jurassic Park*) such as *Microraptor* and *Caudipteryx* from the Liaoning beds of China which had well-developed feathers. Apparently, feathers evolved widely in the theropod dinosaurs originally for functions such as insulation, long before they secondarily became useful for flight. Do the ID creationists mention any of these past 15 years of discoveries? Certainly not in their textbook *Of Pandas and People*, which was last revised in 2004, nor do their websites make any mention of them.

6. Of Integrity and Truth

I could continue to document in much greater detail additional ID creationist falsehoods, distortions, and misstatements, but since space is limited, I welcome the reader to find out the details for themselves in my new book (Prothero, 2007). But the way in which ID creationists approach the issue raises larger questions about intellectual integrity and honesty. The ID creationists have made a great effort to deny in public that their movement is religiously motivated, and claim that the "Intelligent Designer" need not be the Judaeo-Christian God, but in private they admit that their goals are all about pushing their religious viewpoint (documented by many authors, e.g., Shermer, 2006). They are mostly members of right-wing evangelical Christian churches, and their Discovery Institute in Seattle is funded almost entirely by religious organizations and conservative foundations. They have tried to hide their religious motivations to get around the separation of church and state enshrined in the U.S. Constitution, but in 2005 they lost badly in the federal trial of *Kitzmiller et alia vs. Dover*

Pennsylvania School Board. Judge John Jones, a conservative President George W. Bush appointee, ruled against the ID creationists, and pointed out that their ideas were another thinly disguised effort to inject narrow sectarian religious views into the public-school science classroom. He even called their ideas "breathtaking inanity." Judge Jones was particularly irritated by the hypocrisy of the ID creationists, who attempt to sound secular when the Constitution is involved, but crowed about their religious motives when not in court. "The citizens of the Dover area were poorly served by the members of the board who voted for the intelligent design policy. It is ironic that several of these individuals who so staunchly and proudly touted their religious convictions in public would time and again lie to cover their tracks and disguise the real purpose behind the intelligent design policy."

ID creationism is not about science, but about political power, and about dictating the agenda for schools and textbooks now, and eventually exerting control over society. They have described their movement as a "Wedge Strategy" to squeeze their religious beliefs (disguised as science) into the public school science classroom. ID creationists play by whatever rules (dishonest if necessary) they need to in order to win. I showed how they ignore, distort or deny the evidence, quote out of context, and do many other unethical things—all in the name of winning their crusade. As someone who was raised in a Presbyterian church, and learned *Bible* verses every Sunday, it appalls me to see how unethically these supposed "Christian" men and women will act in their battle against their perceived foes. It makes you wonder whether they have second thoughts about violating the word and spirit of many parts of the Scripture with their lies and deceptions, all to accomplish their goals at the price of their souls.

How do they reconcile this un-Christian behavior with their Christian beliefs? Apparently, to the ID creationists lying and deception are lesser sins that Darwinism, and they are willing to sacrifice their integrity in their crusade against what they believe to be the source of all evils in the world. Their intellectual blinders are so strong that they see only what they want to see, and read only what they want to read in a quotation, all in the name of their religious beliefs. To ID creationists, pushing their beliefs about the *Bible* is essential to their religious salvation, and everything else (including science) must be sacrificed so their souls can go to heaven.

The ID creationists reveal their true motivations when they speak among themselves (documented by Shermer, 2006). On February 6, 2000, William Dembski told the National Religious Broadcasters:

> "Intelligent Design opens the whole possibility of us being created in the image of a benevolent God… . The job of apologetics is to clear the ground, to clear obstacles that prevent people from coming to the knowledge of Christ.… And if there's anything that I think has blocked the growth of Christ as the free reign of the Spirit and people accepting the Scripture and Jesus Christ, it is the Darwinian naturalistic view."

At the same conference, Phillip Johnson said:

> "Christians in the twentieth century have been playing defense. They've been fighting a defensive war to defend what they have, to defend as much of it as they can. It never turns the tide. What we're trying to do is something entirely different. We're trying to go into enemy territory, their very center, and blow up the ammunition dump. What is their ammunition dump in this metaphor? It is their version of creation."

In 1996, Johnson said: "This isn't really, and never has been, a debate about science… . It's about religion and philosophy." One of the ID creationist authors, Jonathan Wells, is a follower of the Reverend Sun-Myung Moon and his Unification Church cult (which is vehemently anti-evolutionary). As Wells wrote: "When Father chose me (along with about a dozen other seminary graduates) to enter a Ph.D. program in 1978, I welcomed the opportunity to prepare myself for battle."

Perhaps they should go back to their *Bibles*, where Proverbs 12:22 says: *Lying lips are an abomination to the Lord.*

References

Ahlberg, P. and Clack, J. (2006) A firm step from water to land. *Nature* **440**, 747-749.

Behe, M. (1996) *Darwin's Black Box*. The Free Press, New York.

Davis, P. and Kenyon, D. (2004) *Of Pandas and People: The Central Question of Biological Origins* (2nd ed.). Haughton Publishing, Dallas, Texas.

Eldredge, N. and Gould, S.J. (1972) Punctuated equilibria: an alternative to phyletic gradualism, **In**: T.J.M. Schopf (ed.) *Models in Paleobiology*. Freeman Cooper, San Francisco, pp. 82-115.

Knoll, A. (2004) *Life on a Young Planet: The First Three Billion Years of Evolution on Earth*. Princeton University Press, Princeton.

Prothero, D.R. (1994) Mammalian evolution, **In**: D.R. Prothero and R.M. Schoch (eds.), *Major Features of Vertebrate Evolution*. Paleontological Society Short Courses in Paleontology **7**, 238-270.

Prothero, D.R. (2004) *Bringing Fossils to Life: An Introduction to Paleobiology* (2nd ed.). WCB/McGraw-Hill, New York.

Prothero, D.R. (2007) *Evolution: What the Fossils Say and Why it Matters*. Columbia University Press, New York.

Prothero, D.R. and Dott, R.H., Jr. (2004) *Evolution of the Earth* (7th ed.). WCB/McGraw-Hill, New York.

Prothero, D.R., and Shubin, N. (1989) The evolution of Oligocene horses, **In**: D.R. Prothero, D.R. and R.M. Schoch (eds.), *The Evolution of Perissodactyls*. Oxford University Press, New York, pp. 142-175.

Schopf, J.W. (1999) *Cradle of Life*. Princeton University Press, Princeton.

Shermer, M. (2006) *Why Darwin Matters: Evolution and the Case Against Intelligent Design*. Henry Holt/ Times Books, New York.

Zimmer, C. (1998) *At the Water's Edge: Macroevolution and the Transformation of Life*. The Free Press, New York.

Part 5

A Dialogue between Faith and Reason

Biography

Dr. Malcolm E. Schrader was born in New York City in 1925. After serving in the US Army during World War II, with award of the Purple Heart and Combat Infantry Badge, he received his PhD (1956) in Physical Chemistry from Brooklyn Polytechnic Institute. He subsequently performed fundamental and applied research for the US Navy, largely in the field of surface chemistry and nano-surface science. He pioneered in the application of radioisotopes to the study of the interface between fiber and matrix in composite materials, and was the first to use ultra-high vacuum techniques in the study of wettability. Some highlights of the achievements of his experimental research are the solving of the famous controversy on the wettability of gold by water, obtaining of a previously elusive result for water-wettability of graphite, and determination of the nano-locus of hydrolytically induced failure of the glass-resin bond in glass-reinforced plastics.

Dr. Schrader served for four years as Meeting Secretary (program manager) of the Division of Colloid and Surface Chemistry of the American Chemical Society, and as long-range coordinator of symposia on fluid-solid surface interaction of that Division. He served on the Board of Directors of the Society for Plastic Industries, and was involved with surface chemistry applications as a member of the American Vacuum Society and the American Society for Testing Materials.

On retiring as senior scientist from the U.S. Navy, he was appointed professor on the research staff of the Hebrew University of Jerusalem. In the theoretical area, he subsequently revised the thermodynamic approach to wettability, and showed that, contrary to previously accepted concepts, the adsorption of a nano-layer of its own vapor does not necessarily impede spreading of a given liquid on a solid surface. He is presently pursuing an interest in chemical evolution as Guest Scientist with the Department of Inorganic and Analytical Chemistry at Hebrew University.

25
Evolution and the Bible
The Secular Approach

MALCOLM E. SCHRADER

Department of Inorganic and Analytical Chemistry, Institute of Chemistry
The Hebrew University of Jerusalem 91904, Jerusalem, Israel
schrader@vms.huji.ac.il

1. Secular Approach to *Bible*

The subtitle of this volume is "*Questions of science and faith in biological evolution*". Most of the public controversy on this question has centered on the *Bible*. However, the debate on this subject, as it refers to the *Bible*, or, specifically, the first section of the book Genesis of the Pentateuch (five books of Moses), suffers from a peculiar, rarely mentioned asymmetry. While the *occurrence* of biological evolution (as distinct from its proposed *mechanism*) is regarded as a fact by all "secularists", and many people of faith as well (Schrader, 2006), this recognition has itself resulted from a cultural evolution in western civilized thought, in which the "scientific method" has become perhaps the most powerful tool of inquiry into various areas of knowledge. The secular approach regards this method as supreme, and thus regards contradictory results from other approaches as false. An example of course is secular rejection of the fundamentalist religious view, which accepts information obtained by direct communication with God.

The latter religious view regards direct divine communication as an unchallengeable truth. Thus, any contradictions to it, even if based on direct observation, must be false. The secular approach, on the other hand, regards the scientific approach as supreme, and thus unchallengeable by any other.

Moreover, it challenges, on principle, the existence of any direct communication with God whatsoever.

However, it is not only the understanding of biological evolution that has come about through the culturally evolved scientific method. Understanding and interpretation of the *Bible* has also undergone evolution, and has been subjected to a critical approach which attempts, to whatever extent possible, to apply the scientific approach to *Bible* understanding.

The secularists, while applying modern scientific method to observation of biological phenomena, have chosen to ignore similar advances in *Bible* study. They use ancient approaches to *Bible* interpretation for comparison with modern approaches to biology. This sets up an easy mark for debate, but hardly yields meaningful results.

A proper secular approach to the *Bible* and evolution should use modern historical and other techniques for *Bible* interpretation, alongside the modern historical knowledge of other ancient ideas on evolution.

2. The Two Accounts of Creation

The first important point to be made in a secular approach to the biblical treatment known as "creation" is the fact that there are two accounts of creation in the first part of Genesis. The existence of these has occasionally been treated by traditional authorities, see for example, Soloveitchik (1974). He points out "Sefer Breishit" (Genesis), as is well known, contains two accounts of the creation of man The first chapter's description of man's creation pictures man as almost an afterthought. His 'birth' is mentioned in one breath with that of the animals. The first chapter talks of man and woman (a biological description) while the second talks of husband and wife (a social or ethical description)." It can be seen that Soloveitchik's recognition of two separate accounts extends to defining different purposes for each account, but not different origins for each account. From the secular point of view, however, any forcing of the two different physical descriptions of each account into one is neither necessary nor desirable. The secular approaches accept, and often require, the existence of separate documents, or at least separate traditions (Cassuto, 1961) for each of these two accounts, as well as for many other sections of the Pentateuch. A widely accepted documentary approach resolves the Pentateuch into sections with four origins, called the J, E, P, and D (with the J and E often overlapping). This approach assigns the first chapter of Genesis, containing the first account, to the P section, and the second chapter, containing the second account, to the JE section.

3. The First Account

The first account (FA) opens the book of Genesis with the declaration that in the beginning, the heavens and earth were created. The state of the earth when first created is described as chaotic, and a type of order is then brought about. First, light is introduced (it is neither created nor crafted, but simply comes into existence on command). Then it is seen that the light is good, and the light is then separated from the darkness, and the two named day and night. This brings to an end what is described as the first "day" of creation. On the second day, a firmament (atmosphere) is placed in the water, and the water under the firmament is then separated from the water above the firmament (to form clouds above the liquid water covering the earth). Then, on the first part of the third day, water covering the land is withdrawn to its own locality, thus exposing dry land, and the two are named land and ocean, and it is seen that this is good. This sets the stage for the appearance of life. The command is then given for the earth to give forth vegetation, including grasses and trees, which reproduce themselves. The earth then obeys the command and produces the vegetation as called for, and it is seen that this is good, ending the third day.

After an interlude on the fourth day for the creation of the sun, moon and stars, the history of life is then resumed with the crafting (Note 1) of the animal kingdom on the fifth "day", starting with the command for the water to produce insect-like creatures and for birds to fly above the earth on the face of the sky. In the execution of the command, large sea animals are created, as are all crawling sea animals, and all species of birds, mentioned in that order. Both for the case of the fifth "day" command and for its execution, mention of the crafting or creation of underwater animals is followed immediately by mention of the "coming into existence", or creation, of birds. Following this, on the sixth "day", land dwelling cattle, land dwelling creeping things, and land dwelling beasts are brought forth from the earth and crafted, and finally, man is crafted and created.

4. Discussion of First Account

The first thing we may notice is that there is a gradual but relentless buildup of matter from a virgin, disorganized inanimate state to the highly organized state of living matter (starting with vegetation), followed by increased organization of the living matter to higher and higher complexity and functionality of mobile living forms (now called animal kingdom) climaxed by the crafting and creation of the *Homo sapien* species, called "the Adam" (as distinguished from Adam, the name of the first living human in the second account of creation).

The second thing noticeable is that there is some detail given regarding the order of appearance of the various species, even in situations where it is not required to represent a continuous increase in complexity. For example, bird life is presented along with life under water, with the life under water mentioned first (both on fifth "day"). The main line of succession following sea life and birds is then continued with animal and insect life on land (sixth "day"), including, and climaxed by, the crafting and creation of *Homo sapiens*.

5. The Present Day Scientific Account: Some Highlights

It is of course now an obvious exercise to compare this first biblical account of the crafting of life with what is now known as the scientific account of evolution, especially for further comparison with ancient historical recording of Greek ideas which are credited with being fore-runners of the now accepted "Darwinian theory of evolution".

The present scientific description of evolution is vague regarding the classification of earliest life as plant, animal, or other. The first life forms are considered to have appeared between 3.5 to 3.9 GYa (billion years ago) (Kasting, 1993). It has been speculated (Margulis, 1992) that unicellular organisms, such as for example bacteria or algae plants, were the earliest life forms. Due to the lack of bony structure, fossils of these forms and knowledge of them is obtained by indirect means, such as molds in rock or carbon-envelope residues. Approximately 600 MYa (million years ago), fossils of invertebrate underwater crawling animals, such as trilobites and brachiopods, suddenly (in the geologic time sense) began to appear. Approximately 500 MYa the first corals and vertebrates, including first jawless fish, appeared. Roughly 400 MYa the first jawed fish appeared, followed by insects and amphibians. Also, at this time, land plants appeared and evolved. There were still however, virtually no land animals. At roughly 300 MYa the first reptiles and land insects appeared. About 230 MYa mammal like reptiles, and at 180 MYa early bird types, appeared. At 135 MYa small and very primitive mammals were present, and at 65 MYa primitive mammals developed and proliferated. At 35 MYa monkeys and apes, at 25 MYa man-like apes and at 10 MYa apelike men. At some time after 0.6 MYa prehistoric men.

Of particular interest is the lineage. Crossopterygian ganoid fishes (vertebrates) are thought to be the immediate ancestors of labyrinthrodont amphibians, which are the immediate ancestors of cotylosaur reptiles (Moody, 1953; Dobzhanski, 1953). These reptiles (land animals) then form various ancestral branches, of which one is to birds and dinosaurs, and another to mammals. The mammals, of course, consist not only of cattle, horses, lions and tigers, but of monkeys, apes and modern humans.

6. Comparison of First Biblical with Scientific Account

We first compare the present day science-based picture with the biblical first chapter, and start with a listing of the order of appearance of the different life forms, and separately, with the ancestry (descent) of the life categories. The scientific account of bioevolution of the Cambrian (600 MYa) and post-Cambrian forms starts with appearance of invertebrate underwater crawlers, followed by vertebrates, followed by jawless fishes, followed by jawed fishes, followed by amphibians, followed by reptiles. The reptiles branched out to birds and dinosaurs on one branch of the evolutionary tree, and to mammals (which include humans) on another branch.

The biblical first account has crawling sea creatures, all birds, and large sea animals appearing in one time period (Note 2, 2006), described as the fifth "day". In the next period, described as the sixth "day", the land dwelling cattle, creeping things, beasts, and man were all crafted.

It can be seen, then, that both the biblical and biological accounts state underwater creatures of the "animal kingdom" as having been crafted first, and all land dwelling creatures of the animal kingdom as being crafted in a subsequent period of time. Thus, the crafting of underwater crawlers and swimming fish are assigned to the same period by both the scientific and first biblical accounts. In the scientific account, amphibians and reptiles, which are not mentioned separately in the biblical account, bridge the time of appearance between water animals and mammals.

Birds are given a somewhat different position in the evolutionary tree in the *Bible* as compared to the present day speculations of the scientific approach. The *Bible* has them appearing directly after underwater animals or fish, while the scientific account has them descending indirectly from the underwater animals and fish, via amphibians, reptiles, then dinosaurs according to today's mainline theory, or from four-legged reptiles preceding the dinosaurs according to another view.

With respect to insects, both the biblical and modern scientific accounts report their origin in water. The modern scientific account in fact specifies fresh water as their origin, from arthropods, in the Devonian period, 300 to 350 MYa.

It is of course obvious that the first account in Genesis resembles the present day scientific account far more closely than it resembles the second account in Genesis. It is of interest then to compare various ancient ideas which have been proposed as the earliest statements of the basic idea of evolution.

7. Ancient "Ideas of Evolution"

At this point, we pause to examine historical research on the concepts of biological evolution. They all focus on the "idea" of evolution, its origin, and degree of resemblance to the concept as we now know it. Speculations and authoritative statements on the origin of the life forms we now know, are as old as history. Fothergill (1952), in describing ideas relating to evolution from the "obscure" ancient period, states: "As early as the sixth century before Christ, Confucius tried to show that 'complexity was derived from a single source' and that 'things were originated from a single simple source through gradual unfolding and branching'."

The early Greek scientists were Thales, Anaximander and Anaximenes, in that order (Burnet, 1930; Zeller, 1931). Thales is considered to have lived from 624 to 565 BCE. He taught that the earth floated on water and that water is the primary substance from which all else is derived. Anaximander (611-547 BCE) was possibly Thales' pupil. His primary substance was called the "boundless", out of which came heat, cold, water, air, earth and fire. The earth was an amorphous mud, from which arose plants, animals, and man, in that order. Furthermore, the first humans were born on the inside of a fish, then ejected after some maturing. Anaximenes (585-525 BCE) proposed air as the one basic substance, identifying it with the "boundless" of Anaximander. Fothergill quotes Newman as stating that Anaximenes introduced the idea of spontaneous generation, from the mud, of all living forms. It is thought that Anaximander also held this view.

Among the later scientists, Heraclitus (540-475 BCE) and Parmenides (540-470 BCE) did not teach any evolutionary changes whatsoever. Empedocles (495-435 BCE), however, is sometimes called the father of the evolution idea. He taught that plants and animals arose from the earth (Zeller, 1931). At first only separate limbs of animals arose. They then combined to form products of monstrous shapes. When the present life forms, i.e., animals and men, were crafted, they were amorphous masses which "only with time achieved their structure" (Zeller, 1931). Zeller states further "It is in itself *not probable*, nor is it affirmed by Aristotle, that Empedocles explained the efficient structure of organisms by the theory that only those of these chance creations which were fitted for life could have survived". Empedocles further believed in the existence of a community of blessed spirits, from which exiled sinners wandered through plant, animal and human bodies. "Aristotle is often described by some writers as holding evolutionary opinions, while others deny this" (Fothergill, 1952).

8. Discussion of Empedocles "Idea of Evolution"

What can be regarded as a precursor to the evolutionary concept is: the human form did not suddenly appear on earth as is. There were originally simpler biological entities which combined in various ways.

Having extracted these abstract essences from the Empedocles model, the historians proclaim this to be the earliest statement of the evolution idea. The following should be noted. Indeed, it proposes that the human form did not appear suddenly and complete. However, the combination of preexisting limbs is certainly not how it came about. There were no preexisting limbs nor was there ever any such combination, correct or otherwise. Furthermore, Zeller points out that it is most unlikely "that Empedocles explained the efficient structure of organisms by the theory that only those of these chance creations which were fitted for life could have survived".

9. Comparison of Empedocles with Biblical First Account as an Evolutionary Model

Now let us look at the biblical model. Here is a statement of the coming into existence of complicated life forms, not only human but mammals, including the carnivorous wild beasts (e.g. lions, tigers), and herbivorous horses, cows, and dogs, etc. and, furthermore, not only mammals but reptiles and birds, and fish and underwater crawlers, and the insect world. They are listed in an order of appearance which progresses from the simplest to the most complicated. Now, why, however, must it be concluded that this description of appearance of species in FA (first account) is essentially an evolutionary account? In answer, there is first, as mentioned above, the not insignificant matter of the order in time, of crafting, or creation, of the species. It starts from the simple and relentlessly works its way up to the more and more complex forms of life. Why? Is it a natural thing to list the crafting of the species in increasing order of complexity, without any reason? If so, why is this the only ancient document where this is done? One convenient example of listing the creation of the species is the second account in Genesis, found in chapter 2. Crafting there starts with the most complex of all species, man himself. Following this is vegetation (Garden of Eden), and following that is a random listing of creatures of the animal kingdom. The Empedocles picture does not deal with interspecies complexity at all, or with species as such, but rather conjures up parts of an individual for assembly, a process of even less scientific significance than the biblical second account's crafting of woman from man's rib. The Babylonian accounts of "creation", despite occasional touting otherwise, do not contain any systematic description of biological crafting of the species in increasing order of complexity remotely resembling the contents of FA. If a religious reason is chosen for the order of listing in the FA, as many choose to suppose, then we are presented with the notion that God crafted each species separately from the same raw materials, but did not have the confidence to proceed to a more complicated sculpture until a simpler one worked. This does not conform even to simple ancient concepts of an omniscient monotheist God. It must be accepted that the order of appearance of the species is based on a scientific concept as to how it physically occurred, to which is added a theological cover, with the purpose of the latter being to emphasize that all this occurred under the initiative and supervision of the monotheistic God, rather than with the involvement of the pagan gods of other nations. Selection of a physical narrative was then either drawn from prevailing wisdom of an existing culture available to the P author or authors, or invented by the P authors themselves. It is hard to imagine any plausible reason for arranging the order of crafting of these species according to their complexity, other than a dependence of each step of crafting on the existence of the previous one, which means that each succeeding step derived from the previous one in some physical manner. Thus FA is the true precursor of knowledge of the facts of evolution that have been accepted by much of the world since the days of Wallace and Darwin.

The mechanism of transition in the step-wise advance to increased complexity is not known, but, in the Greek version, mechanism is not only not known but the identity, even approximate, of the creatures undergoing evolution, for which a mechanism is required, is not remotely guessed at. There is no hint of transition of lower to higher species, just an impossible assembly of impossibly preexisting limbs. With respect to the circular proposal of survival of the "fittest" (what are the fittest? those which survive), this concept, contrary to popular notion, is explicitly included in the biblical first account, but in a more sophisticated, noncircular manner. After the crafting of each listed classification, the *Bible* pauses to state "and God (Elohim) saw that it was good". The last quote is not necessary to the theological interpretation of the overall creation process. It results from an understanding, based on observation of nature, on the part of the author, that many variations were possible in the evolutionary process, but the ones which occurred and survived did so as a result of being *fit* to survive in the environment. Not necessarily as a result of competition, or of being the fitt*est*, but of being quite fit, and altogether, as a final ecology, *very fit* as a whole on completion in the sixth day. It was only after each class proved itself by at least the ability to survive, and possibly other unstated positive evolutionary achievements, that the evolutionary process was permitted to proceed to the next step, each one occurring in successive geological periods (of length in time probably unknown to the author of the account).

10. Mechanism of Inheritance

The question arises, of course, how can this be considered evolution when there is no mechanism of inheritance and when the text repeats "and the earth gave forth", for many of the species. Of course, the mechanism of evolution by inheritance, such as it may be, did not come along until Darwin, and even then was presented in most vague form. It is even today highly speculative in many respects. There is no suggestion that this type of knowledge was possessed by the ancients. It is consequently not unreasonable that the FA would be vague on this matter. It is supposed by some that the phrase "the earth gave forth" or that "the water (produced)" represents a statement of "spontaneous generation". This supposition is flawed. First, given that the FA must be vague about the unknown details of evolutionary transition, a statement that it was produced from the earth, in an unstated manner, is not at all unreasonable. In any event the earth plays a part. Second, this device is used in some of the cases of evolutionary transition, but by no means in all. For example, for the case of the bird genus, the command simply states "and birds will fly on the land, on the face of the firmament of the sky", without stating the source or method of production. When the command is carried out, the word "create" is used. When man is brought about, the command uses the word "craft", without indicating from what. (Note: FA, unlike the second account, does *not* create man from the dust of the earth). When the command is carried out, the word "create" is used. Furthermore, even with the underwater crawlers, which are commanded to be brought forth from the water, when the command is carried out the word "create" is used (without indicating from what). Thus, the argument that earth or water bringing forth is meant to specify spontaneous generation, is a rather weak one. The FA simply does not commit itself on the question of mechanism of inheritance.

11. Historical Dating of the Biblical First Account of Creation

The foregoing, in terms of the history of the idea of evolution, invites a most intriguing question. What is the date of original recording of the biblical FA, especially as compared to that assigned to the Empedocles proposal. A bit of discussion on the history of the P document is in order here.

The documentary theory originally proposed by Wellhausen has received considerable acceptance. However, Wellhausen went further and proposed time periods, as well, for the final writing or editing of these individual documents. His approach was that the religion evolved gradually from pagan worship to monotheism, more or less in linear fashion. This evolution took place during the time of the literary prophets. The most advanced stage of the evolution, complete monotheism, was the last to take place. Therefore, the law in the books of Moses, clearly contained largely in P, had to be written after the literary prophets. He presents some evidence in support of this thesis. However it is now recognized that his basic assumption is not correct. The religion did not evolve linearly from paganism to pure monotheism. Like all social evolution, advancement is cyclical, with regressions as well as advancements along the way. Wellhausen (1878) had proposed that J and E (now called JE) were combined and edited in the 9th and 8th centuries BCE, that D was composed in the last quarter of the 7th century, and P in the 6th to 5th centuries.

Kaufmann (1960) however, in his 7 volume treatise "*The Religion of Israel: From its Beginnings to the Babylonian Exile*", disagrees and presents overwhelming evidence that the P document is independent of, and probably older than, the 7th century D book. Noting that the cosmology of Empedocles is the chief support for the claim that the idea of evolution stems from the Greek philosophers, it is obvious that the biblical FA is by far the first and most complete account of biological evolution known to history. As mentioned above, FA is part of the P document, which according to Kaufmann would precede Empedocles by about 300 years. Even if the Wellhausen time schedule is accepted, the FA version, to which, in any event, the Empedocles myth cannot be compared as an evolutionary document, will have been written before Empedocles. Thus, the FA history of species evolution preceded modern day speculations, by roughly two thousand four hundred years.

12. The Second Account

In the foregoing, we have concentrated exclusively on the systematic account of evolution given in the FA (first account) of evolution in Genesis. We point out here, however, that although the second account (the story of Adam and Eve) is not a systematic account of evolution, there is interesting evidence that it, as well as the first account, was privy to some facts of evolution known to the advanced culture, or cultures of that day. In this case, moreover, it is not interspecies evolution that is portrayed, but rather evolution within a species itself. The account implies that the snake is, at the time of the story, walking on legs, and then clearly states that in the future it will have to crawl on its belly. It is a known fact of evolution history that there are snakes which once had legs, but then lost them through evolution (Green, 2000; Rieppel, 2003). The theory is that they adapted themselves to long periods of dwelling in tube-like shelters where legs were a hindrance. The author or authors of the second account, then, were aware of scattered evolutionary occurrences, but did not or could not put them together into an overall system as was projected in the first account. This knowledge of evolution upon which both were able to draw probably came about from examination of bodies and cadavers of then-existing species, and possibly also of some fossils.

13. Conclusions

The historical origin of the idea of biological evolution occurs in the *Bible* as the first account of "creation" in the first chapter of Genesis.

The "objectively" estimated date of this account is given as roughly 700 BCE on the basis of strong arguments by the scholar Y. Kaufmann. Previous "objective" estimates assigned the date of approximately 400 BCE.

Historians of evolution have often credited the Greek philosopher Empedocles with first proposing the idea of evolution.

There is little substance behind this claim, however, either in terms of the date proposed or its content. Empedocles makes no suggestion of a succession of real biological forms starting from the simple and proceeding to the complex.

The first chapter of Genesis presents an account of "creation" that is consistent in broad scope and considerable detail with the present-day concept of the facts of evolution. The second chapter presents an alternative version of the physical events which is not compatible with the evolutionary system. The common feature to both is the real message the Bible is transmitting, i.e., regardless of what the physical events were it was God who planned and executed the entire process.

References and Notes

Burnet, J. (1930) *Early Greek Philosophy,* Macmillan, New York.

Cassuto, U. (1961) A *Commentary on the book of Genesis, vol. 1*, I. Abraham, trans. Magnus Press, Hebrew University, Jerusalem.

Darwin, C.E. (1859) *The Origin of Species*, John Murray, Albemarle Street, London.

Dobzhansky, T. (1955) *Evolution, Genetics, and Man*, Wiley, New York.

Fothergill, P.G. (1952) *Historical Aspects of Organic Evolution*, Hollis and Carter, London.

Green, H.W. and Cundall, D. (2000) Evolutionary Biology: Limbless Tetrapods and Snakes With Legs. *Science* 287 (5460), 1939-1941.

Kasting, J.F. (1993) Earth's Early Atmosphere. *Science* 229, 920-926.

Kaufmann, Y. (1960) *The Religion of Israel: from Its Beginnings to the Babylonian Exile*, Translated and Abridged by Moshe Greenberg, University of Chicago press.

Margulis, L. (1992) Biodiversity. Molecular biological domains, symbiosis and kingdom origins. *Biosystems* 27, 39-51.

Moody, P.A. (1953) *Introduction to Evolution*, Harper, New York.

Rieppel, O., Zaker, H., Tchernov, E. and Polcyn, M.J. (2003) The Anatomy and Relationships of Haasiephis Terrasanctus, a Fossil Snake With Well developed Hind Limbs, from the Mid-Cretaceous of the Middle East, *Journal of Paleontology* 77, 536-558.

Schrader, M.E. (2006) *Is Life on Earth Accidental? Natural Selection and the Second Law of Thermodynamics,*In: J. Seckbach (ed.) "Life As We Know It", Springer , Dordrecht, pp. 163-178.

Soloveitchik, J.B. (1974) *Sheeurei Harav*, In: J. Epstein (ed.) Hamevaser, Yeshiva University, New York, N.Y.

Wallace, A.R. (1855) On the Law Which Has Regulated the Introduction of New Species. *The Annals and Magazine of Natural History,* ser.2,16, 184.

Wellhausen, J. (1878) *Prolegomena to the History of Israel.* Eng. Trans. 1965.

Zeller, E. (1931) *Outlines of the History of Greek Philosophy,* 13th ed., revised by W. Nestle. L. R. Palmer, trans. Routledge and Kegan Paul, London.

Note 1 – "Crafted" is the author's translation of the Hebrew word "VaYa-as", as used in this chapter of Genesis.

Note 2 – Possibly Genesis meant the "day" to be taken literally. On the other hand, that it refers to a unit of time equaling a day seems more likely, in view of the fact that there could not have been day and night before the appearance of the sun and moon on the fourth day. Also, not at all unlikely is a third possibility that the word "day" refers to an epoch of unknown length.

Biography

Dr. Taner Edis is associate professor of physics at Truman State University, Kirksville, MO, USA. He received his Ph.D. from The Johns Hopkins University in 1994. Trained as a theoretical condensed matter physicist, his research interests have been diverse, from atmospheric modeling with collaborators at Lawrence Livermore National Laboratory to the philosophy of machine intelligence. Lately his focus has been questions relating to science and religion, and he has written a number of books on such topics. *The Ghost in The Universe: God in Light of Modern Science* (2002), presents a comprehensive naturalistic picture of the world, based on the best of modern knowledge. *Why Intelligent Design Fails: A Scientific Critique of the New Creationism* (2004), co-edited with Matt Young, presents the scientific case against intelligent design. *Science and Nonbelief* (2006) explores the relationship between modern science and religious nonbelief. *An Illusion of Harmony: Science and Religion in Islam* appeared in 2007.

26
Muslim Resistance to Darwinian Evolution

TANER EDIS

Truman State University, Department of Physics
Kirksville MO 63501, USA
edis@truman.edu

1. A Range of Options

Ever since Darwin, the theory of evolution has come into conflict with popular religious views about the nature and history of life. And since evolutionary biology arose in scientifically advanced countries where the predominant religion was Christianity, Christian views of evolution have generated the most extensive literature. Most Christian responses to evolution fall into broad categories of theologically conservative denials of evolution, efforts to reach some accommodation, or a more radical reinterpretation of doctrine to avoid conflict with scientific views. Judaism and Islam, however, are also Abrahamic religions that share many beliefs about divine action with Christianity. They face similar tensions between evolution and traditional beliefs concerning the accuracy of scripture and the divine role in creating life. Hence responses to evolution within Judaism and Islam are very similar to the better known Christian examples.

Very often, the first option for religious thinkers has been to preserve orthodox doctrines in their traditional forms. Believers who understand scripture fairly literally and who take their holy texts to be authoritative tend to reject evolutionary biology. If such believers also have positive attitudes toward science and technology, they will be attracted to claims that science, when properly done, actually refutes evolution and confirms a creationist view of life instead. The classic example is the way many conservative Protestants, especially in the United States, endorse a "scientific creationism" that is considered a pseudoscience by the scientific community (Scott, 2005).

Another common response is to seek a middle ground. Religious people will often read scriptural creation stories in a more metaphorical sense while preserving an active role for their God in the history of life. Many Christians and Jews, for example, agree that the universe is billions of years old and that life forms have changed and developed throughout time. They accept that evolution has taken place, in the sense that the forms of life that we see today all share a common ancestry. This does not, however, mean that they fully accept Darwinian evolution. The consensus view of modern biologists is that undirected natural mechanisms are sufficient to explain the history of life on Earth. Darwinian variation-and-selection, in particular, is the mechanism responsible for the functional complexity displayed by life forms. Such a view denies any explicitly noticeable divine influence on biology. Hence even liberal-minded believers are often uncomfortable with full-blown Darwinian evolution. They tend to prefer the notion that evolution has been divinely guided in order to produce human life. Typically they conceive of evolution as an inherently progressive process rather the blind exploration of biological possibilities envisioned by modern scientists.

There also are continuing attempts to fully reconcile Darwinian evolution and some usually very liberal versions of the Abrahamic religions. These attempts tend to be limited to academic theologians and religious scientists; they have little direct effect on the general public. Furthermore, there are many secular people in the advanced Western countries, particularly Western Europe, who might be indifferent to supernatural claims but who identify with the culture and moral tradition associated with their religious background.

All of these options are available for Muslims as well as Christians and Jews. Forms of creationism, reinterpretation combined with guided evolution, and a secularized acceptance of evolutionary science are all common in Muslim lands. There are, however, also noteworthy differences. In the Muslim context, all these broad forms of response to evolution naturally take on a more Islamic coloring. And the Muslim world is more similar to the United States than Europe, in that the cultural and political strength of religious populism encourages the more overtly anti-evolutionary responses. In fact, compared to the United States, liberal religious options are much less developed in Muslim lands (Edis, 2007). And so today, the strongest and most successful versions of creationism in the world are inspired by Islam rather than Christianity.

2. Denying Evolution

Almost all devout Muslims believe that the Quran is the direct, unaltered word of God, untainted by human mediation. Such a belief does not always require a literal reading of the text, and there are competing interpretations of the Quran and the other sacred sources of traditional Islam. Most interpretations, however, keep close to the traditional understanding of Islam as developed by religious scholars over the centuries. So by and large, Muslims take claims made in the Quran at face value. This has changed little with modernization. Social and political changes have reduced the status and power of the class of religious scholars in all Muslim lands, and increased literacy rates have favored an emphasis on how guidance from the sacred sources is directly available to every Muslim. As a consequence, modern Muslims are often drawn to a straightforwardly literalist reading of the sacred sources.

The Quran does not explicitly include the Genesis story of creation, though it has references to the six days of creation scattered throughout. The Quran also does not have as detailed a story about Adam and Eve, but it portrays the first human couple as being created directly, by a special act of God. The Quran, in other words, is most naturally read as a creationist text. It treats life, especially human life, as an immediate divine creation, both in explicit statements and in the way its allusions make sense in the context of common Abrahamic religious beliefs about special creation. Since the overwhelming majority of observant Muslims take Quranic claims at face value, Muslims tend to have a creationist view of life.

In the most traditionally religious communities, evolution is rarely a point of contention. Some awareness of evolution may exist—perhaps conceived vaguely as a Western materialist point of view—but religious scholars point to the sacred sources and denounce evolution because it goes against traditional Islamic beliefs. In such communities, a short and dismissive ruling settles the matter.

The more devout segments of Muslim populations generally take a similar position against evolution. There is a vast amount of conservative Muslim literature, from books, pamphlets, and newspaper columns to web sites. Much of this literature is devoted to answering practical questions by consulting the sacred sources and the traditions of Islamic Law. Attacks on Darwin or evolution are not very frequent, mainly because evolution is not usually a matter of practical concern. Evidently, evolution rarely comes to the attention of conservative Muslims. One reason is that in Muslim countries, students do not often get thoroughly exposed to evolutionary ideas. As the creationist movement in the United States demonstrates, perhaps the most reliable way to provoke an anti-evolutionary reaction is to impose secular biology education on a religiously conservative population. In much of the Muslim world, this has not happened.

In the late nineteenth century, some Muslim intellectuals showed interest in European debates over Darwin's theory, largely because of the religious and philosophical implications of evolution rather than any deep interest in biology. Indeed, a handful of Muslim westernizers and secularists became very enthusiastic about evolution. But more devout thinkers, even among prominent modernists, quickly rejected evolution as an impious, materialistic myth. They argued that while importing Western technology had become imperative, Muslims had to be vigilant against adopting aspects of Western thought that could cast doubt on religious convictions (Bezirgan, 1988; Ziadat, 1986). By and large, they succeeded. Devoutly Muslim intellectuals never had any extensive debate over creation and evolution, largely because the most enthusiastic secular evolutionists remained marginal figures who did not even merit much denunciation. In the twentieth century, military leaders who favored a more secular politics attained political power in many Muslim countries. But religious resistance to military secularism rarely included opposition to evolution as a central concern, since devout Muslims typically faced far more severe challenges to the traditional religious social order.

So one of the most common Muslim responses to evolution has been to ignore it. Many Muslims are quiet creationists who feel no need to actively oppose evolution because the cultural penetration of Darwinian evolutionary ideas remains very low.

More recently, this state of affairs has been changing. Islam has always provided the dominant cultural frame of reference in Muslim countries. Recent revivalist currents have only strengthened the public role of religion. Indeed, Islamist political movements have been very successful in the past few decades. In this more publicly religious environment, conservative Muslims, including political Islamists, have begun to put more emphasis on opposing Darwinian evolution. Furthermore, with continuing modernization and Muslim economies becoming more integrated into the current global order, constituencies for a new kind of creationism have arisen. There are now a significant number of people who depend on advanced technologies for their livelihood, who live modern lives in most respects, but still rely on traditional religious beliefs as the main source of moral and political legitimacy. Such people tend to be more aware of scientific influences on modern culture, are very positive in their attitudes toward technology, and appreciate the role of mass education and the media in shaping modern societies. A quick dismissal of evolution that only invokes the authority of the sacred sources is not sufficient for such an audience, as they keenly feel the cognitive authority of science as well as religion. An attractive resolution of this tension comes from a more ostensibly scientific form of creationism. If *real* science actually supports creation, and if evolution is merely a falsehood promoted by cultural forces hostile to Islam, then the tension disappears.

Turkey, which has been the most Western-oriented among Muslim states, also provides the best examples of the recent, more aggressively pseudoscientific form of creationism. Since the 1920s, the Republic of Turkey has been committed to a program of westernization and military-supported secularism, in the face of much resistance from devout and rural populations. Since the 1970s, however, political Islam has become very influential, and there has been a considerable re-Islamization of public culture. As a result, official secularism has been eroding, and a Muslim version of "scientific" creationism has become increasingly popular.

Turkish creationism used to be confined to conservative religious circles, plus some modernizing religious movements that promoted the view that modern science confirmed Muslim beliefs. The Nur movement, based on the teachings of Said Nursi, was often at the forefront of attacks on evolution. Until an Islamist party took part in coalition governments in the 1970s, however, Turkish creationists usually did not confront the educational establishment, continuing their more quietist political stance of passive resistance to official secularism. The 1970s saw infrequent objections to evolution in textbooks expressed in parliament, but creationism largely remained confined to traditionalist and

Islamist constituencies (Atay, 2004). There was no large creationist literature, and opposition to evolution did not attract significant public attention.

Turkish creationism made its first breakthrough in the early 1980s, under a right-wing military government. Although the Turkish military has historically supported an strongly secular state, the military government of 1980-1983 decided to exploit Islam as part of a cultural policy to promote national unity and combat left-wing political tendencies. The following civilian government, which included an Islamist faction that controlled the Ministry of Education, continued these policies. In the mid-1980s, the Ministry of Education consulted with some devout Muslim intellectuals with links to the Nur movement, asking them what could be done about combating Darwinian evolution in Turkish education. Furthermore, concerned that the younger generation was coming under immoral materialist influences, the Ministry made translations of American Protestant creationist literature. These were distributed to secondary schools, intended as supplemental reading (Edis, 1994).

Government support, and a source of fresh ideas from Christian creationism, invigorated Turkish creationists. They borrowed wholesale from the Protestant-produced "scientific" creationist literature, adapting it to Muslim needs. Since they did not care about defending the literal accuracy of the Genesis story, Muslim creationists did not emphasize questions about the age of the Earth. American creationism is dominated by a young-earth position, and American creation-scientists devote much effort to "flood geology" in order to fit Earth history into a few thousand years. While some Muslim creationists have expressed sympathy to these young-earth arguments, this is entirely due to the influence of the Christian creationist literature. Muslims tend to be old-earth creationists—they usually interpret Noah's flood as a local event and think that the days of creation mentioned in the Quran could just as well be long ages. Nevertheless, with these few differences, Turkish creationists of the 1980s found Protestant creation-science remarkably useful, adopting much of it with only small changes (Edis, 1994).

The beginning of creationist influence on the educational establishment was a significant victory for religious conservatives in Turkey. Together with translated creationist literature, paragraphs expressing skepticism about Darwinian evolution began to appear in textbooks used in high school biology and in some occasions, even medical schools. This success, however, was limited, and was noteworthy only in the context of the strictly secularist tradition of the Turkish Republic. After all, creationist sentiments in textbooks are common throughout the Muslim world. Moreover, creationism imposed from the top by the government is vulnerable to changes in the political winds. Indeed, recently, when more secularist political parties took power in Turkey, they removed the traces of creationism from the curriculum, while more Islamic-colored parties reintroduced creationism when back in office. Government-sponsored creationism has not been much more than another tool in the Turkish culture wars between Islamists and secular elites.

Beginning in the mid-1990s, however, Turkish creationism reached its most popular and mature expression, and started to influence the rest of the Muslim world. This new wave of creationism has been distinguished by its popular appeal and its broad reach due to its extensive use of modern media technologies. It has found a more modern audience, beyond traditional, conservatively religious circles.

The new Turkish creationism is mainly associated with literature presented under the name of Harun Yahya (Edis, 1999). Harun Yahya is a pseudonym claimed by Adnan Oktar, a leader of a Turkish sect. But since Oktar has no appreciable scientific or other academic background, and since hundreds of books, articles, web sites, and other media productions have appeared under the Harun Yahya name in only a decade, it is evident that Harun Yahya should be thought of as a brand name for particular style of creationism rather than as a single author.

Yahya's brand of creationism is very similar to its predecessors in content. He also borrows his distorted "science" from Christian creationist sources, again adapting them and giving them an Islamic emphasis. Indeed, the Yahya material is typically opportunistic. In making unsubstantiated charges of fraud and gross incompetence against evolutionary scientists, Yahya quotes young-earth creationists as if they were scientific authorities, uses the words of legitimate scientists out of context, or uses arguments from anti-evolutionary thinkers who are not associated with fundamentalist Protestant views (Yahya 1997). He argues that evolution is a scientific mistake, often by presenting examples of functional complexity in biology and declaring that it is obvious that chance variation and selection could never achieve such intricate structures. In doing so, he relies on a traditional and still very common Muslim perception that the world very obviously has to be a result of divine design, and that it is a strange failure of reason to think otherwise. The anti-evolutionary arguments that were originally produced by Christian sources only enhance what is a very intuitive conclusion for a Muslim audience.

Attacking evolution is the centerpiece of the Harun Yahya brand of popular Islamic apologetics, but the Yahya corpus also addresses many other topics of concern to Muslims today. The Harun Yahya name has also appeared on endorsements of anti-Masonic conspiracy theories, literature denying that the Holocaust happened, appeals to unity among the Abrahamic faiths against the common threat of atheism and secularism, praise of Said Nursi, and arguments that terror and violence has no place in authentic Islam. Yahya has even invited some theological controversy by arguing for an idealistic view of existence where only the mental is real and matter is only an illusion. Nevertheless, he keeps coming back to the claim that the universe is obviously designed. Resisting evolution is a task that carries much moral urgency, not just because evolution denies God's creative role and can also cause a more gradual slide toward materialism, but also because evolution leads to ideologies such as communism and fascism. Even terrorism, Yahya argues, is rooted in the false belief in evolution (Yahya, 2002).

Yahya's message is easily available; indeed, it is difficult to avoid. Yahya's creationist literature includes many attractively illustrated, high-quality but also low-cost books, plus magazines, DVDs, CD-ROM's, a profusion of web sites and internet resources. These are sold in mainstream bookstores, not just Islamic outlets patronized by religious conservatives. Yahya-affiliated organizations sponsor creationist presentations worldwide, often giving a platform to creationists who hold scientific credentials. In 2006, Yahya's associates even began opening "creation museums" to the Turkish public. Clearly, the Harun Yahya operation is backed by considerable financial resources, especially since almost all of Yahya's creationist material is available at no cost or at artificially low prices. The source of this funding remains unclear.

Besides its popular appeal and independence of government support, the most important characteristic of Yahya's creationism is its growing international influence. Yahya material has been translated to just about every language in use in the Muslim world. Significantly, this includes most European languages—Yahya's books are easily available in London's Islamic bookstores as well as all over Turkey. Harun Yahya's works have found a large audience in Pakistan, Indonesia, and many other Muslim-majority countries. Even mass-market introductory books about Islam published in the United States refer to Yahya as a Muslim scientist who has made a solid case against evolution. The modern, well-packaged, media-savvy, technology-affirming Islamic creationism presented under the Harun Yahya brand has clearly tapped into a worldwide market (Edis, 2003a; Edis, 2003b).

3. Beyond Popular Creationism

Strict creationism, whether of the traditionalist sacred source-based variety or a Yahya-style pseudoscience, is not the only option available to Muslims. More secularized Muslims, for example, tend to accept evolution due to their general trust of modern education and science as a cognitive authority. As always, where the average educated person is concerned, this acceptance rarely involves more than a superficial knowledge either of the science or of the religious worries about evolution articulated by more conservative Muslims. Still, there is definitely a constituency for efforts to interpret Islam in a way that is compatible with evolution.

As with many Christians, the favored view among liberal Muslims is to downplay Darwinian explanations of the evolutionary process, affirming common descent while portraying biological evolution as a divinely guided progression toward higher forms of life (Ateş, 1991). A particular concern for Muslims, naturally, is to try and interpret the Quran in such a way as to allow for a degree of evolution. Some theologians, for example, read verses that say God created all animals from water as a statement that life emerged from the oceans, just as the scientific history of life on Earth has it. Verses that deal with the special creation of humans call for more strenuous attempts at reinterpretation. Indeed, many Muslims accept considerable evolution under divine guidance for all non-human forms of life, but consider humanity to be a totally separate creation.

Another way to make evolution more acceptable to Muslims is to try and find evolutionary ideas in the Muslim philosophical tradition, and to revive them as an "evolutionary creation theory" (Bayrakdar, 1987; Altaytas, 2001). Indeed, some medieval Muslim philosophers elaborated on the Hellenistic idea of a Great Chain of Being and speculated on how different species were related to one another, although they never came up with a truly evolutionary explanation. In a cultural climate that favors ideas that seem more Islamically authentic, such historical connections can make guided evolution a more attractive view. In any case, efforts to find a suitable compromise between Darwinian evolution and strict creationism are common. If the religious experimentation going on in the Muslim world today ever develops a more liberal trend that can appeal to more than a westernized elite, guided evolution will become a more popular option.

Many influential Muslim intellectuals also remain skeptical of Darwinian, naturalistic evolution while avoiding Quranic literalism and strict creationism. Their main concern appears to be that God should continue to directly shape the observable world, and that the common Muslim intuition that the world is obviously a product of divine design should be sustained. Prominent examples are the views promoted by figures such as Seyyed Hossein Nasr and Osman Bakar, well known academics who propound a specifically Islamic philosophy of science. Both Bakar and Nasr reject Darwinian evolution, though they allow for limited, non-creative changes in species over time. Like popular creationists, they charge evolutionary theory with being a materialist philosophy rather than a true empirically-based science. Unlike Harun Yahya and similar creationists, however, they try and contrast the Darwinian view of life with a more substantial perspective than simple acceptance of a revealed text. Bakar and Nasr want to revive the classical, God-centered conception of knowledge in Islam. Revelation and the Islamic religious sciences are supposed to be restored their position of preeminence, providing the framework for all knowledge claims including investigations of the natural world (Bakar, 1987, 1999; Nasr, 1987).

With proposals to "Islamize science" or otherwise reconstruct modern knowledge in a more Islamic fashion attracting much attention in Muslim academic and intellectual circles, Bakar and Nasr's views continue to resonate among Muslim thinkers concerned about science and Islam. Indeed, there is an important difference between Western and Muslim intellectual cultures in regard to evolution. In the scientifically advanced West, there is very little creationism in the academic and mainstream intellectual environments, and even broader anti-Darwinian views tend to be muted. The Muslim intellectual world is much more hospitable to ideas hostile to Darwinian evolution. The view that complex structures must be the result of intelligent design remains deeply embedded in Muslim intellectual culture (Edis, 2004).

Therefore, it is not surprising that the latest version of antievolutionary thought in the United States, promoted by the "intelligent design" movement, has begun to attract attention in the Muslim world. In Turkey, where the debate over evolution has been most public, the major books defending intelligent design have been translated and have been favorably reviewed in the Islamic press. Indeed, Turkey has begun to develop its own intelligent design proponents (Akyol, 2005). The intelligent design approach, which ignores questions about scripture or the age of the earth and concentrates on claiming that Darwinian processes cannot create the information-rich structures seen in biology, fits well with common Muslim intuitions. It is likely to continue to receive sympathetic attention from devout Muslims (Edis, 2007).

4. Defending Evolution

In Muslim countries, as in the United States, scientists and science educators are the groups most interested in promoting a better understanding of evolutionary science. And in both cases, the debate is complicated not only by the conservative religiosity of much of the general population but by the overall political context.

In Turkey, scientists are naturally concerned about Harun Yahya's creationism. And they have made some halting efforts to respond; for example, by producing popular scientific literature explaining and defending evolution. The strongly religious nature of creationism, however, means that aside from scientists, the natural constituency in favor of evolution is secularists. Portraying attacks on evolution as further attempts to erode the secular nature of the Turkish Republic is the best way to appeal to a wider audience and to give the issue some urgency (Sayın and Kence, 1999). But such a strategy might not help make the case for evolution in the long run, since secular elites in Turkey no longer enjoy as privileged a position as they used to. As religiously conservative and provincial populations begin to enjoy more economic and political power, their cultural demands become harder to ignore or suppress.

Similar currents affect other Muslim countries. Increasing tendencies toward political democracy means that the strongest political movements in Muslim countries will have a moderate Islamist identity. Religious populist movements typically attach great importance to symbolic and cultural politics, and therefore creationism will likely enjoy a significant advantage. As Muslim populations continue to modernize, and as varieties of Islam adapt to new circumstances, the Muslim world will continue to be a stage for a struggle between evolution and creation. Popular creationism will remain strong for the foreseeable future, and its main competition will come from superficial views of guided evolution that will often consider humanity to be a special creation outside of the natural biological order. Darwinian evolution will continue to be affirmed by a small group among scientists and intellectuals, but even the general intellectual culture in Muslim lands will most likely remain cool toward evolution.

References

Akyol, M. (2005). Intelligent Design could be a bridge between civilizations. *National Review Online*, December 2.

Altaytas, M. (2001). *Hangi Din?* İstanbul: Eylül Yayınları.

Atay, T. (2004). *Din Hayattan Çıkar: Antropolojik Denemeler.* İstanbul: İletişim Yayınları, pp. 136-37.

Ateş, S. (1991). *Gerçek Din Bu,* Volume 1. İstanbul: Yeni Ufuklar Neşriyat.

Bakar, O. (1987). *Critique of Evolutionary Theory: A Collection of Essays.* Kuala Lumpur: The Islamic Academy of Science.

Bakar, O. (1999). *The History and Philosophy of Islamic Science.* Cambridge, UK: The Islamic Texts Society.

Bayrakdar, M. (1987). *İslam'da Evrimci Yaradılış Teorisi.* İstanbul: İnsan Yayınları.

Bezirgan, N. A. (1988). The Islamic world, In: T. F. Glick (ed.) *The Comparative Reception of Darwinism.* Chicago: The University of Chicago Press.

Edis, T. (1994). Islamic creationism in Turkey. Creation/Evolution **34**: 1-12.

Edis, T. (1999). Cloning creationism in Turkey. Reports of the National Center for Science Education **19(6)**: 30-35.

Edis, T. (2003a). Harun Yahya and Islamic creationism, In: A. Chesworth et al. (eds.) *Darwin Day Collection One.* Albuquerque, NM: Tangled Bank.

Edis, T. (2003b). A world designed by God: science and creationism in contemporary Islam, In: P. Kurtz (ed.) *Science and Religion: Are They Compatible?* Amherst, NY: Prometheus Books.

Edis, T. (2004). Grand themes, narrow constituency, In: M. Young and T. Edis (eds.) *Why Intelligent Design Fails: A Scientific Critique of the New Creationism.* New Brunswick, NJ: Rutgers University Press.

Edis, T. (2007). *An Illusion of Harmony: Science and Religion in Islam.* Amherst, NY: Prometheus Books.

Nasr, S.H. (1987). *Knowledge and the Sacred.* Albany: State University of New York Press.

Sayın, Ü and Kence, A. (1999). Islamic scientific creationism. Reports of the National Center for Science Education **19(6)**: 18-20, 25-29.

Scott, E.C. (2005). *Evolution vs. Creationism: An Introduction.* Berkeley: University of California Press.

Yahya, H. (1997). *Evrim Aldatmacası: Evrim Teorisi'nin Bilimsel Çöküşü ve Teorinin İdeolojik Arka Planı.* İstanbul: Vural Yayıncılık.

Yahya, H. (2002). *Islam Denounces Terrorism*, 3rd edition. Bristol, UK: Amal Press.

Ziadat, A.A. (1986). *Western Science in the Arab World: The Impact of Darwinism.* London: McMillan.

Dialogue

Peter Bahn

I read your chapter for this book and I liked it very much because it emhasizes the importance of teaching a scientific view of evolution in the modern Islamic country of Turkey.

The questions that I was curious about as I read your chapter were these:

1. How did Kemal Atatürk, the Father of Modern Turkey view Darwinian evolution?
2. I assume that Turkey is predominently of the Sunni branch of Islam. However, I wondered: Are there any differences in the how Sunnis versus Shiites view Darwinian evolution?

Taner Edis

My answers to the questions:

1. Atatürk was known to be interested in Darwinian evolution during his years as a student in the Ottoman War Academy. And after the 1920's, when he and like-minded westernizers took power in Turkey, they westernized the curriculum as well, including topics such as evolution.
2. Turkey is predominantly Sunni, with perhaps 20% of the population (estimates vary) belonging to the Alevi sect. Alevis have some Shiite affinities, but they are in many ways a very liberal sect that have little in common with mainstream Shiite belief. I don't know of mainstream Shia such as those in Iran being all that different from Sunni intellectuals in terms of their response to Darwinian evolution. I would be surprised if there was much of difference.

Oktar Babuna

Below you may find my views on Taner Edis' article titled "Muslim Resistance to Darwinian Evolution," which has been brought to my attention as a referee. Before answering the false accusations made in the article, it is useful to remind about the well-known Muslim intellectual identity and works of Harun Yahya.

As is known, works of Harun Yahya enjoy a wide readership in many countries, from India to America, England to Indonesia, Poland to Bosnia, Spain to Brazil, Malaysia to Italy, and France to Russia. Besides many of his books refuting the theory of evolution by means of scientific evidence, many others on such subjects as faith in God, facts of creation, the destruction caused by irreligious ideologies on the world are translated into 57 languages and greatly appreciated all around the world. His books' wise, genuine and sincere style that's easy to understand, together with the rational and scientific approach, directly affect anyone who reads them. Web sites based on the author's works, the number of which is more than a hundred, are visited everyday by tens of thousands of people.

These works are characterized by conclusiveness and irrefutability. This is why those who seriously consider these books, can no longer truly advocate materialism, atheism or any other perverted ideology or philosophy. All godless movements of our age are intellectually defeated by Harun Yahya's collection for a total of more than 45,000 pages.

However, some atheist and communist circles still argue for their philosophies out of emotional stubbornness. It is an outcome of this defeat that they consider Harun Yahya's success as "worrisome" and try to slander it by means of groundless claims. Facts about the false accusations made in the article by Taner Edis are as follows:

Adnan Oktar, who authors his books under the pen name Harun Yahya, is certainly not a leader of a sect, as opposed to what is claimed. He is the "Honorary President" of Bilim Arastirma Vakfi (Science Research Foundation) and Milli Degerleri Koruma Vakfi (Foundation on the Preservation of National Values). These foundations are faithful followers of ideas and principles of Atatürk—the founder of the Republic of Turkey—and the organizer of over 1,500 national and international conferences. They make use of Mr. Adnan Oktar's broad range of works in each and every service they perform.

The claim that Mr. Oktar does not have an academic background is false. Mr. Oktar studied at Mimar Sinan University's Academy of Fine Arts and at the Philosophy Department of Istanbul University, two of Turkey's established institutions. It is acknowledged by hundreds of thousands of readers that his books are written in accordance with contemporary scientific advances, and ideas presented in these books are supported by totally scientific evidence.

Besides, if an author is a researcher, he or she does not necessarily have to be academic. What matters is the accuracy of what the author writes. That one is from an academic circle does not mean that he or she tells the truth. The same applies to the contrary; that one is not from an academic circle does not mean that what he or she tells is not true. Still, books by Mr. Oktar are regarded as true by academic circles as well.

It needs mention to say that Mr. Adnan Oktar's works are not the type of works that can be prepared solely on academic information. His works require Divine gift, knowledge from Allah's sight. If they could be accomplished through academic information, it would have by now been done in many universities throughout the world. Thanks to this distinction of this special knowledge, Mr. Oktar's works affect people so deeply. Indeed, had 50 scientists—each one of them graduated from more than one university—come together, they would still not be able to produce this series of works, for the effect of these works cannot be acquired through an academic career. Allah's special guidance and power of influence is clearly seen in Mr. Oktar's works. That these works become a means to inspire faith in millions of people and make them feel excitement of faith are their distinctive features. The fact that these works have such a high power of inculcation and persuasion is the main factor why atheist, Darwinist and materialist circles are so uneasy about them.

The claim that books by Harun Yahya are written by more than a single person is nothing but an aspersion. Mr. Oktar has dedicated all his life to these works. His books are the product of research and study he has been engaged in since his university years, and an effort of more than 30 years. Furthermore, these books are before the eyes of the millions; their official printing house and contracts are indisputable. The fact that no other person pretends to the authorship of these books reveals how baseless this accusation is. This applies to the books of any other person: so long as there is no evidence against, the author of a book is the person who has his or her name on it. Besides, Mr. Oktar has assistants helping him with translation, research of foreign material, layout and proofreading during preparation and publication of the books. He works together with his technical team, and does not assert otherwise on any occasion. Yet, after all, it is him determining the research subject and images, interpreting, planning and eventually putting them into writing.

It is true that creationist Christian scientists are used as source material for the books. However, a great majority of the quotations are from scientists who hold evolutionary views, admitting that they cannot come up with evolutionary explanations for various matters. Therefore this claim is a misleading account put forward merely by demagogic motives. Besides, it's important to pick the accurate information and discard the wrong. As a matter of fact, regardless of religion, school of thought, or belief, Mr. Oktar criticizes wrong accounts and makes use of accurate ones in his books. Moreover, statements of scientists are cited sticking to the original and are referenced, therefore there is no use of quotes out of their context.

Mr. Oktar provides evidence for all the accusations of forgery he brings against evolutionist scientists. One should encourage, not be disturbed by, disclosing those who went down in scientific literature as forgers or frauds that were exposed in the subsequent years, and letting people know of these realities by means of books.

The article states that Harun Yahya "attacks evolution." In his books, Mr. Oktar scientifically criticizes and proves the theory of evolution false and invalid. This is not an attack, but a scientific and intellectual effort. Mr. Oktar's strong writing, and respected and upright style of good standing are distinctive. It is now accepted by the entire world that Adnan Oktar's works play a major role in the realization of the evolutionary theory as a "scientific forgery." The impact of his works have been analyzed in various prestigious publications:

November 1999 issue of the *Reports of the National Center for Science Education*, or *RNCSE*, devoted its cover and 30 pages to ***The Evolution Deceit*** by Harun Yahya and his works refuting the theory of evolution and revealing the facts of creation.

New Scientist noted in an article called "Burning Darwin" in its 22 April 2000 issue that the author's books played an important role in the anti-evolutionary intellectual campaign globally: ***"Harun Yahya is an international hero. His books have spread everywhere in the Islamic world.*** *Science*, a leading periodical of the scientific community, observed in the article "Creationism Takes Root Where Europe, Asia Meet," dated 18 May 2001, that ***Harun Yahya's books have become more influential than textbooks in many places***.

In an article posted on http://www.pitch.com, a US based web site, evolutionist professor Umit Sayin admitted the intellectual defeat of evolutionists, stating that ***"There is no fight against the creationists now. They have won the war...***

In 1998, I was able to motivate six members of the Turkish Academy of Sciences to speak out against the creationist movement. Today, it's impossible to motivate anyone." The article went on to refer to Turkey as a country where proponents of the theory of evolution were almost completely defeated. It was also stated in the article that the Science Research Foundation ***"sparked a revolution in its own country and is now so successful that it's been asked to send an emissary"*** to Kansas to help the American creationists.

The article asserts that Harun Yahya denies that the Holocaust happened. The author's book *The Holocaust Violence*, in which he describes the Holocaust as "violence," is the best answer to this groundless claim. If requested, there is an official answer to such claims given through a court of law.

Mr. Oktar's aiming to unite Divine religions against irreligion and suggesting that terror and violence has no place in true Islam should be commended. It is simply not understandable why defending peace against war, announcing that true Islam is the religion of peace, and encouraging a peaceful, tolerant and understanding concept of Islam bothers one and why these are portrayed as bad. This, in fact, proves Mr. Oktar right in his correlating the real foundation of terrorism with the evolutionist mindset.

The article in question mentions, in a criticizing tone, that the author praises Bediuzzaman Said Nursi in his books. It is most natural to praise a prominent 20th-century Islamic scholar whose writings became a means whereby millions of people gained deeper insights into their faith. To find this approach odd can result merely from failure to appreciate Said Nursi's works and holding a materialist prejudice.

It is claimed that the author is engaged in theological controversy by suggesting that matter is only an illusion and what is experienced in mind is real. It should be noted before all else that the "secret beyond matter," as the author discusses in several books, is a scientific explanation taught at primary and high schools today. It is a technical piece of information studied even in the fundamental biology textbooks that all external images are perceived in the brain. Contrary to what is claimed, the author does not maintain that matter does not exist, but states that the matter existing outside is perceived in the brain as an illusion. In his discussion of this topic, he makes use of the works of respectable scientists. On the other hand, references to noted Islamic scholars and their views on the subject is evidence that these ideas do not constitute a matter of Islamic dispute.

The author is right when he declares that opposing the theory of evolution is an urgent moral duty, and that this theory forms the basis for the materialist philosophy as well as such ideologies as communism and fascism. He presents his argument in his books with proofs and past instances. The most striking accounts on the subject are in his books *The Disasters Darwinism Brought to Humanity* and *Darwinism's Social Weapon*. These are truths that all readers would acknowledge to be right.

It is suggested in the article that Harun Yahya's books are given away for free and that the financial resources of the Harun Yahya operation are unclear. First of all, the so-called given away works are no-cost materials which anyone can download from the Internet merely by pressing a couple of buttons. The books of the author, on the other hand, sell in millions all around the world. Conferences, which are referred to as the Harun Yahya operation, are legal activities to which Science Research Foundation members are invited as guest speakers.

It is a grave error how the article portrays attacks on evolution as attempts to weaken the secular nature of the Turkish Republic. On the contrary, through Mr. Oktar's anti-evolutionary works, it is aimed to protect the secularity of the Republic, and how Darwin's evolutionary claims pose a threat to the Turkish State is revealed in his *Darwin's Hostility to Turks* with proofs.

Moreover, it is significant that not a single scientific evidence that could be a proof for the theory of evolution is presented in Edis' article. Apparently, the entire content of the article is characterized by hostility to religion and a blind advocacy of evolution. Unfounded accusations against Muslims, Christians and Jews—all of whom are members of the three Abrahamic religions—and a style unobservant of the reverence due to our holy book Qur'an and Almighty Lord Allah will cast a shadow over the value and quality of your publication. The article, which is by no means scientific, contains accusations and insults without any proof or document.

Taner Edis
I clearly lack the "Divine gift" to guarantee that what I say is accurate. I should, however, clarify one point. In the late 1990's, the Harun Yahya corpus included a book called *Soykırım Yalanı,* or *The Holocaust Hoax* in English translation. This was derived from American holocaust denial literature. This book soon

became an embarrassment and vanished from Yahya catalogues and websites, to be replaced with *Soykırım Vahseti*, or *The Holocaust Violence*, which acknowledges the holocaust and blames Nazi policy on evolution.

Biography

Dr. Josef Svoboda is a Professor Emeritus at the Department of Biology, University of Toronto at Mississauga, Canada. He studied Philosophy and Biology in Czechoslovakia but in 1949 he was arrested and sentenced by the communistic regime. He spent almost 9 years in prisons and uranium mines' labour camps. After immigrating to Canada in 1968, he graduated at the University of Western Ontario, Canada. In 1974 he obtained his PhD from the University of Alberta, Canada for a study of productivity processes in polar semi-deserts. For more than 30 years, together with his students, he conducted research in the High Arctic focusing on plant colonization following deglaciation of polar landscapes and responses of arctic plant communities to global warming. He also published papers on the origin of life and evolution. Honours: Northern Science Award by the Government of Canada and Honorary doctorate from the Masaryk University, the Czech Republic.

27

To Creation *Via* Evolution?

Creationists' About-Face or a Belated Insight of Reflective Evolutionists?

JOSEF SVOBODA

Department of Biology, University of Toronto at Mississauga
Mississauga, Ontario, L5L 1C6, Canada
Josef.Svoboda@utoronto.ca

1. A preliminary

How to make a meaningful contribution to a discussion where positions on both sides are deeply entrenched and communication between the belligerent sides is almost non-existent? The situation reminds me of a movie about the Western Allies invasion into Nazi-occupied Europe. When in a certain combat development a German general demanded the surrender of a contingent of the invading forces under German siege, the allied command replied by a single word: "Nuts!" No discussion, just a digging deeper into the trenches.

The following book titles demonstrate the uncompromising polarization of the **Divine Action and Natural Selection** camps: *The Science of God* (Schroeder, 1997), *God and the Cosmologists* (Jaki, 1998), *The Case for a Creator* (Strobel, 2004) — with *The Blind Watchmaker* (Dawkins, 1991), *Creationism's Trojan Horse* (Forest & Gross, 2004), *The Counter-Creationism Handbook* (Isaak, 2005), to name just a few, on the other side. All of these are eloquent treatises, written by esteemed experts in their field, published by first-rate publishers, yet standing immutably opposed to each other. One wonders how it is possible!

2. In the Beginning

A new-born baby, which Bronowski (1976) calls "a mosaic of animal and angel", learns first to distinguish between itself and the external world. Similarly, the primeval hominid learned *subconsciously* to distance and to dissociate its own kind from the surrounding "wilderness". Our early ancestors were well aware of the various kinds of living beings — from the most simple (edible worms, insects) to the most advanced (dog, horse, monkey) to their knowledge — and judged them as to their usefulness or threat. Significantly, they considered themselves a different breed, superior to all other creatures. This radical separation between humans and the rest of the animated world is common to all ancient cultures, and geographically from the Pigmies in Africa to the Eskimo in the Arctic.

When this fast evolving hominid became fully conscious of this quintessential distinction, he became fully human — *Homo sapiens sapiens* we call him in retrospect. It is still held by many that this self-realization, that is being conscious of his consciousness, occurred through a moment of sudden awakening. A more plausible explanation, however, suggests that the state of self-consciousness had been achieved through successional stages, involving brain and language development, over a longer period of time. It culminated some 50,000 years ago in the "mind's Big Bang" event (Calvin, 2004) which catapulted humanity into "cultural evolution".

> "It took at least two million years for man to change from the little dark creature with the stone in hand … to the modern form *H. sapiens* … But it has taken much less than twenty thousand years for him to become the creatures that you and I aspire to be: artists and scientists, city builders and planners for the future…" (Bronowski, 1976).

The author sees the pace of cultural evolution/revolution to be hundred times faster than biological evolution. Kurzweil (2005) says it even more succintly:

> "It took two billion years from the origin of life to cells; fourteen years from the PC to the World Wide Web."

Cultural evolution got into an exponential spin during the recent scientific and technological era which became the modern flagship of human progress.

3. The Divine was First

"Divine" voices talked to primeval humans long before they became conscious of them and started asking questions about God's existence. Gods and demons, and belief in them, came first. More precisely, "the divine" was not merely believed, it was real and present, entrenched in the depth of consciousness (Jaynes, 1990).

People, like other intelligent living beings, are curious creatures. Our ancient ancestors had been exploring the world around them in meticulous detail and at a certain stage began asking abstract questions. The first sentence of Aristotle's Metaphysics states: "All people, based on their very nature desire to know...", and this regardless of practical application or apart from personal gain. In every culture there have been individuals who wondered and asked impractical, imaginative questions. Aristotle's queen of sciences, *Sophia* (wisdom) was primarily concerned with questions of "causes" and "origins".

Not surprisingly, the *Bible* starts with an answer to the ultimate, and clearly burning question, also that of origins: "In the beginning God created heaven and earth..." (Genesis 1,1). For millennia this scenario of creation was accepted word for word. Similar references in other sacred texts and oral traditions (e.g. Hindu, native Africans and Americans, etc.) illustrate that questions of origins (*cosmogonies*) and also of the individual and his social conduct, of death and afterlife, have been a matter of fundamental importance. From the beginning humanity embraced the creationist's explanation of the world and its own birth. That deeply rooted frame of mind cannot be referred to as mere belief. It formed an intrinsic component of the human intellectual and spiritual makeup which has held firm until its crisis in modern times. Joseph Campbell (1991) in his acclaimed book *The Power of Myth* demonstrates the role of mythology in the formation of human social consciousness, and as a reference base from which to judge and assess the world. Naturally, these mythological roots were in the "divine".

4. From Divine to Human

In Europe this period of cultural transition from rootedness in the divine to the human is known as the Renaissance, and the expression of its specific worldview as Humanism. Not God *per se* anymore, but Man emerges as the measure of truth and values. Everything was laid on the table to be questioned anew.

In pre-Darwinian times, a great many people, including biologists, had journeyed to exotic parts of the world and gathered plentiful information on floral and faunal abundance, and anthropological diversity. Yet it took Darwin, a keen observant naturalist, to realize that there are connections between closely related species (e.g. Darwinian finches) and, at a greater evolutionary distance also between less closely related groups such as monkeys, apes and Man. After much searching and hesitation Darwin formulated a revolutionary, but very plausible theory of evolution which attempted to explain the *con-generic* relationship of all living forms.

Apples were always falling down from the tree and no-one wondered about it. Yet it required the bliss of genius to ask *why* do they fall down? The answer, leading to the formulation of the laws of gravitation continues to be mind-boggling and far-reaching in its consequences. By proposing that higher organisms have *evolved* from their lower forms, Darwin (with due credit to his fore-runners) just stated, what *now* seems to be glaringly evident, yet with repercussions equivalent to greatest discoveries of Newton, Einstein and the like.

5. Clash of Ideas

As could be expected, Darwin's idea challenged the old biblical belief that in the beginning God created all living things in their variety and diversity, albeit in stages, "days". The reaction was vicious and forceful, exemplified by the famous 'Oxford meeting of British Association' in June 1860. At this gathering summoned to defy Darwin's "blasphemous" theory of the origin of species, men of religion and science clashed in a hardly academic style. Bishop Samuel Wilberforce represented the traditional belief while T.H. Huxley and J. Hooker argued in defence of Darwin's theory. Darwin was absent due to illness (Moorhead 1978). There were deeply hurt feelings but no winners at the meeting or in the many subsequent ardent disputes over the issue during the following 150 years.

As more convincing evidence in favour of evolution has accumulated, Darwin's revolutionary hypothesis became generally adopted as the only plausible theory.

No biologist in good standing would dare to challenge the basic idea of biological evolution any more. This fact may have prompted Chaisson (2001) to remark:

> "The power of this quintessential Darwinian concept is so pervasive in biology and related fields that many so called ultra-Darwinists subscribe to it wholly and without limit, indeed with a kind of fervor matched only by the fundamentalists whom they abhor."

6. Big Bang and Before?

Fundamental questions of origins aim at events prior to biological evolution, even prior to the origin of life. As to how the world itself came into being, or is the universe eternal?

The discovery of the "spectral red shift" by Hubble in mid-1920's which led to the formulation of an astounding Big Bang (B-B) hypothesis seemed to provide the answer: the universe exploded into being 13.7 billion years ago (Lemonick, 2003). Although, humanly speaking, such an initial scenario is hard to imagine, the concept is still within the realm of our qualified educated comprehension.

As if the idea, that the visible universe and its even larger invisible components (dark energy, dark matter) were once cramped in an infinitesimally small volume, has not been mind-boggling enough, a new question emerged: What was before Big Bang? To this logical question Stephan Hawking was heard to reply: "This is, as if you asked what is North of the North Pole?!" There was no 'before'. Time itself was born, and the reality we call the cosmos or the universe arose from nothing at the point of B-B.

Hawking doesn't subscribe to any pre-Big Bang existence. He maintains that our universe is completely self-contained, has no boundary, is self-regenerating and eternal (White and Gribbin, 1992). He is skeptical that constructing mathematical models would ever answer the question why the universe exists and whether it needs a creator. "Who created the creator?" (Hawking and Mlodinow, 2005).

For others, however, accepting the Big Bang as an ultimate and universal starting point of the extant physical reality, its time and space, would herald an end for further scientific inquiry. To avoid the dead-end trap, some theoretical physicists started experimenting with a fresh approach.

Alice, in a pursuit of a White Rabbit dove into a rabbit hole and emerged — in Wonderland. Similarly these cosmic magicians rammed through the Big Bang "STOP" signpost fashioning a *virtual window* into a *virtual reality*. Vistas to new, time-exempt horizons charmingly opened and a world of 'multiple' and 'parallel' universes emerged.

Theoretical physicists like to resort to substitutes for real things. In quantum field theory, 'virtual' particles (e.g. virtual photon, antiproton, virtual particle pairs: electron and positron) play an essential role, although "they don't have an independent existence as actual particles" (Penrose 2005).

Ideas, such as the various versions of parallel universes and self-reproducing inflationary models have become favored themes of many theoretical physicists. These theories are derived from advanced principles of quantum mechanics, especially from the surreal, not yet fully understood M-theory of a super-cosmic membrane of 11 dimensions (Seife, 2003). The seemingly absurd question "what is North of the North Pole?" lingers in their collective consciousness as the background radiation lingers in the cosmos around us. However, from the heuristic attitude that 'no stone be left unturned' the search is quite legitimate. It is an expression of the unceasing intellectual drive, so characteristic of the inquisitive human mind.

The meaning of these new theories, although mathematically brilliant and elegant, is even less comprehensible to an untrained mind, and at this stage cannot be supported by any evidence. Yet, for an equation "to have beauty" is more important "than to fit experiment" (Dirac, 1963, Farmelo, 2002). "It (the equation) gave me just the properties one needed for an electron. That was really an unexpected bonus for me." (Dirac cited in Wilczek, 2002). To be elegant is more appealing than to be "true" in an applied sense. Interpretation in physical terms may come later — or may never come, which is, with respect to the validity of the equation, not essential. Dickerson's overriding rule demands that scientists work "to explain the behavior of the physical and material universe in terms of purely physical and material causes, without invoking the supernatural" (Dickerson, 1992, cited from Behe, 1996). The scientist is not barred from believing in the supernatural, only to operating with it as a causal factor in the realm of his discipline.

These daring theories are trying to advance our understanding of our universe's permanence by defining the underlying condition extant *prior* to Big Bang, and thus bypassing the "out of nothing quandary". No time dimension is needed or applicable here. The sublime superstring matrix, out of which universes pop up randomly, is eternal. There is or was no absolute *beginning*!

These conclusions, however, are conditional upon the degree of certainty that the Big Bang did really occur the way our models describe it. At present, the B-B theory is favoured by most cosmologists and the evidence for it is growing (Hawking, 2001, Singh, 2004). Is this a final word on the origin of our Cosmos or will there be, in the weird world of quantum mechanics, another yet unanticipated explanation for its existence? Already we've got used to thinking about the relativity of time and space, about the super-dense matter of black holes, the simultaneous existence and non-existence of particles. Nevertheless, we shall be reminded that the science of cosmology is still in its turbulent stage. New theories will be popping up, as the parallel and bubble universes are popping up out of Guth's (1997) inflationary ideas now (see also Seife, 2003).

The searchers are well aware of the paralyzing uncertainty principle, considered a universal rule. It precludes any definitive statement about the state of the universe, even about its existence and non-existence. Paradoxically, "nature itself must obey the dictates of Heisenberg's principle" (Seife, 2003).

7. Yet, Parallel Universes Do Exist

It depends on whom do one asks. The above mentioned theories do not, and cannot, address the crucial question of the all-encompassing universe's *existential* origin. If not when (B-B has been fairly well dated) then *how* "physical reality", the ultimate matrix, substance or essence came into being? What is the *nature* of the ultimate *essence* of physical reality? The idea of the *First Mover* sparks up in every deeper debate but is immediately extinguished as non-scientific by science puritans. *Vis major*, the Almighty "maker of all that is *seen and unseen*" (Nicene Creed) may surface as a question but, as a rule, cannot be the answer! God is a not an acceptable hypothesis in science. There is the *physical* universe we see, live in and are preoccupied with. The other universe which we may not see but some are able to experience and communicate with is *spiritual*. Mystics would argue that the spiritual universe is more real and less "virtual" than are the hypothetical multi-verses of the mathematicians. But how could its existence be proven? Science is a respectable discipline, but there are other reputable ways to probe reality. While the physical universe is the exclusive domain of the logic supported by empirical verification, the other may be more a path of pure intellect, intuition and heart. Both roads have separate objectives and limitations and may lead to specific discoveries. They are non-competitive and may be equally convincing to their adherents and users.

8. To Creation *via* Evolution

Curiously, something unexpected has happened. Over the span of the last few decades, the more judicious members of the creationist flock switched their strategy of defence and positioned themselves at the very onset of the evolutionary process. The theory of evolution began to make sense to them. However, they argue that the evolutionary processes involved and the resulting physical and enlivened structures of the universe are too complex to emerge by chance in the time available since the universe's conception. Thus the divine factor, freshly renamed "Intelligent Design", has been folded-in as an inevitable prime cause and premise. It was a gradual but a complete about-face: from the initial scorn for any form of evolution, to condoning a micro-evolution here and there, to a wholesale embrace of macro-evolution from Big Bang to Man!

9. The Question of *How* Remains

Since Darwin "there has been an overemphasis on the power of selection as opposed to the generation of diversity" (Beardslay, 1997). Reservations have been raised about the evolutionary mechanism, in particular about the inadequacy of the chance process to *facilitate* species diversification (speciation) and mainly about its power of *fast-forwarding* the true evolutionary advancement. In other words, questions as to *how* organisms evolve have not been convincingly resolved.

Darwin and Wallace proposed "natural selection of fitter populations" as an engine of the evolutionary progression and this driving force is clearly involved. But the skeptic may see the role of natural selection in its rather destructive role, primarily as a clearing house, "a pruning device" (Chaisson 2001) for less competent or competitive populations, even entire species. It was found less proficient, rather incapable, as a working mechanism for more expeditious evolutionary advancement (Davies, 1999; Dembski, 2002).

For a selection to take place, a pool of suitable alternatives *to select from* is needed. Darwin believed that some individuals during their lifespan acquire new characteristics which, if successful, may survive, perpetuate and help establish a new, more advanced generation. He had in mind changes in a phenotype ("usage or non-usage" of a particular organ or faculty) resulting in hereditary fixing. Darwin was not aware of the parallel work of Mendel, nothing was known about genes yet. Luckily, the discovery of spontaneous gene mutability became a welcome generator of modified genetic material for natural selection and fresh ammunition to neo-Darwinists against the Creationists. This awesome heuristic process was crowned with the unravelling of the DNA code by Watson and Crick in 1953. Thus, a new transparent source for natural selection at the very molecular level was identified. The genetic pool of every living individual overlaps in a great number of genes, yet no single genome is absolutely identical with that of anyone else.

10. Many Still Beg to Differ

At present an entire spectrum of views and positions can be identified. From non-compromising biblical fundamentalists taking the story of creation literally — to a guarded pronouncement of Pope John Paul II (1997) on the compatibility of evolutionary theory with belief in God the Creator, — to a broad embrace of 'cosmic evolution' from Alpha to Omega by Teilhard de Chardin (1959), — from emphatic advocates of Intelligent Design such as Johnson (1993, 1995), Behe (1996) and Dembski (2002), — to proponents of the Anthropic principle (Carr and Rees, 1979) — to condoners or at least sympathizers of appearances of *purpose* or *directionality* in nature (Einstein: "God doesn't play dice"; Kauffman's (1995) principles of self-organization; Denton's (1998) "directed evolution", Davies' (1999, p122, 271) "bio-friendly and mind-friendly" universe), — to strict neo-Darwinists and ideologically motivated adherents of natural selection as the only organizing agent in living nature, such as Dawkins (1991), Dennett (1995), de Duve (2002) and many others.

11. Intelligent Design and God's Hypothesis

"Intelligent Design" (ID) is a modern buzzword — with a unique meaning specifically applicable to a *directed* evolution, whereby the entire evolutionary process is not aimless, as if propelled by "blind-law" forces (Ruse, 2003). Evolution is, therefore, a managed and directed course of actions from the initially "simple" to the present "complex". There is a common denominator for authors using this term: Intelligent Design refers to a universal blueprint (and its execution in time) underlying all non-living and living systems.

The Big Bang event released enormous energy and created conditions in which a small proportion of the free energy condensed into rudimental matter. In a micro-instant after the birth of the universe, the process of sub-nuclear *self-structuring* started and nucleosynthesis was accomplished in minutes. The building of atoms and molecules continued, followed by the agglomeration of galaxies with stars and their planetary systems, all this, however, at a rapidly decreasing rate. Several billion years of the universe existence, under extremely rare circumstances, likely on some "life-friendly" celestial bodies, definitely on our planet Earth, the self-organization proceeded up to emergence of the first living cell. (Let's omit here the controversial issue of natural spontaneity or divine intervention in life's emergence). At this stage, 4-5 billion years ago, the universe had already reached a state of relative stability, characterized by a very slow rate of change but with a new accelerated rate of expansion. Curiously, the disproportionately tiny allotment of the enlivened matter began to march in an anti-entropic direction through biological evolution. Most interestingly, in contrast with the explosive speed of events following the B-B, the unfolding of the biological world had been very slow for almost 4 billion years. However, after yet unexplained pre-Cambrian biological explosion of new life forms (~ 550 Mil. y. ago) the rate of self-organization suddenly picked up speed. Since then biological evolution progressed exponentially (although being "punctuated" by catastrophic events) through many stages of complexification, crossing boundaries towards intelligence and rationality. In retrospect, directionality, if not finality of the evolutionary trajectory, is evident. The biocosmos is a self-propelling, self-propagating super-system with no parallel in the inanimate world (Svoboda, 2006) and its evolution will likely proceed through us beyond what we are now (Kurzweil, 2005).

The idea of *self-organization*, advanced by Ashby (1947) and later Kauffman (1995 and 2000), has many admirers but also avid opponents. To some

Kauffman's use of this terminology has a too teleological odour but the author makes no apology. Yet, being a declared agnostic, he does not subscribe to the Designer idea either.

In cosmology and bio-evolution, ID means that the world was envisaged and brought to existence by a higher intelligence, which, to most authors, is synonymous with the "Divine". Design, especially of cosmic dimensions without a Designer, would be a contradiction. The presupposed Designer exists apart, and is more or less (pending on authors' preferences) independent of his designs. The Designer is not a micro-manager as is the biblical Creator. He operates through natural laws established by Him. Up to a point though, as we shall explain below!

Neo-Darwinists seem to have certain systemic problems with explanations of the sudden emergence of structures or functions, which, the critics say, could not evolve gradually. These are Behe's (1996) *irreducible complexity* features or sudden *evolutionary leap advances* occurring in a relatively short time period. The pre-Cambrian biological Big Bang, the rapid evolution of the human cranium (cerebralization, enabling the rise of consciousness), the structural and functional perfection of certain organelles, or the sophistication of some biochemical systems, are often cited as prime examples. In such cases, some of the ID proponents do not hesitate to call for help to the supernatural as to a direct intervener — to the "God of the gaps".

However, implying the divine as a rational postulate and a *virtual non-natural agent* beyond reach and proof is to science purists a crossing of the no–no line which prohibits invoking of the supernatural (cf. Dickerson, 1992). In this respect, however, some opponents of ID seem to adhere to a double standard in judging what is and what is not 'off-side'. They also propose theories short of invoking the supernatural. Penrose (2005) has strong words for Guth's inflationary theory:

> "What reason is there to believe that such an inflationary picture of the universe is likely to be close to the truth? Despite its evident popularity ... I believe that there are powerful reasons for doubting the very basis of inflationary cosmology..."

Penrose (2005) is optimistic about the eventual resolution of problems "which are still beyond our present-day scientific understanding". Yet he has strong reservations as to a 'solution' which "provides a powerful driving force behind the inflationary position". In other words, he has a reservation about ideas, which may seem to be driven by other than purely scientific objectives. The problem of Origins is of such gravity, touching many people intellectually and emotionally, that it should be approached with utmost respect and sensitivity.

The "God hypothesis" of philosophers and the cosmologists' hypothesis of the eternal universe are both equally valid in their particular realm of competence. However, in the strict sense, they could hardly qualify as true hypotheses, since they cannot be tested or verified. They are both based on mere rational arguments and/or circumstantial 'evidence'. "The most incomprehensible thing about the universe is that it is comprehensible" (Albert Einstein: cited from Singh 2004). No discovery in physics may lead conclusively to a Creator, yet a number of modern cosmologists remain puzzled by the weird quantum mechanical paradoxes and continue to look for explanations beyond the horizon of their discipline (Zajonc, 2004).

For a long time classical philosophy was strongly influenced by the dispassionate Aristotle-Aquinas rationalism. Various causality arguments of that period could be summarized as: since physical reality exists (including me, the subject and observer), the ultimate cause of this objective reality must exist. In Thomistic élégant expression: *Quid non est esse ex essentia pended ab esse ex essentia.* What is not a being by its own quintessence, owes its existence to a being by its own quintessence. Everything, except God, receives its existence. God *is* existence.

Modern philosophers draw their conviction about God through the phenomenon of transcendence, intimate existential experience and the finality of human existence. While *esse ex essentia* fits more a deistic concept (and therefore might be more appealing to authors in natural sciences), the anthropological concept of God is personal, relating to Man and therefore theistic. In contrast to a strict and logical classical rationalism this is a softer psycho-anthropological approach.

Unfortunately, there is no undeniable, irrefutable proof of God's existence. The God of reasoning is merely a hypothesis (Ondok, 1998). Human existence is more a journey of faith than of reason, although the two do not contradict each other. A litmus test of intellectual sincerity: Do philosophers pray to their hypothetical God? If they are religious, they probably do, but the source of their faith comes from revelation not syllogism. God is silent, and requires faith to be believed in. The universal principle of uncertainty prevails.

12. No Side Has the Answer

Clearly, there are difficulties with natural selection *and* with the Intelligent Design, both lacking clear explanation about the operation of mechanisms. The neo-Darwinists' selection mechanism *via* "random mutations" does not unequivocally explain the sudden outcome of the emergence of new features. Neither do we read any convincing clarification of *how* ID manifests itself in the cosmic or biological evolutionary process. What agent does the Designer use at various stages of evolution to explain sudden appearances of new complex mechanisms or organs (cf. Behe's, 1996, "all or nothing" phenomenon)? Is *Deus ex machina* always directly involved in new acts of creation?

Yet there is clearly a challenge to our understanding, and reasoning is all we have as a tool. If we argue from the reductionist's position, our mind craves further explanation. If we allow ourselves the luxury of extrapolation beyond the facts, we may feel guilty or be accused of undue speculation. All these doubts have been here before. Things are in constant motion, *panta rhei*. "You cannot step into the same river twice", declared Heraclites, 25 centuries ago. Nothing is possible to know with certainty. What is said today will be disproved tomorrow. Cratylus, a Heraclitean follower and later critic uttered: "You can't step into it once!" By the time you make a statement, the object has already changed. Poignantly, he ended up only pointing his finger at things (Guthrie, 1974).

A provocative question comes to mind: what might the agnostic cosmologists and the present ID folks say about the universe and physical reality 100, 500, 2500 years from now! *Ignoramus et ignorabimus!?* "We may never know the ultimate reality" (Ruse, 2003). Ruse is analytical without taking sides. He enjoys the discussions of great minds as long as they are civilized.

13. In Summary

It seems that there are no winners in the heuristic game in progress between the combined Cosmogonists-Natural selection team and the team of Intelligent Designers.

Natural science has been unable to explain satisfactorily three basic questions:

1. The Origin of the world. Is it eternal, or final in the past and future?
2. The Origin of Life. How and where?
3. The Origin of Man as a *rational* being. In particular, the mysterious and fast transition (Bronowski, 1976), the "quantum leap" which separated Man from other animals (Reid 1985).

It has been long assumed that everything between the first three pillar questions was explainable by gradual evolutionary processes until Guth (1997) proposed the possibility of "The Inflationary Universe" and Behe (1996) formulated his challenging enigma of irreducible complexity systems. Is the game back to square one?

Divine action proponents base their arguments on the inability of natural science in general, and natural selection in particular, to explain the three fundamental questions stated above. Relying heavily on arguments by *reason alone* they *deduce* that the world is too complex and that it has been developing in a clear and meaningful direction. This, without Supreme Intelligence, is unexplainable and makes no sense. However, in our age of science and demand for constant verification, the weakness of these arguments is in their pure rationality. The era of pure thinking is over. The modern skeptic is impassive to reason alone. He demands "proof". Naturalism in the Academy, as well as popular culture prevails (Johnson, 1995). Yet deductions founded in good logic are not irrational. Also in logic, as in mathematical equations, there is an appreciation for elegance. But as in math, the purely logical conclusion may not always correspond to reality — a good reason to be skeptical!

As in any good game emotions run high, often at the expense of objectivity. The fascinating course of discovery and the contest of opposing interpretations of these discoveries will not finish soon. In fact it will never end. Then there is another interfering factor. Per analogy with the frontiers of physics, *the contest shapes and modifies both, the players and the playing field,* which are changing and transforming during the endless game. The universe is not immune to the power of intelligence; it does and will respond to it. Kurzweil (2005) believes that intelligence is "more powerful than physics" since the "smart matter" can manipulate other matter.

14. Personal Reflection

God cannot be a legitimate concept for natural science, since its scope is narrowly defined to physical structures, forces and causalities but may be a "virtual reality" to philosophers and even to some theologians. God is real to believers, including scientists. God is the answer to the transcendental quest for ultimate causes and the meaning of our existence, and a sustainer of peace and order. God is the terminal harbor of our terrestrial journey and a hopeful fulfillment of our desire for eternal Creator-creature union. Thus, on a personal level, the great aspirations of science and faith are mutually compatible, as long as a clear distinction between them is maintained and respected. Many great scientists have been believers. de Duve (2002) argues that, "a naturalistic view of the origin of life does not necessarily exclude belief in a Creator. The notion propagated at the same time, though for opposite reasons, by militant atheistic scientists and by many antiscientific circles, that the findings of science are incompatible with the existence of a Creator is false."

However, personal conversions occur in both directions. The same author, a Nobel laureate and a life-long Catholic, an elite member of Louvain University, has confessed with discernible regret that at the end of his life's journey, he now considers himself an agnostic.

"No one has ever seen God" (Jn.1, 18). We are completely lost with respect to the evidence of God's existence, unless God himself has revealed it to us... Many religions (esp. the Judeo-Christian tradition) believe that God did just that. The great mystics shy away from describing God in any human terms. John of the Cross had one word referring to our ineffability to deal with Divinity: *"Nada, nada, nada"*, nothing, nothing, nothing — in terms of human language and imagination. Yet through a dark night, this "nada", pregnant with the ultimate meaning, is leading to the radiance of invisible stars and profound personal love-union with God (Dubay, 1989). In religious terms God is better defined as Lover than Designer. What purpose would the best-mastermind world serve if there were no rational beings, sentient souls populating it, cherishing their own existence and glorifying him? There is essential distinction between God as a hypothesis, i.e. rational derivative of a deistic or theistic nature, and God dwelling in one's heart. Only a personal God, one who communicates with its creation and who can be addressed in return, can be loved. God's presence through subjective experience does not need proof.

15. The Meaning of Being

This is a legitimate question, perhaps more important than those about the origins: "Why is there something rather than nothing?" (Rees, 2000). Accepting that there is "something", what is its meaning? My personal view is drawn from the following considerations.

The meaning of the universe without God and without Man... A timeless era! The universe must have always existed since there was no agent which would have created it. Such a chaotic ever-changing reality would preclude even the question of meaning. Not only because there would be no one responsible for its existence and no one there to question it. A surreal state of existence!

The meaning of the universe with God but without Man... God to God alone would have to answer such a question. Why would God create the universe without the possibility of being addressed by its creation in return?

The meaning of the universe with Man but without God... An immensely sad situation for Man! On one side he would be able to participate in unraveling the grandeur, beauty and complexity of the universe, enjoying his own presence, intellectually and emotionally being part of the diverse and exuberant life, maturing in personhood — yet bound to die. (Not to mention all of the troubles and suffering that he must often endure in a realm without ultimate meaning.) A crude prank of wild chance?!

The meaning of the universe with God and Man... Being a person starts making sense. The joy of participation within the universe and potential creativity with God carries its rewards. Nothing will be taken from us even when we suffer and die. Man is God's mirror image, like, but not equal to God — still elevated enough to call God his Father. God as Father is a guarantor of Man's essential immortality. Thus God is the meaning of Man's being. This conclusion does not come from the dispute between two contesting sides but from humble submission to the mystery of our existence. Shouldn't every thinking person desire that God exist?

I wish God to exist.
I hope God exists.
I trust God exists.
I trust in God who exists.

References

Ashby, W.R. (1947). Principles of the self-organizing dynamic system. *Journal of General Psychology* 37: 125-128.

Beardslay, T. (1997). Evolution evolving. *Scientific American* (September). 15-16.

Behe, M.J. (1996). *Darwin's Black Box: The Biochemical Challenge to Evolution.* Touchstone. New York. pp. 307.

Bronowski, J. (1976). *The Ascent of Man.* British Broadcasting Corporation. London. pp. 448.

Campbell, J. with B. Moyers (1991). *The Power of Myth.* Anchor Books. New York. pp. 293.

Calvin, W.H. (2004). *A Brief History of Mind. From Apes to Intellect and Beyond.* Oxford University Press. New York. pp. 219.

Chaisson, E.J. (2001). *Cosmic Evolution. The Rise of Complexity in Nature.* Harvard University Press. Cambridge. pp. 275.

Davies, P. (1999). *The Fifth Miracle. The Search for the Origin and Meaning of Life.* Simon & Schuster. New York. pp. 304.

de Duve, C. (2002). *Life Evolving. Molecules, Mind and Meaning.* Oxford University Press. New York. pp. 341.

Dembski, W.A. (2002). *No Free Lunch. Why Specific Complexity Cannot Be Purchased without Intelligence.* Rowman & Littlefield. Lanham, Maryland. pp. 404.

Denton, M. (1998). *Nature's Destiny. How the Laws of Biology Reveal Purpose in the Universe.* The Free Press. New York.

Dawkins, R. (1991). *The Blind Watchmaker. Why the Evidence of Evolution Reveals a Universe Without Design.* Norton. New York.

Dennett, D.C. (1996). *Darwin's Dangerous Idea. Evolution and the Meanings of Life.* A Touchstone Book. New York. pp. 587.

Dickerson, R. (1992). *Journal of Molecular Evolution.* 34: 277.

Dirac, P. (1963). The Evolution of Physicist's Picture of Nature. *Scientific American* (May 1963) Dubay, T., S.M. (1989). *Fire Within.* Ignatius Press. San Francisco. pp. 358.

Farmelo, G. (ed.). (2002). *It Must be Beautiful. Great Equations of Modern Science.* Granta Books. New York. pp. 284.

Forrest, B. and Gross, P.R. (2004). *Creationism's Trojan Horse. The Wedge of Intelligent Design.* Oxford University Press. New York. pp. 401.

Guth, A. (1997). *The Inflationary Universe.* Jonathan Cape. London.

Guthrie, W.K.C. (1974). Flux and Logos in Heraclitus. (p. 1979-213) In: Mourelatos, P. D. (ed.): *The Pre-Socratics.* Anchor Books. Doubleday. New York. pp. 560.

Hawking, S. and Mlodinow, L. (2005). *A Briefer History of Time.* Bantam Books. New York. pp. 162.

Isaak, M. (2005). *Rebut Creationists with Scientific Arguments.* Greenwood Press. New York. pp. 368.

Jaki, S.L. (1998). *God and the Cosmologists.* Real View Books. Fraser, MI. pp. 286.

Jaynes, J. (1990). *The Origins of Consciousness in the Break-down of the Bicameral Mind.* Houghton Mifflin. Boston. pp. 493.

John Paul II (1997). The Pope's Message on Evolution.*Quarterly Review of Biology* 72: 377-383.

Johnson, P.E. (1993). *Darwin on Trial.* InterVarsity Press. Downers Grove, Illinois.

Johnson, P.E. (1995). *Reason in the Balance.The Case Against NATURALISM in Science, Law & Education.* InterVarsity Press. Downers Grove, Illinois. pp. 245.

Kauffman, S. (1995). *At Home in the Universe. The Search for the Laws of Self-Organization and Complexity.* Oxford University Press. New York. pp. 321.

Kauffman, S. (2000). *Investigation.* Oxford University Press. New York. pp. 287.

Kurzweil, R. (2005). *The Singularity is Near. When Humans Transcend Biology.* Viking – Penguin. New York. pp. 652.

Lemonick, M. D. (2003). *Echo of the Big Bang.* Princeton University Press. New York. pp. 232.

Moorhead, A. (1978). *Darwin and the Beagle.* Penguin Books. New York. pp. 224.

Ondok, P. (1998). *The Proof or Hypothesis of God.* Trinitas. Svitavy. The Czech Republic. 155pp. In Czech.

Penrose, R. (2005). *The Road to Reality. A Complete Guide to the Laws of the Universe.* Knopf. Borzoi Book. New York. pp. 1099.

Reid, R.G.B. (1985). *Evolutionary Theory. The Unfinished Synthesis.* Cornell University Press. Ithaca. New York. pp. 405.

Ruse, M. (2003). *Darwin and Design. Does Evolution have a Purpose?* Harvard University Press. London. pp. 371.

Schroeder, G.L. (1998). *The Science of God. The Convergence of Science and Biblical Wisdom* 1st Broadway Books. New York. pp. 226.

Seife, C. (2003). *Alpha & Omega. The Search for the Beginning and the End of the Universe.* Viking, Penguin. New York. pp. 294.

Singh, S. (2004). *Big Bang. The Origin of the Universe.* Fourth Estate (Harper Collins). New York. pp. 532.

Strobel, L. (2004). *The Case for a Creator.* Zondervan. Grand Rapids. MI. pp. 341.

Svoboda, J. (2006). Life as an Unfolding Biocosmos. In: J. Seckbach (ed.). *Life As We Know It.* Springer. New York. pp. 431-444.

Teilhard de Chardin, P. (1959). *The Phenomenon of Man*. Collins, London.

White, M. and Gribbin, J. (1992). *Stephen Hawking: A Life in Science.* Penguin. New York.

Wilczek, F. A Piece of Magic (2002) The Dirac Equation. In: G. Farmelo (ed.). *It Must be Beautiful. Great Equations of Modern Science.* Granta Books. New York. pp. 284.

Zajonc, A. ed. (2004). *The New Physics and Cosmology. Dialogues with the Dalai Lama.* Oxford University Press. New York. pp. 246.

Dialogue

Arie Issar

In general I agree with the description by Prof. Svoboda of the stages through which humans' world-view went from mythology to science to "Creationism". I attribute this agreement to the fact that Prof. Svoboda is a biologist, while I am a geologist, thus both of us have been "indoctrinated" by the evolutionary dogma, which claims that what we observe today, either in the laboratory or in the field or ideas, which we read or hear about in scientific conferences, went through stages of development from a certain primitive to a more sophisticated state. Thus the evolution from the first stages of humans' way of thought to the present can be followed back on the dimension of time.

Yet, I must admit that I could not read Prof. Svoboda's exposition, without being influenced by my own philosophical world view (which I present in this volume in my essay "The evolution of intelligence in the bio-world"). Thus the comments, which I will bring up in the following paragraphs, should be regarded as just my personal perspective once I add the Dimension of Information and Entropy as a field of force.

In this connection I see the internet as the verification of my theory regarding the Dimension of Information. (I challenge the reader to answer the question: On what dimension is google's search machine running?). Thus when reading the following lines:

> "The author sees the pace of cultural evolution/revolution to be hundred times faster than biological evolution. Kurzweil (2005) says it even more succintly: 'It took two billion years from the origin of life to cells; fourteen years from the PC to the World Wide Web.' Cultural evolution got into an exponential spin during the recent scientific and technological era which became the modern flagship of human progress."

I ask myself whether this "exponential spin" does not tell us something about velocities of events happening along the Dimension of Information ????

> "3. The Divine was First Divine" voices talked to primeval humans long before they became conscious of them and started asking questions about God's existence. Gods and demons, and belief in them, came first. More precisely, "the divine" was not merely believed, it was real and present, entrenched in the depth of consciousness (Jaynes, 1990). People, like other intelligent living beings, are curious creatures. Our ancient ancestors had been exploring the world around them in meticulous detail and at a certain stage began asking abstract questions."

Comment: From the evolutionary point of view I explain the motivation for the creation by the human mind of "Gods and demons, and belief in them" as a survival mechanism, once the instinctive existential response to fright of the animal evolved into a conscious "angst". This involved in the first place the invention of the sacrifice as a payment to the over-power, to guard the donor against the powers of evil, i.e. unpredictable natural and human antagonist and catastrophes. As in modern times, the more shrewd individuals were quick to take advantage of this situation and suggested themselves as go-betweens, once a certain fee was paid.

The further evolution along the Dimension of Information, yet due the same basic instinctive existential impulse, the human mind thrived to understand its environment, namely how it was formed and what are the laws governing its forces. The evolutionary process is characterized by a slow de-ego-centralization process, quite similar to what we see in the evolution of the child's understanding of the world. Thus the more primitive stories of creation, like that of the Sumerians, describe the creation as a series of quarrels between gods and human beings as victims or beneficiaries of these quarrels.

> "5. Clash of ideas 'No biologist in good standing would dare to challenge the basic idea of biological evolution any more.'"

Comment: While Darwin investigated the distribution of biological forms along global spatial dimensions, one should not forget the role of the geologist Sir Charles Lyell in giving Darwin the idea of distribution of forms along the dimension of time (Darwin had Lyell's book on his Beagle voyage). Thus the concept of gradual evolution could take place only after accepting Lyell's geological time scale enabling evolution of the species. Moreover, Lyell's theory of Uniformitism, which established that the geological and biological processes, have always been working in the same way and with the same pace, brought to the conclusion that the universe is many millions of years old. Albeit Lyell did not include the evolutionary theory of Darwin in his 1871 edition of his book *Student's Elements of Geology* (I boast an original copy in my library inherited from my father), yet, in the public debate, Lyell supported Darwin's theories and he and Darwin were close friends.

> "No biologist in good standing would dare to challenge the basic idea of biological evolution any more. This fact may have prompted Chaisson (2001) to remark: 'The power of this quintessential Darwinian concept is so pervasive in biology and related fields that many so called ultra-Darwinists subscribe to it wholly and without limit, indeed with a kind of fervor matched only by the fundamentalists whom they abhor.'"

Comment: Needless to say that as a geologist I back up the basics of the theory of evolution in all what regards the "Tree of Life" and also the role of "natural selection". Yet, not being an "Ultra Darwinist" I claim that there exists a difficulty of explaining, through just random mutations and selection by adaptation, the increase in complexity in the bio-world, especially when the evolution of intelligence is considered. In this respect I agree that albeit intelligence is an important tool for survival, yet, the question is whether Darwin's theory, especially the principle of random mutations, could have produced enduring increase in intelligence and not just a variety of species.

> "For others, however, accepting the Big Bang as an ultimate and universal starting point of the extant physical reality, its time and space, would herald an end for further scientific inquiry. To avoid the dead-end trap, some theoretical physicists started experimenting with a fresh approach."

Comment: To be consistent with my conceptual model of our universe as a space-time-information continuum, I am afraid that this will not work. It is like asking what did I think before I existed.

> "The searchers are well aware of the paralyzing uncertainty principle, considered a universal rule. It precludes any definitive statement about the state of the universe, even about its existence and non-existence.

> Paradoxically, 'nature itself must obey the dictates of Heisenberg's principle' (Seife, 2003)."

Comment: Speaking on Heisenberg's Uncertainty Principle, in the quantum size world, this principle, as a matter of fact, ushered into the stage of physics the observer as part of the physical event. Thus the decision whether to observe the location, or the momentum of an electron, as well as to regard it as a particle or as a wave depends on the observer's decision how he would prefer to carry out his experiment or measurement. In this connection I would like to comment that by introducing the Dimension of Information into any space-time event the 'conscious observer' becomes an intrinsic part of the physical event. This means that the observer is not merely influenced by the event he observes, but also influences it in some way. In other words this multi-dimensional model explains the 'uncertainty principle' of Werner Heisenberg which, in the framework of the multi-dimensional model, one will say that whether an event will be regarded as a particle or a wave will depend on the dimensions to which the event is referred. When these are space-information (time is regarded constant) the event will be grasped as a particle. When the event is measured along time-information (space is regarded constant), it will be referred to as a wave. In other words the 'absurd' of the interference of the decision of the observer with the observed is shown to become logical, once Information is introduced as a dimension.

Moreover it is suggested that the conscious observer, who is 'heavily laden' with ordered information, 'influences' the event, which is very 'light' with ordered information, like a heavy mass influences a very light mass. Thus, if one is ready to admit information as a dimension to his physical model, he admits also probability as an intrinsic part of reality, as probability is proportional to the load of information of any system.

In this connection I would like to add that the 'uncertainty principle' also led to the formulation of the 'complementarity principle', which states that two simultaneous views or models of the same event are legitimate from the scientific point of view. Einstein disagreed with the fuzzy or non-deterministic aspects of·this principle and was convinced that once theory is further developed the science of physics will become deterministic again. Here comes the EPR (Einstein, Podolsky and Rosen) thought experiment, which was carried out later by Aspect in practice and showed that indeed one polarized photon 'knows' how its twin photon behaves. Thus if one does not admit super-light-velocity one has to admit that Bohr's inseparability paradigm, rather than of Einstein's world view is right. Davies and Brown (1986) asked leading physicists to explain this experiment. John Wheeler gave the following answer:

> "If we're ever going to find an element of nature that explains space and time, we surely have to find something that is deeper than space and time — something that itself has no localization in space and time.... It is indeed something of a pure knowledge-theoretical character, an atom of information which has no localization in between the point of entry and the point of registration. This is the significance of the delayed-choice experiment."

I dare say that the Dimension of Information answers the challenging forecast of Prof. Wheeler.

"The Question of *How* Remains": Indeed natural selection is "rather incapable, as a working mechanism for more expeditious evolutionary advancement"

Moreover, with all due respect to: "strict neo-Darwinists and ideologically motivated adherents of natural selection as the only organizing agent in living nature, such as Dawkins (1991), Dennett (1995), de Duve (2002) and many others", natural selection is also incapable to explain emerging complexity in the bio-world.

Re: "after yet unexplained pre-Cambrian biological explosion". To explain this trend I suggested that there exists a general field of force, namely that of "entropy", which pulls towards infinite disorder. Yet, like in the case of the field of gravity, once energy is invested there is a "take off" towards higher levels of organization, i.e. ordered complexity.

Although Physicists may not agree to define entropy as a field of force, I still maintain that it is of a less *Deus ex Machina* character than the mechanism explaining emerging complexity by "The idea of *self-organization*...".

> "!4. Personal Reflections... . God is the answer to transcendental quest about ultimate causes and the meaning of our existence, and a sustainer of peace and order."

Comment: I think Prof. Svoboda will be interested in my translation of a few lines written by **Rabbi Abraham Isaac Kook** (1865–1935) (*HaRav Avraham Yitzchak HaCohen Kook*), who was one of the two chief Rabbis of Palestine under the British Mandate and one of the most celebrated and influential Rabbis of the 20th century:

> "The human being has to free himself from his personal frame of mind, which fills up all his existence, and which drives him to the state that his thoughts circle only about his personal fate, which brings him down to the depth of pettiness, causing him physical and spiritual suffering, which are interwoven. His thinking and will and the basis of his ideas should strive to the general, the world in general, the human being, the Israeli people in general, to the whole universe. This will bring to the founding of his private personality in the right way."

Biography

Dr. Attila Grandpierre is currently a senior scientific researcher at the Konkoly Observatory of Hungarian Academy of Sciences. He obtained his degree of University Doctor in 1977 from the Eötvös Lóránd University, his academic degree candidate of physical sciences in 1984 from the Hungarian Academy of Sciences and continued his research at the Konkoly Observatory.

Dr. Grandpierre's scientific interests are in the areas of: origin of solar activity, the Sun as a self-organizing system and as a system showing lifelike activities, non-protein based life forms in the Universe, first principles of physics and biology.

28
Fundamental Complexity Measures of Life

ATTILA GRANDPIERRE
Konkoly Observatory of the Hungarian Academy of Sciences
H-1525 Budapest, P. O. Box 67, Hungary
grandp@iif.hu

1. Introduction

At present, there is a great deal of confusion regarding complexity and its measures (reviews on complexity measures are found in, e.g. Lloyd, 2001 and Shalizi, 2006 and more references therein). Moreover, there is also confusion regarding the nature of life. In this situation, it seems the task of determining the fundamental complexity measures of life is especially difficult. Yet this task is just part of a greater task: obtaining substantial insights into the nature of biological evolution. We think that without a firm quantitative basis characterizing the most fundamental aspects of life, it is impossible to overcome the confusion so as to clarify the nature of biological evolution. The approach we present here offers such quantitative measures of complexity characterizing biological organization and, as we will see, evolution.

Fortunately, some important complexity measures are already established. Two such fundamental complexity measures are the algorithmic complexity of the human brain and the genetic complexity of the human organism. Let us consider how they are obtained.

The complexity measure C of a system consisting from N elements can be characterized by the number of distinctive connections (not including replicas) between the elements $C = Nc_d$ where c_d is the average number of distinctive connections per element (Denbigh, 1975, 99). The term 'distinctive' is necessary, because an airplane is more complex than a watch, but a hundred watches of the same kind are not more complex than a single watch. The complexity measure is stated in terms of information units that estimate the information content i of one element (having an average number of connections) as $I = Ci$. On this basis, the brain's complexity is usually considered in terms of neurons and synaptic connections, e.g. Stripling (2004). For a number of neurons $N_{neurons} = 10^{11}$–10^{13} (Smith, 1997, 921), taking a value for the number of their interconnections as a few thousand per neuron $c_d \sim 10^4$ (Koch and Laurent, 1999), we obtain for the measure of the brain's complexity the number $C_l = cN_{neurons} = 10^{15}$–$10^{17}$. We emphasize that the complexity measure C measures the number of (distinctive) interconnections, and so it is a dimensionless number that corresponds to objective reality. The general view assumes that a connection (synapse) represents $i{\sim}1$ bits of information. In this way, we obtain for the information measure of the brain's complexity a value of $I_l = I$ (human brain) ~ 10^{15}–10^{17} bits. Now since algorithmic complexity may be characterized by the size of the memory, we obtain that $I_{\underline{1}} = I_{algorithmic}$(brain).

Although it may seem that this estimation is overly simplified and evident, we point out a baffling problem that surfaces immediately when comparing $I_{algorithmic}$(brain) with the genetic complexity of the whole human organism. Maynard Smith (2000) noted that the genetic information content of human DNA corresponds to "instructions," and assigns a value of I_2(DNA) ~ 10^9 bits to this information measure. (Notice that pure genetic complexity can be measured by a dimensionless number, characterized by the number of coding base pairs. See Maynard Smith and Szathmáry, 1995, 5). Now it is generally accepted that the DNA controls all the biochemical processes of the organism (e.g. Woski and Smith, 2002, 28). We point out that the comparison of these two firm complexity measures presents a fundamental paradox: How can it be that the genetic complexity of the human organism (including the brain) — $I_2 \sim 10^9$ bits — is smaller than the algorithmic complexity of the human brain, $I_1 \sim 10^{15}$–10^{17} bits? We have to think that the organism as a whole has to be more complex than one of its parts, the brain[1].

---

[1]Actually, a part of brain's activity is governed by self-conscious activity. But its is easy to show that $\dot{I}$(self-conscious activity) < 100 bits s^{-1} (we cannot read more than a few pages per minute; Breuer, 1995), and so the information obtained by self-conscious activity during a lifetime I (self-conscious activity per lifetime) < 10^{11} bits dwarfs in

We propose that the solution of this *brain–DNA paradox* lies in the fact that genetic complexity is related to elementary instructions determining the organization of simple biochemical reactions of the cells, while algorithmic complexity corresponds to simple biochemical reactions organized together into cycles or units of reaction sequences. Genetic complexity controls all the biochemical reactions occurring in cells; and so the gross activity of the brain corresponds to a comparatively small subset of biochemical reactions only, which have a much lower level of complexity than that of the algorithmic complexity of the human brain. In this simple model, the brain works only on some special characteristics of neural cells, such as their firing. Therefore, the resolution of the brain–DNA paradox lies in the recognition that genetic complexity corresponds to a deeper level of complexity than the algorithmic complexity of the brain, corresponding to the memory of the neural network.

We note that we are faced immediately with another fundamental problem. The DNA contains 10^9 bits of information in the sequence of the base pairs, and this is a static form of information. In comparison, the biochemical reactions of cells represent a dynamic flow of information corresponding to their selection, coordination, and timing. If DNA is static, it could not govern the continuously changing reactions. No machine can work without moving components. This is the *DNA–dynamism paradox*. It turns out that with the introduction of the two most firmly established biological complexity measures, we are faced with two paradoxes immediately. In this chapter, we will suggest a simple resolution of these paradoxes.

2. Fundamental Complexity Measures of Life

In mathematics, there are no numbers without mathematical rules determining their interactions. Actually, numbers and mathematical rules are fundamentally different entities. In the phenomenological aspect, the most fundamental entities are numbers, sets, and complex sets of sets. In geometry, the most fundamental entities are the point, the geometrical structure (circle, square, etc.), and complex structures comprised of structures of structures. In the organizational aspect, the most fundamental entities are relations, rules, and axioms.

---

comparison to the information represented by the whole neural network $I_l \sim 10^{15}$–10^{17} bits.

We note that the point and the circle have not only different levels of complexity, but they also express different kinds of complexities. Any circle can be represented as consisting from an infinite number of points, and as the correspondence between the points. This idea can be formulated by a simple concept: A circle consists of a set of points in a plane that are equidistant from a single point which is the centre of the circle. If one were to assign one bit (one yes or no question and its answer) to one point, the information content of the circle would be infinite. In a world in which only points exist, the construction of a circle would require giving the position of the "next" point, consecutively; and so the circle would require an infinite number of yes or no questions and answers, and so an infinite number of bits.

Actually, in the process of constructing the circle from points one by one, still one would need concepts representing deeper level complexities, such as "continuity" and "closed line," which cannot be expressed directly in the language of points. Yet we recognize that the idea of the circle can be characterized by a deeper-level complexity — *algorithmic complexity*. The algorithmic complexity of a circle is a finite and small quantity. The distinction between the two levels of complexities of the circle — it's phenomenal and algorithmic complexity — is fundamental.

In physics, there are no elementary particles without laws of interaction. Elementary particles form structures, and complex structures of structures. Their interactions are described by physical laws. And all the fundamental laws of physical interactions can be derived from the least action principle.

The next step towards finding quantitative measures of biological organization is to consider what a machine is. A machine is a special arrangement of service parts or components put together according to its blueprint. In the organizational aspect, the machine is made up by the assembly of its components, and by the realization of its blueprint, which determines the interactions between its components. The components have to be produced, and the instructions required to their production represent a type of algorithmic complexity: The blueprint determines the relation of the components to each other. Therefore, the specified complexity of the blueprint corresponds to a still deeper level of complexity than the complexity of the production of its components. A machine is governed by its blueprint. Yet not only the blueprint, but the components themselves are static, passive. In the phenomenological aspect, regarding their complexity levels, machines consist of: a) particles, b) components, and c) components put together.

In comparison, living organisms consist first of all of processes of biological interactions. Their complexity levels are a) simple biochemical reactions, b) biochemical cycles and units of reaction sequences, and, at their deepest organizational level, c) biological organization. Ultimately, living organisms are governed by genetic instructions determining which reactions and units of reaction sequences have to occur, and when and where. Genetic instructions change from one timestep to the next, and involve all the components of a living organism simultaneously. This is a fundamental difference between living organisms and machines.

Living organisms are creative beings, as evolution, plants sciences and ethnology show. Regarding their physical level, living organisms can be compared to robots. Living organisms can behave in the next timestep as a new robot, a robot with a new function. They are able to invent new blueprints serving new, unforeseen tasks. Living organisms are dynamic at their deepest level of complexity.

The difference between the machine and the living organism is like the difference between numbers and mathematical rules. The complexity of the machine is phenomenological, static, and passive; while that of the living organism is organizational, dynamic, and active.

Let us approach the distinctions between machines and living organisms in light of the difference between physical "organization" (termed as "self-organization") and biological organization. As the root of the word "organization" ("organ") tells, organization belongs to the realm of biology. Physical "organization" is present in the order of crystals, of magnets, of snowflakes patterns, of convection patterns, of reaction-diffusion patterns, etc. Physical "organization" represents actually not organizational, but ordering processes.

Actually, ordering and organization are two fundamentally different processes (see e.g. Denbigh, 1975, 89-98; Elitzur, 1994). When starting from the living state, the larger is the order, the smaller is the organization, as shown when a living organism becomes frozen. In physical ordering, patterns of elements can be generated, and in man-made machines they follow prescribed rules. In living organisms, biological organization generates new rules from time-step to time-step. *Biological processes are governed by complexity* present in the boundary conditions of physical equations. In contrast, even the most complex *physical orderings* (like reaction-diffusion processes) *are governed by physical equations*, and their boundary conditions are simple. Actually, biological processes are governed by an extraordinarily complex system of time-dependent boundary conditions of the physical equations.

Written sentences are composed of words, and words are composed of letters. Letters represent the fundamental elements or building blocks, corresponding to numbers in mathematics, and particles in physics. Words correspond to structures in mathematics, and patterns in physics. Sentences correspond to machines and organisms. One can compare the difference between machines and organisms to the difference between syntax and semantics. Machines follow the (once-for-all) established syntactical rules only, since physics does not deal with problems of meaning. Sentences written by a computer following merely the rules of syntax will form an incoherent sequence of sentences, most of which will be without any meaning. In contrast, when the sentences follow meaning, the result will be a poem, a novel, or a book on science: one single organism.

A living organism follows the continuously changing internal and external contexts, and reacts to them on the basis of its own principle driving its biological organization towards the optimization of life's conditions. Biological organization is like writing, while physical ordering is mechanical repetition of words following merely syntactical rules, if any. It is these syntactical rules that represent algorithmic complexity. In contrast, the semiotic principles correspond to a deeper, principal level of complexity. This is why machines cannot rebuild themselves from time-step to time-step. At the same time, this is the most fundamental property of organisms.

Denbigh (1975, 96-97) emphasizes that "one cannot speak of an entity as being organized without at once raising the question: What is it organized for? (…) A machine is not explainable by the laws of physics and chemistry (even though the material of which it is composed obeys these laws); machines have always to be understood in terms of their own specific operational principles laid down by those who design them."

The chemical reactions within living organisms are mostly organized into cycles and units of reaction sequences and thus represent instances of algorithmic complexity. Genetic instructions elicit the proper cycle and unit of reaction sequences at the proper place and time. They correspond to a deeper level of complexity than the algorithmic complexity of biochemical cycles and units of reaction sequences. The ultimate level of biological organization occurs at the genetic level. Genetic complexity characterizes a deeper level of complexity than algorithmic complexity. In every timestep, biological organization switches into new algorithms and new contexts. The new perspectives are determined by the new internal and external factors.

Ontologically, there are three fundamental levels of complexity corresponding to the fundamental levels of existence. We observe by our outer senses the phenomenal world W_0. W_0 is governed by the laws of Nature, corresponding to a deeper ontological level we shall call the 'level of laws' or 'lawlike level' — W_1. The laws of Nature do not represent the ultimate level of existence, for they are derived from first principles like the action principle of physics and the Bauer principle (Bauer, 1935/1967, 51) of biology. *By the term "first principle" we mean a principle from which all the fundamental laws of natural sciences (physics, biology, and psychology) can be derived.* Therefore, it appears only three first principles are necessary in the natural sciences: one in physics (this is the action principle), one in biology (this is the Bauer principle; Bauer, 1935/1967), and one in psychology (formulated by Grandpierre, 2005, 76-102).

The Bauer principle states that "*The living and only the living systems are never in equilibrium and they continuously invest work from their free energy resources against the equilibrium that should be reached on the basis of the given initial state and the physico-chemical laws.*" Thus the first principle of biology acts to generate consecutively the initial and boundary conditions that will be the input elements of the physical equations in the next timestep, which in turn correspond to evolving elementary biochemical reactions within the cells. Therefore the phenomenal–algorithmic–genetic complexities correspond to the phenomenal–lawlike–principal ontological levels. These three ontological levels correspond to three complexity levels. We note that the three levels of complexity as interpreted here show a certain similarity to Maynard Smith's (2000) three complexity levels: the morphological, the selection, and the genetic level.

Denbigh (1975, 93) notes that biological organization is basically "the organization of chemical processes each taking place continuously". Now let us estimate the complexity of biological organization of the human organism as expressed in elementary biochemical processes realized within cells. Certainly, the number of chemical reactions per second is larger than the number of ATP molecules produced per second. Kornberg (1989, 65) determined that the average daily intake of about 2500 kcal, corresponding to approximately 100 W, translates into a turnover of a whopping 180 kg of ATP. This number translates into $N_2 = N_{ATP}$(organism) ~ 2×10^{21} ATP molecule production per second in the human body, or $4\text{x}10^7$ ATP molecule production per cell per second. Regarding the fact that ATP is produced in a chain of electron transfer events, and acts through energy coupling that involves the coupling of two reactions occurring at the same time, at the same place, typically utilizing the same enzyme complex, we find it plausible to assume that the rate of ATP production of N_{ATP}(organism) ~ 2×10^{21} reactions per second is smaller than the number of all chemical reactions of the human organism, $N_3 = N_{reactions}$(organism) > 2×10^{21} chemical reactions per second. It is clear that both the production of each ATP molecule together with its reactants has to be timed so that the energy coupling can be effective, and that this timing is not completely pre-programmed because it depends on dynamic, on-going changes of state at the cellular, intercellular, and global organizational levels. One may presume that at least 1 bit is necessary for the proper timing of a chemical reaction. Therefore the flux of biochemical reactions would correspond to a rate of information production $\dot{I}_1 = \dot{I}_{biochem} >$ 2×10^{21} bits/s. With 6×10^{13} cells in the body, we obtain a lower limit $\dot{I}_{lower}$(cell) > 4×10^7 bits/s. When this measure is applied to neurons, we find the dynamic chemical complexity of the brain exceeds by 6 orders of magnitude the complexity we find at the neural level. With the number of neurons in the human brain estimated at $N = 10^{11}$–10^{13}, the rate of flux of biochemical reactions in the human brain would be above 4×10^{18}–4×10^{20} bits s^{-1}. In a period of 10 years (~3×10^8 sec), this flux of biochemical reactions can produce an amount of information exceeding I_{biochem}(brain) ~ 10^{27}–10^{29} bits, a much larger quantity than the quantity of information represented in the neural network of the human brain I_1(human brain) ~ 10^{15}–10^{17} bits obtained above.

In this way, we found a simple solution for resolving the brain–DNA paradox: The genetic complexity of DNA and the algorithmic complexity of the brain are measures that characterize different levels of complexity altogether. Now let us consider the DNA–dynamism paradox.

3. The Working Mechanism of the DNA

The question remains: How are we to understand the physical realization of the genome's activity? Actually, how can DNA regulate 4×10^7 biochemical reactions per second in its host cell? The first problem we need to solve to answer this question is the fact that DNA contains 10^9 bits information in the sequence of the base pairs, and this is a static form of information. It seems clear that static information cannot elicit any processes. Requiring that DNA control all the biochemical processes of the organism (Woski and Smith, 2002, 28), the DNA must be active, not static. This means that *some part of the DNA must change relative to the others and it is these changes that instruct the biochemical reactions*. In order for these internal changes within DNA to elicit biochemical processes, they must be related to activating processes. The fastest means of such activation are light-induced excitations and electron transfer.

Recently it became clear that long-range single electron transport along the DNA as modulated by intervening sequence and sequence-dependent dynamics might help to switch genes that are far apart on and off (Nunez, Hall and Barton, 1999; Coghlan, 1999). Electron transport and proton translocation are intimately connected with metabolic activity (Demetrius, 2003), and so with the elementary biochemical reactions corresponding to $\dot{I}_{\text{biochem}}$. Electronic excited states of complex molecular systems represent the main reservoir of free energy in biologic processes (Korotkov, 2004). Electronic states of complex molecules such as DNA may extend to the whole of the molecule (actually, to the whole of the cell and more); and therefore they are suitable tools to transform sequential static information into a dynamic form. Certainly, not only the static sequential information will play a role, but also the information present in the continuously changing excited states and their biological coupling, which is governed not by the physical laws per se, but by the biological principle. All biochemical reactions can be coupled through electronic states. Our proposal tells that the essence of biological organization is that it couples endergonic to exergonic processes in a suitable manner, preparing the next timestep's input boundary conditions for action by the physical laws by means of these couplings. In this way, DNA becomes able to supply the requirement of timing, determining which chemical reactions should occur in the next timestep. Certainly, DNA cannot do the regulation alone, since its activity must be coupled to cellular organization supplying the necessary chemicals in the necessary places in the right moments, utilizing also a significant part of their thermodynamic capacities. Yet in our model, the dynamic DNA with its active electrons as modulated by sequential information can still maintain its key role of facilitating genetic control over the cellular reactions.

Actually, the timescale for light-induced transfer of electrons (electronic transitions) is $\tau \sim 10^{-12}$ s (Stryer, 1995, 6, Fig. 1–7). The excited electronic states of the cell are modulated by the sequential information of DNA within its collective electronic cloud. This means that biological organization may couple endergonic to exergonic processes that utilize the sequential information of DNA in a way that generates light-induced electronic excitations. The modulated forms of excited electronic states can decay and emit a photon which can activate, e.g., an enzyme, in due time and place. Certainly, the enzymes must be able to act in accordance with instantaneous biological needs. Therefore the whole cell effectively prepares itself as a receptive state in which the activation of a molecule can lead to the realization of the requisite biochemical reaction. Most biochemical reactions of cells are related to enzymes and endergonic–exergonic couplings. This means that the whole cell mobilizes its thermodynamic capacities in accordance with the biological principle.

Now let us estimate the thermodynamic potential of a cell. An average human organism works with an energy flow of ~100 W distributed on $\sim 10^{14}$ cells. Therefore, an average cell consumes $\sim 10^{-12}$ J s^{-1}. This value at ~310°K corresponds to an energy flow $\dot{E} \sim 3\times10^{-14}$ J K^{-1} s^{-1} in entropic units; converting it to units [bits s^{-1}] by the simple formula: Information flow in [bits s^{-1}] can be obtained from the energy flow in [J K^{-1} s^{-1}] when divided by the Boltzmann constant $k \sim 1.38\times10^{-23}$ J K^{-1} (see e.g., Brillouin, 1956, pp. 1–3), we obtain that the dynamic thermodynamic capacity of a cell is $\dot{I}_{TD}$ (cell) $\sim 2\times10^{9}$ bits s^{-1}. In the foregoing we estimated that more than 4×10^{7} reactions occur per cell per second. It seems to be plausible to estimate that inducing one reaction involves at least one photon and at least one bit of information. Such an estimation yields a lower limit for the biologically utilized information flow of a cell $\dot{I}_{biol}$(cell) > 4×10^{7} bits s^{-1}. This estimation shows that cells may utilize a significant part of their thermodynamic potentials for biological aims.

4. DNA as the Central Factor of Cells' Cooperation

Regarding the governing activity of DNA, a second question also arises: How does a multicellular organism like a human being *coordinate* the activities of its cells? If the control of the organism is due to DNA, and the individual cell's biochemical reactions are also controlled by the DNA, then we can conjecture that DNA has a twofold function: It has to control *local* cellular processes as well as organizing all these processes into a unique *global* biological organization. This would mean that DNA controls the activity of all the DNA molecules that in turn control the activity of the individual cells. All the DNA molecules are under the control of all the other DNA molecules such that all biochemical reactions serve biological needs useful for the global multicellular organism. The question then becomes: How does the DNA molecule sitting in one cell know about all the chemical reactions occurring in all the other cells governed by the DNA molecules sitting in all the other cells?

It is apparent that multicellular government should be mediated by a coordinating activity between the DNA molecules themselves. Therefore, the DNA molecule sitting in one molecule should follow all the changes occurring in all the other cells. This is a much more demanding task than controlling one cell's activity. This is a task that would require one DNA molecule to act in concert with all the other DNA molecules. This means that instead of $\dot{I}$(DNA) ~ $\dot{I}_{biol}$(cell) ~ 4×10^7 bits s^{-1}, the changes in the internal states of the DNA molecules must be around $\dot{I}$ (DNA) ~ 10^{21} changes s^{-1}.

Thus the question arises: How is it possible to produce 10^{21} changes s^{-1} in a DNA molecule if the information content in the sequence of its base pairs is only 10^9 bits? We point out that this requirement can be fulfilled if the timescale required to induce a change is 10^{-12} s. Actually, the DNA molecule is able to realize ~10^{21} changes per second by the fastest known means of biochemical processing — by light-induced electronic excitations (Stryer, 1995, 6, Fig. 1-6). This is what one can expect if these changes are generated by biological organization that is itself the manifestation of the first principle of biology (e.g., the Bauer principle). *The action of first principles does not require mechanisms in order to be achieved.* The first principle of physics, the action principle, does not require a computer built into each elementary particle that would compute which way the particle ought to go. Instead, particles work with the action principle because *the action principle is 'built into' the elementary particles*.

In the hypothetical absence of a first principle of biology, an explanation of the governance of electronic excitation states, and the activation of photons, enzymes, and proteins that correspond to biologically optimal trajectories, would be a computationally unsolvable problem. To govern the states of DNA's 10^9 base pairs in a way that each changes in every time step of ~10^{-12} s should occur in the biologically optimal manner requires the presence of a first principle of biology in action. There is no way to solve this enormous computational problem except by positing a first principle of biology — the most economic and (we suspect) the only possible solution.

5. On the Activity of the First Principles

The activity of the action principle is best understood by the Feynman path-integral interpretation. It states that the action principle has a quantum physical origin: Each elementary particle emits virtual particles which map all possible paths of the whole environment, the collective behavior of which summarizes all possible quantum mechanical paths and realizes the extremum of action having a dimension [energy][time]. The extremum usually indicates the minimum in physical situations (Feynman, Hibbs, 1965, 245). When Feynman introduced the path-integral principle, he pointed out that to be able to follow the principle of least action, quanta must 'virtually' go over all the possible histories, and then these add up to the 'actual' shortest route. The precondition of such an adding up is that in the course of surveying all the possible routes, each quantum virtually travels over all the routes — certainly, at a speed much larger than the velocity of light. Therefore, it is usually said that the Feynman path integral method offers only a model. We note that this model works not only as one of the best of physics, the most exact branch of the natural sciences, but it is the core technology of modern theoretical physics (Moore, 1996; Taylor, 2003; Moore, 2004).

The point is that the first principles and the faster-than-light virtual actions governing actual interactions are two sides of the same coin.

The integral form of the action principle contains a non-negligible advantage over its formulation in differential equations. Differential equations need definite initial conditions, while the integral formalism — virtually — includes informative interactions with a large set of the environment. Integral principles are independent from coordinates, and therefore they can cope with time-dependent boundary conditions as well. The apparent teleological behavior of living organisms may correspond to computational processes determined at the organism level, where the organism acts as an agent, following its own interests and biological needs, such as survival. Once the biologically favorable endpoint of a biological process is prescribed by the organism at the organism level, the problem is simplified; and with the help of the action principle of physics it becomes possible to determine the trajectory to be followed, which involves the organism's satisfaction of its biological needs by rearranging its internal physical environment.

Let us consider an extended version of the Galilei experiment. In this example, not only inanimate things, but also living organisms are dropped from the Pisa tower. When a living bird is dropped from a height, it will decide which way to go — i.e., to fall to the ground or fly away — and the biological decision will be realized directly by the action principle since this is the most economic way to organize the physical activities of the bird. Clearly the action principle is subservient to biological organization. The decision of the bird about the optimal endpoint is obtained through switching out into a new context and wider perspectives. If the bird before dropped from a height were in a state of short-period or quasi-instantaneous perspectives, at the moment when dropped it switches into a much wider perspective, the perspective of its life. In this new perspective, it decides about the endpoint of its flight. Once the endpoint is determined, the physical realization of the trajectory best suited to the selected endpoint — survival in this case — can be supplied by the action principle of physics as the most economic solution. This economy extends not only to the minimization of action, but also to the minimization of biological interventions. Absent an integral principle, the bird would need to compute in every time step all the necessary boundary conditions for all the physical processes occurring within it that may lead to the optimal biological solution regarding the selected endpoint of the flight. This task seems to require an overly demanding computational faculty, since every elementary process should result from previous computations, taking into account all the combinatorial possibilities. Without doubt, the action principle is suitable to biological government through endpoint determination. It seems that now it is only a step forward to assume that the action principle is tailored just for biological purposes. Let us consider this hypothesis a bit more closely.

We propose a simple but powerful qualitative idea: that the first principle of biology arises at the other extremum of integrated action [energy][time], which corresponds to the maximum of integrated action instead of its minimum. Biological organization acts to secure the optimal conditions for life. Remarkably, it is the maximum of integrated [energy][time] that corresponds to the optimal quality of life, as is easily seen with the help of an example.

Life's quality is the better, the more the number of years we live and the more free energy we have by which to live our years. Therefore, life's quality can be measured by a quantity: integrated [energy][time]. Now if the first principle of biology acts to realize the maximum of vitality (defined as the distance of the living organism from the deathly thermodynamic equilibrium) not only in the momentary context, but also in the context of our full lifespan, then it appears the Bauer principle likewise would require a maximum of a quantity that also takes the form of integrated action [energy][time].

We note that vitality — the distance of the living organism from thermodynamic equilibrium — is different from Schrödinger's negative entropy, which measures the distance from thermodynamic equilibrium for isolated systems in a static or closed environment. By definition, isolated systems cannot exchange matter and energy with their environment.

We acknowledge a similar idea of Rashevsky (1973, 177), the founder of mathematical biology, who worked out a unified approach to physics, biology, and sociology. In this work one of his main achievements is his Postulate 1 (ibid., 185), which states: "The evolution or time course of change in any organismic set is characterized by the requirement that during the total time of the existence of an organismic set the total number of different relations involved should have a maximum." In comparison, our formulation tells: "*Living organisms invest internal work from their free energy resources against the equilibrium that should be reached on the basis of the given initial state and the physico-chemical laws in order to maximize the total integrated action.*"

We saw that the Bauer principle is a variant of the action principle of physics, since the Bauer principle maximizes the action, while in physics the action is usually minimized. Certainly if DNA works with the Bauer principle, and the Bauer principle is a kind of action principle, then the Bauer principle must act in the same way as the action principle does: by virtual interactions able to map all the universe instantaneously.

The idea that Nature pursues economy in all her workings is one of the oldest principles of theoretical science. In a physical problem, the "action" in the action principle represents a cost. For example, in Fermat's principle the cost expended by a light ray moving along its path is the transit time. In many engineering problems time as well as energy plays the role of a cost, and in such cases the most economic solution is what minimizes the product [energy][time]. This recognition triggered suspicions of the presence of economical aspects.

But it seems a bigger problem is involved here: Such economical aspects are alien to the physicalist world picture and so usually remain undefined and unclear. The action principle has a teleological character that does not fit into the present conceptual scheme of physicalism. Instead of predicting the future from initial conditions, as for instance when working with differential equations, from the more fundamental viewpoint of the action principle, the system starts with a combination of initial conditions and final conditions and the task is to find the path in between them, *as if* the system somehow knows where it wants to go. Actually, it is generally argued that the system does not need the advantage of prior knowledge where it has to go, since the path integral calculates the probability amplitude for any given process. But the real problem is that these probability amplitudes would need (ostensibly) to be calculated for all the possible paths in between.

These problematic and unresolved aspects led to the strange situation in which the highest achievement of physics, its first principle, remained largely ignored in its real impact, and so left without proper interpretation. Additionally, by our proposal action is a basic concept in biology just as it is in physics, since the first principle of biology maximizes it. In this way, an elegant situation arises in which the two known first principles of Nature both take the same quantity — action — to the extreme, each in its own way.

Moreover, since biological organization maximizes action, it is a natural requirement of the Bauer principle that *once the coupling that can lead to the biologically optimal solution is decided and the biological endpoint corresponding to optimal life conditions is determined, the physical realization of decisions should occur with the minimum of action*, since this circumstance is the condition of the requirement that life ever navigates towards its maximum by means of efficient action within the context of the systemic whole. These considerations seem to favor our proposal that the action principle of physics arises as a natural consequence of the Bauer principle.

Returning to the working mechanism of DNA, we can illustrate it with the help of a parable. DNA acts as a watchtower having 10^9 lamps. Each lamp is switched on or off in every time step, at a frequency interval of roughly 10^{-12} s. Collectively the lamps illuminate an enormous biologically useful information flux. Now switching "on or off" are already decisions that one expects would be preceded by information processing. But since the switching on or off occurs at the speed of light, and no physical event can occur at a speed faster than the speed of light, there is no physical possibility for information processing to precede the event characterized by the condition "on or off."

Therefore we realize that DNA works with the help of a factor that is utterly beyond DNA's or any other material system's physical capabilities. On our view, this something is immaterial yet effective and belongs to science — this something we have denoted as a *first principle.* The first principle of biology acts as a deeper intelligence of the "vacuum," in the sense that it virtually maps all the possible histories; summarizes the results of this mapping on its own basis; then decides about the biological endpoint; and from there, "chooses" the optimum physically realizable path.

Let us keep in mind that the action principle *also* acts as a mediated, faster-than-light virtual process that maps the entire universe before the quantum makes its "decision" as to which way to go. Therefore, the strange ability of DNA processing information faster than the speed of light is in good company; namely, in the company of the action principle, the formulation of which by the path-integral formalism has led to some of the greatest achievements of physics. Just as physics resonates to its first principle, the action principle, so DNA acts by the first principle of biology, the life principle (about the life principle see the theoretical biology of Bauer, 1935/1967).

Now if virtual particles correspond to the vacuum, and virtual particles correspond not only to the physical but also to the biological first principle, than *the vacuum must have not only a physical but also a biological nature.* We revealed the existence of the biological vacuum.

Perhaps we are now aware that we have penetrated the realm of ultimate reality. Nevertheless, we feel safe because we find ourselves in the best company, together with the first principles of modern science, which are the safest ground yet achieved by science. Now it is clear that if DNA is governed by the life principle, then the theory of evolution and theoretical physics work with toolkits from partially different conceptual storehouses. We do not require the introduction of new elements into scientific research; we just take into account the most effective tools of science, the first principles, at their real face value. Our solution is the most economic possible: *It extends the action principle just by one step so to include the selection of biological endpoints.* This solution offers the most effective way to integrate the action principle and Ervin Bauer's life principle.

6. Evolution or Divine Action? Complexity Jumps in the History of Life

As Maynard Smith and Eörs Szathmáry show in their Table 1.1 (1995, 5), the coding part of the bacterial genome has N_{bp} (bacterial) ~ 4×10^6 base pairs and the human genome has N_{bp} (human) ~ 6×10^8 base pairs. We find it remarkable that the size of the coding DNA shows a mere hundredfold increase from bacteria to humans, from 4×10^6 base pairs to 6×10^8 base pairs. It is widely thought that terrestrial life was already present within 100 million years after the solidification of the Earth's crust. Now we only point out the obvious, below we will argue for it quantitatively: Evolution is possible only in the presence of life. Chemical abiogenesis cannot produce algorithmic and genetic complexity from morphological complexity, just as emergent phenomena cannot produce the laws of nature.

In this context, it is important to take into account the fundamental fact that the laws of physics have a very low information content, since their algorithmic complexity can be characterized by a computer program less than a thousand characters (Chaitin, 1985). In a personal communication, Chaitin wrote (2004): "My paper on physics was never published, only as an IBM report. In it I took: Newton's laws, Maxwell's laws, the Schrödinger equation, and Einstein's field equations for curved spacetime near a black hole, and solved them numerically, giving 'motion-picture' solutions. The programs, which were written in an obsolete computer programming language APL2 at roughly the level of Mathematica, were all about half a page long, which is amazingly simple."

Now one may estimate the complexity of a page as approximately 2×10^3 bits, since the average rate of information processing in reading is about 50 bits s^{-1} (Breuer, 1995, 13); and so at a reading rate of 1.5 pages per minute, the information content of a page is about 10^3 bits. Taking a page from Chaitin, we thus surmise that the algorithmic complexity of physical equations is surprisingly low, $I_{algorithmic}$(physical equations) ~ 10^3 bits. We think that the low algorithmic complexity of the physical laws is shown also by the fact that present-day physical cosmological models fail to account for such basic phenomena as stellar activity, not to mention protein-based life, with its extremely rich variability.

In contrast to the basic claim of physicalism, the failure of cosmological models with respect to biology is not a practical, but a principal one. Physical models are not only unable to predict biological phenomena, but there is no physical model that can calculate from the positions of particles whether an animal will go left or right from its initial state, nor account for any characteristics of the trajectory of the animal. Moreover, seemingly there are no scientific works in progress in this particular field. No physical models are under construction that could predict such simple phenomena. In contrast, our biological approach is able to work out such a model. Once the main range of the biological endpoint is determined by biological needs and aims, the arising physical trajectory of a bird dropped from a height can be derived (Grandpierre, manuscript).

This suggests that information-producing complexity jumps simply are not possible in the case of physical systems, for such jumps are novel *increases* of complexity: They are ever jumps "up".

Certainly, the observed flow of environmental information is enormous, but it is morphological information. Now since we cannot expect that Big Bang (or recycling) cosmological models obtained initial conditions corresponding to an algorithmic complexity higher than the algorithmic complexity of the physical laws themselves, we can estimate that the complexity measure of physics — initial and boundary conditions and physical equations included — is also about $I(\text{physics}) \sim 10^3$ bits.

The central thesis of physicalism proclaims the causal closure of the physical. Ashby's Law (Ashby, 1962) and Kahre's Law of Diminishing Information (Kahre, 2002) stated that physical systems cannot produce more information at their output than was present at their input. This means that for physical systems, complexity jumps are simply not possible. Therefore the fact that we observe complexity up-jumps here on Earth strongly indicates the presence of life.

The comparison of machines and living organisms can shed light on the nature of biological organization. Once the machine is constructed, its algorithmic complexity is fixed. Even in machines programmed with "learning abilities," only phenomenal data can be involved, and such data cannot increase algorithmic complexity. In contrast, biological organization is able to increase not only algorithmic, but also genetic complexity, as shown by the blossoming of complexity in plants, animals, and in evolution generally.

We point out that there was a much greater complexity up-jump between the early Earth without life and the first bacteria (from 10^3 bits to 4×10^6 bits, a jump of $J_1(10^8$ years) ~ 4×10^3, within about $t_1 \sim 10^8$ years) than between the first bacteria and humans (from $4\text{x}10^6$ bits to 6×10^8 bits, a jump of $J_2(4\times10^9$ years) ~ 150, during $t_2 \sim 4\times10^9$ years). It means that abiogenetic evolution should be more than thousand-fold faster ($J_1/J_2*t_2/t_1 \sim 1067$) than biological evolution. This fact seems strange, since we recognize that chemical abiogenesis appears (in principle) unable to accelerate the evolution of complexity faster than the evolution of life itself. The question inevitably arises: How can we expect chemical evolution to reach a twenty-seven times higher increase in complexity within a forty times shorter time period than life itself managed to do?

In this context, an example may be enlightening. Hoyle (1983, 243) pointed out that to solve the Rubik cube by one random step in every second, it would take 1.35×10^{12} years. The chance against each move producing perfect color matching for all the cube's faces is about 5×10^{19} to 1. Now if an intelligence is present, reporting after each move whether it was successful or not, reckoning 1 minute for each successful move and, say, 120 moves to reach the solution, the solution of the same Rubik cube may be reached within 2 hours. This fact indicates that the presence of life or intelligence can accelerate evolution in a rate higher than $\sim 6\text{v}10^{15}$ (1.35×10^{12} years/2 hours $\sim 5.9*10^{15}$). Certainly, the abiotic processes are not completely random — modifying the success ratio with and without intelligence from about 10^{16} to somewhat lower.

There exists a popular example of monkeys that can type Shakespeare's complete oeuvre on a typewriter. Actually, to type only one sentence from *Hamlet*, consisting of 40 letters, each selected from 30 possibilities, it would be necessary to realize $30^{40} \sim 10^{59}$ trials. Let us assume that we have ten billion monkeys — that is, rather more monkeys than there are currently people in the world. And let us imagine each monkey hits one key per second. Let us further assume that they never stop to sleep or eat or anything else. It will still take more than 10^{49} seconds before one of the monkeys has the luck to hit on the right sequence. Now one year is about 32 million seconds, so it will take our world population of monkeys about 3×10^{41} years to get there.

Now how would it be possible that the absence of monkeys and typewriters — corresponding to the case of chemical abiogenesis — could accelerate the process to write an amount of information corresponding to Shakespeare's whole *Hamlet*? Certainly, one cannot expect that chemical evolution would be able to produce useful amounts of genetic complexity in the absence of agents. Even in the presence of "inanimate agents," it seems highly implausible to expect that the accumulation rate of genetic information by chemical abiogenesis in an assumedly *physical* environment (information accumulation in physical systems is excluded by Ashby's Law, Kahre's Law, and causal closure) could produce a meaningfully higher jump in genetic information than the jump produced by life during its 4×10^9 years of evolution. Why should "inanimate agents," if they exist at all, be more efficient than living agents possessing much higher genetic complexity?

The complexity measures and their analysis presented here argue that the birth of life here on the Earth is counterindicated by two strong constraints. The first constraint arises from Ashby's Law, Kahre's Law, and the hypothesis of the causal closure of the physical. This constraint tells that *there is no free lunch of algorithmic and deeper complexities*. Algorithmic and deeper (genetic, principal) complexities can be generated only in the presence of life. The second constraint arises from the comparison of the complexity jumps J_1 and J_2 estimated above. It remains the task of the believers of abiogenesis theory to show how abiogenetic evolution can be more than 10^3 faster than biological evolution, even if biological evolution is indicated to be faster by a factor $\sim6\times10^{15}$ or higher (see the example of writing Shakespeare's *Hamlet* by abiogenetic ways, without monkeys and typewriters).

Our results quantitatively argue that life at its fundaments acts by almost fully informed light, filled with information produced in the virtual reality of the quantum vacuum fields, governed by the first principle of biology: the Bauer principle. This aspect lends an intelligent character to the dynamic form of genetic complexity. On the other hand, we also acknowledge the role of natural selection in the evolution of genetic complexity. The arguments presented here together seem to point towards the hypothesis that evolution understood as the increase of complexity is possible *only* in the presence of life.

We found strong quantitative indications showing that life in the virtual quantum vacuum belongs to the utmost foundation of the Universe, and it is this cosmic life that is responsible for the evolution of genetic complexity from bacteria to humans here on Earth.

In the picture outlined above, the natural sciences do not exhaust their possibilities in physics and applied physics, but involve biology (and arguably psychology or the science of self-consciousness) as well. The integral and natural approach presented here differs from the intelligent design theories in that it does not refer to an intelligent factor *beyond* Nature, but involves a deeper intelligence represented *within* Nature *in the form of first principles* from which all the fundamental laws of physics and biology can be deduced and tested by empirical experience.

The integral approach opens new perspectives before modeling biological behavior, collecting, grouping and explaining yet unexplained phenomena. Moreover, it is able to offer predictions on the basis of these first principles, which can then be tested by empirical measurements. For example, the Bauer-principle predicts that in living organisms all the processes governed by physical laws (decay of proteins, heat radiated away, entropy increase, etc.) will be accompanied by processes that will practically compensate the decrease of distance from thermodynamic equilibrium. In living organisms, the fundamental coupling is not between spatiotemporal coordinates, but between global thermodynamic variables of the organisms.

More concretely, the integral approach offers an explanatory model for biological phenomena such as a bird dropped from a height (a prototype of all physical processes). All physical phenomena proceed towards equilibrium: a stone dropped from a height, a warm pond cooling at the onset of evening cold, or a sugar cube dissolving in a cup of tea. All these processes have their characteristic timescales. In living organisms, many similar processes occur, but most of them are compensated for by thermodynamically uphill processes. Therefore, the physical approach cannot model biological behavior at the level of global organization, such as the trajectory of a thirsty animal towards a river. In contrast, the integral model allows the existence of biological aspects such as the determination of biological endpoints by the organism itself. Once the existence of biologically determined endpoints is acknowledged, the integral approach facilitates the determination of the physical trajectory corresponding to the given biological end. The integral model offers the simplest and most effective approach; it is able to predict and it can be tested, therefore fulfilling the sharpest criteria of methodological science. The utmost simplicity at the level of seeming utmost complexity is also explained by the biological principle.

7. Summary

We derived quantitative measures of algorithmic complexity of the human brain and of genetic complexity of the human organism. Already these simple complexity measures indicate a paradox of how the brain's complexity can be larger than the complexity of the whole organism. The resolution of this paradox leads us to recognize that genetic information corresponds to a deeper than algorithmic level of complexity. In our consideration of how DNA can regulate the biochemical activities of the organism, we point out that "static" DNA must be complemented by a dynamic complexity that corresponds to its static sequential information content. Numerical estimations show that DNA's changes are regulated by light that is almost fully informed by biologically useful information. Qualitative arguments based on Ashby's Law, Kahre's Law and the hypothesis of causal closure as well as quantitative arguments based on complexity measures of life show strong indications against abiogenesis. We point out that complexity measures show that genetic complexity cannot be produced from environmental effects alone but are governed by the already known, quantitatively formulated first principle of theoretical biology (Bauer, 1935/1967, 51) that is on a similarly firm footing as modern theoretical physics. This means that biological evolution is governed fundamentally by the Bauer principle, and natural selection represents an important, but secondary factor.

Key words: algorithmic complexity, dynamic complexity measures, genetic complexity, working mechanism of DNA.

8. Acknowledgements

It is a pleasure to express my thanks to my friend Jean Drew for inspiring communications and correcting the English.

References

Ashby, W.R. (1962) in: H.V. Foster, and G.W Zopf (eds.): *Principles of Self-Organization*. Pergamon Press, Oxford, 255.

Bauer, E. (1935/1967) *Theoretical Biology* (1935: in Russian; 1967: in Hungarian) Akadémiai Kiadó, Budapest, 51.

Breuer, H. (1995) *Informatik. DTV-Atlas zur Informatik.* Deutscher Taschenbuch Verlag. Kg, München, 13.

Brillouin, L. (1956) *Science and Information Theory*. Academic Press, New York, 1-3.

Chaitin, G.J. (1985) An APL2 gallery of mathematical physics — a course outline. Proc. Japan '85 APL Symposium, Publ. N:GE18-9948-0 IBM Japan, 1-26.

Coghlan, A. (1999) Electric DNA, New Scientist, 13 Feb 1999.

Demetrius, L. (2003) Quantum statistics and allometric scaling of organisms, Physica A 322, 477-490.

Denbigh, K.G. (1975) *An Inventive Universe*. George Braziller, New York, 99.

Elitzur, A.C. (1994) Let There Be Life. Thermodynamic Reflections on Biogenesis and Evolution. *J. theor. Biol.* 168, 429-459.

Feynman, R.P. and Hibbs, A.R. (1965) *Quantum Mechanics and Path Integrals.* McGraw-Hill, New York.

Grandpierre, A. (2005) *Our Life and the All-pervasive Order* (in Hungarian). Barrus Publishers, Budapest.

Hoyle, F. (1983) *The Intelligent Universe*. 243, see also http://en.wikipedia.org/wiki/Fred_Hoyle

Kahre, J. (2002) *The Mathematical Theory of Information*. Kluwer Academic Publishers, Boston/Dordrecht/London, http://www.matheory.info/chapter1.html

Koch, C. and Laurent, G. (1999) Complexity and the Nervous System. Science 284, 96-98.

Kornberg, A. (1989) For the love of enzymes. Harvard University Press, Cambridge, 65.

Korotkov, K. (2004) Assessing Biophysical Energy Transfer Mechanisms in Living Systems: The Basis of Life Processes, Journal of Alternative and Complementary Medicine, 10, 49-57.

Lloyd, S. (2001) Measures of Complexity – a non-exhaustive list, Control Systems Magazine, 21, 7-8, http://web.mit.edu/esd.83/www/notebook/Complexity.PDF

Maynard Smith, J. and Szatmáry, E. (1995) *The Major Transitions in Evolution.* W. H. Freeman – Spektrum, Oxford, 5.

Maynard Smith, J. (2000) Concept of Information in Biology. Philosophy of Science 67, 177-194.

Moore, T.A. (1996) Least Action Principle, entry in Macmillan Encyclopedia of Physics, John Rigden, ed., Simon and Schuster, Macmillan, Vol. 2, 840.

Moore, T.A. (2004) Getting the most out of the least action: A proposal, Amer. J. Phys. 72, 522-527.

Nunez, M.E., Hall, D.B. and Barton, J.K. (1999) Long-range oxidative damage to DNA: effects of distance and sequence, Chemistry and Biology, 6, 85-97.

Rashevsky, N. (1973) A Unified Approach to Physics, Biology and Sociology, Chapter 2C, in: R. Rosen (ed.) *Foundations of Mathematical Biology*, Vol. III, Academic Press, New York, 177-190.

Shalizi, C.R. (2006) Complexity Measures. http://cscs.umich.edu/~crshalizi/notebooks/complexity-measures.html

Smith, T.E. (1997) *Molecular Cell Biology.* In: T. M. Devlin (ed.) *Textbook of Biochemistry.* 4th edition. Wiley Liss, New York, Ch. 22, p. 921.

Stripling, J. (2004) Introduction to Cognitive Science. Neuroscience, http://comp.uark.edu/~jstripli/CogSci-JS-L1-web.pdf

Stryer, L. (1995) *Biochemistry.* Fourth ed. W. H. Freeman and Co., New York, p. 6 & Fig. 1, 7.

Taylor, E.F. (2003) A call to action. Guest editorial. Amer. J. Phys. 71, 423-425.

Woski, S.A. and Smith, F.J. (2002) *DNA and RNA: Composition and Structure.* in: T. M. Devlin (ed.) *Textbook of Biochemistry.* 5th edition. Wiley-Liss, 28.

Dialogue

Philip Ball

Understanding life processes in terms of flows of information seems like a fruitful way to proceed in attempting to answer Erwin Schrödinger's question 'what is life?' But it seems to me that this issue is a subtle one that can be obscured as much as it is elucidated by the contemporary emphasis on DNA as a depository of digital information. The relationship between the genetic sequences in DNA and the molecular processes of life is by no means obvious. Take, for example, the way in which cell behaviour is regulated by the operation of protein ion channels. There is certainly a form of logic to this function: the output, transmembrane potential say, can be regulated by a variety of input signals, such as mechanical forces, temperature, the presence of other ions or ligands. The way this transduction occurs is mediated by physical laws of an analog nature, for example Fickian diffusion, thermal fluctuations, electrical gradients. One might argue that, in defining the shape and structure of the ion channel, genetic information in DNA 'governs' the process — but what does that really mean? Does the static information in the gene encoding the protein somehow dictate the temporal switching behaviour of the gated channel over time? No, surely here DNA is more like the medieval God who lays down the initial laws of the universe before sitting back and simply letting them unfold, without constant intervention. The process unfolds because of the nature of the changing environment, acting on a set of preconditions, and not because of tampering by the agency of those preconditions.

This is why I am somewhat puzzled by Grandpierre's conception of DNA 'controlling' cellular biochemical reactions. It rather sounds as though he requires this control to be constantly 'active': a reaction cannot proceed as it should unless DNA is *doing something* to ensure that, which then seems to demand ultrafast processes involving, say, electron transport. This, perhaps, is where we are led by modern biology's insistence on the primacy of DNA as the 'author' of life, so that things cannot be trusted to unfold of their own accord (which is to say, on the basis of simple physicochemical processes). The problem seems to become even more profound once we consider how individual cells coordinate their activity: as Grandpierre asks, "How does the DNA molecule sitting in one cell know about all the chemical reactions occurring in all the other cells?" It seems to me that the answer is that the DNA does not have to 'know' anything; there are mechanisms for cell-cell communication (some of which, in higher organisms, use the very fast transmission of electrochemical potentials to connect remote regions) which have become adapted in such a way as to permit spontaneous, functionally directed self-organization of the multicellular body. One can see primitive self-organized behaviours of this sort in single-celled organisms that can display some collective behaviour when the conditions dictate, such as the slime mold *Dictyostelium discoideum*. The DNA does not have to act as some kind of godlike molecular overseer in these situations.

With this in mind, it seems no longer obvious why one need invoke the kind of biological first principle that Grandpierre discusses towards the end of his article — one that acts as a kind of faster-than-light 'deeper intelligence of the vacuum', deciding about biological endpoints and then mapping the pathway there. (Incidentally, where does the teleological 'biological endpoint' come from? Since when did we need to invoke any prescience to biology in order that it 'works'? Biology is, from moment to moment, surely quite blind, and it is only evolution that has installed an apparent 'purpose' to it all.)

This insistence on absolute and rigid genetic control of biological processes seems to be what motivates the notion of DNA as a 'watchtower' switching on each of its lamps every 10^{-12} s. To me, this conjures up the image of an over-zealous lighthouse keeper convinced that no sailor can by themselves work out how to navigate the rocks, whatever the weather or the time of day. At any event, why does this switching have to happen 'at the speed of light', demanding some kind of mysterious superluminal agency that dictates the pattern of switching? I don't understand that. Yet it leads to the even stranger notion of a 'biological vacuum', which seems to me to be some agency that gives organisms biological foresight of the kind that evolution ensures they don't actually need.

Another claim for which I can't find sufficient motivation is the idea that biological systems maximize action. This seems to depend on an assumption that organisms 'try' to live for as long as possible. But that isn't so. They live for as long as is evolutionarily convenient. In conventional Neodarwinian terms, the sole imperative is to maximize the prospects of propagating one's genes: survival is generally a concomitant of this, but not survival at any cost — for example, an organism might stand to benefit more from devoting its limited resources to reproduction than to cell repair. In any event, the hypothesis of maximal action seems to be one that is asserted but never proved.

At root, I am perhaps most perplexed by the notion that algorithmic complexity has to be high to account for biological phenomena. Has it not been one of the underpinnings of complexity science that complex behaviours can arise from simple rules? Grandpierre asserts that no physical models can account for the trajectories of biological organisms. But they can! Models of ant motion, driven by simple ideas such as chemotaxis and random searching, can reproduce the behaviour of ant colonies rather well. For simple organisms such as bacteria, it seems even possible in principle that one might measure from moment to moment all the environmental influences acting on a single cell, and thereby predict its motion with great precision. Certainly, it is not clear why there need be anything mysterious or aphysical about this behaviour.

Grandpierre argues that abiogenesis cannot seem to create, in a sufficiently short time, the complexity we see in life: if I understand correctly, he implies that only life (or 'intelligence') can beget life. To my mind, there are two shortcomings with this. First, it assumes that accumulation of complexity is linear, whereas it now seems that many complex systems possess thresholds above which entirely new modes of behaviour — new capabilities — appear. Secondly, I see no explicit role here for evolution: for the quite remarkable efficiency of searching in the landscape of possibilities for effective 'answers' that is permitted by the rather simple algorithm of random mutation and replication in the face of limited resources. Diversification and complexification are, in this respect, boosted by the fact that every evolutionary step broadens and modifies the landscape in which subsequent steps are taken: evolution does not simply have to respond to a preordained landscape, but to itself. To my mind, "intelligence in Nature" here becomes another God of the gaps, an expression for what we do not yet understand (and what therefore astounds us) about the capacity of the physical world to generate richness and complexity.

Attila Grandpierre
I agree with Ball's note that the recent emphasis on DNA as a depository of digital information might be an overstatement. Indeed, as I tried to indicate it in my chapter, perhaps not consequently enough, it is the cell as a whole, with all its constituents and biological couplings, which governs the cell's behavior, and not the DNA alone. For example, I argued that the cell utilizes a significant part of its thermodynamic potential for biological organization. Regarding the

problem I raised in my chapter, namely, that the static information of the DNA in itself is not suitable to govern (or participate in) the time sequence of biochemical reactions. In my point of view, this is a fundamental unsolved problem of modern biology. It seems that Ball approaches only the physical aspect of the cell's behavior. Indeed, enlisting the physically influential parameters of the input and the output of the process regulating the behavior of the cell, he implicitly ignores the biological aspects of the problem. The biological aspects of the cell's behavior are related to thermodynamically uphill reactions made possible by biological couplings between endergonic and exergonic reactions. My point is that the DNA also contributes to the biological coupling processes through spontaneous photon emissions and absorptions, electron transfer and many other ways, in coherence with all the biochemical processes, all of which are governed ultimately by an autonomous biological principle. (continued below).

> *Philip Ball*
> This seems possible, but is there any evidence for it? I'm aware only of, e.g. electron transfer in DNA perhaps playing a role in DNA damage.
>
> *Attila Grandpierre*
> It has been pointed out that the genetic code is useless without the supporting cellular machinery — first of all, without properly functioning proteins, required for DNA/RNA functioning (Ben Jacob, Shapira, Tauber, 2006, Seeking the foundations of cognition in bacteria: from Schrödinger's negative entropy to latent information. *Physica A* 359, pp. 495-524; p. 515). It is clear that static, sequential information is useless in generating the dynamic biological organization (Grandpierre, 2007, *NeuroQuantology* 5, pp. 346-362). As I wrote in my chapter: "Recently it became clear that long-range single electron transport along the DNA as modulated by intervening sequence and sequence-dependent dynamics might help to switch genes that are far apart on and off (Nunez, Hall, and Barton, 1999; Coghlan, 1999). Electron transport and proton translocation are intimately connected with metabolic activity (Demetrius, 2003), and so with the elementary biochemical reaction flux".

Attila Grandpierre (continued)
This means that, in contrast of Ball's opinion, DNA does not act like the medieval (deistic) God who lays down only the initial laws of the universe before sitting back and simply letting them unfold. Instead, DNA, together with

all the subsystems of the cell, continuously changes and these changes add up to the changes of the cell's behavior (continued below).

Philip Ball
I still don't see why this is necessary. Isn't it one of the basic principles of self-organized systems that they do not need a constant 'hand on the tiller'?

Attila Grandpierre
The difference between physical self-organization and biological organization is that physical self-organization does not need a continuous control. In contrast to physical self-organization, in biological organization a continuous flux of information is required to govern biochemical reactions.

Philip Ball
To take a simple example, where is the 'information' that promotes lipid assembly?

Attila Grandpierre
Lipid assembly corresponds to a concrete, fix form of information. In contrast, another type corresponding of biochemical reactions also exists representing continuously changing information corresponding to continuously changing external and internal conditions. Biological organization governs the relation between such prefixed cycles.

Philip Ball
It is not obviously 'in' the genes that encode lipid-synthesis enzymes, but follows from physicochemical principles.

Attila Grandpierre
It seems that the example of lipid assembly within given physical conditions simplifies biology to physics by throwing out the baby with the bath water. Biology is present not within the framework of an already definite physical problem, but, on the contrary, biology prepares the conditions for the physical laws to act. Biology is the control science of physics. The point is that biology starts 'before' physics, preparing the input conditions for the physical laws. Once the physical conditions are suitably prepared by biological organization, the rest is physics, I agree. I am speaking about the biological aspect, while you seem to be

> concerned with the physical aspect of the problem. It seems we are speaking about different aspects of the same subject.

Attila Grandpierre (continued)
I emphasize that not only the outer environment is influential in determining the behavior of the cell, but the internal environment, too, the determination is not absolute, and it does not occur only on the basis of physical laws. Physical laws are the ultimate, instantaneous tools of biological reactions, but the conditions within which the physical laws act are governed by biological couplings in a way that within the continuously changing biological conditions the physical laws result a biological behavior which is at variance with the physical behavior which would arise in the absence of biological couplings. Cells can utilize the significant part of their thermodynamic potential only through couplings of endergonic and exergonic reactions making it possible to compensate the otherwise inevitable approach towards thermodynamic equilibrium due to entropy increasing physical processes by thermodynamically uphill processes like active transport, regeneration of gradients etc.

Attila Grandpierre
In the second paragraph, Ball seems to ignore my point that the whole cell is involved in biological organization when speaking about "Grandpierre's conception of DNA 'controlling' cellular biochemical reactions". My conception is much more modest. I argue only that the DNA must contribute to the government of biological processes.

Philip Ball
It rather sounds as though he requires this control to be constantly 'active': a reaction cannot proceed as it should unless DNA is *doing something* to ensure that, which then seems to demand ultrafast processes involving, say, electron transport.

Attila Grandpierre
Again, the root of the misunderstanding lies in the different approaches. Ball seems to be involved in the physical approach, considering the moment to moment changes of a certain reaction, with all its physical input conditions already prepared. In contrast, I consider biological behavior in a longer timescale, in the biological context, in which these input conditions are influenced by biological couplings related not only to physical laws and physical conditions but to biological needs and ends as well. (continued below)

> *Philip Ball*
> But is this 'broader picture' not an emergent property?

Attila Grandpierre
A property can emerge, but a law cannot. Phenomenal complexity can emerge in a physical process, but algorithmic complexity cannot. Emergence is a process at the morphological/phenomenal level, while laws exist at the level of algorithmic complexity. Physical laws cannot emerge from material properties (like mass, charge, or size). If biological behavior is governed by the Bauer principle, which cannot be derived from the physical principle of the least action, it cannot emerge from physics.

Philip Ball
Shaped by intermolecular interactions but not obviously derivable from them?

Attila Grandpierre
Recently, it has become clear that simple bacteria can exhibit rich behavior, have internal degrees of freedom, informational capabilities, and freedom to respond by altering itself and others via emission of signals in a self-regulated manner (Ben-Jacob, E. 2003, Bacterial self-organization: co-enhancement of complexification and adaptability in a dynamic environment. *Phil. Trans. R. Soc. Lond. A*, 361:1283-1312). Each bacterium is, by itself, a biotic *autonomous* system, having a certain freedom to select its response to the biochemical messages it receives, including self-alteration, self-plasticity, and *decision making*, permitting *purposeful* alteration of its behavior (Ben Jacob, Aharonov and Shapira, 2005, Bacteria harnessing complexity. *Biofilms*, 1, 239-263; http://star.tau.ac.il/~eshel/papers/11.11.04.pdf). Bacteria are able to reverse the spontaneous course of entropy increase and convert high-entropy inorganic substances into low-entropy life-sustaining molecules (*ibid.*). Similarly, di Primio, Müller, and Lengeler (2000, *SAB2000 Proceedings Supplement, International Society for Adaptive Behavior*, 3-12, http://www.ais.fraunhofer.de/~diprimio/publications/diprimio_MinCog.pdf) have demonstrated that bacteria and other unicellular organisms are *autonomous* and social beings showing cognition in the forms of association, remembering, forgetting, learning, etc., activities that are found in all living organisms. It is widely recognized recently that biochemical reactions are regulated by complex conditions involving practically the whole cell, governed possibly by a yet unknown principle (see Ben Jacob, Shapira and Tauber, 2006, Seeking the foundations of cognition in bacteria:

from Schrödinger's negative entropy to latent information. *Physica A* 359, 495–524.).

Philip Ball
To my mind, DNA simply encodes the local rules that enable such large-scale properties to emerge from the biochemical network. In this sense, DNA does not really contain a 'blueprint' of the organism.

Attila Grandpierre
One of the main points in my chapter is that the changes of the DNA that contribute to selecting, timing and localizing biochemical reactions are regulated by the Bauer principle. I agree with Ball's claim that that DNA's instructions are the result of the biochemical network, but only with the reservation that the biological 'network' is in the actual cell an astronomically enormous dynamic flux of biochemical reactions, estimated in my chapter to represent an information flux around 10^7 bits s^{-1} $cell^{-1}$, which is governed by the Bauer principle. In this respect, the term 'blueprint of the organism' is misleading since suggesting that the biochemical processes are determined by a static material structure similar to a 'blueprint'. Instead, my argument tells that the dynamic flux of biochemical reactions is ultimately governed by the Bauer principle; similarly to physics, since physical behavior is governed by the least action principle. Actually, the 'local rules of the DNA' can be at work only when relying on the universal biological principle; therefore, they are based on the Bauer principle.

(continued)
Philip Ball
It seems to me that the answer is that the DNA does not have to 'know' anything; there are mechanisms for cell-cell communication (some of which, in higher organisms, use the very fast transmission of electrochemical potentials to connect remote regions) which have become adapted in such a way as to permit spontaneous, functionally directed self-organization of the multicellular body.

Attila Grandpierre
In contrast, I emphasize that DNA has to be informed about the cellular processes, otherwise it cannot contribute to the biological organization in a biologically useful manner. I suggest that ultimately, the DNA is informed by virtual interactions (continued below).

Philip Ball
What are 'virtual interactions'?

Attila Grandpierre
Virtual interactions are interactions mediated by virtual particles. In the double slit experiment of quantum physics, the action principle is realized by virtual interactions mapping the whole situation and integrating the quantum amplitudes corresponding to all possible trajectories of the quanta. Feynman's path integral approach indicates that quanta explore all possible paths between the initial and end states (Taylor, 2003, *Amer. J. Phys.* 2003; 71: 423–425.; Moore, 2004, *Amer. J. Phys.* 2004, 72 : 522–527.), and the resulting path is the integrated sum of all these paths. Virtual interactions are governed in physics by the action principle (Feynman and Hibbs, 1965, *Quantum Mechanics and Path Integrals*, McGraw-Hill.

(continued) that determine the biological couplings, which govern biological processes like cell-to-cell communication etc. It seems that the conflict between the physical and the biological approach is manifest when Ball speaks about permitting "spontaneous, functionally directed self-organization of the multicellular body". The point is that physical self-organization is a process which is governed by physical laws. In contrast, biological or functional organization serves a biological need or end, like a biological function, a concept alien to physics. Therefore functionally directed self-organization is not a physical process. It is possible only with the assistance of biological couplings which most fundamentally determine the input conditions for the physical laws (continued below).

Philip Ball
Aha! Maybe this is really where our views diverge. It seems to me that all one needs to obtain function is physical self-organization coupled to selective pressure.

Attila Grandpierre
As I pointed out above, physical self-organization occurs only occasionally, while life requires continuous modifications of the input conditions of physical laws. Function is clearly a biological concept. In general, I define biological function as consisting from processes solving a biologically useful task. In cases of physical self-organization like formation of snowflake patterns, or Benard convection cells, we cannot speak about biological functions. Certainly, we cannot speak about 'selection pressure' in case of

physical self-organization processes. Therefore, it is possible to couple physical self-organization to selective pressure only in case of living organisms. Now living organisms are governed by the Bauer principle, and so it is not allowed to exclude the biological principle from the picture by substituting it with physical self-organization plus selective processes.

Philip Ball
Certainly, that seems to work in *in vitro* chemical evolution.

Attila Grandpierre
Certainly, one cannot speak about biological functions in case of chemical evolution. I note that this point requires ramifications in the context of the problem of continuity of life with the apparently inanimate world, as many scientists suggested; e.g. Editorial, 2007, The meaning of 'life', *Nature* 447, pp. 1031-1032.

(continued)
Philip Ball
"With this in mind, it seems no longer obvious why one need invoke the kind of biological first principle that Grandpierre discusses towards the end of his article,...deciding about biological endpoints and then mapping the pathway there".

Attila Grandpierre
Certainly, with this in mind, i.e. with the physical approach in mind and overlooking the fundamental difference between the physical and biological self-organization, it seems no longer obvious the need for an autonomous biological organization.

Philip Ball
Incidentally, where does the teleological 'biological endpoint' come from? Since when did we need to invoke any prescience to biology in order that it 'works'?

Attila Grandpierre
Yes, the teleological biological endpoint comes from the biological principle, the most action principle, as well as from the autonomous selection from the range of all possible biologically prescribed biological endpoints by the organism. For example, when a bird is dropped from a height from the Pisa tower, in the first moment it moves exactly like a dead bird or a stone, i.e. following the law of free fall. But as moment comes after moment, slowly the bird starts to modify its internal structure with the help of biological couplings, and with the help of the energy resources arising from the exergonic reactions

make it possible to initiate biologically useful endergonic reactions. With the help of self-initiated internal changes, modifying its external shape, the bird extends its wings and modifies its trajectory. In contrast to widespread opinions, teleology is not alien in science, see e.g. Thomas Nagel (1979, *Teleology Revisited and Other Essays in the Philosophy and History of Science*. New York, Columbia University Press, p. 278.). Teleology is directly related to functions (continued below).

> *Philip Ball*
> But evolutionary theory surely shows that function does not imply teleology?
>
> *Attila Grandpierre*
> Evolutionary theory does not explain the origin and nature of biological functions. It only indicates that systems of biological functions, once they exist, can evolve. Machines receive functions only by human activity. Machine's functions represent human teleology, or purpose. Biological functions serve biological purposes (see also Buller, D. J. 2002, Function and Teleology, in: *Encyclopedia of Life Sciences*, p. 393), since ultimately they represent biological endpoint selection as input for the action principle. Without biological endpoint selection serving the most action principle, only the least action principle would be at work, and so only physical processes could occur, driving the organism towards thermodynamic equilibrium and death. Therefore, biological functions inevitably represent biological teleology.

(continued)
Attila Grandpierre
...and biology is the science of mechanisms and functions. "Biology, the scientific study of living organisms, is concerned with both mechanistic explanations and with the study of function." (Purves, Orians and Heller, 1992, Life. The Science of Biology, 1) Therefore, it would be unscientific to ignore the directly teleological aspects of living organisms. The characteristic difference between the physical approach and the biological one is present again in the sentence of Ball:

Philip Ball
"Biology is, from moment to moment, surely quite blind".

Attila Grandpierre
In physics, the differential equations plus completely blind random fluctuations govern the behavior from moment to moment. In biology, the case is very

similar for changes occurring from moment to moment, but very different for changes occurring from the initial state to the final state. Living organisms are able to mobilize their internal energy resources, and, through biological couplings, they can initiate investment of biologically useful work (to recharge their potentials, or to fly away from the free fall trajectory). While the flight of the fallen bird can be regarded from moment to moment as determined by physical conditions, biological couplings act in order to modify the input conditions of the physical equations continuously and systematically, deteriorating the bird from the physical trajectory and making it able to follow any trajectory consistent with the biological principle. (continued below).

Philip Ball
To my mind, the bird has been given a 'flying instinct' by evolution, which it exercises. This 'flying instinct' is a consequence of blind exploration of evolutionary space, coupled to selection.

Attila Grandpierre
In the first chapter "Gaps and Inconsistencies in Modern Evolutionary Thought" of the 728 pages book of West-Eberhard (2003, *Developmental Plasticity and Evolution*, Oxford University Press) a whole list of basic problems of evolutionary theory are presented. It becomes more and more clear that Darwinian theory is so logically flabby it can "explain" anything by subtly changing the terms of the debate. Evolutionary theory can show only that systems of functions may evolve in a changing environment, but does not explain how an individual cell selects from the astronomically large domain of biological possibilities. Evolutionary theory concerns only the historical life forms appearing on the earth. It considers only a part of biological phenomena, instead of working out the general theory of biological processes and deriving the more special phenomena from the more general laws as it is possible in physics. In contrast, the theoretical biology of Ervin Bauer established the most universal law of biology in an exact manner which is quite compatible with the exactness of physics. These arguments indicate that selection is not the cause but the result of biological organization. Therefore, ultimately, the flying instinct, together with the phenomenon of evolution, is based on the Bauer principle.

Philip Ball
I think the example is clearer if we take a less emotive case: a bacterium released near a source of nutrient, to which it can find its way by chemotaxis.

Attila Grandpierre
I repeat my answer given in the previous round: chemotaxis is already a biological input.

Philip Ball
One can design a robot, or most probably soon an artificial cell, to do that. It is all purely mechanical.

Attila Grandpierre
No, it is not mechanical, even not physical; ultimately, it is always the decision of the organism to decide to go for the food. Deciding to go for the food is prescribed not in a coercive manner like the laws of physics but depend on the autonomy of the organism, too. Biological decisions correspond to the most action principle. Once the autonomous organism decided to go for the food, the rest is physics, because the endpoint is decided, and so the most action principle requires the work on the basis of the least action principle in order to secure the most action possible for the organism. Therefore, the first principle of physics is derivable from the most action principle of biology.

Philip Ball
In nature, it's surely now clear why the 'watchmaker' involved in 'design' can be blind.

Attila Grandpierre
I repeat: physics comes only after the biological endpoint is determined. One can close the eye only if it already selected what to do.

(continued)
Philip Ball
"It is only evolution that has installed an apparent 'purpose' to it all."

Attila Grandpierre
Evolution acts only on the species and not on the individual. The trajectory of a fallen bird is not prescribed by evolution in all its details. Evolution cannot play the role of an absolute and rigid control of biological processes.

Philip Ball
Evolution nevertheless shaped all individuals in a species. And I'm not sure it's true to say that evolution can't explain the bird's trajectory.

Attila Grandpierre
Definitely, evolution acts only at the level of species, and cannot determine the trajectory of a concrete bird in all its details. Even in the same environment, the bird dropped from a height can fly to many directions and can select many different trajectories.

Philip Ball
Evolution can explain how the bird gets the instincts and apparatus for flying; the rest is mechanics.

Attila Grandpierre
This claim can be stated only when ignoring the most fundamental aspect of biological organization, namely, biological endpoint selection.

Philip Ball
Yet it leads to the even stranger notion of a 'biological vacuum', which seems to me to be some agency that gives organisms biological foresight of the kind that evolution ensures they don't actually need.

Attila Grandpierre
Evolution is a special aspect of biology. The theory of evolution has quite a different character from theoretical physics. The fundamental laws of physics are the most general laws of physical phenomena, while the theory of evolution considers only a special phenomenon, the evolution of species, and only within special conditions present in the Earth, and so it fundamentally lacks due generality. Biology will reach the position of a mature natural science only if it finds the most general laws of biological behavior. Once theoretical biology will develop, finding its first principle, it will be a more fundamental theory than the theory of evolution. Therefore, rejecting the possibility to consider the first principle of biology on the basis of sticking to a special theory has a corollary hindering the development of natural science (continued below).

Philip Ball
I'm not sure I agree that evolutionary theory need be seen as a 'special case' in biology, or that it precludes other views on biological behaviour, such as a systems view of the cell or explanations for group coordination.

Attila Grandpierre
The fundamental equations of physics determine the most general laws of all possible physical objects. In contrast, evolutionary theory is not the theory of the most general laws of motion of all living organisms possible. It is only the theory of evolution of species of the biosphere present on the Earth. As it is not correct to claim that the semiempirical theory of the growth of the tree is the ultimate theory of the tree, it is also not correct to claim that the semiempirical theory of evolution is the ultimate theory of biology.

(continued)
Philip Ball
Another claim for which I can't find sufficient motivation is the idea that biological systems maximize action. This seems to depend on an assumption that organisms 'try' to live for as long as possible. But that isn't so. They live for as long as is evolutionarily convenient.

Attila Grandpierre
Again, the subject of evolution is the species, and not the individual. In most cases it is an irresistible biological urge for the individual to survive, independently of its "evolutionary convenient" age. Even if *Homo sapiens* would have an evolutionary convenient life span, let us take 80 years, most older people still want to survive. (continued).

Philip Ball
Ultimately the average lifespan is determined by evolutionary factors, regardless of the fact that we have a survival instinct.

Attila Grandpierre
Again, we have to distinguish between the level of biosphere, the level of a certain species and the level of a certain individual. The average lifespan of a species is determined at the level of the biosphere, which can be regarded also as a living organism; see Shapiro, R. 1998, *Planetary Dreams*. Therefore, the average lifetime of a species corresponds to the most action principle applied to the biosphere, to the longest lifespan of the biosphere. The same is true on the level of the individual, which also tries to

live in accordance with the most action principle applied to the individual. Apparently, Ball denies the aim of the individual to live as long and as healthy as possible on the basis of a consideration switching implicitly to a different context, from the individual to the species and the relations of species within the biosphere.

Philip Ball
Of course, humans might one day be able to subvert that technologically (probably to our cost), but we're a special case!

Attila Grandpierre
I hope it is more clear now that I mean not a technological but a biological wish of maximizing survival and health, at the level of the individual.

(continued)
Moreover, the aim of survival only one of the essential aspects maximizing action, since the other factor is to reach the highest amount of available free energy (action=energy*time). Indeed, most living organisms strive not only for survival but for the healthy state in which the available free energy is maximal (continued below).

Philip Ball
Sounds rather like both stem from the same survival instinct.

Attila Grandpierre
Not necessarily. There are two variables: maximal distance from equilibrium (E) and maximal lifespan (T). Action is roughly the product E*T (more precisely, $\int E(t)dt$). Moreover, my chapter presents not a qualitative argument but a quantitative one formulated mathematically. This can be regarded as a definite achievement.

(continued)
Moreover, Ball considers only extreme cases in which the species and the individual are in conflict (continued below).

Philip Ball
In any event, the hypothesis of maximal action seems to be one that is asserted but never proved.

Attila Grandpierre
As I stated in my chapter, Ervin Bauer (1920, 1935/1967; refs. in the chapter) already established the first principle of biology, proving that all the fundamental life phenomena like metabolism, growth, reproduction, regeneration and death can be derived from it. He also formulated the first principle of biology in a quantitative mathematical form, and made it suitable for calculations of biological processes. Moreover, my re-formulation of the Bauer principle in the form of most action principle is proven in Grandpierre, 2007, *NeuroQuantology* 5, 346-362.

Philip Ball
I confess that I'm not familiar with these works – I will need to take a look.

(continued)
Philip Ball
At root, I am perhaps most perplexed by the notion that algorithmic complexity has to be high to account for biological phenomena. Has it not been one of the underpinnings of complexity science that complex behaviours can arise from simple rules?

Attila Grandpierre
This is a very good question, related to a main aspect of my chapter. Indeed, today it is a dominant view that complexity can arise from simple rules. But one must distinguish between the above weak form of such a statement and a stronger form claiming that all complexities found in nature must be derived from simple rules regulating only the physical properties of the systems and organisms. As I enlightened it in my chapter, algorithmic complexity has a fundamentally different nature from the morphological or phenomenological complexity. Let us take an example. The circle has a low algorithmic complexity (cca. 100 bits), and an extremely high morphological complexity (infinite points, infinite bits). Therefore, it is apparent that a small algorithmic complexity is able to produce an extremely high amount of morphological complexity. In this sense, algorithmic complexity is more fundamental than morphological one. Does it follow from the fact that the circle has a low algorithmic complexity that we must think that all the mathematical functions can be derived from simple rules? No, because, for example, there are many mathematical objects that cannot be given in algebraically closed form. Let us take another example. There are simple machines like a watch having a low algorithmic complexity. Does it follow that we must accept that all machines must have low algorithmic complexity? No, because a computer with higher algorithmic complexity can solve more tasks and more easily than a smaller computer. Moreover, once the

machine is ready, its functions are specified, its algorithmic complexity is given. But there are tasks for living organisms requiring revealing a problem, to realize the existence of an unexpected task. Living organisms must continuously solve new and new problems, and problem solving is by definition corresponds to the production of algorithmic complexity. Production of algorithmic complexity is possible only if a still deeper level of complexity (generative complexity) exists which can produce algorithmic complexity on the basis of a unified context corresponding to the generative principle. One of the main points of my chapter is to show that we must realize that algorithmic complexity and generative complexity can be regarded as full members of the conceptual framework of science and they are fundamental aspects of nature. My answer to Ball's problem is that we have to consider systems and organisms with high algorithmic complexity. The algorithmic complexity of a circle or a fractal is low, a watch has a higher algorithmic complexity, a computer still higher, and a living organism still much higher. Indeed, the algorithmic complexity of a living organism must be high to account for biological phenomena (continued below).

> *Philip Ball*
> I need to give this more thought! All I'd say now is that our point of reference in such questions should in my view be a bacterium rather than any higher organism: to my mind, once you have the former, the latter follow…
>
> *Attila Grandpierre*
> There are more and more evidence indicating the existence of autonomous bacterial purpose and intelligence, like e.g. Mathieu and Sonea 1996, Time to drastically change the century-old concept about bacteria, *Science Tribune*, August 1996; Ben-Jacob, 2003, Bacteria harnessing complexity. *Phil. Trans. R. Soc. Lond. A*, 361, pp. 1283-1312; Ben Jacob, Aharonov and Shapira, 2005, Bacterial self-organization: co-enhancement of complexification and adaptability in a dynamic environment. *Biofilms*, 1, 239-263; http://star.tau.ac.il/~eshel/papers/11.11.04.pdf; di Primio, Müller, and Lengeler, 2000, Minimal Cognition in Unicellular Organism. *SAB2000 Proceedings Supplement, International Society for Adaptive Behavior*, 3-12,
> http://www.ais.fraunhofer.de/~diprimio/publications/diprimio_MinCog.pdf; Ben Jacob, E., Shapira, Y. and Tauber, A. I. 2006, Seeking the foundations of cognition in bacteria: from Schrödinger's negative entropy to latent information. *Physica A* 359, 495–524; Shapiro, J. A. 2007, Bacteria are small but not stupid: cognition, natural genetic engineering and socio-bacteriology. *Stud. Hist. Phil. Biol. & Biomed. Sci.* 38, 807-809; B.

J. Ford, 2004, Are cells ingenious? *Microscope* 52, 135-144; B. J. Ford, 2006, Revealing the ingenuity of the living cell. *Biologist* 53, 221-224.

(continued)
Philip Ball
Simple models of ant motion, driven by simple ideas such as chemotaxis and random searching, can reproduce the behaviour of ant colonies rather well.

Attila Grandpierre
Again, it seems that Ball thinks in a physical approach without realizing the difference represented in biological behavior. Definitely, when introducing biological concepts like chemotaxis, it is already possible at least in some simple cases to reproduce the observable behavior of ant colonies (but not so well of each individual ant, I guess). My biological approach intends to establish the idea that it is necessary to add suitable biological concepts to our physical vocabulary when attempting to determine biological behavior (see Grandpierre, 2007, *NeuroQuantology* 5, 346-362.).

Philip Ball
For simple organisms such as bacteria, it seems even possible in principle that one might sense from moment to moment all the environmental influences acting on a single cell, and thereby predict its motion with great precision.

Attila Grandpierre
Again, the difference arises in the approaches. Ball seems to stick to the physical approach corresponding to extremely short timescale changes "from moment to moment". In the long run, in a long enough timescale, it is not possible to determine the behavior of a fallen bird in the absence of suitable biological endpoints (like chemotaxis, reaching the food, escaping from a danger etc.). Therefore, even if it would be possible to predict the behavior of a cell from moment to moment (like the position of a fallen bird in the next instant) approximately, it is not possible to determine the biological behavior of the cell on a purely physical basis on a biological timescale (the trajectory of a fallen bird deteriorating itself from the free fall path) (continued below).

Philip Ball
Hmm… I'm not sure I agree with the way the problem is posed, as though a complex biological entity like a bird can be compared with an inanimate particle.

Attila Grandpierre
The problem is posed in the framework of the Galileo experiment, considering the fall of different objects from the Pisa tower and watching their behavior. The behavior of the bird is compared not to that of an inanimate particle but of an inanimate thing. The problem is posed in accordance to the central task of science, namely, finding the laws of nature governing observable behavior. The laws of physics explain physical behavior. We must consider biological behavior in order to understand the nature of the biological principle. To understand the difference of biological principle from the physical one, we must compare the behavior of inanimate objects and living organisms.

Philip Ball
It is clear that organisms possess 'motivations', which means that their trajectories, while not violating Newtonian mechanics, can't easily be deduced from that.

Attila Grandpierre
It is not only the case that biological behavior "can't easily be deduced from Newtonian mechanics". The plain fact is that the most action principle cannot be derived from the least action principle, independently of trying easily or more systematically. The first principle of biology simply cannot be derived from the physical principle. The fundamental complexity measures I considered in my chapter underpin this fact quantitatively. Since the algorithmic complexity cannot be produced by physical processes, and because the laws of physics has an algorithmic complexity around 1 000 bits (see in my chapter), therefore the algorithmic complexity of biological organisms, being much higher than the 1 000 bits of physical laws, cannot be derived from physics.

Philip Ball
But I'm not clear why this should be a mystery that requires any principles beyond the ones we have already.

Attila Grandpierre
On the contrary. Biological behavior is a mystery at present only when narrowing down our considerations to the framework of physics. This apparent mystery can be resolved on the basis of a principle which is published more than eighty years ago by Ervin

Bauer, which is, unfortunately, still ignored. So we do not require any principles beyond the ones we have already.

Philip Ball
Why doesn't evolution alone suffice to provide the imperatives and mechanisms, which are then acted out in particular situations and contexts by particular organisms?

Attila Grandpierre
The evolution of species is merely a historical process of special, namely, earthly life forms and it should not be taken as the most fundamental biological process. The most fundamental biological processes are the ones corresponding to the most action principle, which are the virtual interactions generating biological couplings between exergonic, energy liberating and endergonic, energy requiring processes. Evolution is like the growth of the tree. Similarly, the growth of the tree is not the fundamental life process of the tree. Cellular biochemical activity, the biological couplings governed by the most action principle, and the virtual interactions manifesting the most action principle are more fundamental biological processes. Growth of the tree is the result of cellular activity and not the other way around. One cannot explain cellular biochemical reactions in terms of growth of the tree. The "biological imperatives" arise ultimately not from evolution and selection pressure, but from biological organization governed by the Bauer principle.

(continued)
Philip Ball
Grandpierre argues that abiogenesis cannot seem to create, in a sufficiently short time, the complexity we see in life: if I understand correctly, he implies that only life (or 'intelligence') can beget life. To my mind, there are two shortcomings with this. First, it assumes that accumulation of complexity is linear, whereas it now seems that many complex systems possess thresholds above which entirely new modes of behaviour – new capabilities – appear.

Attila Grandpierre
Actually, I argued in my chapter that the accumulation of complexity can be faster in living organisms than in abiotic systems. Moreover, one of the main results of my paper is that I determined quantitatively the rate of complexity upjumps from abiotic systems to the first simple living organisms, and the complexity upjumps from the smallest bacteria to humans. The results tell that the complexity upjump assumed in the hypothetical abiogenesis during a period

less than hundred million years is larger than the complexity upjump during the four billion years of biological evolution. I argued that this result makes the assumption of abiogenesis improbable.

Philip Ball
As I understand it, this is consistent with my claim that, once you have a bacterium, all the rest (up to humans) follows. I agree that the origin of life is a huge step in increase of complexity. But I'm not sure that we should necessarily have any expectations about whether life, once begun, would maintain a comparable rate of increase in complexity. Bacteria and other single-celled organisms are extremely successful; humans are anomalies. Work like that of Stuart Kauffman at least claims to show that, once you have autocatalytic cycles, you have the basic ingredients of 'life' in place.

Attila Grandpierre
Autocatalytic cycles when coupled together into an integrated unit form an automaton. Organisms are much more than automata since they are able to reorganize themselves by biological organization. It follows that biological organization is not the result of autocatalytic cycles but is the cause of organizing such cycles, together with ingenious reactions, into a living organism. Indeed, bacterial autonomy, motivations and significant achievements (see the references indicated above) show the profound significance of non-mechanically repeated biological reactions.

Philip Ball
In any event, I'm not sure we understand the processes that lead to a living organism well enough yet to be able to make convincing claims about how 'unlikely' abiogenesis is.

Attila Grandpierre
The argument presented in my chapter is a quantitative one which is more close to the norms of established science than opinions which did not reach the phase of quantitative arguments.

Philip Ball
Hence my comment below: it seems you are introducing a new idea/process largely because we don't yet have enough understanding to bridge the gap on the basis of established ideas, rather than because there is any clear empirical demand for it.

Attila Grandpierre
There are arguments indicating that complexity science is the next frontier of natural sciences as well as arguments telling that the 21st century is the century

of biology. Unfortunately, it seems that both complexity science and theoretical biology lacks its exact base which could be comparable to the exact base of physics. In my chapter I tried to introduce fundamental measures of complexity. I found that besides the complexity we observe with our eyes (morphologic or phenomenal complexity) there exist a fundamentally different level of complexity which is already known as algorithmic complexity. I tried to show that genetic complexity, which is a static thing, represents a still deeper level of complexity. I argued that biological organization has a complexity corresponding to the genetic level of complexity, but, at variance with genetic complexity characterized by the number of non-coding base pairs, biological organization represents a form of activity. With the help of these fundamental complexity measures I tried to obtain new light on the difference between machines and organisms. I hope I succeeded to make some steps towards a quantitative theory of biological organization and show the possibility of putting complexity sciences and theoretical biology to a more firm basis. Actually, there is a whole list of experimental data and theoretical underpinnings indicating the increasingly vital need for a realistic complexity science and theoretical biology (given in Grandpierre, 2007, *NeuroQuantology* 5, 346-362.). Therefore, I did not introduce a mystic new factor to fill the gaps of our present knowledge. Instead, my approach explains biological behavior on the most fundamental, exact and elegant way possible, since the most action principle of biology and the least action principle of physics form a natural union: the general action principle.

Philip Ball
To my mind, "intelligence in Nature" here becomes another God of the gaps.

Attila Grandpierre
To my mind, the first principle of physics is not another God of the gaps. It is quantitative, predictive, and consistent with our best and broadest theories and the widest range of established empirical facts. Similarly, the biological first principle is also quantitative, predictive, it is the best and broadest biological theory, consistent with the theoretical biology of Ervin Bauer, and with a large body of yet unexplained facts that it can explain. This is science and not religion. In the twenty-first century more and more biological data are accumulated. In the absence of a general theoretical biology, there is an increasing frustration between millions of biologists (Brent and Bruck, 2006, *Nature* 440, 416). Recently, biological physics became a new frontier of natural sciences (Phillips and Quake 2006, *Physics Today* 2006; 59(5): 38-43.; Sung, 2006, *Crossroads: Journal of Asia Pacific Center for Theoretical Physics* 2006; 4: 1-3.). Biological physics is the interdisciplinary effort to cross the barriers between physics and biology from the biology side (Sung, 2006). The US National Science Foundation allocates billions of dollars year by year for initiating the birth of

theoretical biological physics (Ladik, 2004, *Journal of Molecular Structure* 2004; 673: 59-64.). "Promoting research that encourages a holistic perspective to "understand complex systems" is a long-term investment priority in the strategic plan of the National Science Foundation of the United States" (Hübler, 2007, *Complexity* 12, No. 5, 9). The fundamental complexity measures of nature introduced in our chapter serves both needs simultaneously.

I think it is time to realize the need for a realistic theoretical biology and complexity science.

Philip Ball
"I think it is time to realize the need for a realistic theoretical biology and complexity science." I completely agree!

Biography

Rabbi Natan Slifkin directs "**Zoo Torah**," http://www.zootorah.com/, an educational enterprise that teaches about the relationship between Judaism and the animal kingdom. He is the author of numerous works on this theme, most recently *The Challenge of Creation: Judaism's Encounter with Science, Cosmology and Evolution*.

29
Evolutionary Mechanisms and Intelligent Design

NATAN SLIFKIN
2a Nachal Raziel, Ramat Bet Shemesh 99632, Israel
zoorabbi@zootorah.com

1. Science and Religion

One of the primary religious objections to there being *any* sort of naturalistic mechanism that would explain how one species changes into another is that "Scientists say that they can explain the natural world with evolution, but we know that God did it." People feel that science, in giving a naturalistic explanation for how bacteria turned into the diverse creatures of the world—and one based on randomness, no less—is painting God out of the picture. But is it indeed doing so?

In order to understand God's role in biology, let us first consider astronomy. The *Bible* considers that the celestial phenomena are a primary means of perceiving God as a Creator:

> Lift up your eyes upon high, and perceive Who created these! (Isaiah 40:26)
> The heavens speak of God's glory, and the sky tells of His handiwork! (Psalms 19:2)

And yet today, the science of astronomy is highly advanced. We have built up a comprehensive picture of the universe. We understand how stars form. We understand the laws of gravity that govern their motion. So should religion oppose the science of astronomy? Does it mean that the prophet Isaiah and King David were speaking out of ignorance when they told us to look up at the cosmos and perceive God?

The answer is that of course it doesn't. A natural, scientific explanation of things in no way contradicts the concept of God as a Designer. Instead, it provides a new tapestry for Him to have designed. As Einstein said:

> "You find it surprising that I think of the comprehensibility of the world to the degree that we may speak of such comprehensibility as a miracle or eternal mystery. But surely, *a priori*, one should expect the world to be chaotic, and not to be grasped by thought in any way.... Even if the axioms of the theory are posited by a human being, the success of such a procedure supposes in the objective world a high degree of order, which we are in no way entitled to expect *a priori*. Therein lies the 'miracle' which becomes more and more evident as our knowledge develops. And here is the weak point of positivists and professional atheists, who feel happy because they think that they have pre-empted not only the world of the divine but also of the miraculous" (Einstein, 1956).

Nobel Prize-winning physicist Eugene Wigner (1902-1995) made the same point:

> "It is, as Schrödinger has remarked, a miracle that in spite of the baffling complexity of the world, certain regularities in the events could be discovered.... It is not at all natural that "laws of nature" exist, much less that man is able to discover them" (Wigner, 1960).

Why, then, should biology be any different from astronomy? Certainly there is no basis in classical religion for saying that God is perceived only in the animal kingdom and not elsewhere in the universe. In the same way as we do not see the science of astronomy as negating this, the science of evolutionary biology does not paint God out of the picture of the development of life.

Judaism sees the laws of science as a description of the mechanism that God uses in creating and operating the world. As Rabbi Samson Raphael Hirsch (1808-1888) puts it:

> "All the laws of mechanics, physics, chemistry and physiology can no more do without the One Who governs and guides the course of the universe (and, according to Jewish teaching, also the life of the individual in accordance with His purposes) than a living body, breathing in accordance with the laws of physiology, can do without the unfettered guidance of a human intelligence, or than a steamship, operating in accordance with the laws of mechanics, can do without the helmsman who guides it in accordance with his own free will" (Hirsch, 1984).

Everyone appreciates that science is very good at explaining the daily functioning and operation of the world. And yet religious people have no difficulty in seeing God's hand in the day-to-day running of the world notwithstanding the scientific explanation. Since science does not prevent us from perceiving God in the operation of the world, it should certainly not prevent us from perceiving God in its design.

At the end of the day, just as our understanding of astronomy does not prevent us from saying that God makes the sun shine, so too an explanation for the development of life based on seemingly random mutations and natural selection would not prevent us from saying that God designed animal life. Darwin (1859) himself, in later editions of his book, pointed this out:

> "I see no good reason why the views given in this volume should shock the religious feelings of any one. It is satisfactory, as showing how transient such impressions are, to remember that the greatest discovery ever made by man, namely, the law of the attraction of gravity, was also attacked by Leibnitz, "as subversive of natural, and inferentially of revealed, religion." A celebrated author and divine has written to me that "he has gradually learned to see that it is just as noble a conception of the Deity to believe that He created a few original forms capable of self-development into other and needful forms, as to believe that He required a fresh act of creation to supply the voids caused by the action of His laws."

In 1873, fourteen years after the publication of Darwin's *Origin of Species*, Rabbi Samson Raphael Hirsch addressed the topic of evolution. Although he considered at that time that evolution was "a vague hypothesis still unsupported by fact," he declared that while certain stated implications of the theory were wrong, the essence of it is by no means incompatible with Judaism:

> "Even if this notion were ever to gain complete acceptance by the scientific world, Jewish thought, unlike the reasoning of the high priest of that nation [probably a reference to Thomas Huxley, who advocated Darwinism with missionary fervor—N.S.], would nonetheless never summon us to revere a still extant representative of this primal form [an ape—N.S.] as the supposed ancestor of us all. Rather, Judaism in that case would call upon its adherents to give even greater reverence than ever before to the one, sole God Who, in His boundless creative wisdom and eternal omnipotence, needed to bring into existence no more than one single, amorphous nucleus, and one single law of 'adaptation and heredity' in order to bring forth, from what seemed chaos but was in fact a very definite order, the infinite variety of species we know today, each with its unique characteristics that sets it apart from all other creatures" (Hirsch, 1984).

Rabbi Hirsch makes it clear that a naturalistic explanation does not paint God out of the picture. On the contrary—in this case, it testifies all the more to the genius of the Creator, to His "creative wisdom."

2. The Intelligent Design Movement

Intelligent Design advocates state that it is apparent from nature that there is an intelligent Designer Who made the world with a certain end goal. The point of the theory is that even if all life forms developed from earlier forms, this cannot be explained in purely naturalistic terms. Mathematician and philosopher William Dembski, a primary figure in the Intelligent Design movement, writes as follows:

> "...Logically speaking, Intelligent Design is compatible with everything from the starkest creationism (i.e., God intervening at every point to create new species) to the most subtle and far-ranging evolution (i.e., God seamlessly melding all organisms together in a great tree of life). For Intelligent Design the first question is not how organisms came to be (though this is a research question that needs to be addressed), but whether they demonstrate clear, empirically

> detectable marks of being intelligently caused. In principle, an evolutionary process can exhibit such 'marks of intelligence' as much as any act of special creation" (Dembski, 1998).

In practice, however, ID does not refer to either of the extremes described by Dembski. Although some religious people trumpet various statements by ID spokesmen in order to attack evolution and implicitly support special creation, these ID spokesmen generally accept that all life-forms did indeed develop from a common ancestor over billions of years. And, at the other extreme, although some ID advocates may speak of perceiving God within a fully naturalistic framework, this is not at all what the ID movement presents.

The Intelligent Design movement presents certain things in particular that are considered to show evidence of intelligent design. One of the spokesmen for ID, biochemist Michael Behe, wrote a book called *Darwin's Black Box* which was a flagship work for the ID movement. Behe claims that certain features of biological organisms exhibit what he calls "irreducible complexity." This means that such organisms require various components to work in unison and would not work if any of these components were faulty. The claim is that such organisms could not have evolved without guidance, since there is no intermediate stage which is functional. The prime examples offered are the bacterial flagellum—a tiny whip-like structure that some bacteria use to swim—and the structure of proteins in the human blood-clotting system.

Whether these things do indeed demonstrate irreducible complexity is debated; the arguments do not concern us here. The problem is that the ID movement involves some extremely problematic theological aspects of which many people are unaware. Many well-meaning religious people assume that the ID movement is a friend of religion and their cause should be supported. But this is not the case at all.

The ID movement claims that we should look for signs of intelligent design in biological phenomena, and that schools should teach this perspective in their biology classes. But nobody claims that we should do that in a class studying (secular) history, or physics, or geology! A bizarre aspect of ID is the place that it allocates for God showing His hand in nature. It tells us that while God's hand is not to be looked for in the movement of the planets, in the formation of the world, or in the harmony of nature, it can be seen in the bacterial flagellum and the blood-clotting system. These are odd choices of places for God to show Himself.

The ID movement effectively tells us that the prophets of Scripture who perceived God in the cosmos and the natural world were all speaking out of ignorance. ID claims that, since we have scientific explanations for these things, God cannot be seen there. Instead, God is to be found in the bacterial flagellum and the blood-clotting system!

Aside from the theological absurdity of such a belief, the danger of intelligent design theory is that it risks forming a "God-of-the-gaps"—a Creator who is invoked to account for phenomena that science cannot explain, but whose existence is unnecessary if science is eventually able to explain such phenomena in naturalistic terms. Phillip Johnson admits to this peril, and presents his response:

> "There is a risk in undertaking such a project, of course, as the theistic evolutionists constantly remind us by referring to the need to avoid resorting to a 'God of the gaps.' If the naturalistic understanding of reality is truly correct and complete, then God will have to retreat out of the cosmos altogether. I do not think the risk is very great, but in any case I do not think theists should meet it with a preemptive surrender" (Johnson, 1993).

Johnson's statement that "if the naturalistic understanding of reality is truly correct and complete, then God will have to retreat out of the cosmos altogether," is by no means true. A complete explanation of the celestial bodies by astronomy, or an explanation of the formation of mountains by geology, or of rain via meteorology, does not paint God out of the picture, but instead means that He works through science. But Johnson denies such a role for God.

An ID proponent might take a different position from Johnson and claim that they indeed perceive God in those aspects of organisms that can be explained by evolutionary mechanisms, but that they see Him all the more powerfully in aspects which they believe cannot be described by such processes. Yet this would appear to be a slight to God's creative abilities. Was He incapable of designing laws that could accomplish all His objectives, and therefore had to interfere to bring about the results that He wanted?

The ID movement, in its usual manifestation, is no friend to Judaism. It denies the role of God in 99% of the universe, relegating Him to being little more than the designer of bacterial flagella and blood-clotting systems; or it implies that He was only able to engineer processes that would accomplish 99% of His objectives, but not all of them. Postulating the intelligent design of organisms means postulating the unintelligent design of natural laws.

3. Theistic Darwinian Evolution

On the other hand, some argue that Darwinian evolution is uniquely opposed to religion in a way that other sciences are not, due to the "blind watchmaker" aspect of Darwinian evolution; that which Richard Dawkins describes as "the blind, unconscious, automatic process which Darwin discovered, and which we now know is the explanation for the existence and apparent purposeful form of all life, has no purpose in mind." It is this which is regarded by some as the aspect of evolution most dangerous to religion. Proponents of ID consider this ideology to be inextricably linked to Darwinian evolution. It would certainly appear at first glance that the blind-watchmaker thesis, which is the essential feature of Darwinian evolution, is incompatible with the concept of a seeing-watchmaker. We shall have to analyze this carefully.

One point to note is that it is impossible to ever determine whether something is truly random. One could take a string of a hundred seemingly random numbers and perform every conceivable test, and discover no pattern. Yet it could be that those numbers were actually the numbers preceding the one hundred digits that appear after the millionth digit of Pi. So, they were not actually random at all.

Another point to note is that the words "blind," "chance," and "random," when used in the context of Darwinian evolution, refer to the combination of chance genetic mutation together with arbitrary deterministic forces of circumstance such as environmental change, and so on; they are only "blind" in that they are not visibly geared towards the eventual structure of the creature. Thus, while the term "chance" causes many religious people to instinctively recoil, a careful analysis of its usage in describing evolution reveals that there is no reason for this.

It cannot be adequately stressed is that apparent "chance" is fully consistent with the classical religious view of how God runs the world. This is spelled out explicitly in Scripture:

> [When] the lot is cast in the lap, its entire verdict has been decided by God. (Proverbs 16:33)

The nineteenth-century Biblical commentator Rabbi Meir Leibush elaborates:

> "There are things that appear given to chance but are actually providentially determined by God... . 'the lot is cast in the lap,' hidden from the eye of man, handed over to chance, but nevertheless the eye of God's providence is displayed in it, and the verdict that the lot brings up is not chance but is from God; just as with the apportioning of the land and so on, where the lot was under God's providence" (Leibush, 2000).

The example that Rabbi Leibush gives is the dividing of the Land of Israel amongst the twelve tribes, which was done via a lottery:

> God spoke to Moses, saying, To these the land shall be divided for an inheritance according to the number of names. To the more numerous you shall give a larger inheritance, and to the fewer you shall give a smaller inheritance; to every one shall his inheritance be given according to those who were counted by him. However, the land shall be divided by lot; according to the names of the tribes of their fathers they shall inherit. According to the lot its possession shall be divided between many and few. (Numbers 26:52-56)

Despite the seeming randomness of the process, Jewish tradition is emphatic that the apportionment was not truly random. Rather, each tribe received the portion of land that was predestined for it. The seemingly random process of the lottery was merely a guise for the Divine decision.

It is abundantly clear from all this that the randomness of Darwinian evolution poses no theological problem whatsoever. Religion has no problem with processes that *appear* to be random, and in fact it sees them as an ideal means via which God dynamically exerts His will. While the evolutionary process may well *appear* to be random from our perspective, it can simultaneously be directed from God's perspective.

References

Darwin, C. (1859). On the Origin of Species.

Dembski, W. (1998). "The Intelligent Design Movement," Cosmic Pursuit, Spring 1998.

Einstein, A. (1956). Lettres a Maurice Solovine.

Hirsch, S.R. (1984) Collected Writings (Jerusalem: Feldheim Publishers).

Johnson, P. (1993). Darwin on Trial, 2nd edition (IL: InterVarsity Press).

Leibush, M. (2000). Commentary to the Bible (HaTorah VeHaMitzvah) (Bnei Brak: Mosdos Chassidei Alexander).

Wigner, E. (1960). "The Unreasonable Effectiveness of Mathematics in the Natural Sciences," Communications in Pure and Applied Mathematics, **13**:1.

Part 6

Science Curricula in Schools of Various Countries

Biography

Galina Leonidovna Muravnik is a Vice-Rector in St.Andrew's Biblical theological institute. She graduated the Biological faculty of the Moscow State University (Russia) in 1979 and specialized at the Department of Genetics. In 1985 she finished a post-graduate course at the Institute of Genetics and Selection of Microorganisms (Moscow). At the same time she attended lectures in various fields of theology at Father Alexandr Men's Open Orthodox University. In 1998 she graduated the course on orthodox pedagogy and pedagogical skill. In 1992 Muravnik turned to teaching at an Orthodox Lyceum of Moscow. She created a program with the purpose of implementing a harmonious synthesis of modern evolutionary biology and Christian divinity. As a result she has been teaching a course called "Introduction to Christian natural sciences" all these years. Nowadays she leads the Educational Department of St. Andrew's Biblical Theological Institute and gives there a number of courses — "Science and religion", "Theology of evolutional biology", "Bioethics" — as well.

30

The Divine Creation and Biological Evolution

The Experience of the Teaching of Evolution in Russian Secondary School

GALINA L. MURAVNIK

St.Andrew's Biblical theological institute 3, Jerusalem St., 109316 Moscow, Russia
muravnik-gl@mail.ru

1. The Problems of Teaching Evolution in Russian Secondary Schools

For many years I have tried to find the answer to a question, whether science, in particular — evolutionary biology, is capable of contributing to the Christian education and formation of religious world outlook in schoolboys. Whether the study of mysteries of the development of living nature contributes to the most mystical of processes — the Knowledge of God? And if so, what should the relationship between the secular and religious in a school course of biology be?

In Russia at the beginning of the 1990s the official ban on religious education was removed, and soon a lot of orthodox schools were created. Now in Moscow alone there are about 30 comprehensive schools of this kind. Similar schools exist in many towns of Russia. The teachers of biology working at orthodox schools of Russia, whether they want it or not, at each lesson have to search for their own answer to the question on the relationship between Christian religious doctrine and modern natural sciences. The situation is exacerbated by the fact that for the time being there are no special school programs created for orthodox schools. The same also applies to biology as a subject mandatory for all Russian schools.

But the current version of modern education, which is offered in natural sciences, can be called the "school of atheism", as it radically excludes a Christian view on creation. And it is impossible to solve this problem, which requires urgent revision, by simple and quick methods like reprinting old, pre-revolutionary school textbooks and programs. A new approach based on achievements of sciences of the 20th century is necessary.

On the relationship between science and divinity there is rather a wide spectrum of opinions, from a complete denial of science as the product of mankinds' inherently sinful mind, "fallen of man's mind", up to a complete denial of the religious view on the world order and calling science "the new religion". However, there is another point of view, and I support it too. Its essence is that religion and science, belief and reason, are not in confrontation with each other, but are accepted as two ontologically interconnected paths leading to knowledge of the Creator. Their interaction contributes to enrichment and development of each other. Anglican theologian and scientist John Polkinghorne has called it "friendship of science and religion" (Polkinghorne, 1997).

I am quite sure it is exactly on this beneficial course that both science and divinity will make the most considerable discoveries since this course widening the boundaries of our world outlook, implies the ascension to the summit of knowledge.

2. The Situation at Orthodox Schools

When we consider the situation at orthodox schools, the teacher has to maneuver skillfully between the program offered by the Russian Ministry of Education, and his/her own religious views of the world. As a matter of fact such practice brings into the minds of schoolchildren the basis for an inevitable spiritual crisis and an inner split. Some of them as they grow up label science as a "demonic occupation", and reject it. Others gradually drift apart from the Church, losing live, sincere belief. At best they remain the nominal Christians, who only occasionally come into a Church. In the worst case skepticism born in their souls brings them to the camp of atheists.

Such a situation cannot be acceptable, that is why it needs to be changed. But how? In my view, for a long time there has been the necessity for developing a modern concept of Christian natural sciences, however unusual this term may sound today.

3. In Search of a New Conceptual Space

What kind of a conceptual space is needed in which the search for and birth of modern Christian natural sciences is possible? And, above all, what are the problems of Christian natural sciences, which should be solved? They can be formulated as follows:

- to show convincingly, on the basis of modern scientific achievement, actual conformity of scientific deductions to what is disclosed in The Holy Writ and developed in the Christian divinity;
- to show that the Divine Will in the world is carried out through natural processes and laws which are a subject of scientific research;
- to facilitate the formation of a Christian world outlook in the pupils, and, on this basis, of the system of harmonic relations with God's world.

This kind of a pedagogical method has a number of advantages. First of all, it enables not only acquainting schoolchildren with recent achievements of biology, but also allows them to get in touch with the most mystical secret, by feeling the actual presence of God in the world. Thus the overlapping of scientific and religious views of the world should inevitably take place. We may say that biology gives us a perfect opportunity to speak with schoolchildren about the most sacred things on the basis of our belief, contributing to the formation of a Christian world outlook.

4. The New Course — "Introduction to Christian Natural Sciences" and its Ideas

I have created a special program for teaching in the Secondary School — the course "Introduction to Christian natural sciences". It is intended for teaching in the10th and 11th forms of general educational orthodox schools of Russia. The basic purpose of the course is formation of the Christian world outlook in the students on the basis of the data of modern sciences, withdrawal from a far-fetched "conflict of a science and religion" and, on the contrary, an effort to make them to think that scientific and religious views of the world can and should be indivisible, as is the world created by the God. This ontological integrity of Faith and Science (if we address summits of both) is caused by their common basis, i.e. "*one and the same Spirit*" (1Cor., 12.11).

The course, which I offer, is based on a principle of teleology (from Greek "Telos" — extremity, end, purpose), that ultimately leads to theology. The main idea is that the whole course is formed in such a way that it constantly shows that the origin, development and formation of the world is the result of Divine creativity, that the creation was from the very beginning carried out according to some purpose — God's Design. The laws formulated by science, in particular, the so-called the anthropic principle, proves the expediency of the world. Thus, the aim of science wholly and evolutionary biology in particular, is not just the study of the world for the sake of making life comfortable for mankind or for the satisfaction of natural curiosity, but the comprehension of wisdom of the Creator through His traces in the world. The Apostle Paul states this idea in his Message to Romans: "… Wherefore His invisible attributes, His eternal power and divine nature, since the creation of the world through examining of creations visual…" (1.20).

Thus, the task of the course, developed on the basis of teleology, is a consecutive continuous formation of a Christian world outlook in the pupil on the basis of the data of modern evolutionary biology, perceived through a prism of Christian divinity. Actually the course of Christian natural sciences becomes a sort of the natural science of apologetics, one of the aims of which is the protection of the foundations of Christian dogmas against perversion and assaults. But the natural science of apologetic is a child of its time. It rates with those data, which are offered to it by modern science. Therefore it is forced to be updated constantly, involving in the circle of its interests all new discoveries which are made in science. It is gratifying that lately both scientists and theologians realized the productivity and necessity of a dialogue between science and religion. Logically and inevitably it is leading to interaction of belief and knowledge by a principle of complementarity. They, as two wings, are lifting mankind to the heights of knowledge of Truth.

I believe that the whole course "Introduction to Christian natural sciences" should be built around the text of chapters 1-2 of the Book of Genesis. It is possible to say, that Six Days of Creation is some kind of "roll of the world".

And the aim of science is to unroll and to read it. But it is not necessary to try in a sense "to find" literally, in the text of chapters 1-2 of the Book of Genesis, scientific laws or theories. This text is just a field of open questions and an indication of the direction of scientific search.

At the same time, the whole course should be formed around the above mentioned anthropic principle as a core, which enables us to see and to feel an excellently "tuned" universe, created not randomly, i.e. the only world, in which God-like man, could appear and live.

Christian science, recognizing the world as cognizable, by studying its laws, ultimately approaches the comprehension of the reasonable design of the world, studying the world not in general, but comprehending it as God's creation, as His creative workshop. It is not surprising that many scientists, not being formally members of the Church, through their scientific investigations came to the idea of God. As A. Einstein said: "...the belief in the sensibility of the world inspires the explorers". This sensible world creation, more precisely the creation contemplated by the Creator and embodied in the material world according to the laws set by Him, is the very subject of study in the course "Introduction to Christian natural sciences".

So, the basic pedagogical aim is to equip children with knowledge of the evolution of living beings, which will allow them not only to work in science professionally if they wish, but also to carry the light of Christian belief into the world.

5. Feature and Principles of the Course

The course is based on a historical-hierarchical principle. The program of the 10th form begins with the study of the molecular level of organization of living beings, (acquaintance with biological chemistry, molecular biology and molecular genetics). Further it goes step-by-step through higher organizational levels providing acquaintance with the biology of a cell, embryology, and also different fields of genetics (general genetics, genetics of the man, cytogenetics, genetics of populations, mutagenesis, genetical bases of selection, genetic engineering).

All these sections of biology form the theoretical base, which enables us to study, on a serious scientific level in the 11th form, such complex problems as the origin of life, biological evolution of species, and the origin of the man, by no means indifferent for Christian consciousness.

My choice of necessary scientific material is rather different from that generally studied at such lessons. I use a historical principle. It seems to me extremely important not just to inform the pupils about this or that scientific fact of modern evolutionary biology, but to show them how long and difficult was the search for it. In this way the pupils get acquainted with a history of the origin of scientific discoveries including not only successes, but retreats at times, delusions, dead-ends and search for the way out, as well as further advance step by step approaching the knowledge of secrets of nature. A lot of attention is also given to personalities of scientists, their lives, religious and philosophical views. It makes the conversation about a history of discovery emotional, stirring, invoking active interest in schoolchildren, and, as a consequence — better comprehension and mastering of the material.

Besides I try to present problems at a modern scientific level, and to include all new discoveries, taking into account that biology is an extremely dynamically developing science. Certainly, this approach considerably expands the volume of the studied material. But in this case much depends on the pedagogical skill of the teacher.

When introducing this or that scientific problem, I try to present it in a wide historical prospective. For this purpose four time periods were selected: the Ancient world, the Middle Ages, New Time, Present Time. At every period we observe the development of scientific research, all interesting and important events that took place in science; answers which were found, as well as questions left without the answers for scientists of the following periods.

However, the Present Time in which we live still does not know the answers to many questions. Within the framework of the course no attempt is intentionally made to present a modern pattern of the evolution absolutely completed. The main conclusion, which can be made is that in modern evolutionary biology there are considerably more questions than answers given to them. But it is with a correctly formulated question that the way leading to a discovery starts. Therefore we should not be discouraged that many questions have yet no precise, universally recognized answers. They form an information field of open questions, where further search is possible, and where the pupils are given the opportunity to reflect and to propose hypotheses. Thus, they can participate in a fascinating process of scientific creativity and test their abilities. It can be a start of their way to serious science.

6. The Basic Methodological Principles of the Course

The basic methodological principles used in studying the course "Introduction to Christian natural sciences" can be formulated as follows:

- Comparative historical approach, i.e. the problem of evolution is traced from the moment of its origination and realization by mankind;
- Acquaintance with a wide spectrum of ideas, theories, concepts, hypotheses offered by various scientific schools;
- The use of original author's texts, instead of their popular adaptations;
- Critical analysis of the theories, revealing disputable and problematic points;
- Freedom of wide discussion after finishing study of the problem of evolution;
- Harmonization of the studied scientific concepts of evolution with what is disclosed in the Holy Writ, the theological analysis and understanding of the problem on the basis of Holy Father's and modern theology.

The basic theological idea of the course is its so-called conceptual dominant combining all educational material, and which this educational material should illustrate is the following: "Religion and science on the path to knowledge of Truth". The word Truth is not casually written with a capital letter, as Christ Himself said to us: "I am the way, and the truth, and the life" (John, 14.6). Therefore the offered course, besides other things, is centered on Christ. The Light of Christ is the landmark, which helps us to not be mistaken in the search of Truth.

The studying in the 11th form of evolutionary idea — the central problem of modern biology — is based on these principles.

7. Argument of the Course: The History of the Idea of Evolution

Schoolchildren study different points of view about evolution as deeply as possible. The formation and development of the idea of evolution is traced from the moment of its origin in the minds of ancient-Greek philosophers, like as Anaximander, Heraclitus, Empedocles, Aristotle (Voronin, 1999).

The result of this acquaintance is understanding that the philosophers of the Ancient world created and presented to mankind a pattern of the world for the first time. They conveyed to scientists of modern times, as a testimony, a very deep idea of the gradual, expedient, historical development of nature. Of course, this pattern was very naïve, but the central place was lead off of God–Creator, Initial Chaos, rules of the all creations. There is no conflict between scientific and religious views of the world in this period. But on the contrary, they form an organic whole, developing in a common spiritual space. The science of nature hadn't left at that time it's "cradle" — religious philosophy. The schism of science and religion hadn't happened yet.

Later schoolchildren study the next historical period — the Middle Ages, when the attempt of Christianization of heathen wisdom was undertaken. Finally it led to the creation of the so-called rational theology. The goal of rational theology is the theological comprehension of nature. So the natural science of this period developed as one of the spheres of theology. Therefore the creators of Middle-Ages science were not scientists, but the great Christian theologians — John Chrysostom, Basil the Great, Gregory of Nyssa, Gregory the Theologian, Augustine of Hippo, Thomas Aquinas, Albert the Great (Basil the Great, 1900; Gregory of Nyssa, 1995; John Chrysostom, 1993). They left their meditations — commentaries of the chapters 1-2 of the Book of Genesis. So developed the school of symbolic interpretation of the *Bible*. These commentaries on the Six Days of Creation lie in adjoining spheres, because the subject of theological comprehension is the material world, and its study is the goal of the natural science.

But from a certain time the figurative, poetic style of the Book of Genesis was considered as strictly scientific, and the Six Days of Creation — as an absolutely exact, literal description of the pattern of creation. So developed the school of literal interpretation of the *Bible*. This school excludes any symbolycal or allegorical interpretation of the chapters 1-2 of the Book of Genesis.

These two theological schools produced two trends in natural science:

- Transformism (from Latin "transformis") — the idea of historical development of nature. This development is sent by the Creator.

- Creationism (from Latin "creatio") — the idea that God created every species of plant and animal during the 3rd, 5th and the 6th Days. Every Day of Creation continued 24 hours sharp. All species remain invariable from the period of their creation. One species doesn't produce another. All these events happened about 7.5 thousand years ago.

I think it is necessary to tell schoolchildren about the ideas of creationism, since it is an outstanding example of pseudoscientific and pseudotheological concept. It has very deep contradictions both with science, and with orthodox theology.

Schoolchildren must learn the history of the separation of science and theology (the doctrine of Thomas Aquinas about "Double Truth") in the 13th century.

The analysis of the theological (patrological) commentaries leads us to an important conclusion: the idea about historical development of nature obtains a new sense. This sense is founded on *Bible* Revelation and Christian divinity.

Saint Gregory of Nyssa in the treatise "*On the Arrangement of the Man*" says: "Nature makes the way of the ascent from the smallest to the perfect" (Gregory of Nyssa, 1995). This idea of the gradual ascent of nature from simple forms to more complicated links the Holy Fathers' theology and the evolution.

The main result of this period's study is the following: science develops in the bosom of religion; aspiration to perceive God's world leads to forming rational theology — a religious comprehension of nature. God–Creator has the central place in the pattern of the world. All pattern of the world is pierced by feeling, that nature is a beautiful and harmonious creation of God; it is inflated with His Wisdom and Beauty.

But the first signs of "the conflict" between science and religion appear in the later Middle Ages. "Knowledge is power" — said Francis Bacon in the XVII century. This idea became the scientific motto of the New Time. The scientists came to the idea of the unlimited power of science in the perception and transformation of the world. But it was a "bitter fruit of knowledge". Deism was born, and this philosophic school took away the Creator out the frame of the universe. This universe without God became the object of study of the new science of the time.

Schoolchildren learn the proceedings of many scientists of this period: Sh. Bonnet, G. de Buffon, E. Darwin (E. Darwin, 1960; Lamarck, 1955; Wallace, 1911) and of course, Charles Darwin (1926). We try to understand what is necessary for a theory to be accepted and which of these theories should be rejected; we also try realize what we may take from this scientific heritage, and what not. However, a review of either one or another theory and even a fundamental difference of opinion keeps us from a simple trip of blame. At that we try to understand what we may take from this scientific heritage, or not. This travel along the annals of the history of evolutionary biology allows understanding the logic of the development of a problem, to find mistakes, and to realize the causes of their appearance. It is necessary, if we want to move forward on the way to Truth.

8. Argument of the Course: The Current Status of the Idea of Evolution

Thus after a deep study of history of the evolution idea we proceed to the study of its current status. At the lessons we start to discuss the evolution theories of the Present Time. We try to ask ourselves a principal question: does the Divine Creation refute biological evolution? Are Divine Creation and the biological evolution strong alternatives to each other or not?

To find the right answer it is necessary to understand clearly the meaning of the term "evolution" today. This is a Latin word that comes from the verb "evolvo". As a scientific term it was for the first time used by English theologian M. Heil (1609-1676). The modern scientific definition of the term "evolution" is the following: "Evolution is an irreversible, fairly directional process of historical change (modification) of living nature" ("*Biological Encyclopedia*", 1989).

The schoolchildren learn next the doctrines of evolution of the Present Time: Neo-Darwinism and "Punctuated equilibria" (Eldredge & Gould, 1972) — so-called "Tychogenetics" theories (from Greek "τύχη" — fortuity), i.e. these theories are based on a fortuitous events.

Then the schoolchildren learn so-called "Teleological" theories (from Greek "τέλος" — aim, end): nomogenesis by L.S. Berg (1977), Christian evolutionism by P. Teilhard de Chardin (1987), Diatropica by S.V. Meyen (1974), the concept of A.A. Lubitschev (1982) and others.

Evolution is the fact of the biological history, but it can't go on accidentally. This is the fundamental difference between tychogenetics and teleological theories. There is an issue: has evolution an aim or not? Both theories answer this question in different ways. Tychogenetics theories can't be united with the idea of Divine Creation, because all evolutionary development results from fortuitous events. In opposition to this, teleological theories permit such opportunity and even require a Creative Intelligent Power. This Power, i.e. God-Creator, leads an evolutionary process. And we come back to our issue: can we unite scientific and religious points of view, religious pattern of the creation and the dates of the contemporary doctrine of evolution? Once more we appeal to the text of chapter 1 of Genesis and the theological commentaries.

9. Theological Analysis of the Idea of Evolution

In verse 20 we read: "Then God commanded: Let the water be filled with many kinds of living beings, and let the air be filled with birds". How does the creation of living beings occur? God doesn't act as a master, demiurge. The Creator orders the water: "Let the water be filled..." And water generates "living beings". And further: "So God created the great sea monsters, all kinds of creatures that made the water..." John Chrysostom comments on that verse: "The word of God became the action, and the command of God gave rise to the living beings and explained a way of their origin" (John Chrysostom, 1993).

In such a way, according to the pattern of Divine Creation, living beings appear in water and then came to the land. "Then God commanded: Let the Earth produce all kinds of animal life..." (1.24). But science tells the same. Thus the idea of Divine Creation doesn't contradict the scientific principles.

It is important to note that the Divine Creation is carried out by stages, i.e. expansion or evolution of the species takes place. But evolution is not equal to the creation, because only things that exist already can evolve in some form, for example in the form of design (logos).

Surely, the Divine Creation can't have any analogies in our world. Possibly, there is the reason of the impossibility to create a living organism *in vitro*. This is direct evident of the existence of a principal cause of the being, and this cause is inaccessible for scientific research. Yet the presence of such a principal Cause (God's Law) doesn't contradict the scientific laws.

Attentive reading of the Story of Creation and the commentary for it allows a feeling for the fundamental scheme of the evolutionary scale of rank — the Design of God develops step-by-step in the world. The Scriptural Six days of Creation are the steps of the ascension, adopted by Hebrew word-symbol — "יום" (day). The Afflatus about evolution of the world is delivered not as a scientific treatise, but in a special form called a "Divine protocol" (Shifman, 1993). Each word, each phrase in this text is a formula, having several dimensions.

This point of view allows connecting the Divine Creation and evolution noncontradictorily. But such evolution will not happen by Darwin's mechanisms. Darwinism and Neo-Darwinism don't need the Creator, since all mystical reasons are eliminated from the pattern of evolution.

Despite all the contentiousness of Darwin's theory, we must remember: Charles Darwin didn't exclude the initial act of Divine creation (see the concluding paragraph of "*The Origin of Species*") (Darwin, 1926).

The main conclusion is that evolution has a clear trend, an aim. But if there is an aim it is necessary that there is the presence of a purpose, and we must recognize the presence of Higher Intellect.

10. Results of Study of the Idea of Evolution

The conclusion of these lessons-reflections is that the universe is the "creative laboratory" of God; in this laboratory the Design of God opens gradually. The motional power of such an evolution is the love of the creation for the Creator. This idea is present in the writings of the French theologian and scientist P.T. de Chardin. This is the pure essence of evolution. Such evolution doesn't enter into conflict with the idea of Divine Creation; it can't be realized without God. The Russian theologian N. Fioletov wrote: "The idea of evolution, the development of species, can't be examined as conflicting with the Christian doctrine about the Divine creation of world" (Fioletov, 1992).

Therefore evolution as the development of the Design of God, as its realization in the material world, as the movement to God's aim — such evolution forms the harmonic unity with Divine Afflatus.

I believe the study of the evolutional biology gives the schoolchildren a wonder of greatness of God's Design and allows them to understand that scientific truth may and must serve the Truth of Afflatus. So evolutionary biology becomes one of the ways of the Knowledge of God.

I am sure, the course leads to a complete comprehension of Revelation, to the truth of God, to realization of the joyful meeting of scientific and divine thinking. Thus, the course is an attempt to find of a compounded point of view on a pattern of the evolving world.

11. Summary

In this article I have tried to find the answer to the question, whether science, in particular — evolutionary biology, is capable of contributing to Christian education and formation of a religious world outlook of pupils. I have created a special program for teaching in the Secondary School — the course "Introduction to Christian natural sciences". It intended for teaching in the 10th and 11th forms of general educational orthodox schools of Russia. The basic purpose of the course is the formation of the Christian world outlook of school students on the basis of the data of modern sciences, withdrawal of a far-fetched "conflict of a science and religion" and, on the contrary, a statement getting them to think, that a scientific and religious pattern of the world can and should be indivisible, as the world was created by God.

The course is based on a principle of teleology, which leads to theology. The main idea of the course is to show that the origin, development and formation of the world are the result of Divine creativity. The creation was from the very beginning carried out according to some purpose — The Design of God. The basic theological idea of the course, its so-called conceptual dominant, combining all educational material, and which this educational material should illustrate, is the following: "Religion and science on the way to knowledge of Truth". The teaching in the 11th form of the evolutionary idea – the central problem of modern biology – is based on these principles.

Thus, the aim of evolutionary biology is the comprehension of wisdom of the Creator through His traces in the world.

So the course's development on a basis of teleology is a consecutive continuous formation of the Christian world outlook of the pupils on the basis of the data of the modern evolutionary biology, perceived through a prism of Christian divinity.

The pupils study the different points of view about evolution as deeply as possible. The formation and the development of the idea of evolution is traced from the moment of its origin in the mind of the ancient-Greek philosophers (Anaximander, Heraclitus, Empedocles, Aristotle and others). Further, schoolchildren learn the writings of many scientists: Sh. Bonnet, G. de Buffon, E. Darwin, J. Lamarck, A.R. Wallace and Charles Darwin. Later the schoolchildren learn so-called "teleological" theories: nomogenesis by L.S. Berg, Christian evolutionism by P. Teilhard de Chardin, Diatropica by S.V. Meyen, the concept by A.A. Lubitschev and others.

We try to ask ourselves the principal question: does the Divine Creation refute biological evolution? Are Divine Creation and biological evolution strong alternatives to each other or not? We appeal to the text of Chapter 1 of Genesis and the theological commentaries. Attentive reading of the Story of Creation and the comments for it allows a feeling of the fundamental scheme of the evolutionary scale of rank — the Design of God develops step-by-step in the world. The Scriptural Six days of Creation are the steps of the ascension. This point of view allows connecting the Divine Creation and the evolution noncontradictorily. But such evolution will happen by non-Darwinian mechanisms.

The main conclusion is that evolution has a clear trend, the aim. But if there is the aim it is necessary that there is the presence of purpose, and we must recognize the presence of Higher Intellect.

Therefore evolution as the development of the Design of God, as its realization in the material world, as the movement to God's aim — such evolution forms the harmonic unity with Divine Afflatus.

References

Basil the Great, (1900). *The Hexaemeron.* Moscow.
Berg, L.S. (1977). *Works on the Theory of Evolution*, Leningrad.

Biological Encyclopedia (1989), Moscow.

Darwin, Ch. (1926). *The Origin of Species*, Collected Works, **v.1**, book 2. Moscow –Leningrad.

Darwin, E. (1960). *Temple of Nature*, Moscow – Leningrad.

de Chardin, P. Teilhard (1987). *Phenomenon of Human Being*, Moscow.

Eldredge, N., Gould, S.J. (1972). Punctuated equilibria: an alternative to phyletic gradualism, In: Schopf, *Models in Paleobiology*, San Francisco, p. 82.

Fioletov, N.N. (1992). *Essay of Christian Apologetics*, Moscow.

Good News Study Bible (1986). Thomas Nelson Publishers, New York.

Gregory of Nyssa (1995). *On the Arrangement of Man*, St. Petersburg.

John Chrysostom (1993). *Homilies on Genesis*, Moscow.

Lamarck, J.-B. (1955). *Philosophy of Zoology*, Moscow.

Lubitschev, A.A. (1982). *Problems of Form, Systematics and Evolution of Organisms*, Moscow, p. 353-364.

Polkinghorne, J. (1997). *Science and Christian Belief*, Moscow.

S.V. Meyen. (1974). About correspondence between nomogenetics and tychogenetics aspects of evolution, *Journal of General Biology*, **v.35, №**3.

Shifman, I.Sh. (1993) *Doctrine. Pentateuch*, Moscow.

Voronin, N.N. (1999). *Development of Ideas of Evolution in Biology*, Moscow.

Wallace, A.R. (1911). *Darwinism.* Moscow.

Dialogue

Peter Bahn

Galina Muravnik traces her pioneering work to integrate true science into a religiously Christian based high school in Russia. The author describes the course that she has designed for this purpose. This manuscript is fascinating reading. I have actually learned something reading it, namely that religion and science are really not contradictory. Science and religion actually approach the same source of wonder — creation — from two different directions.

In particular, what I like about the author's work, among other things, are the following:

- I like this sentence: "Christian science, recognizing the world as cognizable, by studying its laws, ultimately approaches the comprehension of the reasonable design of the world, studying the world not in general, but comprehending it as God's creation, as his creative workshop." This is inspiring.
- "So, the basic pedagogical aim is to arm the children with knowledge of evolution of living beings, which will allow them not only to work in science professionally, if they wish, but also to carry the light of Christian belief into the world." OK, I, the reviewer, am actually Jewish, but when I hear Christianity talked about in this way, I am almost ready to go out and convert myself to Christianity. This is how I would like to think about Christianity.
- "It seems to me extremely important not just to inform the pupils about this or that scientific fact of modern evolutionary biology, but to show them how long and difficult the search for it was. In this way the pupils get aquainted with a history of the origin of scientific discoveries including not only successes, but retreats at times, delusions, deadends and search for the way out as well as further advances in the step by step approaching of the knowledge of the secrets of nature". This statements wants me to become a student in the author's class because this author knows what science is really like. Extraordinary! In the next paragraph, the author says "biology is an extremely dynamically developing science." This makes me want to become a biologist. Excellent!

- "The main conclusion, which can be made after studying the given course is that in modern evolutionary biology there are considerably more questions than answers. But it is with a correctly formulated question that the way leading to a discovery starts. Therefore, we should not be discouraged that many questions have yet no precise, universally recognized answers. They form an information field of open questions, where further search is possible, and where the pupils are given the opportunity to reflect, to propose hypotheses. Thus, they can participate in a fascinating process of scientific creativity and test their abilities. It can be a start of their way to serious science." The author speaks with much wisdom and beauty here. It points out very strongly the beauty of science.
- "There is no conflict between scientific and religious pattersn of the world in this period. But on the contrary, they form an organic whole, developing in a common spiritual space. The science view of nature hasn't left that 'cradle' of religious philosophy. The schism of science and religion hasn't happened yet." This is an author after my own heart. I was just saying similar things to the Editor [J.S.] just the other day. What a wonderful knowledge of the ancient philosophers this author has! The author says: "science develops in the bosom of religion." This is quite extraordinary, and it is true, and IDers and Creationists are not even aware of it, because such pseudoscientists never do any heavy reading! This paper is enough to make Michael Behe faint!
- "...we must remember: Charles Darwin didn't exclude the initial act of Divine creation (see the concluding paragraph of '*The Origin of Species*')." Wow! In my own contribution to DINA [this book], I have actually quoted this famous last paragraph of Darwin's *Origin of Species*!

Taner Edis

Galina Muravnik equates the exclusion of a Christian view on creation with atheism. But many scientists in the United States, including many who are personally religious, disagree. They say that evolution does not deny supernatural realities, that it is religiously neutral. Excluding religious interpretations merely keeps science neutral and focused on scientific matters. How exactly does Muravnik think such an view is mistaken?

It also strikes me that Muravnik's primary concern is maintaining the religious convictions of her students, rather than improving their understanding of science. In the context of a specifically Orthodox Christian education, this is perhaps as it should be. But there is a danger of misrepresenting science in the process. Indeed, I think Muravnik does just that.

For example, Muravnik appears to have the impression that the “anthropic principle” is a scientific result presenting us with a universe fine-tuned to achieve certain human-like purposes. Whatever the religious attractions of the anthropic principle, it certainly is not useful for physical cosmologists. Physicists do, on occasion, use a variety of anthropic reasoning, but this is strictly to highlight the fact that we do not occupy a representative corner of the universe. It is unfortunate that this so regularly gets misunderstood outside of scientific circles.

In the context of a science course, Muravnik’s admiration of Teilhard de Chardin is also troubling. I am not suggesting that theological responses to evolution are intellectually uninteresting. But Teilhardian ideas of progressive evolution have had no influence on biology as a science. Indeed, to the extent that such progressive evolutionary views make contact with scientific tests, they are almost certainly mistaken. The notion that evolution is directed by intrinsically progressive forces is a common misconception that biologists take care to avoid when training their students.

In the end, this is perhaps the most serious misunderstanding Moravnik propagates. It seems her view of evolution is close to those Intelligent Design advocates who insist that Darwinian, naturalistic processes cannot drive evolution. The overwhelming consensus of scientists stand against such a claim.

I understand that religious schools will be specially concerned with theological thinking about science. This is perfectly legitimate. But surely any educational establishment should also be concerned about the accuracy of what they present as scientific views. Selectively focusing on ideas on the fringes of the scientific mainstream, or worse, emphasizing ideas that are currently rejected by the scientific community, strikes me as a disservice to students, no matter what their cultural background.

Biography

Dr. Hermann Josef Roth, is a Studiendirektor, biologist, Roman-Catholic Priest and Cistercian-Monk; manager of the "Naturhistorischer Verein der Rheinlande und Westfalens." He has a seat at the University of Bonn, whose foundation was assisted by Carl Fuhlrott (1843). Dr. Roth obtained his Ph.D. from the University of Nijmegen (Netherlands) in 1990. Dr. Roth published many books and articles on natural history of the Rhineland, and history of botany and zoology in Middle-Ages and during the $18^{th}/19^{th}$ centuries. His main interest is in Monastic Medicine including ethnobotany in the Portuguese world. He worked on the project "Klostermedizin" at the University of Würzburg.

31

Message of the *Bible* or Theory of Darwin? A Pseudo Problem

An Interdisciplinary Statement on the Current Controversy in Germany

HERMANN JOSEF ROTH

Universitätsbibliothek, Naturhistorischer Verein
Nußalle 15a, D - 53115 Bonn, Germany
Hermannjroth@aol.com

1. Introduction

"God versus Darwin" is the headline of the *Spiegel*'s feature article, which reports on the "war of beliefs surrounding evolution". The article is written from the liberal point of view, usually supported by the magazine (Spiegel 2005). The Catholic magazine "*Herder-Korrespondenz*" attacks the issue from a religious position, summarizing the key arguments of the discussion under the heading "intelligent design" (Herder 2005). Only two voices have so far been quoted, albeit from widespread and influential areas of the German media. The debate around this topic has erupted again, last but not least due to the controversial remarks of the Archbishop of Vienna, Cardinal von Schönborn, a debate which appeared to be settled publicly long ago. A similar evolutionary article was published in *TIME Magazine* (2005).

One of the reasons for this newly evoked interest can be described in the controversy which originated in the USA and then quickly spread to Europe as well. Additionally, especially in Germany, there has been a noticeable reorientation to values and belief rooted in Christianity, which last but not least can also be attributed to both the Catholic "World Youth Day" and the official visit of the "German Pope". Finally, the review of the discovery of the fossils in Neanderthal 150 years ago by C. Fuhlrott (Fuhlrott 1859) has caused an avid response in the popular media (Archiv-NHV 2006) and further triggered discussion on this matter.

As a consequence a new variation of the old debate about the teaching of evolution has erupted which is taking place in North America under use of two recently formulated neologisms. Creationism (not to be confused with 'Kreatianismus' — a theory from the Middle Ages) claims to explain Creation in a fundamentalist-biblical manner. This theory clearly and provocatively contradicts the claim of evolutionism, which postulates the notion of evolution to be a principle of all realities. Both stories of Creation in the *Bible* (Gen. 1 to 3, 24) are controversially handled and thus often misunderstood within the debate.

It is unnecessary and would be tiresome to reiterate the well known positions taken on this side of the argument. Instead, this report will attempt to describe a loophole, which exposes the debate as a pseudoproblem, so long as creationism is taken to refer to the wording of the *Bible*. However, the same methodological error is also committed by evolutionism if, as in the course of opposing standpoint, it treats the biblical texts as outdated.

The viewpoint articulated here was developed in the course of an interdisciplinary forum, which was organized by the "Natural History Foundation" in 2005 at Drachenburg in Königswinter, near Bonn. As is perhaps noticeable, this text resulted from the forum. The essay is, therefore, to be read as a spontaneous contribution to the discussion and not as purely scientific work. In order to retain a free reading-style, *Bible* quotations will be given from memory. Quotes originally written in Hebrew will be offered in simplified Latin, instead of using the official phonetics of conventional dictionaries found in Germany. Some additional subheadings have been added to make the train of thought easier to follow.

Some of the conclusive points featured in most of the recent publications on the debate between Biblical beliefs and the teachings of evolution which have appeared in Germany require critical revision and amendment if further understanding is to be truly attained. Such a critique deals with:

1. The exegetical methods,
2. The key ideas, their faithful translation and retention of the sense of the original text so that it remains recognizable, and
3. An interpretation that is acceptable for us today.

The aim is to examine those statements repeatedly used in debates and rounds of discussion with reference to the controversy between creationism and evolutionism (or whatever the two positions wish to call themselves).

2. Sources

Fundamentalists attempt to understand the *BIBLE* word for word. The opposite standpoint advises against this. Instead of hastily choosing a position, the reader should simply let himself/herself be affected by the text, preferably through reading aloud. The Hebrew text is written in verse form; it is therefore a matter of lyrics! These passages must be read and understood as such. These aspects often fail to come across in the brittle German translation. The translation by M. Buber and F. Rosenzweig (1925-1992), both Jewish scholars, attempts to emulate the Semitic verse-form in the German version.

Incidentally, it can be asked who actually decided to criticize from an astronomical perspective a poem that praises the sunrise. Let us allow the Semitic verse to retain its aesthetic magic! This magnificent poetry does indeed convey clear messages. The verse is a medium for expressing a certain view of the world, namely, a religious one that has essentially been accepted by millions of people to this day.

Pinchas de Lapide (1982), a Jew, makes it clear that: Faith seeks God. Science, on the other hand, seeks the work of God. Or, to put it another way: Religion deals with the beginning and the end of time — in Christian symbolism, with Alpha and Omega. And once again: God is unfathomable (Ezekiel 20:3; *New Jerusalem Bible*). God's work on the other hand is fathomable. Humans question where they came from and where they are going; that which lies in between, man can and does find out for himself.

In order to appreciate listening to or reading the answers expressed in the poetry of Genesis, the literary character of the text has to be given as much consideration as the historical background from which it originated. A grave error of Christian script analysis is, in my opinion, to forget to which extent our thinking is already influenced by fundamental principles derived from ancient Greek and Roman philosophy.

Table 1: Genesis Verses for Comparison.

German Bible	Buber	Hebrew
Die Erde aber war	*Erde ward*	*V' haarez hajtā*
wüst und leer	*Irrsal und Wirrsal*	*tohu wa bohu*
Und Gott sprach:	*ER sprach:*	*Vayomer Elohijm:*
Es werde Licht,	*Licht werde,*	*Yehi Or,*
und es ward Licht.	*Licht ward.*	*Va yehi or*
Und es war Abend,	*Abend ward,*	*Va jehi ereb,*
und es war Morgen,	*Morgen ward,*	*va jehi boker,*
der erste Tag.	*erster Tag.*	*Yom echad.*

3. Answers

Both "stories of Creation" already hold important answers to questions about where we come from, where we are going and why. Good literature — which the *Bible* is — tends to place the answers at the beginning or the end of the text.

And there it is, in the first words of the *Bible*: (*B'reschith baráh elohim*) — in the beginning HE (Jews are not suppose to pronounce the sacred name of God) created Heaven and Earth. Three statements are contained within this sentence:

HE exists.
HE created.
HE created "heaven and earth".

4. "In the beginning"

4.1. *God Was and Above All is*

In his razor-sharp logic, Thomas Aquinas (1225-1274) defines the word "God" (*quod loquatur deus*) as existence per se. HE exists from himself *(deus a se)*, whilst what we know from our own experience has its origin from something else (*ens ab alio*).

4.2. *God Creates or Makes*

Both translations (Gr.: έν αρχή εποίσεν ο θεός τόν ουρανόν και τήν γήν = *en archē epoísen ho theós tón ouranón kai tēn gēn*; Lat. *In principio fecit Deus* ... equally ... *creavit* ... (= In the beginning God made ...) are valid. The equivalent Hebrew verb is used emphatically in Genesis.

Only HE can bring something into being. That is at the same time a radical denial of all polytheistic mythology and thus surely also a denial of scientifically embellished fantasies of today, which theological and scientific statements try to bring together, instead of leaving them as they were originally intended and written.

"Heaven and Earth", which is the material world, whether visible or not! Put another way, everything that we can experience through our senses or that which is ascertainable through science is attributed to HIM — he who we call God, as Thomas (1225-1274) would say — due to his existence.

5. "From Nothing"

Teacher: "What did God create the world from?" Pupil: "From nothing." Joker: "What does that look like?" This daft school joke reveals a misunderstanding that has led to a widespread academic error. There is no mention of this in Genesis. In fact, this belief derives from church teaching, which (if my memory serves me correctly) has its origin in ancient philosophy (Plato: τό μη ον ειvάι = *to mē on einai* = "non-existence").

Verse 2 of Genesis: delivers the decisive quote: *v' ruah Elohim m'rachéphet al p'nei hamayim.* Commonly used German Bibles, including the renowned *Jerusalem Bible* translate: "…and the Spirit of God moved upon the face of the waters". However, in Buber and Rosenzweig it is written: "…and the Spirit of God moved upon the face of the first breeze". Previously we find the statement already quoted about the current state of the world: *Ve-haárez hayetá tóhu wa bóhu.* The saying has found its way into everyday German language, just as the analogous τό χάος, from the foreign Greek word ("chaos"), has been adopted.

The closely related assumption that the Creator acts like the demiurge of Ovid, who creates a shapeless lump mass, is knowingly refused by theologians and doctors of the church. Indeed, if one were to accept this, as in the naïve portrayal of ancient poetry and mythology, it would result in a belittlement of the biblical message. The arbitrary random element of ancient cosmology, beginning with the "gaping abyss" of Hesiod, has complicated the discussion and led to the breakdown of dialogue with religious theology.

If we were to set the scientific structural model theory aside and understand "chaos" simply as a lack of all order, then a justifiable analogy to "utter chaos" could, in my opinion, be found in Genesis. It was therefore HE (= God), who created order (Greek 'ο κόσμος). The act of Creation was said to be the very first Creation of order. To fathom and formulate this within the realms of the law of nature is a matter of concern for science.

Incidentally, the "Spirit of God" is also a misunderstood translation, because, to Christian ears, it is reminiscent of the teachings of the Trinity: the meaning of this word is something which is completely unbelievable for the Jews. The word *rúach* can have a variety of meanings in Hebrew ranging from "breath" to "spirit".

Yet it is only with great ambiguity that the wording of Genesis seems to refer to the spirit, because "the Spirit of God" does not appear in Verse 3 nor in the following verses, as in the Greek demiurge, instead it is always written: *vayómer elohim* = "HE spoke".

Did the Creator create the world "from nothing" or are we going to have to humbly leave the question open? For believers, HE can unquestionably do everything. But, in my opinion, we should not dare to say whether he actually does everything that he can. In this case I find the thought that HE gave us the laws of nature more acceptable. Whether matter or energy (which are in any case equivalent) were present "at the beginning" is largely irrelevant. As a scientist, I would like to agree with the words of Klopstock (1974) from his "*Messiah*" (slightly changing the emphasis): "Schöner ein froh Gesicht, das den großen Gedanken deiner Schöpfung noch einmal denkt." ("More beautiful is a merry face, which reflects upon the great idea of your creation once again.")

6. "God made"

Further development of both stories of Creation could perhaps shatter the interpretation given above, if the individuality of literature is not given enough consideration. After "God said" we read "God made". The actual reason for this is the change of authorship (so-called Elohist ↔ Yahwist). As our discussion focuses on meaning, questions related to this issue seem to be irrelevant.

In the first account, during the six days work of creation (Hexaemeron), HE constructed a garden: Eden. The abstract plan follows a theological aim, namely that of the typical Jewish sanctification of the Sabbath (= *Schabbát*), the seventh day, on which God "rests". It seems relevant for our discussion that "HE said" is not used throughout the description of the story, but that instead, this is surprisingly expanded upon. Genesis 1:11-12 speaks of the self-Creation of nature: "Let the earth bring forth grass...". Even in our basic text, therefore, there is a kind of "development, an independence of nature. This is repeated in 1:24 in relation to animals. God says only: "Let the earth bring forth living creatures according to their kinds...".

Namely, I would like to boldly suggest that the Creator instigated the events, which in turn caused Creation to occur autonomously. Darwin could probably have happily lived with this theory. The scholastics also appear to see it in this way. If my memory serves me correctly, Thomas Aquinas, (1225-74) once said: *Deus non facit per se, quod facere potest per creaturas* (God does not do himself, what he can do through his creations).

The so-called second account of Creation describes God's actions in a powerful poetic language, in which the author of the *Bible* shows God working conscientiously. First of all, HE planted the Garden of Eden (= gan-*'edén* = delight), which was later given the Persian word "paradise" (= pardēs, gr. Παράδείσος) and can be understood as a park or garden. Regardless of which description may be correct: In the eyes of the sons of the desert it was wonderful, for it reminds them of the vision of the rescuing oasis. In the framework of our discussion I cannot hold back from recalling the phrase "national park". If one wanted, one could almost talk about it as a prehistoric document on conservation.

The second part describes the creation of humans using the image of a potter, surely borrowed from Mesopotamia, where ceramic art flourished. When it comes to creating humans, the Creator therefore uses the most modern technology of his time (Message 1). By creating humans from the same material ("rib"), the text confirms the equality of the sexes. With this, the final thought of the first story is confirmed: "HE created man … male and female HE created them" (Gen. 1:27). The claim that women are inferior has no support in the biblical text. The Hebrew expresses this fact by using a play on words, which does not talk of man and woman as such, but rather talks of "Man and Manness" ("*isch*" ↔ "*isch-scha*").

7. "It was good"

With almost monotonous insistence, the first narrator rounds off the day's work of the Creator with the conclusion: "And God saw that it was good" (= *v'hinéh-tov*). Thus, along the way HE created and ordered the world to be good. We shall hurry along in the story but remember that it was man himself who, through an improper intervention in the existing order, gambled Eden away (Gen. 3). Here, thoughts of "ecological balance" or "fouling (of waters)" spring to mind.

8. Conclusion

The approach to reading the stories of Creation from Genesis (b^{e}re'šīt), which has been suggested here, allows for a more flexible interaction with the text than some people in their enthusiasm, might allow. This flexible reading of the text is achieved by reading the text literally.

However, what follows is always true: verifiable facts are just one side of human experience; the other involves philosophical classification, religious meaning and aesthetic acquisition of knowledge. Each side should complement the other. However, this can only be achieved by careful, methodical separation. One must always be aware of what is being discussed. We should bear this in mind without having to declare anything new: an oil painting can be judged both from a chemical viewpoint as well as from the point of view of art and the history of art, just as a piece of music can be analysed according to criteria of physics or through artistic interpretation and music history.

Similarly, the stories of Creation given in the *Bible* allow for different approaches. Just as theological exegesis and philosophical reflection develop different aspects of these texts, so do literary history, stylistics, mythology, political history, as well as cultural history of art of the ancient world, archaeology, applied geography and natural history. In order to ensure that the texts are correctly understood, the literary genre must be considered as the most important basic requirement.

The passages quoted appear to be Semitic fiction with a religious message (Kerygma) and as such, correspond to the context of their date of origin. The imparted facts are not part of the content, but rather should be considered as stylistic devices which solely serve to convey a message. Attempting to use the devices to prove the accuracy of the theory of evolution means that they lose their literary style and is, therefore, methodically unacceptable.

Thinkers of the middle Ages most probably allowed for the free interpretation demanded by the Creationists today. To cite an example from a maxim of Siger von Brabant († 1281/84): *Nec de Dei miraculis, sed de naturalibus naturaliter disseramus* ("We want not to speak of the wonders of God, rather, we want to talk of natural things"). Someone who is unfamiliar with the free thinkers of the Middle Ages may instead consider Thomas Aquinas, who also understands natural events as being independent from the direct intervention of God: *Deus non facit per se, quod facere potest per creaturam* ("God does nothing with his own hands, that he can better achieve through his Creation [probably better: creatures]").

In the same way that one would not normally quote "The Moon has risen" by Matthias Claudius (1998) in an astronomical dispute, texts from Genesis should not be invoked as arguments for or against specific views of evolutionary biology. *Bible* or evolution? — The question in reality is a pseudoproblem.

Table 2: The view of Creationists vs. the written text.

"Creationist"	According to the text
Bible seen as specialist book	Genesis texts are Semitic verses
In the beginning: God "moved over the abyss"	In the beginning: God "on the first breeze"
God created things "from nothing"	God delivered chaos in the cosmos
God created everything himself	Literally: "The earth brought forth grass, herbs yielding seed after their kind" (Gen 1:13); "The land generates all kinds of living creatures" (Gen 1: 24) — quoted. NJB

9. Alternative Interpretations of Creationism

Creationism "is a fundamentalist and biblically justified concept of Creation, which expressly fights against a widely applied evolutionism" (LThK 1997).

Creationism's criticism fails to address the core arguments of the scientifically justified theory of evolution, as it misconceives the technical methodology, its power and its limits.

Creationism seeks to take the *Bible* word for word and uses this approach to infer unjustified conclusions.

It is exactly through use of this method, by taking the *BIBLE* text in its original and word for word form, that the fundamentalist Creationist interpretation can be most swiftly laid aside.

Creationism represents a pseudoproblem. Relaxed treatment of the *Bible* was anticipated in the beginning by thinkers of the European Middle Ages.

References

Archiv Naturhistorischer Verein, Bonn (2006): Biologie in unserer Zeit 1/2006, pp. 61-62; Bonner General-Anzeiger vom 19. 1. 2006, pp. 12; Damals. Das Geschichtsmagazin 5/2006, pp. 71-75; Bild der Wissenschaft 5/2006, pp.

18-27; Natur und Museum **136**, 5/6, 2006, pp. 97-106; National Geographic 1/2006, pp. 128-136.

Buber, M. and Rosenzweig, F. (1925-29, 1956/57, 1992): Die Schrift, 1-5. Köln; rev. ed. Heidelberg, Stuttgart.

Claudius, Matthias (1778): Abendlied. Der Mond ist aufgegangen. Last edition by Reinhard Görisch. Frankfurt 1998 (= itb 2178).

Fuhlrott, C. (1859; First published in: "Verhandlungen des naturhistorischen Vereins der preußischen Rheinlande und Westphalens" **16**, pp. 131-153. *Herder-Korrespondenz* **59**, 10, 2005 pp. 497-500.

Klopstock, Friedrich Gottlieb (1750): Der Zürchersee. New ed. in: Werke und Briefe, by Elisabeth Höpker-Herberg et al. Berlin, New York 1974 ff.

Lapide, Pinchas (1982): Mit einem Juden die Bibel lesen. Stuttgart.

LThK = Lexikon für Theologie und Kirche, 1-11. [3]Freiburg, Basel, Rom, Wien 1993-2001.

NJB = Neue Jerusalemer Bibel. Freiburg, Basel, Wien 1985, p. 1217: "Ich lasse mich von euch nicht befragen."

Spiegel, Der, Nr. 52, 24. 12. 2005, pp. 136-147; Interview with the philosopher Daniel Dennet (pp. 148-150).

Thomas von Aquino: Opera, 1-7 (1225-74). Ed.: R. Busa. Stuttgart-Bad Canstatt, 1980; compare with: Die deutsche Thomasausgabe, 1-36. 1933-1962.

Time (2005) *Evolution Wars*, August 15, volume 166 (7). 26-35.

Biography

Karol Sabath was recently popularizing evolution in two major museums of Warsaw: the Geological Museum of the Polish Geological Institute and the Museum of Evolution of the Institute of Paleobiology, Polish Academy of Sciences. He died October 10, 2007, so this is a posthumous chapter, perhaps his last. He was a member of the Committee of Evolutionary and Theoretical Biology, Polish Academy of Sciences. He coauthored biology textbooks for secondary schools and was a member of the editorial board of *Biologia w Szkole* (a journal for biology teachers). As an active popular science writer he engaged in debates with creationists since the late 1980s, and later co-founded the website http://www.ewolucja.org/ intended to provide resources addressing creationist claims. His research interests were in vertebrate paleontology: dinosaur eggs, tyrannosaurid theropods and dinosaur tracks; he was assistant editor of the *Acta Palaeontologica Polonica*. As an amateur paleoartist, he depicted extinct animals (now using digital media).

See:

Kielan-Jaworowska, Z. (2007). In memoriam: Karol Sabath (1963–2007). *Acta Palaeontologica Polonica* **52**(4), 828-829.

http://www.app.pan.pl/acta52/app52-828.pdf/

32

Teaching Evolution in Poland

The Education System and the Creation–Evolution Debate

KAROL SABATH

Museum of Evolution, Institute of Paleobiology, Polish Academy of Sciences Twarda 51/55, 00-818 Warsaw, Poland

1. Introduction

Teaching evolutionary biology has not been seriously contested in Poland since the country regained independence in 1918, when after more than half century since the "*Origin of Species*" the debates among scientists over the fact of evolution settled down. Later the scope of evolutionary topics in school curricula varied, depending on the trends in education, school reforms or political pressures. However, various shades of religiously motivated creationism persisted among the predominantly Christian nation. Therefore, recent campaigns of "scientific creationists" or recently Intelligent Design (ID) proponents have appealed to a substantial fraction of the society, even though less vocal than in the United States. I shall briefly recapitulate below the development of the Polish creation-evolution debate, and present the current state of biological education in the national school curricula, as well as the recent threats to the integrity of science education in Poland.

It seems appropriate to begin with a short account of reception of the theory of evolution in Poland and of the development of the national education system.

2. Historical Background

In Charles Darwin's lifetime the territory of Poland was partitioned between three neighboring empires: Prussia, Russia and Austro-Hungary. The foreign powers did not encourage development of Polish culture and science, and to various degrees attempted to minimize the extent of Polish-language education, especially at higher levels. Nevertheless, new ideas and trends were assimilated by the national elites striving to keep in touch with current intellectual developments abroad.

2.1. *Pioneers of Darwinism in Poland*

2.1.1. *Warsaw*

Within the divided Polish territory, one of the major scientific and cultural centers was Warsaw, within the Russian Empire throughout the 19th century. This former (and later) Polish capital became the main center of evolutionary thought in Poland.

Darwinian theory was first introduced to Polish students already in 1862 by the famous zoologist Benedykt Dybowski, then teaching in Warsaw's "Main School" (Szkoła Główna). A year later, however, this Polish-language college was closed down by the Tsarist authorities, and Dybowski sentenced to death for his involvement in a national uprising against the Russian rule (later his sentence was changed to deportation to Siberia, where he studied Lake Baikal fauna).

The Imperial Warsaw University, set up in place of the Main School in 1870, was a Russian-language school, where, however, some Polish professors remained, like August Wrześniowski and Henryk Hoyer, who continued to teach evolution.

School curricula in conservative monarchies occupying Poland contained no mention of "Darwin's dangerous idea". Evolutionary theory was then popularized via scientific societies and newly established popular science magazines, like "*Przyroda i Przemysł*" (*Nature and Industry*, published in 1872-1881) and "*Wszechświat*" (*Universe*, since 1882), where Darwinian theory was presented by August Wrześniowski, Mieczysław Kowalewski, Antoni Ślusarski and others (see Nusbaum, 1952: 648).

Evolutionary ideas spread among wider circles of Polish society thanks to magazines addressed to general audience like "*Przegląd Tygodniowy*" (*Weekly Review*), "*Prawda*" (*Truth*) or "*Ateneum*". Besides journals, the first Polish books on Darwinism were published, like "*Teoria Darwina rozwinięta przez Haeckla*" (*Darwinian Theory, Expanded by Haeckel*) by Bronisław Rejchman (1873), countered by anonymous pamphlets like "*Homo* versus Darwin" (Dr. Nemo 1874). The anti-Darwinian attitudes were common among clergy and aristocracy, and the Tsarist secret police also treated Darwinian literature as "subversive".

Nevertheless, the Polish translation of the "*Origin of Species*", started by Szymon Dickstein, graduate of the Faculty of Nature, Warsaw University in 1877, and after his premature death completed by Józef Nusbaum (graduate of 1881) was first published in 1884. Also several other books by Darwin and Huxley were translated and published in the last quarter of the 19th century, mainly by the influential Warsaw-based "*Przegląd Tygodniowy*" weekly.

2.1.2. *Cracow*

Another center of Polish national culture and science was Cracow, formerly a royal capital of Poland. In the 19th century, Cracow belonged to the Austro-Hungarian province of Galicia, together with Lwów (now Lviv, Ukraine), another important university town. The province had substantial autonomy, and the Habsburg regime was more tolerant than the Tsarist one, thus Polish culture thrived there more freely, and Cracow surpassed Warsaw as a refuge for national art and humanities. Natural sciences were also cultivated there, even though the elites were more conservative there and new ideas gained ground more slowly than in Warsaw.

For example, Darwin's "*Descent of Man*" and "*Sexual Selection*" were translated into Polish by Ludwik Masłowski already in 1874-1876 in Kraków, but still twenty years later, lectures on the origin of man by Tadeusz Garbowski (Privatdocent who arrived from Vienna) were cancelled by the President of the Jagiellonian University Stanisław Tarnowski at the request of Cracovian archbishop Jan Puzyna.

2.1.3. *The early controversies*

The religiously motivated opposition to Darwinism has been recently discussed by Rev. Zbigniew Kępa (2004), who devotes more attention to an apologetic book by Heinrich Reusch, translated as "*Biblia i natura*" by Michał Nowodworski, in 1872. Rev. Nowodworski, editor of "*Przegląd Katolicki*" (*Catholic Review*) and lecturer at the Warsaw's Akademia Duchowna (Clerical Academy), replaced the author's text in many chapters or added footnotes attacking Darwinism as a false and godless doctrine, reducing humans to mere animals and contradicting the divine revelation recorded in the narrative of Genesis.

Despite such attacks, by the early 20th century, Polish scientists were already mostly evolutionists, as were the secular elites. J. Nusbaum-Hilarowicz, for some time working as a school teacher, popularized evolutionary biology in his own writings, including the "*Idea of Evolution in Biology*", first published in 1910 (Nusbaum, 1952), as did many other Polish authors, mainly naturalists and medical doctors by training. Some of the books, like the novel "*Wotchłaniach czasu*" (*In the Depths of Time*, 1896) by Erazm Majewski, were addressed to younger readers, presenting the development of life on Earth.

2.2. *Brief History of Teaching Evolution in Polish Schools*

Below is a brief overview of the history of teaching evolution in Poland, divided into major periods used in the Polish historiography: the "Second Republic" (the first was the Polish-Lithuanian Commonwealth, dubbed "Noblemen Republic", before the partitions), the Communist-ruled "People's Republic", and the current "Third Republic" (the numbering indicates that the Communist Poland was not fully independent).

2.2.1. *The Second Republic (1918-1939)*

When Poland regained independence in 1918, it was obvious that one of the principal goals of the country was to develop national industry and infrastructure, and that one of the top priorities was national education. During the partitions of Poland, preserving the language, national culture and art was regarded as the crucial patriotic obligation. In the independent Republic, science and technology seemed at least equally crucial for the prosperity of the nation. The two decades between the world wars brought quick development of Polish scientific institutions, with rapidly growing universities, and improvements in educational curricula, including biology. Among others, brief accounts of evolutionary history of life and the geological past of the country became part of standard college-level education. The regular textbooks were supplemented with additional materials on evolution, such as selections of excerpts from classic works by Lamarck, Darwin and Wallace, like that compiled by Kazimierz Czerwiński (1927). Biology schoolbooks of that period were, however, mostly concentrating on traditional descriptive botany and zoology.

Moreover, only a small fraction of the population received more than elementary education. After World War I more than half of the population was illiterate, due to the policies of the partitioning powers. In 1919, mandatory grammar education was introduced for children aged 7 to 14. In 1932, a reform of education was introduced, dividing primary schools into three categories: I (2 teachers per 120 pupils; 4-grade curriculum extended into 7 years), II (4 teachers per 210 pupils, 6-grade curriculum, with the last grade extended into 2 school years), and III (full 7-grade curriculum). Those who continued their education went to a 4-year gymnasium (ending with a state exam, called "mała matura") and then could enroll in a 2-year lyceum, ending with a graduation exam ("duża matura"). The curriculum in these secondary schools depended on a selected profile: classical, humanities, mathematics-physics or natural sciences. Because only a small percentage of the population attended secondary schools, and most of them followed the first three profiles, actually only very few students were learning biology at more than a very basic level, and confronted with the basic tenets of evolutionary theory. On the other hand, religion was taught at public and private schools of all levels, so that most of the population was exposed to the Genesis story, whether in Catholic, Protestant, Orthodox or Jewish religion classes.

Besides schools, also natural history museums are important institutions serving public education, and reaching wider audiences, including adults. In Poland there were only few university cabinets of natural history and private collections. One of the first decisions of the new Parliament (Sejm), was establishing the Polish Geological Institute in 1919; in the 1920s, a new building was erected for the institute, with a large exhibition hall in the center. The hall was inspired by the architecture of great 19th century museums of natural history in London and other cities of Western Europe. Due to financial shortages during the economic crisis of 1929, only shortly before World War II the exhibition on geological history and structure of Poland, with focus on the country's mineral resources was completed in the Geological Museum. Along with illustrating the geological processes of the past, the exhibition showcased fossils from various epochs (Mizerski and Urban, 2006).

In Cracow, since 1865, there existed the Physiographic Museum of the Cracow Scientific Society, but it was mostly devoted to regional fauna, flora and mineralogy. After World War I, it was incorporated into the Academy of Arts and Sciences, and enriched in spectacular specimens, such as the woolly rhinoceros preserved with soft parts, found in Starunia (now Ukraine) in 1929. However, the Museum had a very limited exhibition space until 1992, when it was transferred to a renovated building of former baths. It is now known as the Natural History Museum of the Institute of Systematics and Evolution of Animals, Polish Academy of Sciences (and has recently presented a large exhibition on evolution of life, to commemorate the Xth Congress of the European Society for Evolutionary Biology, held in Cracow in 2005).

A fine example of Polish private natural history collections was that of the Count Włodzimierz Dzieduszycki, who established a private museum in 1855 and in 1880, after moving it to a new large building, donated it to the Polish nation, providing also financial basis for its development. The Dzieduszycki Museum in Lwów contained a large natural history division, including the largest ornithological collection in the country, as well as paleontological and geological exhibitions. The museum, lacking government's support, suffered a deep crisis in the 1930s, resulting in its closing to the public and damages to the collections due to insufficient care (the staff was drastically reduced). After World War II, the museum became the State Natural History Museum of the National Academy of Sciences of Ukraine in Lviv.

Unfortunately, the economic crisis delayed the project of establishing the National Museum of Natural History in Warsaw. Formally, the Minister of Religious Confessions and Public Education, Jan Łukasiewicz, issued a decree in September 1919, a few months later that the Geological Institute, but contrary to the latter, only a Zoological Museum has been actually organized, located provisionally (till now!) in a tenement house. Finally, after the crisis of the early 1930s, the government decided to build a modern state-of-the-art natural history museum in Warsaw, next to the National Museum (of art), being constructed in a great downtown location. The construction of the Natural History Museum was to begin in 1940. The war interfered with the plans. The museum has never been built, despite several initiatives in the past decades, and Poland remains one of the few countries without a National Museum of Natural History (Dzik, 2004).

2.2.2. *The People's Republic (1945-1989)*

After World War II universities, museums, and schools were ruined. Polish intellectual elites were decimated. At the same time, the Communists declared access to education for everybody. A spectacular campaign against illiteracy (in 1960 the illiteracy dropped to about 2 percent) was accompanied by publishing great amounts of very cheap books. Since the new political system was advertised as stemming from Marxism, presented as a synonym of "the scientific worldview", natural sciences became promoted more than ever. There was, however a dark side of this "enlightment of the masses".

One of the central tenets of Marxism was the inevitable historical progress. Obviously, also evolution became laden with ideology, as the biological equivalent of social progress and as an argument against the "reactionary" religious worldview. Moreover, the Stalinist doctrine of the first post-War decade favored "creative Darwinism", aka Lysenkoism. Political pressure on science affected also education. School textbooks of that period praised Michurin and his neo-Lamarckian achievements (and generally giving the impression that modern biology was developed mostly by Russian and Soviet scientists), while criticizing Mendelian-Morganian genetics and other theories essential for the modern neo-Darwinian synthesis as "bourgeois pseudo-science" (the more so that Gregor Mendel was a priest). This was also accompanied by rewriting the history of biology. Also the Polish pioneers of evolutionary biology, such as Nusbaum-Hilarowicz, were judged according to their political sympathies or views on Lamarckian heredity (Nusbaum, 1952, a reedition of his 1910 book, was so heavy edited, that the authorship actually belongs to the editorial committee). The usage of "creative Darwinism" as a tool of antireligious propaganda was reviewed in 1998 by Rev. Zbigniew Kępa in his Ph.D. thesis (published as Kępa, 1999). He also presented the reactions of Catholic philosophers and theologians to this campaign.

Among the damages done to biology in the public opinion at that time was associating evolutionary biology with the Marxist doctrine, so that many people are still prone to Creationist propaganda claiming that the collapse of

Communism means also the demise of its companion materialist ideology: "Darwinism". In fact, ideological biases of the Communist rulers hampered the development of many branches of natural sciences, such as evolutionary genetics, in the Eastern block (and cost the careers and lives of many brilliant biologists). Absurd experiments in agriculture not only failed spectacularly, but also raised well-founded skepticism towards the claims of "true Darwinism" (as Lysenkoism was promoted) among farmers and society in general.

Nevertheless, the Lysenkoism ended within a few years after Stalin's death and Polish biologists could join the global community of natural sciences. Since the 1960s, school textbooks provided a fairly accurate, if simplified, account of evolution.

At that time the education system was reformed. Since 1947 it consisted of 7-grade primary school (attendance was mandatory, beginning at the age of 7 years), and various secondary schools, such as 4-grade lyceum, 3-grade technical schools or 2-grade vocational schools. The Parliamentary Act of 1961 on the Education System extended the primary school from 7 to 8 years (the change was effective since 1967). Only a minority attended lyceums (where in biological-chemical profile biology was taught more extensively; other profiles had limited biological education, concentrating on human biology and hygiene, as well as brief accounts of systematic biology, ecology and evolution), while most young people went to technical or vocational schools. This reform also extended pedagogical studies and led to establishment of new university faculties, etc., to provide enough qualified teachers for the new demands.

In 1978, a new reform began: the school education was to begin at the age of 6, and consist of mandatory 10-grade primary school for all (with curriculum divided into 3 years of elementary education followed by 7 years of specialized education divided into particular subjects), supplemented by 2-grade secondary schools of various kinds, including those preparing for university enrollment. The prolonged 10-year primary education along a common curriculum was intended to improve the chances of children from rural areas and from families with worse education background. However, after a few years this reform was aborted, preserving only the earlier start of school education (with six-years-olds attending "zero" grade preparatory classes).

The education system was reinforced by the national TV (a large part of the schedule was devoted to education or popular scientific films and programs, dramas, etc.). There were also many widely accessible popular science journals, explaining various aspects of science and technology at a fairly advanced level — though with rather poor visual quality.

Due to historical circumstances, natural history (and science in general) was always given less public attention than "truly patriotic" activities of national heroes, such as humanities and arts — poetry, historical paintings, etc. — or military traditions. As a result, amateur science is not very popular. Still, in the 1960s and 1970s a wave of interest in paleontology (and thus evolution) rose in the wake of very successful Polish-Mongolian Paleontological Expeditions. Dinosaur mania erupted for the first time in Poland, fuelled by the dinosaur exhibition in the Warsaw's Palace of Culture and Science (open in 1968, then relocated to form the core of the Museum of Evolution, established in 1984). A dinosaur park was open in Chorzów, Silesia, in 1975, and young readers followed the vivid accounts of the desert digs written by the expedition leaders, Prof. Zofia Kielan-Jaworowska (1969, 1973) and Maciej Kuczyński (1977). Vertebrate paleontology remained one of the few disciplines, where Polish scientists achieved remarkable success, becoming leading international authorities in their field (see, e.g., Kielan-Jaworowska et al., 2004 or Dodson et al., 2004). However, the public interest in paleontology faded, only to be briefly revitalized by the *Jurassic Park* movies and then TV series, such like the BBC *Walking with Dinosaurs* in the following decades.

2.2.3. *The Third Republic: after 1989*

After the political system changed in 1989, the education became again the subject of debates and reforms. Of course, curricula and handbooks of history were changed first. Other curricula followed, with less modifications. An important feature of the developing new education system was its pluralism. Non-public schools (basically nonexistent for past half a century) were organized, and the school curricula were developed by various teams of authors, with textbooks and supplementary teaching materials produced by competing publishers. They are all to follow the national education standards published by the Ministry of National Education, and are subject to approval by the Ministry experts. Otherwise, teachers are free to choose particular detailed course curricula and associated handbooks for their classes. The competition between publishers resulted in quick improvement of at least visual attractiveness of the new schoolbooks, compared to times, when a single state publisher

(Wydawnictwa Szkolne i Pedagogiczne) practically had a monopolist position in the textbooks market for primary and secondary schools.

The major reform in recent years (announced in 1991 and now completed) introduced tripartite division of the Polish education system into primary schools (6 years preceded by an obligatory year in preschool grade "0"), gymnasium (3 years) and lyceum (3 years). The reform also affected school curricula, with the declared general goal being a reduction of the burden of "encyclopedic" knowledge, and increase of practical abilities (including finding and utilizing various information sources, self-learning, integrating multidisciplinary data, social skills, etc.). Recently, the Ministry of Education suggested the gymnasium might be reunited with grammar schools; the status of matura exams as achievement tests vs. university admission exams is being debated.

Among the successes of Polish education in the past decade was the increase of the proportion of students entering lyceums and then universities or other college level schools. The critics point to devaluation of high school diplomas, and poor overall quality of education (partly explainable by understaffing of schools of all levels by underpaid teachers); despite the statistical improvement of average formal education, secondary illiteracy (difficulties in understanding moderately difficult written text) seems to be increasing. Unfortunately, the same is true for scientific illiteracy.

There are only few popular science magazines left in the market (including Polish editions of *Scientific American* and *National Geographic*), and since literature generally became more expensive relative to buying power, popular science books usually sell only in few thousands (or even hundreds) copies (the population of Poland is about 38 million), orders of magnitude less than a few decades ago. In bookstores, the "Science" sections are often dominated by pseudoscientific or esoteric literature. And more than half of the adult population has not read any book in the preceding year.

On the other hand, cable and satellite TV enables many people to view popular scientific channels (e.g., Discovery or National Geographic). The Internet allowed both teachers and students easy access to great information resources, including several web portals in Polish addressed specifically to schools, and covering, among others, biological topics — like

- http://www.eduseek.interklasa.pl,

- http://www.gimnazjum.pl,
- http://www.liceum.pl,
- http://www.biolog.pl,
- http://www.profesor.pl,
- http://www.ids.edu.pl/wwwbio,
- http://www.wiw.pl/biologia

(in many of these websites the official education standards, curricula and lists of approved textbooks can be found).

3. Current Biology Curricula in Poland

3.1. *Primary Schools*

Under the current system, primary schools are divided into two stages. The elementary teaching (grades 1 to 3) integrates grammar education (reading, writing, arithmetic) with basic information about the surrounding world. The next stage (grades 4 to 6) involves teaching specialized topics. Among the subjects is one named "Nature". The national education standards ("Podstawa programowa" — literally: curricular basis) list its aims such as "understanding relationships in the natural environment", "acquiring observation and description skills regarding natural phenomena", "knowing interdependence between humans and environment", "performing simple experiments and interpreting their results". The topics include elements of astronomy, geography, ecology, physiology, biodiversity and rudimentary taxonomy, environment protection, etc. The curriculum does not include evolution *per se* but should build the foundations of empirical thinking and provide basic knowledge necessary during further education.

3.2. *Secondary Schools — Gymnasium*

At the gymnasium level (3 grades), Biology is a separate subject. Among the expected achievements of students are more general ones, like "Formulating hypotheses. Analyzing and interpreting the results of observations and experiments, and assessing their reliability", or more specific, like "Interpreting correlation between the organism's environment and its structure and functions". Among the topics listed in the national education standards are intra- and interspecific interactions in nature, cycling of matter and energy flow in various natural systems, hereditary information (characters determined by genes and by environment), biodiversity, etc. This knowledge is fundamental to understanding evolution, and — if learned well — should make young people less prone to standard creationist "arguments" bases on ignorance of genetics, thermodynamics, etc.

There are currently more than thirty available biology schoolbooks for gymnasium, with about half a dozen publishers having the lion's share of the market. Generally, they invoke evolution explicitly in several contexts: history of the life on Earth (with inevitable reference to dinosaurs), species extinctions (also due to human-induced environmental changes), and natural classification of organisms (higher rank taxa as products of adaptive radiation of ancestral forms). Nevertheless, evolution of life on Earth is only briefly mentioned (and in many cases not mentioned at all, because not all teachers do teach these issues, as a brief enquiry in a web forum of young dinosaur enthusiasts recently revealed).

3.3. *Secondary Schools — Lyceum*

The 3-grade lyceum is now chosen by more than half of the gymnasium graduates. They can enroll in a particular profile. Depending on profile, some subjects are taught in a basic version, while others are taught in expanded version.

In most profiles (except the biological-chemical), biology is offered in the "light" version. It focuses mainly on human biology (anatomy, physiology, health), with elements of genetics and ecology (addressing biodiversity issues, environmental awareness, GMOs [genetically modified organisms], etc.). Even this reduced scope explicitly includes evolution, namely evolution of the Earth's biodiversity and human origins; the elements of genetics and ecology also include basic ideas necessary to understand natural selection.

The expanded biology curriculum (used in a minority of classes) is based on national standards including elements of evolutionary biology dispersed among various sections (e.g., systematic botany and zoology), but also specifically listed under the heading "Evolution" (section 5 of the curricular basis): (1) the concept and evidences of evolution; (2) mechanisms of evolution; Charles

Darwin and the theory of natural selection; (3) population genetics; (4) speciation; (5) basic regularities in evolution; (6) origin and history of life on Earth; (7) anthropogenesis.

Actual presentation of the topics and their sequence in particular curricula depends on their authors (for example, some textbooks, e.g., Bartnik and Lewiński, 2004) include a brief presentation of early, non-evolutionary concepts in science, from Aristotle, through Christian concepts to Cuvierian catastrophism). Typically, the teachers are expected to cover the above evolutionary topics within about 20 hours (e.g., Lewiński et al. 2003).

The regular lyceum textbooks are often accompanied with supplementary books by the same publishers, covering selected topics in more detail. In the case of evolutionary biology, such additional books include, e.g., Sabath and Balerstet (2001). Sometimes, parts of regular textbooks are printed separately (e.g., Kaszycka and Ryszkiewicz, 2005), allowing them to be used independently (e.g., by students using another textbook for the main curricular course of biology). That knowledge is required at the graduation exam (if someone chooses biology among the exam subjects) and by universities and medical academies testing the candidates.

4. The Creation-Evolution Controversy in Poland

4.1. *Antiscience Attacks*

During the Communist rule in Poland, the state (and thus the Communist party) had control over most media (the radio and TV was state-owned, as were most publishing houses and newspapers). Some journals and books were published, e.g., by churches, but were subject to censorship. Theoretically, the official ideology was based upon the "scientific worldview", but even the major national media sometimes leaned toward pseudoscience, entertaining various speculations on psychic forces, telekinesis, UFOs [unidentified flying objects], etc., perhaps as a secular substitute for religious miracles and mysteries. Even though American-style creationism was virtually nonexistent publicly (and occasionally mentioned as an exotic curiosity, along with the flat-Earthers), conspiracy theories contesting the mainstream view on the past were represented by Erich von Daeniken's claims about "paleoastronauts". His books became bestsellers in the 1970s, and the movie "Chariots of the Gods" was shown in

Polish cinemas attracting crowds. Perhaps the Communist censors believed that Daeniken may help seed doubts about the *Bible* (with his claims that "Gods" and "angels" were in fact ancient astronauts).This indicated how great is the demand for such persuasively told stories undermining scientific rationality.

After 1989, the irrational triumphed. In the early 1990s, Russian psychic healer Anatoly Kashpirovsky held regular hypnotising sessions on public TV, watched by millions. Healers and alternative medicine kept gaining popularity as the national health care system became more and more inefficient. Horoscopes abound in the press and the internet. A couple of years ago, crop circles began to appear in Poland. There are several magazines and regular TV programs devoted entirely to unexplained phenomena, "other dimensions", UFOs, government conspiracies, etc. One of the guests was Michael Cremo, of the "Forbidden Archeology" fame. The translated book itself (Cremo and Thompson, 1998) sold in huge number of copies and was enthusiastically endorsed by creationists, even though the Vedic interpretation contradicts the Christian scenario of Genesis. "Forbidden Archaeology" was, however, a much more scientific-looking attack on mainstream paleoanthropology than the creationist publications available in Poland (usually overtly religiously motivated, while not all readers were aware of the Hare Krishna affiliations of the "Forbidden Archeology" authors), so it has been acclaimed as an evolution-bashing tool.

A poll conducted by the Pentor Polling Agency for the "*Wprost*" weekly (Górecki, 2002) was summarized in the lead of the article: "What the Poles and Americans have in common is often blatant ignorance". The poll was inspired by the publication of a report "*Science and Engineering Indicators 2002*" by the National Science Foundation in the USA. The results shown that in Poland 61 percent of the people believed that "astrology is a reliable source of information about the world" (including 41 percent of those declaring university level education), compared to 25 percent of Americans (an additional 43 percent of Americans occasionally consulted horoscopes). As for psychic forces, there were 49 percent of believers in Poland, as opposed to 60 percent in the USA; 26 percent of Poles (and 30 percent of Americans) believed that extraterrestrial civilizations visited our planet in the past; 39 percent of Poles (and 48 percent of Americans) claimed that people lived alongside dinosaurs; 52 percent of Poles (and 47 percent of Americans) disagreed with the statement that humans appeared on earth as a result of evolution of earlier organisms.

Even science teachers seem disoriented. When the Polish Geological Institute announced a geological contest for secondary schools, among those participants who chose the "Cambrian Explosion" as their topic, a substantial minority pasted whole passages of an anti-evolution essay, together with figures, from a creationist webpage. And all the entrants were supervised by teachers, who signed letters of support for them.

Concerned scientists try to counteract the surge of antiscience and pseudoscience: for the past ten years a Festival of Science has been organized each September in Warsaw (http://www.icm.edu.pl/festiwal), with many cities joining the initiative in recent years. For the past four years Polish Radio has helped organize a one day Science Picnic in June in the historical part of Warsaw. The interest of people attending these events shows that science can be presented in an attractive way, appealing to people of all ages and professions. A modern science museum, called the Copernicus Science Center is planned in Warsaw along the Vistula River — to allow year-round popularizing of current science.

However, the task of preventing further degradation of scientific awareness in Poland is difficult. Despite occasional political declarations praising science as the key factor of national economic growth and competitiveness in the 21st century world, spendings on science relative to GNP are steadily dwindling: from 1.2 percent in 1991 down to about 0.3 percent in the current budget. Young scientists, with monthly salaries equivalent to less than 500 euros are prime targets for brain drain.

4.2. *The Antievolution Movement in Poland*

Until the collapse of Communism, creationist literature was marginal, largely published by Christian denominations with little influence in Poland (like Seventh Day Adventists). Articles appeared in low-impact religious publications, translated pamphlets or comics, often distributed by American missionaries. Probably most widely known was an elegantly printed color book by the Jehovah's Witnesses (Anonymous, 1989), whose missionary activity in Poland has been noticeable.

Young Earth creationism (YEC) received wider attention, when Prof. Maciej Giertych, a dendrologist, translated and published a book "*The Crumbling*

Theory of Evolution" by J.W.G. Johnson (1989). Not only did Giertych have scientific credentials, but also the book was printed by a Catholic publisher (and excerpts were earlier published by "*Rycerz Niepokalanej*", a periodical associated with the Franciscan Order). Giertych was also known in the political arena, as the member of Consulting Board to General Wojciech Jaruzelski, who proclaimed martial law in 1981 and banned "Solidarity". He has also opposed reforms in the Catholic Church after the Vaticanum II Council. After he openly supported YEC, several Catholic priests, professors at the Academy of Catholic Theology (later renamed to Cardinal Stefan Wyszyński University in Warsaw), Papal Theological Academy in Cracow and Catholic University of Lublin wrote a letter of protest to the chairman of the Scientific Council of the Polish Episcopate, Bishop Alfons Nossol (partly because the book was published with an official ecclesial imprimatur). Also secular scientists wrote letters to Catholic media, pointing to the entirely pseudoscientific nature of the book. The Committee of Evolutionary and Theoretical Biology, Polish Academy of Sciences, issued a special brochure (Łomnicki, 1994), exposing the errors in a VHS movie featuring M. Giertych and other creationists. Voices in the creation-evolution debate from that time were collected by Moczydłowski (1994). Attempts to reconcile evolution and creation from a Catholic (Theillardist or St. Augustine-based) perspective were published (McMullin, 1993; Kloskowski, 1994).

In 1995, the Polish Anti-Macroevolutionary Society (Polskie Towarzystwo Kreacjonistyczne, PTKr) was established, that became a hub for various initiatives and created a website (http://www.creationism.org) providing a growing collection of antievolution resources (mostly translated American creationist materials). The Society is ecumenical — among its prominent members are a Jesuit Professor Piotr Lenartowicz (philosopher and anthropologist), an Adventist theologian, Professor Zachariasz Łyko, as well as an evangelical apologist Mirosław Pajewski (pseudonym).

In many online discussion forums even remotely related to science, creationists keep posting antievolutionary claims, often excerpts from that website. There are also lectures by creationist supporters from Poland or visiting speakers, often organized in churches. They accuse the "evolution indoctrination" (up to 20 hours in biology classes at the lyceum in the extended programme, usually a couple of hours throughout the whole secondary school curriculum and nothing in primary schools) of depraving youth, prevailing over religious teaching to which Polish youth is exposed a couple of hours per week for at least twelve years. Rarely they go beyond the most typical rhetoric (immorality of Darwinism, thermodynamic impossibility or statistical improbability of evolution, Piltdown man forgery, etc.), borrowed from translations of creationist literature (e.g., McLean et al., 1999; Wieland, 2000, Zillmer, 2003).

In 1998, a philosopher from the Pedagogical Academy in Zielona Góra (later Zielona Góra University), Kazimierz Jodkowski published a large monograph on "*Methodological Controversies between Evolutionism and Creationism*" (Jodkowski, 1998), including numerous translations of creationist publications. He concluded that methodologically scientific creationism is just another paradigm, adding to the pluralism of science, and the choice between the creationist and evolutionary explanation is merely a matter of different arbitrary assumptions (because both are actually ideologies). Jodkowski published also extensively in creationist journals, such as "*Na początku...*" ("*In the Beginning*"), published under the auspices of the Polish Anti-Macro-Evolutionary Society, where also large portions of the monograph were published on-line. In 1999, his monograph was awarded the Individual Prize by the Minister of Education Mirosław Handke (representing a right-wing party). Jodkowski is now the editor of an on-line journal "*Filozoficzne Aspekty Genezy*" ("*Philosophical Aspects of Genesis*", http://www.nauka-a-religia.uz.zgora.pl/index.php?action=czasopismo). Recently, however, Prof. Jodkowski seems more skeptical about the scientific merits and perspectives of creationism in explaining and predicting natural phenomena.

On the other hand, there are available popular science books and textbooks on evolution, including many translations, public TV has repeatedly broadcast popular series on biology or paleontology, teachers are being educated in current developments of evolutionary theory (e.g., by the "*Biologia w Szkole*" quarterly), there are websites like http://www.ewolucja.org and several discussion boards that debunk creationism. Still, the awareness of creationist tactics is low, and many scientists tend to overlook the scale of the problem, regarding it as a purely American phenomenon, despite alarming similarities in the overall education levels and religiousness of both societies, and the fact that the Polish society is still less prepared to cope with modern political marketing methods. Especially Intelligent Design is often treated as a *bona fide* scientific hypothesis, due to lack of understanding of the legal reasons why the less overtly religious concept has been chosen for the Wedge Strategy in the USA (see Forrest and Gross, 2004). I have heard educated non-biologists who judged "*Darwin on Trial*" by Phillip E. Johnson (Polish edition: 1997) as a competent and balanced treatment of scientific evidence (and not a one-sided prosecutor's speech using the whole arsenal of creationist rhetoric to support the claim that evolution is merely an ideological weapon of a secular naturalistic attack on true faith).

The Catholic Church remained rather distanced from radical creationism. Archbishop Józef Życiński is even a member of the Committee of Evolutionary and Theoretical Biology, Polish Academy of Sciences. He published several books on evolution and theology, e.g., "*God and Evolution. Fundamental Questions of Christian Evolutionism*" (Życiński, 2002), where he regards fundamentalist ideas rejecting evolution as damaging to the authority of the Church in modern society.

In August 2005, when about 1500 scientists attended the Xth Congress of the European Society for Evolutionary Biology in Cracow, one of the biggest scientific symposia in the country, the media were completely silent about the event. At the same time, the "*Ozon*" weekly, in one of the first issues published by the new editorial team, sympathizing with the Catholic right, published a pro-ID essay "Dogma of Evolution?" (Terlikowski, 2005), soon reprinted at the website of the Polish Anti-Macroevolutionary Society.

Most church authorities and Catholic scientists (probably a majority), used to embrace the notion of non-conflicting truths of science and faith. Often, they invoked the famous address to the Papal Academy of Sciences by Pope John Paul II on October 22, 1996, proclaiming evolution "more than a hypothesis", corroborated by independent lines of evidence. A recent example was a letter by Professor Jan Kozłowski (2005), evolutionary ecologist from the Jagiellonian University in Cracow, commenting upon Cardinal Christoph Schonborn's famous anti-Darwinian statement; the commentary was titled: "God could create man via Darwinian evolution".

However, negating the existence of conflict between biology and theology with claims that Catholic doctrine accepted evolutionary theory as it is now understood by working scientists is becoming difficult, since Pope Benedict XVI himself in his inaugural sermon clearly spoke against the neo-Darwinian model of evolution, and stressed that humans are not a product of accidental naturalistic processes.

While scientists and philosophers (and in Poland the majority of professional philosophers are Christian philosophers, largely Catholic priests) debate on the relationships between natural sciences and religious faith, exchanging sophisticated arguments and witty remarks during numerous symposia and conferences devoted to philosophical aspects of evolutionary biology, origin of humans, etc., the future of evolutionary science and education in Poland may be changed by less subtle politicians.

4.3. *Towards the Fourth Republic?*

After the 2005 elections, the perspectives for science and the citizens hoping for reducing the civilization distance between Poland and the old European Union look dim, and more and more young educated people declare willingness to emigrate.

The new ruling coalition consists of Euro-skeptical parties, known from populist and anti-intellectual attitudes. The deputy Parliament speaker (wicemarszałek Sejmu) Ms. Genowefa Wiśniowska has a bachelor degree in psychic phenomena received from a Russian diploma mill. The government is praising the nationalist Catholic Radio Maryja and TV Trwam, as the only free information media in Poland. These media also oppose evolution (when filming a program at the Cracow natural history museum, the TV Trwam crew explicitly demanded the interviewed staff not to mention the e-word).

The new Minister of Education, appointed in May 2006, is Mr. Roman Giertych, lawyer by profession, leader of the nationalist-Catholic party League of Polish Families (LPR). Both his widely criticized past as a political right-wing extremist and his lack of academic or pedagogical qualifications inspired a wave of protests ranging from street demonstrations in many cities to a petition to the Prime Minister, signed by more than 60 thousand people within two days (http://www.bezgiertycha.pl). In the context of creation-evolution debate, it is noteworthy that Roman Giertych is the son of Maciej Giertych, member of European Parliament and unsuccessful candidate of LPR in the last presidential election (about 1 percent support), but also the most famous Polish young earth creationist. Roman Giertych is known to share his father's worldview and thus it is possible that the new Minister of Education will try to remove evolution from school curricula or insert creationism/ID. This would be the first serious challenge to teaching evolution in Polish schools. Teaching religion (mostly

Catholic) is already assigned more time than teaching the whole biology curriculum. Removal of biological evolution from the curricula would leave young Poles with only the knowledge of Genesis creation stories. The new coalition suggested they wish to establish the Fourth Republic, with a more centralized, authoritarian political system, and if accompanied by official adoption of the religious version of origins and philosophy by the public education system — it might be a step towards theocracy in the middle of Europe.

Paradoxically, the Catholic Church may become the mitigating factor to some extent, because many of its influential leaders (including the Pope) oppose greater involvement of the Church in the secular realm, remembering bitter lessons from the past in many countries.

Note added in proof: Indeed, in October 2006, Maciej Giertych organized an antievolutionary conference in the European Parliament, and soon his son publicly stated that it's time for a serious debate on the biology curricula, and the vice-Minister of Education Mirosław Orzechowski (LPR) called the theory of evolution "a legalized lie". This led to a series of protesting letters by the Academy of Sciences, leading universities, scientific societies and individual scientists – more than a thousand signed a letter demanding resignation of Orzechowski. Also Catholic scholars and the Scientific Council of the Polish Episcopate issued statements critical of creationism and ID. The news about the antievolution campaign was commented on in "*Nature*" (Graebsch, 2006; Giertych, 2006; and letters in vol. 444: 679-680). A selection of materials in Polish is available at http://www.ewolucja.org/d3/d37-6a.html.

5. Summary

Teaching evolutionary biology has not been seriously contested in Poland since the country regained independence in 1918. Later the scope of evolutionary topics in school curricula varied, depending on the general trends in education, school reforms, and sometimes political pressures (like the Lysenkoism of the 1950s). Current national education standards place evolution mainly in the biology curriculum of secondary schools, covering topics from genetic mechanisms of evolution through history of life to human origins. The coverage is, however, fairly limited, and evolutionary biology is taught in a more comprehensive way only to a small minority of each generation (the students who have chosen the expanded biology curriculum in Lyceum-type schools).

Prevalence of antievolutionary attitudes (and scientific illiteracy) in Poland is similar to that in American society. Therefore, more than a decade of active political marketing by "scientific creationists" or recently ID proponents, found a substantial resonance among the less educated and strongly religious sectors of the society. Since the current government coalition stems from that electorate, attempts to introduce creationism or downplay evolution in the public education system are possible.

References

Anonymous (1989). *Jak powstało życie? Przez ewolucje czy przez stwarzanie?* Watch Tower Bible and Tract Society of Pennsylvania, 256 pp.

Bartnik, E. and Lewiński, W. (eds.) (2004). *Biologia LO 3. Zakres rozszerzony. Podręcznik.* Operon, Gdynia, 320 pp.

Balerstet, J. and Sabath, K. (2001, 2nd ed, 2004). *Podstawy ewolucjonizmu. Książka pomocnicza dla uczniów liceów (kandydatów na akademie medyczne i uniwersytety).* Operon, Gdynia, 152 pp.

Cremo, M.A. and Thompson, R.L. (1998). *Zakazana archeologia.* Arche, Wrocław, 354 pp.

Czerwińki, K. (1927). *Wypisy z zakresu teorji ewolucji (Lamarck, Darwin, Wallace).* Warszawa.

Dzik, J. (2004). Narodowe Muzeum Przyrodnicze — historia i marzenia. *Ewolucja* **2**: 47-57. http://www.paleo.pan.pl/ewolucja/Ewolucja2.pdf

Forrest, B. and Gross P.R. (2004). *Creationism's Trojan Horse: The Wedge of Intelligent Design.* Oxford University Press, New York, 401 pp.

Giertych, M. (2006). Creationism, evolution: nothing has been proved. *Nature* **444**: 265.

Górecki, P. (2002). Magia zamiast nauki. *Wprost,* **1017** (26 V 2002): http://www.wprost.pl/ar/?O=13068.

Graebsch, A. (2006) .Polish scientists fight creationism. *Nature* **443**: 890-891.

Jodkowski, K. (1998). Metodologiczne aspekty kontrowersji ewolucjonizm-kreacjonizm. *Realizm, Racjonalność. Relatywizm, t.* **35**, Wyd. UMCS, Lublin, 540 pp.

Johnson, J.W.G. (1989). *Na bezdrożach teorii ewolucji.* Michalineum, Warszawa-Struga, 199 pp.

Johnson, P.E. (1997). *Sąd nad Darwinem*. Vocatio, Warszawa, 239 pp.

Kaszycka, K.A. and Ryszkiewicz, M. (2005) *Biologia. Ewolucjonizm. Podręcznik dla liceum ogólnokształcącego. Zakres rozszerzony*. Wydawnictwa Szkolne PWN, Poznań, 97 pp.

Kępa Z. (1999). *Marksizm i ewolucja. "Twórczy darwinizm" jako narzędzie propagandy antyreligijnej w latach 1948-1956*. Wydawnictwo Diecezji Tarnowskiej Biblos, Tarnów, 199 pp.

Kępa Z. (2004). *Recepcja darwinizmu na ziemiach polskich w latach od 1859 do 1884*. http://www.opoka.org.pl/biblioteka/F/FB/darwinizm_polska.html

Kielan-Jaworowska, Z. (1969). *Polowanie na dinozaury*, Wydawnictwa Geologiczne, Warszawa, 96 pp.

Kielan-Jaworowska, Z. (1973): *Przygody w skamieniałym świecie*. Nasza Księgarnia, Warszawa, 166 pp.

Kielan-Jaworowska,Z., Ciefelli, R. L., and Luo, Z.-X. (2004). *Mammals from the Age of Dinosaurs: Origins, Evolution, and Structure*. Columbia University Press, New York, 630 pp.

Kloskowski, K. (1994). *Między ewolucją a kreacją*. Wyd. Akademii Teologii Katolickiej, Warszawa, 208 pp.

Kozłowski, J. (2005). Pan Bóg mógł stworzyć człowieka drogą ewolucji darwinowskiej. *Tygodnik Powszechny* **30**; http://tygodnik.onet.pl/1546,1238283,0,355785,dzial.html

Kuczyński, M. (1977). *Zwrotnik Dinozaura*. Iskry, Warszawa, 241 pp.

Lewiński, W., Prokop, J., and Skirmuntt, G. (2003). *Biologia. Liceum ogólnokształcące. 1. Zakres rozszerzony. Przewodnik dla nauczyciela*. Operon, Gdynia, 135 pp.

Łomnicki, A. (ed.) (1994). *Opinia o filmie video "Ewolucja: rzeczywistość czy domniemanie"*. Universitas, Kraków, 68 pp.

McLean, G.S., Oakland, R., and McLean, L. (1999). *W poszukiwaniu prawdy o początkach. Ewolucja i kreacjonizm w świetle dowodów*. Pojednanie, Lublin, 160 pp.

McMullin, E. (1993). *Ewolucja i stworzenie*. Ośrodek Badań Interdyscyplinarnych PAT, Kraków, 95 pp.

Mizerski, W. and Urban, H. (2006). Muzeum Geologiczne Państwowego Instytutu Geologicznego W Warszawie W latach 1919–1939. *Przegląd Geologiczny* **54** (5): 388-391. http://www.pgi.gov.pl/pdf/pg_2006_05_15.pdf

Moczydłowski, E. (ed.) (1994). *Pan Bóg czy dobór naturalny*. Megas, Białystok, 145 pp.

Nusbaum J. (1952). *Idea ewolucji w biologii. T. 1. Zarys dziejów ewolucjonizmu do końca XIX wieku*. PWRiL, Warszawa, 793 pp.

Terlikowski, T.P. (2005). Dogmat ewolucji? *Ozon*, **17:** 33-36. http://www.ozon.pl/a_tygodnikozon_2_5_432_2005_17_1.html

Weishampel, D.B., Dodson, P., and Osmólska, H. (eds.) (2004). *The Dinosauria*. 2nd edition. University of California Press, Berkeley, 833 pp.

Wieland, C. (2000). *Kamienie i kości*. Megas, Warszawa, 45 pp.

Zillmer, H.-J. (2003). *Pomyłka Darwina. Zadziwiające dowody podważające teorię ewolucji*. Amber, Warszawa, 192 pp.

Życiński, J, (2002). *Bóg i ewolucja. Podstawowe pytania ewolucjonizmu chrześcijańskiego*. Towarzystwo Naukowe Katolickiego Uniwersytetu Lubelskiego, Lublin, 212 pp.

Dialogue

Taner Edis
I found this chapter fascinating, especially since it deals with Roman Catholic opposition to evolution.

There is some conservative Catholic anti-evolutionary sentiment in the United States, but it attracts much less attention compared to highly popular Protestant versions of creationism. Indeed, Catholicism here has a reputation of comparative liberality concerning evolution. Most visible Catholic views tend toward guided, non-Darwinian evolution. This has some affinities to Intelligent Design, and to that extent is scientifically dubious. But guided evolutionary views have much less potential to interfere with secondary science education, so they create no friction with the scientific community.

I was saddened to hear of Dr. Sabath's death, as I would have been very interested to get his views on some questions. I would have asked Sabath if he would make any more general comments about antievolutionary thought in a Catholic context. I would also have asked about how Cardinal Schoenborn's pronouncements critical of Darwinian evolution (to the extent of suggesting some Intelligent Design sympathies) played in Poland. We will never get to know how Sabath would have replied, but this remains a very intriguing review of creationism in Poland nonetheless.

Part 7

Are there Possible Avenues Towards Convergence?

Biography

Dr. Doron Aurbach is a Professor in the Department of Chemistry in Bar Ilan University (BIU), Ramat Gan, Israel. He obtained his Ph.D. in physical-organic chemistry from Bar Ilan University in 1983 and received also B.Sc in chemical engineering from the Israel Institute of Technology (Technion), Haifa, Israel in 1981. Dr. Aurbach is the founder of the electrochemistry group of BIU at the beginning of 1986. He teaches basic and advanced courses in physical chemistry and electrochemistry and works in the field of energy storage and conversion. He has more than 320 publications in leading journals of electrochemistry, physical chemistry, materials and surface science, and 11chapters in different books, 12 patents, and a book ("*Nonaqueous Electrochemistry*", 1999 published by Maecel Dekker N.Y.). Dr. Aurbach is an associate editor of the *Electrochemical & Solid State Letters Journal* and the *Journal of Solid State Electrochemistry*. Currently he is leading one of the biggest research groups in Israel (working in electrochemistry, energy, water desalination and materials science).

33
Intelligent Design *vs.* Evolution Theory

DORON AURBACH

Department of Chemistry, Bar-Ilan University
Ramat-Gan 52900, Israel
aurbach@mail.biu.ac.il

1. Introduction

When we examine what is universally known as evolution theory, we need to distinguish between two distinct aspects of the theory, namely, micro- and macro-evolution. Micro-evolution is a reality that is experienced in life and documented in reliable scientific experiments and observations (Freeman and Herron 2004). It includes observation of speciation in some kinds of birds and insects, due to the interrelationship between genetics, environment, and adaptation. In fact, the genetic code for all living species allows for certain degrees of freedom and change for any property that is inherent in these species. These degrees of freedom are in part intrinsic, that is, they are part of the genetic code itself, and may be caused by mutations, or, in layman's terms, by sporadic and occasional changes in the genetic code. This freedom of the genetic code is of critical importance because it allows for the adaptation of life to environmental changes and for the wonderful diversity that is seen in life on earth, enabling human beings, for example, to be distinguished from one another, by appearance, as well as by unique identities and characteristics. This adaptive ability of genetic codes, as will be argued below, is nothing more than a fine tuning of basic properties.

In contrast to the micro-evolutionary approach, there is the macro-evolutionary philosophy, which proposes that everything that we see around us — reality, life, and faith — is accidental and could simply have occurred naturally. This approach asserts that, for some reason, matter starts to live by spontaneous and possibly inevitable processes. Thus, "living" molecular clusters crystallize into living cells, after which these primitive cells develop internal, well-defined organs (e.g., the so-called prokaryotes-eukaryotes transformation). Single cells then spontaneously join together to form primitive multi-cell creatures, which continue to develop — again, spontaneously-into increasingly more complex entities. The various cells in these primitive multi-cell communities then develop specific functionalities over the years and become true organs. Thus, as macro-evolution posits, over a period of millions of years, creatures have gradually developed to possess functional systems that we recognize today as the various organs of living species. What is the driving force behind this amazing development? The magic words — 'selection rules'. The environment is thus acknowledged as a dynamic force that causes adaptation and improvement of the species, with the better functions persisting and the worst functions or responses to the environment disappearing.

In brief, this is the essence of macro-evolutionary philosophy — the evolution of matter into a living, breathing human being. To put it differently, the mere existence of water, the earth's elements, and a flux of energy from geo-thermal sources and radiation from the sun leads inexorably to the creation and development of life, from primitive to complex forms.

While this approach is described in detail in such books as C. de Duve's *Life Evolving* (de Duve 2002), it is, in effect, nothing more than a belief — and a fanatic belief, at that — in accidents and in the power of serial, random occasions to bring matter to life. In point of fact, in all books of this type, including that of de Duve, the authors make constant use of the verb "believe" in describing the various processes involved in the evolution of life (i.e., they "believe" that this or that happened ...). And surprisingly, at a first, superficial, glance, it appears that this approach is widely accepted by most scientists.

To cite one instance, the last issue of the prestigious journal "*Science*" in 2005, focuses on "a breakthrough of the year" (2005), discoveries of "evolution in action" (see the cover picture and the related editorial note) (Culotta et al., 2005), demonstrating that the above macro-evolution theory has conquered the human conscience and its general understanding of life. This "belief," as it were, appears in encyclopedias, in educational programs in many schools, and has become a vital matter of research validated in most leading academic and research institutions throughout the world. The questions here are thus how and why, and in order to understand them we need to have an historical perspective, as briefly outlined below.

2. An Historical Perspective

During the last few centuries, the human race has experienced an astounding scientific and technological revolution. The beginning of what is generally defined as history itself, approximately 6,000 years ago, is marked by the appearance of civilization, mostly in China, Mesopotamia, Egypt, and some areas in the Middle East. These civilizations were characterized by impressive skills in organization, building, art, agriculture, and communications. While from time to time archaeological discoveries leave us amazed at the abilities of ancient people in medicine, navigation, astronomy, and the like, it can be said that most of these ancient cultures remained stagnant for thousands of years at a consistent level of knowledge in the realms of life, reality, and the universe. Furthermore, it can also be said that since the dawn of civilization — a watershed period marked by man's acquisition of the ability to read and write, leaving a written heritage and testimonial for ensuing generations — and up until the end of the Middle Ages (about 300 years ago), there were no pronounced changes in the human race's level of knowledge, both as regards reality and the universe. In some cases, in fact, there was appreciably more regression than there was progress.

All religions were born in ancient times to early civilizations whose level of knowledge was very low compared to that which exists today. Unfortunately, many religions created their own bad reputations and negative associations due to the cruel work of missionaries, expansion via terrible wars, the corruption of the church and religious institutions, and — far more serious — their markedly narrow-minded approaches towards scientific discoveries at the beginning of the scientific revolution.

Since most religions regard the human race as the most important entity on earth — in fact, the real reason for and goal of the creation of the world — it was logical for them to adopt the geocentric approach, namely, that our earth is the center of the universe and that all the planets revolve around it. Galileo,

Copernicus and Kepler, however, presented much better models to explain the trajectories of the planets around us, namely, a heliocentric description that is the basis of astronomic observations today. They theorized that our planet rotates around its axis, causing the motion of planets and stars, with the sun at the center, and the various planets surrounding it in elliptic trajectories. The Christian church felt threatened by these new theories, expressed an overwhelmingly negative attitude to the new models of modern astronomy, and even endangered the lives of Galileo and Copernicus (Atkins, 2003). Hence, a very natural antagonism seemed to develop between religions that reflect ancient, primitive knowledge, and modern life and civilization, which experience explosive revelations in the perception and knowledge of life, matter, reality, and the universe. Moreover, because of the lack of machines and power sources that we have today, life in ancient times and in the Middle Ages was generally hard. The various constraints that religions demanded from their followers could be adopted relatively easily by people whose daily lives were, in any case, already difficult.

For modern generations, whose lives have become easier due to the technological revolution of the last two centuries, religious burdens may seem to be more annoying and irritating than they were for people of ancient times and the Middle Ages, whose lives were so hard. Another important point that should be raised is that all religions relate to the unknown — the unquantifiable, the mysterious, and the apparently transcendental aspects of human life. Most religions emphasize the fact that the human being is an interface between a physical body that can be touched and measured, and a spiritual entity that is transcendental and may be immortal. Hence, religion as a concept may be very attractive from a psychological point of view, as it is expected to touch upon critical aspects of reality: the meaning of life and death, faith, and the possibility that human beings may have immortal elements that remain after death. The modern age and its scientific revolution, a revolution to which we all bear witness, has considerably increased our ability to measure and sense reality. Many mysterious aspects of life that could formerly be dealt with purely in terms of spiritual, miraculous or mystic approaches, and which therefore encouraged people to accept religious explanations, are well understood today because of modern science. Today we comprehend many peculiar phenomena that a few hundred years ago appeared mysterious, supernatural or enigmatic, and, above all, challenging. Thus, it was eminently natural and logical that modern science would lead people to develop a materialistic approach to human life on earth. Why should we believe in spirits and in other mystical phenomena if we develop increasingly more effective tools that enable us to measure reality precisely, and to decipher, based on these scientific tools, more riddles related to our existence? Hence, it is apparently very difficult today to take too seriously those religions that were born at a moment in history when human society's level of knowledge was so low, as compared to that which exists today.

All of the above remarks are only part of the logical arguments against any central role that ancient religions and philosophies should play in modern life. Hence, in light of modern, scientifically explosive developments, with their resultant new perspectives of reality, it is quite natural that modern societies would be driven to divest themselves of old religions. However, the main point of strength for all religions is that they claim to represent omnipotent forces. For instance, the monotheistic religions, Judaism, Christianity, and Islam, all speak in the name of a single God Who created the entire universe, including intelligent creatures such as the human race, to whom He revealed Himself, and from whom He made demands. The most elegant way to free the human race from beliefs in an omnipotent creator who created the universe for some specific purpose that relates to divine exigencies, is simply to show that the world does not need a creator. Enlisting modern science to 'show' and 'prove' that life is spontaneous, accidental, and, quite simply, happens, evolving with no purpose or demands from the human race, seems to be a straightforward way of eliminating religious approaches from our lives.

3. More on Macro-Evolution

Darwin's theory appeared at the right time in the evolution of the scientific/technological revolution. With his acumen and life experience, Darwin analyzed processes in which living species could ostensibly adapt themselves to their environment by gradual processes, led by the driving force of better adaptation yielding better existence, later defined as natural selection. Darwin's analysis was further extended by him and his successors to become the theory of evolution, namely, a so-called scientific explanation of how life generally developed from primitive to complex forms by adaptation to the environment, driven by the rules of natural selection (de Duve, 2002). This theory has been promulgated by its proponents to take a central role in our understanding of the history of life and of faith. Vital and ongoing research revolving around this theory is being carried out, as is evident from the number of scientific journals, as well as books, encyclopedia entries, and exhibitions in museums, dedicated to the central premises of evolution.

I would argue that the macro-evolution theory has no real scientific background or basis. Maybe it is widely accepted only because it serves a purpose, freeing intelligent human beings from the need to believe in a demanding God and to obey His agents on earth, namely, various existing religious institutions. Despite its popularity, as described above, and despite the fact that it is so deeply seated in the consciousness of many people, including top scientists, it must be stated clearly that macro-evolution and its attendant scenarios about the accidental and inevitable development of life on earth have no solid scientific basis and are purely speculative. Below I summarize — and will later refute — major arguments that are used by evolutionists as evidence that macro-evolution is the ultimate story of life on earth, the fundamental basis to an understanding of biology (Culotta et al., 2005):

1. As mentioned above, there are the well established and solid observations of micro-evolution.
2. The many remnants of ancient animals, fossils, petrified plants and animals, as evaluated by appropriate dating methods, such as radioactive methodologies, appear to show that life appeared on earth in stages. It is eminently logical to assume that the initial formation of primitive forms of life, soon after the appearance of stable liquid water on earth, was when these primitive life forms became organized as single cell microbes. The next stages were the appearance of multi-cell creatures, plants, marine vertebrates, terrestrial animals — reptiles, insects, birds, mammals, various types of hominids, and finally, human beings, or *Homo sapiens*.
3. There are remains of creatures that can be considered as intermediate between marine and terrestrial life (Culotta et al., 2005), reptiles and birds (Leinfelder, 2006), primates and hominids (Horgan, (1992), etc. So why should we not connect all of this history of life by common denominators: evolution from primitive to complex, with speciation on the way, which forms the divergent life forms that we know? In embryological observations, as a further instance, it seems that we see a patterning of evolutionary processes. Initially, in their early stages, different embryo species look very similar. Specific shapes and morphologies appear only in later stages of embryonic development, which may show that so many distinct forms of life may well have the same origin.

4. Recent studies of the genomes of a large trajectory of living species — from microbes to insects, developed animals, including monkeys, and finally, to human beings (as surveyed in a special issue of "*Science*", 2001) — show interesting similarities and diversities that apparently seem to be connected to a development process from common ancestors that separated into the various species at some point in the history of life here on earth (Culotta et al. 2005). However, scientifically speaking, these new discoveries are not necessarily related to any of the genealogical trees drawn up by the evolutionists (de Duve, 2002), who describe how certain species evolved from each other.

On the basis of real science that connects conclusions to solid facts based on reliable measurements, all of the above four aspects do not necessarily dictate that life evolved spontaneously, accidentally, or inevitably. However, what is definitely true is that for people who think *a priori* in terms of the evolution theory, it is very easy to enlist the above observations or aspects as "convincing" arguments. As an example of how such arguments can be misleading, the above note in "*Science*" (Culotta et al., 2005) mentions that the difference in DNA between *Homo sapiens* and chimpanzees is only 1%, which is, of course, astounding. But from this note one can also learn that about 40,000,000 genetic changes are needed to transform the latter into a human being and vice versa (provided that many of these changes are coherent!).

Below I raise very briefly some problems related to the belief that life evolves accidentally and spontaneously.

4. Remarks on the Origin of Life

It is far beyond the scope of this article to describe the subject of the amazing chemistry of life, even briefly. However, some pertinent examples are necessary. At the heart of life as a reality, there are synthetic processes. Plants absorb simple molecules such as water, nitrogen, carbon dioxide, and some elements such as sulfur, phosphorous, and various metallic cations (of sodium, potassium, magnesium, iron, and more). From these simple moieties they compose highly complex macromolecules and polymers, including proteins, DNA, RNA, sugars, fats, polymers based on sugar molecules, and much more (Voet et al., 2004). One of the most basic laws of nature is known as the second law of thermodynamics, which determines, among other things, to what extent a chemical reaction is spontaneous. In fact, this law ranks all possible chemical reactions by their level of spontaneity. In light of this law, all synthetic reactions in nature, whereby macromolecules are formed from small molecules, atoms, and ions, are not spontaneous (Atkins, 1998). Continuous biosynthesis in nature is possible only because of the radiation flux that comes from the sun. This energy flux, which periodically floods the earth, enables the photosynthesis of biomaterials to take place despite the fact that spontaneous processes perpetually decompose the various biomaterials into the small molecules from which they were naturally synthesized. However, photosynthesis can take place only because there are highly complex molecules available, which absorb radiation from the sun and transform it into chemical energy. Photosynthesis and all other chemical processes of life are possible only because there are enzymes available that can conduct the chemical processes in the highly efficient and specific way that is crucial for the chemistry of life to proceed. However, enzymes are complex proteins that can only be produced by complicated photosynthetic processes and, moreover, photosynthesis can take place only when all of the components, namely, the different macromolecules involved, are available in the right structure, which we find in plants. Hence, it is impossible to see how photosynthesis could ever be started spontaneously. What we have here is the famous riddle of the chicken and the egg, and in fact this riddle applies to almost every cycle of the chemistry of life. Proteins, for instance, can only be synthesized in nature by processes that begin with genetic codes, namely, DNA and RNA, and they are synthesized based only on the chemical information that is stored in these macromolecules by the intensive and critical involvement of enzymes. However, DNA and RNA only function in nature when they are structured in complex proteins and are surrounded by enzymes that are responsible for their activity and information transfer. Hence, the synthesis of proteins from DNA and RNA is yet another form of the chicken and the egg conundrum. Scientists who work in the field of chemistry and explore the actual chemistry of life are the people who have the real view of its genius.

We, the scientists, struggle on a daily basis with chemical reactions and experience their complexity, the difficulties in reaching the desirable efficiency, and the unexpected diversity of the products that chemical reactions may produce. Based on our daily experience with these reactions, we perceive the chemistry of life to be simply astonishing. Researchers have received Nobel Prizes for their success in producing natural products, where they had to design highly complex, multistage reactions, starting with kilograms of starting materials, and, at the end of a long and complex process, demonstrating the achievement of their goal with milligrams of pure materials. To return to the example of photosynthesis, each leaf of a plant conducts hundreds of simultaneous processes, all occurring at 100% efficiency, and milligrams of starting materials are converted by multistage processes to milligrams of desired products with no undesirable residues.

Many polymers that comprise living tissues such as proteins are composed of building blocks, or molecules, and, in the case of proteins, amino acids, which have the lowest symmetry possible. Such molecules are termed chiral and possess a property that is called optical activity (related to their interaction with polarized light). This low level of symmetry is of signal importance in order to have proteins with very specific structures and active sites that can act as enzymes. However, all the usual syntheses of low symmetry compounds produce mixtures of molecules that are called racemic mixtures, which possess **no** optical activity. The production of optically-active molecules requires the so called asymmetric synthesis in which chiral, optically-active molecules **must** be involved (Solomon et al. 2000). Here we have another chicken and egg situation: how did nature begin to produce chiral, optically-active materials that could then continue to direct most of the natural syntheses to produce chiral, optically-active materials?

Clearly, even as we better understand the chemistry of life, it does not enable us to suggest solid and sound routes for its spontaneous beginning. All this invokes our increasing wonder at the ingenious design of amazing, highly complicated and simultaneous processes that make life possible. It should be emphasized that the simultaneous manner of all the multistage chemical processes of life is crucial to its existence. The failure of a single system or a single reaction may, in most cases, lead to the failure of the entire living system.

In conclusion, the origin of life cannot be explained by spontaneous, sporadic, accidental reactions that, by chance, crystallized into the amazing and inspiring chemistry of life that we are struggling to decipher. This conclusion is freely admitted by leading scientists who have tried to investigate the origins of life, among them the Nobel Laureate, Prof. F. Crick (Horgan, 1992).

5. From Single Cells to Developed Creatures?

Now, let us move on to the next stage, provided that, for some reason and by some mysterious mechanisms, living cells exist, namely, well-organized structures that include the necessary synthetic mechanisms for multiplying these cells, based on their genetic codes (i.e., DNA-> RNA-> enzymes and proteins, all different kinds of polymers needed to construct cell envelopes such as poly-carbohydrates, lipids and the building block molecules such as amino acids, fats, sugars, etc.). According to the theory of evolution, sporadic mutations take place and lead to the formation of many possible diverse bio-structures, morphologies, and functionalities. Now, on an accidental basis, some structures are favorable and some are not. The environment attenuates the "weak" structures and selects the better ones, and in this way, over exceedingly protracted periods of time, the sporadic changes crystallize into well-functioning organs, such as eyes, ears, muscles, nerves, and the brain, to cite a few instances, all reflecting amazing design and the highly judicious use of all the laws of nature. These assumptions, however, are nothing but beliefs, and there is no single scientific evidence that this can really take place spontaneously, without ingenious, intelligent design.

These improvement processes assumed by the macro-evolution theory are, by definition, gradual in nature, and hence, should take place through a host of intermediate forms. However, what we really find on earth is evidence of "jumps." For instance, the most ancient fossils that we find relate to protozoa (i.e., single cell life) and to the remains of creatures that possessed all the organs of well-developed life, with no evidence of the expected gradual processes from single cells to well-organized, multifunctional creatures, whose remnants we find in fossils. The suggestion put forward by the theory of evolution that living species produce all possible biological structures in a sporadic diversity which is then attenuated by the environment through so-called "selection" rules, is, in fact, absurd. It is no less absurd than a claim that a tornado hitting a junk yard has the chance of producing well-functioning cars — why not? All the ingredients are there, including winds at a speed of hundreds of miles per hour — or the possibility that well-defined paintings were formed by artists' sporadic attempts to throw paint on canvases, on the assumption that having been given sufficient time for trials, the desired pictures would, by chance, be formed. Attempts by the disciples of evolution to prove the validity of macro-evolution through the well-established observations of micro-evolution are nothing more than the promotion of their beliefs by mixing — or in fact confusing — the fine and the coarse tuning of life on earth.

Micro-evolution relates to the well-designed mechanisms of adaptation that each creature may have to some degree. It should be emphasized that micro-evolution really works when the impact of the environment on a genetic change is immediate and fatal. For instance, mutation in microbes that are attacked by antibiotic poisons, which change the cells' envelope so that it does not allow transfer of the antibiotic compounds into the cell, has an immediate impact. Hence, only microbes that have undergone such a mutation can survive, while the rest should die. This is indeed the type of mutation that works positively and leads to a preferred selection of species by the environment. In fact, most of the mutations that we observe in genetic codes have a negative impact on living creatures. Giving sporadic mutations (in series) the ability to form functional organs is nothing but an absurd belief in the power of accidents to evolve into ingenious design. The evolutionists enlist on their side the apparent natural history of the earth, which may indeed contain many pieces of information about the past. In their excavations, archaeologists have uncovered fossils and remnants of creatures that lived many years ago and have disappeared. They have also found the remains of ancient creatures that possessed the intermediate functioning of both birds and reptiles, for example (Leinfelder, 2006), or of both marine and terrestrial creatures (Pennisi, 2006). Evolutionists have attributed these findings to missing links in the evolution of birds from reptiles, and from marine creatures to terrestrial ones. However, when they existed, these creatures were highly developed, as compared to primitive forms of life. Hence, if the macro-evolution theory were valid, then many intermediate forms of life would have developed during the course of the evolution process, for each "intermediate"/"missing link" creature whose remains have been found. Thus, there are many "missing links" on the way to the development of any "intermediate" creature, whose remains have been discovered thus far.

Evolutionists list findings related to fossils, remnants of ancient animals, and the like in order to prove their theories. There is no question that our world has a history that is reflected by ruins and fossils, and it is clear from these relics, which have been found almost everywhere on earth, that there were species that lived during certain epochs and then disappeared. There have been many events of mass extinction of the flora and fauna of our planet. For instance, we have the disappearance of the entire dinosaur species, as well as many species of ancient mammals, including some species of hominoids. There is strong evidence that the earth's climate changed over the years, and that our planet underwent pronounced geological and geographical changes since its inception. However, do any of these findings really prove that there was macro-evolution, namely, processes whereby species developed and became increasingly more complex and sophisticated as the result of random, accidental genetic changes and mutations that were attenuated and selected by environmental constraints? Definitely **not**! Real science, true science, involves cycles of induction and deduction and can only be fully conclusive in relation to the present, not to the past. Scientific progress is achieved by a series of experiments in which phenomena are explored at present, and based on these results scientists try to develop theories that explain them. These theories must then be proven by predicting results that can be obtained experimentally. Hence, real science relates to chemistry, physics, astronomy, biology to some extent, and psychology, for example, in which experiments can be carried out, and from which conclusions can be drawn and further validated. Any conclusions related to the past, especially to the prehistoric past from which we have no written documentation, are, by definition, speculative. Therefore, any scenario suggested as the history of our planet cannot really be proven.

6. On Intelligent Design

The only alternative to understanding complex life as a result of a series of genetic sporadic accidents, is the assumption that there is intelligent design behind the reality that we know, feel and live. It should be emphasized that modern science has no tools or mandate to help us decide on this matter. Science has a mandate to answer questions of "how," but not "why." Modern science provides us with nature's most beautiful inspirational tools for seeing and understanding the amazing complexity and ingenious design of everything around us. I previously mentioned the chemistry of life. However, astronomy, the nuclear physics of elementary particles, and the delicate balance that we measure among all the forces of nature that maintain an existence which is always metastable, are likewise inspiring. In fact, the leading theory in astrophysics, the big-bang theory, describes the history of the universe in terms of the creation of matter from nothing, from an infinite vacuum.

In the second part of this article, I will try to present an alternative approach to the above described belief in the power of a series of accidents to crystallize complicated forms of life, which include intelligent creatures.

What I describe below is also a belief. However, I am certain that this belief, which is based on very deep historic roots and the intellectual heritage of hundreds of generations and many thousands of geniuses over the years, has a much more solid basis than the macro-evolution theory.

The key to solving the riddle of our existence and its purpose, can only be found through understanding the history of the human race, something that in fact forced the "invention" of the religion called Judaism. The materialistic approach (out of which the evolution theory was born) suggests an explanation as to how and why religions appeared. Generally, religions are described as the manifestation of the psychological needs of human beings to face limitations in understanding the natural phenomena which surround them and the inevitability and inexorability of death. This, however, is simply not true, at least in one case — that of Judaism. Judaism is the source of all monotheistic religions, of which Christianity and Islam are the dominant ones today. Judaism in fact holds the key to understanding what we human beings are doing here on earth, as well as the purpose and meaning of life. Throughout thousands of years of human history, dedicated, idealistic and highly intelligent religious Jews have conveyed a critically important spiritual and intellectual heritage from one generation to the next, which can be summarized, as follows, in the next section. One can ask what is the place of religious matters in a scientific paper? At a first glance, religions are based on belief, on feelings and arguments that for them may be all contested. However, Judaism is different than all other religions because it is based on solid historical events that were experienced by millions of people. The author of this paper is fully convinced that upon having all the necessary information at hand: the literature, the bibliography of the main spiritual Jewish leaders along with the Jewish history and all the well documented details of 3300 years of Jewish history, any unbiased scientist can reach the conclusion that Judaism is much more than a belief. Coming to such a conclusion however, needs an extensive, systematic learning of facts and literature. In such a paper it is impossible to discuss all the many necessary arguments needed, in order to reach the above conclusion. Some points for thinking however, are described in the next sections.

7. Some Main Messages of Judaism

Several important messages of Judaism, related to the discussion about intelligent design *vs.* accidental development, are listed below.

1. The world was not developed randomly or accidentally. It was created with a purpose, which relates to the existence of intelligent creatures on earth, i.e., the human race.
2. The world has an omnipotent Creator who revealed Himself to millions of people. Hence, we know about the Creator and His creation through direct communication between Him and an entire nation, the Jewish people. Communication by the people of Israel with the Creator was not a single event, but in fact lasted for approximately 1,000 years and relates to an amazing phenomenon called prophecy.

Prophecy was a mode of communication between human beings and the Creator, the Lord, which was relevant mostly to people who had prepared themselves for this. Prophecy disappeared about 2,400 years ago because the Jewish people were no longer deserving of this privilege because they had betrayed their covenant.

The compulsory relationship between the Jewish people and the Creator began with a series of great events that affected millions of people. Millions of Jews were inevitably witness to the fact that all the laws of nature are nothing more than the will of the Creator who revealed Himself to them as the Lord. These events included the Exodus, namely, the release of the Jewish nation from bondage in Egypt, after it was struck by 10 plagues that related to all aspects of nature, the reality of life, and human existence. Then, seven days later, the Red Sea parted, allowing the people of Israel to cross, and then closed, drowning the pursuing Egyptian army. Forty three days later, the entire Jewish nation participated in an amazing and inspiring event near Mount Sinai (probably somewhere in the southern part of the Sinai Peninsula), where they were all involved in direct communication with the Creator and experienced prophecy. For forty years after this, the people of Israel wandered in the Sinai Desert, experiencing miracles on a daily basis, and finally, about 3,278 years ago, they defeated the Canaanites and conquered the Land of Israel. The Canaanites were much stronger than they were and could not be defeated on a natural basis by a population of shepherds, who only a few decades previously had been slaves and, since their release from bondage, had mostly concentrated on spiritual studies and not on military training.

The period of prophecy, which lasted about 1,000 years, has been fully documented in the 24 books of the Jewish bible. The reality of communication with the Lord and the great events that began the relationship between Him and the Jewish people are conveyed from generation to generation, as an essential part of Jewish life.

The Lord gave the Jews remarkable instructions, commandments that include extensive demands from them that are related to all aspects of life, as well as other instructions to convey to the entire human race. However, these instructions regarding general human life are much less demanding than those related to Jewish life.

The essence of Jewish life is learning. The Lord's instructions, codified in the "*Torah*," as it is termed in Hebrew, are very sophisticated. They were given in a complicated way that requires intensive study by each individual in every generation, in order to understand and use them as a way of life.

All communication with the Lord, and His instructions, while compulsory, still allow for latitude and for freedom of choice and free will. In fact, one of the most important privileges that a human being possesses is that of choice regarding the management of his life, despite the existence of the Lord's compulsory instructions.

For further reading, there are several fundamental books that summarize Jewish philosophy and its main messages concisely and beautifully. Of especial importance are "*The King of Kuzar*" ("*Hakuzari*", by Rabbi Yehuda Halevi, translation by Korobkin, 1998); "*The Way of the Lord*" ("*Derech Hashem*" by Rabbi Moshe Haim Luzato, translation by Kaplan, 1997); and "*The Duty of Hearts*" ("*Hovat Halevavot*" by Rabbi Bachya, translation by Feldman, 1996).

8. On the First Part of the *Bible*, the *Torah*

The first five books of the Bible were canonized upon the death of Moses, about 3,278 years ago. They contain prophecies that describe complex and sometimes terrible events that would happen to the Jews in the future, events that in fact took place many hundreds of years after their outline was initially delineated in the *Torah* (see for instance Leviticus, Chapter 26 and Deuteronomy, Chapters 28-32). When we read those sections of the *Torah* carefully, comparing them to actual Jewish history, which is so well documented, and taking into account that after the canonization of the *Torah* it was impossible to add even a single word to its text, one can conclude that the *Torah* was composed by a supreme entity that has unlimited control over the faith of our world.

As an introduction, in the first chapter of its first book, "B'resheit" or "Genesis," the *Torah* includes a very brief description of the creation of the universe, our planet, its flora and fauna, and finally, the human race. The period of creation is defined as six days of sequential steps: energy, matter, planets, the appearance of water, separation between land and sea, flora, marine fauna, birds, terrestrial fauna, and finally, the first human family. It should be emphasized that the *Torah* does not intend to provide us with a detailed history of the universe. What is described in this first book is just the framework of creation. We know from Einstein's Theory of Relativity that time is not invariant, but rather depends on the speed of the system in which it is measured, and on its mass. It is clear that when the *Torah* talks about the six days of creation, it does not refer to today's time scale of 24 hours. It most definitely leaves room for investigating the past, and it cannot contradict, by definition, any physical findings and conclusions obtained by any well-established scientific methodology. In fact, the *Torah* expresses its appreciation of people who try to understand the world, based on true, objective, and unbiased observations. There are several books that have appeared in recent years, which demonstrate very nicely the correlation between the data appearing in the *Torah* that describes the creation of the world and major findings from modern astrophysics and prehistoric studies, including a discussion of the time scales (e.g. Aviezer, 1990 and Schroeder, 1997). These two books also show very nicely that the so called evolution theory does not meet at all standards of valid scientific theories, as usually accepted in exact and life sciences.

Science is definitely an indirect, but highly important mode of communication between the human race and its Creator. All of the Laws of Nature, which are so complex, obey, without exception, the laws of mathematics, which the human brain has been so well equipped to invent. Moreover, in several cases, strange mathematical approaches which were developed in the past and regarded, at the time, as useless, were later found to describe precisely laws of nature that were only discovered subsequently. This correlation between our computational ability and the behavior of the Laws of Nature is a special gift from the Creator that enables us to exploit these laws in order to make our life on earth easier and more comfortable. All the technological advances from which we benefit so much nowadays have only come about because nature is governed by rules that can be precisely and mathematically described. Hence, it is a great pity that science has been enlisted improperly by some groups to support invalid theories that describe reality as accidental.

The *Torah* is not a source for the study of natural history because its focus is on the history of the free choice of the human race. The main theme of the first book of the *Bible*, "B'resheit," is to show how the human race used the privilege of free choice badly at the beginning of its existence, bringing upon itself mass extinction in the form of the biblical Flood. However, 3,700 years ago some people proved that the privilege of free choice can be used to promote good, and that, by unbiased observation, it is possible to understand that the world was created by the Lord for a purpose. These people were Abraham and his wife, Sarah, followed by their son Isaac and his wife, Rebecca, and by their son Jacob and his entire family. Most of this first book of the *Bible* is dedicated to detailed descriptions of the lives of these first three generations of Jews who proved that free choice can be used positively. Through descriptions of their lives, this book conveys important educational messages to the entire human race. The Jewish nation descended from the family of Jacob. The 24 books of the *Bible* were intended to guide the entire human race, based on 1,000 years of prophecy. Unfortunately, too many Jews misused their free choice during critical periods in history, violating the Lord's written Covenant (see for instance, the book of "V'ayikra," or "Leviticus," Chapter 26, 3rd verse and following), and hence the Lord had to punish the Jewish nation (as was promised so clearly within the framework of this covenant). All of these penalties, including the destruction of the Holy Temples, the destruction of the entire Jewish existence in the Land of Israel, and the long and terrible exile of the Jews, who were dispersed throughout the world, are part of the Covenant and its attendant consequences, if the Jewish nation violates it. This is clearly outlined in the *Torah* (e.g., "Leviticus," Chapter 26, Deuteronomy, Chapters 4 and 28-32).

From its inception, the Jewish nation developed a culture of study, a culture in which illiteracy is the greatest shame. This culture of study was enabled by the fact that the Hebrew alphabet contains only 22 letters, which was revolutionary at the time of the birth of the Jewish nation about 3,000 years ago. It should be noted that at that time, writing skills in the dominant cultures and civilizations of Egypt, Mesopotamia and China, were very complicated and included thousands of letters, pictures, or symbols. Therefore, most of the population of ancient civilizations was illiterate, and the ability to read and write was limited to small groups of people, such as priests and nobles in the courts of the king. Thus, it was easy for them to rewrite history in a way that could serve their own political needs and aspirations.

The Jewish *Bible*, which reflects about 1,000 years of history during a period in which the Jews were privileged to have prophets, differs from any other historical document. By following the course of Jewish history and *Torah* studies since the birth of the nation, it can be concluded that never, in any generation or period, was it possible, for any reason at all, to deceive an entire nation whose people have been literate for successive generations, nor to plant in their minds accounts of events that never took place at all, and, based on this, force them to accept the demanding Jewish way of life from an imaginary creator with whom they never had any connection.

What I have presented above is, of course, very brief and superficial, because the Jewish *Torah* is an ocean of wisdom and information. Its systematic study enables us to conclude that the Jewish people convey a remarkable message from an omnipotent Creator who created the universe in general, and the human race in particular, for a purpose. In order to understand this purpose, one has to study the *Torah*, a discussion of which is far beyond the scope of this article.

9. Concluding Remarks

In conclusion, intelligent design is not an alternative to macro-evolution, which is a false theory without any real scientific basis. Intelligent design is, in actuality, the sole and only explanation for the reality of life on earth, because communication with the intelligent, omnipotent designer is an essential part of human history. The understanding that we are all the result of intelligent design should not prevent us from exploring the past and learning more about natural history. Our universe has an amazing past that, based on physical findings, can be speculated on and explored, for better or worse. The history of our planet, I would argue, cannot be understood in terms of a series of accidents, but rather as a series of designed events that enable the existence of an intelligent human race, where every individual should and can be a reflection of the supreme Creator, the Lord.

References and Further Reading

Atkins, P.W. (1998). *Physical Chemistry*, 6th Edition, Oxford University Press, Oxford, Chapters 4-6.

Atkins, P.W. (2003). *Galileo's Finger: The Ten Great Ideas of Science*, Oxford University Press, Oxford.

Aviezer, N. (1990). *In the Beginning--: Biblical Creation and Science*, Ktav, Hoboken, N.J.

Culotta, E. and Pennisi, E. (2005). Breakthrough of the year: evolution in action. *Science* **310**: 1878-9.

de Duve, C. (2002). *Life Evolving*, Oxford University Press, Oxford.

Freeman, S. and Herron J.C. (2004). *Evolutionary Analysis*, 3rd Edition, Pearson/Prentice Hall, Upper Saddle River.

Halevi, Y. (1998). Translated and annotated by N. Daniel Korobkin, "*The Kuzari": In Defense of the Despised Faith*, Aronson, Northyale, N.J.

Horgan, J. (1992). D2 or not D2. A barroom brawl over an "alcoholism gene". *Scientific American* **266**(4): 32.

Horgan, J. (1992). The new challenges. *Scientific American* **267**(6): 16.

Leinfelder, R. (2006). *Archaeopteryx*: the lost evidence. *Science* **312**: 197-8.

Luzzatto, M.C. (1997). Translated and annotated by Aryeh Kaplan, reviewed by Gershon Robinson. *"Derech haShem": The Way of God, Ma'amar Haikkarim: an Essay on the Fundamentals*, Feldheim, Jerusalem.

Pennisi, E. (2006). *Science* **312**: 33.

Schroeder, G. (1997). *The Science of God: The Convergence of Scientific and Biblical Wisdom*, Broadway Books, N.Y.

Solomon, G. and Fryhle, C. (2000). *Organic Chemistry*, 7th Edition, J. Wiley & Sons, N.Y., Chapter 5.

The Duties of the Heart (1996), "*Bachya ibn Pakuda*", translated and commentary by Yaakov Feldman, Aronson, Northyale, N.J.

The Human Genome (2001). Special Issue: *Science* **291**: 1145.

Voet, D. and Voet, J.G. (2004). *Biochemistry*, Wiley & Sons, N.Y.

Dialogue

Harun Yahya

I was honored to read Prof. Doron Aurbach's article demolishing, with full supporting scientific evidence, the theory of evolution, and supporting Creationist belief. As a Muslim who believes that all members of the Divine faiths should combine together in an intellectual struggle against Darwinism and materialism, I was delighted to see that our believing Jewish brothers are performing such a service in that same awareness.

As we all know, atheist, materialist, and other anti-religion power centers are waging an intensive war against religious morality. They have been working systematically across the world to divert people from Allah's path, have them deny Allah's existence and unity, and, in their own eyes at least, annihilate religious morality, and they seek to portray Darwinism as the supposed "scientific" foundation for this.

Darwinism portrays the world as an arena in which only the fittest survive and seeks to distance people from religious morality, and thereby turn them into selfish and merciless animals. In reality, many sciences (e.g., paleontology, biochemistry, genetics, comparative anatomy, and biophysics) prove that the origin of life cannot be explained through natural processes and coincidences, as evolutionists claim. In short, evolution is defended not by science, but in spite of science.

The leading positions of power in the scientific world and in the media are still filled by people who assert that the theory of evolution is absolute truth. Seeking to keep this theory alive at all costs, they have no qualms about concealing or twisting those scientific discoveries that refute it, or belittling or disparaging those discoveries that expose its falsity or those scientists who criticize it. This inquisition causes many scientists and researchers to conceal their true views so that they will not be excommunicated by having their funds cut off or their reputations destroyed.

All people who believe in Allah must join forces to resist and, ultimately, overthrow the Darwinist dictatorship ruling the world of science and thought.

Therefore, it is of vital importance that we unite in using the scientific and technological means of the twenty-first century to tell the world that Darwinism is nothing but a sham, and that the true source of life is creation. The eradication of Darwinism, using scientific evidence, will permit people to see the existence of Allah by encouraging them to think about the purpose behind creation. In this way, virtues such as peace, love, affection, justice and brotherhood, which are commanded by all the Abrahamic faiths, will, by Allah's leave, rule the whole world.

Biography

Professor Michael J. Behe is in the Department of Biological Sciences at Lehigh University in Bethlehem, Pennsylvania. He obtained his Ph.D. in biochemistry from the University of Pennsylvania in 1978. He is the author of *Darwin's Black Box: The Biochemical Challenge to Evolution* (Free Press 1996), which argues that the functional complexity discovered in the molecular machinery of the cell is best explained as the result of deliberate intelligent design.

34
When Science Renounces a Facet of Reason
Biology's Choice Between Solipsism and Intelligent Design

MICHAEL J. BEHE

Department of Biological Sciences, Lehigh University
Bethlehem, Pennsylvania 18015 USA
mjb1@lehigh.edu

1. Introduction

Centuries ago nascent modern science took a crucial step. Breaking at last with the old Aristotelian thinking, it would no longer consider nebulous "final causes" in its explanations. Whether the ultimate purpose of, say, a mountain was to display the grandeur of God or something else could not be decided by an investigation of nature. Henceforth such questions would be relegated to philosophy or theology. Science would deliberately confine itself to issues about the mechanics of nature, and ignore issues of purpose. What a horse or river or star is "for" would trouble science no longer.

At the time it seemed like a prudent course of action. Yet the simplistic division of labor was doomed from the start, because some parts of nature quite clearly are "for" identifiable things. Science wished to explain the mechanics of nature apart from purpose, but especially in biology the mechanics themselves often cannot be understood apart from purpose. The "function" of a mountain may not be decidable, but the function of a wing surely is. The purpose of a horse might be obscure, but the purpose of a horse's eye is not.

Biologists of course realized this. Only when Darwin seemed to provide a non-purposive explanation for apparent purpose could life itself be accommodated within the new framework. Biology was the last scientific discipline to come fully under the sway of non-Aristotelian, mechanistic thinking. Subsequently biologists continued to think and speak of aspects of life in terms of purpose, but only with the (usually implicit) understanding that it was merely apparent purpose. Although over the years many scientists acting as individuals concluded that purpose was real, "official" biology self-consciously restricted itself to considering only non-purposive explanations for the origin of the apparently purposeful systems of life.

I argue that this state of affairs is no longer tenable. Although for convenience humans may carve up intellectual pursuits into various academic disciplines, reality is not obliged to respect such boundaries. In order to come to grips with the real world—in order to arrive at true explanations—rationality demands that we abjure artificial distinctions when their usefulness appears to have reached its limit. Rather, in proposing explanations for nature we must use all of the data and all of the experience we have available. The end result of stubborn adherence to a simplistic division of nature into discoverable mechanics and undiscoverable purpose, I say, is nothing less than the official divorce of science from reason.

2. The Appearance of Design in Life is Overwhelming

Although science is supposed to "officially" spurn purpose, the proposition that aspects of life appear to be purposely designed is readily affirmed even by scientists who most strongly argue that the appearance is deceiving. For example, Oxford biologist Richard Dawkins, who is certainly the most active ambassador of Darwinian theory to the public, on the first page of *The Blind Watchmaker*, his classic defense of Darwinian evolution, defines biology thus: "Biology is the study of complicated things that give the appearance of having been designed for a purpose." (Dawkins 1986, 1) Let me stress Dawkins' remark. He says that is the *very definition* of biology—the study of things that display apparent purpose. Purpose, or its appearance, says Dawkins, is at the *very heart* of life. Well, is the appearance of purpose subtle? Is it easily missed or mistaken? Is it akin to seeing faces in the clouds or shapes in the constellations? Not according to himself. Although as a Darwinist Dawkins ascribes the appearance of design to natural selection, he nonetheless insists that the appearance of design is *overpowering*:

> "Natural selection is the blind watchmaker, blind because it does not see ahead, does not plan consequences, has no purpose in view. Yet the living results of natural selection *overwhelmingly* impress us with the appearance of design as if by a master watchmaker, impress us with the illusion of design and planning" (Dawkins 1986, 21, my italics).

Is the overwhelming appearance of design a matter of personal taste? An aesthetic judgment, akin to being overwhelmed at the beauty of a pretty sunset? Or is it some sort of moral judgment, that the way things are just seems to be right, seems to be fitting? Not according to Dawkins. According to Darwin's foremost modern advocate, the judgment of design is neither an aesthetic nor a moral judgment—it is an *engineering* judgment:

> "We may say that a living body or organ is well designed if it has attributes that an intelligent and knowledgeable engineer might have built into it in order to achieve some sensible purpose, such as flying, swimming, seeing ... [A]ny engineer can recognize an object that has been designed, even poorly designed, for a purpose, and he can usually work out what that purpose is just by looking at the structure of the object" (Dawkins 1986, 21).

Thus the way in which we decide that a biological system appears to have been designed is the same way in which an engineer decides that a nonbiological system appears to have been designed: by the way in which the parts of the system fit its apparent function.

Dawkins is not the only Darwinian to readily admit the strong appearance of purpose in life. Another public defender of Darwin's theory is State University of New York biologist Douglas Futuyma. Futuyma is the author of a widely-used textbook of evolutionary biology and the author of a 1982 trade book entitled *Science on Trial: The Case for Evolution*, which defends Darwin's theory for a wider audience. In an online interview with ActionBioscience.org Futuyma explained why Darwin's theory is important: "The reason that natural selection is important is that it's the central idea ... that *explains design in nature*." (my italics). He continues:

> "[Natural selection] is a very simple natural mechanism that explains the appearance of design in living things. Before Darwin, the adaptations and exquisite complexity of organisms were ascribed to creation by an omnipotent, beneficent designer, namely God, and indeed were among the major arguments for the existence of such a designer. Darwin's (and Wallace's) concept of natural selection made this 'argument from design' completely superfluous."

So according to Futuyma, that's exactly what Darwin's theory of evolution by natural selection does—it explains the appearance of design, the exquisite complexity of organisms. In other words, if it weren't for the overpowering appearance of design in life, Darwin's theory would have little to explain.

The appearance of design in life—especially at its molecular foundations—has grown enormously stronger than it was when Darwin lived, its resemblance to mechanical systems much greater than when William Paley first wrote his watchmaker argument. Consider the remarks of former President of the National Academy of Sciences Bruce Alberts in his introduction to a special issue of the journal *Cell* he was editing on the topic of Macromolecular Machines:

> "We have always underestimated cells. Undoubtedly we still do today. But at least we are no longer as naive as we were when I was a graduate student in the 1960s. ... But, as it turns out, we can walk and we can talk because the chemistry that makes life possible is much more elaborate and sophisticated than anything we students had ever considered. ... [W]e now know that nearly every major process in a cell is carried out by assemblies of 10 or more protein molecules. And, as it carries out its biological functions, each of these protein assemblies interacts with several other large complexes of proteins. Indeed, the entire cell can be viewed as a factory that contains an elaborate network of interlocking assembly lines, each of which is composed of a set of large protein machines.
>
> "Consider, as an example, the cell cycle–dependent degradation of specific proteins that helps to drive a cell through mitosis. First a large complex of about 10 proteins, the anaphase-promoting complex (APC),

> selects out a specific protein for polyubiquitination; this protein is then targeted to the proteasome's 19S cap complex formed from about 20 different subunits; and the cap complex then transfers the targeted protein into the barrel of the large 20S proteasome itself, where it is finally converted to small peptides" (Alberts 1998).

The special issue carried articles with titles such as: "The Cell as a Collection of Protein Machines: Preparing the Next Generation of Molecular Biologists"; "Polymerases and the Replisome: Machines within Machines"; and "Mechanical Devices of the Spliceosome: Motors, Clocks, Springs, and Things". Alberts emphasized why these biomolecular systems are called machines: "Why do we call the large protein assemblies that underlie cell function protein machines? Precisely because, like the machines invented by humans to deal efficiently with the macroscopic world, these protein assemblies contain highly coordinated moving parts."

The incapacity of biology to understand parts of life apart from their purpose was emphasized recently by developmental biologists Michael Levine and Eric Davidson, who have spent much of their careers studying gene regulatory networks that govern the development of embryos.

> "Gene regulatory networks (GRNs) are *logic maps* that state in detail the inputs into each *cis*-regulatory module, so that one can see how a given gene is fired off at a given time and place. ... *The architecture reveals features that can never be appreciated at any other level of analysis but that turn out to embody distinguishing and deeply significant properties of each control system.* These properties are composed of linkages of multiple genes that together perform specific operations, such as positive feedback loops, which drive stable circuits of cell differentiation" (Levine and Davidson 2005, my italics).

If the appearance of design in the abilities of animals to fly or swim is, as Dawkins affirms, overwhelming, then the appearance of design in molecular systems such as Alberts, Levine, and Davidson describe—which self-assemble multiple complex components in order to interact specifically with other molecular systems to accomplish precise functions upon which life depends—well, then the appearance of design in them is undeniable.

3. There Exists No Nondesign Explanation for Life

Since the appearance of design in life is far beyond overwhelming, then in order to rationally ignore the explanation of real design and purpose, we must have solid reasons for thinking a nondesign explanation is likely. But no such explanation is anywhere in sight. For example, just last year the Harvard biologist Marc Kirschner and Berkeley biologist John Gerhart published *The Plausibility of Life: Resolving Darwin's Dilemma.* The point of the book is that Darwin missed a very large piece of the evolutionary puzzle (the "Dilemma" of the title). Although his idea of natural selection was on the mark, neither Darwin nor his contemporaries knew what could make life give rise to the variations from which nature could select. The tiny, gradual changes Darwin envisioned would have been grossly inadequate, think Kirschner and Gerhart. Only now, with their idea of "facilitated variation" can we see that evolution could work. (Facilitated variation, roughly, is the idea that, with the developmental genetic programs discovered in the past several decades, we can supposedly see that organisms are predisposed to vary in ways that would be evolutionarily helpful.) Toward the end of their book Kirschner and Gerhart ask coyly:

> "Can evolution be imagined without facilitated variation? What capacity to evolve would a hypothetical organism have if it did *not* have facilitated variation? If animals did not use and reuse conserved processes, they would, we think, have to evolve by way of total novelty—completely new components, processes, development, and functions for each new trait" (Kirschner and Gerhart 2005, 242).

Their clear implication is that without facilitated variation, Darwin's theory *could not work.* Yet reviewers have been unimpressed with the new idea. Reviewing the book for *Science*, evolutionary biologist Brian Charlesworth arches an eyebrow:

> "In *The Plausibility of Life* ... Marc Kirschner and John Gerhart argue that the Darwinian explanation is incomplete and that the results of recent discoveries in cell and developmental biology can be used to remedy this defect. Are they right, or does their effort represent the latest entry in the catalogue of failed attempts by developmental biologists to supplement or replace neo-Darwinian evolutionary biology?" (Charlesworth 2005).

and waives away facilitated variation:

> "Kirschner and Gerhart do not present any detailed examples of how the properties of developmental systems have actually contributed to the evolution of a major evolutionary novelty. Nor have they shown that alternative properties would have prevented such evolution. Although *The Plausibility of Life* contains many interesting facts and arguments, its major thesis is only weakly supported by the evidence" (Charlesworth 2005).

Writing for Harvard's alumni magazine, Harvard biologist and Kirschner-colleague Daniel Hartl says many nice things about the authors and the book, but even he hedges about the theory:

> "Although Kirschner and Gerhart make a strong case for facilitated variation, it remains at this point only a hypothesis whose predictions need to be tested… . At this early stage, facilitated variation merely suggests that the origin of novelty and complexity might be rather more straightforward than previously appreciated" (Hartl 2005).

Another academic who thinks Darwin needs rescuing is University of Southampton Lecturer Richard Watson. As a computer scientist interested in evolutionary algorithms, he comes at the topic from a different angle than Kirschner and Gerhart, but arrives at the same conclusion. In his monograph *Compositional Evolution*—published by MIT Press in its Vienna Series in Theoretical Biology—Watson writes bluntly:

> "In computer science we recognize the algorithmic principle described by Darwin—the linear accumulation of small changes through random variation and selection—as *hill climbing*, more specifically *random mutation hill climbing*. However, we also recognize that hill climbing is the simplest possible form of optimization and is known to work well only on a limited class of problems" (Watson 2006, 272).

Those problems include very simple ones that can be solved by changing just one or a few variables—as in the evolution of drug resistance. "Darwin's masterful contribution was to show that there was *some* principle of optimization that could be implemented in biological systems" (Watson 2006, 278), grants Watson—just not the right one for complex systems, such as those that fill the cell. To save Darwin, Watson proposes his idea of compositional evolution, which is roughly the same as symbiosis. Without compositional evolution, thinks Watson, evolution by unintelligent processes would be unattainable. But what is the biological evidence for compositional evolution? "The existence of appropriate modularity in nature is an empirical matter that is only now becoming testable, but there are a number of ideas about modularity discussed below that suggest that appropriate modularity in natural systems is not an entirely unreasonable possibility" (Watson 2006, 280). In other words, maybe in the future we'll find some evidence.

Reviewing *The Plausibility of Life* for the journal *Cell*, Smithsonian paleontologist Douglas Erwin (Erwin 2005) is unenthusiastic: "[T]he authors' limited view of the evolutionary literature on variation and constraint undercuts most of their own arguments in favor of their theory of facilitated variation." But he sees the book as part of a trend: "Kirschner and Gerhart's book must be placed in the context of a number of other recent contributions to evolutionary thought, *all of which* argue that the current model of evolution is incomplete." [emphasis added] Well, *is* the current model incomplete? "Is there reason to think that our view of evolution needs to change? The answer is almost certainly yes," affirms Erwin. He quickly adds "although not, as the purveyors of creationism/intelligent design would have it, because the reality of evolution is under question." Apparently by "the reality of evolution" Erwin means common descent, although he does not explain.

Thus, as a spate of recent books by researchers at the highest levels attests, Darwinian theory has not explained, and likely can't explain, what we have found in biology. But as the unenthusiastic reviews of those books also attest, the newly proposed additions to Darwinism don't help. So at the end of the day we're left with the overpowering appearance of design but no explanation for it other than real design.

4. Rationality versus Science

Rigorous definitions of science are hard to come by. So for purposes of discussion I'm going to take "science" to indicate the activity of investigating nature by means of systematic observation, measurement, and classification. This is an intentionally simplistic definition, because I just want to highlight the fact that to a large extent science depends on modes of reasoning outside of itself. In other words, reason and science are not co-extensive; rationality encompasses more than natural science. As an easy example, consider mathematics. Mathematics is a distinct branch of knowledge from science. Nonetheless, science of course routinely uses mathematical reasoning to arrive at conclusions about nature; in fact, without abstract mathematical reasoning science would be greatly diminished. To put a point on it, the example of mathematics demonstrates that science depends critically on nonscientific reasoning. As a more difficult example, consider classification. Scientific classification is the grouping of natural objects into existing conceptual frameworks. Yet the prior judgment of which patterns exist in nature, which are important, and which frameworks are likely to be helpful for classification, although it is a rational activity, is not itself science. The construction of classifications is more basic than science.

Another, much more fundamental, example gets us closer to the nub of the problem I want to highlight. The perception of existence itself is not a scientific conclusion. The cognizance that the external world exists—that what we call reality is not the construct of our own minds, that it has an existence separate from our own minds—is also not a scientific conclusion. Since science can perform no experiment, make no measurement, construct no classification that would show that the world is real (without begging the question), then the conclusion that it is real is not a scientific one. Nonetheless, we properly regard the conclusion that the world really exists as a rational one—in fact the most basic rational conclusion—and we lock up in padded cells those folks who act as if they really deny it (as opposed to pretending to). Again, my point is that science and reason are not co-extensive, that rationality is prior to science, and that without non-scientific modes of reasoning, science itself would be severely hobbled, if not outright abolished.

Finally, here's the crux of the matter. Very closely related to our ability to perceive reality external to ourselves is our ability to recognize the existence of other minds. Not only do we understand that our own minds are not all that exists—that there is a world external to our own consciousness—but we also perceive that the external world also contains other minds capable of rational activity. That need not have been true. The only mind in the world may have been ours. But from a combination of our understanding that the external world exists and our perception of patterns in that world, we can come to the conclusion that the best explanation for some patterns is the existence of other minds. Here is the point I wish to emphasize. *The perception of other minds is a rational activity*. Like the use of mathematics or the perception of a world external to ourselves, our awareness of other minds is a rational activity that is more basic than science.

5. The Consequences of Spurning Reason in Physics

Solipsism is the categorical denial either that other minds exist or that we can perceive their existence. Solipsism can come in stronger or weaker forms. In its pure form, it holds that the world and everything it contains, including other seeming minds, are just manifestations of the solipsist's own mind. They don't really exist. A less extreme form of solipsism, I think, but solipsism nonetheless, is when we rule out a priori other minds as an explanation for events. For example, suppose we took it as a fundamental, nonnegotiable principle that only humans had minds. If human explorers came upon an undiscovered group of animals in a remote region that looked nonhuman, then, if the humans stayed true to their assumption, absolutely no evidence could persuade them of the animals' intelligence. If the animals did amazing tricks, like making fine artwork, that might be ascribed to instinct. If they seemed to communicate with each other, again, it must be brute instinct, or a trick by one of the explorers' party, or even a hallucination by the explorers, since the explanation of intelligence was ruled out from the start.

I contend that the express elimination of purpose, final cause, and mind as explanations in science was concomitantly the embrace of solipsism by science. Although first banished for what seemed to be good reasons, over time the rejection of real purpose in nature has grown pernicious and corrosive of science. To the extent that mind and purpose, or their effects, may really subsist in nature, then to that extent non-Aristotelian science has forsworn reason itself. To the extent that mind and purpose, or their effects, may really subsist in nature, the solipsistic bent of modern science will lead it deeper and deeper into irrationality.

As an example, I'll point to a recent book by the Oxford University philosopher Nick Bostrom, *Anthropic Bias: Observation Selection Effects in Science and Philosophy*. The chief problem that interests Bostrom is the apparent fine-tuning of the universe; that is, the fact that the laws and physical constants of the universe seem to exist in the narrow range that is compatible with life. Bostrom writes, "Some philosophers and physicists take fine-tuning to be an explanandum that cries out for an explanans. Two possible explanations are usually envisioned: the design hypothesis and the ensemble hypothesis." (Bostrom 2002, 11) In other words, design is seen by some—those who don't rule it out a priori—as a rational explanation based on the observable evidence. What is "the design hypothesis"? Bostrom defines it as follows:

> "The design hypothesis states that our universe is the result of purposeful design. The 'agent' doing the designing need not be a theistic God, although that is of course one archetypal version of the design hypothesis.... We can take 'purposeful designer' in a very broad sense to refer to any being, principle or mechanism external to our universe responsible for selecting its properties, or responsible for making it in some sense probable that our universe should be fine-tuned for intelligent life" (Bostrom 2002, 11-12).

He explains that not only will some theists conclude that the design hypothesis is reasonable, but also some persons who are not theists.

> "For people who are already convinced that there is a God, however, the design hypothesis is likely to appear as an attractive explanation of why our universe is fine-tuned. And if one is not already convinced about the existence of a Designer, but thinks that it is a coherent possibility, one may be tempted to regard fine-tuning as a reason for increasing one's credence in that hypothesis" (Bostrom 2002, 12).

Nonetheless, Bostrom restricts his analysis throughout most of the book to the selection effects that would be expected if ours were one of the lucky universes that happened to have gotten the laws and constants just right in a huge multiverse, where most universes got them wrong.

But in an infinite multiverse, Bostrom points out, a universe would not have to be finely tuned to contain a mind.

> "When black holes evaporate, they do so in a random manner such that for any given physical object there is a finite (although, typically, astronomically small) probability that it will be emitted by any given black hole in a given time interval. ... [T]here is thus a finite probability that a black hole will produce a brain in a state of making any given observation. Some of the observations made by such brains will be

> illusory, and some will be veridical. For example, some brains produced by black holes will have the illusory of [*sic*] experience of reading a measurement device that does not exist.
>
> "The probability of *anything* macroscopic and organized appearing from a black hole is, of course, minuscule.... Yet even a low-probability outcome has a high probability of occurring if the random process is repeated often enough. And that is precisely what happens in our world, if the cosmos is very vast. In the limiting case where the cosmos contains an infinite number of black holes, the probability of any given observation being made is one.
>
> "...It isn't true that we couldn't have observed a universe that wasn't fine-tuned for life. For even 'uninhabitable' universes can contain the odd, spontaneously materialized 'freak observer', and if they are big enough or if there are sufficiently many such universes, then it is indeed highly likely that they contain infinitely many freak observers making all possible human observations. It is even logically consistent with all our evidence that *we* are such freak observers" (Bostrom 2002, 52-53, 55).

In other words, any mind, such as, say, your own, along with all its beliefs about the past and about nature, is certain to spontaneously pop into existence in an infinite universe, utterly disconnected from reality. This is not the typical way in which solipsism is presented, but it amounts to the same thing: a solitary mind that can't know anything outside itself. Design is ruled out, and irrationality rushes in.

Of course the argument is self-defeating. How can there be "good grounds for believing" anything at all if minds can pop into existence with detailed false beliefs about the world? Bostrom tries to save the day by assuming that the universe is very large, but not infinite. So while "freak observers" (that is, minds produced by black holes or something) will exist, they are supposedly vastly outnumbered by ones who "emerged via ordinary evolutionary processes", presumably like us. But the move is a dodge. In order to explain apparent fine-tuning Bostrom delicately maneuvers to enjoy the benefits of infinity, but he shrugs off its drawbacks.

Here's one big drawback. It is a strange property of infinity that things of the same "cardinality" all exist in equal quantities. For example, consider the infinitely-many positive integers. There are as many integers divisible by two as total integers; that is, there are as many even integers as even-plus-odd integers. There aren't half as many even integers, because half of infinity is still infinity. In the same way, there are as many integers that are evenly divisible by a trillion—or by a trillion to the trillionth power—as there are total integers. With infinity, there is no such thing as rarity. This means that in an infinite universe there would be the *same number* of "freak" observers as observers produced under whatever we might be pleased to call "ordinary" circumstances. (This just scratches the surface of problems with the freak-observer scenario. If minds can pop into existence filled with false beliefs, how does one know that such concepts as "black hole", "infinity", "universe" and so on have any connection to reality?)

So are you actually a solitary mind in a barren, life-hostile universe that has just popped out of a black hole (or something) along with false memories of an earlier life? Setting aside Bostrom's dodge, but keeping his other arguments, you have at least an even chance of it if design is ruled out. In fact, you may have a much better than even chance of at least being a mind filled with false beliefs about yourself and the world. In a paper in *Philosophical Quarterly* (Bostrom 2003) he argues that if humans advanced to a level where they could run extremely sophisticated computer simulations of people, then by far most minds would exist in the simulations, filled with false beliefs about the world. This is scarcely different from being a brain in a vat! He even considers minds within simulations that run their own simulations. What about simulations run by freak observers who pop out of a black hole with an advanced computer and the knowledge to run it? Stay tuned.

6. The Consequences of Spurning Reason in Biology

Physics thinks big; biology, smaller. Nonetheless, one can also see there the baleful effects of radical skepticism concerning our ability to recognize the effects of other minds—to recognize real design. In biology this currently is manifested mainly in the inability to see when we have discovered actual reasons for an apparently purposeful phenomenon or not. In other words, when we can't recognize our own non-rationality. Here's an example. In a review article discussing cellular protein folding, the author offhandedly remarks:

"Another important constraint is the inability of a cell to tolerate significant amounts of unfolded, non-functional protein. As a result, every cell has *evolved* mechanisms that identify and eliminate misfolded and unassembled proteins" (Johnson 2005, my italics). But in fact as recent books show we actually have no knowledge of how such sophisticated mechanisms could have come about by random mutation and natural selection. Now re-read the quoted sentence, this time leaving out the word *evolved*. What knowledge has been lost?—None at all. If one is sensitive to it like I am, one sees that the science literature is filled with such utterly gratuitous—that is, non-rational—uses of the word evolution. Here's a second example from the same article: "To accomplish a timely recognition of nascent membrane proteins by the translocon-bound RNC during the integration process, the system has therefore evolved a mechanism that utilizes ribosome-induced folding of selected nascent chain sequences within the ribosomal tunnel" (Johnson 2005). Re-read the sentence, leaving out the words "therefore evolved." Notice that no information was lost.

Although most often the non-rational invocation of random evolution is unthinking, as it apparently was above, in other cases it reflects a deliberate rejection of the possibility of design. A review article in *Nature*, ironically on the intentional design of biological features by humans, asserts: "Synthetic biology is the logical corollary of the realization that cells, like mechanical or electronic devices, are exquisitely 'designed'—albeit by evolution rather than on the drawing board." (Ball 2004) Yet as recent books show we have no knowledge that non-purposive processes can produce exquisite design. And the habit of reflexively excluding design from consideration is taught to students. For example, in a section on protein structure in the textbook *Biochemistry* by Voet and Voet we read: "Keep in mind that only a small fraction of the myriads of possible peptide sequences are likely to have unique stable conformations. Evolution has, of course, selected such sequences for use in biological systems" (Voet and Voet 2004, 279). Notice that jaunty phrase, "*of course*"! Since we do not actually have sufficient evidence for the conclusion the book asserts, the basis of our confidence is not knowledge, but exclusion of purpose. In other words, the conclusion is non-rational.

Although the biological examples and language don't strike our ears as quite so strange as the physics examples of Nick Bostrom's monograph, the trajectory of thought is the same. If push came to shove, this line of thinking must claim that the appearance of functional biochemical complexity is the delusion of a freak observer from a black hole, rather than consider forbidden design. Reason is all of a piece, and when a part of rationality is rejected, eventually it all unravels.

7. Final Comment

Centuries ago science renounced consideration of purpose and final causes, shifting them to philosophy and theology. But the compelling appearance of design in biology remained a lingering rebuke to the agreement. Darwin finally broke the contract. Darwin himself addressed final causes and purpose, arguing that they were illusory. Following Darwin, modern biology now routinely spins nonpurposive stories to attempt to account for the overwhelming appearance of purpose in biology. Yet if the appearance of purpose is fair game for science, then one can't arbitrarily rule out the possible explanation that the appearance is due to real purpose. Arbitrarily ruling out real purpose, while simultaneously allowing nonpurposive explanations, does violence to rationality itself, one component of which is our ability to recognize the existence of other minds. The end point of the rejection of one facet of reason is the inability to recognize reality.

8. Abstract

In the past half century biology has discovered aspects of life that overwhelmingly appear designed. Darwinian claims notwithstanding, no explanation for them that invokes only undirected causes currently exists. Although genuine teleology was supposedly banished from science to philosophy centuries ago, the rigid separation is no longer intellectually tenable. Because a facet of rationality is our ability to recognize the existence of other minds, the adamant refusal to admit the possibility of real design in biology, I argue, opens the door to solipsism, which eventually leads to a radical skepticism that corrodes reason.

References

Alberts, B. (1998). The cell as a collection of protein machines: preparing the next generation of molecular biologists. *Cell* **92**: 291-294.

Ball, P. (2004). Synthetic biology: starting from scratch. *Nature* **431**: 624-626.

Bostrom, N. (2002). *Anthropic Bias: Observation Selection Effects in Science and Philosophy*. Routledge: New York.

Bostrom, N. (2003). Are we living in a computer simulation? *The Philosophical Quarterly* **53**: 243-255.

Charlesworth, B. (2005). On the origins of novelty and variation. *Science* **310**: 1619-1620.

Dawkins, R. (1986). *The Blind Watchmaker*. Norton: New York.

Erwin, D.H. (2005). A variable look at evolution. *Cell* **123**: 177-179.

Hartl, D.L. Better living through evolution. *Harvard Magazine* [November-December], 22-27. 2005.

Johnson, A.E. (2005). The co-translational folding and interactions of nascent protein chains: a new approach using fluorescence resonance energy transfer. *FEBS Letters* **579**: 916-920.

Kirschner, M. and Gerhart, J. (2005). *The Plausibility of Life: Resolving Darwin's Dilemma:* Yale University Press: New Haven, CT.

Levine, M. and Davidson, E.H. (2005). Gene regulatory networks for development. *Proceedings of the National Academy of Sciences USA* **102**: 4936-4942.

Voet, D. and Voet, J.G. (2004). *Biochemistry*, 3rd ed. J. Wiley & Sons: New York.

Watson, R.A. (2006). *Compositional Evolution: the Impact of Sex, Symbiosis, and Modularity on the Gradualist Framework of Evolution*. MIT Press: Cambridge, Mass.

Dialogue

Charles H. Lineweaver

I too have problems with Darwinism. For example, I think that the unit of selection is not obvious. Can group selection explain any genetic components of altruism or homosexuality? Can the entire biosphere be considered alive? Could Lovelock's Gaia have evolved? I think that Kirschner and Gerhart's notion of "facilitated variation" is a potentially very important explanation of how biological diversity is generated. One should also appreciate the large role that lateral gene transfer and endosymbiosis have played in generating biological diversity. Like you, I also think that purpose should not be taboo to biology. I think that the purpose of life, just like the purpose of any far from equilibrium dissipative system, such as a hurricane, a convection cell or a star, is to reduce the gradients in free energy—whether those gradients are gravitational, chemical, thermal or pressure (Schneider and Sagan 2005). Finding evidence for this hypothesis may be difficult. These are all interesting subjects of debate, motivated by attempts to understand the universe as scientists. All good scientific theories are subject to revision and reinterpretation. When there are flaws or interesting holes, we jump in and try to fix them or, as in the case of Darwinism, see if we can verify that no evolutionary pathway could have led to some extant complex structure. I agree that biologists have become very complacent about invoking evolution to explain some complexity or apparent design. The solution to this is to try to reconstruct the phylogenetic relationships of the complex proteins. Maybe you can show that none exist. There are many missing pieces to our understanding of Darwinian evolution and neo-Darwinism—that is why it is such an active field—but calling the non-recognition of your god's role in evolution, "solipsism" is not rational or scientific. It is a religious response to the ubiquitous uncertainties of scientific theories.

Michael Behe

I think that purpose in biology cannot be thought of in the same way as "purpose" in non-biological systems. Perhaps the "purpose" of a hurricane is indeed to dissipate a gradient (or maybe that's just what it does). In any case, the purpose of an eye is not to dissipate a gradient; the purpose of an eye is to see. One can easily tell that the purpose of an eye is to see by noting that the components of the eye are fitted to each other to allow that very complex process to occur. I'm much less sanguine than you about newer theories, such as Gerhart and Kirschner's, which promise to do the job that they claim Darwin's theory cannot. I see no evidence of that. Nor do I see much likelihood that mere phylogenetics will settle the question of the mechanism of evolution. On the other hand, the adamant refusal to recognize the existence of a mind at work, no matter what the evidence, is indeed solipsism.

Biography

Dr. Gerald Schroeder currently teaches at the Aish HaTorah College of Jewish Studies in Jerusalem and does research in environmental radioactivity. He obtained his PhD (in 1964) at the Massachusetts Institute of Technology in two fields (in Earth and Planetary Science, and in Nuclear Physics). He served seven years on the staff of the M.I.T. Physics Department prior to moving to Israel where he joined the Weizmann Institute of Science, and then the Volcani Research Institute, and the Hebrew University of Jerusalem Isotope Separation Mass Spectrometer Facility. He also witnessed the detonation of six atomic bombs.

His formal theological training includes twenty years of study under the late Rabbi Herman Pollack, Rabbi Chaim Brovender and Rabbi Noah Weinberg in Jerusalem.

Dr. Schroeder has approximately 60 publications in peer-reviewed scientific journals. He is the author of *Genesis and the Big Bang* (Bantam Doubleday); *The Science of God* (Free Press of Simon & Schuster, and Broadway Books of Bantam Doubleday); and *The Hidden Face of God* (Free Press). In the first full year of its publication, *The Science of God* was on the Barnes & Noble list of non-fiction best sellers and was Amazon.com's best selling book in the field of physics/cosmology for that entire year. His books appear in nine languages.

35
The Right and the Wrong of Intelligent Design

GERALD SCHROEDER

Aish HaTorah College of Jewish Studies, HaHish 5,
Jerusalem, 93223 Israel
Gerald_Schroeder@alum.mit.edu

1. Introduction

Courts in the United States of America have adjudicated that it is wrong to teach Intelligent Design (ID) in public school science classes. This in no way implies that the theory of ID is flawed or false. The decision is a legal one. The first amendment of the Constitution of the USA forbids the government establishing of a religion. ID assumes a metaphysical force active in the processes of the physical world. One such process would be the evolution of life. Though ID advocates do not necessarily include God in their statements, the putative metaphysical interaction assumed by ID is so similar to that which is posited for the biblical description God as to make ID a theory essentially couched within a God-based religion. In that sense the courts feel it is wrong to teach ID in a public school science class even though ID may be correct and relevant to life's evolution.

The irony of ID being disavowed as a scientific theory due to its metaphysical implications finds root in the fact that modern physics has about it much that mirrors the metaphysical. The physics of the 20th century discovered actions in the world that are totally outside human logic or intuition. Unlike special and much of general relativity which were deduced by logical extension of acts observed or envisioned within the physical world, the basics of quantum mechanics were discovered only after observations forced acceptance of explanations of events that were outside the realms of a materialist logic. That an item can be a wave and a particle simultaneously is illogical. Particles have discrete dimensions, edges. Waves are extended. Yet Einstein earned his Nobel Prize when he interpreted certain observations as only being compatible with light exhibiting a wave/particle duality, a concept totally illogical to our human concept of logic. Then de Broglie theorized and it was later demonstrated that all particles, not merely ethereal light beams, also have a wave aspect. And not only a wave/particle duality, but within this composite, there seems to be an aspect of mind.

Such a bizarre supposition stems largely from the famous double slit experiment. Light beams (photons) or sub-atomic particles are directed toward an opaque barrier in which are two narrow, widely spaced openings (slits), A and B. Most of the particles strike the solid barrier, but those that pass through the slits display the most illogical behavior. The path the particles (photons and sub-atomic) take after passing through the slits depends upon whether or not both or only one of the slits is open. If for example B is closed and A open, a particle passing though A senses that condition at B, even though in atomic terms, B is vastly distant from A. The path taken by particle A in this case is totally different from the path taken when both A and B are open. It is as if there is a form of knowledge at a distance. Freeman Dyson, physicist at The Institute For Advanced Study, Princeton, interprets this eerie behavior as follows. "Atoms are weird stuff, behaving like active agents rather than inert substances. They make unpredictable choices between alternative possibilities according to the laws of quantum mechanics. It appears that mind, as manifested by the capacity to make choices, is to some extent inherent in every atom." Mind in matter? Very strange and yet quite in accord with the physics of matter as we know it. Dyson is not alone in this opinion.

Sir James Jeans, knighted mathematician, physicist, and astronomer who helped develop our understanding of the evolution of stars; in his book, *The Mysterious Universe*, writes: "There is a wide measure of agreement *which, on the physical side of science approaches almost unanimity*, that the stream of knowledge is heading towards a non-mechanical reality; the universe begins to look more like a great thought than a great machine. Mind no longer appears as an accidental intruder into the realm of matter. We are beginning to suspect that we ought rather to hail mind as the creator and governor of the realm of matter [emphasis added]."

Materialist theories of existence need make room for the illogical, that is, for aspects of reality that lie outside the defines of human logic. ID being disavowed as a scientific theory simply because it calls upon phenomena that might be termed metaphysical, as in outside the physical, stands in opposition to observations of the scientific community itself. However, there are aspects of the ID theory that overstep the limits of that which may be surmised from observations in the physical world. These fall strongly within the domain of the evolution of life.

2. The Wrong of ID

The strict version of ID posits that there are aspects of life so complex and components so interwoven that they must have been formulated as a single unit, rather than having evolved sequentially over time. The logic being that if any one of the components were removed the entire organ or function would fail. In this sense the organ or system under study would be irreducibly complex. An example brought is the clotting of blood in humans. There are twelve factors that work in a fixed sequence to produce the clotting of blood when exposed to air and hopefully not when still within the body. The highly orchestrated progression implies that removing any one of the factors would disrupt the development, resulting a lack of clot formation. At least one research paper discusses the clotting of blood in a different mammal, the dolphin. Here one of the twelve factors is absent and still the dolphin blood clots. In humans the result would be hemophilia.

We see that a varied arrangement retains functionality, although in a different mammal. Less sophisticated vertebrates such a jawless hagfish have a simpler form of clotting based on only three of the factors present in the mammalian system. These data imply that although in its present form, human clotting may require all twelve factors to function effectively, this intricate system may have been built up to its present complexity in stages, and not necessarily as a single unit. The concept of irreducibly complexity suggests that such a sequential development is not possible. Nature has demonstrated this not to be true. Sequential development is indeed possible.

A more modest version of ID is less open to criticism. That is the concept of specified complexity. The complexity observed in even the simplest forms of life is so extreme that it may exceed the possibility of being the product of random mutations that are then screened by the local environment for survivability. Image forming visual systems are found in three phyla, arthropods, mollusks, and chordates. In all, the light sensitive molecule contains the protein, opsin, bound to a form of vitamin A. The discovery of this union and its function earned Professor George Wald of Harvard University the Nobel Prize. Discovery of the complexity and rapidity of the reactions involved in forming and re-forming the images (in excess of ten times per second) impressed Wald to the extent that he changed his world view to include the metaphysical within the physical. But more on that later. Interestingly, the opsin protein is also found in bacteria, and homologs are abundant is aquatic microbes. Again we see that although the finished product has a very highly specified complexity, simpler and varied combinations of the components are also functional. Thus specified complexity may not be a valid argument demanding ID for its origin.

These objections to ID not withstanding, basic problems do confront neo-Darwinian concepts of evolution. Can randomness be the only driving force for nature discovering the mutations that produce evolutionary change. If so, statistically it its hard to envision how nature could have arrived by random mutations at the few viable forms from among what might be termed the vast bio-hyperspace of non-viable forms of life. Some 530 million years ago, the Cambrian explosion of animal life marked the appearance of all animal phyla extant today. Approximately 50 phyla appeared but within a geologically short period of time, a third became extinct. Some 32 phyla remain. The others apparently were not viable. Prior to the Cambrian profusion of animal life, a complex form of proterozoic life (the Ediacaran fauna) appeared in the fossil record. It failed as an experiment of life. The surviving phyla of the Cambrian developed, radiating and in doing so, formed different classes of animals, each retaining the basic structure of its parent phylum. Then 250 million years ago, disaster of unknown origin struck, decimation in the full meaning of the word occurred. Some 90% of all life disappeared from the fossil record. The environment was wide open for new phyla to appear. No new phyla developed. These events imply that only certain forms of life are able to succeed within the constraints of the earth's environment. Other forms are not functional. The question arises as to how nature arrived at the winning combinations that allowed for survivability.

3. Problems with Randomness

The basic building blocks of life are proteins, long chain-like molecules consisting of varied combinations of 20 different amino acids. The molecules coil into a highly specified shapes dependent upon the exact sequence of the amino acids within the chain. A mutation will insert a wrong amino acid into the chain, altering the final shape of the coil, making that protein useless, and possibly lethal to the animal within which it resides. In all of life, from bacteria to humans, there are fewer than a million different viable proteins. A typical protein has some 200 or more of these 20 amino acids joined together to form its chain. The number of possible combinations of amino acids in the chain is 20 (that is the number of possible different amino acids) to the exponent power of 200 (the number of locations within the chain filled by one of the 20 different amino acids). In the more usual ten-based system, this number equals 10 to the power of 260. That number is a one with 260 zeros after it. To gain a sense of the magnitude of this truly gargantuan number, in words this number of possible amino acid combinations in a chain 200 amino acids long is a billion. If purely random processes were the source that led to the million viable proteins, nature would have had to discover by chance from among this vast non-viable biological wasteland, the fewer than one million combinations that allow for life. The Laplacian probability of that happening by random chance is one chance out of a thousand billion. In other words, there is no way that nature could have happened upon the viable combinations by chance.

Simon Conway Morris, professor of evolutionary paleobiology at the University of Cambridge, Fellow of the Royal Society of England, is probably the world's leading paleontologist. His research revealed that the Cambrian explosion of animal life, some 530 million years ago, produced *simultaneously* every phylum of animal life that exists today. His research changed the concept of evolution, forcing the abandonment of the catechism of an evolutionary tree of life, with simple phyla evolving into more complex phyla. The history of life became the bush of life, with all animal phyla branching simultaneously out from the microscopic cell clusters of life that pre-dated the Cambrian explosion.

In his seminal book, *Life's Solutions*, Morris describes how he understands the almost illogical success of life set within the vast biological hyperspace of failure. But he does so in a manner that ties together the physical with the metaphysical:

> "If you happen to be a 'creation scientist' (or something of that ilk), may I politely suggest that you put this book back on the shelf. It will do you no good. Evolution is true, it happens, it is the way the world is, and we too are one of its products. This does not mean that evolution does not have metaphysical implications; I remain convinced that this is the case. …[E]ven more mysterious is the attempt to explain the origins of sentience, such that the product of ultimately inanimate processes can come to understand both itself, its world, and as I have already noted its and thus our strange sense of purpose. … Life is simply too complex to be assembled on any believable time scale. … evolution's uncanny ability to find the short cuts across the multidimensional hyperspace of biological reality. It is my suspicion that research might reveal a deeper fabric to biology in which Darwinian evolution remains central as the agency, but the nodes of occupation are effectively predetermined from the big bang. One such node is of course that of the humanoid. To reiterate: life may be a universal principle but we can still be alone."

Morris, who knows with intimate detail the tale the fossil record tells, realizes that these very data endorse the supposition of "metaphysical implications." This suspicion arises not merely from the need "to explain the origins of sentience, such that the product of ultimately inanimate processes can come to understand itself," but equally from "evolution's uncanny ability to find the short cuts across the multidimensional hyperspace of biological reality." That hyperspace of biological reality includes the 10 to the power of 260 possible combinations of amino acids out of which fewer than 10 to the power of 6, that is one million, are viable. How did nature stumble upon the survivors?

In his book, *Tour of a Living Cell*, Nobel laureate, organic chemist de Duve, summarizes this incomprehensibly complex and unlikely flow of non-living matter into life. "If you equate the probability of the birth of a bacteria cell to chance assembly of its atoms, eternity will not suffice to produce one. … Faced with the enormous sum of lucky draws behind the success of the evolutionary game, one may legitimately wonder to what extent this success is actually written into the fabric of the universe." Written into the fabric of the universe, by what mechanism? The Declaration of Independence of the USA refers to such a force as "Nature's God ... the Supreme Judge of the world." This, of course, is exactly the claim of the legally disavowed theory of ID.

ID identifies the directing force as a metaphysical intelligence. Conventionally referred to as God, this intelligence must have been actively involved in the process of life's origins and development. Exactly how God's involvement in nature might manifest itself is debated among even the staunchest supporters of ID. The difficulty in discerning God's methods lies in our human limitations. All our thoughts arise from within the physical aspects of time space matter. There is no way we can think outside the physical. The greatest of poets, philosophers, scientists and theologians all face this same limitation. How a metaphysical Designer, God, interacts with the physical universe is not limited to mechanisms we can conceive from within the confines of our physical world. It would be folly to confine God's power to that which we can imagine.

The entire concept of ID is based on the acknowledged reality that life is marvelously complex. While the debate remains as to how that complexity arose, there are aspects of how life arose from non-living matter about which there is broad agreement. Rather than contesting how a fish may or may not have become a frog, or how a pre-human hominid became a human, let's look at those portions of the flow from the big bang to sentient life about which there is little or no debate.

4. Teaching the Wonder and Discovering the Intelligence

Science classes, in public or private schools, should start at the very beginning, which is a very good place to start. The first topic of discussion would do well to consider the beginning of existence, the big bang creation of the universe. Certainly the big bang is an acceptable topic for teaching in public school science classes. We may take the existence of existence for granted, as the fish might take water for granted, but the fact that there is an 'is' within which life might or might not have evolved is a wonder in itself. In fact it is probably the greatest wonder of all. Why is there existence? Why is there a universe? Existence in any form with or without life is not mandated by any scientific principle. And yet we are living proof not only of the fact of existence, but equally phenomenal, of the marvelously, and from first principles, unexpected, complexity of this existence we take for granted.

It comes as a surprise to the younger generation that the fact of our universe having had a beginning is quite recent knowledge. A survey of scientists taken in 1959, the results of which were republished in *Scientific American*, revealed that 2/3 of those scientists queried about the age of the universe opted for an eternal universe, no big bang, no beginning. Then in 1965 two scientists at Bell Labs in New Jersey, USA, Arno Penzias and Robert Wilson, discovered that all of space is filled with a very long wave radiation, as if all space were at 3 degrees Kelvin. Penzias and Wilson had discovered the echo of the big bang, the radiation once unknowably intense now stretched by the million million-fold expansion of space to microwave proportions. They had discovered the echo of the big bang. Since then many refined measurements have confirmed these data, validating that the distribution is that of black body radiation, the expected profile if indeed this is the radiation remnant of the big bang.

Next we might study the laws of nature. We'd discern that, scientifically, the laws of nature are exactly correct for sustaining life. These include facts so commonly a part of our lives that we do not even marvel at them, such as there being three spatial dimensions; only one time dimension. The mass of an electron is 1837 times less than the mass of a proton and yet they have identical but opposite charges. Had the charge relationship been similar to gravity, then the charge of the electron would have been minute compared with that of a proton. To balance the charge of the nucleus, for each proton there would have to be 1837 orbiting electrons. Vast clouds of electrons would encompass each nucleus. The slightest input of energy would expel outer electrons, continuously ionizing the atom. The result: no molecular stability, no stable chemistry. No complex structures and hence essentially no chance for life.

Scientific American, the most widely read scientific journal, avidly materialist in its approach, acknowledges in its May 2003 issue, that if our universe is the only universe, then we live in a designer universe. Since this is patently unacceptable for a materialist view of existence, the authors tell us that there must be an infinite or near infinite number of universes, most of which will have less a favorable nature. With a vast number of putative universes, one of them had to be just right. That's our world. Of course there are no data that support the concept of there being vast numbers of other universes, only the necessary assumption to avoid the designer image. Of course having a vast number of universes greatly increases the conundrum of why there is anything instead of nothing.

The first material product of the Big Bang creation of our world was not matter as we know it and not the sub-atomic particles that combine to form the atoms of the 92 naturally occurring elements. The first material product of the Big Bang was pure exquisitely powerful energy. In a simplistic characterization, this initial burst of energy might be referred to as super powerful light beams. Observations taken of light reaching us from deep space, from galaxies billions of light years away, indicate that the universe is expanding, space stretching. Running the time line back into the past, the expansion data reverse to become contraction. Based on these observations, it appears that at the first instance of existence, the entire universe was compressed within a minuscule speck of space. All that was eventually to produce a hundred billion galaxies each with some one hundred billion stars was packed into a volume not larger than the black of an eye. Albert Einstein's discovery of the bizarre fact that solid matter is a form of condensed energy allowed for nature to accomplish this miracle. Energy is ultimately compressible into ever higher compaction as the wave length of that energy gets ever shorter. No solid form of matter can accomplish such compaction.

Matter makes its appearance in our world as a dot of superbly powerful energy. And every item about us from the tiniest grain of sand to the mass of a galaxy is the product of that primordial burst of energy. For fourteen billion years the universe has been coasting on the initial burst. And we were part of that beginning, not as our bodies are today, but as that energy. We are literally made of the light of the big bang creation. An analogy may help to internalize the significance of the cosmological concept that there was only one physical creation from which all that we see is constructed.

Let's assume that in one had I hold a clear glass jar filled with the gas, oxygen. In my other hand I have a similar jar filled with the gas hydrogen. I study the properties of each gas and discover that by mixing them at a given temperature they chemically react. Cooling the mix after the reaction gives me liquid water. Water does not look or feel like oxygen or hydrogen gases. Yet we all know very well that water is indeed a combination of hydrogen and oxygen. These two elements do not stop being them selves when they combine. They merely change form.

The parallel I seek is that we may not look like the light beams of the creation but we are. We are the condensed energy of the big bang. This fact may read as if some new age fantasy is at work. But it is as old as our universe. Our cosmic genesis started with the creation. Then as the universe expanded and in doing so, cooled, a small fraction of that initial energy condensed into the two lightest of elements, hydrogen and helium. And we were part of that condensate. Not as spectators, but actually gases in space. The forces of gravity then drew these gases together into galactic sized clouds. Within those titanic structures, smaller clusters formed. These became stellar systems. The pressures at the cores of the stars, and then in the supernovae as the stars exploded, forced the lighter nuclei to fuse, forming the heavier elements. The youngest of stars were composed initially of hydrogen and helium. The stardust of those exploded stars became the material for second and third generation stars, as that stardust mixed with the primordial clouds of hydrogen. Pressures within the cores of a star ignited nuclear reactions, as hydrogen nuclei fused and metamorphosed into helium. The products of the reactions are several per cent less massive than the in-going material The lost mass has been converted into energy, the light of a star. A star is an on-going nuclear fusion reactor, a steady state hydrogen bomb.

Our sun is a second or third generation star. We know this because the solar system is so rich in the heavier elements, elements formed within the previous generation or generations of stellar activity. The wonder of this cosmic flow is that we were a part of it, part of the stars, of the supernovae, of the stardust in space and finally of the rocks and water that condensed to form the planet earth, our home in the universe. We witnessed it all. There's no deed to speculate about whether or not intelligent design was the driving force in selecting the laws of nature that allowed this cosmic novel to unfold. Just teach the facts. They speak for themselves.

And then just under four billion years ago, the once molten earth cooled sufficiently to have water condense into its liquid form. Three decades ago, paleontology brought a revolution in thought to studies of the origin of life. Since the time of Darwin, it had been assumed that vast amounts of time, measured in the billions of years, passed between the origin of liquid water and the origin of life on earth. The idea was that random reactions over these eons of time led to the first cells of life. Nobel laureates waxed poetic over the process that Darwin had hinted at as a warm little pond brimming with chemicals and electricity to form the first of life. George Wald, Noble laureate, professor of biology at Harvard University described the putative random process in an article, "The origins of life", published in *Scientific American*, August 1954:

> "However improbable we regard this event [the start of life], or any of the steps which it involves, given enough time it will [happen] almost at least once. And for life as we know it ... once may be enough. Time is in fact the hero of the plot. The time with which we have to deal is of the order of two billion years. What we regard as impossible on the basis of human experience is meaningless here. Given so much time, the 'impossible' becomes the possible, the possible probable and the probable virtually certain. One has only to wait: time itself performs the miracles."

This is certainly a definitive statement, and why not? How else could life have started on a once hot and sterile earth? But the nasty facts spoiled the theory. Twenty five years later *Scientific American*, in a special publication, "Life: origin and evolution," retracted the article that had become a classic in evolutionary theory:

> "Although stimulating, this article probably represents one of the few times in his professional life when Wald has been wrong. Examine his main thesis and see. Can we really form a biological cell by waiting for chance combinations of organic compounds? Harold Morowitz, in his book "*Energy Flow and Biology*," computed that merely to create a bacterium would require more time than the Universe might ever see if chance combinations of its molecules were the only driving force."

The retraction was unequivocal. Random reactions ("chance combinations of organic compounds") could not have yielded life in the time available. But the retraction was not based merely on some calculations of probability by physical chemist, Morowitz, at Yale and Princeton Universities. Darwin's concept of a pool brimming with chemicals was dispelled because there were not billions of years between liquid water forming on earth and the first cells of life. As cited in that issue of *Scientific American*, Professor Elso Barghoorn, also at Harvard, had realized that the search for the earliest forms of life must be done with a microscope. The first forms of life would be microbes. Barghoorn discovered that the oldest of rocks that can bear fossils already have fully formed fossils. Life started immediately on the cooled earth. This bit of knowledge, usually glossed over in the debate over evolution, is a major challenge in origin of life studies. The crucial point here is that there were not billions of years for life to have evolved from the inorganic substrates present on the pre-biotic earth. The path that led from a few very simple molecules into self-replicating molecules and then to the first cellular form of life occurred in a geological snap of time.

Theories for the origin of life abound. All require special conditions, special catalysts, none of which were obviously present on the young, sterile, earth 3.8 billion years ago. The concept of a purely unguided randomness as the engine that produced life, once so popular, has been abandoned.

None of these facts proves that there is an intelligent force acting to accomplish life's origin and development, but it certainly does call out for an explanation. This may be that to which Nobel laureate, organic chemist de Duve, alluded when he wrote that "Faced with the enormous sum of lucky draws behind the success of the evolutionary game, one may legitimately wonder to what extent this success is actually written into the fabric of the universe."

As quoted above, professor of evolutionary paleobiology at the University of Cambridge, Simon Conway Morris, seconds these thoughts and goes a step further toward identifying the origins of life's wonder of complexity. "Evolution is true, it happens, it is the way the world is, and we too are one of its products. This does not mean that evolution does not have metaphysical implications; I remain convinced that this is the case."

One of the basic difficulties in theories of the origin of life is that the first form of life had already to have within its structure the ability to reproduce. Although there may have been vast numbers of molecules forming and disintegrating on the early sterile earth, one fecund molecule had, by itself, to discover the process of reproduction, and not only reproduction but a reproductive process that was not with total fidelity. Full fidelity would produce changeless stagnation. And this progressive molecule then had to continue to mutate till it formed the workings of a viable cell. If it failed along the way of this bio-genesis, then the experiment for life would have to start again, de novo. The deception of the fossil record is that it shows only the most superficial aspects of the putative evolution of life. What took place within those now fossilized organisms, the molecular mechanisms, the molecular biology, defined the flow from a mix of rocks, water and a few simple molecules through the tortuous path that reached viable life. Neglecting the complexity of DNA and the amazing organelles that are able to read and translate the DNA information into proteins, it is instructive merely to consider the end product of this evolution of the cell, mitosis, the replication of a cell. It is essentially the same for all forms of eukaryotic life.

In simplest of terms, first the cell has to replicate its genetic material, the DNA. For this, thousands of readers move along the DNA helix, opening it and then copying and joining to form the new strand. Cables of molecular dimension then line the DNA chromosomes along the equator of the cell in opposing pairs. Molecular cables extending from the opposite poles of the cell then reel in the DNA-bearing chromosomes to their respective poles. A molecular cable forms about the cell's equator and tightens and tightens forming the cell into a figure eight. This molecular motor molecule continues to constrict until the cell is literally cut in two. Mitosis is complete. Reproduction is nothing less than an exquisite ballet set in dimensions of molecules. For a more complete description of the details of this biological wonder, I suggest passing a few hours reading Alberts, *Essential Cell Biology* or Klug and Cummings, *Essentials of Genetics*.

With the first viable cell in place, the debate rages as to how it metamorphosed into the diverse plethora of life extant today. Very intelligent persons argue that random processes and the three billon years since the advent of life on earth are adequate to explain the bio-world we see about us. And very intelligent persons argue the opposite view, that random mutations could not have accomplished the task. There are data relevant to both sides of that equation.

The fossil record has problems. Darwin insisted that *natura non facit saltum*, Nature does not make jumps. Yet in Darwin's time, 1859 at the publication of *The Origin of Species*, there were overwhelming lacunae in the fossil record. Darwin instructed his readers to ignore the fossil record and instead use imagination as to how the record will look when it is eventually filled out. Unfortunately the quandary has become more severe. In Darwin's time, there were no known fossils older (deeper) than those of the Cambrian era, today dated to have been some 530 million years ago. The blank fossil record was good news. A blank record merely means that the older fossils simply have not yet been found. What was a blank fossil record in Darwin's day has now been filled through the discoveries of Harvard professor Elso Barghoorn. The record extends back to the origin of liquid water on earth, between 3.8 and 3.6 billion years ago. Life logically appears in microscopic proportion. The surprise is that it remained microscopic for over three billion years, and in that multi-billion year period developed no discernable external structures. And then the Cambrian layer: animals measured in lengths of centimeters and bearing jointed limbs, sectioned bodies, mouths, and perhaps most amazingly of all, eyes. Obviously all fossils have not been found. But when teaching of the fossil evidence for the evolution of life, it is unjustly prejudicial to skim of the difficulties raised by the data from the Cambrian. It was Simon Conway Morris, who in 1985, 120 years after the publication of '*Origin*,' demonstrated that the fossils of the Cambrian represent every animal phyla extant today.

Several years ago, the Natural History museum in London had an entire wing devoted to showing evolution in progress. Upon entering the exhibit, a series of daisies were displayed, documenting the mutations and breeding that led red daisies to become blue daises. The next exhibit, in similar fashion showed the evolution of big dogs into small dogs. And the final work was most impressive. In this, we learned that in a mere 12,000 years, documented in the layers of sediment of Lake Victoria, some ten species of cichlid fish had developed into 500 species of cichlid fish.

Note the problem with this exhibit. Daisies remained daises, dogs remained dogs and cichlid fish remained cichlid fish remained exactly that, cichlid fish. This is referred to as micro-evolution. The museum did not, or perhaps more accurately, could not provide a single case of macro-evolution in which there were major morphological changes. That does not prove that it never happened. But if it did, the data have been hard to discover.

5. Summary

When searching for intelligence within the design of life, there is no need to fall into the debate of whether randomness or guidance produced the changes that allowed aquatic animals to become amphibians and then fully terrestrial. Compelling data are cited on both sides of the aisle. But there are manifold pieces of information upon which we all agree. And those in themselves call out for an explanation. In searching for the source of the design, it is folly to limit the manner by which a metaphysical reality, call it God, might interact within the physical world. We form all our thoughts from within the confines of the physical aspects of time space matter, the parameters of our physical world. There is no way we can think outside those boundaries. The greatest of poets, philosophers, scientists, theologians all face this limitation. How a metaphysical designer, God perhaps, might or might not interact with the physical universe is not limited to the mechanisms that we can conceive from within our physical existence. Let's not define God's power by that which we can envision.

The great value of the ID controversy is that it has forced upon the public an awareness of the magnificent wonder of life and the fact that not all knowledgeable persons agree on how that wonder arose.

As Carter, in her book, *Mapping the Mind*, points out, our brains are programmed to seek causes for that which we experience. We developed in a world where survival meant being able to discern the source of a sound: was it wind moving branch, or the paw of a threatening beast doing the same. We've left the jungle, but the programmed need to discern causes has remained. Perhaps all events need not have causes. Yet human logic argues for the opposite, even if we cannot sense what that cause might be. The basic problem in teaching evolution is that we get so involved with the minutiae that we neglect the really crucial questions. When teaching the sciences, there's no need to posture a divine intelligence in action as the foundation and cause of the wonder of our existence. However, at the beginning of a study of life and its development, an overview is essential if the sessions that follow are to be understood within the context of a world that had a beginning. All that need be presented are the facts upon which all agree, all the facts including those for which there are no facile explanations.

These might include discussions as to why there is existence rather than nothing at all, and, how life filled with consciousness and the ability to feel love, joy, wonder might have emerged from non-living matter, rocks and water and a few simple molecules, and even more alarmingly from the energy of the big bang creation. When all the facts are given, especially those for which there are no facile explanations, the students, young and old alike, will ask the questions for which there are no facile answers.

References

Alberts, B., (1998) *Essential Cell Biology*, Garland Publishing Co., New York.

Carter, R. (1998) *Mapping the Mind*, Weidenfeld and Nicolson Publishers, London.

de Duve, C. (1984) *Tour of a Living Cell*, Scientific American Books, New York.

de Duve, C. (1991) *Blueprint for a Cell*, Neil Patterson Publishers, Burlington, NC.

Folsome, C. (1979) Life: Origin and Evolution, Special Publication, Scientific American, New York.

Heisenberg, W. (1971) *Physics and Beyond*, Harper and Row, New York.

Jeans, J. (1931) *The Mysterious Universe*, Cambridge University Press, Cambridge, UK.

Klug, W. and Cummings, M. (1999) *Essentials of Genetics*, Prentice Hall, Upper Saddle River, NJ.

Morris, S. (2003) *Life's Solutions*, Cambridge University Press, Cambridge, UK.

Peebles, P. (1993) *Principles of Physical Cosmology*, Princeton University Press, Princeton, NJ.

Tegmark, M. (2003) Parallel universes. Not just a staple of science fiction, other universes are a direct implication of cosmological observations. Scientific American **288**(5), 31-41.

Dialogue

Steve Grand

This is my first response to your chapter. Phew! You cover so many topics that it's difficult for me to know where to start. I can't respond to everything or my email will be longer than your essay, so I'll reluctantly skip the Big Bang, quantum uncertainty, educational theory and so on, for now at least, and stick to your objections to neo-Darwinism itself.

It seems to me as if you might have become disillusioned with certain aspects of the contemporary paradigm in physical science and sought a better explanation. I feel very much the same way, but my own explanation is very different from yours. I find no need for metaphysics, just a retreat from the dogma of reductionism. I'm actually quite sad that you haven't touched on information, which I know interests you, because it's where we follow connected but divergent tracks. Anyway, for your own information, my fields of expertise are artificial life, artificial intelligence, computational neuroscience and computer science, although all these are really just aspects of my interest in self-organization.

I feel obliged to say that I'm troubled by your repeated use of Argument from Authority in this chapter: it smacks a little of sophistry. You emphasize certain scholars' credentials and then quote their mystification or conviction about things that actually lie outside their expertise, making their qualifications largely irrelevant. I wouldn't dream of arguing with Simon Conway-Morris about Ediacaran fossils, nor with Christian de Duve about cytology, but it seems neither is sufficiently familiar with complexity theory and the nature of self-organizing systems. If they were, they wouldn't be nearly so mystified.

In a number of places in your article you actually form an equation between the inexplicable and metaphysics, as if the non-existence of an explanation is equivalent to the existence of a metaphysical cause. But to do that is to fall into the same trap that made people in Mediaeval times resort to "magic" as the explanation for things that are now known to be straightforwardly mechanistic. It's misleading to argue in such a way and Occam's Razor cautions against it. Indeed you use our (present) human inability to conceive of certain things in your own defense of God: you say that God's influence shouldn't be discounted simply because we're incapable of understanding it. Well you can't have it both ways! Nature's ability to produce phenomena that we currently find inexplicable or improbable shouldn't be discounted either. In my view the problem lies in the mathematical foundation of Physics, not the physical world itself.

But let's get to the meat of your argument: the improbability of life arising, or of life's features emerging, by purely random means. For instance, you say that "statistically it is hard to envision how nature could have arrived by random mutations at the few viable forms from among what might be termed the vast bio-hyperspace of non-viable forms of life." You have two objections here: the insufficiency of randomness and the improbability of finding what is supposedly a tiny fraction of "viable" proteins or life-forms. Let's start with the latter.

You point out that there are 10^{260} possible proteins of length 200, and marvel at how nature was supposed to stumble upon the mere 10^6 that actually work for sustaining self-replicating entities. But that's like saying there are an almost infinite number of possible symbol systems and yet humanity somehow managed to stumble upon the few alphabets capable of sustaining language. There are only a few alphabets because most of them are descended from the first one or two that were successfully used. It doesn't mean that none of the other comparable symbol systems are "viable" as sources of written language — pretty well all of them would do, but only a few actually got used and those are the ones we derived our modern ones from, that's all. Every single protein that could exist has many physical properties that could make it a useful part of an autocatalytic network of some kind, and hence of a living thing; it's just that the first successful organisms happened to use a certain subset of those proteins and all existing life is descended from those ancestors.

Much the same applies to the number of possible life-forms on Earth, although with interesting differences, but I won't repeat myself here. I would just like to pick up on a few things before moving on: Firstly, Conway-Morris notwithstanding, the animal phyla did not emerge simultaneously. A short time in geology is still a very long time in biology. Anyway this was all very recent compared to the origin of life. What's more, there's nothing special about phyla (or even kingdoms) — they were just different but closely related species at the time of their divergence. Just as we humans are all descended from one man and one woman (neither of whom lived in Eden), yet there was nothing special or fundamentally different about them compared to their relatives and friends. All branches in a tree are qualitatively similar, it's just that the early ones necessarily dictate the possibilities open to the later ones, so we accord them a higher status.

Now to randomness. I don't know why people persist in believing that evolution by natural selection is a random process. That's nonsense, and Richard Dawkins has done a sterling job in explaining why. Natural selection is a decidedly NON-random process. The only random bit is mutation (and crossover, which is a powerful factor that mustn't be disregarded). All those people who have been mystified by the minute probability that a sufficiently complex network of chemical species should spontaneously assemble itself out of nothing have been missing the point completely and their statistics are worthless. In general, as I'm sure you know perfectly well, natural selection proceeds by making small improvements to a perfectly viable predecessor. The path taken through the search space is therefore very non-random indeed, and as Dawkins' biomorphs or "methinks it is a weasel" illustrations show, any desirable outcome can be reached through a counter-intuitively small number of steps. The process is many orders of magnitude more efficient than classical statistics (with its false assumption of independent variables) would predict.

As you suggest, the tricky part is creating the very first self-reproducing organism. After that point, everything is comparatively easy. But we don't yet know what a minimal self-replicating entity consists of. It may not require very much at all. One thing is for sure: it doesn't have to look remotely like a modern eukaryotic or even prokaryotic cell, with its complex DNA mechanism. In fact the only initial requirement is that a network (or loosely coupled group of networks) of chemical reactions arises, which is capable of making more of each of its own molecular components (at least, those that don't come freely from the outside as "food"). It doesn't even have to be a cell or contain a recipe for its structure in the form of DNA, RNA or anything else. Stuart Kauffman's work suggests that the spontaneous emergence of such networks is actually highly probable. It seems to me that the hardest thing to explain about the origin of life is therefore not why it occurred but why it apparently only occurred once! And even that may not be so mysterious.

The later emergence of the now very complex process of DNA replication and protein synthesis is still a mystery, but not one that requires a supernatural explanation. The important thing to recognize is that these later, more complex mechanisms are probably heavily ***streamlined*** descendants of once far messier but more robust systems. You touch on this in your objection to the evolution of the blood clotting process. Sure, as it stands it makes no sense that the clotting mechanism could have arisen step by step. But there's no reason to suppose that the existing steps are the only ones that ever existed. The present blood clotting mechanism is, I suggest, the minimal, most elegant system that evolution could extract from something that was originally much messier and less efficient. It didn't emerge all at once NOR step by step; it condensed out of something much more fluid. Imagine building a bridge over a river: how do you do it step by step when it won't stay up until all the pieces are in place? Well, you could try dumping a huge heap of stones in the river and then remove all the rocks that don't contribute much to the structure you want.

Anyone who works with artificial forms of evolution is invariably amazed by how efficient and creative it is (as long as they don't miss out important aspects, especially embryological ones). It really doesn't require billions of years; only millions. It's not trying to search a vast space at random; it's an extremely efficient search algorithm. Sudden jumps in the "fossil record" are not only common but entirely to be expected, once evolution discovers the capacity to produce switches (such as homeobox sequences), and given the information multiplication implicit in embryogenesis (e.g. it's very easy to get from a two-legged creature to a 100-legged creature; all you have to do is mutate the "repeat count", not evolve new legs over and over again from scratch). The linear mindset of classical mechanics (that you perhaps dislike as much as I do) may have a hard time coming to terms with these things because of its mathematical heritage, but I think you only have to try it out for yourself in simulation to find that natural selection is perfectly up to the task of creating life, just as a similar principle of selection gave us galaxies and protons.

Gerald Schroeder

Your argument was going along nicely more or less till you brought Dawkins' biomorphs or "methinks it is a weasel" illustrations show. Good grief. Look at the program he used. He gives just enough information in the book to show the total non-random nature of the flow. I am certain that you can figure out the problem. Using his approach and giving a system with 10^{18} places [Seimens' estimate of all the info in all libraries in the world], in 100 trials I can produce with greater than 95% probability all the information in all the libraries in the world. Dawkins showed one thing only. That his computer was working as a random number generator. But the letter selection, regardless of the order, locked in each correct 'choice' never to mutate again though all the others were. Each "slot" was totally independent of all the others.

You take the property of self-organization as a given, as that's just the way nature is. I take the property of self-organization as clear evidence of skewing nature toward the complexity of life. The 10^{260} still remains. The mutations that hit the repeater box had a lot of other potential choices to make. Once the pattern of life types has been established, then only a confined number of proteins are useful when evolving a fin into a foot even though the bone pattern remains the same. Those evolutionary changes have to be by random mutations on the genome as you so well know. Using American terminology, life developed, not evolved, from the simple to the complex. What drove that development was more than a mere bumbling along of random genetic events.

Steve Grand
Richard Dawkin's biomorphs and weasels are illustrations, not scientific models, and you have to make allowances for that. Nevertheless, the point is that every newly conceived creature starts out with a completely viable design, and mutation only makes very minor adjustments to it. Creatures therefore do NOT have to shuffle all the cards completely from scratch every time, so the probability of a viable result (and hence the probability of an improvement in fitness) is vastly higher than would be predicted by classical statistics.

Given the robustness of the more critical chemical pathways, there's a high probability that mutation will not fatally damage the genotype of the offspring and it will survive (the truth of this is self-evident, because we do all contain mutations and yet a substantial fraction of us do survive from conception to maturity). Given that the mutations are small, the offspring will usually (although not always) be very similar to its parents and hence have a similar fitness level. It will either be slightly less fit or slightly more, and if it is more then it is more likely to have offspring and hand down its acquired advantage. This is utterly unlike the situation you are talking about, in which every trial starts with a completely fresh throw of all the dice. Evolution doesn't keep starting from scratch.

The same applies to mutating a fin into a foot. It is NOT a matter of replacing one protein with another chosen at random. If it was then you'd be right — the chances of finding one of the few suitable proteins out of the many irrelevant ones would be miniscule. Mutations take an existing, useful protein and change a few of its amino acids. The result is almost the same protein as before, but usually with slightly different physical properties. It is still very likely to be useful. Structural proteins and enzymes that influence the length of a long bone, its curvature or the number of radiations, are likely to mutate into almost identical proteins that result in a slightly longer bone, or a small change in curvature. Such changes distinguish a foot from a fin, and every step along the way (if you'll forgive the pun) is viable and has a fair chance of being more useful than its predecessor. At no point do we have to throw out a useful protein and replace it with a wholly new one drawn at random from 10^{260} choices – the new one is extremely likely to be similar in properties to the one it replaces (because it is in the same region of phase space), and that is a very NON-random affair.

In fact if, for the sake of argument, we treat a mutation as a single amino acid alteration to your example protein of length 200, then evolution is choosing one new protein out of only 200 × 20 = 4,000 possibilities, not 10^{260}. And most of those 4,000 proteins are guaranteed to be roughly as useful as the one they replace, because they're all similar. (In practice the choice would be even smaller, since a single-codon mutation in DNA is only capable of changing any given amino acid into a small subset of the other 19). Your figures only apply if EVERY amino acid in the chain is mutated at once.

Natural selection IS more than a "mere bumbling along of random genetic events". It contains a ratchet mechanism, and that makes all the difference.

Gerald Schroeder
Steve, you wrote "your example protein of length 200, then evolution is choosing one new protein out of only 200 × 20 = 4,000 possibilities". I beg to differ. The 10^{260} is correct I believe. There are 200 spaces for amino acids and into each space any one of the 20 AA's can be placed. The number is indeed 20 times itself 200 times or 20^{200} which is approx 10^{260} To get a scale to that number it is worth recalling that the estimated number of basic particles in the observable universe is 10^{81}.

You wrote:

> "makes very minor adjustments to it.... every newly conceived creature starts out with a completely viable design [how'd we get to the viable design – it is a huge leap of faith that it all randomly came out of rocks

> & water and a few simple molecules even with nature's built in cleverness of self-organization], and mutation only makes very minor adjustments to it.... the mutations are small".

Small or big, other than the types of mutations that merely repeat already present organs, the mutations are either random or skewed.

Steve Grand

I don't want to get bogged down on this but it's crucially important to what's wrong with your main conclusions. You're guilty of thinking like a physicist, and abstracting the problem too far from its mechanism.

10^{260} is, as you say, the number of possible proteins, and therefore 1 in 10^{260} is the probability of obtaining any given protein on a single "shot-in-the-dark" trial by assembling 200 amino acids at random. But evolution does NOT make shots in the dark. Any single-point mutation to a protein can only change it in 4,000 different ways (this is a simplification, but it's in the right order of magnitude). In other words, there's a random chance of which one of the 200 amino acids will get replaced, and a random chance of which of the 20 (in practice far less than 20) amino acids it might be replaced by. 20×200 is 4,000. QED.

If more than one amino acid happens to mutate at once, then the number of possibilities certainly multiplies, but it reaches 10^{260} only if a mutation (which is a single event, remember, occurring during the transcription of a single DNA molecule) should affect every single amino acid in the chain. This simply couldn't happen in practice — it would require a strand of DNA to be replicated in such a way as to produce a second strand that bore no resemblance whatsoever to the first. The whole point of DNA is that it can copy itself with almost 100% accuracy. The probabilities you're insisting upon are equivalent to asking a mediaeval monk to copy a manuscript and finding that he made an error in every single word, and the copy bears no resemblance to the original. What really happens is more like the monk making an accurate copy with a few small mistakes. The copy is very similar to the original and therefore isn't a jumble of words chosen at random from the myriads of possibilities. In all probability the copy is readable and tells a similar story to the original. It's also highly possible that the new story is slightly "better" than the original. What it won't be is complete gibberish, as the vast majority of random word combinations are.

How we got to a sufficiently complex system to be self-replicating in the first place is indeed a leap of faith at present, but it's not an unjustified one by any means, and is a separate question. First you have to accept that once such an organism exists, evolution searches the design space extremely efficiently and all this talk of "mere randomness" is borne of applying the wrong kind of mathematics to the problem.

Gerald Schroeder
Steve, I agree with your logic completely one we get a sufficiently complex system to be self-replicating in the first place.

Steve Grand
Ok, that sounds like progress to me (and I really appreciate your having the guts to admit it). I think much of the "unbelievability" of evolution is a result of the linear mode of thought that arose around the time of Newton and has become deeply embedded in the human psyche. So much of the physical world can be successfully reduced to averages (for example we don't need to know the trajectory of every particle to work out the behavior of a gas as its temperature rises). But this just isn't true for biology, which is a fundamentally non-linear science. The tools of physics don't apply here.

Speaking of physics, I'd like to make a small point: You said (as have many physicists) that it's bizarre and illogical "that an item can be a wave and a particle simultaneously". Well it would be if it were true, but it isn't. These are just useful analogies, but like all analogies they break down eventually. Photons are just photons — they aren't waves OR particles. The two-slit experiment seems baffling because of a failure of our imagination. If we'd grown up in a quantum-scale world we'd probably find the macro world equally counter-intuitive. There's no "knowledge" required in the two-slit experiment — it just requires that a photon has a distribution in space. Dyson's (and Jeans') connection between this supposed "choice-making" and the (equally suppositional) belief in human free will is absurd – another example of applying the conceptual framework of physics where it doesn't belong. Physics is heavily soaked in the assumption that any property of the whole must also be a property of the parts (despite ample evidence to the contrary), and so Mind must be reducible to some smaller, subatomic "mindlets". But the universe is an inventive place. Completely new phenomena become possible once other, simpler phenomena exist to form a substrate. There is NO reason to suppose that all the phenomena that will ever be (including consciousness) must necessarily be present from the beginning, any more than we should assume that each word in a joke contains a little bit of humor.

Anyway, you mostly use this to explain why we shouldn't disparage ID, just because it is counter-intuitive, since we already accept that many things in nature are counter-intuitive. That's fair enough — we shouldn't. There are plenty of better reasons to disparage it! For instance it just pushes the problems back, instead of solving them. You may be unable to conceive of how lifeless molecules could end up with behaviors that we describe as living and/or conscious, but to invoke a universal intelligence or a creator just pushes the question back one step: how did that intelligence arise? How does it work? Of course you can always sidestep that by saying "God has always been there", or "it's beyond our capacity to know", but those are cop-outs. Whether we can understand it or not; whether God has or hasn't always been present; neither fact excuses him from HAVING an explanation. For instance, what guides God's intelligence? Does he (or did he, as prime mover) make his decisions for a reason or not? There are only two options. If he had no reason for acting as he did then his actions are, by definition, random. If he had reasons, then he was acting lawfully — there was a repeatable, explicable relationship between his actions and their causes. Either way he's not much of a god: he's either a random number generator or a mechanism, just like you and me. If he had reasons for acting as he did, then there must have been prior causes, so what, or who, caused them?

The idea of the universe being put together or guided by an intelligence doesn't make sense — in a far more profound way than the way that the double-slit experiment appears not to make sense. It is fraught with deep paradoxes and absurdities, and we DON'T NEED IT. To explain the universe this way makes the explanation more complex, not less. Explaining how life came to be is a very difficult problem, but because we lack the conceptual skills, not because we lack some magic ingredient. It's all part of the single most basic principle in the universe: things that persist tend to persist, while things that don't, won't. ANY universe will have properties that make some disturbances to its basic fabric more stable and persistent than others. In our universe, waves persist on water because of the way they propagate themselves, while other kinds of splash don't last very long. Photons do something similar in the electromagnetic field. Some shapes of electromagnetic disturbance resonate and hence persist for long periods in one place — we call them atoms. Some of those atoms engage in a dance which itself persists, even when the individual atoms move elsewhere —

we call this chemistry. Some of these dances are very persistent because they're autocatalytic. Some autocatalytic dances have a property that enables them to adjust over time and hence maintain themselves even in a changing environment — we call them alive. Some can adjust before an environmental stress actually arrives, because their mechanism predicts its occurrence — we call them intelligent. As the universe cools, more and more ways to persist become possible (cultures are one minor latecomer, for instance). Maybe one day a universal intelligence will also arise, in which all other intelligences are mere actors, just as our neurons become actors in a mind that they're not even aware of, or humans give rise to a mob with its own agenda. In which case, God could be said to have emerged. But to put him there first doesn't make any sense at all. It's the wrong way round.

Michael Behe

In his essay "The Right and Wrong of Intelligent Design", Professor Gerald Schroeder seems sympathetic to the idea of intelligent design, but finds fault with several of its current formulations, particularly with regard to biology. What's more, he appears to think that discussing some amazing aspects of nature, especially the physics of the universe, will allow most people to apprehend the profound design behind nature, without the need to explicitly and controversially declare that design is present, which one minor judge (a former director of a state Liquor Control Board), representing the central district of Pennsylvania in the United States, has disallowed in public schools in his jurisdiction. I think Schroeder's irenic stance is laudable, but unrealistic. Additionally, I think he misunderstands aspects of the intelligent design argument in biology, which lead him to overlook its similarity to his own views about physics. I will briefly address each of these points.

As have many persons before him, Professor Schroeder points out that many features of our universe are very finely tuned to allow for the existence of life:

> "Next we might study the laws of nature. We'd discern that, scientifically, the laws of nature are exactly correct for sustaining life. These include facts so commonly a part of our lives that we do not even marvel at them, such as there being three spatial dimensions; only one time dimension. The mass of an electron is 1837 times less than the mass of a proton and yet they have identical but opposite charges. Had the charge relationship been similar to gravity, then the charge of the electron would have been minute compared with that of a proton. To

> balance the charge of the nucleus, for each proton there would have to be 1837 orbiting electrons. Vast clouds of electrons would encompass each nucleus. The slightest input of energy would expel outer electrons, continuously ionizing the atom. The result: no molecular stability, no stable chemistry. No complex structures and hence essentially no chance for life."

Yet after citing these particular facts about the universe, out of very many that could be listed which are required for life, Schroeder immediately tells us how materialists respond:

> "*Scientific American*, the most widely read scientific journal, avidly materialist in its approach, acknowledges in its May 2003 issue, that if our universe is the only universe, then we live in a designer universe. Since this is patently unacceptable for a materialist view of existence, the authors tell us that there must be an infinite or near infinite number of universes, most of which will have less a favorable nature. With a vast number of putative universes, one of them had to be just right."

Yes, of course a materialist gloss will be put on whatever data science produces, especially if it would otherwise make us think we lived in a "designer universe." For example, in an article entitled "Our Universe: outrageous fortune" in a recent issue of *Nature*, the possibility that there exists 10^{500} universes was seen as an attractive alternative to fend off the hordes of design supporters:

> "Talk of a Universe fine-tuned for life has already attracted supporters of intelligent design, who claim that an intelligent force shaped evolution. If there's no way to tell whether the values of scientific constants are a coincidence, the movement's followers argue, then why not also consider them evidence of God's handiwork?" (Brumfiel 2006).

But the crux of the problem is this. To anyone but an expert in these fields, it is impossible to tell whether a "multiverse" is any more outrageous than a "Big Bang" in which all the material for 'just' our own observable universe was concentrated in a tiny space eons ago. If it takes only an infinitesimal space for one universe to begin, then, a layman might wonder, why not many, many universes? Since this materialistic point of view will be propounded by authority figures in science, and since it will strike lay-judges as "scientific" and not "religious", it will easily find its way into high school and college science textbooks. On the other hand, the idea that perhaps the universe reflects real design, will most likely not survive current judicial scrutiny, and so be precluded. Thus I am far less sanguine than Professor Schroeder that simply discussing the wonderful laws of nature will lead people to any particular conclusion. Rather, different interpretations of the data must be actively discussed, or most students will adopt the textbook attitude.

Examples from biology as evidence for intelligent design are, I think, more accessible to many more people than is esoteric physics. Anyone watching an eagle in flight or a whale spouting will directly apprehend the purposiveness that these creatures exemplify in their lives. As more is learned of biology even in high school classes, such examples multiply and become more compelling — more so, in my view, than examples from non-biological sciences. Thus discussions of design can start in biology with systems that, in the words of Richard Dawkins, "overwhelmingly impress us with the appearance of design, as if by a master watchmaker... ." (Dawkins 1986, p. 21). Students will more easily see why there is a question of design in biology than in physics.

Yet Professor Schroeder thinks there are problems with current design arguments in biology. He cites the fact that, while humans require a certain number of protein factors to clot their blood, dolphins get by with one less, and thinks this counts against the concept of irreducible complexity. He makes, however, the mistake of assuming that the protein "factors" that are present in humans and dolphins are identical. They aren't. And just as, say, two different bicycles may be made out of similar but nonidentical parts (different size nuts, bolts, screws, etc.) and each still be irreducible, so may a biochemical system such as the blood clotting cascade. Schroeder also thinks that because simpler forms of some systems exist in nature, it means that the more complex forms can be built gradually. But intelligent design does not preclude systems being built gradually; rather it rules out systems being built gradually *by unintelligent processes*. If the formation of a more complex system from a less complex system were guided by an intelligent agent, then that would be no more of a problem in biology than it would in mechanics for an engineer to deliberately make a series of more complex systems (say, bicycle, motorcycle, automobile, and so forth) using modified parts of less complex systems. But in neither case do the systems get built by chance.

In fact, in section 3 of his interesting essay Schroeder identifies *randomness* as the most suspect culprit in the Darwinian story of the unfolding of life on earth. In fact, that is the essence of intelligent design, that complex biological systems arose not randomly, but somehow through the planning of an intelligent agent. The target of the design argument is not this or that evolutionary story, but the limited role of randomness in the unfolding of life (I make this explicitly clear in my new book *The Edge of Evolution: The Search for the Limits of Darwinism.*) I think Professor Schroeder may have more in common with ID proponents than he realizes.

References

Brumfiel, G. (2006). Our Universe: outrageous fortune. Nature **439**:10-12.

Dawkins, R. (1986). *The Blind Watchmaker*. Norton: New York, p. 21.

Biography

Dr. Elia M. Leibowitz is a Professor of Astronomy at the Tel Aviv University; he obtained his Ph.D. in 1971 in Astronomy at Harvard University, Cambridge, MA. After a year as a post-doc researcher at Kitt Peak National Observatory in Tucson, Arizona, he joined the School of Physics and Astronomy of the Tel Aviv University in 1972, where he has been a faculty member ever since. Dr. Leibowitz was for many years the director of the Wise Observatory of the Tel Aviv University and served as the Chairman of the Department of Astronomy and Astrophysics of the University. He was a visiting associate at the University of California at Berkeley, visiting astronomer at the European Southern Observatory and a visiting lecturer in the University of Munster, Germany. Dr. Leibowitz has over two hundred publications in the literature in astronomy, as well as in other subject matters.

36
Is "Intelligent Design" Science? or, What Enables Human Beings to be Intelligent Designers?

ELIA M. LEIBOWITZ

School of Physics and Astronomy, Tel Aviv University
Tel Aviv 69978, Israel
elia@wise.tau.ac.il

1. The Court Ruling in Pennsylvania

On December 20, 2005, a federal judge in Pennsylvania, John E. Jones, made a mark in American culture history by ruling that it was unconstitutional for a Pennsylvania school district to present intelligent design (ID) as an alternative to evolution in high school biology courses. The Memorandum Opinion (MO) issued by the Judge says: "... while ID arguments may be true, a proposition on which the Court takes no position, ID is not science". Though legally binding within the district of his jurisdiction, the ruling itself cannot be considered an argument in the philosophical-conceptual debate over what Science is, and whether or not the basic ID claim is part of it. One may, of course, be influenced by the Judge's reasoning provided in the MO, but any judgment in that debate must rest on the merits of the argumentation and not on the legal status of the document. Consider another (hypothetical) legal case, in which a judge rules that a cellular telephone company should remove one of its antennas from a certain residential area, on the grounds that the intense radio waves emanating from the antenna constitute a health hazard. The phone company will have to remove the antenna, but no physicist or biologist will consider the ruling by itself as evidence that should be taken into account in the study of the very important and difficult field of the interaction of electromagnetic radiation with biological systems. Similarly, the fact that in the eyes of the law enforced in the Dover district of Pennsylvania ID is not Science, is not an argument for rejecting the philosophical claims of ID proponents that ID is Science.

In this paper I shall present a few arguments that in my opinion constitute the main reason why ID is not Science. Alternatively, this paper may be regarded as an attempt to deal with the first part of the judge's statement "… ID arguments may be true". "Truth" is an evasive concept which has escaped many attempts to characterize it with a universally accepted definition. Thus I do not wish to claim that ID arguments are not true but rather that they do not contribute anything of value to the useful knowledge of mankind. In particular they do not add anything to the capability of human beings to be intelligent designers within this world. They probably even hinder this remarkable human capability.

2. Einstein on Science

The question what is Science occupied the thoughts of many people throughout the ages. Today it constitutes a considerable part of the discipline of Philosophy of Science. Yet the meaning of the word "Science", and that of the related term "scientific investigation" remain fuzzy, allowing a wide range of freedom in attaching specific meaning to these words.

One way to approach the question of what is, or what should be considered Science, is to try to learn how individuals, unanimously accepted as outstanding scientists, viewed their own work as scientists or that of fellow scientists. This by itself is also not necessarily a compelling argument in the debate over the nature of Science, but it might help people make their own subjective choice, regarding the meaning of this term.

It may therefore be instructive to try to find out what Albert Einstein, a man recognized by most people as one of the greatest scientists of all times, thought about this question. One may learn how Einstein grasped the concept of Science from his evaluation of the work of another giant of science, Galileo Galilei. Here is what Einstein said about Galileo:

> "Pure logical thinking cannot yield us any knowledge of the empirical world; all ***knowledge of reality*** *starts from experience and ends in it.*

> Propositions arrived at by purely logical means are completely empty as regards reality. Because Galileo saw this, and particularly because he drummed it into the scientific world, he is the father of modern physics — indeed of modern science altogether" (Einstein, 1933).

Let us try to understand to some depth why Einstein saw in Galileo the father of modern science. This may give us important clues on what he considered as the fundamental characteristics of "modern science".

Our starting point is the fact that Einstein sees in Galileo the ***father*** of modern science. This means, among other things, that Galileo begot it. In other words, modern science is something that did not exist in the world until Galileo gave birth to it. Now, even if there are differences of opinion about the exact definition of Science, everybody agrees that Science is the endeavor of human beings to understand the real world and to gather knowledge about it. Human beings have been providing interpretations and explanations of observed natural phenomena from prehistoric times, preceding Galileo by thousands of years, and a great deal of knowledge about the world had been collected by mankind much before Galileo's birth in the 16th century. What is then the new element in Galileo's work or ideas that justifies entitling him the father of modern science?

3. The Meaning of the Galilean Revolution

The word Knowing (or the verb to know) denotes a certain state of mind of a person. The content to which that state of mind refers is knowledge. Knowing something is a subjective feeling that the person who has the knowledge is experiencing. It is associated with a certain configuration of the neural network in the brain of that person, which I shall call a brain pattern. When the pattern dissolves, the person looses the knowledge, i.e. forgets it.

Human beings are exposed to countless stimuli that reach their brains through the senses, excite neural activity and create there all sorts of patterns that manifest themselves as various subjective states of mind, e.g. anger, hope, suspicion, faith, fear, love, belief, etc. Like knowing, most other states of mind are concerned with an object, physical or abstract, to which the state of mind refers. One is angry with someone, fears something, loves somebody, etc.

The formation of a brain pattern that is associated with the subjective experience of knowing is in some sense a second order process. It involves a certain degree of selection on the part of the brain owner, by which he chooses which of the things within his awareness should be "elevated" to the rank of knowledge. Thus upon hearing a broadcast on a radio station, announcing that green Martians have landed in Baha California, one has to decide whether or not he now knows that Martian invasion of Earth has in fact commenced. This selection process is often being performed, perhaps most of the times (but not always), unconsciously.

Consider for example the knowledge of generations of people in the Western world about free fall. For more than 1500 years, most people knew that heavy objects fall faster than light ones. Today we may say that they so believed, but in fact their mental attitude had all the attributes of knowledge. If a student in 12th century Bologna had written otherwise in his physics exam he would have flunked. All people of that time possessed the brain pattern associated with the knowledge of that "fact". A prime generator of this universal brain pattern could be traced back to a few assertions made by Aristotle concerning free fall. For example, he said (Aristotle, 350 BC): "…the downward movement of a mass of gold or lead, or any other body endowed with weight, is quicker in proportion to its size". During the Middle Ages, Aristotle commanded an almost unlimited authority in the Western culture world. Due to this compelling authority, the mere fact that so said Aristotle was probably enough for people to adopt the content of his assertions as their own knowledge, i.e. for the creation in their brain the patterns of knowing it as reality.

There are many other ways by which patterns of knowledge are being created in the brains of people. The process is complicated and its precise nature and physical and biological mechanisms are unknown. But on a macroscopic psychological level we recognize that patterns of knowledge are often being formed out of fear or as an outcome of strong wish. Human beings make that conscious or unconscious selection of what they adopt as knowledge about reality on the basis of what they read in holy scriptures or in the newspaper, or based on what they hear from a respectable or a fearsome person. Spontaneous thoughts fueled by faith, tradition, or conditioning are also common inducers of brain patterns of knowledge. Continuous pouring of propaganda through the electronic media and the mass communication systems is probably becoming the major generator of brain patterns of knowledge of our time.

According to Einstein, Galileo's revolutionary contribution to human civilization was his realization of what are the ingredients of the process that brings about the creation of brain patterns of "***knowledge of reality***", which in this paper I shall call also "***useful***" or "***congruent***" knowledge, for reasons explained below.

4. The Grammar of Useful Knowledge

The ability to speak is one of the fundamental characteristics of the human intellect. It is based on the ability of the human species to produce a peculiar set of sounds from which units of syllables and words can be created. This ability, crucial for the emergence of a spoken language, is by no means sufficient. Language consists of sequences of syllables, but not every string of syllables, or even words, constitutes a meaningful utterance. To be meaningful, a string of words must belong to a certain subset of formations that are called grammar. According to some schools of thought, grammar can be formulated as a set of rules that distinguish between meaningful sentences and sequences of words that are devoid of any content. According to others, grammar takes the form of statistical distribution of different combinations of words and of chains of words. Be it as it may, there is hardly any doubt that for an utterance of words to be useful (e.g. as a means of communication) it must adhere to some sort of grammar.

One does not have to know explicitly the rules of the grammar of his or her language. One does not even have to know that grammar exists at all. In fact all people speak fluently their mother tongue long before they are aware that there are rules that they are following when they say or hear something meaningful. Many people remain in the state of not being aware of the existence of grammar for all of their lives, with no hindrance on their ability to speak.

Thus knowing the grammar, or that there is grammar, is not a necessary condition for speaking a language. However, the awareness that grammar exists and the understanding of its rules are of great intrinsic and practical value. There is hardly any doubt that the discovery that grammar underlies the meaning of sentences, and the formulation of its rules, brought about and is still bringing a great deal of benefit for the well being of mankind. This is especially so in the current computer era.

Another fundamental characteristic of the human species, akin perhaps but not identical to its speech ability, is the capacity to accumulate knowledge about the world, individually and collectively. Man has also the ability to transmit the brain patterns of knowledge from one generation to another. Human beings have been doing this for thousands of years, since prehistoric times when *Homo sapiens* started walking on this planet. However, up until Galileo's time, human beings did not realize that "useful" knowledge, or "knowledge of reality" as Einstein calls it, is what is associated with only a certain subset of all knowledge patterns in the brain. Galileo is the father of modern science because it was his discovery that there is a "grammar" of the creation of brain patterns of "useful" knowledge. Galileo was the first person who realized that for knowledge to be "**of reality**", the formation of the corresponding brain pattern must follow a certain procedure that adheres to a set of a few "grammatical" rules. Modern science, according to Einstein, is that grammar of useful knowledge. The basic rules of that grammar, summed up by Einstein as "*start[ing] from experience and end[ing] in it*", were also characterized, at least implicitly, by the example set by Galileo's research investigations.

5. "Knowledge of Reality"

By "reality" Einstein is no doubt referring to the physical world. If the subjective experience associated with a brain pattern of knowledge is to be considered knowledge of the physical world or some part of it, it must be "congruent" with the world. By this I mean that there must be a certain relation between what is known and some sensory signals that the knowing person is receiving from the world. In particular, knowledge is congruent with reality if it is predictive — raising in the mind expectations of future signals from the world, that are eventually being received. This relation of congruence can also be used for characterizing knowledge of past reality.

Consider, for example, the following two examples in which an attempt is being made to interpret events of the past, thus increasing our knowledge of past reality:

1. Napoleon invaded Russia, an unreasonable strategic move that was indeed a major factor in his eventual downfall. Why did he take that action that was clearly against his own interest? Answer: because his much beloved horse, Marengo, nagged him to do so for three full months.

2. Samson, the Biblical Hebrew Judge, in an utterly unreasonable move, revealed the secret of his unusual strength to his woman Delila, an act that brought about his capture and death. Why did he do such a thing that was clearly against his own interest? Answer: because his much beloved woman Delila nagged him to do so for three full months.

Most people would be willing to accept, at least in principle, the Delila story, as a satisfactory explanation of Samson's astonishingly unwise move. On the other hand, any person in his right mind will reject on the spot the Marengo explanation for Napoleon's unwise move.

The difference is obvious. The Delila story is acceptable since it is congruent with reality. For one, we know from observations that women can speak. Furthermore, there are other cases, not related to Samson, where women drove men to crazy actions, often leading to self destruction. No other example of a horse nagging his master to do something has been recoded in written history. Even if one has grounds to dismiss all other known explanations of Napoleon's move, with the result that Napoleon's action remains a mystery, one would not accept the horse story as an explanation. Note that this is not because we have any proof that Marengo did not used to have chats with Napoleon. The immediate dismissal of this explanation is an example of an application of the scientific grammar, which, according to Einstein, was Galileo's great discovery. One does not adopt the Marengo explanation into the bulk of his historical knowledge because it is not congruent with reality. No signal from the observable universe was ever received by anybody, that could be considered implied by this story.

In some cases, when, based on existing knowledge, we expect to detect some future sensory signals, our congruent knowledge has a predictive power.

All subjective experiences of knowing are associated with brain patterns. However, only knowledge that is the subject of patterns that are formed by the grammatical procedure can be considered "knowledge of reality".

6. Congruent Knowledge and Design Capability

Knowledge of reality in the Einstein/Galileo sense, the congruent knowledge discussed above, is the relevant type of knowledge for being able to design intelligently within this world. This is why it may also be called useful knowledge.

The adjective Intelligent in the idiom ID, indeed the noun design as well, indicate purposeful non-arbitrary activity. Human ID is intended to perform certain operations that will leave some marks in the physical world that serve some specific goal. We emphasize in particular that the plan is of operations to be performed on objects in the physical world. If the designer has a goal that he wants to achieve with his design, as is the case for all human designs, the designer must know ahead of the actual execution of his design how his operations affect the world, and how his actions will fulfill the goals that he has set upon himself. The designer must therefore possess brain patterns that raise the expectations for specific signals that will come from the world following specific operations executed on the world, prior to their execution. If the person does not have such knowledge one would call whatever he has on his mind neither a design nor intelligent. Any brain pattern of knowledge that does not increase the human ability to predict events in the physical world is entirely useless for a designer.

The differentiation between useful and useless knowledge, and the prescription for creating brain patterns of useful knowledge, are the essence of the Galilean scientific revolution. The consequences of this new recognition in the 17th century on the evolution of mankind were indeed profound. This is evident, or at least suggested, in the attached Figure 1, presenting a graph of the human world population as a function of time, from the year 10,000 B.C. up to present day (US Census, 2006). There is a dramatic change in the slope of the curve shown

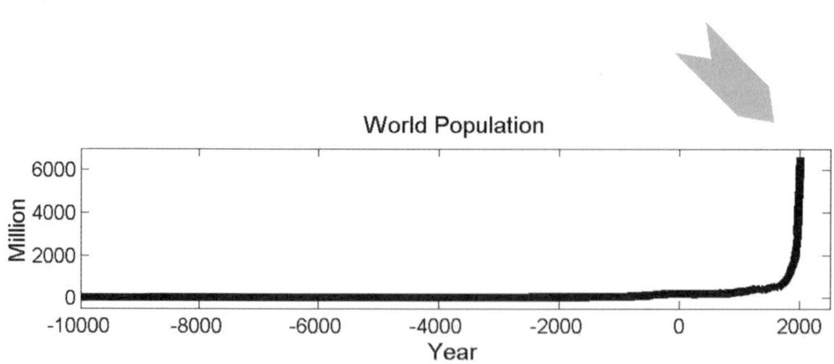

Figure 1: World (human) population from the year 10000 B.C. up to present time [from US and UN sources]

in the figure, from a nearly horizontal line, indicating very small population increase for many centuries, to a nearly vertical line denoting a sudden change to a huge growth rate. On the time scale of the curve shown in the figure, this unprecedented, revolutionary change was almost instantaneous. It occurred around the 18th century. The correlation with the time of the discovery by mankind of the grammar of useful knowledge, personified according to Einstein by the man Galileo Galilei, is striking. It strongly suggests a causal relation between these two historical developments.

7. Does ID "Start from experience and end in it"?

ID proponents, notable among them is Michael Behe, speak about chasms, gaps in the sequences of evolved biological systems, particular at the molecular level of the structure of some systems, that are, in principle, unbridgeable by the Darwinian mechanism of natural selection (Behe, 1999). ID proponents further say that the existence in nature of biological systems that apparently crossed these unbridgeable gaps is the mark of a design by some unspecified intelligent being. They argue that the "irreducible complexity" of such systems is an empirical fact from which the existence of a designer can be inferred (Behe, 1996).

It is this inference that according to Einstein does not belong to modern science, as fathered by Galileo. Even if one accepts Behe's claim that " there are no non-design explanations for the molecular machinery of life" (Behe, 2006), which is in fact rejected by most experts in the field, this cannot be regarded as positive identification of design in nature. The ID claim which is based solely on the lack of explanations for certain natural phenomena, does not "start from experience and end in it". According to Einstein at least, this very fact, in itself, disqualifies ID from being considered part of modern science.

The ID advocates make a leap from a brain pattern corresponding to non-comprehension of a natural phenomenon, to the brain pattern of knowledge. Such a leap is a direct continuation of the mental mechanism at the basis of the traditional deistic worldview that dominated human cultures for centuries. Throughout history, however, there were occasionally people who consciously or unconsciously used the Galilean grammar. They did not accept as useful knowledge an awareness associated with a brain pattern that rests solely on the inability to provide an explanation for a natural phenomenon. If it were not for these people, adhering to the Galilean grammar even without knowing it, mankind would still be ***knowing*** that the sun is the fiery chariot of the god Helios, and that Earth is resting on the back of the sacred cosmic tortoise.

But ID proponents such as William Dembski insist that the inference made by the ID movement (IDM) is indeed based on experience and observations. Dembski argues that the ID inference is part of science no less than the inference that the astronomers engaged in the SETI (Search for Extra Terrestrial Intelligence) program are hoping to make, if and when the signals from outer space that they are after are recorded by their radio telescopes (Dembski, 2005a).

Though the SETI project is deemed controversial by some scientists (Crichton, 2003), we assume here that SETI is part of "modern science". The inference made by SETI that should lead to knowledge of reality has been dramatized in the motion picture "Contact". In that movie, based on Carl Sagan's science fiction story, the heroine, a radio astronomer, deduces the existence of an extra-terrestrial intelligent civilization by carefully analyzing a radio signal recorded by her telescope. The signal is identified as a repeating sequence of all prime numbers between 2 and 101. Since no natural process known to us is producing such a sequence, the heroine concludes correctly, so goes the story, that it must be a message sent out by an intelligent being. This is in fact what the SETI project is set to discover (Korpela & Howard, 2008). Its proponents are hoping to intercept a radio signal that has a structure of similar "specific complexity", namely a pattern that is unexplainable as a product of a known natural process, and is both specified and complex. It is Dembski's claim that in a similar fashion, it is a valid scientific inference to regard terrestrial natural structures manifesting specific complexity as intelligently designed.

There are, however, two crucial differences between SETI and IDM that leave the first within the scientific domain, in the Einstein/Galileo's sense, and keep IDM out. The two are concerned with the reliance of the SETI inference on experience and the lack of it in the IDM case.

The starting point of SETI is our knowledge from experience that there exists in the universe a physical system, the human being, that produces prime numbers. We even know from experience how this intelligent being can represent these abstract concepts using known laws of physics. It does it by encoding them into some modulation of electromagnetic radio waves. SETI is looking for other systems in the universe that have these similar characteristics.

A second point of departure of SETI from experience, as required by the Galilean grammar, is some further knowledge that SETI researchers have about that human being. This includes also knowledge of the inner psychological world of the SETI researchers themselves. From personal introspection, or from the course in number theory that they took in the university, they know that prime numbers have a meaning. Without knowing beforehand that prime numbers have that special mathematical meaning of being primes, SETI astronomers would not be able to arrive at their hoped for conclusion even when faced with a prime number sequence.

During the decades of its operation, SETI has recorded myriads of sequences of radio signals. They consist of variations in the measured radio intensity from thousands of sources. The detailed causes of most of these variations are usually unknown to us. This lack of explanation might justifiably be termed "gaps" in our astronomical comprehension of the world. Nonetheless, nobody claims that these sequences of varying intensity are signatures of intelligence in space. The reason is that the human mind finds no meaning in these signals. In order for a sequence to become a candidate carrier of a mark of intelligence, someone has to identify within the observed sequence either the series of prime numbers again, or the first verses of the American national anthem, or some other structure that has some meaning for humans, ciphered with a simple numerical code. Meaning per se is not something that has an objective existence in nature. It is rather in the eyes of the beholder. Specification of humanistic meaning is therefore a necessary condition for regarding an object or a phenomenon as possessing "specified complexity" that is a mark of intelligent design.

Dembski (2005b) is attempting to convince us that the IDM identification of ID is essentially the same type of deduction made by every normal person who believes that he is not the only intelligent being on Earth, and that other human beings are intelligent as well. Though no one has a direct contact or access to another mind, we all know, or so we believe, that other persons are intelligent beings. We necessarily gain this knowledge only by observing the impact of these intelligences on the physical world.

One would hardly consider as a religious claim or as residing outside the scientific domain Tom's conclusion that his neighbor Jerry is an intelligent being. This conclusion is based on Tom's observations of the kennel that is being built in his neighbor's yard just before the rainy season. According to Dembski this inference demonstrates that deriving the existence of ID from observations in the physical world is an ordinary scientific procedure. The IDM derivation of the existence of an intelligent designer from observations in the physical world, must likewise not be dismissed as nonscientific.

However, here again the ID inference is lacking two fundamental components of Tom's inference that make it a scientific deduction. For one, in addition to Tom's observation of the kennel that is being built in the neighbor's yard, he has further information, gathered through observations in other parts of the world, in particular of himself as a physical entity, as well as from mental introspection. Tom observes, for example, that it is a physical body of an external structure, very similar to his, that is performing the act of building the kennel. Incidentally, the process of building the kennel by that other intelligent being seems to be obeying the same laws of nature that Tom knows to be effective in his backyard. Furthermore, Tom understands perfectly well what is the purpose of the physical action in the neighbor's yard. This is why he considers the growing kennel as an outcome of an intelligent design.

These two crucial elements of Tom's derivation, being acquainted with a physical system which Tom knows to be harboring intelligence, e.g. himself, and his understanding the purpose of the modification that he observes taking place in his neighbor's yard, are entirely missing from the IDM inference. Without any one of them, Tom would be unable to infer that there is an intelligent designer in his neighbor's yard. Indeed, if Tom saw his neighbor, the very same physical body, using the very same hammer that was used in the kennel construction, hitting with a very similar motion of his hand not the last nails in the doghouse but his 2 years old daughter, he would immediately call the police. Chances are that after due process Jerry would be sent to a mental hospital. No person in his right mind, including IDM proponents, will consider this last impact of Jerry on the physical world as a signal of ID. Note that if Jerry broke with his hammer the safe of the nearby bank he would have been sent to jail, not to a mental hospital. The difference is only in **our** understanding of the goals of the actions. We do not consider the hitting of the child a mark of intelligent design only because **we** don't see the purpose of it.

Thus it not true that intelligent design per se could be recognized without presupposing the purpose of the design. Therefore, whoever claims that he recognizes in the molecular structure of the cell a design, must provide us with information about what the design is for. No suggestion has yet been made by

ID advocates explaining in human terms the purpose of the design that they claim is governing evolution in the terrestrial biological world.

An answer that the goal of that designer is to enable the emergence of life, and/or of human beings would not be counted as satisfactory. It begs the question of what is the purpose of life. The SETI people who are looking for prime numbers or some other meaningful signal are exempted from answering the question of what is the purpose of life. They are playing inside the human intelligence domain. They can easily give a simple answer to the question of what is the purpose of the ET who is sending prime numbers out to space. The answer that they give is actually that the message is sent out intentionally as a search for and/or for the establishment of a Galactic communication network. Alternatively they will say that the message that they hope to intercept is a spillover of a working communication network, either within one and the same civilization or of an inter-Galactic chat that goes on between different civilizations.

SETI is not required to go any further, i.e. to explain why civilizations will want to communicate with one another. Here they are anchored in experience as required by the Galilean grammar. They recognize such a wish in the one intelligent civilization that they know by experience, namely our own. I suppose that nobody, including an ID proponent, will take as a serious answer the suggestion that the molecular structure within the cell serves a similar purpose of establishing a channel of communication between the designer and the human race. Thus the gap in the ID hypothesis, of not being able to suggest a purpose of the design work that has a meaning within human culture, seems to be wider and much less bridgeable than the alleged unbridgeable chasms in Darwin's theory of evolution.

8. Subjectivity vs. Objectivity

There is one other distinction between the scientific worldview and the one underlying the ID hypothesis. It has to do with the investigator's psychological attitude towards natural phenomena. Science is an attempt to understand the world in terms that are as objective as possible. In line with the Galilean, and here we might add also Copernican grammar, personal subjective impressions are removed from scientific explanations. Not so in the ID movement. Here, personal impressions and subjective feelings seem to be the driving forces of the ID hypothesis.

This distinction could hardly be expressed more succinctly than in Michael Behe's rebuttal of the MO of Judge John E. Jones, in the ID Dover case (Behe, 2006). In his comment No. 19 Behe quotes Richard Dawkins, described by him as an "adamantly Darwinian evolutionary biologist", saying that the appearance of design (in the biological world) is "overwhelming", and that "Biology is the study of complicated things that give the appearance of having been designed for a purpose". He quotes also other sources, including Darwin himself, proclaiming that there is a strong appearance of design in biology.

Behe then asks: "if .. the appearance [of design in biology] is just [a] 'completely subjective proposition' what is Darwin's theory explaining?" And the answer is plain and simple. Darwinism is explaining just that, namely, that the appearance of biological systems of having been designed, is in fact a completely subjective human impression. This is the essence of Darwin's great contribution. He suggested a mechanism by which the same natural forces and processes that create physical systems such as rocks, stars and volcanoes, can also generate systems that seem to the subjective human eye as being designed for purpose. Behe's question is not much different from a similar question that one may ask: if the rising and setting of the sun and stars are completely subjective propositions, what is the theory of the rotation of Earth explaining?

Thus Behe is right in saying that "Strong opponents and proponents of design both agree that biology appears designed." The difference between opponents and proponents is that the former, in line with the grammar of useful knowledge, explain this appearance out of biology. ID adherents adopt the subjective appearance and incorporate it as an important element in their interpretation of the world.

9. A Biblical View of ID

As a final point I cannot refrain from making a comment about an attitude towards IDM that can be found in the *Bible* itself. In the media it is often stated that the creationism movement and IDM are allies, ideologically as well as politically. This was also strongly echoed in the Memorandum Opinion of the Harrisburg, Pennsylvania court. Although the IDM proponents claim that their arguments and philosophy rest on scientific grounds, many of them do not deny that as human beings they do hold religious beliefs, which presumably include, at least for some of them, a belief in the sanctity of the *Bible*. It may therefore be of some interest to find out whether or not the *Holy Bible* itself expresses an opinion about the ID idea.

For this purpose one should read carefully the biblical creation story as told in Chapter 1 of the book of Genesis, ending in verses 1-3 of Chapter 2 of that book. According to King James Version, these 3 verses read:

> "1 Thus the heavens and the earth were **finished**, and **all the host of them**. 2 And on the seventh day God **ended his work which he had made**; and he **rested** on the seventh day from **all his work which he had made**. 3 And God blessed the seventh day, and sanctified it: because that in it he had **rested** from **all his work which God created and made**."

One cannot be blind to the insistent repetitions in these mere 3 verses of the script, of entire sentences as well as of key terms, expressing a complete termination. This becomes even more apparent when the abundance of seemingly synonymous wording in the description of the seventh day of creation, as marked in the above quotation, is contrasted with the frugality, bordering on miserliness, in the usage of words in the previous chapter, in which the creation of no less than the entire world is being described. There can hardly be any doubt that in these 3 verses the scripture is making an effort to send a message to the reader, indeed, to borrow Einstein's expression, to 'drum in' that message into the reader's mind. And the content of that message is also very clear. The sixth day of creation is the last day in which God is dealing with the world. The seventh day is the day at which God's entire activity of creation has come to a complete end. The day of Sabbath marks the permanent cessation of God's involvement with the physical world. From that day on God is not occupied anymore with making, creating or in doing anything concerning this world. After that day, no metaphysical entity is governing the evolution of the world in its entirety.

The message of these 3 verses should be heard loud and clear by any English speaking person. I should add in parenthesis that in the original Hebrew wording of the 3 verses, this message is even clearer. The last verse that underlines once more the absolute rest that God took from all physical doing, an explicit reference is made also towards the future. A somewhat more accurate translation of the Hebrew last verse would read: "because that in it he had rested from all his work which God would create", or alternatively "which God created and set to operate".

To be sure, there are many references in the *Bible* to deep involvements of God with earthly matters, the story of Noah's Great Flood is an immediate example. However, the writer of the Biblical creation story, that close literary unit of Chapter 1 and the first 3 verses of Chapter 2 in the book of Genesis, leaves very little room for doubt that he or she is not a proponent of the ID idea.

10. Summary

Humans are the only beings known to us in the universe capable of intelligent design. A few possible exceptions are some animals that do show traces of such capability, but all are manifesting orders of magnitude lower level of intelligence. For the ID movement, human intelligent design is the proto-type of this capacity in this world, and it serves as the model for the intelligence of the being which according to the ID hypothesis designed some of the phenomena of the terrestrial biosphere, if not the entire physical universe. ID proponents *know* that such a designer existed in the past and presumably still exists at present. Nobody can take this knowledge away from them against their will. However, from all that we know about the interaction of the human mind with the observable universe, this knowledge is devoid of any use.

It is ironic that the human design capability, which is a fundamental pillar supporting the ID hypothesis, is itself based on the refusal of humans to accept the ID hypothesis as an explanation of the world. Human beings are intelligent designers only because they do not accept as knowledge of reality any brain pattern associated with a wonder, i.e. a craving for a missing explanation for some natural phenomena. Mankind is a species of intelligent designers due to its members who accept as knowledge of reality only the subjects of brain patterns that have been established by "modern science", by the process that adheres to the Galilean grammar of knowledge, namely, "start from experience and end with it".

References

Aristotle, ~350 B.C. *De Caelo, Book IV* ch 2, 309^b 15", translated by R. P. Hardie and R. K. Gaye.

Behe M., 1996, *"Darwin's Black Box"*, The Free Press.

Behe M., 1999, "Experimental Support for Regarding Functional Classes of Proteins to be Highly Isolated from Each Other", The Weekly Standard, June 7, 1999, Books & Arts; Pg. 35 http://www.arn.org/docs/behe/mb_smu1992.htm

Behe M., 2006, "Whether Intelligent Design is Science", Center for Science and Culture, Discovery Institute http://www.discovery.org/scripts/viewDB/index.php?command=view&id=3218

Crichton M., 2003, "Aliens Cause Global Warming", Caltech Michelin Lecture http://www.crichton-official.com/speeches/index.html

Dembski W., 2005a, "In defense if ID", in: P. Clayton (ed.) *Oxford Handbook of Religious and Science.* http://www.designinference.com/

Dembski W., 2005b, *The Design Revolution*, Ch. 32 http://www.designinference.com/

Einstein A., 1933, "On the Method of Theoretical Physics", Herbert Spencer Lecture, Oxford University.

Korpela, E. and A. Howard, 2008. "Future SETI: technologies, techniques, and strategies." *Astrobiology* **8**(2): 384-388.

U.S. Census Bureau, 2006, "Historical Estimates of World Population", and "Total Midyear Population for the World: 1950-2050". http://www.census.gov/ipc/www/world.html

Dialogue

Arie Issar

The questions by Professor Elia Leibowitz's (PEL) of What is "truth"? and especially What is "science"? not to speak about his answers, bring us to be thankful to the initiators of the ID hypothesis for triggering such questions and answers and to editors of the present volume for giving us a stage to discuss them.

The daily occupation with the particular problems, which emerge in the course of personal research projects, of each of us, caused, naturally, the neglect of questioning established truths, and querying into the assertions of axioms. These "truths" and axioms are taken by many of us to be self evident.

In order to utilize to its optimum this opportunity of questioning basic truths and doubt established concepts, I would like to play the Intelligent Designer Devil's Advocate (IDDA), and elaborate further, on the question: What is science?*

I will play the IDA by following in the footsteps of PEL in his examination of this question, to the point where he concludes that the ID hypothesis is not science. More or less at this point I will have to abandon the defense of the ID side. The fact that I arrive at the end at the same conclusions as that as of PEL, but due to different reasons, may be because I am prejudiced by my own "creational" hypothesis, but this I will leave for the reader to decide (and at the same time give him an additional reason to read my article in this volume).

PEL's argument that ID is not science starts by quoting Galileo's maxim that: "all knowledge of reality starts from experience and ends in it. Propositions arrived at by purely logical means are completely empty as regards reality".

In my opinion, in this connection ID supporters may claim that their propositions are based on observations, i.e. experience. The books and articles, which were written by Professor Michael Behe, including that in the present

---

* I refrain from discussing the question: What is truth, because as PEL points: "'Truth' is an evasive concept which has escaped many attempts to characterize it with a universally accepted definition.".

volume, include many observations. A similar claim may be brought up when it comes to the knowledge of reality, which according to PEL is a process that brings about the creation of brain patterns of knowledge of reality. The supporters of ID will undoubtedly claim that they possess a brain pattern, which reflects reality according to their observations. Indeed William Dembski insists that the inference made by the ID movement (IDM) is based on experience and observations.[†]

Another reason of PEL for disqualifying IDM as being regarded a scientific theory is their argument that "there exist gaps in the sequences of evolved biological systems, particular at the molecular level of the structure of some systems, that are, in principle, unbridgeable by the Darwinian mechanism of natural selection". In this case I would have gone even further than IDM proponents and say that the appearance of life on our planet needs a crossing of an "unbridgeable gap", for which, there is no theory up to date, which has been proven by an experiment, and thus, especially in this case, the help of the hand of a supreme intelligent entity would have been very much appreciated. Once involved, His support in the more advanced stages of evolution cannot be offhand rejected.

As an IDA I also see a difficulty to explain phenomena that Michael Behe describes as "irreducible complexity", which according to his definition refers to a system (in our case a living system) "composed of several well-matched, interacting parts that contribute to the basic function, wherein the removal of any one of the parts causes the system to effectively cease functioning".

For myself, as a geologist I had a difficulty in accepting the opposing theory, i.e. that of the neo-Darwinists, who explain the formation of such complex systems by the conceptual model of "Pure chance, absolutely free but blind, at the very root of the stupendous edifice of evolution (Jacques Monod 1972)[‡], not to speak of the intuitional difficulty I had, of accepting that this complex system was assembled by a "blind watchmaker, blind because it does not see ahead, does not plan consequences, has no purpose in view" (Richard Dawkins 1986)[§].

[†] Dembski W., 2005a, "In defense of ID", in: P. Clayton (ed.) *Oxford Handbook of Religious and Science.* http://www.designinference.com

[‡] Monod, J. (1972) *Chance and Necessity*, Vintage Books, New York. p. 112.

[§] Dawkins, G. (1986) *The Blind Watchmaker*, Longman Scientific and Technical Publications p. 21.

Having said this, I would like, at this stage of discussion, to remove the robe of the advocate and put on that of the prosecutor, but not against ID people but against the ND, i.e. Neo-Darwinists, physicists and astronomers, who, to my opinion, failed to suggest a general theory to answer the questions Prof. Michael Behe and his group have been putting forward.

As a matter of fact my accusing finger is pointed towards all Darwinists, starting from Darwin and Wallace, to the most recent evolutionists, of having taken life, its survival and its evolution into more complex, and thus fragile systems, for granted. Reading quite a few of the comments, which were written during the one and half centuries since Darwin wrote "*On the Origin of Species by Means of Natural Selection, or The Preservation of Favored Races in the Struggle for Life*", seldom if at all, one finds the question: Why 'preservation' and why "the struggle for life" in the first place? The world of science took it for granted that once the physical conditions existed for the formation of life it would come into existence and start its 'struggle' with no need of some kind of force to keep this struggle for existence going on. I am not speaking of the supply of energy, which such a struggle needs. In the case of life on earth it is the conversion of solar energy via photosynthesis into oxygen and food, but what about the question of survival itself against all odds in a physical environment containing more anti than pro reasons for life to survive? No physical explanation, to date, suggests, nor was any experiment carried out, to prove that life is an inevitable phenomenon and must evolve into more complex systems in the physical process of evolution following the Big Bang.

Thus, starting with the question, already touched upon, about the appearance of the most primitive forms of life upon earth, one would expect the scientists investigating this question to come up with a description of a physical force, which in the first place was involved in this process and in the second place pushed it to evolve. In other words, after the Big Bang theory was accepted by the majority of the scientists, one would have expected a physical-mathematical conceptual model, making a divine command redundant, and to suggest in its place a Little Bang, which brought the RNA and DNA molecules into existence, to survive and to evolve into more complex forms of life, up to those participating in writing the present volume.

As mentioned above a mere 'random walk of mutations' explanation is difficult for me to accept, taking into account that even without asking the support of a Chimp to use a typewriter to write Othello or even a sonnet by Shakespeare, I am quite convinced that my PC would just produce a "mishmash" of signs rather than a simple sentence, not to speak an article, or for a change a book, by my just hitting the keyboard during all my lifetime. Thus 'mere random walk' would have brought a directionless or 'neutral evolution' (Kimura 1983)[**], rather than the progressive evolution which brought the emergence of the intelligent *Homo sapiens*.

Moreover, the neo-Darwinists who enthusiastically defend the 'throwing of dice' philosophy by arguing that evolution from 'primitive' to 'progressive' living beings gave an advantage in competition, must also admit that once a game of competition is going on, then one has to admit that there is a goal to reach and when there is a goal then there is a direction. It is claimed, thus, that neo-Darwinists are not aware of the fact that 'competition' admits some kind of a 'demon', which pushes living organisms to compete for survival, while letting material bodies comply with the Second Law of Thermodynamics (SLT) and decompose. There is no claim that living organisms violate the SLT, as they are open systems and export entropy, but why should they evolve beyond the level of the most primitive multiplying molecule, or at best, a microbe, or even a bluegreen alga? In the words of the physicist Paul Davies, my arguments, concerning the universe in general and the bio-world especially, appear in the following form:

> "I refer to the fact that the universe is progressing — through the steady growth of structure, organization and complexity — to ever more developed and elaborate states of matter and energy. This unidirectional advance we might call the optimistic arrow, as opposed to the pessimistic arrow of the Second Law" (Davies 1988).[††]

That a new non-conventional answer (of course not necessarily ID) is required to answer my question was suggested by (Schrödinger, 1969)[‡‡], who wrote:

[**]Kimura, M.(1983), *The Neutral Theory of Molecular Evolution*, Cambridge University Press.

[††]Davies, P.C.W., (1988), The Cosmic Blueprint, New Discoveries in Nature's Creative Ability to Order the Universe, Simon and Schuster, New York p. 20

[‡‡]Schrödinger, E. (1967), *My View of the World*, Cambridge Press, Cambridge.

> "We must therefore not be discouraged by the difficulty of interpreting life by the ordinary laws of physics, … We must be prepared to find a new type of physical law prevailing in it."

In his times it was the argument against movement requiring the help of super-physical forces like 'elan vital' [vital force], phrased by Henry Bergson, while today these are movements backing up the Intelligent Designer hypothesis. My interpretation of Schrödinger's suggestion is the urge to look for a non-conventional, scientific law to answer the questions Michael Behe and has put forward.

My claim is that while physicists came up with the idea of an electromagnetic field of force to explain bonds between atoms and their components, the nuclear strong force and the weak force to explain the survival and processes of decay inside atoms, why did they fail to answer Schrodinger's challenge to suggest a physical force coming into play when more complex self-reproducing molecules are observed?

In order to emphasize the failure of conventional science to answer the questions that ID scientists (I insist on calling them scientists, until I claim the contrary) make, I would like again to point my prosecutor finger at the '*The Blind Watchmaker*', by Dawkins and the failure of its author to look for a new field of force to explain progressing evolution.

While explaining how only chance and nothing but chance, brought evolution from the first molecule of RNA to *Homo sapiens*, Dawkins gives the following example:

> "If you walk up and down a pebbly beach, you will notice that the pebbles are not arranged at random. The smaller pebbles typically tend to be found in segregated zones running along the length of the beach, the larger ones in different zones or stripes. The pebbles have been sorted, arranged selected."

Dawkins goes on telling us that a primitive tribe living there will attribute the act of sorting to a Great Spirit in the sky, while we, with a smile of superiority on our lips know "…that the arranging was really done by the blind forces of physics, in this case the action of waves. The waves have no forces and no intentions, no tidy minds, no mind at all… . A small amount of order has come out of disorder, and no mind planned it."

What Dawkins has failed to observe is that the arrangement of the pebbles is due to the combination between the force of impact by the waves, and the mass of the pebbles which is a function of the field of gravity. That this field is involved also in creating order in living systems, like in the case of the development of a fertilized egg to an embryo, one can learn from the development of the bilateral symmetry of chick embryos (Kochav and Eyal-Giladi 1971)[§§].

At this stage of discussion I can not refrain from offering my answer to the question of evolving complexity in the bio-world, by suggesting entropy and negentropy as two opposing poles of a field of force, which pulls every structure to become disordered, while once energy is invested, either solar or hydrothermal (as in the case of the submarine hot vents) it pushes composite life forms to survive and even evolve. I will not go on with defending my model, which I brought up here only in order to demonstrate the failure of conventional science to adopt Schrodinger's advice and look for "a new type of physical law". The reader who is interested in more details is referred to my article 'The evolution of intelligence in the bio-world' in this volume as well as to my book "*From Primeval Chaos to Infinite Intelligence*".

Mentioning intelligence, I am afraid I may join again the ID proponents, in their dissatisfaction with Darwin's theory but from another point of view. My claim is that the theory of Darwin failed to consider the evolution of intelligence, and thus restricted itself only to the evolution of form, which requires only space-time dimensions. This restriction is to be regretted, because Darwin was the courageous pioneer who questioned the Judeo-Christian doctrine that humans had been created by God in His image and thus are totally different from the animal kingdom and cannot be its descendents. My claim, which I defend in the above mentioned article in the present book, is that in addition to the evolution of form there is also evolution of intelligence in the bio-world. But in order to define intelligence and measure its evolution a new dimension, i.e. that of Information, has to be added to space-time.

[§§] Kochav, S. and Eyal-Giladi, H. (1971), Bilateral symmetry in chick embryo, determination by gravity, *Science*, 171, pp. 1027-1029.

Thus my suggestion to answer Schrodinger's quest for "a physical law" is by suggesting a new dimension, i.e. that of Information, along which an already well known law, i.e. the SLT, acts. In other words I suggest that the SLT pronounces a general field of force pulling (or pushing) towards a negentropy pole, in a space-time-information continuum. Such a force makes an Intelligent Designer redundant and leads me to agree with PEL in disqualifying ID from being considered part of modern science.

In this connection I join his argument that: "The ID advocates make a leap from a brain pattern corresponding to noncomprehension of a natural phenomenon, to the brain pattern of knowledge". My agreement with PEL is because I maintain that the evolution of science should be regarded as the continuous and endless construction of an algorithm, which tries to interconnect observations by logical arguments, either grammatical or mathematical, in order to avoid the need of having a non-natural force (i.e. Divine), involved. Once such a force is allowed it is the end of science and the beginning of religion. As I can see it the ID hypothesis indeed cuts short this algorithm and ends it once for all with the "Divine Intelligent Designer". I thus find PEL justified in saying that: "The ID explanation is a direct continuation of the mental mechanism at the basis of the traditional deistic worldview that dominated human cultures for centuries".

I hope nobody will use the same argument, namely of cutting short the scientific algorithm, by claiming that my suggestion that "Information is a dimension and entropy is a field of force" does the same cutting short. On the contrary, my claim is that the introduction of a new dimension opens new vista of research and for this purpose I allow myself to quote a physicist, namely John Archibald Wheeler saying:

"If we're ever going to find an element of nature that explains space and time, we surely have to find something that is deeper than space and time — something that itself has no localization in space and time. The amazing feature of the elementary quantum phenomenon — the Great Smoky Dragon — is exactly this. It is indeed something of a pure knowledge-theoretical character, an atom of information which has no localization in between the point of entry and the point of registration. This is the significance of the delayed-choice experiment" (John Archibald Wheeler in Davies and Brown, 1986)[***].

[***] Davies, P.C.W. and Brown, J.R. (1986), The Ghost in the Atom, Cambridge University Press, Cambridge. p. 66.

Biography

Dr. Cihat Gündoğdu is a medical doctor specialized in family medicine. His main interest is in "Creation Science". He graduated from The School of Medicine, University of Marmara in Istanbul in 1992. He completed the Family Medicine Residency Program in Haydarpasa Numune Educational and Research State Hospital in Istanbul in 1999. As a lecturer, he has given numerous lectures in several universities and high schools in USA (Columbia University), Canada (McGill University, Concordia University), England (10 universities including Cambridge, Oxford and London School of Economics), Wales, Scotland, Northern Ireland, Ireland, Malaysia (International University of Islam), Holland (Erasmus University Faculty of Medicine), Belgium, Japan (6 different campuses including School of Medicine, University of Tokyo), Australia and more than 150 conferences in Turkey, mainly based on the scientific evidence invalidating Darwinism. He had many articles published on several journals dealing with the scientific facts of creation and has been on TV programs as well. He works with The Science Research Foundation of Turkey. He says "it is the mission of scientists to tell people that what we observe in the scientific field is the clear evidence of 'Creation'".

37
The Collapse of the Theory of Evolution

CIHAT GÜNDOĞDU

Marmara University School of Medicine, Department of Family Medicine
Haydarpasa Numune Educational and Research State Hospital, Istanbul, Turkey
CGundogdu@scienceresearchfoundation.org

1. Introduction

How did life emerge? There are only two possible explanations. One of them is the theory of evolution, which claims that life on Earth began by chance, and that therefore, all the species we see around us are the results of mere coincidence. The other explanation is creation. No other, third explanation or proposal can be put forward. This fact is now accepted by even the leading evolutionists in the world, who state this quite frankly. Douglas Futuyma, a prominent American biologist admits as much:

> "Organisms either appeared on the earth fully developed or they did not. If they did not, they must have developed from preexisting species by some process of modification. If they did appear in a fully developed state, they must indeed have been created by some omnipotent intelligence" (Futuyma, 1983, p. 197).

In fact, Futuyma's words underline a very important truth. He writes that when we look at life on Earth, if we see that life emerges all of a sudden, in its complete and perfect forms, then we have to admit that life was created, and is not a result of chance. Even Futuyma—who is himself known as an evolutionist—accepts this as a fact. So as soon as evolutionary theory is proven to be invalid, then creation is the only explanation left.

We will now evaluate the suggestions regarding evolution and test them scientifically. To account for life as the result of an evolutionary process, evolutionists must first answer the following questions:

1. How did first living thing on Earth come to be? How did the first living cell form?
2. What are the mechanisms responsible for living things evolving into one another?
3. If all the species on Earth are the result of evolution, then there must be a vast amount of fossil evidence for this. Are there any such fossils?

Let's evaluate these questions one by one. The first is: How did the first living thing come into being? Or in other words, how can a single cell come into existence by itself?

2. Incredible Complexity of the Cell

The living cell's complex structure was unknown in Darwin's day. At the time, evolutionists thought that ascribing life to coincidences and natural conditions was convincing enough. Charles Darwin had postulated that the first cell could easily have formed "in some warm little pond" (Darwin, 1959, pp. 202-203). One of Darwin's great supporters, Thomas Henry Huxley, examined under the microscope some sludge dredged up from the ocean floor and claimed that this was an inanimate matter that turned into living matter. This so-called "mud that comes to life," known as *Bathybius haeckelii* (named after the German Darwinist biologist Ernst Haeckel), is an indication of just how simple a process the founders of the theory of evolution thought life to be.

But the technology of the 20th century delved into the tiniest particles of life, revealing that the cell is easily the most complex system mankind has yet confronted. Today we know that any single cell contains miniature power stations that produce the energy for the cell to use, tiny chemical factories manufacturing the enzymes and hormones essential for life, a databank where all the necessary information about all the substances to be produced is genetically recorded, complex transport systems and pipelines for carrying raw materials and products from one place to another, advanced laboratories and refineries for breaking down external raw materials into their useable components, and specialized cell membrane proteins to control the inflowing and outgoing materials. And these constitute only a small part of this incredibly complex system.

Even the simplest of cells is so sophisticated that even the high levels of technology attained today cannot produce one. None of the efforts to create a cell artificially has ever met with success; and indeed, all attempts to do so have been abandoned.

The theory of evolution claims that this system—which science, with all the intelligence, knowledge and technology at its disposal, cannot succeed in reproducing—came into existence by chance under the conditions of the primordial Earth. Actually, the probability of forming a cell by chance is about the same as that of an explosion in a printing plant producing a perfect copy of a book.

The English mathematician and astronomer Sir Fred Hoyle made a similar comparison. Although himself an evolutionist, Hoyle wrote in his *The Intelligent Universe* that the chance that higher life forms might have emerged in this way is comparable to the chance that a tornado might assemble a Boeing 747 from the materials in a junkyard (Hoyle, 1983, p. 19). This analogy dramatizes that the cell cannot have come into being by chance and therefore, must definitely have been created.

A living cell maintains itself with the harmonious cooperation of the many organelles within it. If only one of these organelles fails to function, the cell cannot remain alive. The cell does not have the opportunity to wait for unconscious mechanisms like "natural selection" or "mutation" to permit it to develop organelles. Thus, the first cell on Earth was necessarily a complete one that possessed all the required organelles and functions, and this cell must definitely had to have been "pre-assembled"—in short, created.

So much for the cell as a whole, but evolution even fails to account for its building blocks. Under natural conditions, the formation of just one single protein out of the thousands of complex protein molecules making up the cell is impossible.

For instance, an average-sized protein molecule is composed of 288 amino acids. It contains twelve different types of amino acids that can be arranged in 10^{300} different ways. (This is an astronomically huge number, consisting of 1 followed by 300 zeros.) Of all of these possible sequences, only one forms the desired protein molecule. The rest of the possible amino-acid chains are either totally useless, or else potentially harmful to a living cell. In other words, the probability of only one viable protein molecule forming is only 1 in 10^{300}. The probability of this actually occurring is practically nil. (In practice, probabilities smaller than 1 over 10^{50} are thought to be of zero probability).

In the 1970s Hoyle calculated the mathematical probability of the coincidental formation of only the 2,000 types of proteins found in a single amoeba. (There are some 200,000 different types of proteins in a human cell.) The figure that he calculated was 1 over $10^{40,000}$—an incredible number obtained by putting 40,000 zeros after the 1 (Eastman&Missler, 1996, p. 61).

If the coincidental formation of even one of these proteins is impossible, it is billions of times "more impossible" for some one million of those proteins to assemble themselves by chance to form a complete human cell. Moreover, by no means does a cell consist of a mere assembly of proteins. In addition to those proteins, a cell also includes nucleic acids, carbohydrates, lipids, vitamins, and many other chemicals such as electrolytes arranged in specific proportions, and equilibrium, which all reveal that it is created perfectly—in terms of both structure and function. Each of these elements functions as a building block or co-molecule in various organelles.

On to our second question: Are there any mechanisms in nature that could cause living things to evolve?

3. Imaginary Mechanisms of Evolution

The neo-Darwinist model, generally taken as the mainstream theory of evolution today, argues that life must have evolved through the two mechanisms of natural selection and mutation. The theory basically assumes that natural selection and mutation are complementary. The origin of evolutionary modifications lies in random mutations taking place in the genetic structures of living things. Anatomical and behavioral traits determined by mutations are then selected by the mechanism of natural selection, and by this means, living things—and eventually, genetically unique species—evolve. To test out such a hypothesis, let's see how these mechanisms work.

3.1. *Natural Selection*

The process of natural selection was familiar to biologists before Darwin and defined as a mechanism that kept species unchanging without being corrupted. Darwin was the first to assert that this process wielded evolutionary power, and he then erected his entire theory on the foundation of this assertion. The title he gave to his book indicates the basis of his theory: *The Origin of Species, by means of Natural Selection...*

Natural selection is founded on the assumption that there is a constant struggle for survival in nature. It favors organisms with traits that best enable them to cope with the challenges and pressures exerted by their environment. At the end of this winnowing struggle, the strongest and those most suited to natural conditions survive.

For example, in a locale where winters are long and the ground remains covered with snow for extended periods, white-furred rabbits will be better camouflaged and fall prey to predators less often than darker-colored ones. This gives them a greater chance of surviving to reproduce. This being the case, the proportion of white-furred individuals in the rabbit population will keep rising, while the relative number of darker-furred rabbits will decline.

To give another example, in a herd of zebras that must constantly flee from leopards, those which run fastest will survive while the others will be caught and killed. Since fast runners in every generation will survive, in time the herd will come to consist of only the fastest runners. However, no matter how long this process continues, it will not transform those zebras into another species. The weaker zebras will be eliminated, and the stronger will survive. But since no alterations in their genetic data take place, no transformation of the species occurs. Despite the continuous processes of selection, zebras continue to exist—as zebras.

Since Darwin's time, not a single shred of evidence has been put forward to show that natural selection causes living things to evolve. Colin Patterson, the senior paleontologist of the British Museum of Natural History in London and a prominent evolutionist, stresses that natural selection has never been observed to make living things evolve:

> "No one has ever produced a species by mechanisms of natural selection. No one has ever gotten near it and most of the current argument in neo-Darwinism is about this question: how a species originates" (Patterson, 1982).

The greatest problem for the theory of evolution is that natural selection cannot enable new organs or traits to emerge in living things. Natural selection cannot develop new genetic data; therefore, it cannot be used to account for the emergence of distinct new species. Stephen Jay Gould, one of the theory of evolution's most prominent supporters, said that evolutionists are seeking from natural selection a power it does not possess:

> "The essence of Darwinism lies in a single phrase: natural selection is the creative force of evolutionary change. No one denies that natural selection will play a negative role in eliminating the unfit. Darwinian theories require that it create the fit as well" (Gould, 1977, p. 28).

Evolutionists also accept that natural selection is not a conscious mechanism capable of planning or foresight, but an unconscious, undirected process. In his book *The Blind Watchmaker*, Richard Dawkins, one of the most passionate proponents of the theory of evolution, defines natural selection in these terms:

> "Natural selection, the blind, unconscious, automatic process which Darwin discovered, and which we now know is the explanation for the existence and apparently purposeful form of all life, has no purpose in mind. It has no mind and no mind's eye. It does not plan for the future. It has no vision, no foresight, no sight at all. If it can be said to play the role of watchmaker in nature, it is the blind watchmaker" (Dawkins, 1996, p. 5).

Natural selection has no power to improve genetic material. It cannot change DNA by any means. It can only eliminate the defective, weak or sick individuals in a given population. It cannot produce new genetic information, new organs, or new species. That is, it cannot make anything evolve. Darwin accepted this reality by admitting that "Natural selection can do nothing until favourable variations chance to occur" (Darwin, 2004, p. 150).

This is why neo-Darwinism has had to add "mutations" alongside natural selection as the cause of favorable variations. However, as we shall see, mutations can only be the cause of "unfavorable" changes.

3.2. *Mutations*

Mutations are defined as breaks or replacements occurring in the DNA molecule, found in the nuclei of the cells of every living organism and which contains all its genetic information. Such breaks or replacements are the result of external effects such as radiation or chemical action. Therefore, every mutation is an "accident," and either damages the nucleotides making up the DNA or changes their locations. Most of the time, they cause so much damage and modification that the cell cannot repair them.

Mutation acts on the complex information in the DNA in a random and unconscious manner, harming the DNA and therefore, the organism as a whole. At best, it may have no effect at all. However, mutations can never add any new information to DNA, and do not improve the organism in any way. Not a single instance of this has ever been observed.

The direct effect of mutations is harmful. The changes effected by mutations can only be like those experienced by the survivors of Hiroshima, Nagasaki, and Chernobyl: that is, death, disability, and birth defects. The reason for this is very simple: DNA has a very complex structure, and random effects can only damage it. As biologist B.G. Ranganathan states:

> "First, genuine mutations are very rare in nature. Secondly, most mutations are harmful since they are random, rather than orderly changes in the structure of genes; any random change in a highly ordered system will be for the worse, not for the better. For example, if an earthquake were to shake a highly ordered structure such as a building, there would be a random change in the framework of the building, which, in all probability, would not be an improvement" (Ranganathan, 1988).

In recent years, thousands of diseases have been found to be caused by genetic mutations. Genetics textbooks list some 4,500 different genetic diseases, including Down's syndrome, sickle-cell anemia, dwarfism, mental retardation, cystic fibrosis, and certain forms of cancer.

Not surprisingly, no useful mutation has ever been observed. All genetic mutations have proved to be harmful. All efforts to generate a "useful mutation" have resulted in failure. For decades, evolutionists carried out many experiments on fruit flies, hoping to observe any single beneficial mutation. These insects reproduce very rapidly, so any genetic changes would show up quickly. Generations upon generation of these flies were mutated, yet no useful mutation was ever observed. In all the thousands of fly-breeding experiments carried out for more than fifty years, no distinctly new species was ever seen to emerge… nor even a new enzyme.

Pierre-Paul Grassé, former president of the French Academy of Sciences and editor of the 28-volume *Traité de Zoologie*, likened mutations to spelling mistakes, and added that they could never give rise to evolution:

> "Mutations, in time, occur incoherently. They are not complementary to one another, nor are they cumulative in successive generations toward a given direction. They modify what preexists, but they do so in disorder, no matter how… . As soon as some disorder, even slight, appears in an organized being, sickness, then death follows. There is no possible compromise between the phenomenon of life and anarchy" (Grassé, 1977, pp. 97-98).

The evolutionist science writer Gordon Taylor writes thus:

> "It is a striking, but not much mentioned fact that, though geneticists have been breeding fruit flies for sixty years or more in labs all round the world—flies which produce a new generation every eleven days—they have never yet seen the emergence of a new species or even a new enzyme" (Taylor, 1983, p. 48).

Another researcher, Michael Pitman, comments on the failure of the experiments carried out on fruit flies:

> "Morgan, Goldschmidt, Muller, and other geneticists have subjected generations of fruit flies to extreme conditions of heat, cold, light, dark, and treatment by chemicals and radiation. All sorts of mutations,

> practically all trivial or positively deleterious, have been produced. Man-made evolution? Not really: Few of the geneticists' monsters could have survived outside the bottles they were bred in. In practice mutants die, are sterile, or tend to revert to the wild type. (Pitman, 1984, p. 70)

To summarize, there are two main properties of mutations:

1) The direct effect of mutations is harmful: Since they occur as accidents, they almost always damage the living organism.
2) Mutations add no new information to an organism's DNA: The genetic information is either destroyed, or its position in the sequence is shuffled or changed. Thus, mutations cannot lead to a new organ or a new trait, much less to the emergence of a new species. They only cause structural abnormalities like a leg sticking out of the head, or an ear out of the abdomen.

In order for a mutation to be passed along to the subsequent generation, it must have taken place in the organism's reproductive germ cells: A random change that occurs in a cell or organ of the body only cannot be transferred to the next generation. I.L. Cohen, a member of the New York Academy of Sciences, writes that:

> "To propose and argue that mutations even in tandem with 'natural selection' are the root-causes for 6,000,000 viable, enormously complex species, is to mock logic, deny the weight of evidence, and reject the fundamentals of mathematical probability" (Cohen, 1984, p. 81).

His statement unmasks the absurdity of those who believe that all life forms are the work of mutation and natural selection. Briefly, it is impossible for living beings to have evolved, because there exists no mechanism in nature that might cause them to do so. This agrees with the evidence of the fossil record, which demonstrates that this scenario is far removed from reality.

The third question is whether any fossil evidence can prove the evolutionary process.

4. The True Origin of Species

The theory of evolution claims that all living things descend from a common ancestor. According to the theory, the origination of such diverse living species took place through minor and successive changes over a very long period of time.

When Darwin's *The Origin of Species* was published in 1859, it was widely believed that he had advanced a theory that could account for the extraordinary variety of living things. He had observed that there were different variations within the same species.

However, Darwin's assumption about "the origin of species" was not actually able to explain their origin at all. Thanks to more recent developments in genetics, it is now understood that increases in variety within one species can never lead to the emergence of another, entirely new species. What Darwin believed to be "evolution," was actually variation.

5. Variation

Variation, a term used in genetics, refers to a genetic phenomenon that causes the individuals or groups of a certain type or species to possess different characteristics from one another. Variation does not constitute evidence for evolution, because variations are simply the outcomes of different combinations of already existing genetic information, and they do not represent any new characteristics added to the genetic information. The important question for the theory of evolution, however, is how brand-new genetic information to make a brand-new species with markedly different characteristics could come about.

Variation always takes place within the limits of existing genetic information. In the science of genetics, this limit is called the "gene pool." For example, as a result of variation, varieties that have relatively longer tails or shorter legs may appear in a certain species of reptile, since information for both long-legged and short-legged forms may exist in that species' gene pool. However, variations do not transform reptiles into birds by adding wings or feathers to them, or by changing their metabolism from cold-blooded to warm. Such a change would require an increase in the living thing's genetic information, which is certainly not possible through variations.

If a geographic obstacle arises between members of a species—in other words, if they are isolated from one another—then it is very probable that different variations will begin to predominate in the two groups that are now separated from each other. Despite being originally of the same species, such variations with specific morphological differences between them are called "sub-species." (For convenience's sake, we can call them as "variation A" and "variation B.")

Evolutionary biologists were forced to distinguish between variation within species and the formation of new ones, and to propose separate mechanisms for these entirely different phenomena.

6. The "Microevolution" Misconception

The theory of evolution claims that living species evolved from one another, from the earlier and simpler to the later and more complex, completely by chance. For the theory to be taken seriously, therefore, it needs to posit a mechanism for increasing genetic information. It must be able to explain how living things lacking eyes, ears, hearts, lungs, wings, feet, and other organs and systems came to acquire them, and where the genetic information to encode such systems and organs came from. A mechanism that divides an already-existing species into two groups, each of which undergoes a loss of genetic information, clearly has nothing to do with this scenario.

This point is, in fact, accepted by evolutionists. For this reason, they define variations within a species and instances of speciation by division of a population into two as "microevolution"—a term used to describe variations that develop within an already existing species. Yet the inclusion of the term "evolution" in this description is a deliberate deception. There is no evolutionary process here at all, not even a "micro" one. This process merely distributes genetic information already existing within the genetic pool among a different combination of individuals.

The theory of evolution proposes that living things can develop and accumulate new genetic data by the mechanisms of mutation and natural selection. However, since variations can never create new genetic information, they're thus unable to bring about "evolution." Giving variations the name of "microevolution" betrays an ideological preference on the part of evolutionary biologists.

In fact, even evolutionist experts now accept that the variations they call "microevolution" cannot lead to new classes of living things—in other words, to "macroevolution." In a 1996 article in the leading journal *Developmental Biology*, the evolutionary biologists S.F. Gilbert, J.M. Opitz, and R.A. Raff explained the matter this way:

> "The Modern Synthesis is a remarkable achievement. However, starting in the 1970s, many biologists began questioning its adequacy in explaining evolution. Genetics might be adequate for explaining microevolution, but microevolutionary changes in gene frequency were not seen as able to turn a reptile into a mammal or to convert a fish into an amphibian. Microevolution looks at adaptations that concern only the survival of the fittest, not the arrival of the fittest. As Goodwin (1995) points out, 'the origin of species—Darwin's problem—remains unsolved'" (Gilbert and et al., 1996, p. 361).

We can sum up the situation like this: variations—which Darwinism has seen as "evidence" of evolution for roughly a hundred years, actually have nothing to do with "the origin of species." Cattle can be bred for millions of years, and different breeds of cattle may well emerge. But cows and bulls can never turn into a different species—giraffes or elephants, for instance.

In the same way, the differently-billed finches that Darwin saw on the Galapagos Islands are another example of variation that is no evidence for "evolution." Recent observations have revealed that those finches did not undergo an unlimited variation, as Darwin's theory presupposed. Moreover, most of the types of finches that Darwin thought represented 14 distinct species actually interbred with one another, which means that they were only variations belonging to the same species. Scientific observation shows that the finches' beaks, which have been mythicized in almost all evolutionist texts, are in fact just an example of variation and therefore, constitute no evidence for the theory of evolution. Peter and Rosemary Grant, who spent years observing the finch varieties in the Galapagos Islands looking for evidence for Darwinistic evolution, were forced to conclude that "the population, subjected to natural selection, is oscillating back and forth," a fact which implied that no "evolution" leading to the emergence of new traits ever takes place there (Grant, 1991, pp. 82-87).

7. The Origin of Species in the Fossil Record

Evolutionists assert that each species on Earth arose from a single common ancestor through minor changes. In other words, their theory considers life as a continuous phenomenon, without any preordained or fixed categories. However, the observation of nature clearly does not reveal any such continuous picture. What emerges from the living world is that life forms are very distinctly separated in very distinct categories. Robert Carroll, an evolutionist authority, admits this fact in his *Patterns and Processes of Vertebrate Evolution:*

> "Although an almost incomprehensible number of species inhabit Earth today, they do not form a continuous spectrum of barely distinguishable intermediates. Instead, nearly all species can be recognized as belonging to a relatively limited number of clearly distinct major groups..." (Carroll, 1997, p. 9).

According to the theory of evolution, every species has emerged from a predecessor. A species that existed previously developed into other species over time, and every new species has come into being in this way. According to the theory, this transformation proceeds gradually—over millions of years. But if this were the case, then countless intermediate species should have existed during the immense period of time during which these transformations were supposedly occurring.

In the past, for instance, there should have lived some half-fish/half-reptile creatures that had acquired some reptilian traits in addition to the piscine features they already had. And there should have existed some reptile/bird creatures, which had acquired some avian traits in addition to the reptilian traits they already possessed. Evolutionists refer to these imaginary creatures, which they suppose must have lived in the past, as "transitional forms."

But if such creatures had ever really existed, there would have been millions, even billions, of them. More importantly, the remains of these creatures should be present in the fossil record. The sheer number of these transitional forms should have been far greater than that of present animal species. And necessarily, their remains should be found all over the world. In *The Origin of Species*, Darwin accepted this fact:

> "... if my theory be true, numberless intermediate varieties, linking most closely all the species of the same group together, must assuredly have existed... Consequently evidence of their former existence could be found only amongst fossil remains..." (Darwin, 2004, p. 151).

Darwin himself was aware of the absence of such transitional forms, but hoped that they would be found in the future. Despite his optimism, he realized that these missing intermediate forms presented the biggest stumbling block for his theory. That is why he wrote the following in the chapter of the *The Origin of Species* entitled "Difficulties on Theory":

> "Why, if species have descended from other species by insensibly fine gradations, do we not everywhere see innumerable transitional forms? Why is not all nature in confusion instead of the species being, as we see them, well defined?... But, as by this theory innumerable transitional forms must have existed, why do we not find them embedded in countless numbers in the crust of the earth?... Why then is not every geological formation and every stratum full of such intermediate links? Geology assuredly does not reveal any such finely graduated organic chain; and this, perhaps, is the most obvious and gravest objection which can be urged against my theory" (Darwin, 2004, pp. 145-146, 227).

To counter this objection, the only explanation Darwin could come up with was the argument that the fossil record uncovered so far was inadequate. When more specimens were unearthed and the fossil record had been studied in detail, he asserted, the missing links would be found.

When we examine the paleontological findings today, we come across an abundance of fossils. Billions of examples have been uncovered all around the world. (Gish, 1995, p. 41) Based on them, some 250,000 distinct species have been identified, which bear striking similarities to the 1.5 million identified species currently living on Earth (Day, 1989). (Of these 1.5 million species, 1 million are insects.) Yet despite the abundance of fossil sources, not a single transitional form has been uncovered. It seems increasingly unlikely that any transitional forms will be found as a result of new excavations. A professor of geology from the University of Glasgow, T. Neville George, admitted this fact years ago:

> "There is no need to apologize any longer for the poverty of the fossil record. In some ways it has become almost unmanageably rich, and discovery is outpacing integration ... The fossil record nevertheless continues to be composed mainly of gaps" (George, 1960, pp. 1, 3).

8. Stasis in the Fossil Record

If evolution had really taken place, then living things should have emerged by gradual changes, and have continued to change over time, under varying environmental pressures. But the fossil record shows the exact opposite, revealing that species emerged suddenly, each with entirely different anatomical structures, and remained exactly the same throughout even the longest geological periods. The evolutionist paleontologist Stephen Jay Gould admitted this fact first in the late '70s:

> "The history of most fossil species includes two features particularly inconsistent with gradualism: 1. Stasis. Most species exhibit no directional change during their tenure on earth. They appear in the fossil record looking much the same as when they disappear; morphological change is usually limited and directionless. 2. Sudden appearance. In any local area, a species does not arise gradually by the steady transformation of its ancestors; it appears all at once and 'fully formed'" (Gould, 1977, p. 14).

The fossil record is now rich and extensive enough for us to understand the origins of life. It explicitly reveals that distinct species came into existence on Earth all of a sudden, each with all its own distinct forms and features.

9. The Invalidity of Punctuated Equilibrium

As we have mentioned above, the different living groups in the fossil record emerged suddenly, and stayed fixed for millions of years without undergoing any changes. This fact was ignored for many years by paleontologists, who kept hoping that imaginary intermediates would one day be found. In the 1970s, some paleontologists accepted that finding "intermediate forms" was an unfounded hope and that the "gaps" in the fossil record had to be accepted as a reality. However, because these paleontologists were unable to relinquish the theory of evolution, they tried to explain this reality by modifying the theory. And so was born the "punctuated equilibrium" model of evolution, which differs from neo-Darwinism in a number of respects.

This model began to be vigorously promoted at the start of the 1970s by the paleontologists Stephen Jay Gould of Harvard University and Niles Eldredge of the American Museum of Natural History. They proposed that living species came about not through a series of small changes, as Darwin had maintained, but by sudden, large ones.

This theory was actually a modified form of the "Hopeful Monster" theory put forward by the German paleontologist Otto Schindewolf in the 1930s. Schindewolf suggested that living things evolved not, as neo-Darwinism had proposed, gradually over time through small mutations, but suddenly through giant ones. When giving examples of his theory, Schindewolf claimed that the first bird in history had emerged from a reptile egg by a huge mutation—in other words, through a giant, coincidental change in genetic structure (Stanley, 1979, pp. 35, 159).

According to this theory, some land animals might have suddenly turned into giant whales through a comprehensive change that they underwent. This fantastic theory of Schindewolf's was taken up and defended by the geneticist Richard Goldschmidt at the University of California at Berkeley. But the theory was so inconsistent that it was quickly abandoned. The factor that obliged Gould and Eldredge to embrace this theory again was that the fossil record is at odds with the Darwinistic notion of step by step evolution through minor changes. The fact of stasis and sudden emergence in the record was so empirically well supported that they had to resort to a more refined version of the "hopeful monster" theory again to explain the situation. Gould's famous article "Return of the Hopeful Monster" was a statement of this obligatory step back (Gould, 1980, pp. 186-193).

Gould and Eldredge did not just repeat Schindewolf's fantastic theory, of course. In order to give the theory a "scientific" appearance, they tried to develop some kind of mechanism for these sudden evolutionary leaps with the interesting term, "punctuated equilibrium." In the years that followed, Gould and Eldredge's theory was taken up and expanded by some other paleontologists. However, the punctuated equilibrium theory of evolution was based on even more contradictions and inconsistencies than the neo-Darwinist theory of evolution.

The punctuated equilibrium theory of evolution, in its present state, holds that living populations show no changes over long periods of time, but stay in a kind of equilibrium. According to this viewpoint, evolutionary changes take place in short time frames and in very restricted populations—that is, the equilibrium is divided into separate periods or, in other words, "punctuated." Because the population is very small, large mutations are chosen by natural selection and thus enable a new species to emerge. For instance, according to this theory, a species of reptile survives for millions of years, undergoing no changes. But one small group of reptiles somehow leaves this species and undergoes a series of major mutations, the reason for which is not made clear. Those mutations which are advantageous quickly take root in this restricted group. The group evolves rapidly, and in a short time turns into another species of reptile, or even a mammal. Because this process happens very quickly, and in a small population, there are very few fossils of intermediate forms left behind, or maybe none. On close examination, this theory was actually proposed to develop an answer to the question, "How can one imagine an evolutionary period so rapid as not to leave any fossils behind it?" Two basic hypotheses are accepted while developing this answer:

1. that macromutations—wide-ranging mutations leading to large changes in living creatures' genetic make-up—bring advantages and produce new genetic information;
2. that small animal populations have greater potential for genetic change.

However, both of these hypotheses are clearly at odds with scientific knowledge.

10. The Misconception about Macromutations

The first hypothesis—that macromutations occur in large numbers, making the emergence of new species possible—conflicts with known facts of genetics. One rule, put forward by R.A. Fisher, one of the last century's best known geneticists, and based on observations, clearly invalidates this hypothesis. Fisher states in his book *The Genetical Theory of Natural Selection* that the likelihood that a particular mutation will become fixed in a population is inversely proportional to its effect on the phenotype (Fisher, 1930).

Or, to put it another way, the bigger the mutation, the less chance it has of becoming a permanent trait within the group. It is not hard to see the reason for this. The more an individual is affected by mutation, the less chance it has of surviving. Ernst Mayr, the doyen of Darwinism, makes this comment on the subject:

> "The occurrence of genetic monstrosities by mutation ... is well substantiated, but they are such evident freaks that these monsters can be designated only as 'hopeless'. They are so utterly unbalanced that they would not have the slightest chance of escaping elimination through stabilizing selection... the more drastically a mutation affects the phenotype, the more likely it is to reduce fitness. To believe that such a drastic mutation would produce a viable new type, capable of occupying a new adaptive zone, is equivalent to believing in miracles... The finding of a suitable mate for the 'hopeless monster' and the establishment of reproductive isolation from the normal members of the parental population seem to me insurmountable difficulties" (Mayr, 1970, p. 235).

It is obvious that mutations cannot bring about evolutionary development and this fact places both neo-Darwinism and the punctuated equilibrium theory of evolution in a terrible difficulty. Since mutation is a destructive mechanism, the macromutations that proponents of the punctuated equilibrium theory talk about must have "macro" destructive effects. Some evolutionists place their hopes in mutations in the regulatory genes in DNA. But the feature of destructiveness which applies to other mutations, applies to these, as well. The problem is that mutation is a random change: any kind of random change in a structure as complex as genetic data will lead to harmful results. In their book *The Natural Limits to Biological Change*, the geneticist Lane Lester and the population biologist Raymond Bohlin describe the blind alley represented by the notion of macromutation:

> "The overall factor that has come up again and again is that mutation remains the ultimate source of all genetic variation in any evolutionary model. Being unsatisfied with the prospects of accumulating small point mutations, many are turning to macromutations to explain the origin of evolutionary novelties. Goldschmidt's hopeful monsters have indeed returned. However, though macromutations of many varieties produce drastic changes, the vast majority will be incapable of survival,

> let alone show the marks of increasing complexity. If structural gene mutations are inadequate because of their inability to produce significant enough changes, then regulatory and developmental mutations appear even less useful because of the greater likelihood of nonadaptive or even destructive consequences… . But one thing seems certain: at present, the thesis that mutations, whether great or small, are capable of producing limitless biological change is more an article of faith than fact" (Lester and Bohlin, 1989, pp. 141-142).

Observation and experiment both show that mutations do not enhance genetic data, but rather damage living things. Therefore, it is clearly irrational for proponents of the punctuated equilibrium theory to expect greater success from "mutations" than the mainstream neo-Darwinists have found.

11. The Misconception about Restricted Populations

The second concept stressed by the proponents of punctuated equilibrium theory is that of "restricted populations." By this, they mean that the emergence of new species comes about in communities containing very small numbers of plants or animals. According to this claim, large populations of animals show no evolutionary development and maintain their "stasis." But small groups sometimes become separated from these communities, and these "isolated" groups mate only amongst themselves. (It is hypothesized that this usually stems from geographical conditions.) Macromutations are supposed to be most effective within such small, inbreeding groups, and that is how rapid "speciation" can take place. But why do proponents of the punctuated equilibrium theory insist so much on the concept of restricted populations? The reason is clear: Their aim is provide an explanation for the absence of intermediate forms in the fossil record.

However, scientific experiments and observations carried out in recent years have revealed that being in a restricted population is not an advantage from the genetic point of view, but rather a disadvantage. Far from developing in such a way as to give rise to new species, small populations give rise to serious genetic defects. The reason for this is that in restricted populations individuals must continually mate within a narrow genetic pool. For this reason, normally heterozygous individuals become increasingly homozygous. This means that defective genes which are normally recessive become dominant, with the result that genetic defects and sickness increase within the population (Soulé and Mills, 1998, p. 1658). This reveals that the claim by the proponents of punctuated equilibrium theory that small populations are the source of evolution has no scientific validity.

12. Conclusion

Evolution is a theory that collapses at its very first step, because evolutionists are unable to explain even the formation of a single protein. Neither the laws of probability nor the laws of physics and chemistry offer any chance for the fortuitous formation of life. And there is no mechanism in nature that can cause species to evolve into another. And finally, the fossil record indicates that living things did not evolve from primitive to the advanced forms, but instead emerged individually, suddenly, and in a perfect state. In short, both extinct species and those we observe today originated suddenly, with their body structures in complete working order. This shows that they did not come into existence by evolution, they were created by God, the Omnipotent and Omniscent.

References

Carroll, R.L. (1997) *Patterns and Processes of Vertebrate Evolution*, Cambridge University Press, p. 9.

Cohen, I.L. (1984) *Darwin Was Wrong: A Study in Probabilities*, New Research Publications, New York, p. 81.

Darwin, C. (2004) *The Origin of Species*, Barnes and Noble, New York, pp. 145-146, 150-151, 227.

Darwin, F. (1959) *The Life and Letters of Charles Darwin*, Basic Books, New York, vol. II, pp. 202-203.

Dawkins, R. (1996) *The Blind Watchmaker*, W.W. Norton, New York, p. 5.

Day, D. (1989) *Vanished Species*, Gallery Books, New York.

Eastman, M., Missler, C. (1996) *The Creator Beyond Time and Space*, Costa Mesa, CA: TWFT, p. 61.

Fisher, R.A. (1930) *The Genetical Theory of Natural Selection*, Oxford University Press, Oxford.

Futuyma, D.J. (1983) *Science on Trial*, Pantheon Books, New York, p. 197.

George, T.N. (Jan 1960) "Fossils in Evolutionary Perspective," *Science Progress*, **48**: 1, 3.

Gilbert, S., Opitz, J., Raff, R. (1996) "Resynthesizing Evolutionary and Developmental Biology," *Developmental Biology*, article no. 0032, **173**: 361.

Gish, D. (1995) *Evolution: Fossils Still Say No*, Institute for Creation Research, CA, p. 41.

Goodwin, B. (1995). Neo-Darwinism has failed as an evolutionary theory. *The Times*(May 19, Higher Education Supplement), http://www.timeshighereducation.co.uk/story.asp?storyCode=98152&sectioncode=98126.

Gould, S.J. (May 1977) "Evolution's Erratic Pace," *Natural History*, **86** (5): 14.

Gould, S.J. (June/July 1977) "The Return of Hopeful Monsters," *Natural History*, **86**: 28.

Gould, S.J. (1980) "Return of the Hopeful Monster," *The Panda's Thumb*, W. W. Norton Co., New York, pp. 186-193.

Grant, P.R. (Oct 1991) "Natural Selection and Darwin's Finches," *Scientific American*, **265**: 82-87.

Grassé, P.P. (1977) *Evolution of Living Organisms*, Academic Press, New York, pp. 97-98.

Hoyle, F. (1983) *The Intelligent Universe*, Holt, Rinehart and Winston, New York, p. 19.

Lester, L.P., Bohlin, R.G. (1989) *The Natural Limits to Biological Change*, Probe Books, Dallas, pp. 141-142.

Mayr, E. (1970) *Populations, Species, and Evolution*, Belknap Press, Cambridge, p. 235.

Patterson, C. (March 4, 1982) *Cladistics*, Interview on BBC by Peter Franz.

Pitman, M. (1984) *Adam and Evolution*, River Publishing, London, p. 70.

Ranganathan, B.G. (1988) *Origins?*, The Banner of Truth Trust, Pennsylvania.

Soulé M.E., Mills, L.S. (1998) "Enhanced: No need to isolate genetics," *Science*, **282**: 1658.

Stanley, S.M. (1979) *Macroevolution: Pattern and Process*, W. H. Freeman, San Francisco, pp. 35, 159.

Taylor, G.R. (1983) *The Great Evolution Mystery*, Harper and Row, New York, p. 48.

Dialogue

Taner Edis

My main question to Dr. Gündoğdu is whether he really expects his article and similar creationist efforts to be taken seriously within the scientific community. Reading his contribution, it is striking how many errors of basic scholarship stand out, beyond specifically biological mistakes. In my conversations with creationists, I have found that they often resent the offhand dismissal they encounter from mainstream scientists. If, however, they regularly produce material of such appallingly poor quality, is this any surprise?

Consider just a few outstanding problems, obvious even without any biological expertise.

Gündoğdu heavily relies on quotations to make his case. Creationists are notorious for quote-mining and pulling statements out of context. One example here is from Robert Carroll. Gündoğdu's use of Carroll is a wild misrepresentation. And then there is the way Gündoğdu quotes from creationist literature as if it had the same standing as the peer-reviewed biological literature. For example, Cohen (1984) is a classic crank book, entirely worthless except as a minor curiosity. Lester & Bohlin (1989) is a creationist text put out by a conservative Christian publisher. And so on.

There is a deeper problem here. Even if Gündoğdu did not misrepresent his biological sources and slip disreputable sources alongside them, how could a series of quotations add up to an argument? This kind of substitution of authority for substantive argument might be legitimate in certain religious contexts, but it is far from how science operates.

Gündoğdu's sources are also curiously outdated. Proper scholarship demands that a critic should address the best, most current views of his opponents. But with few exceptions, his references date from the 1970's and 1980's. For example, he refers to the debate over punctuated equilibrium from the 1970's. But mainstream biology has moved on; it has long been clear that punctuated equilibrium ideas harbor no challenge to Darwinian evolutionary mechanisms.

Moreover, none of the debate among biologists, even in the 1970's, called common descent into question.

If Gündoğdu wants to claim that no transitional forms exist, the proper approach would be to examine cases where biologists do claim that a reasonable (if still partial) fossil record of transition is available. I see no mention of examples such as whales and the transition between mammal-like-reptiles and mammals. Proper criticism addresses the stronger cases; this is just a matter of basic scholarly ethics. The selective picture Gündoğdu presents, of "a theory that collapses at its very first step" is more typical of propaganda.

And then there are examples of the simply ridiculous. Gündoğdu cannot resist standard creationist themes such as the immense improbability of complex biological and biochemical structures. Invariably, such calculations are based on naive equiprobability and single-step assembly assumptions. As variation-and-selection is an entirely different process, such results are completely irrelevant. Gündoğdu should address actual claims, not fantastic scenarios that have nothing to do with evolution. Indeed, Gündoğdu seems to understand very little of what he is ostensibly criticizing. Another example is his insistence that natural selection cannot cause things to evolve. Presumably this incoherent statement is trying to say that selection cannot create novelties. Indeed, it cannot, but biologists have never claimed that it could. Blind variation, due to mutations and other errors, are the source of novelty. Gündoğdu then proceeds to an inept discussion of mutation that shows no awareness of the relevant professional literature.

Again, there is a question of basic scholarly competence, perhaps even ethics, here. Surely Gündoğdu cannot be unaware of the rather extensive literature wherein mainstream scientists address common creationist and intelligent design claims in considerable detail. If there is any real dialogue to take place, creationists have to begin to address such responses, rather than repeat arguments that have changed very little over the last few decades.

Cihat Gündoğdu
How would an individual confront scientific evidence? If it is against one's belief certainly he would get upset. Here, Mr. Edis is upset with the scientific evidence against Darwinism just like other evolutionists who believe in Darwinism as *a priori*. Edis and other Darwinists in fact are the members of an

ancient religion which they do not verbalize, but serve it under the name of Darwinism.

Darwinism is a Shamanistic religion that has adopted nature as a deity. Darwinists' present-day beliefs are just as odd and irrational as those of people who once worshipped crocodiles. Darwinists regard chance and inanimate, unconscious atoms as a creative force, and are as devoted to that belief as if to a religion.

Shamanism is a belief system based on the worship of forces such as rain, snow, lightning, storms, wind, and the Sun. Darwinism is also a religion of nature-worship; it describes nature as an entity with "mythical and mysterious powers." It anticipates the belief that stone, earth, the Sun, lightning and wind combined to give rise to life.

Shamans claim to be the physicians, sages, leaders and administrators of their tribes and societies. Darwinists describe themselves in the same way. Shamans maintain that they understand the secrets of nature and can foretell the future. Similarly, Darwinists maintain that they know the secrets of matter, the atom and the Earth, and seek to describe imaginary changes that human beings and nature will be subject to in the future.

Astronomy, biology, paleontology, physics, geology, chemistry, geophysics, embryology are all sciences. Darwinism, however, is not a science, but a primitive shamanistic religion.

Since the theory of evolution was first put forward, advances in a great many branches of science, have demolished the theory's claims, one by one. Nonetheless, Darwinism still has its adherents. Typically, when a scientific theory is disproved, it is shelved, and all debate and discussion comes to an end. But not so with Darwinism. No matter how powerful and indisputable the evidence against their theory may be, evolutionists ignore it and continue to defend their beliefs in a fervent manner.

Worshipping fire, the stars or the Sun, believing that the Pyramids were built by aliens, or venerating certain animals as sacred, are not scientific. The same applies for Darwinism because just like other superstitious beliefs, Darwinism is also a religion of idols and false deities. Darwinism's foremost idol is the "idol of chance." Whatever Darwinist text you read, you will see claims about the so-called power and limitless abilities of "natural selection," the lifeblood and essence of Darwinism. Evolutionists claim that everything performed by the "idol of chance" is actually based on pragmatic calculations. In their view, this idol is able to consider everything and calculate in advance every step it will take.

Evolutionists believe in one very strange force. They ascribe divine status to matter. They have belief that matter once assembled itself into a living cell and that one organism can give rise to another, entirely different one. Science has refuted these ideas. But for Darwinists, they are irrefutable facts that everyone must believe.

Any rational mind is quite able to conceive that no complex entity can come into being spontaneously, by chance, but absolutely must be the product of a conscious plan. However, just like pagans who worship the idols they have crafted with their own hands, Darwinists believe in false deities.

Just like pagans believe that inanimate idols created all things, so do evolutionists and materialists believe that inanimate matter created all living things. (God is surely beyond these.) They claim that even their own human bodies are the sum total of various coincidences.

We are exhibiting thousands of fossils in different countries to the public which are evidence of no change in species in time. Species are stable and did not exhibit primitive features when compared to specimens of the same species living today. They shout out that they didn't evolve but were created. They have appeared all of a sudden with their fully formed organs and systems and stayed as they are for hundred milions of years. That is why they are called as "stable species". But are there any "unstable species"? None at all!

The public laughs at the ruses resorted to by evolutionists, because they cannot prove evolution by way of slander and aggression. Any evidence they have, they should put forward, whereupon everyone can easily distinguish between truth and error. For months now, the exhibitions of fossils taking place in the UK, the US or Turkey have helped the people realize that living things never underwent evolution.

It is expected that evolutionists exhibit any transitional fossils if they have any. At the very least, they should display a few of them in the centers of the cities. If they cannot do that, then they should cease their defense of evolution. They have no intermediate-form fossils to show, because such creatures never existed.

However, each fossil that evolutionists have put forward as a proof of evolution has turned out to be a fake or else misinterpreted. "Piltdown Man," for example, turned out to be a hoax. The tooth of Nebraska "Man" turned out to be from a fossil boar. The coelacanth has been caught alive—and unchanged—since 1938. All the skull and bone fragments that are claimed to show the alleged human evolution have been proven to be either those of present-day human beings, vanished human races or else of species of ape that have become extinct.

I recommend Edis and other Darwinists to rely on positive sciences — paleontology, biochemistry, physiology, biology, etc. Superstitions of the past are falsified by modern-day science and that is why even the leading figure amongst atheists has converted recently. Anthony Flew defended atheism as a teacher at the universities of Oxford, Aberdeen, Keele and Reading, at many American and Canadian universities he visited, in debates, books, lecture halls and articles. In recent days, however, Flew has announced that he has abandoned this grave error and accepts that the universe was created.

Flew realized, in the face of the information-based complexity of life, that the true origin of life is an intelligence and that the atheism he had espoused for 66 years was a discredited philosophy. Flew announced the scientific reasons underlying this change in belief in these terms:

> "Biologists' investigation of DNA has shown, by the almost unbelievable complexity of the arrangements which are needed to produce [life], that intelligence must have been involved."[*]
>
> "It has become inordinately difficult even to begin to think about constructing a naturalistic theory of the evolution of that first reproducing organism."[†]

---

[*]Richard N. Ostling, "Lifelong atheist changes mind about divine creator," The Washington Times 10 December 2004; http://washingtontimes.com/national/20041209-113212-2782r.htm

[†]Antony Flew, "Letter from Antony Flew on Darwinism and Theology," Philosophy Now; http://www.philosophynow.org/issue47/47flew.htm

> "I have been persuaded that it is simply out of the question that the first living matter evolved out of dead matter and then developed into an extraordinarily complicated creature."[‡]

The DNA research which Flew cites as a fundamental reason for his change of opinion has indeed revealed striking facts about creation. The helix shape of the DNA molecule, its possession of the genetic code, the nucleotide strings that refute blind chance, the storage of encyclopaedic quantities of information and many other striking findings have revealed that the structure and functions of this molecule were arranged for life with a special creation. Comments by scientists concerned with DNA research bear witness to this fact. Francis Crick, for instance, one of the scientists who revealed the helix shape of DNA admitted in the face of the findings regarding DNA that the origin of life indicated a miracle:

> "An honest man, armed with all the knowledge available to us now, could only state that in some sense, the origin of life appears at the moment to be almost a miracle, so many are the conditions which would have had to have been satisfied to get it going."[§]
>
> Based on his calculations, Led Adleman of the University of Southern California in Los Angeles has stated that one gram of DNA can store as much information as a trillion compact discs.[**]

The most striking fact about DNA is that the existence of the coded genetic information can definitely not be explained in terms of matter and energy or natural laws. Dr. Werner Gitt, a professor at the German Federal Institute of Physics and Technology, has said this on the subject:

> "A code system is always the result of a mental process... . It should be emphasized that matter as such is unable to generate any code. All experiences indicate that a thinking being voluntarily exercising his

---

[‡]Stuart Wavell and Will Iredale, "Sorry, says atheist-in-chief, I do believe in God after all," The Sunday Times, 12 December 2004;
http://www.timesonline.co.uk/article/0,,2087-1400368,00.html

[§]Francis Crick, Life Itself: Its Origin and Nature, New York: Simon & Schuster, 1981, p. 88.

[**]John Whitfield, "Physicists plunder life's tool chest", 24 April 2003;
http://www.nature.com/nsu/030421/030421-6.html

> own free will, cognition, and creativity, is required… . There is no known natural law through which matter can give rise to information, neither is any physical process or material phenomenon known that can do this."[††]

Creationist scientists and philosophers played a major role in Flew's acceptance of creation, backed up by all these findings. In recent times Flew participated in debates with scientists and philosophers who were proponents of creation, and exchanged ideas with them. The final turning point in that process was a discussion organized by the Institute for Metascientific Research in Texas in May, 2003. Flew participated together with author Roy Abraham Varghese, Israeli physicist and molecular biologist Gerald Schroeder, and Roman Catholic philosopher John Haldane. Flew was impressed by the weight of the scientific evidence in favor of creation and by the convincing nature of his opponents' arguments, and abandoned atheism as an idea in the period following that discussion. In a letter he wrote for the August-September, 2003, edition of the British magazine *Philosophy Now*, he recommended Schroeder's book *The Hidden Face of God: Science Reveals the Ultimate Truth* and Varghese's book *The Wonderful World*.[‡‡] During an interview with the professor of philosophy and theology Gary R. Habermas, who also played a major role in his change of mind[§§], and also on the video "Has Science Discovered God?," he openly stated that he believed in a Supreme Being with creative force.

In the face of all the scientific developments outlined above, the acceptance of an Omnipotent creative force by Antony Flew, famous for defending atheism for many years, reflects a final scene in the process of collapse being undergone by Darwinism. Modern–day science has revealed the existence of a "wisdom pervading the universe," thus leaving atheism out of the equation.

In his book *The Hidden Face of God*, Gerald Schroeder, one of the creationist scientists who influenced Flew, writes:

[††]Werner Gitt, In the Beginning Was Information, CLV, Bielenfeld, Germany, pp. 64-7, 79.

[‡‡]Antony Flew, "Letter from Antony Flew on Darwinism and Theology," Philosophy Now; http://www.philosophynow.org/issue47/47flew.htm

[§§]"Atheist Becomes Theist: Exclusive Interview with Former Atheist Antony Flew;" http://www.biola.edu/antonyflew/index.cfm

> "A single consciousness, a universal wisdom, pervades the universe. The discoveries of science, those that search the quantum nature of subatomic matter, have moved us to the brink of a startling realization: all existence is the expression of this wisdom. In the laboratories we experience it as information that first physically articulated as energy and then condensed into the form of matter. Every particle, every being, from atom to human, appears to represent a level of information, of wisdom."[***]

Scientific research into both the functioning of the cell and the subatomic particles of matter has revealed this fact in an indisputable manner: Life and the universe were brought into being from nothing by the will of an Entity possessed of a superior mind and wisdom. There is no doubt that the possessor of that knowledge and mind that pervade the universe at all levels is Almighty Allah (God).

Darwinism is not science and this is being admitted by the evolutionists like Henry Gee, who is the editor of *Nature* magazine. In his book *In Search of Deep Time*, published in 1999, Gee pointed out that conventional theories of the origin and development of human beings are "a completely human invention created after the fact, shaped to accord with human prejudices," and adds:

> "To take a line of fossils and claim that they represent a lineage is not a scientific hypothesis that can be tested, but an assertion that carries the same validity as a bedtime story—amusing, perhaps even instructive, but not scientific."[†††]

Darwinism is nothing more than a bedtime story. A frog may change into a prince but only in fairy tales. Neither scientific world nor the public needs amusements on topics which shape our lives. Humanity should rely on facts and what scientific observations and findings show is the fact of creation.

[***]Gerald Schroeder, *The Hidden Face of God*, Touchstone, New York, 2001, p. xi.
[†††]Henry Gee, *In Search of Deep Time*, New York: The Free Press, 1999, pp. 32, 116, 117, 202.

Biography

Dr. Lee M. Spetner is currently Director of Research at Redoxia Israel, Ltd., of which he is a founder, where he is engaged in cancer research. He received the BS degree in mechanical engineering from Washington University in 1945 and the PhD degree in physics from MIT in 1950. He then joined the Applied Physics Laboratory of The Johns Hopkins University, where he was Principal Physicist engaged in development of guided-missile systems. In 1970 he joined Eljim, Ltd. in Israel as Technical Director, and later as manager of Elbit, Nes Ziona where he supervised development work in military electronic systems including electronic countermeasures, and a military electronic navigation system. He has taught courses at the Johns Hopkins University, Howard University and the Weizman Institute. These courses include classical mechanics, electromagnetic theory, real-variable theory, probability theory, and statistical communication theory.

38
The Evolution Controversy and Randomness

LEE M. SPETNER

Redoxia Israel Ltd.
27 Hakablan St., Jerusalem 93874, Israel
LSpetner@alum.mit.edu

1. Introduction

The public arguments currently raging about evolution revolve principally around the origin of life: Is life the result of divine action or of purely natural phenomena? The evolutionist advocates that life arose by chance, following natural laws, while the creationist advocates that life is the result of divine creation. The creationist acknowledges that his position is faith based, while the evolutionist invokes Science. Yet both positions stem from choices made outside of Science—both are based on a faith. Observational data and logic alone do not force one to accept either of the two positions.

Before the publication of *The Origin of Species*, conventional wisdom held that life was divinely created. Darwin's theory opened the possibility that life could have arisen by purely natural means, without any need for divine intervention. While Darwin offered no convincing suggestion about how life could have got started, he did offer what seemed to be a plausible outline of how complex life might have developed after some simple, self-reproducing, life form had come into existence. His outline served as a possible alternative to divine creation, and was appealing to some for that reason (Thomson 2005). He seemed to have convinced himself that his theory was indeed the way life had developed, and that the necessity of divine creation as an explanation could be dispensed with.

Darwin was long on speculation and short on convincing data, but his idea nevertheless had enough of an appeal to be eventually embraced by scientists and other intellectuals. As discoveries were made in paleontology, in the new science of genetics, and later in the even newer science of molecular biology, new data were fit into the Darwinian speculations, and hailed as support for his theory, when in reality they also required speculation to make the fit. What Darwin really offered was not a theory of how new species came into being, but a speculative outline as to how it *might* have happened. Life *could* have developed along the lines he suggested. As his speculation gained adherents, what started out as how life *might* have developed, progressed to being how life *must* have developed, and finally into how life *did indeed* develop. This progression was made in spite of the failure to find any really convincing evidence of how life *might* have evolved from a simple beginning and the lack of even an acceptable theory of how that simple beginning might have occurred.

So after nearly 150 years of Darwinian theory, we are left with nothing more than a speculation, no better than that offered by Darwin, of how life *might* have developed rather than having been divinely created. The major philosophical impact of evolutionary theory on the lay public is its claim to offer a scientific basis for the natural emergence and development of life from *molecules-to-man* (M-to-M). But the theory does not fulfill that claim.

2. The Nature of the Evidence for Evolution

Examples of evolution are often cited as evidence for the speculations of M-to-M, but none provides the kind of support usually required for a scientific theory. Some of these examples even refute, rather than support, the Darwinian theory, which in its modern form says that evolution occurs through natural selection acting on random genetic changes.*

How evolution in fact works can only be elucidated by direct evidence of evolution in action. Darwin had to rely on indirect evidence such as the fossils and comparative anatomy. Indirect evidence may have convinced Darwin that evolution had occurred, but it could not show *how* it did. Fortunately, today, we have evidence of evolution in action, showing *how* evolution works, and it is not

*A genetic change is usually considered random in this context if its occurrence is independent of its benefit to the organism.

at all the way Darwin thought or the way it is described by the modern neo-Darwinian theory.

Direct observations of evolution can be divided into two sets. One set comprises examples of evolution that, by their very nature, cannot serve as prototypes of what could lead to M-to-M. An illustration of this is the often proffered example of the evolution of bacterial resistance to antibiotics.

The acquisition of antibiotic resistance can also be divided into two types. The first is the gaining of resistance by horizontal transfer from other species of microorganisms that already possess the genes for resistance. Clearly, since horizontal transfer does not create any novelty in the biosphere, it cannot by itself be a source of novelty and it therefore cannot serve as a prototype step in a process illustrating M-to-M. That is to say, that any sequence of horizontal transfers, no matter how long, could not by itself produce M-to-M.

The second type of antibiotic-resistance evolution, which also cannot serve as a prototype step of M-to-M evolution, is the gaining of resistance by degrading or destroying the match between the antibiotic molecule and the microorganism. An antibiotic molecule attacks a microorganism by obstructing its life processes, and to do this it usually has to attach itself to the microorganism at a matching site on its surface. If the site should become sufficiently degraded through a mutation, the antibiotic cannot attach well and its antibiotic effect will be defeated. Here again, a sequence of such mutations, no matter how long, could not lead to M-to-M. I have dealt with this type of example elsewhere (Spetner 1997).

The second set of direct observation of evolution produces rapid and significant changes, which may look as if they could lead to M-to-M, but the rapidity with which they occur is not compatible with current Darwinian theory. These examples entail the activation of a capability latent in the organism.[†] Many of these examples are of sympatric speciation, which is the rise of a new species propinquant with the extant species through the appearance of an isolating mechanism that prevents the former from being diluted by the latter. The concept of sympatric speciation has gone in and out of favor among students of evolution over the past century and a half. It was thought to be an important

---

[†] Such a capability could, by the way, be a third way for pathogens to acquire antibiotic resistance, although I am not aware of any examples.

method of speciation until Mayr (1963) gave cogent theoretical arguments against it. He argued that sympatric speciation could occur only with the simultaneous appearance in the nascent species of a preference for a new niche and cross sterility with the extant species, or at least a preference for mating with its own kind, an unlikely pair of occurrences. In addition, he held that sympatric speciation was not supported by irrefutable data.

3. Some Evolutionary Examples

By the end of the 20th century, however, enough irrefutable examples of sympatric speciation had been accumulated that Mayr began to acknowledge the existence and importance of sympatric speciation (Mayr 2004). Thoday & Gibson (1962) claimed to show the evolution of sympatric speciation in the laboratory. Shortly thereafter, Maynard Smith (1966) proposed a theoretical model of sympatry. In the following decades, more examples of sympatry were reported.

None of these models, however, help to account for M-to-M. Some of the models postulated the immigration into an existing population of a few individuals already sexually isolated from the extant population (Doebli 1996). Others postulated the appearance of a few reproductively isolated mutants. Only the latter would seem to have a chance of representing a component in a sequence leading to M-to-M.

One example of sympatric speciation is that of the apple maggot fly *Rhagoletis pomonella*. This fly originally bred on hawthorn, but in the 19th century it began to infect apple trees. It has now spread to cherries, pears, and roses. For these flies to take on a new host, they must undergo several simultaneous changes. Their behavior has to evolve to prefer to feed and to mate on the new host, which often ripens either earlier or later than the old host. They have to evolve new mating preferences of both the males for the females and the females for the males. Their mating procedures also change, and serve to isolate the new from the old population and keep them from interbreeding. They must also change their maturation time to match the ripening time of the new host fruit. All these changes must occur simultaneously. The changes are, moreover, genetically based (Barton, et al. 1988, Feder et al. 1988, McPheron et al. 1988, Smith 1988). All these genetic changes occur together rather quickly, making it impossible to account for them by random point mutations and natural selection. This example, and others like it, cannot be prototypes of evolutionary events leading to M-to-M, nor do they support the neo-Darwinian hypothesis.

How could *R. pomonella* have evolved so rapidly? It could, perhaps, be said there was enough genetic diversity in the population that a recombination could have led rapidly to all those required changes. It is not clear at all how so many changes could be effected by one recombination. If many recombinations would have been required, the changes would be unlikely since the probability of getting many specific events by chance is low and they are unlikely to happen. Whatever mechanism is suggested to account for these changes, it will not be one that could support M-to-M.

Character displacement is a speciation phenomenon in which two species that tend to be similar when the species are separated, are divergent in one or more characters when they occupy the same territory (Brown & Wilson. 1956). The differences are thought to be genetically based. A striking example of rapid character displacement in which the divergence appeared after residing with another species for only 22 years, and in which the character displacement occurred rapidly—in just one year—was recently reported by Peter and Rosemary Grant (Grant & Grant 2006). The medium ground finch (*Geospiza fortis*) was the sole finch on an undisturbed Galápagos island until 1982 when two females and three males of the large ground finch (*Geospiza magnirostris*) arrived. As these new arrivals bred, the size of the bills of the *G. fortis* were monitored. The mean bill size remained constant[‡] until 2004, when it suddenly underwent a substantial decrease in a single year.[§] The change was adaptive in that it permitted the *G. fortis* to feed on small seeds, avoiding competition with the now well-established *G. magnirostris*. In this example, the competitive stimulus built up very slowly, starting from 1982. Until 1997 the number of *G. magnirostris* were too small to estimate. From that time they built up gradually, and by 2004 they were suddenly almost as numerous as the *G. fortis*, and that is when the abrupt change in mean bill size of the latter underwent its sudden change. Here the evolutionary phenomenon points toward a nonrandom genetic change triggered by an environmental cue. As in many other such examples, there is no support here for M-to-M.

---

‡ Except for a temporary increase, which started before the arrival of *G. magnirostris*.

§ The decrease was four times the one-sided 95% confidence limits on the estimate of the mean in 1973.

Sympatric speciation of anole lizards has been observed on islands of the Greater Antilles (Losos 2001, Losos & Schluter 2000). Each of the islands, Puerto Rico, Cuba, Hispaniola, and Jamaica, has species adapted to a variety of habitats. There are about 110 species of these lizards on those islands that have apparently evolved to fill most conceivable niches such as on tree canopies, tree trunks, twigs, and grass.

How do we know that these species evolved sympatrically (together in the same locality) rather than allopatrically (each in a separate locality) and then came together? There are two strong indications of sympatric speciation among several of the species. First of all, in the 1970s the brown anole, *Anolis sagrei*, was introduced to 20 islands, each of which offers a variety of habitats. About 20 years later it was found that, on many of the islands, lizards of different morphologies were adapted to the various habitats and were reproductively isolated from one another. Since each island had several adapted species, one can conclude that in the time between the introduction of the lizards and when the resultant population was observed, the new species diverged sympatrically from the original *A. sagrei*.

Second, DNA analysis indicates that the various species on an island evolved on that island and did not immigrate from elsewhere. How is this shown? The anoles on each of the islands of Cuba, Jamaica, Hispaniola, and Puerto Rico, have the same types of adaptations. Species on different islands adapted to the same habitats have nearly the same morphology. For example, all four islands have a species of anoles adapted to living at the base of a tree and one adapted to the tree canopy. Three of the islands have species adapted to the tree trunks and a species adapted to living in grass.

If these various species evolved their separate adaptations together in one place and then dispersed, one would expect the DNA of a lizard on one island to more closely match the DNA of a lizard having the same adaptation on another island than to a differently adapted anole on the same island. DNA analysis, however, has shown the reverse—that the variously adapted species on any one island are more closely related to each other than each is to similarly adapted species on the other islands (Losos 2001, Losos et al. 1998, Jackman et al. 1997). The implication is then strong that the first lizard population to colonize each island diverged over time to fill the various niches.

4. Nonrandom Genetic Changes

How could the lizard genotypes have changed to adapt their corresponding phenotypes to the niches they found on the island? To say that random point mutations could have by chance produced similar adaptation over and over again stretches the limits of credulity. To say that genetic recombinations by chance put together pieces of the genome to produce the adaptation is likewise incredulous. On the other hand, to say that the genome somehow changes itself to adapt to the needs of the organism is to invoke a mysterious Lamarckian effect for which there is no known mechanism.

It seems clear that the molecular mechanism of the adaptation is nonrandom. Losos (2001) has suggested that there may be a deterministic process that drives the lizards' evolution to produce appropriate adaptations over and over again. This implies, as I have previously suggested (Spetner 1997), that the genotype can have within it the capability of adapting the organism to a limited range of environments, and that cues from the current environment stimulate the genome to trigger an adaptive genetic choice among predetermined potential adaptations of the phenotype. Environmental cues are often manifest as stress, and stress is known to stimulate hormonal secretions that can have wide-ranging effects on the organism.

Losos has noted that the environment can influence the phenotype during development. The process is called *phenotypic plasticity*, which is the ability of an organism to adapt by responding to an environmental cue with an adaptive change in the phenotype. Adaptation to a new environment by means of phenotypic plasticity, without genetic change, is known to occur (West-Eberhard 1986, 1989, 2005).

West-Eberhard (1998, 2005) suggested that phenotypic plasticity plays an important role in evolutionary change. The change can occur during development or even afterwards. This phenomenon is familiar in the increase in size and strength of a muscle with use. The evolutionary significance of such adaptations has usually been rejected by students of evolution because it is without genetic effect. Yet sequences of fossils are often accepted as evidence of evolution, in spite of the lack of evidence that their differences are of genetic origin. An example of this phenomenon is the long sequence of snail fossils found and reported by Williamson (1981) in the Turkana basin in Northern Kenya. He reported the changes in shell forms through a 400-meter thickness of sedimentary rock to consist of sequences of no change in shell shape punctuated by abrupt changes. Williamson, a student of S.J. Gould at Harvard, presented his find as evidence of punctuated equilibrium. Others, however, pointed out later that the changes were more likely brought about by changes in the chemistry of the water in which they lived (Fryer et al. 1983).

Losos suggested that phenotypic plasticity may be just the beginning of the change in the phenotype, which later can be fixed through genomic change (Losos 2001). West-Eberhard (2005) also makes this point, noting that the phenomenon may be widespread. One of the earliest reported examples of evolution beginning with an adaptive change in the phenotype induced by the environment and proceeding to fixation by genetic change is that of the callosities on the rump of the ostrich. These hardened calluses serve as protection to its skin where the ostrich sits on rocky or sandy ground. Such calluses would normally form as a response to the skin irritation stemming from repeated sitting. But these calluses are of genetic origin as evidenced by their appearance in the ostrich embryo, before there is any opportunity for stimulation by the environment (Duerdon 1920). These calluses, then, serve as an example of the phenomenon that West-Eberhard describes of phenotypic plastic effect fixed by genetic change.

It is far from clear how the Darwinian theory of natural selection on random mutations can account for this kind of evolution. We don't know what genomic changes are necessary to cause the callus formation, but whatever they are, natural selection has nothing to work on since the calluses would form anyway without genomic change. This is a problem with any of the examples of an adaptive phenotypic-plastic change induced by an environmental cue, which later gets fixed through a genomic change.

5. Mutations Cued by Environment

The above examples of evolution do not occur through random mutations, but rather through nonrandom genetic rearrangements that seem to be responsive to the need of the organism to adapt to a changing environment. This kind of evolution does not correspond to Darwinian theory, which invokes genetic changes that are unrelated to the benefit they may confer on the organism. These genetic changes that are observed are statistically biased toward the adaptive.

That is not to say, however, that Lamarckism has returned with a mysterious drive toward adaptation.

While there is evidence that such evolution occurs, natural selection on random mutations cannot account for it. There must be another explanation. James Shapiro (2002) has noted that the genomic changes that lead to evolution must be nonrandom, echoing on the genotypic level Losos's suggestion on the phenotypic level that the lizard evolution must be driven by a deterministic process. Shapiro notes that random point mutations are only copying errors, which are largely repaired by normal cellular mechanisms. Such errors cannot account for the above examples of evolution. It is the nonrandom genetic changes that appear to drive the evolution that has been observed. One is then led to consider evolution as driven by normal biological processes of the genome rather than by accidental genomic changes feeding natural selection.

Nonrandom genetic changes have, in the last three decades, been studied in single-celled organisms—particularly in bacteria—and adaptive genomic changes have been found to occur. The discovery of what have come to be called *adaptive mutations* was at first received with much skepticism, but after the experiments had been repeated at many laboratories, the phenomenon has been acknowledged to exist (Drake 1991). All of these adaptive mutations are effected by a sophisticated genetic information-processing mechanism that Shapiro calls natural genetic engineering (Shapiro 1999a, b, 2002).

An early report of a manifestation of an adaptive mutation induced by an environmental cue was made by James Shapiro (1984). He inserted a *Mu* prophage into one of the genes in the arabinose codon of *E. coli*. The insertion blocked the lactose genes downstream of it, preventing the bacterium from metabolizing lactose. If the *Mu* element is deleted, the lactose genes become fused to the control gene of the arabinose codon, providing ON/OFF control for the lactose gene. The lactose gene will then be expressed when both lactose and arabinose are present. Under normal circumstances, the *Mu* is not deleted and the fusion does not occur. But when both lactose and arabinose are present the fusion occurs enabling the bacteria to metabolize lactose. The *Mu* deletion and gene fusion is then a genetic change that occurs when it is adaptive. Somehow, the presence of arabinose and lactose, which will permit lactose metabolism when the *Mu* element is excised, induces the excision of the *Mu*. Shapiro's *Mu* modified *E. coli* exhibit an adaptive genetic change when the environment is favorable for it.

About 25 years ago, Barry Hall prepared a strain of *E. coli* bacteria that could not metabolize lactose (Hall 1982). For this set of experiments Hall prepared a strain that lacked the gene encoding the first enzyme of the array of lactose enzymes, which prevented the bacteria from metabolizing lactose. When the bacteria grew and multiplied on another nutrient, but in the presence of lactose, *two mutations* were found to appear in the same bacterium. One of these mutations was in a hitherto unknown structural gene and the other was in its control gene. The mutated structural gene encodes an enzyme that can perform the missing first step in lactose metabolism. Because the mutant bacteria activated a substitute gene, it was able to live on lactose. The gene that mutated had been present all along, but had been dormant. Its normal function is, however, unknown.

Neither of the above two mutations is of any use by itself to the bacterium. For the bacterium to metabolize lactose, both mutations have to occur. In the absence of lactose, these two mutations occur independently and with low probability. They will occur together only by chance, and will do so with a probability of only 10^{-18} per replication. In the presence of lactose, Hall found about 40 of them in just a few days. The lactose in the environment evidently induced the mutations.

Hall (1988) did another experiment with a similar result. He cultured bacteria in the presence of salicin, which is a nutrient the bacteria cannot normally metabolize. But the surprise was that the bacteria have another latent gene, normally repressed by a regulatory gene, encoding an enzyme that can break down salicin. The latent gene will become active if one of a few mutations would occur in the regulatory gene. One is a specific nucleotide substitution. Another is the insertion of either one of two transposable elements.

Hall used a strain of bacteria whose latent gene itself did not work. His strain had in its genome an extra piece of DNA, an insertion sequence known as IS103, having 1,400 nucleotides. It sits upstream of the latent gene, and keeps it from being expressed because it shifts the coding frame and garbles the transcription from DNA to mRNA.

For Hall's strain to metabolize salicin, two genetic changes had to occur. The sequence IS103 had to be precisely deleted. Then the right nucleotide had to be changed, or else a sequence called IS1, or another one called IS5, had to be inserted into the latent regulatory gene. Hall tried to measure the spontaneous rate of the precise deletion of IS103, but found it to be too low to measure. He could say only that its probability is less than 2×10^{-12} per replication. In the absence of salicin these two mutations would occur in the same cell with a chance of less than 10^{-19} per replication.

The Darwinian paradigm says that mutations are random in the sense that they are independent of the value of the mutation to the organism. If they are random, then the chance of the right double mutation in at least one cell of the population in two weeks is about one in thirty million. Hall would have had to wait about a million years to see one of them in his culture. Yet he found that within two weeks about *60 per cent* of his colonies underwent both mutations and could metabolize salicin. He investigated in particular the occurrence of the precise deletion of IS103 in the presence of salicin. (Remember that in the absence of salicin the rate of this deletion is too low to be measured.) In the presence of salicin, he found that within 8 to 12 days as many as 89 per cent of the cells in a colony underwent the deletion. These mutations occur only in the presence of the nutrient the cell needs and whose metabolism these mutations facilitate. The important point here is that elaborate genetic machinery had to be in place for these nonrandom, adaptive, mutations to occur.

John Cairns and his colleagues described a similar experiment (Cairns et al. 1988). They used a strain of bacteria with a defect in the gene encoding the first enzyme for metabolizing lactose. Their bacteria could therefore not metabolize lactose. Cairns's team fed their bacteria lactose and looked for cells to appear that could live on it. They found such cells, and they interpreted their appearance to be the result of mutations. They found that these mutations, which were occurring in a gene encoding an enzyme that breaks down lactose, occurred only in the presence of lactose. They concluded that:

"The cells may have mechanisms for choosing which mutations will occur... . Bacteria apparently have an extensive armory of such latent genes that can be called upon for the metabolism of unusual substrates. The mechanism of activation varies... . *E. coli* turns out to have a cryptic gene that it can call upon to hydrolyse lactose if the usual gene for this purpose has been deleted. The activation of (this cryptic gene) requires at least two mutations... . That such events ever occur seems almost unbelievable..." (Cairns et al. 1988).

The mutations that occurred in these bacteria occurred only when they would be adaptive. They afford the bacteria a method of adapting quickly to certain new environments. They are effected by a variety of genetic rearrangements that turn ON latent, or cryptic, genes. These changes include precise deletions, frame shifts, and insertions. A common thread running through these genetic changes is that a genetic infrastructure has to be in place, which is the elaborate genetic mechanism giving the organism the capability of what Shapiro has called genetic engineering.

Of course, even if we do not yet understand the molecular mechanism of exactly how the environment triggers these adaptive mutations, there is no reason to believe that it is mysterious. The phenomenon seems strange only because it does not fit easily into the paradigm that biology has been working under during the past century. Research efforts have been underway in the last few decades to elucidate the sequence of signaling events leading from the environmental cue to the adaptive mutation.

The several examples cited above indicate that the phenomenon may be widespread in bacteria. Although we do not have as detailed evidence for the phenomenon in animals as we have for bacteria, there is evidence, such as that cited above, that the phenomenon does exist in animals—animals can heritably adapt rapidly to a changing environment. Moreover, animals are known to have the same kinds of genetic infrastructure of mobile or transposable elements that are known in bacteria to effect adaptive genetic rearrangements. In some animals, for example, nearly half the genome consists of these elements, and up to 90% in some plants (Kazazian 2004, SanMiguel et al. 1996).

6. Conclusions

Although one can infer from indirect evidence that evolution has occurred, direct evidence is needed to discover how it in fact works. There is now abundant direct evidence that was not available half a century ago. But all this evidence points more toward nonrandom genetic changes than the random mutations postulated by neo-Darwinian theory. These nonrandom events, moreover, require a sophisticated genetic and biochemical infrastructure for their operation.

We arrive then at the following view of a naturalistic origin of life, which necessarily falls into three successive stages. In stage 3 self-replicating organisms have the genetic infrastructure enabling them to adapt quickly to a changing environment. This is the stage for which we have direct evidence of the mechanism by which it is effected. The infrastructure for this stage had to evolve in the preceding stage.

In stage 2, the genetic-engineering mechanisms required for the third stage had to evolve, and this stage had to begin with a self-replicating organism, which in turn had to result from Stage 1. Since there is no direct observation of the evolutionary mechanism of stage 2, we can only hypothesize how it must have occurred. The neo-Darwinian paradigm suggests that this stage evolved through random heritable changes filtered through natural selection.

Stage 1 had to start from an abiotic origin and arrive at a self replicating organism from which stage 2 might start. There is, as yet, no satisfactory theory that can explain the emergence of life from inert chemicals, but presumably random molecular collisions had to play a major, if not *the* major, role.

The time window into which stage 1 must fit begins with the end of the late heavy meteorite bombardment of the earth, which presumably sterilized the earth in case any life happened to have started before that. The end of this window is the time of the actual appearance of life as revealed in the fossil record. The latest estimate of the origin of life is about 3.85 billion years ago (Holland 1997). The end of the heavy bombardment can be determined from dating the rocks that were delivered by the meteoric bombardment. Erosion on the earth has removed any trace of these rocks, but the moon, which presumably suffered the same bombardment as the earth, has not undergone such erosion. Rocks recovered from the moon by Apollo have been dated by measurements of ^{40}Ar, showing that most of them were formed between 3.75 and 3.9 billion years ago (Duncan et al 2004). The origin of life is thus put approximately concurrent with the cooling of the earth and the time frame for stage 1 is too short to be measurable.

The time frame for stage 2 would date from the first appearance of life until the earliest appearance of the mechanisms of genetic rearrangements. The appearance of mobile genetic elements has been dated to the emergence of the eukaryotes (Ostertag et al. 1999). The eukaryotes are estimated to have arisen between 1.6 and 2.1 billion years ago (Vellai & Vida 1999).

Thus the time for stage 2 is about 2 billion years, and the past 1.8 billion years or so has been that of stage 3. These considerations then lead to the conclusion that from well before the Cambrian, which occurred about 500 million years ago, evolution has been that of adaptive nonrandom genetic rearrangements, triggered by the environment.

This strange picture is what one must accept to believe in a Science-based, naturalistic, emergence and development of life. It is not an impossible picture to accept, but it does not result in a hands-down win for naturalism over divine creation. Just as the acceptance of divine creation requires faith, so the acceptance of naturalism requires faith. Tolerance on both sides of the issue would benefit the debate.

References

Barton, N.H., Jones, J.S., & Mallet, J., (1988). No barriers to speciation, *Nature*, **336**, 13-14.

Brown, W.L. Jr. & E.O. Wilson (1956) Character displacement. *Systematic Zoology*, **5**, 49-64.

Cairns, J., Overbaugh, J., & Miller, S., (1988). The origin of mutants, *Nature*, **335**, 142-145.

Doebeli, M. (1996) A quantitative genetic competition model for sympatric speciation. *Journal of Evolutionary Biology*, **9,** 893-909.

Drake, J.W. 1991. Spontaneous mutation. *Annual Review of Genetics* **25,** 125-46.

Duerden, J.E., (1920). Inheritance of callosities in the ostrich, *The American Naturalist*, **54**, 289-312.

Duncan, R.A. et al. (2004) Identifying impact events within the lunar cataclysm from ^{40}Ar-^{39}Ar ages of Apollo 16 impact melt rocks. *Lunar & Planetary Science XXXV*.

Feder, J.L., Chilcote, C.A., & Bush, G.L. (1988). Genetic differentiation between sympatric host races of the apple maggot fly *Rhagoletis pomonella*. *Nature*, **336**, 61-64.

Fryer, G., Greenwood, P.H., & Peake, J.F., (1983). Punctuated equilibria, morphological stasis and the palaeontological documentation of speciation: a biological appraisal of a case history in an African lake, *Biol. J. Linnean Society*, **20**, 195-205.

Grant, Peter R. & B. Rosemary Grant (2006) Evolution of character displacement in Darwin's finches, *Science* **313** (5784), 224-226.

Hall, B.G. (1982). Evolution in a Petri dish, the evolved β-galactosidase system as a model for studying acquisitive evolution in the laboratory, *Evolutionary Biology*, **15**, 85-150.

Hall, B.G. (1988). Adaptive evolution that requires multiple spontaneous mutations. I. Mutations involving an insertion sequence, *Genetics*, **120**, 887-897.

Holland, H.D. (1997) Evidence for life on earth more than 3850 million years ago. *Science*, **275**, 38-39.

Jackman, T., Losos, J.B., Larson, A. & de Queiroz, K. (1997). Phylogenetic studies of convergent adaptive radiations in Caribbean *Anolis* lizards. p. 535-557 in T. Givnish & K. Systma, Eds., *Molecular Evolution and Adaptive Radiation*. Cambridge Univ. Press.

Kazazian, H.H. Jr. (2004) Mobile elements: Drivers of genome evolution. *Science*, **303**, 1626-1632.

Losos, J.B. & Schluter, D. (2000) Analysis of an evolutionary species-area relationship. *Nature*, **408**, 847-850.

Losos, J.B. (2001). Evolution: A lizard's tale. *Scientific American*, **284**(3), 64-69.

Losos, J.B., Jackman, T.R., Larson, A., de Queiroz, K., & Rodríguez-Schettino, L., (1998). Historical contingency and determinism in replicated adaptive radiations of island lizards. *Science,* **279**, 2115-2118.

Maynard Smith, J. (1966). Sympatric speciation. *American Naturalist* **100**, 637-650.

Mayr, E., (1963). *Animal Species and Evolution*, Belknap Press, Harvard University.

Mayr, E. (2004) 80 years of watching the evolutionary scenery. *Science,* **305**, 46-47.

McPheron, B.A., Smith, D.C., & Berlocher, S.H. (1988). Genetic differences between *Rhagoletis pomonella* host races. *Nature,* **336**, 64-66.

Ostertag, E.M. & Kazazian, H.H. Jr. (2001) Biology of mammalian L1 retrotransposons. *Annual Review of Genetics*. **35**, 501-538.

SanMiguel, P. *et al.* (1996) Nested retrotransposons in the intergenic regions of the maize genome. *Science* **274**, 765-768.

Shapiro, J.A. (1999a). Genome system architecture and natural genetic engineering in evolution. In *Molecular Strategies in Biological Evolution,* L. H. Caporale (ed.) Annals N. Y. Acad. Sciences, **870**, 23-35.

Shapiro, J.A. (1999b) Transposable elements as the key to a 21st-century view of evolution. *Genetica*, **10**, 171-179.

Shapiro, J.A. (2002) Genome organization and reorganization in evolution, Formatting for computation and function. In *From Epigenesis to*

Epigenetics, The Genome in Context. L. Van Speybroeck, G. Van de Vijver & D. De Waele (eds.) Annals N. Y. Acad. Sciences, **981**, 111-134.

Smith, D.C. (1988). Heritable divergence of *Rhagoletis pomonella* host races by seasonal asynchrony. *Nature*, **336**, 66-67.

Spetner, L.M. (1997). *Not by chance!* Judaica Press, Brooklyn.

Thoday, J.M. & J.B. Gibson (1962). Isolation by disruptive selection. *Nature* **193**(4821), 1164-1166.

Thomson, K. (2005) *Before Darwin: Reconciling God and Nature*, Yale University Press, New Haven, CT.

Vellai, T. and Vida, G. (1999) The origin of eukaryotes: the difference between prokaryotic and eukaryotic cells *Proceedings of the Royal Society B: Biological Sciences*, **266**, 1571-1577.

West-Eberhard, M.J. (1986). Alternative adaptations, speciation, and phylogeny (a review), *Proceedings National Academy of Science USA*, **83**, 1388-1392.

West-Eberhard, M.J., (1989). Phenotypic plasticity and the origins of diversity, *Annual Review of Ecological Systems*, **20**, 249-278.

West-Eberhard, M.J., (1998). Evolution in the light of developmental and cell biology, and *vice versa. Proceedings National Academy of Sciences, USA* **95**, 8417-8419.

West-Eberhard, M.J. (2005) Developmental plasticity and the origin of species differences. *Proceedings National Academy of Science USA*, **102, suppl. 1**, 6543-6549.

Williamson, P.G., (1981). Palaeontological documentation of speciation in Cenozoic molluscs from Turkana Basin, *Nature*, **293**, 437-443.

Dialogue

Steve McGrew

My background is in physics, electrical engineering, optics and biology. In 1993 I developed and still often use a genetic algorithm software package. Here is my critique.

In your chapter, "The evolution controversy and randomness", you begin with the thesis that both the creationist position and the evolutionist positions are based on a faith. In Section 2 you introduce a second thesis (which I would say is really the most important thesis in your chapter) to the effect that evolutionary experiments and the fossil record give some reason to suspect that evolution is driven by something more directed than simple random mutations and natural selection.

The examples you cite indicate that adaptively useful mutations occur much more rapidly than a straightforward Darwinian model would allow, and that multiple mutations sometimes occur that, together constitute an adaptive mutation but individually confer no advantage. It appears that you have used an implicit assumption that the probabilities of synergistic multiple point mutations would be found by multiplying the probabilities of the individual mutations—which would make such multiple mutations very improbable.

The reasoning you present seems to boil down to this: Because evolutionary change seems almost always to be adaptive to a changing environment, and because evolutionary change appears to occur much more rapidly than possible according to a straightforward use of point mutation probabilities, you conclude that there must be an unknown process or inbuilt mechanism that drives the evolutionary process *toward* adaptation. Almost as a side note you mention a bottom-line conclusion (in effect) that because that process or mechanism is unknown, we cannot rule out the possibility of divine creation. In the context of your chapter, it seems that divine creation in that case would be in the creation and introduction of that mechanism, or even in ongoing divine direct guidance of the evolutionary process.

Though I agree with the general course of your arguments, and even agree with the idea that all reasonable possibilities should be considered when looking for explanations of observed phenomena like rapid evolutionary adaptation and punctuated equilibria in the fossil record, I think you may have overlooked some important background evidence. There are several entirely natural mechanisms that can dramatically increase the expected rate of evolutionary adaptation.

One class of such mechanisms is the lateral transfer of genes between species, especially between bacteria, such as through the action of retroviruses. In organisms that reproduce sexually there is a dramatically increased rate at which favorable mutations are accumulated into the genome of an individual.

Another class of mechanisms that clearly exists has to do with the algorithmic structure of the processes driving embryonic growth—that is, of the process by which instructions encoded in DNA are executed to build a mature organism. There is plenty of reason to expect that what are commonly called "macromutations" can occur when one or a few genetic factors restructure parts of the process, resulting in dramatic changes such as the number of body segments, the number and arrangement of digits, the density of blood vessels, limb proportions, routing of nerves, etc. In fact, it seems that most of the really significant adaptive differences between species are more in the timing and sequence of processes occuring in embryonic growth than in the proteins encoded by genes.

At least in many "higher" organisms, there is yet another very important mechanism: storage and recycling of obsolete genetic information. Within a single human genome, there are vast stretches of DNA of unknown function, some portions of which appear to be non-functional fragments of old genes. Recombination can easily drop those fragments into functioning genes, resulting in mutations much greater than single point mutations.

The Baldwin effect provides a plausible mechanism for sharply focusing natural selection in an adaptive direction [Sterelny, K. (2004). *A review of Evolution and Learning: the Baldwin Effect Reconsidered* edited by Bruce Weber and David Depew. *Evol Dev* **6**(4), 295-300].

All of that said, I agree with you that life probably has, somewhere along its evolutionary path, acquired mechanisms for enhancing the efficiency of the evolutionary process. To whatever (unknown) extent it is possible for our DNA to code for chemistries, physical structures, or control loops that respond to specific environmental stresses by increasing mutation rates in genes that are overworked by stress, it is reasonable to expect that evolution has explored the adaptive value of that coding. Sure, there are open questions, lots of them. We don't know nearly enough about the inner workings of the cell, nor of the genetic programs that control embryo growth, nor of the factors that influence mutation rates at different locations in the genome, to guess the actual rates and directions of mutation occurring in nature.

However, none of this really has anything to do with the question of whether or not divine intervention has occurred at any stage in the evolution of living things. Like an aerodynamics engineer whose model shows that a bumblebee cannot fly, we should question our model rather than rush to the conclusion that bumblebees are levitated by divine intervention. As scientists (and engineers) we are obligated to keep reaching for natural explanations even if natural explanations seem to be unreachable at our current level of knowledge.

Massimo Negrotti

The arguments and the empirical findings proposed by Prof. Spetner open a new way of looking at evolution without any deliberate objective of defending aprioristically a creationist point of view. Instead of limiting himself to supporting a thesis adopting more or less persuasive theoretical issues (as do neo-creationists such as W.A. Dembski and M. Behe), Spetner's main argument points directly towards supporting the plausibility of a view according to which mutations, and consequent evolution, may often be the result of non-random processes. Instead of resorting to the usual themes of the very low probability of almost all chains of mutation/selection, or the 'irreducible complexity' of almost all biological systems, Prof. Spetner, maintains that biological systems show "an adaptive phenotypic-plastic change induced by an environmental cue, which later gets fixed through a genomic change", along with other researchers he quotes. This observation is based on a wide range of empirical evidence, at least with regard to bacteria, wherein cryptic or latent genes switch on, thus initiating a long-term stratagem of molecular processes culminating in the change, and the survival, of the organism.

As I am a methodologist and not a biologist, I especially appreciate, in Spetner's exposition, the rigorous attempt to draw the best hypotheses from experimental findings, without thereby claiming to solve the problem of evolution in its entirety. In fact, Prof. Spetner does not deny the action of chance in all cases: he simply wishes to suggest, if I understand him correctly, that it is not the only means that biological systems have at their disposal in order to evolve and survive.

In 2006, independently of Prof. Spetner's account, I published a paper in an Italian journal under the title "Intelligent Design or Design of Intelligence?" (*Prometeo*, Milan, 24, 93), which supported the idea that the random-based theory of evolution—although very simple and persuasive, not only at a biological level, but also at many other microscopic and macroscopic observation levels—would not be able to cover all circumstances, and should therefore be integrated with some other hypothesis, particularly when we are dealing with micro-biological systems.

The most suitable tool, in my opinion, was, and remains, the cybernetic model, and I'm very happy to learn that Spetner agrees with some highly specialized scientists in the field, such as J.A. Shapiro, when they speak of 'genetic engineering' as an ability of the micro-organism itself. Thanks to this model, cells and bacteria could be attributed with the ability to resort to a cybernetic self-regulation process with which they could develop biochemical strategies to respond to various environmental challenges. Incidentally, such processes could explain why many systems, although very distant from one another in space or time, evolve equifinally—that is to say, towards the same goal—as emphasized by proponents of so-called *Convergent Evolution*, as if a given class of problems could be faced only by means of a given class of solutions. In fact, bacteria or cells often seem to find an effective and sufficiently rapid solution to environmental challenges, which suggests the triggering of some form of 'calculation' that quickly bypasses the long road that a classical random-based procedure would need to follow.

The only problem with a cybernetic approach—and also, I think, with a genetic-engineering one—is that a self-regulation loop, in order to be a well-ordered and meaningful process, should, it would seem, be teleologically established, because the system must somehow *know* that specific values of its parameters must be preserved. When Prof. Spetner speaks of an 'environmental cue', he understandably avoids other, more generic expressions, such as, for example, 'environmental *events*, or *forces*, or *changes*', because he is aware that not all environmental facts can be 'cues'. Indeed, most environmental changes will carry little or no relevant information for a given system. An environmental fact thus becomes a *cue* for an organism if, and only if, that organism is made *aware*, in some sense, that things are developing against its preferential state, thereby inducing, as a deterministic consequence, the initiation of a process of regulation or correction.

A cybernetic process may place itself at two different levels: a purely energetic one and an informational one. In the former situation, the concept of feedback deals only with the energetic adjustment that follows some event to be controlled (e.g., Watt's regulator). In these systems there are no choices nor decisions or calculations to be made, as their very structure provides the appropriate adjustments.

In the latter situation, by contrast, a feedback is established by some form of decision-making structure, which *evaluates* (i.e., transduces or translates, recognizes, calculates, compares) the inputs with respect to the relevant parameters to be preserved. The correct measurements are stored in the system's memory, thereby describing its *preferential state*. I believe that, in a biological system, this is the only meaning we can assign to the term 'determinism' as used by Losos—that is to say, to a teleology that forces the selection or activation of strategies capable of guaranteeing the homeostasis of the system.

It seems clear to me, at this point, that the principal problem amounts to answering the question: What evidence do we need in order to scientifically persuade ourselves that within bacteria and other cells there are sub-systems dedicated to storing the 'preferential state' and the corresponding responses to be effected when that state is challenged by different environmental cues? Furthermore, why could an evolutionist not claim that even this 'instinct for survival', stored in the genome, is one of the main outcomes of a random-based evolution? We could suggest, as a mere conjecture, that surviving species – including those that subsequently die out for one reason or another – are the only ones, among the many created, that have developed the aptitude to preserve themselves both phenotypically and genotypically. The only obvious alternative is to believe that all species are given such an aptitude on the basis of a universal design, which is a faith-based view and not a scientific theory.

An interesting situation arises when the system does not find, in its memory, any stored response—not even an analogical one—to deal with a dangerous input coming from the environment. In these situations, a truly cybernetic system can: a) die, or b) withdraw itself to a state of stand-by behaviour, waiting for better times, or c) try a random response. It is clear that the third possibility, very frequent in higher species, would require, in order to be accepted as a valid hypothesis, some plausible insight regarding the 'random engine' capable of generating a random-based response within the system.

In any case, the engineering, or cybernetic, hypothesis has several intellectual and methodological strengths. In my opinion, such a hypothesis is even compatible with the evidently random-based dynamics of systems at every observation level: the trial-and-error behaviour of many animals and plants, or the so-called *random walking* of bacteria, may belong to sets of responses to environmental changes stored along with others, of a non-random-basis, within all systems. The resulting diversity and complexity of the biosphere is also compatible with a non-random hypothesis, at least according to the classical Aristotelian definition of chance as the result of two (or more) events independent of one another, or according to Laplace's view of chance as ignorance, on our part, of what is actually going on. On the other hand, the reference to the concept of *random walking*, introduced by R. Feynman in the field of quantum physics in relation to particle movement, reminds us that this expression—or something like it—could be involved in the discussion, along with the possible reduction of all phenomena to a random-based origin at a sub-microscopic level. Also this alternative, however, would require some thesis, or conjecture, regarding just what might constitute a 'random engine' which generates events without antecedent.

Nevertheless, we should not forget that Darwin himself, at the beginning of the 5th chapter of *The Origin of Species*, admits having "hitherto sometimes spoken as if the variations so common and multiform in organic beings under domestication, and in a lesser degree in those in a state of nature, had been due to chance". He continues: "This, of course, is a wholly incorrect expression, but it serves to acknowledge plainly our ignorance of the cause of each particular variation."

Steve McGrew

Although "operons" per se occur only in bacteria, the concept applies under different names and in a somewhat more convoluted way to eukaryotes like yeasts and humans. So, we can make use of the concept of "operon" as meaning a group of genes, along with control mechanisms, in which the control mechanisms affect the expression of the genes. The control mechanisms can in principle be affected by such things as the relative amounts of various nutrients, metabolic waste products, proteins expressed by other genes outside the operon, the amount of sunlight, and even mechanical forces or emotional stress.

There is evidence that in some cases the state of activation of a gene can be inherited. For example, so-called "epigenetic inheritance" effects have been observed such as inheritance of a gene's methylation state. We could imagine that a "directed mutation" could occur in two steps: first, environmental cues put a gene into an inheritable but temporary deactivated state in which it is not expressed; second, if in its deactivated state the gene is less likely to be accurately copied during DNA replication, the gene is more likely to be permanently deactivated due to random mutations. According to that scenario, epigenetic inheritance provides a way for a line of descent to "test" a mutation before committing to it irreversibly.

In the scenario you suggest, latent operons would exist that, if activated in germ cells, could alter the phenotype via embryonic development in ways that could not be accomplished by activating the same operons in non-germ cells. *If* there were a mechanism by which those latent operons could be activated in germ cells by environmental cues; and *if* the organism and its progeny needed to survive and reproduce in an environment whose character changed on a time scale comparable to the generational time for the organism; then it seems that natural selection should support the continued existence of those operons and the mechanisms that can activate them. Once established, the operons and mechanisms should be stable.

The biggest problem I see with either of those two scenarios is that they do not explain genomic differences between species. The first scenario could trim deleterious genes from the genome fairly efficiently, but it could not create novel genes. The second scenario would not change the genome at all unless somehow aided by effects like those in the first scenario. But I don't yet see how either of the scenarios would speed up the creation of new genes tailored to meet new needs. If the mechanisms I keep pointing out in these communications can be demonstrated to be inadequate to explain the rate of adaptive evolution, then we need to look for mechanisms that *create new genes in an adaptively directed way*. However, I don't think you, Lee, have answered my arguments regarding the adequacy of known evolutionary mechanisms.

Lee Spetner
I cannot understand why you, Steve, misrepresent my position on divine creation (or intervention). The default position of the divine creation of life that I mentioned in my previous reply is historically the philosophical position held almost unanimously by Western scientists before Darwinian theory took hold. That is just the way it was—a historical fact. The default is not arbitrary and that is why neither pagan creation myths nor magic, which you want to suggest, can qualify as the default. The creation position was displaced only because Darwinian theory appeared to offer a naturalistic, and therefore scientific, alternative to the philosophical position of creation. Consequently, if our theory of evolution cannot account for the origin of life, then it has not displaced the creation position. That is not because creation is a scientific position—it is not. But it was the previously held dominant position.

Naturalism as a working assumption is fine. It is the only way science can go. That does not mean, however, that creationism is an invalid philosophical position, as many claim. Surely, creationism is not scientifically testable. Some critics of evolutionary theory hold that evolutionary theory also is not testable. I, however, hold that it is testable, but it has failed the test. Testability alone, therefore, does not give evolutionism an advantage over creationism as a philosophical position.

I think an experimental program can be defined to test my hypothesis that some organisms can adapt to a changing environment through triggers from environmental cues. There is already some indication of this possibility in the references I cited of the works of Jonathan Losos (who works with phenotypes) and of James Shapiro (who works with genotypes). But no matter how far such research would go, one cannot expect it to shed any light on the creation position, because the latter is not subject to scientific test, as you have noted.

You ask what I mean by *random genetic change*. By that I mean changes that occur as accidental events, independent of their any usefulness they may have. These are precisely the changes on which current evolutionary theory is based. James Shapiro contends that evolution is driven, not by random, but by nonrandom mutations, which are part of cellular biochemistry rather than stochastic events. These nonrandom events involve high specificity and feedback. The organism, moreover, has a complex built-in mechanism that effects these events. On this basis I have suggested the possibility that the triggers for these events may be environmental cues. Environmental cues are usually manifest as stress, and we know that stress can, through hormonal secretions, affect every cell in a body.

You write, "… it is very wrong to think that variation and selection have been proven incapable of providing a complete explanation of the origins and evolution of life." It is, of course, usually difficult and often impossible to prove a negative. But if life is to be explained as a natural phenomenon, then the burden of such explanation lies on the advocate of evolution to produce a theory to account for it, not on the opponent to show it is impossible. Since evolution is a historical theory, and since we cannot go back and examine how life originated, to support evolution as an explanation of life's origin we must have at least a theory that can account for it. A good deal of support claimed for evolution consists of stories of how, in a very superficial way, evolution *might* have happened. None of these stories ever comes with a plausible examination of the details of just how that could have happened. For example, no one has ever displayed a sequence of possible mutations, each more adaptive than the previous, that could account for at least a portion of M-to-M evolution. Moreover, no random mutations have been observed that could possibly form the basis of the mutations that would be required for M-to-M evolution. In this meaningful sense, evolutionary theory has failed the test.

Steve McGrew
I certainly have not intended to misrepresent your position. I think I have misunderstood your meaning of "default". If I understand correctly now, by "default" you mean a return to a position that was previously widely held; and you do not intend "default position" to necessarily be a valid or even scientifically plausible position. Would Daoist philosophy, or other such philosophies without conscious divinities, be an even more basic default position?

You write, "For example, no one has ever displayed a sequence of possible mutations, each more adaptive than the previous, that could account for at least a portion of M-to-M evolution." By "M-to-M" evolution I assume you mean "molecule-to-man".

I think it is a common mistake to think that evolution proceeds in a linear and one-directional fashion, as would be indicated by "a sequence of mutations, each more adaptive than the previous". In fact, there is no reason to believe that A) evolution has proceeded in a linear fashion, nor that B) every mutation in an evolutionary process must be "more adaptive than the previous". Animals are very good at surviving injuries as serious as a broken leg or a punctured lung; and they are also good at surviving genetic defects. Consider the vast range of genetic variations within the human population (or the population of any other species), and the fact that the vast majority of human beings are not only fertile but cross-fertile with the vast majority of other human beings of the opposite sex. Presumably according to any single measure of "adaptive" fitness, there is one and only one male individual out of six billion whose genome is "more adaptive" than that of all others. But we can be very, very sure that in a million years that one individual will *not* be the single common ancestor of all surviving descendants of today's generation. Instead, we can be very, very sure that virtually all of those descendants will have ancestors in our current generation whose adaptive fitness is less—probably much less—than that one individual. In "M-to-M" evolution, we can be equally sure that there needn't have been, and almost certainly was not, any one sequence of mutations, each more adaptive than the previous.

This is fully consistent with a modern, informed, theory of evolution based on "random" variation and natural selection. The fact is that *there is no single measure of adaptive fitness* that can be applied to an individual at the time it is living. Nobody can say that this genome or that one is more fit than another. Fitness of an ancestor in the context of evolution should only be judged by whether or not an unbroken line of descent stretches from the ancestor to fertile members of the present population.

By the way, I like your definition of "random genetic change". Some might argue about the meaning of "accidental", but to nail down a testable meaning for that word would require more effort than it's worth!

Lee Spetner
Although it is often said that evolution does not proceed in a straight line, you, Steve, have misapplied that mantra. Evolution going forward is indeed said to be many branched, but if you follow any one of those branches backwards, there is only one path—it is not branched, it is unique. Following such a path backwards, there *must* be, according to neo-Darwinian theory, a succession of mutations where the selective values (each within its own environment) going backwards must be monotonic nonincreasing. My point is that no one has shown such a possible sequence of mutations.

You wrote, "… we can be very, very sure that in a million years that one individual will *not* be the single common ancestor of all surviving descendants of today's generation. Instead, we can be very, very sure that virtually all of those descendants will have ancestors in our current generation whose adaptive fitness is less—probably much less—than that one individual." I do not know what you mean by "our current generation," but in any case, this is not what the prevalent theory of evolution says. I don't think evolution happened this way, but this is the theory, against which I am arguing.

The prevalent theory of evolution is that a random genetic change occurs in a single individual, granting that individual a selective advantage. If that individual is lucky enough to survive and produce offspring, and if it is lucky enough to have its selective advantage expressed in its having more viable progeny than other individuals, the new mutated genome increases its numbers in the population. If that process is continued over several generations the mutated genome will be expressed in a larger and larger fraction of the population. To make it simple, suppose the population size remains the same, so that the increase of the subpopulation of the mutated genome will come at the expense of a decrease in the subpopulation of the original genome. As the mutated subpopulation increases, less and less luck will be required for the process to continue. Eventually the mutation with its selective advantage may take over the population. This scenario is a random walk of the number of mutants in the population, for which there are only two stable states—either the mutation disappears or it takes over the population. If the mutation was sufficiently rare that it did not occur a second time in the population, then the original mutated individual will certainly be the ancestor of the entire new population, contrary to what you wrote. In a sufficiently large population of bacteria in the presence of an antibiotic, an adaptive mutation could occur in several individuals in each generation, and then no one of them would be the ancestor of the entire population. I don't think that a long process of successively more adaptive mutations will occur at all, for a few reasons. One of which is that I doubt that there are many single random mutations that can grant a selective advantage strong enough to make a difference.

I think that all the examples of rapid evolution that have been observed (and we can observe only rapid evolution) are effected by genetic changes triggered by environmental cues, what I have called the *nonrandom evolutionary hypothesis* (NREH). When I proposed the NREH about a dozen years ago, several steps in such a process were already known. Additional support for this hypothesis has recently appeared in Lema, Sean C. (2008) The phenotypic plasticity of Death Valley's pupfish. *American Scientist* **96**: 28-36.

Steve McGrew
Dear Massimo, though Dick Gordon did not quite invite you, me, and Professor Spetner to engage in a 3-way dialog, he did not say we shouldn't—so why not? Accordingly, here are my thoughts concerning your response.

You make several key points, in effect that: a) we should ask what physical mechanisms in an organism might be able to calculate appropriate responses to environmental cues and express those responses in the form of directed genetic changes; b) some sort of "random engine" would be required to respond to environmental changes never encountered before; and c) a capacity to calculate and implement appropriate responses might be a natural outcome of Darwinian evolution.

Every homeostatic system, whether it be the climate of a moist planet like the Earth or the metabolic system of a cell, is composed of feedback loops that automatically provide responses supporting survival of the system. The real challenge posed by Professor Spetner is to seek mechanisms that a feedback loop might use to direct mutations into useful directions that support homeostasis not for the organism but for its line of descent. Note that in the Darwinian view, "homeostasis" equates to "producing viable offspring". One highly plausible mechanism would simply be increased mutation rates in response to stress. Higher mutation rates can result in more rapid evolution. Another highly plausible mechanism is natural selection. It is the very nature of natural selection that higher-fitness variants are amplified in the population and lower-fitness variants are de-amplified so that, as long as the environment changes slowly enough with respect to the rate of mutation, the mean phenotype in the population automatically adapts. Sexual recombination coupled with natural selection provides "targeted" variants, in that it greatly increases the likelihood that independent high-fitness characteristics of different individuals in any generation will be combined in their descendants. I suspect there are other such randomly driven but structurally constrained and entirely natural mechanisms that "channel" evolution toward more effective adaptation. Such mechanisms could very plausibly arise in a way analogous to the way that hibernation, food storage, sporulation, nest building and other such "forward-looking" behaviors can arise via natural evolution. It is obvious that random events provide a vast supply of variants; and it is also obvious that reproduction and mutation are constrained by the structures and circumstances in which they occur. Plausible mechanisms for genetic adaptation exist. What we lack are computational models adequate to tell us how fast and how effectively a cascade of descendants can adapt when driven by those mechanisms.

Massimo Negrotti
Dear Steve, let me answer your comments. Even if Dick would not agree to add this 'two-ways' discussion, I wish to underline that your remarks have been very useful for me.

From my viewpoint, the several 'mechanisms' you refer to, are particular specifications of the more general self-regulation model.

Nevertheless, as far as a discussion on the 'beginning status' of matter and life is concerned, the main point is that both feed-forward and feed-back loops have to base themselves on some pre-condition, namely that of an intrinsic teleological feature towards survival.

Given such a pre-condition, the list of the mechanisms that can be adopted by the system (both for surviving itself or for allowing the descents to survive) may be as long as we want, according to the level of 'intelligence' or 'rationality' of the considered system.

We may also conceive the above pre-condition as a sort of 'attractor' (or, if you prefer, as a 'rejector') which qualifies or classifies the environmental cues according to their relevance for the survival of the system, but, in any case, it seems to me that some kind of a 'control value' is needed in order to give to living systems their typical behaviour.

To be clear, it is not my intention to be renewing philosophical doctrines like the 'élan vital' by H. Bergson or the vitalism by H. Driesch. I'm only persuaded that living systems do not work on the basis of simple energy-based feed-backs but on information-based ones. Therefore, the homeostasis of living systems cannot be described as pure physical recovery of a certain state.

You are right when you say that "The real challenge posed by Professorr Spetner is to seek mechanisms a feedback loop might use to direct mutations into useful directions that support homeostasis not for the organism but for its line of descent." But let me remark that the extension of the cybernetic models to the offspring evokes an even higher level of self-regulation and, therefore, of teleology.

Really I do not have knowledge enough to follow you in your detailed descriptions of what happens or can happen with the many strategies that living systems follow to survive.

I would only underline that randomness could be conceived not only in a passive way but also in an active one.

In other words, a system in trouble with environmental cues, can resort to randomness as a source of responses of both 'waiting' that, by chance, the things change by themselves or activating some 'mechanism' able to generate random (and then filtered) responses. This seems to happen with all high level systems—humans included. Why we should exclude that this could happen even in the simplest cell?

I think that the discussion—but this could be the theme of another book, maybe!—could gain further interest if we should place at its core just the theme of randomness. Not only in its classical terms (both philosophical and mathematical) but also, and mainly, in its most fascinating aspect (at least for me): are there living systems able to generate 'true' random sequences or whatever sequence has a random configuration only in the classical sense of our ignorance of the micro events that are acting at some lower level on the basis of some deterministic rule? By the way, the relevance of such a question seems to me crucial also considering the wide adoption of expressions like 'intrinsic randomness' by quantum physicists.

Thank you very much for your patience in reading the above conjectures!

Steve McGrew
Dear Massimo, in your original reply to Lee's chapter, you said something about "teleology" that I thought particularly interesting. You said, "…a self-regulation loop, in order to be a well-ordered and meaningful process, should, it would seem, be teleologically established, because the system must somehow *know* that specific values of its parameters must be preserved".

Then you went on to say, "The correct measurements are stored in the system's memory, thereby describing its *preferential state*. I believe that, in a biological system, this is the only meaning we can assign to the term 'determinism' as used by Losos—that is to say, to a teleology that forces the selection or activation of strategies capable of guaranteeing the homeostasis of the system".

If I understand correctly, you are thus using "teleology" to mean processes that detect deviations from some optimal (preferential) state and respond to deviations by taking actions that can correct those deviations.

Normally we would think of the "optimal state" as being a static condition. But it can also be a dynamic condition. "Balance" is one such condition, and "alive" is another such condition. In evolution, the real optimal state is a dynamic condition of *continued existence of the line of descent*. Natural selection and limited resources guarantee that only a few lines of descent will continue on to each succeeding generation. What distinguishes today's living species from those that have gone extinct over the billions of years of life's history is simply that today's living species are the product of unbroken lines of descent.

It is entirely appropriate to ask what mechanisms function in living organisms to help ensure continuity of line of descent. We know of several such mechanisms, including:

1. Ordinary homeostatic mechanisms that improve odds of survival of an individual until it can produce offspring, including "forward-looking" behaviors that prepare individuals for changes in the environment. (Engineers often design "feedforward" control loops that forecast environmental changes and prepare a system for the changes in advance. A computer is not needed for this!)
2. Exploratory behaviors that enable individuals to find new niches or strategies that reduce competition from other members of their own species, thus increasing survival odds for their own offspring.
3. Behaviors that support survival of offspring, sometimes to the detriment of the parent.
4. Behaviors that modify the environment to a state that promotes survival of the individual and its offspring (e.g., beaver dams).
5. Mate selection behaviors that enhance odds that an individual's offspring will have favorable combinations of inheritable traits.
6. Mechanisms such as that ensure wide genetic variation but channel that variation into directions having high likelihood of viability (e.g., sexual reproduction, crossover).

7. Mechanisms that allow genetic mutations but control the mutation rates (e.g., DNA repair mechanisms, increased mutation rates in presence of environmental stress).
8. Mechanisms that store something like a genetic history of an individual's line of descent, providing a "database" of genetic factors that were useful to the individual's ancestors (e.g., genetic variation within a sexually reproducing population, "junk" DNA).

Lee Spetner is convinced that mechanisms 5, 6, 7 and 8 are not sufficient to explain how quickly a species' genome can change to adapt to changing environment, but he does not propose any additional mechanisms (other than perhaps divine intervention) that might explain it better.

You propose that there may be a cybernetic process, built into the cellular machinery or into the genome itself, that analyzes environmental cues and makes decisions based on stored information about how to change the genome to adapt most effectively.

I would suggest we do not have computational models yet that can tell us whether or not mechanisms 5, 6, 7 and 8 are sufficient to explain the rate of adaptation. But more germane to the perspective of our current discussion, I would suggest that mechanisms 5, 6, 7 and 8 constitute the cybernetic process you hypothesize and, indeed, provide the "intelligence" behind design in Nature.

Massimo Negrotti

Dear Steve, your list of 'mechanisms' that allow the descendents to survive is very fascinating and persuading.

Cybernetic loops surely apply also to the dynamics of systems: positive feedbacks are, partially, means for explaining the development of a species. In this frame, the survival of descendents is only a particular case of the general ability of cybernetic loops to explain the survival of an individual (as a stationary system) and of a species (as as a self-amplifying process).

The problem of randomness remains, anyway. In my opinion—that is based on a pure theoretical view in terms of general cybernetics—the capacity of generating random sequences of 'trials' could be one of the 'mechanisms' that systems can enable when they have no possibilty to put at work other procedures (like the ones you list).

If this would be true—as it seems to be true in high level species, or, at least, in many human activities—then the main problem becomes: what enables a system to generate such random sequences?

It is sure that, for doing so, living systems do not act like computers as we know them—whose random generation is correctly qualified as 'pseudo'—but some form of procedure, not necessarily of a mathematical type, is anyway needed.

In other terms, a generalized 'random walk' seems to be a rather diffuse model of action both of particles and for living systems, but we need to understand how it is generated just to understand the nature of randomness which is not only in the intimate structure of the matter, but even inside us as a resource.

Steve McGrew
Dear Massimo, actually, randomness is automatically generated in the process of reproduction. Reducing that randomness to a safe level is a much more important challenge for living organisms, than generating the randomness in the first place.

I'll give you several examples of how randomness is automatically generated.

1. Point mutations
Point mutations are "mistakes" in copying DNA. The rate of mistakes is always nonzero because of the quantum mechanical nature of the molecular processes involved. Organisms minimize the rate of point mutations via several mechanisms. One important method is to use "DNA repair enzymes" which detect differences between the original and the complementary DNA copy and cause the error to be repaired. These enzymes are not perfect, but they work pretty well.

2. Sexual recombination
The human genome is parcelled out into 23 pairs of chromosomes. Sperm (and ova) contain just one member of each pair. There is apparently no mechanism to bias selection of one member relative to the other member, so a given member has a 50:50 chance of being selected (by the way, one member of each chromosome pair is from my mother and one member of each pair is from my father). So, if I had 2^{23} = 8,388,608 children, the chances are roughly 50:50 that my genetic contribution to one child would be exactly duplicated in any other of my 8 million-plus children. The same would be true with respect to my

wife's genetic contribution, so my wife and I could have $(2^{23})^2 = 2^{46}$ = roughly 70 *trillion* children without necessarily duplicating our respective contributions in any two of our children. By the way, sexual recombination makes the entire species (plus any species it is cross-fertile with) a repository of potential genetic variations. Although my wife and I carry different variants (alleles) of many genes, the whole human population carries many more variants of any given gene than my wife and I carry together. Through mate selection, all of those variants become available to the next generation in an unthinkably large number of possible combinations.

3. Crossover

Crossover occurs when the two members of a pair of chromosomes come into contact. The two chromosomes in a pair are both organized the same way, so they can be laid parallel to each other and connected at two points, then the DNA sequences between those points can be switched between the two chromosomes. That way, the genes end up in the right locations on the chromosomes, even though they have been exchanged between the chromosomes. The exact probability of a crossover point occuring at each possible location on a chromosome is not known, but it probably depends on many factors. Crossover "blends" the genes of two parents more intimately that sexual recombination does. We have something on the order of 25,000 genes, plus an unknown number of DNA segments that influence gene expression. Crossover that simply exchanges genes can increase the number of possible genetically different children by a huge factor: from the 2^{46} in #2 above, to $2^{25,000}$. That number is so huge that there isn't even a name for it. And, the crossover points are basically random without any special mechanism to 'create' the randomness. Note that crossover is 'not' a rare event.

4. Gene duplication

Sometimes in the (random) crossover process, or in the process of DNA replication, a gene is duplicated in a child. This does not affect the child much. However, it provides a surplus gene that can be modified without damaging the original necessary gene. Genetic studies have shown that many important genes have almost certainly arisen through gene duplication followed by mutations in the duplicate gene.

My reason for detailing these mechanisms is to show that randomness is easy and occurs automatically I think Lee Spetner's question is really asking if any mechanisms (such as divine intervention) exist to "channel" the random genetic changes caused by these mechanisms, into adaptive directions. I would suggest that there *are* such "channeling" mechanisms, that they are entirely natural, and that they have been visible ever since we first understood the chemistry of reproduction—but that we have not realized what we were seeing.

Massimo Negrotti
Dear Steve, the points 2 and 3 you list seem to me to belong to the realm of random phenomena due to the huge number of interactions that happen within all observation levels of reality: even a crowd in a public square behaves in this way, with analogous consequences in terms of exchange of information.

In contrast, point 1 seems to imply some true regulatory behaviour since if "DNA repair enzymes [...] detect differences" they have to know the right parameters at stake. I would be very interested in knowing the ways enzymes do their job. The errors in the copying process of DNA can be due to quantum phenomena but I assume that the 'recognition' by the enzymes is a fact that places itself at a higher level. Therefore, it implies a sophisticated model of behaviour.

Anyway, randomness seems to be present at all levels of reality, though in different forms. Significantly, the genetist Langaney wrote that in the moment of fecundation a sexual cell is taken out by chance in each of the parents.

What I maintain is that living species 'know', so to say, that randomness is always acting on their own structure (maybe due to different decay times of particles?). As a rule, living systems control the effects of randomness, but, in some circumstances, they resort right to it in order to give responses to environmental cues. This means that, having to control some unknown input, they act *as if* they would allow their internal random generators to work freely. This would start subsequent selection of the most suitable action, possibly, but not necessarily, involving the death of the parents, preserving the descendents.

Steve McGrew
Dear Massimo, repair enzymes work by increasing the energy or entropy difference between an exactly complementary double-strand of DNA and double-strand with a "bubble" of non-complementarity. It is, indeed, regulatory behavior, but it doesn't really require very complicated information processing.

By analogy, a nut and bolt only screw together if they are complementary, and no information processing is required to determine their complementarity. DNA is like a string of nuts and bolts of two different kinds side by side. Nut "A" can only screw onto bolt "B", and nut "C" can only thread onto bolt "D". So, two strings of nuts and bolts are laid next to each other, the nuts from one can screw onto the bolts from the other and form a double string, but only if the two strings are complementary (nut "A" opposite to bold "B" and nut "C" opposite to bold "D"). But if one nut on one of the strings is wrong, there will be a "bubble" in the double string because that nut will not be able to screw onto the bolt in the corresponding position on the other string. A "bolt string repair enzyme" can simply bind to a "bubble" and induce random changes until the "bubble" goes away. It does not need to know which of the two strings is the original one. It only needs to keep changing things randomly at the location of the bubble until the nut and bolt fit each other. There is some probability that the result will correct the copying error, and some probability that the result will be a point mutation. (I haven't calculated the actual probabilities.)

The above analogy is very rough. Actually there are several different kinds of DNA damage, and there are different repair mechanisms for each kind of damage. A good summary appears at
http://en.wikipedia.org/wiki/DNA_repair#DNA_repair_mechanisms.

One way that environmental stress could increase mutation rates is via accumulation of metabolic by-products that interfere with the production or effectiveness of repair enzymes. "Environmental stress" would occur whenever the organism's normal homeostatic functions approach their limits. In that case, the sensing of a need to explore new genetic territory in search of appropriate adaptations would be a simple deterministic process; and the exploratory response (i.e., decreased repair rate -> increased mutation rate) would also be simple and deterministic.

Massimo Negrotti
Dear Steve, yes, I undertsand that many micro-bio phenomena are regulated by 'shape-matching' procedures. Of course, in these circumstances there are no information processes. In this case, the 'preferential state' is, so to say, enbedded in the architecture itself of the living systems (LiS, from now on).

Nevertheless, it is sure that many LiS are able to perform self-regulation by means of information processing loops. The problem is to see if at the micro level there are or not such models of regulation.

Anyway, my deepest interest is in random-based processes as means to respond to environmental cues. I think that LiS, in addition to high-level feedbacks, can resort to random as a means for widening their range of possible responses. In order to do so, they have to enable some special feature of their biostructure, or, maybe, they have to allow external entropy to enter their structure 'analyzing' (accepting or refusing) the flow of combinations that are generated this way. In all cases, LiS of high-level seem to perform comparisons and this implies some level of 'calculation'.

There is a book, now in press in Italian, edited by me and by Prof. G. Lanzavecchia (who is a theoretical physicist) whose title is "*The Chance Enigma*". In this book, we try to clarify not the solution of this ancient enigma, but the new boundaries within which the problem could be placed.

The strange thing is that all the 20 authors speak, as you do, of randomness without defining it and, especially, without any idea on the way LiS can produce it. The only two authors who give some definition are philosophers (one is a colleague from one of the Pope universities), almost in terms of 'free will'.

Does this mean that we use everywhere a concept that cannot be studied in its empirical features? Maybe. The 'randomness wall' seems to be absolutely insurmountable in the realm of particles, but what about for LiS?

In my opinion randomness is not only an epistemological expedient due to our ignorance on what is going on. Rather, as the theory of evolution by Darwin only exemplifies, randomness is a real dynamic condition of empirical reality that can be found at every level. As such, randomness is, at the same time, a permanent menace and a potential resource for LiS. Provided that LiS have at their disposal not only energy—this can be enough for responding to only a part of low-level environmental cues—but also some more or less high-level ability to process information.

In some way, information and randomness are the two faces of the same coin. In fact, when information processing is not enough, many LiS resort—

paradoxically—just to randomness for trying to survive (both as individuals and as species).

In this sense, biological evolution—and even more the cultural one—appears to be driven by a teleology which has to face randomness or to use it according to need.

By the way, the last sentence makes clear why, from my viewpoint, creationism and evolution are quite compatible provided that the object of Creation is not a finished system but a teleology introduced in the matter.

Steve McGrew
Massimo, you bring up a good question: 'What is the definition of "random"?'

I have explored that question extensively, and have reached the same conclusion as you have: randomness cannot be defined. There really is no test that can determine whether or not a sequence of numbers was generated by a "random" process or by a deterministic process; and any definition that does not provide a test is not scientifically useful.

I have developed (and often use) genetic algorithms. These are driven by deterministically generated pseudorandom number sequences such as the digits in the value of "pi", or by any of the more common pseudorandom number generators used in computers. I have not observed any differences between the performance of genetic algorithms attributable to their use of different pseudorandom number generators. It seems that the only thing that matters is that there be no simple correlation between the GA, the problem being solved, and the pseudorandom number string.

So, I must conclude that in all discussions using the word "random" in the context of evolution, the word "pseudorandom" can be substituted without any loss of meaning. Moreover, I must conclude that "pseudorandom" variations in genes are indistinguishable from "random" variations.

Lee Spetner
Massimo, ciao, Thank you for your contribution to the dialogue, which I guess is now a triologue. I apologize for taking so much time to respond.

I also think that randomness could contribute to the mechanism in what I have called in my book the *nonrandom evolutionary hypothesis* (NREH). The hypothesis is that of a latent adaptive capability triggered by environmental cues in the following way. An organism has a set of latent operons in the genome, each with the potential ability to grant the organism adaptation to one of a set of alternative environments. The environmental cue will activate the latent operon in the germ cells that is appropriate to the environment. Such a genetic change (or mutation) will grant the progeny adaptation to that environment. The adaptation of the population would occur in many members of the population, making it rapid—occurring in just one or a few generations. A mechanism of this sort in multicelled organisms could account for the rapid evolution observed in several species, where adaptation to a new environment has been observed to occur in only a few years. The mechanism in microorganisms may be simpler because they could rely on their short generation time and rapid exponential increase to experiment with alternatives among the various latent operons.

Losos has suggested an alternative—an environmental cue could lead to adaptive phenotypic plastic changes, which could then lead to genetic changes that would improve the adaptation. I cannot, however, think of a mechanism for the phenotypic changes leading to genetic changes. Losos implies that random mutations would produce such a genetic change. He is not alone in this—others have made such a suggestion before. But I fail to see how, even if appropriate mutations occur, natural selection could have much of an effect, since the extant phenotypic adaptation, which is presumably the same as the genetic adaptation, is already present. Your quote from my chapter, "an adaptive phenotypic-plastic change induced by an environmental cue, which later gets fixed through a genomic change", is not my opinion, but a paraphrase of the opinion expressed by Losos. I don't see how that could work.

Steve McGrew

Lee, your arguments appear to be attacking a straw man rather than the informed modern theory of evolution! Perhaps you have been misinformed about modern evolutionary theory on several counts:

1. It is a drastic and seriously misleading oversimplification—or a mistake—to say that every line of descent follows a single, unbranched path when read backwards in time. This is neither a tenet of, nor a logical consequence of the tenets of, modern evolutionary theory. Moreover, it is not true. You have two parents, four grandparents, eight great-grandparents and so on. Each

member of each of your paired chromosomes has converged with the other member of its pair after traveling its own independent line of descent, distinct from the line of descent traveled by your other chromosomes. For example, let's pretend that you have only two paired chromosomes "C1a" and "C1b", and "C2a" and "C2b". You inherited C1a and C2a from your mother, and C1b and C2b from your father. But C1a may have come from your maternal grandmother *or* your maternal grandfather. The same is true for C2a, which may have come from your paternal grandmother *or* your paternal grandfather. The probability that all of your chromosomes have the same ancestral line of descent is zero. The probability that even half of them have the same ancestral line of descent is infinitesimally small. If you were to retreat to stating that any given chromosome, say C1a, can be traced backwards in time along a single path, you still would be wrong. Crossover ensures that if you trace your ancestry back far enough, you will find that each of your genes has arrived in your chromosomes via its own unique path, passing through largely unrelated individuals scattered all over your genetic ancestral watershed. Even individual genes are subject to crossover, and can consist of components contributed by multiple individuals. The term "evolutionary path" is a misnomer; its simplistic meaning is only a crude approximation to that which it attempts to name.

2. You wrote: "If the mutation was sufficiently rare that it did not occur a second time in the population, then the original mutated individual will certainly be the ancestor of the entire new population ...". Unfortunately, that statement is very wrong too. If you had said, "... then the original mutated individual *could eventually* be *an* ancestor of the entire new population", I would agree with you.

   I'll demonstrate in a very straightforward way that your statement is wrong. Suppose two males in a population carry different unique mutations, resulting in different beneficial phenotypic changes. Male A carries mutation Alpha, while male B carries mutation Beta. Assume that the two mutations are on different chromosomes. Eventually, perhaps, all members of the entire population many generations later will carry *both* mutations, Alpha and Beta. Now, answer this: Who is *the* ancestor of all the members of that new population?

   The correct answer is that *both* male A and male B are ancestors of each member of the population. Nobody is *the* ancestor. This argument, by the way, deflates the concept of an "ancestral Eve": a concept suggesting that all human beings descended from a single early human female based on the near-universality of certain components of mitochondrial DNA among humans. Even if one genetic component such as mitochondrial DNA in all living humans could be traced to a single early human female, it would certainly not imply that all chromosomes and all genes in today's human beings could be traced to that "Eve" and her mate (or mates). Yes, it might imply a genetic bottleneck, but *only* with respect to the one component being analyzed. If two different genetic components were to reveal such bottlenecks, it is overwhelmingly probable that those bottlenecks occurred in different individuals in ancestral populations at different times. There could be a single ancestral mitochondrial Eve, a different ancestral bloodtype B Eve, and yet a different ancestral curly haired Eve, all within a single population of modern-day human beings.

So, your statement might be nearly correct when applied to a single advantageous gene mutation, but it is grossly incorrect when applied to the genetic attributes of members of a population, wherein each member's genome comprises a collection of hundreds or thousands of genes.

3. The course of evolution in bacteria and other asexually reproducing organisms is different from that of sexually reproducing organisms, for many reasons. However, if you were to retreat to a position where your statement is to be applied only to bacteria, your statement at the beginning of #2 above would still be wrong but for different reasons. Numerous mechanisms have been demonstrated to exist that cause genetic information to be exchanged between bacteria, both as entire genomes and as small portions of genomes. Like crossover in eukaryotes, those mechanisms provide ways for genes from disparate bacterial individuals to accumulate into single individuals. So, even the genes in a bacterium have arrived via disparate paths.

It is important to understand that an analysis of evolution from the perspective of a single gene is grossly incomplete and leads to dramatically incorrect conclusions such as those expressed in your statements quoted above.

However, all of this should be aside from the real point I think you want to make: that NREH mechanisms might exist and could help explain rapid adaptive genetic change.

In the context of this dialog it will not work to build a plausibility argument for NREH by attacking a straw man. It would be far better to concentrate on trying to find and demonstrate the existence of NREH mechanisms (which by the way are *not* excluded by modern evolutionary theory; they're merely assumed to be unnecessary). If they don't exist, they will not be found. If they do exist, hopefully they will be found.

You stated, "I doubt that there are many single random mutations that can grant a selective advantage strong enough to make a difference." I am curious to know your opinion on how big a selective advantage (in bacteria) needs to be to make a difference. Would a reproductive advantage of 0.1% per generation be enough to make a difference? How about 1%? How about 0.01% or 0.0001%?

Lee Spetner
Dear Steve, in your last two letters you have merely been reciting textbook-type snippets about evolution that are quite irrelevant to our discussion. Perhaps you forgot what the discussion is about. The erroneous path you are on began with your comments on my stating, "no one has ever displayed a sequence of possible mutations, each more adaptive than the previous, that could account for at least a portion of M-to-M evolution. Moreover, no random mutations have been observed that could possibly form the basis of the mutations that would be required for M-to-M evolution. In this meaningful sense, evolutionary theory has failed the test."

If M-to-M evolution has indeed occurred, then a large quantity of information had to have been inserted into the genome of the later life forms that did not exist in the earlier forms. Evolutionary theory purports to account for this information buildup through successive random mutations, each fixed into the population by natural selection. If you disagree with this, then please state how you think evolutionary theory accounts for the information buildup.

Any character of a later life form, for example an organ, which contains a lot of information, stands at the end of a process in which information was built up in a sequence of many small steps. Each of these steps was a consequence of a random genetic mutation. Each of those mutations occurred first *in an ancestor of all later life forms possessing that character* (barring the case of prokaryotes, in which horizontal transfer of genetic material is known to occur widely, and barring possible viral DNA transfer in higher forms of life, which is not generally considered to be important for evolution).

It is in principle possible that two different information-bearing characters may not be traceable to the same ancestor in that they may have arrived in the same individual through hybridization, although I consider this unlikely. Even more unlikely is that a genetic recombination could be one of the random mutations in the above sequence wherein the two segments that recombined came from entirely different ancestral lines. For this to occur, extreme parallel evolution would have to have occurred, which is extremely improbable. But even if such an unlikely event would have occurred, no one has ever exhibited the possibility of such sequences either, and my statement quoted above still stands.

The important point here is that any given information-bearing character should, in principle, be *traceable backwards through the successive mutations that represent the steps in the information buildup.*

Steve McGrew
Dear Lee, I have not forgotten what the discussion is about; I have simply been pointing out errors in the arguments you've made to support your assertion that evolutionary theory cannot explain the observed rate of adaptation.

I can't take seriously your complaint that "no one has ever displayed a sequence of possible mutations, each more adaptive than the previous, that could account for at least a portion of M-to-M evolution." Nobody can be expected at this stage of the game to be able to accurately predict the adaptive value of any mutation. We simply do not know enough yet about how genotype determines phenotype. And, the underlying premise that evolution occurs in a linear, monotonic fashion is incorrect. Your next statement, "Moreover, no random mutations have been observed that could possibly form the basis of the mutations that would be required for M-to-M evolution," is made without any support at all.

I agree that there is new information in today's genomes that did not exist in earlier forms; but I hesitate to accept your term "inserted into the genome". I have extensive experience with genetic algorithms, which build up high-quality, information-rich solutions to complex problems with zero input of information other than information implicit in the definition of fitness. That is, no information needs to be "inserted" for information buildup to occur. Natural selection alone is sufficient.

Evolutionary theorists account for genomic information buildup in essentially the same way that laser engineers account for the buildup of a highly structured coherent beam in laser: there is a source of random signals (spontaneous emission of photons), an amplifier (stimulated emission of photons in an inverted population), and a filter (the laser cavity with feedback). Interestingly, what emerges from a laser is an assortment of different beam structures, each optimal for the "environment" they must "survive" in. This kind of behavior is exhibited in myriads of nonlinear physical processes including biological evolution. In natural evolution the random signals are generated by point mutations, crossover, sexual recombination, gene duplication and other such mechanisms. Amplification is provided when individuals produce multiple offspring. Natural selection is the filter.

You would be right to say that each mutation occurs first in an ancestor of all later life forms possessing a "character" requiring that mutation. But it seems that you have made the (false) extrapolation that, therefore, the *character* occurs first in an ancestor of all later life forms possessing that character. Did I misunderstand you? Your example of a character is an organ, which is of course defined by a large cluster of genes and gene expression controls.

The evolutionary process does *not* build up information in a "sequence of many small steps", if by that you mean a *linear* sequence: one step, then another, then another. In fact, the evolutionary process largely builds up information through a combinatorial process. All the points you have presented in your chapter and in our dialog indicate a deep misunderstanding on your part about this fact.

Maybe we're simply not communicating here. I assume that by "hybridization" you mean hybridization between species. But if (for example) your mother had blue eyes and black hair, and your father had brown eyes and blond hair, and you have blue eyes and blond hair, then you are a living example of an individual whose information-bearing genetic structures arrived from different sources without cross-species hybridization.

Here is a crucial point I think you are probably overlooking. Phenotypic features in creatures like us are rarely controlled by a single gene, or even by several genes all on a single chromosome. Instead, they are *influenced* by multiple genes and by multiple gene expression controls that occur at multiple places in our genome. Even if we had the complete molecular-level evolutionary history of life laid out in front of us, it would be impossible to trace a phenotypic feature such as an organ backward in time the way you describe, along a one-dimensional, unbranched series of ancestors. The genetic components of each phenotypic feature involving multiple chromosomes or multiple genes on a single chromosome have a branched ancestral tree—*without any need for "hybridization"*—with only a need for sex.

You seem to share a widespread misconception that, according to evolutionary theory, there must have been a single first human being. That could conceivably be true if "human-ness" could be attributed to a single gene, but it cannot. There never was a "first" human being, any more than there was a "first" trickle of water that became the Mississippi River.

Your arguments seem to be based on an unconscious assumption that we reproduce by budding! In fact, every "information-bearing character" should, as you say, be traceable backwards through mutations representing the steps in the information buildup, but *not in a linear sequence.* Because we reproduce sexually, each character's components can only be traceable along multiple paths that diverge exponentially into more and more branches as we trace further back in time. In the forward-time view, complex "information-bearing characters" are assembled by convergence of separate evolutionary paths of independently evolved genetic components, combined via sex or horizontal gene flow.

Evolution is usually described as a tree that branches more and more as time progresses. That description is woefully incomplete and leads to the incorrect understanding you have expressed. Evolution would be more accurately described as a forest of trees in which a root system feeds each tree trunk, then each trunk splits into diverging branches which become root tendrils for trees in the succeeding forest.

Linear sequential evolution (which can occur only in clonal organisms) is very different from the combinatorial (converging-paths) evolution that actually operates among sexually reproducing organisms. Very roughly speaking, what takes N steps in the linear process can take as few as $\log_2(N)$ steps in the combinatorial process. For a very simplistic example, let's say that 500 mutations need to come together to make a modern human brain starting from the brain of a common human/ape ancestor. Let's also be generous and say that one in 10,000 offspring will have a particular desired point mutation. Then, in the linear analysis, roughly 5,000,000 generations (of one offspring per individual) are required to produce the 500 sequential mutations. On the other hand, if the order of mutations is not crucial, the combinatorial process in a population of 10,000 allows the 500 point mutations to occur independently and be brought together in an individual in just 9 generations. Evidently your only argument against this combinatorial process would be to insist (incorrectly) that point mutations must occur in a very specific sequence (i.e., producing monotonically increasing fitness) or they cannot produce viable individuals.

I am not saying evolution actually occurs that fast; I am only saying that well-known natural mechanisms exist that make it possible for evolution to occur that rapidly. Your estimate of the rate of "neo-Darwinan" evolution does not take these mechanisms into account and is consequently wrong by many orders of magnitude.

Lee Spetner
Dear Steve, you suggested that we now summarize our dialogue. From your latest reply to me, I see that we are a long way from summarizing. I have been trying to carry on a dialogue while you have been carrying on a monologue. I see now what the problem is that you do not understand what you read. I think I write clearly, but you are not reading clearly.

For example, I wrote: "If M-to-M evolution has indeed occurred, then a large quantity of information had to have been inserted into the genome of the later life forms that did not exist in the earlier forms. Evolutionary theory purports to account for this information buildup through successive random mutations, each fixed into the population by natural selection. If you disagree with this, then please state how you think evolutionary theory accounts for the information buildup."

You object here to my use of the word "inserted"—you seem to think it means something magical. The word "insert" means "put in". In the context of my discussion, it means "put in" by natural selection. I said explicitly, that evolutionary theory accounts for the buildup of this information via natural selection, but you apparently read something else. I think my meaning was clear that the information was *inserted* by the process of random mutation and natural selection. In your comments on this you engage in irrelevancies and make the silly assertion that "no information has to be 'inserted' for information to buildup to occur. Natural selection is sufficient." You do not understand that I wrote that the information gets in by natural selection. This is a serious misunderstanding of my position, and accounts for some of your irrelevant comments on it.

Another example.—You objected to my statement that "no one has ever displayed a sequence of possible mutations, each more adaptive than the previous, that could account for at least a portion of M-to-M evolution." Your comment was, "Nobody can be expected at this stage of the game to be able to *accurately predict* the adaptive value of any mutation" (my emphasis). Apparently you think it would be easier for you to argue against my position if you radicalize my statements. A careful reading of the above passage of mine will show that I did not require an *accurate prediction* of adaptive values, nor did I require *any prediction* at all. I said that no one has ever *displayed* such a sequence. That display should be of actual mutations in the laboratory. This is a serious objection to evolutionary theory because these mutations together with natural selection are the mechanism that is supposed to make the whole thing work. If the theory were valid, there should have been, and still should be, vast numbers of such mutations; yet none have been observed. You will notice that I did not even require the demonstration of natural selection, just a display of the mutations. The assertion of M-to-M evolution is based on circumstantial evidence alone and the alleged existence of this mechanism is a critical component of the evidence needed to support the theory.

Another example is your misunderstanding of my tracing backwards to its origin a genetic configuration that is an essential component of some character of an organism. What I wrote was, "Any character of a later life form, for example an organ, which contains a lot of information, stands at the end of a process in which information was built up in a sequence of many small steps." It is beyond me how you could possibly misconstrue this passage to mean" the *character* occurs first in an ancestor of all later life forms possessing that character." But aside from your mistaken interpretation of "character", I see you now agree with me that one can indeed trace back a genetic configuration to its origin.

Your misunderstanding of what I write has led you to make the host of irrelevant comments. I asked, if you disagree with my understanding of how evolutionary theory accounts for the information buildup in M-to-M evolution, to please tell me how you think the theory accounts for it. Your description of how you think the theory accounts for the buildup is vague, but I shall examine here what you wrote and try to extract your meaning.

If I understand your laser analogy correctly, then what corresponds in evolutionary theory to the amplification in your laser is the takeover of the population by a new adaptive mutant. This would be in opposition to what you wrote, namely that "amplification is provided when individuals produce multiple offspring." This is incorrect. Multiple offspring alone will not do it. Before the second mutation can be in a position to have its information added to the first by natural selection, the first must already be present in most, if not all, of the population. If you think otherwise, you would be wrong because then you would be advocating simultaneous selection of independent mutations and it is generally agreed that such an occurrence is much more improbable than the selection of a single adaptive mutation, and too improbable to be an important contribution to evolution. You may want to suggest that even such an improbable event can still occur, but you certainly cannot claim it is the dominant mechanism of evolution. My picture of how information is added by two successive mutations is that the first adaptive mutation occurs in an individual in a population and is lucky enough to take over the population after many generations. Then the second mutation occurs in an individual in this population and this mutation too is lucky enough to take over the population after many generations. All the individuals in the population now contain both mutations. This scenario is my picture of a sequence (which you call "linear" and which you mistakenly reject) of two mutations through each of which a small amount of information is added to (i.e., inserted into) the genome.

You have a naïve and incorrect picture of the way evolution branches back to the past. You seem to have the misconception that your ancestors, starting from your two parents, to four grandparents and so on, continue to diverge into the past. This is patently false because if it were true then just your ancestors alone would, only a thousand years ago, would have numbered more than the world's total population. You seem to be oblivious to the *convergence* that traces back into the past together with the divergence. For example, it is conceivable that your father's great-great grandfather is the same person as your mother's great-great grandfather.

Now let's look at your claim that a "combinatorial (converging-paths) evolution that actually operates among sexually reproducing organisms" is the dominant process that produces evolution. Here again your description is somewhat vague, but I shall do my best to extract your meaning. Rather than two independent adaptive mutations occurring and becoming incorporated into the genome successively, you claim they occur independently, and then come together in sexual reproduction. I can show you why this cannot be an important factor in evolution, even it could occur on extremely rare occasions. Let me first point out that you seem to have the fallacious notion that evolution occurs in individuals. It is universally accepted in the theory that evolution occurs not in individuals, but in *populations*. If two independent mutations are to come together for the first time in the population, then at least one of them is rare in the population. Where do you think these mutations originated? If you think they both originated in the same population before takeover by either one, then at their origin you must be dealing with a case in which both of them are rare and each exists in only a single individual. In such a case, their coming together is a very rare occurrence and cannot therefore be the general method of evolution. If they originated in different populations, then how did they come together? Did the two populations merge? Or did the rare mutant in one population somehow wander over to enter the other population? No one claims this as a dominant method of evolution, for the very good reason that it is much rarer than the two mutations occurring successively in the same population, with the second mutation occurring only after the first has taken over the population (the "amplification" in your laser analogy) — it can be only your idea, and it is incorrect.

So much for your criticism of my "linear, monotonic" picture of evolution. You have no valid alternative explanation of how evolutionary theory accounts for the information buildup in M-to-M evolution.

You complain that I gave no support for my statement that "no random mutations have been observed that could possibly form the basis of the mutations that would be required for M-to-M evolution." The support, which I have given elsewhere, is that I have scanned the literature for many years and have found no such observations. That is the best that can be done to support a negative. I challenge you to show counter examples. I look forward to your reply.

Steve McGrew
Dear Lee, before I respond please answer a few questions so I can be sure I understand your position clearly.

1. Do you believe the ancestry of every living organism today can be traced backwards along a single line of descent, such that every genetic feature in any given organism in that line of descent is either a new mutation or was

present in a single parent of that given organism, and that each ancestor in that line of descent had an adaptive advantage over that single parent?
2. Do you believe that the modern theory of evolution requires #1 to be true?
3. Do you believe that the modern theory of evolution does not provide adequate mechanisms to drive the genetic change required to make #1 possible, but that the environmentally-triggered genetic change mechanisms you propose would make #1 possible?

If I have understood your letters and your chapter correctly, your answers to all three questions would be "yes".

Lee Spetner
Dear Steve, your questions presuppose a misconception of my position, so let me try to state my position in a way that will address what I think is behind the questions.

First of all, I do not believe evolution took place (or takes place) the way the theory says it did (or does). You ask what I think the theory says. The dominant theory says that evolution occurs through long sequences of random mutations, each followed by natural selection. Again, I emphasize that I do not believe evolution happens this way, but I am now discussing only what the theory says. The concept of what constitutes a mutation has changed over the years. As now understood, it is any heritable change in the genome. This includes base-pair substitutions, recombinations, deletions, insertions, and transpositions. The important point about mutations in the theory is that they are random in the sense that their occurrence is independent on how that mutation affects the phenotype. It is an activity that is confined to the genome alone. I look upon the random-mutation-and-natural-selection mechanism as a method of information transmission. In fact I have published 3 papers on this subject: One in 1964 (*Journal of Theoretical Biology*) entitled *Natural selection: an information transmission mechanism for evolution*, a second in 1968 (*IEEE Transactions on Information Theory*) entitled *Information transmission in evolution*, and a third in 1970 (*Nature*) entitled *Natural selection versus gene uniqueness*.

In our present discussion I have focused on the manner in which the theory accounts for the information buildup in the genome of, let's say, present-day life forms. I think we agree that the mechanism of random mutation and natural selection correctly characterizes the theory. Now let me address the question of how this mechanism could possibly account for the buildup of information.

Random mutations occur all the time, and the one that is most obviously random is an error in copying a base pair during DNA replication. (There is some serious discussion in the literature indicating that the other types of mutation are not random.) These random mutations are subsequently filtered by natural selection.

You asked if I believe that "the ancestry of every living organism today can be traced backwards along a single line of descent, such that every genetic feature in any given organism in that line of descent is either a new mutation or was present in a single parent of that given organism, and that each ancestor in that line of descent had an adaptive advantage over that single parent?" As I said above, I don't believe that , but I do think a much less extreme form of this statement is indeed a consequence of the theory. According to the theory, there must be at least large stretches of ancestry that can be traced back that way. Rare exceptions would include the following:

1. A sibling species merges with a given species. The merger could be just a few individuals or even a large population.
2. Horizontal transfer of DNA, possibly through viral infection.
3. Offspring of two rare mutants or descendants of mutants.

Your 3rd question is the result of a misconception, so it does not have an answer as it stands. My nonrandom evolutionary hypothesis was not proposed to account for long chains of mutations, nor was it proposed to account for the buildup of information in evolution. It was proposed to account for the many observations for which normal evolutionary theory fails to account. These observations are principally those of rapid evolution where an entire population evolves in just a few generations—in some cases of fish, in a year or two. I hope this answers your questions.

Steve McGrew

Dear Lee, I am happy to continue the discussion, but I think it is important to summarize points of agreement and disagreement in order to focus our discussion on the key issues. If I have any of the following points wrong, please correct me.

1. We agree that the fundamental assumption of evolutionary theory is that genetic information accumulates via heritable genetic change and natural selection.

2. We disagree on how evolutionary theory says that accumulation occurs. You insist that evolutionary theory requires heritable genetic changes to accumulate in linear succession. I insist that modern, informed evolutionary theory acknowledges that heritable genetic changes accumulate through a combinatorial process. Note that this is separate from any agreements or disagreements we might have regarding how mutations *actually* accumulate.
3. You have stated that you understand "mutation" to mean any heritable change in the genome including base-pair substitutions, recombinations, deletions, insertions and transpositions. However, sexual recombination is normally *not* considered to be a type of mutation, so we would be broadening the meaning of "mutation" to use it as you propose. So, I prefer to use the term, "genetic change", or "heritable genetic change". I will use "HGC" as an acronym for heritable genetic change.
4. We agree that evolutionary theory considers the *initial* occurrence of HGC to be independent of how the HGC affects the phenotype. However, I think we also agree that within a population, the rate of increase of the proportion of individuals having a given HGC is directly affected by the "adaptive advantage" of that HGC.
5. We haven't discussed this, but I hope we agree that "adaptive advantage" according to evolutionary theory equates to "reproductive advantage"—and "reproductive advantage" relates to the probable number of viable offspring an individual will have.
6. You mentioned errors in copying base pairs as being the most obviously random mutation. However, I keep mentioning sexual recombination and crossover as being even more important sources of random mutation (again, your definition). Is this a point of disagreement between us?
7. It appears that your conviction that evolutionary theory cannot account for the observed rates of HGC is based on your understanding of how HGCs accumulate according to evolutionary theory. Accumulation occurring per your understanding of evolutionary theory would take far, far longer than accumulation according to my understanding of evolutionary theory. This point is a disagreement about the assumptions of evolutionary theory, not about what actually occurs in nature.
8. You and I agree that there might exist nonrandom sources of HGCs, but our reasons are very different. Because of your understanding (which I think is mistaken) of evolutionary theory, and in particular because of #7 above and because of observations of rapid evolution (e.g., fish—by which I guess you may be referring to the cichlids in African lakes), you propose that there may be a nonrandom source of HGCs. I, on the other hand, think that there are

well-known *biased-random* sources (e.g., the grouping of genes into chromosomes subject to sexual recombination and the arrangements of genes on chromosomes subject to crossover) of HGCs whose meaning and importance are often overlooked. Evidently you think I am mistaken.

9. You and I disagree on how HGCs *actually* accumulate. You think they primarily accumulate in a linear fashion, while I think they primarily accumulate in a combinatorial fashion.

Are there any other important points of agreement or disagreement that you think should be included in this list?

I would like to focus our discussion on point #9 rather than on #7. We agree that evolutionary theory is at best incomplete. With that agreed, we can abandon further arguments about what the assumptions and consequences of evolutionary theory are, and maybe help make some progress toward deciding what the assumptions of a more complete evolutionary theory *should be.*

Massimo Negrotti
Dear Lee and Steve, after having read your post, trying to understand, in some way, your technical language and the actual issues of your disagreement, the following three points seem to me to be clear enough. Please, let me know if you find them acceptable within your lines of thought, forgiving my being outside your specialized field.

1. You both agree on the main role of random processes, but, in your letters, I have found no specifications on the nature of chance triggering biological phenomena, for instance recombination. What is your main conception of chance? Is it a sort of background noise or an autonomous change of something at the level of particles? Is it a mere expedient for defining processes that are, as yet, not describable in full detail, or is it a true and originating phenomenon in nature?

2. You both speak of information, but Lee appears much more unsatisfied with its current definition, which goes back to Shannon, Wiener and so on. I am also inclined to think that, in biological systems, information is not what it is in a computer or in any other system capable of distinguishing logical states or strings of logical states. Steve rightly gives the example of an information system with a well-ordered library, but it is clear that the amount of information is not a property of the books nor of the library in itself. Indeed, information

always requires some agent that is capable of capturing it from the background and its multi-level organization, randomness included. Thus, in a microbiological system, who, or rather what, is the actor? It seems to me that *at such a level there is only a sort of mutual matching among parts whose topology allows them to be complementary to each other,* and not an example of true information processing—at least, surely not of the kind that is at work inside a computer, nor even of the kind that a user or a librarian of a book library adopt, because they outside of the system.

3. I found the idea of 'environmental cues', able to trigger responses, very interesting. Actually, it does not appear to me incompatible with the Darwinian view on evolution. Indeed, the silent database of responses to which Lee refers could be conceived as the result of the accumulation, within genes, of randomly generated changes. The system embodies them—provided they have no lethal effects—building up a sort of depot of properties, or a depot of variety, that, in the face of some environmental change, could be enabled for giving the right response.

The idea of the silent responses is, in some measure, analogous to the innate ideas in Descartes, and also to rationalism, while the pure reference to the power of the environment, in selecting, at a population-wide level, the best fitting systems, could be the analogue of the empiricist tradition. Do we, perhaps, still lack the synthesis proposed by Immanuel Kant?

Steve McGrew
Dear Massimo, thank you for your letter! We are snow-bound here in Spokane, so I have had time to put this reply together.

1. The processes that cause DNA replication errors and determine the particular combination of chromosomes passed on to offspring during sexual reproduction are random in that they are not controlled by any deterministic mechanisms inside or outside the cell. Lee has hypothesized the existence of mechanisms that at least strongly bias those random processes in favour of changes that are adaptive. I agree with him, although not in a way that he likes. The arrangement of genes on chromosomes, for example, strongly biases the random changes in favour of *combining* changes that have been adaptive in the past. But I think Lee has something more "intelligent" in mind: a mechanism that somehow "decides" which genetic changes will be adaptive, and implements them, based on environmental cues.

2. As in any constructive argument, agreeing on the meanings of key terms is a crucial first step! Lee and I are still working on reaching an agreement regarding the meaning of various terms he uses which include the word, “information”. You are right: information processing inside a microbiological system is not strictly analogous to the sort of information processing a digital computer does. Computer designers tend to segregate “hardware” from “software” as much as possible, mostly to make it easier to think about; but in a microbiological system like a cell, every component of the system behaves like both hardware and software.

3. The “gene pool” in a population provides a depot of HGC “packets” that have been tested by natural selection and are available to be incorporated into new genes and into new genomes via sexual recombination and crossover.

Lee is careful to say that he is not proposing a Lamarckian evolutionary process, but so far it is not clear to me what distinguishes Lee’s hypothetical mechanisms from a Lamarckian process. Current evolutionary theory assumes that heritable genetic changes (HGCs as Lee and I have come to call them) occur randomly, without any information from by the environment; and that the environment only *selects* against the resulting phenotypes by modulating the reproductive success of individual phenotypes. If (per Lamarck’s idea) giraffes grew long necks because stretching to reach leaves from high branches makes offspring tend to have long necks, and (per Lee’s idea) that occurs because “something” in the reproductive mechanism senses that stretching and consequently fiddles with reproductive cells to preferentially mutate genes that control neck length, it sounds Lamarckian to me. If that “something” is actually found, Lamarck will be back in favour alongside Darwin, and Lee rightfully will be hailed for having foreseen the mechanism that redeemed Lamarck.

Lee and I are wrestling over one key issue: differences between our understandings of what modern evolutionary theory (MET) says about how multiple HGCs come to be incorporated together into a genome. His understanding leads him to conclude that evolutionary mechanisms allowed by MET are too slow by many orders of magnitude to produce observed rates of evolutionary adaptation. My understanding is the opposite: that MET-allowed mechanisms operate many orders of magnitude faster than Lee’s calculations indicate, because his understanding requires linear accumulation of individual HGCs in a specific order between episodes of total genetic takeover of a population; whereas my understanding allows combinatorial accumulation of independently occurring individual HGCs, *not* in a specific order. His understanding leads to evolutionary "steps" (amounting to the accumulation of some given set of HGCs) taking an amount of time proportional to the number of individual HGCs; while my understanding leads to evolutionary "steps" taking an amount of time more nearly proportional to the logarithm of the number of individual HGCs.

So, the debate is really more about what MET *says*, than about what actually happens in nature. Lee, reasonably enough, does not want to discuss what actually happens in nature until we have settled this issue. That is why I've proposed that we formulate the key question and pose it to the authors of a recent MET text.

Steve McGrew
The debate between Lee and myself turned out to be over the assumptions of modern evolutionary theory, rather than over what actually happens in evolution. It would have been nice to go through one more exchange, in which Lee and I submit a concise question to the authors of a textbook on modern evolutionary theory: *"Does modern evolutionary theory include sexual recombination and crossover as major contributing processes in the generation of new heritable genetic characteristics?"* Naturally, I have no doubts about what the answer would be!

Lee Spetner
Dear Steve [summarizing dialogue that has been edited out],

Steve earlier wrote: I have never thought or said that individuals evolve. However, as noted, new HGCs do occur first in an individual before they spread out into more individuals in the population via sexual recombination.

Lee then replied: If you are not saying that single individuals evolve, then you are at least saying that several individuals evolve, and this is still not true. Only whole populations evolve.

Steve counter-replied: Of course individuals don't evolve. Evolution requires change of genotype. An individual has a fixed genotype. Individuals are the products of evolution; populations are collections of individuals.

Lee's current reply: But you seem to think that an adaptive HGC that appears in one individual and is passed on to its progeny is evolution. It is not. The adaptation has to be taken up by the population. Populations are not just collections of individuals. A population in evolution is an isolated group of individuals who interbreed. Of course evolution requires change of genotype. But that is not sufficient. An evolutionary change is a change in a population; it is not a change in only an individual.

Steve wrote: Can you give an example of what such an increment [of information] might be? For you to consider it to be an increment of information, must it *first* spread to an entire population? If your term "evolutionary information" only applies to an *increment* of information, what terms would you use to refer to the total information in an individual genome or in a collection of individual genomes?

Lee: An example of such an increment is the adaptive change of beak size and shape and body size in finches such as those that have been observed in the Galápagos Islands that occurred between 1971 and 2001 as reported by Grant & Grant. Although we don't know what the genomic changes were, there clearly was a heritable information increment (evolution) to account for the appearance of these phenotypic changes that were adapted to a changed environment. A clear genetic change, that is an example of an addition of information, would be the horizontal transfer of a battery of genes for antibiotic resistance from a microorganism that has the resistance to one that did not have it. In a laboratory environment, examples of each of these types of information increments have been observed to achieve population takeover. For the record, however, let me say that although these examples add information to the population, they cannot serve as prototypes of M-to-M evolution.

The total information in a population is the information that is common to all members of the population. This information is supposed to have been built up by evolution. The buildup of this information is what evolutionary theory must account for.

Steve wrote: It is not correct to say, "... the buildup of information in evolution does not occur until a new HGC has taken over the population". Why do you think this is true?

Lee: It is true because the characteristics we associate with a population are those that are common to all of its members. Thus all humans have two arms, legs eyes, etc. The information in these common characteristics is what is supposed to have been built up by evolution. The fact that these characteristics are common to all members of the population indicates, according to the evolutionary story, that those are adaptations that have taken over the population. Since this is true in the populations we observe today, then according to the principle of uniformitarianism, which evolutionary theory has invoked from the time of Darwin until today, it should have been true as well throughout the entire timeline of the evolutionary process.

It is true because of the following considerations. Consider a single rare HGC that can confer an adaptive advantage on the organism containing it. Then natural selection begins to operate. The adaptive advantage means that the organism will, on the average, produce more offspring than the other individuals in the population. For simplicity, let this HGC be dominant so that the adaptive trait will appear in its offspring whenever it mates. The number of individuals having the HGC will increase as in a random walk with two absorbing barriers—one barrier is extinction and the other is population takeover. These two barriers are the only stable states of the random walk. When there is only one, or just a few, individuals having the adaptive HGC, the walk is very likely to go to extinction, with the probability depending on the selection coefficient. Extinction becomes less likely as the numbers grow. When the numbers become a sufficiently large fraction of the population, they are more likely to go to population takeover.

Steve counter-replied: A dominant gene will *not* appear in *all* offspring of an individual who carries that gene. It will appear in only half of the offspring. Maybe you made this misstatement accidentally; but if not, it is a misconception that could be the basis of your opinion that linear buildup and genetic takeover are required by evolutionary theory.

Lee's current reply: This is a nit pick. Of course it will occur in approximately half. This has no effect on my picture and on the argument I have given. Since this is your only comment, you have evidently accepted the substance of my above statement.

Steve: Why do you say multiplication of an adaptive mutation (HGC) is rare? The likelihood of multiplication depends strongly on the particular way in which the mutation is adaptive. Suppose, for example, the only consequence of a particular dominant HGC is that a female fish produces twice as many eggs than her cousins. In that case, she will have twice as many offspring as her cousins. That is a very large amplification factor. Of course she could be eaten by a bigger fish before she ever spawns—

Lee's current reply: Of course! The likelihood of multiplication depends on the selection coefficient of the adaptation, which is a quantitative way of saying that it depends on the adaptivity of the mutation. As I have noted previously, Fisher has shown that the probability of population takeover is proportional to the selection coefficient, and all evolutionists agree that in nature the selection coefficient is small, and therefore the probability of amplification (whose ultimate result is population takeover) is small.

Lee earlier wrote: Thus, this evolutionary event needs all the help it can get to succeed, and large numbers will be a great deal of help. It is very improbable for another adaptive HGC (call it HGC_2) to appear in an individual already having HGC_1 (which is what I shall now call the first HGC) when there is only one or only a small number of individuals having HGC_1. The larger the number of individuals having HGC_1 the more likely it is for this second event to happen. Although it can happen when there are only a small number of HGC_1's, it is improbable. Letting the growth in numbers of HGC_1 continue for enough generations to reach takeover will increase the probability that an HGC_2 would occur in the subpopulation of HGC_1. Thus, suggesting that HGC_2 occurs and builds up before population takeover by HGC_1 is suggesting a less probable mechanism than one in which HGC_1 goes to takeover first.

Steve then replied: The reasoning in your paragraph above would make sense except for the fact that its fundamental assumption is wrong. You assume that HGC_1 and HGC_2 *cannot* end up together in an individual unless HGC_2 occurs *after* HGC_1 occurs in that individual's ancestry. In fact, HGC_2 can occur entirely independently of HGC_1. This is well understood, and it is one of the key aspects of modern evolutionary theory. Because HGC_1 and HGC_2 can occur independently and then come together through sexual recombination, crossover and horizontal gene transfer, it is not necessary for HGC_1 to take over the entire population in order for there to be a good probability that HGC_2 will be combined with HGC_1.

Lee's current reply: First of all, I did not assume here that the HGC's have to occur in any particular order. You made that up. Furthermore, what you are incorrectly suggesting as "one of the key aspects of modern evolutionary theory" is much less probable than what I am suggesting as the accepted mechanism of evolutionary theory. Let me show you why what you are suggesting is not true.

I shall compare here two scenarios: Scenario I is the standard mechanism for the working of evolution, which is the appearance of a random adaptive genetic change followed by natural selection, which will eventually bring the new change to take over the population. This is one step in the evolutionary process. Another adaptive random genetic change can then appear in the population in which all members already possess the first. When the second takes over the population, the combination of the two adaptations will be characteristic of the population.

Scenario II is what you are advocating, namely one adaptive HGC appears in the population, and without it taking over the population, the second appears. The two adaptations are in each of two different individuals. These two individuals, through random mating, sexually combine, in one individual.

I compute here the probability of two HGC's coming together according to each scenario

Let p = probability of one of the HGC's appearing in a particular individual in one generation.

I am taking the probability of appearance of the two HGC's to be the same. The calculation can be done just as well with two different probabilities, but making them the same will save me from having to use subscripts.

Let s = the selection coefficient of either one of the HGC's in the absence of the other.

Here again I could just as easily make the s's different, but for the same reason as above I am making them the same.

Let N = the number of individuals in the population.

I am assuming that N remains constant through the generations. This is typically the case. I am further assuming that p is very small compared to 1, and that N is very large, but that Np is very small. This again is typical.

Then the probability of getting at least one of the two HGC's in at least one member of the population is approximately $2Np$.

According to scenario I, after one HGC appears, natural selection will operate and the HGC that appeared will eventually take over the population with probability approximately $2s$. Thus the probability of at least one of the two appearing in at least one individual in a particular generation and then becoming, in many subsequent generations a characteristic of the population, is approximately $4Nps$.

The probability that the second HGC would appear in at least one individual of the population in a particular generation after the first has taken over the population is Np so that the probability of the two appearing and coming together according to scenario I is $4N^2p^2s$.

Now let me look at scenario II. Consider the probability of both HGC's appearing in the population together, which then combine, perhaps by the mating of two individual bearers of the two different HGC's.

As before, the probability of getting either of the two HGC's in at least one individual in one generation is $2Np$. Let's say the second one appears perhaps some few generations after the first appeared. The probability that it appears in at least one individual is Np. Since the appearances of the two HGC's are independent events, the probability of both happening is the product of the two probabilities, or $2N^2p^2$. We must now consider the probability of both of these HGC's coming together in a mating. Since there are N individuals in the population, and assuming that half are male and half are female, the probability that one would mate with the other (assuming random mating) is $2/N$. Hence, the probability that the two HGC's will occur and come together in the same individual is $4Np^2$.
That is, the probability of the two HGC's coming together is

$4N^2p^2s$ according to scenario I,

and

$4Np^2$ according to scenario II.

The ratio of these probabilities is

Scenario I/Scenario II = Mine/yours = Ns.

A typical value of N for large vertebrates is 10^5, and a typical value of s is 0.01. With these typical numbers, the probability of my scenario is 1000 times that of yours. As you are an engineer and are used to thinking in dB (decibels), you see that your idea of the prevailing evolutionary theory has a 30 dB disadvantage over what I claim is the prevailing theory. It is therefore absurd to suppose that anyone doing any serious study of the way information is built up by evolution would claim your version as a better theory.

Lee earlier wrote: In any case, the new information represented by the new adaptive HGC is not stably in the population until it is present in all members of the population.

Steve then replied: If this is your conclusion, you must have made some crucial unstated assumptions. Is the sickle cell gene stably in the human population? Or the gene for red hair? What HGC can you point at that *is* in *all* members of the human population? And, why must an HGC be stable in an entire population in order for it to amount to an "evolutionary step"? Why can't it be stable in just a subpopulation?

Lee's current reply: No, Steve, the examples you bring are not counter examples to my statement. My statement is correct. Stability is determined over long evolutionary time periods. The fraction of sickle-cell anemia phenotypes in a population has been observed only over a time very short compared to evolutionary times. This fraction is observed in an unstable state. As time approaches infinity, the fraction must go either to zero or 1. Compare this to a random walk with two absorbing barriers. A random walk with two absorbing barriers, that, at each unit of time, moves one unit in the positive direction with probability p and in the negative direction with probability $(1 - p)$. This phenomenon has been shown mathematically to have only two stable states, namely at the two barriers.

Steve wrote: Is the information you refer to here the same as "evolutionary information" as you defined it above? If so, would you apply the term "information buildup" to linear accumulation of multiple HGCs in a single genome (followed by amplification and ultimately takeover of the whole population)?

Lee: When an adaptive HGC (multiple HGC's also qualify as an HGC) takes over a population, evolutionary information has been added.

Steve wrote: Genetic evidence indicates that the vertebrate eye was preceded by eyes of one sort or another in pre-vertebrate organisms. I'm not sure how literally I should read your words above. I can't imagine that you really think that a *single* mutation (let's call it an HGC) resulted in development of a "final eye" from a "proto-eye"? Construction of a salmon's eye requires the orchestrated action of many, many genes.

Lee: I cannot believe that you are questioning this. Do you really think that a proto-eye, which preceded the present eye by one evolutionary event, was achieved by an "orchestrated action of many, many genes all at once"? Do you have any idea of how improbable such an event is? No serious evolutionary biologist could even contemplate such an event. In any case, even in your scenario, the last event before the appearance of the present eye, no matter how complex that event, was the *last event*. That is tautologically true. You cannot argue with that.

Lee wrote earlier: Since all salmon now have eyes, that last mutation clearly took over the population. The salmon population is genetically diverse, but they all have eyes. According to modern evolutionary theory, there is no reason to believe that the present stage of evolution of the salmon eye is a special case of evolution, different from all other evolutionary stages. Just as *all* salmon have eyes, there is no reason to suppose that *all* proto(-1)-salmon did not have proto(-1)-eyes. That means that the mutation that converted the proto(-2)-eye into a proto(-1)-eye had also taken over the population. The population undoubtedly was diverse in many characteristics, but they *all* had proto(-1)-eyes.

Steve then replied: This is a *non-sequitur*. You are taking it as a given that all salmon today are exclusively the descendants of one individual proto-salmon. It simply is not true! Perhaps any one gene contributing to the eyes of a salmon today arose in a single individual ancestor, but because many genes contribute to the construction of a salmon eye, the whole gene complex certainly did not arise in a single individual, or even through a linear succession of individuals. It arose combinatorially.

Lee's current reply: The fact that, according to current evolutionary theory, all salmon today are descendants of a single individual follows from the evolutionary scenario I described above, and which I have shown is a much more reasonable scenario than yours.

Steve then wrote: We may be stumbling over language here. An eye is constructed during embryo development through the orchestrated action of many, many genes all at once. The DNA sequences constituting the genes and orchestrating their actions are constructed over hundreds of millions of years via the accumulation of myriads of HGCs. Neither an eye nor a proto-eye is defined by a single genetic DNA base pair or a single gene. If we were to choose an individual salmon and take the appropriate parts of its DNA as defining the set of genetic factors defining the present salmon eye, then we might be able to trace that DNA's ancestry back to a recent event where a single small genetic change occurred. The parents of the individual in which that HGC occurred would still have eyes, but we'll call them proto-eyes since the determining DNA is a little bit different from our representative present salmon's DNA. Those proto-eyes probably would be nearly indistinguishable from the present salmon's eye. Follow backwards in time the whole watershed of events that constructed the genetic instructions for building our present salmon's eye, and there might be a few phenotypically abrupt changes in the branching sequence of proto-eyes and proto-proto-eyes, such as when two individuals with distinctly different eye structure mate and produced hybrid offspring, or when one of the highly influential gene control factors is subject to an HGC. However, in tracing the evolution of the salmon's eye all the way back to a patch of light-sensitive cells on the skin of a wormlike creature half a billion years or so ago, we would find only a few (if any) such abrupt changes.

Lee's current reply: Modern evolutionary theory does not rely on abrupt (large) changes. Let me show you how Richard Dawkins (who is warmly embraced by the evolutionist community as their spokesman) described, in his famous book *The Blind Watchmaker*, the hypothetical evolution of the eye. He asked rhetorically,

> "Could the human eye have arisen directly from something slightly different from itself, something that we may call X? The answer … is clearly *yes*, provided only that the difference between the modern eye and its immediate predecessor X is sufficiently small. Provided, in other words, that they are sufficiently close to one another in the space of all possible structures. If the answer … for any particular degree of difference is no, all we have to do is repeat the question for a smaller degree of difference. Carry on doing this until we find a degree of difference sufficiently small to give us a 'yes' answer to [the] question.
>
> "X is *defined* as something very like a human eye, sufficiently similar that the human eye could plausibly have arisen by a single alteration in X. …
>
> "Now, having found an X such that the answer to [the] question is yes, we apply the same question to X itself. By the same reasoning, we must conclude that X could plausibly have arisen, directly by a single change, from something slightly different again, which we may call X′. Obviously we can then trace X′ back to something else slightly different from it, X″, and so on."

This quote from an acknowledged spokesman for evolution, which describes the current thinking on how evolution is supposed to work, explains how the eye is thought to have evolved through a series of single small HGC's. Moreover, this description of how evolution is thought to work is very different from your ideas.

Steve wrote: Lee, you read into Dawkins something that simply is not there. In our discussion, you have based all of your arguments on *your* belief that modern evolutionary theory assumes only linear accumulation of HGCs; and I continue to insist that modern evolutionary theory instead assumes that combinatorial accumulation of HGCs is the dominant mechanism.

It is useless to discuss the consequences of the two conflicting positions when the fundamental disagreement is about what the tenets of modern evolutionary theory are– and not the rightness or wrongness of modern evolutionary theory.

So, please consider the following excerpts from a definitive textbook on modern evolutionary theory: *Evolution*, by Barton, Briggs, Eisen, Goldstein and Patel (2007, Cold Spring Harbor Press):

1. P. 364: "Even within bacterial species, where recombination occurs only sporadically, it may be frequent enough over the history of the sample to make it impossible to assume a single genealogy... With recombination, different stretches of the sequence have different genealogies.

2. P. 427: "...In a strictly asexual population, every gene shares the same ancestry simply because all genes are passed on together from mother to daughter. However, if there is any kind of sexual reproduction (e.g., the occasional transfer of genes between bacteria or regular meiotic sex in eukaryotes,) then ancestry will vary from gene to gene. We can see this most clearly by thinking of the ancestry of the human mitochondrial DNA or the human Y chromosome. In mammals, because mitochondria are always inherited from the mother, they all must trace back to a single ancestral female—the so-called "mitochondrial Eve." Similarly, all Y chromosomes descend from a single male, the "Y chromosome Adam." Now, "mitochondrial Eve" and "Y chromosome Adam" lived at different times and places. Moreover, there is nothing special about these two individuals, except that by chance they contributed a small part of their genome to the future human population. Indeed, it is possible that they contributed nothing else to future generations. Conversely, if we look back in time, we see that the present-day human genome is divided into many blocks that trace back to a large number of different ancestors (Fig. 15.11).

3. P. 429: "We see, then, that the genome of a sexually reproducing organism consists of a mosaic of blocks, each with a different ancestry and a different fate (figs. 15.11 and 15.13). So far, we have argued forward in time. When we analyze samples from present-day populations, however, we need to trace their ancestry back through time. This can be done by extending the coalescent process to include recombination (Box 15.4). We will now see how this process determines the patterns of genetic variation that a re shaped by mutation, recombination, and drift." [Figures 15.11, 15.12, and 15.13 all illustrate genes from *multiple ancestors* coalescing and recombining to form a present-day genome]

4. P. 430 (Box 15.4) "Neighboring sections of genome come to have a different ancestry because of recombination."

5. P. 662: "…As we shall see, mutation is much less effective than recombination as a means of generating useful variation."

6. P. 662"There are several reasons why the rate of adaptation may not be limited by mutation rates. First, in a large population all possible single-base changes may occur many times in each generation. … Finally, in asexual populations we will see that mutations must be fixed in series. If many different favorable mutations occur at about the same time, they will compete with each other so that only one can fix.(this is called clonal interference). ***In sexual populations, recombination can bring different mutations together, which allows much faster adaptation.*** [my emphasis]

7. P. 723: "For the reasons explained in Chapter 23, sexual reproduction improves performance; most importantly, it allows advantageous variants what arose in different lineages to be brought together by recombination."

Lee replied:
You wrote that I read into Dawkins something that isn't there, but you did not support that statement. You did not say how you would interpret his words differently. Your criticism therefore falls flat.

The basic disagreement between us, and the one you have just addressed, is whether modern evolutionary theory considers evolution to be, as you have described it, a linear or a combinatorial process. This disagreement we have encapsulated in, and it actually originated with, the question as to whether in general a population can be said to have a common ancestor or not and whether or not it accumulates adaptations successively. I hold that a consequence of evolutionary theory is that in general a population has a common ancestor (barring rare exception's as I have noted earlier), and you hold that in general it does not. Having a common ancestor means that if we trace the lineage of each individual in the population we will arrive at a common ancestor for all individuals in the population. There may be several common ancestors, but all will share this property. Note that there is no disagreement about an <u>individual</u> having multiple ancestors not all shared by other individuals. For example, I have a great many ancestors because I have two parents, 4 grandparents, 8 great grandparents, and so on. Most of these, particularly the later ones, are not shared by you. Having multiple ancestors implies, of course, that different genes in my genome come from different ancestors. There is no disagreement about that. You contend, however, that there is no such thing as a common ancestor to a population. All of the citations you bring address only the multiple ancestors of an individual genome. They do not address the common-ancestor issue. None of your citations says, or even implies, that a population does not, in general, have a common ancestor. You misread and misconstrue what they have written.

Please note that your second citation says explicitly that there was an ancestral female – the so-called "mitochondrial Eve" from which **all** present females are descended. This is in direct contradiction to what you think, as you have written

> "This argument, by the way, deflates the concept of an "ancestral Eve": a concept suggesting that all human beings descended from a single early human female based on the near-universality of certain components of mitochondrial DNA among humans."

The same applies to the so-called "Y-chromosome Adam". Just because the cited authors write that these so-called "Adam" and "Eve" may have lived in different places and at different times in no way implies that a population does not have at least one common ancestor, and in no way implies that adaptations are incorporated into the population in any way other than successively. Note that "Eve" and "Adam" are both common ancestors of the entire population.

Let us see how a population could arise with all males descended from a common "Adam" and all females descended from a common "Eve." At some time in the distant past a woman in the population (let's call her "Eve") was born with an adaptive mutation in the germ cell from which she developed that gave her the ability to have more descendents than others in the population. Moreover, these descendents passed on the same adaptation, and the females of that line passed on her mitochondrial DNA as well, to their descendents and eventually they took over the population. After that happened, all females in the population were descended from "Eve." Then, an adaptive mutation appeared in the germ cell of a man (let's call him "Adam") and he also passed on his adaptive genome (together with his Y chromosome to his male descendents) to his progeny and eventually they took over the population. Now all the males in the population are descended from "Adam", and all females are descended from "Eve." Note that "Eve" and "Adam" arose at different times, but nevertheless the entire population has at least two common ancestors, and may have many more.

Your second citation also says, "if we look back in time, we see that the present-day human genome is divided into many blocks that trace back to a large number of different ancestors." You seem to think that this implies that the population had no common ancestor. That is not true. Of course the theory says that different sections of the genome have different ancestors, as I have noted. But that in no way implies that there was not at least one common ancestor of the population.

You have presented, and emphasized, the following quote, "In sexual populations, recombination can bring different mutations together, which allows much faster adaptation." You seem to think that this implies a process opposed to a "linear" one in which adaptive genetic changes are fixed into the population successively. That is not so. If a recombination happens to produce an adaptive phenotype, that phenotype still has to be fixed in the population and the successive adaptations will have to be fixed in the population successively. I have shown earlier that a scenario of different adaptive recombinations combining before either of them has taken over the population is much more improbable than a scenario where a second adaptive mutation does not get incorporated until after the first has taken over the population. Therefore the theory says that the more likely scenario is the linear, successive one.

I have shown that the interpretation you have been projecting throughout this discussion is not contained in the words of the authors you have cited. Your misreading and misconstruing of the passages you have cited, and perhaps others like them, may well be the source of your deep misunderstanding of modern evolutionary theory.

Steve replied: Lee, having "a common ancestor" simply means that *at least one gene* present in a population can be traced back to a single individual. That has never been a point of contention between us. Your position has been that modern evolutionary theory assumes that *all fixed genes* (genes that have largely "taken over a population") can be traced back to a linear succession of mutations in a single line of descent, while mine has been that modern evolutionary theory recognizes that the majority of such fixed genes originate on separate lines of descent that converge via recombination. Do you now agree with me?

Lee replied: You have defined "common ancestor" in a strange and unconventional way. Having a common ancestor is not ordinarily defined in terms of genes. Simply put, a population has a common ancestor if all individuals in the population are descended from a single individual. You have denied that modern evolutionary theory holds that in general a population has a common ancestor. You have written earlier,

> "You seem to share a widespread misconception that, according to evolutionary theory, there must have been a single first human being.

Yes I do, and that "misconception" is also shared by all scientists engaged in the study of evolution. You also wrote,

> "You are taking it as a given that all salmon today are exclusively the descendants of one individual proto-salmon. It simply is not true!"

True or not, it is a notion commonly held by those working in evolution, and is even implicit in the citations you have presented. It is, in other words, part of modern evolutionary theory.

You are still confusing the generation of genetic diversity with the generation of genetic information in the process of evolution; you seem to think that because genetic recombination plays an important role in evolutionary theory that evolution proceeds in a "combinatorial" way. You are mistaken. The theory of evolution is obliged to account for the appearance of the vast amount of information in the genome, and it attempts to account for it as a step-by-step process. Genetic diversity is the raw material on which natural selection is said to work. The theory indeed holds that genetic recombination is an important contributor to genetic diversity. That does not mean, however, that evolution proceeds in any fashion other than what you call linear, step by step building one adaptive genetic change on another.

Steve wrote: The whole point in saying that evolution occurs in populations (not individuals) is to highlight the fact that the whole population is a reservoir of HGCs that are recombined via sex to produce new combinations—new HGCs—in individuals.

Lee: That is not the *whole* point. In fact that is not the point at all. The point is simply what the statement says, that evolution occurs only in populations. Changes that are only in individuals are not yet considered evolution. A population is said to have undergone an evolutionary step when the population as a whole has assimilated within it the new adaptive feature.

Steve wrote: You have introduced a new term, "evolutionary step", and defined it in an odd way. Can you cite an authoritative reference that defines it that way? It seems that, per your definition, our ancestors could never have taken any "evolutionary steps" on the way from an ancestral population that split into apes and humans—because for the split to occur, the ancestral population could never have been completely taken over by any HGC that differentiates humans from other primates. If it had, there would be no apes today!

Lee: You do not seem to understand that there were many independent populations of apes that did not interbreed. If one of those populations evolved into humans, the other populations would remain as apes.

Steve earlier wrote: As you point out, "… in population genetics we start with a simple model and don't extend it unless we need greater accuracy". Your calculations, based on Sir Ronald Fisher's work (which pre-dates modern genetics by a half century), do not give correct results, so it is time to extend your model—or at least incorporate more of what the theory actually says. A simplified model based on parts of a theory is not the theory.

Lee then replied: This appears to me to be pure bluff. It would not be if you could estimate the size of the correction a more "correct" model would make to Fisher's calculation and justify that estimate.

Steve's counter-reply: No bluff. Unfortunately you reject the corrections I recommend because you do not believe that they are either realistic or allowed by modern evolutionary theory! I've already pointed out that sexual recombination and crossover would allow HGCs to accumulate in a time on the order of log(N) instead of (N). And, I've already pointed out that adaptive genes whose double-recessive and double-dominant combinations are maladaptive cannot take over a population at all.

Lee's current reply: No Steve, you have now demonstrated that what you wrote is a bluff. I wrote that it would not be a bluff if you could *estimate the size of the correction* to Fisher's result a more "correct" model would make. You did not attempt to make such an estimate. I therefore suspect that you don't even know how Fisher's calculation was made. It would then follow *a fortiori* that you have no idea what correction your more "correct" model would make. If you don't know what you are talking about in suggesting that Fisher's result is wrong and if you cannot calculate what magnitude correction you are talking about, then your contention that it is wrong is pure bluff.

I also see here that you are still confusing generating diversity with generating evolutionary information. The accumulation of HGC's that you like to talk about is only about diversity. I think that this confusion is the seat of your problem in this discussion. Perhaps that confusion prevents you from understanding how information is supposed to be generated in evolution.

Steve then replied: No bluff, and the confusion is yours. I will be glad to point out fatal flaws in your calculations, and will offer calculations of my own, after we submit our basic disagreement about what MET says to authorities who do the saying. Please take a look at the draft letter. It is you who are bluffing, unless you are willing to join me in submitting these questions and accepting the answers as definitive.

Lee's current reply: I have addressed your proposed letter in a separate note, which please see.

If you indeed have any calculations, please present them now to show what you think is the error in Fisher's result. You have had three opportunities to show those calculations and you have not done so. But I see now that the reason you have not presented them is that you cannot. And the reason you cannot has become apparent, namely that you don't even know what the probability of extinction is all about. And yet you felt yourself qualified to criticize Fisher's calculation. I wrote you that your simulation left out the probability of extinction of the HGC's in that they would leave no progeny. The probability of each of the HGC's surviving until your 200 generations is (assuming, say, that each individual gives rise to 10 progeny) about 0.064. That means that the probability that *both* of the HGC's you postulated will survive 200 generations is about 0.004 Thus your figure of 98% probability of the two HGC's combining by the 200th generation must be reduced by this factor. This is not a minor correction. You replied:

> The simulation leaves out lots of important features. However, to first order, random disappearance of HGCs merely changes the amplification factor slightly. The rate of disappearance is on the order of the mutation rate, which is on the order of 1 in 10^6 to 1 in 10^8. This is entirely swamped by the amplification factor.

The correction of extinction is not a "slight change". It is a major one. The mutation rate you mention is irrelevant to the extinction of the HGC. I had explained to you what the extinction problem is by writing,

> ... each individual will, on the average, give rise to one reproducing offspring (or each couple will give rise to two reproducing offspring, but for simplicity let me talk in terms of a single individual). This means that if an individual will give birth to k offspring, then on the average (k-1) of them will not survive to reproduce due to random causes having nothing to do with their fitness -- each newborn will have a probability of survival of only 1/k. Thus if k=10, then each newborn has a survival probability of 0.10.

but I guess you didn't get it. You tried to explain it away with your irrelevant remark on mutation rate. Since the extinction of a mutant is an essential feature of Fisher's calculation, it has become clear that you are simply not qualified to make any believable comment on it, and certainly not to criticize it as not giving the correct results and having been superceded by more modern knowledge of genetics. You simply don't know what it is all about.

Part 8

Intelligent Life in the Universe and Divine Action

Biography

Professor Juan G. Roederer is Professor of Physics Emeritus at the University of Alaska-Fairbanks. Doctor of Science from the University of Buenos Aires in 1952; between 1959 and 1993 Professor of Physics at the universities of Buenos Aires, Denver and Alaska. He was director of the Geophysical Institute of the University of Alaska (1977-1986), chairman of the United States Arctic Research Commission (1986-1992) and senior advisor of the International Centre for Theoretical Physics, Trieste (1997-2003). His research interests are space physics, psychoacoustics and theory of information; his most recent book, on which the present article is based, is *Information and its Role in Nature* (2005), Springer-Verlag, Heidelberg. He is also an organist.

39

Neurobiological Foundations of Religion and Science

JUAN G. ROEDERER

Geophysical Institute
University of Alaska-Fairbanks
Fairbanks, AK 99775, USA
http://www.gi.alaska.edu/~Roederer

1. Religion and Science: Conflict, Coexistence or Cooperation?

Ever since science emerged during the history of mankind, even in its most primitive form as natural philosophy in the times of antiquity, conflicts with established religious beliefs abounded. Nature, when observed systematically and objectively, often seemed at odds with the "absolute truth" attributed to statements of deities and passed down from generation to generation by high priests, prophets, holy men, and their writings. Yet as we shall examine in this chapter, the predisposition for religious beliefs and the search for scientific knowledge have a common evolutionary origin: the human brain's conception of future time and its unique capability of making long–term predictions, the innate urge to do so and a feeling of satisfaction when it is done. Hence it is only natural that one should understand as clearly as possible relevant aspects of human brain function before embarking in a study of the emergence of faith and science. This is a tall order: the human brain is recognized as the most complex system in the Universe as we know it, and a quantitative and objective exploration of its functions has begun only recently. Therefore it is quite difficult to present a summary picture without gross oversimplifications, and then to relate it to specific aspects of religion and science. Yet this is precisely the aim of the present chapter.

We recognize that most religions are focused on the human being, with the Universe relegated to playing a supporting role, whereas science is centered on the Universe, with the human being just one of its many components. In religion, spatial and temporal dimensions are those familiar to human experience; information about the Universe is that which can be acquired through our senses; and relevant causal interconnections between events are those whose consequences directly affect us. It is then quite natural that phenomena extending over a time-scale beyond a few hundred human lifetimes, images like the Earth orbiting around the Sun (which we never actually see happening), or experimental results showing that white light is composed of many different colors, were met with resistance for a long time. The same is happening today with the realization that order can emerge out of chaos and purposeful behavior out of chance events without any outside intervention except for some available energy and a few universal physical laws—and yet many such processes happen in familiar phenomena (e.g., water boiling in a pot, the development of a hurricane, a memory recall).

But there is no need for conflict. Already in the 13th century, St. Thomas Aquinas dealt with the contradictions between faith and reason and tried to find venues for peaceful coexistence between the two; this, indeed, was the approach later taken by Luther and the Reformation. In Aquinas' times, however, natural philosophy was built mainly on observation and pure reasoning, and it was not difficult to accommodate the latter (for instance, the interpretation of Aristotelian philosophy) to fit current religious doctrine. The foundations of the scientific method, with its rigorous protocol of hypothesis, measurement, theory, prediction and experimental verification, were laid only three centuries later by Galileo, and consolidated and formalized by Newton. The emerging "scientific truth" with its reproducible and verifiable statements brought many results that flew in the face of the "absolute truth" as extracted from the Scriptures through interpretation by mortal human brains.

It took six centuries after Aquinas for Vatican doctrinarians to recognize that coexistence between faith and science, even cooperation, was possible. It does require some compromises, though. Religious faith should recognize that one cannot challenge scientific truth with ideas alone and that many more unforeseen natural phenomena revolutionizing previously held world views may still be discovered. Science, in turn, should recognize that there will always be people—including scientists—who need religion for spiritual guidance and comfort, and that there will always be questions concerning the "why of things" to which the scientific method cannot provide answers. Religions should turn away from a literal-sense interpretation of their sacred scriptures by recognizing when they were written, by whom, for whom and for what purpose (e.g., would Jesus have talked about quarks or DNA to his disciples?) Science should turn away from the "easy way out" offered by the anthropic principle ("if it wasn't this way, we wouldn't be here to ask the question") and recognize the existence of possible "points of contact" with theology such as questions concerning the values of the universal constants, the actual form of physical laws and the key fluctuations which gave rise to, inter alia, the Big Bang, the appearance of the first living organism and the emergence of self-consciousness. Pope John Paul II stated (Russell et al., 1988, pp 14 and 29): "... Science can purify religion from error and superstition and religion can purify science from idolatry and false absolutes", and "The *Bible* itself speaks to us of the origin of the universe and its makeup, not in order to provide us with a scientific treatise but in order to state the correct relationship of man with God and with the universe". And Heisenberg wrote (Heisenberg, 1972): "Science deals with the material, objective world, religion deals with the world of values. ...Science deals with 'what is' and 'what we are', religion deals with 'what ought to be' and 'what we ought to do' ...In science, we are concerned to discover what is true or false, in religion with what is good or evil, noble or base. Science is the basis of technology; religion is the basis of ethics".

Scientific thought and religious faith are what a quantum physicist might call "basis states of the human brain". They are mutually orthogonal, but at any given time the actual state of the brain can be a superposition of the two without violating the principles of either. Any attempt to force a brain into one or the other basis state, like the so-called scientific creationists[1] and some scientific agnostics would wish to do, goes counter to the very nature of human brain function.

Why is that so? And why is it so difficult to achieve coexistence or cooperation between the religious and scientific views of the world? The answer lies in some peculiarities of how the brain works. The description of the most relevant aspects of animal and human brain function and their role in the emergence of religion and science will be the specific purpose of this chapter.

---

[1] In this chapter I will not pay any attention to "scientific creationism", which is neither science nor religion and belongs in books on contemporary American politics.

2. The Information-processing Animal Brain[2]

The vertebrate brain is viewed as the "central processor" carrying out the fundamental operations of monitoring the state of the body and the surrounding environment, committing to memory and predicting relevant events, and executing life-preserving and species-preserving somatic and behavioral responses. There are two fundamental ways in which information is represented in the brain. One is dynamic, expressed in the form of a rapidly changing pattern of neural activity, specifically, the spatial and temporal distribution of electrical impulses which individual neurons send to other neurons, representing the operating state of the neural network. The other is quasi-static, given by the spatial distribution and efficacies of inter-neuron connections (the synapses), representing the internal state or "hardware" of the neural network. The first form varies on a time-scale of a few to hundreds of milliseconds and usually involves millions of neurons even for the simplest information-processing tasks, requiring a substantial supply of energy to be maintained (it is the increased vascular blood flow and oxygen consumption that appear mapped as images in functional magnetic resonance imaging (fMRI) and positron emission tomography (PET), respectively). The second type comprises the pattern of synaptic interconnections in the so-called "pre-wired", i.e., genetically inherited, networks, and the so-called "plastic" networks, mostly in the cerebral cortex, which have learning capability and can change slowly on a time-scale of minutes or hours. There is no equivalent to computer "software" in the neural system—the "programs" are embedded in the changing configuration of the neural hardware (also called the synaptic architecture). Finally, there is a chemical information transmission system: certain substances (neurotransmitters, hormones) injected into the blood stream, which play the role of a temporary modulator (stimulator or inhibitor) of the overall neural activity in specific brain regions; they determine the affective responses of brain activity (see next section) and control many internal organ functions.

Unfortunately, none of the neural patterns, dynamic or quasi-static, can be expressed adequately in a mathematical form. The task would be totally hopeless: in the human brain there are over a hundred billion (10^{11}) neurons in the cortex, each one connected to thousands of others, which means that there are over 10^{14} synaptic connections that would have to be described in terms of their position and efficacies. Besides, there is no continuity: two neighboring

[2] The next three sections are based in part on Chapter 6 of Roederer (2005).

neurons may have totally different firing rates. What is important to understand is that the spatio-temporal distribution of neural activity and the spatial distribution of synapses taken together represent the global state of the functioning brain at any instant of time.

There are well-defined stages of information-processing in the neural circuitry of the brain from the sensory receiving areas of the cortex to the frontal lobes, and from there to the motor areas that command the muscles, as well as back to the sensory areas. How a specific spatio-temporal neural activity distribution elicited by the sight of an object or by listening to a sound becomes a specific mental image is an old question that has puzzled biologists and philosophers alike. Today neurobiology provides a radical answer: the pattern doesn't "become" anything—it is the image! Let me restate this with an (oversimplified) example for the human brain. When you see a "shiny red apple"; when you close your eyes and imagine a "shiny red apple"; when somebody says the words "shiny red apple"; or when you are reading these very lines, there appears a (extraordinarily complex) spatio-temporal distribution of neural activity in several specific regions of your brain, part of which is nearly the same in all cases. That common part represents the cognition of "shiny red apple", and is your mental image—your neural correlate—of the concept "shiny red apple". It is yours only; physically/physiologically it would be very different from the one that forms in my brain or in anybody else's under the same circumstances (only the participating regions would likely be the same)—but still these patterns are all expressions of the same information[3]. What counts is the univocal character of the correspondence "object → neural activity distribution"[4]. This aspect of brain function is of fundamental importance but it takes time and effort to clearly comprehend!

One fundamental brain function is memory. The short-term or working memory is dynamic, represented in the form of neural activity kept in some "holding pattern", which has a storage lifetime of the order of tens of seconds. The synaptic architecture per se of the information-processing networks in many

---

[3] Meant here is pragmatic information, i.e., that which represents a univocal correspondence between the pattern in a "sender" and an ultimate change in a "receiver" (e.g., Roederer, 2005).

[4] Patients recovering from brain lesions (e.g., stroke) can, under certain circumstances, learn to establish a new correspondence "old object : new neural activity"; the triggered activity distribution is new, but it represents the same image from the informational point of view.

parts of the brain represents the long-term or structural memory—both inherited genetic memory (in pre-wired networks) and memory acquired through learning experiences during the lifetime of an organism. To retrieve information in a memory recall, the brain must recreate the activity distribution that prevailed during storage, i.e., the neural activity pattern that is specific to the object, event or concept that is being recalled. This is accomplished by feeding into the relevant network a partial pattern or key, which may then trigger the full activity distribution if some minimum information threshold is surpassed. This process is called associative recall (for example, the acoustic perception of the name of a person triggering the visual image of the person). The external key may be just part of the image to be recalled, in which case we have an auto-associative recall (e.g., recognizing the person in a partially obliterated photograph).

The act of long-term memory storage consists of appropriate changes of the synaptic architecture during a learning process. When two or more near-simultaneous sensory input events (e.g., the name and the photograph of a previously unknown person) are presented repeatedly, parts of the relevant neural circuitry change in such a way that, in the future, only one of the input events can trigger the neural representation of the other(s). Note that the brain's mechanism of distributed memory storage and associative recall is fundamentally different from the familiar modes of addressed memory storage and retrieval. Since the mechanism is equivalent to what happens in optical (Fresnel) holography, it is also called a hologic mode of information processing.

Memory storage is hierarchical; images of concepts with lower information content[5] (for example, the concept "apple") are retrieved faster than a subclass with higher information content (for example, the concept "shiny red apple"). An important feature is that the information of millions of images can be stored in the same neural network; another feature is the fact that associative recalls are triggered processes: once launched, they run their course until the replay of neural activity is completed. These processes can all be simulated on a computer using simplified numerical models of neural circuits (e.g., see Kohonen, 1988).

The re-elicitation of neural patterns that represent the images stored in memory requires feedback mechanisms—reverse information-driven interactions in which

---

[5] Meant here is algorithmic information, i.e., the minimum number of binary steps needed to identify the object in question from among a total set of possible ones (e.g., Roederer, 2005).

higher level patterns trigger associated patterns at lower levels. A consequence of this brain mechanism is that the expectation or anticipation of a sensory input will trigger neural activity in relevant primary sensory areas of the cortex even before an actually occurring external feature can elicit the corresponding response; if the expected feature is missing, the corresponding neural pattern appears anyway! This has been verified in many experiments in which the response of feature-detecting neurons in the visual primary cortex is measured while the laboratory animal is exposed to stimuli in which the expected feature is sometimes present, sometimes missing. Much of sensory perception (e.g., speech, music) is indeed based on a process of confirmation and, if necessary, interpolation of missing information or, if the actual input a moment later does not match the expectation, correction of the anticipated image. This leads to affective responses and plays an important role in art appreciation.

3. The Feeling Brain

So far we have focused mainly on input. The neural output of the brain controls the organism's striate musculature for posture and both voluntary and stereotyped movement, some of the smooth muscles of internal organs, and the chemical endocrine system. The more advanced an animal species, the more options it will have for the response to a momentary constellation of the environment. This requires decision-making based on some priorities. Unfortunately, here we tend to think in terms of our own experience with "reasoned" decisions. Animals, however, can neither reason nor engage in long-term planning like humans do (see next section); to reach a goal (hunting, feeding, finding a mate, building a shelter) they follow instructions based both on instincts (genetically acquired information) and experience (learned information). Such instructions are coordinated by brain structures which include a group of subcortical nuclei located near the midline of the brain, historically called the limbic system (brain scientists are quite reluctant to use such "umbrella" designations for processing centers that interact in different ways depending on context), the hypothalamus, the amygdalas, the hippocampi and the basal forebrain. Their functions are to check on the state of the environment and the organism, direct the animal's attention and motivation, and make sure that the output—the behavioral response—is beneficial to the survival of the organism and the propagation of the species in conformity with evolutionary and ontogenetic experience (see sketch in Fig. 1). For a recent review see Dolan (2002).

It is important for the purpose of this chapter to point out some characteristics of the information processing modes coordinated by the limbic system. (i) Since the amount of information input is utterly unmanageable, only relevant information must be stored in the brain for future use. This implies that a selective filtering process, based on genetic and learned information, must first take place before new information is cleared for long-term memory storage. In this selective storage process, "subjective relevance" ranks higher than "objective truth". (ii) Since expediency of environmental sensing operations is a determining factor in maximizing the survival probability, the brain cannot afford accurate statistics in the detection of cause-and-effect relationships: simultaneity is often a sufficient criterion for assuming the existence of a cause-effect relationship among subjectively relevant events. (iii) A fundamental drive compels the brain into exploratory information-acquisition operations even if the current state of the environment does not require that such operations be carried out—the curiosity and playfulness exhibited by many animals are an example.

The limbic system works in a curious "binary" way by dispensing sensations of "reward or punishment": hope or anxiety, boldness or fear, love or rage, satisfaction or disappointment, happiness or sadness, and so on. These are the emotional states of the brain (controlled by the deeper subcortical nuclei). They evoke the anticipation of pleasure or pain, whenever certain environmental events are expected to lead to something favorable or detrimental to the organism, respectively. Since this anticipation comes before any actual benefit or harm could arise, the emotional state helps guide the animal's motivation (controlled by the anterior cingulate cortex) to respond into a direction of maximum chance for survival or procreation, dictated by information acquired during evolution and experience—the so called instincts and drives. Of course, only human beings can report to each other on emotional states or feelings, but higher vertebrates also experience such a "digital" repertoire. The pleasantness or unpleasantness of a feeling is controlled by the chemical information system of the brain (e.g., opioids, monoamines).

Just as the cortical activity is being continuously monitored by the limbic structures, so is the state of the organism; the resulting information-processing operations in turn are mapped onto certain cortical regions (Fig. 1), especially in the prefrontal areas. In this interplay a balance must be achieved, or the organism could succumb to conflicting behavioral instructions. Venturing for just a moment into a computer analogy, there is only one main program operating at any given time in the brain—surely, with many subservient parallel

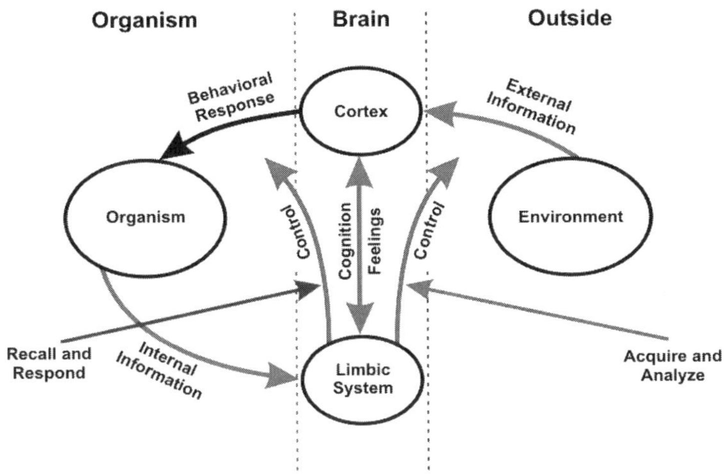

Figure 1: Brain regions which assure that cognitive functions and behavioral response are beneficial for the organism and the propagation of the species. The limbic system generates emotion, dispenses feelings and communicates interactively with higher processing levels of the cortex, particularly the prefrontal regions, relating everything the brain perceives and plans to the needs of the organism and vice versa. In higher species, the coherent operational mode of the cortico-limbic interplay gives rise to consciousness. The human brain is able to map and coordinate this interplay at an even higher level, as well as to overrule limbic dictates; this leads to self-consciousness—and religion and science. After Roederer (2005).

subroutines. In other words, we have total parallel processing under tight central management. Based on real-time input, the memory of experienced events and the instincts, out of this balance or compromise emerges just one primary goal for action at a time—this is the essence of animal core consciousness (Damasio, 1999) and mental singleness. Note the specific epochs of time involved: (i) information from the past (the distant past built up over a long time in Darwinian evolution, collectively representing the instincts, and the individual's past as it has impacted the organism); (ii) information from the present on the state of the body and the state of the environment; and (iii) a narrow window of time for short-term predictions of perhaps a few tens of seconds (probably defined by the capacity of the short-term memory). It is concerning this last window of time that the most striking differences between infra-human and human brains are found.

4. The Information-generating Human Brain

So far we have dealt with the functions of the brains of higher vertebrates in general. From the neurophysiologic and neuroanatomical points of view the human brain is not particularly different from that of a chimpanzee. It does have a cortex with more neurons and some of the cortico-cortical fascicles have more fibers, but this difference is of barely a factor of 2 or 3. More significant is the total number of synapses in the adult human brain. Is the difference in the resulting information-processing capabilities only one of degree but not one of substance?

Aristotle already recognized that "animals have memory and are able of instruction, but no other animal except man can recall the past at will". More specifically, the most fundamentally distinct operation that the human, and only the human, brain can perform is to recall stored information as images or representations, manipulate them, and re-store modified or amended versions thereof without any concurrent external sensory input—this is information generation par excellence (Roederer, 1978). In other words, the human brain has internal control over its own feedback information flow; an animal can anticipate some event on a short-term basis (seconds), but only in the context of some real-time somatic and/or sensory input, i.e., triggered by "automatic" associative recall processes. The act of information recall, alteration and re-storage without any external input represents the human thinking process or reasoning. J. Z. Young (1987) stated this in the following terms: "Humans have capacity to rearrange the 'facts' that have been learned so as to show their relations and relevance to many aspects of events in the world with which they seem at first to have no connection". And Bickerton (1995) wrote: "...only humans can assemble fragments of information to form a pattern that they can later act upon without having to wait on... experience"[6].

---

[6] Categorical statements like these, which trace a boundary, or "ontological discontinuity", between human and animal brain capabilities, are disputed by many animal psychologists. They point out the fact that some apes and birds use tools, construct nests, assemble food caches and exhibit a highly sophisticated social behavior, thus demonstrating ability of complex cognition and mental representation of time (e.g., Emery and Clayton (2004)). However, such behavioral activities are not based on the knowledge of how a tool works or why this or that feature of a nest is better suited for future needs.

The capability of recalling information without any concurrent input had vast consequences for human evolution. In particular, the capability of re-examining, rearranging and altering stored images led to the discovery of previously overlooked cause-and-effect relationships, to a quantitative concept of elapsed time and to the awareness of future time. In animals, the time interval within which causal correlations can be established (trace conditioning) is of the order of tens of seconds and decreases rapidly if other stimuli are present (Han et al., 2003); in humans it extends over the long-term past and the long-term future. Along with the ability of ordering events in time came the possibility of long-term prediction and planning ("information about the future"; Squires (1990), Roederer (2000)), i.e., the mental representation of events that have not yet occurred.

Quite generally, the human thinking process involves the creation of new images, i.e., spatio-temporal distributions of neural activity that do not correspond to any previously sensed or experienced information input. Patterns or objects can be crafted and changes in the environment can be effected that did not exist before and which would never be a deterministic consequence of physical laws and natural initial conditions. This should not be confused with the capacity of higher vertebrates to anticipate the course of current events on a short-term basis of tens of seconds or with animal behavior like tool-making and shelter-building. The latter looks very sophisticated to us but it follows "blueprints" handed down in the genome which differ substantially from the kind of "blueprints" followed by a person building a hut or designing a mansion. The latter are conceived "on the spur of the moment" based on "information from the future"—i.e., as a result of long-term planning—and they can be modified radically in real time not only to take into account unforeseen environmental circumstances but also because of changing ideas, i.e., changing mental images of the goal itself.

Concomitantly with this development came the postponement of behavioral goals and, more specifically, the capacity to overrule the dictates of the limbic system (e.g., sticking to a diet even when you are hungry) and also to willfully stimulate the limbic system, without external input (e.g., getting enraged by thinking about a certain political leader). In short, the body started serving the brain instead of the other way around! Mental images and emotional feelings can thus be created that have no relationship with momentary sensory input—the human brain can indeed go "off-line" (Bickerton, 1995).

In parallel with this development came the ability to encode complex mental images into simple acoustic signals and the emergence of human language. This was of such decisive importance for the development of human intelligence that certain parts of the auditory and motor cortices began to specialize in verbal image coding and decoding, and the human thinking process began to be influenced and sometimes controlled by the language networks (e.g., Premack (2004)) (but it does not mean that we always think in words).

It is important to point out that the capabilities of recalling and rearranging stored information without external input, making long-term predictions, planning and having the concept of future time, stimulating or overruling limbic drives, and developing language, most likely evolved hand-in-hand as neural expressions of human intelligence. At the root of it all from the informational point of view lays the human capability of making "one representation of all representations". There is a higher-order level of representation in the human brain which has cognizance of consciousness and which can manipulate independently the "main program" or primary neural representation of current brain activity both in terms of cognitive acts and feelings. It can construct an image of the act of forming images of environment and organism, as well as of the reactions to them. In other words, in human beings the informational processes have an integral representation at a higher level, and top-down propagation allows those higher-level patterns to influence the neural activity distribution at the lower levels. Collectively we call this human self-consciousness (close but not equal to the "extended consciousness" defined by Damasio (1999)). Self-consciousness is far more than just a feeling—it represents the capacity for some very unique information processing actions. A useful metaphor for this process would be the following: while consciousness is "watching the movie that is running in the brain", self-consciousness is the capacity of human brains "to splice and edit that movie and even to replace it with another one" (Roederer, 2000)[7].

There is no need to assume the existence of a separate neural network to accomplish all this; there is enough information-handling capacity in the human

[7] It may well be that higher primates have "bursts" of self-consciousness during which internally recalled images are manipulated and a longer-term future is briefly "illuminated". But there is no clear and convincing evidence that any outcome is stored in memory for later use. There is a contentious debate on this issue between animal psychologists and brain scientists.

cortical networks that can be shared—except the need for a greater involvement of the prefrontal cortex and the possibility that language networks of the dominant hemisphere may participate in important ways. As a matter of fact, in this sharing there is a give-and-take (Goldberg et al., 2006) between tasks involving pure introspection (the "going off-line", Bickerton, 1995) and the processing of sensory information. For an up-to-date discussion of animal and human consciousness, see Damasio (1999), Koch (2004) and Seth and Baars (2005).

5. The Believing and Knowing Brain

We shall now systematically revisit sections 2-4 almost paragraph by paragraph, in order to point out the relevance of characteristic modes of brain function for the emergence of religious beliefs and science.

We begin with the brain's particular hologic mode of information storage and recall (Section 2). We stated that the recall process runs its course once it is triggered by an input pattern, the "key". When this key is incomplete or in any way distorted or noisy, the triggered neural pattern (the remembered image) may be "wrong", which means not in correspondence with an undisturbed key. When the processing neural network itself is disturbed (chemically, by an abnormal concentration of certain neurotransmitters or other substances in the blood stream, or physically by a lesion), the triggered pattern may not be wrong but distorted or noisy. The result of either case could be an illusion if the stored image corresponds to a static concept (object, word or symbol), or a hallucination, a time-sequence of images that have no correspondence with sensory input of the moment. These are typical responses of a dysfunctional network of distributed, non-addressed memory; they, too, can be simulated by computers.

Illusions are quite common and include ambiguous or erroneous representations of sensory input (e.g., Necker's cube or "I could swear I saw him in the audience", respectively). Hallucinations have played an influential role in the history of human culture, building the framework for many superstitious beliefs—perhaps even religious beliefs. Caused by metabolic disorders, fever, drugs, dementia, damage to neural tissue or by more indirect circumstances such as sensory deprivation and hypnotic trance, hallucinations can be triggered externally by a normal perceptive experience, by illusions or internally by normal cognitive activity; they can also be elicited by localized electrical or magnetic stimulation of the brain. Giants, dwarfs, monsters, ghosts, and flying witches all pertain to an almost standard and, we should point out, surprisingly limited repertoire of characters in different mythologies, and may have originated in rather well-defined dysfunctions of neural processing mechanisms (for example, impairment of the vestibular system by hyoscyamine, found in witches' "brews", giving a sensation of "flying"; disfigurations in optic tract processing by bufotenin, found in the skin of venomous toads, causing moving objects to appear of enlarged or reduced size; perceptual distortions by ingestion of ergot-contaminated cereal grains; e.g., Siegel and West, (1975)). The "voices" heard by schizophrenics are neural images triggered in the speech processing centers by back-propagating activity from the left frontal regions. And if we generalize and define hallucinations as any triggered sequence of systematically misplaced or out-of-context, but otherwise episodically self-consistent neural activity patterns (images) disconnected from sensory input, we may even include normal dreams among them.

Next on our list are consequences of the active intervention of the limbic system (Section 3). The subjectivity and self-centrism imposed by limbic functions pervades human behavior in all its aspects. Consider astrology as an example: when an astrologer (or the daily horoscope in our newspaper) makes a series of (totally random!) predictions, we will remember those that by pure chance did come true, report them to our friends, write them down in our diary—and forget all others. The average value of the data (in this case, the true/false value of each prediction) is biased away from zero—information that was true (just by chance) survived and propagated, false information died off! The self-centered, non-statistical nature of memory storage of events and the consideration of simultaneity as a sufficient criterion to validate a cause-and-effect relationship lie at the basis of many religious rituals and the wealth of personal superstitions and fetishes which pervade the lives of most individuals ("if I wear this magnetic bracelet, I'll win the race").

We mentioned the fundamental drive to engage the brain in exploratory information-acquisition operations even if they are not necessary at the moment. Indeed this drive—the anticipation of a feeling of pleasure when such information is obtained—has a profound impact on human behavior (humans have recently been designated as "infovores" (Biederman and Vessel, 2006)). It represents the evolutionary origin of artistic appreciation and today represents the very basis of scientific curiosity.

Finally, the "binary" mode of the affective response controlled by the limbic system transcended into the behavior of human society since its very beginning, and is responsible for the mutually antagonistic pairs of good/bad, heaven/hell, peace/war, compromise/winner-takes-all, etc. On the other hand, we also referred to the capability of overruling the dictates of the limbic system; this makes it possible to suppress behavior judged detrimental to society (e.g., limit sexual drive and predatory instincts) or to oneself (e.g., hygiene, diets). All this leads to the roots of moral codes and the moral content of religions.

Long-term prediction is the single most important exclusive characteristic of the human brain. As stated in Section 4, the ability of recalling stored information without external input leads to the ability of ordering events in time and discovering causal relationships long after they have occurred, or correlations between current events that are separated in time by more than the short-term memory span of just a few tens of seconds. This ability in turn endowed humans with the capacity of long-term prediction based on observations. For some subjectively relevant complex events the brain builds a repertoire of possible outcomes and their respective likelihoods on which to base behavioral decisions—we may call this process "building a knowledge base"[8]. Organized tribal warfare, weather prediction, rudimentary agriculture and animal breeding were probably the first "practical applications" of this ability in the societal realm—and principal drivers in the early Darwinian evolution of human intelligence.

The chain "observations→information→knowledge" is intimately tied to the characteristics and functions of the human sensory system and the neural networks that evolved to process incoming information. When information is not obtainable through the senses, the human brain can still build images based on self-generated information (Section 4). In this case we have the chain "thoughts→information→belief". The belief in divine beings who govern uncontrollable aspects of environmental events, human life and the after-death may have been the first "practical application" in the spiritual realm. In particular, evidence of attempts to predict what happens after death is found in the earliest burial sites, which included food and artifacts for the deceased's soul. Even today, the continuity of our self throughout and beyond death is perhaps the most basic subject of every religion on earth. As intelligence

[8] Animals build memory banks but not knowledge bases: a lion cannot decide today which of several hunting strategies it is going to use tomorrow.

evolved and early hominids became more and more preoccupied with this kind of after-death prediction-making, peace of mind (controlled by the limbic system) became an important factor for survival.

Finally, let us turn to the emergence of science. The brain works with models that are only approximations of the so-called "reality" outside. These models are built during learning experiences and amended in later experiences, in which errors can be corrected and details can be expanded. In animals these amendments occur mostly by chance through repeated exposure to external happenings and/or limbic (instinctive) guidance; humans, as "infovores", have the ability to seek out improvements for the mental images they already possess. This led to a drive toward a systematization of observation and representation, to the observation of events that were not immediately relevant, and to a truly "statistical" approach to information acquisition. Much later, long past the latest stage of human brain evolution, came (i) the introduction of systematic measurements and the development of instruments to extract information from domains not directly accessible to the sensory system; (ii) the sharing of information through linguistic discourse and documentation of environmental information in more permanent and unbiased memory banks than the brain; and (iii) the mathematical idealization of regularities, symmetries and trends in cause-and-effect relationships in the form of laws of nature.

Science, like the brain, also works with models that are only approximations of reality. They are just better approximations and extend over domains that are not accessible to our senses. And it is precisely here that collisions with religious dogma have appeared and will continue to appear. Understanding scientific facts like the existence of self-organizing systems, the utter complexity of life systems, Darwinian evolution, the operating modes of our brain and the strange behavior of the quantum world is nearly impossible for someone who is not scientifically trained and thus at the mercy of naïve intuition—on which most religions are built. There is no easy solution to this dilemma. Better public education helps, but one cannot pretend to teach fundaments of science to every human being. On the other hand, scientists themselves must also recognize that there are intrinsic limitations to science. Indeed, the evolutionary expansion of the higher-level processing networks of the human brain did not take place in order to satisfy the intricate requirements for scientific thought, but, rather, was shaped by an explosive expansion of requirements for planning tasks to assure survival in a rapidly changing environment under increasing competition from peers (and the consequent need to outsmart, not just out-power, groups of competitors).

The systematization of observation and prediction had to content itself with the use of "old" information-processing machinery whose evolution was driven by more mundane factors. Up to this very day, all models in science and their mental (neural) representations are tied to the classical macroscopic world with which we interact (even in quantum physics we use mental strategies such as "following a given particle on a given path", which doesn't really exist in the quantum domain, and then wonder why we are running into paradoxes; e.g., Roederer (2005)). Scientific intuition is thus condemned to be based on what the brain machinery has originally evolved for, namely creating biologically useful neural images from sensory exposure to the classical macroscopic world—which is also the world as viewed by religion!

In the above, the almighty limbic system (Section 3) is deeply involved: we are motivated to know and to believe, and we are rewarded by knowing and by believing. The coordinated interactions between cortical and limbic functions shown in Fig. 1 in fact constitute the neural basis which makes humans build science—and they are the "antenna" through which human beings relate with their God and the universe.

6. Abstract

Predisposition for religious beliefs and the search for scientific knowledge have a common evolutionary origin: the human brain's conception of time and its unique capability of making long–term predictions, the innate urge to do so and a feeling of satisfaction when it is done. It is therefore useful to understand the relevant cognitive and affective mechanisms of brain function if one wishes to analyze the reasons for age-old real or apparent conflicts between religion and science. We describe current knowledge concerning key characteristic modes of cerebral information-processing, memory storage and recall, the role of the limbic structures in affective and cognitive functions, and the unique capabilities of the human brain. We then link these modes to questions relating to the emergence of faith and science.

References

Bickerton, D. (1995). *Language and Human Behavior.* University of Washington Press, Seattle.

Biederman, I. and Vessel, E. A. (2006). Perceptual pleasure and the brain. *Am. Scientist*, May-June 2006, 247-253.

Damasio, A. (1999). *The Feeling of What Happens: Body and Emotion in the Making of Consciousness.* Harcourt, Inc., San Diego, New York, Boston.

Dolan, R. J. (2002). Emotion, cognition, and behavior. *Science*, 298, 1191-1194.

Emery, N. J. and Clayton, N. S. (2004). The mentality of crows: convergent evolution of intelligence in corvids and apes. *Science*, 306, 1903-1907.

Goldberg, I. I., Harel, M. and Malach, R. (2006). When the brain loses its self: prefrontal inactivation during sensorimotor processing. *Neuron* 50, 329-338.

Han, C. J., O'Tuathaigh, C. M., van Trigt, L., Quinn, J. J., Fanselau, M. S., Mongeau, R., Koch, C. and Anderson, D. J. (2003). Trace but not delay fear conditioning requires attention and the anterior cingulated cortex. *Proc. Natl. Acad. Sci. USA*, 100, 13087-13092.

Heisenberg, W. (1972). *Physics and Beyond: Encounters and Conversations*, Harper & Row, Publ., p. 82.

Koch, C. (2004). *The Quest for Consciousness, a Neurobiological Approach.* Roberts and Co., Englewood, CO.

Kohonen, T. (1988). *Self-Organization and Associative Memory*, Springer-Verlag, Berlin, Heidelberg, New York.

Premack, D. (2004). Is language the key to human intelligence? *Science*, 303, 318-320.

Roederer, J. G. (1978). On the relationship between human brain functions and the foundations of physics. *Found. of Phys.*, 8, 423-438.

Roederer, J. G. (1995). *Physics and Psychophysics of Music.* Springer-Verlag, New York.

Roederer, J. G. (2000). Information, life and brains, In J. Chela-Flores, G. Lemarchand, & J. Oró (eds) (2000) *Astrobiology: Origins from the Big-Bang to Civilization.* Kluwer, Dordrecht, 179-194.

Roederer, J. G. (2005). *Information and its Role in Nature.* Springer Verlag, Heidelberg, New York.

Russell, R. J., Stoeger, W. R., S.J. and Coyne, G. V., S.J. (eds) (1988). *Physics, Philosophy and Theology: a Common Quest for Understanding.* Vatican City State, Vatican Observatory, p. 14 and p. 29.

Seth, A. K. and Baars, B. J. (2005). Neural Darwinism and Consciousness. *Consciousness and Cognition*, 14, 140-168.

Siegel, R. K. and West, L. J. (eds) (1975). *Hallucinations, Behavior, Experience and Theory*. John Wiley and Sons, New York, London, Sydney, Toronto.

Squires, E. (1990). *Conscious Mind in the Physical World.* Adam Hilger, Bristol and New York.

Young, J. Z. (1987). *Philosophy and the Brain.* Oxford Univ. Press, Oxford, New York.

Dialogue

Arie S. Issar

When the editors asked me to start a dialogue with Prof. Juan Roederer's and sent me his CV, (from which I learnt that he is a physicist) and the title of his article, *Neurobiological Foundations of Religion and Science*, I was not too optimistic, to say the least, about the possibility of a dialogue between us. The main reason was my experience, up to date, gained from my discussions with physicists about my conceptual model, that we live in a world in which information is an additional dimension. Intrinsic in the space-time continuum and entropy is a cosmic field of force pulling all complex ordered systems towards disorder. From these discussions I learnt that the physicists view the universe as a spatio-temporal-energy complex and thus any event which takes place in it should be measurable by meters, clocks, weights, thermometer and voltmeters. All that is not measurable by these instruments, like knowledge, and any field of force that does not answer the classical definitions of such a field, does not fall into the category of the physical sciences, and thus is not worthwhile spending time examining.

Yet, after reading the full article I changed my mind, not so much with regard to the above mentioned constraints of measuring our world, but despite these. It occurred to me that the fact that Prof. Roederer was brave enough to enter an arena which, although it is measurable by the above mentioned instruments, may still hide the secret of what many people regard as the dwelling place of the "mind", gives me a chance to discuss this issue with him. I thus decided to accept the challenge and also enter the arena (from the opposite side). My arguments will be based, in the first place, on the fact that as a geologist I studied paleontology, i.e. the evolution of the bio-world, and this study included the evolution of the intelligence of the hominids as judged from the evolution of their tools. Yet, despite many efforts I was not able to explain the evolution of intelligence in the bio-world, without adding an additional dimension, i.e. that of Information (see my article in this volume).

I will start my discussion with quoting a statement of my opponent: "What is important to understand is that the spatio-temporal distribution of neural activity and the spatial distribution of synapses taken together represent the global state of the functioning brain at any instant of time".

The attention of the reader is drawn to the fact that the only dimensions by which Prof. Roederer's measures the neural activity of the brain are just space and time (namely: "spatio-temporal distribution of neural activity and the spatial distribution of synapses").

Here I come to another statement of my opponent, namely: "There are two fundamental ways in which information is represented in the brain. One is dynamic, expressed in the form of a rapidly changing pattern of neural activity, specifically, the spatial and temporal distribution of electrical impulses which individual neurons send to other neurons, representing the operating state of the neural network. The other is quasi-static, given by the spatial distribution and efficacies of inter-neuron connections (the synapses), representing the internal state or 'hardware' of the neural network."

My question is: Fine, this tells us only about the spatio-temporal representation of information but does not touch the question: What is information?

In other words: I do not oppose Prof. Roederer's statement as a partial one, yet I do not think that it answers the question: What is the difference between the processing of information as a function of outside stimulations and the building of a "theory", which connects these stimulations in a logical (not to say mathematical) algorithm. Is it just "electrical impulses", or just "spatial distribution and efficacies of inter-neuron connections"? I will put this question in another form: Is the difference between the theory of creation as brought in the Book of Genesis and that of the Big Bang, just a matter of "spatio-temporal distribution of neural activity etc".? In this connection, in sub-chapter 2, "The Information-processing Animal Brain", Prof Roederer writes:

> "What counts is the univocal character of the correspondence 'object→neural activity distribution' (4). This aspect of brain function is of fundamental importance but it takes time and effort to clearly comprehend!"

If I comprehend it well then there is a "spatio-temporal distribution of neural activity", which corresponds to each observed object. In other words the "spatio-temporal distribution of neural activity" is the "image" of the object. What about "abstract ideas". I guess they are also presented by typical "spatio-temporal distribution of neural activity". How is this turned into knowledge? For example, consider the knowledge of "danger" in the case of "fire". It may be assumed that there must be a spatio-temporal distribution of neural activity which associates with different views of fire. Thus, a certain configuration for the same fire may mean warmth, another configuration, danger.

My claim is that the association of the notion of fire and the "meaning" of it must each have a different: "spatio-temporal distribution of neural activity and the spatial distribution of synapses", yet, these include also the logical connection, namely: *"if...then"*. Thus, if the fire is in the stove then it is warmth but if it is in the carpet *then* it is danger.

I thus suggest that *"If... then"* is an "information component additional to the spatio-temporal distribution of neural activity". In other words, an additional dimensional entity, which enables the interaction between different spatio-temporal distributions of neural activity. This means that the link between two different spatio-temporal distributions of neural activities to form a logical sentence of: *If* the fire is in the stove *then* it is safe and warm, etc., is not only a spatio-temporal arrangement but also informational, namely a configuration on an additional dimension. This involves also the adding of a few spatio-temporal distributions of neural activities to form one comprehensive spatio-temporal-informational neural activity.

Thus I suggest that Prof. Roederer changes his former quoted sentence:

> "What is important to understand is that the spatio-temporal distribution of neural activity and the spatial distribution of synapses taken together represent the global state of the functioning brain at any instant of time".

to the following sentence:

> What is important to understand is that the spatio-temporal-information distribution of neural activity and the spatial distribution of synapses taken together represent the global state of the functioning brain at any instant of time along the dimension of information.

In this connection I will add the observation that all brains evolve, and are able to think only by the investment of energy, and decay at older age to become senile, and later decay into organic and then simple molecules. These phenomena I attribute to the (scalar?) field of entropy. In this field energy has to be invested in order to take off and keep flying against the pull of the universal arrow of entropy.

Now for a few additional remarks of secondary importance (relative to that concerning the Dimension of Information). I will start with the opening chapter: "Religion and Science: Conflict, Coexistence or Cooperation?"

A. I liked the categorization of religion and science, namely: "that most religions are focused on the human being, with the Universe relegated to playing a supporting role, whereas science is centered on the Universe, with the human being just one of its many components". Indeed this is true for the religions based on the *Bible*, where creation of the world is just for the purpose of creating the living conditions for Adam and Eve. I wonder however whether this applies also to the Hindu and its branch-off religion, i.e. the Buddhist faith. Although these religions deal with the human being, they claim that life is just a stage to *Nirvana*. Thus life should be devoted to exercises for the fading out of the self in relation to the infinite dimensions of the universe. I think thus that a more general categorization of religion and science would be: *The world view of all religions is based on the wisdom of the ancients (the more ancient the more significant), while science is thriving constantly to build a new world view, while searching for new observations and new explanations.*

B. I agree with the author "that there will always be people——who need religion for spiritual guidance and comfort". I wonder, however, whether a scientist has to agree that these people should "include scientists". In this connection I disagree from the *scientific- philosophical* point of view with the author's conviction "that there will always be questions concerning the 'why of things' to which the scientific method cannot provide answers". *I claim that scientists should start from the assumption that ALL questions should be asked and for all questions there might be found a scientific answer! Yet, to be aware of the fact that, once the questions, which exist today are answered, new questions will crop up, which makes the occupation and life of a scientist so interesting and challenging.*

C. These are the reasons that I can not share the author's optimism about finding "'points of contact' with theology such as questions concerning the values of the universal constants, the actual form of physical laws and the key fluctuations which gave rise to, inter alia, the Big Bang, the appearance of the first living organism and the emergence of self-consciousness." All are legitimate scientific fields of investigation, and an answer will be found in the framework of science, with no need for theologians' help.

In this connection I disagree with Heisenberg that "Science is the basis of technology; religion is the basis of ethics". *From a scientific point of view, I claim that ethics are the fundamentals of the existence of the human society and one can see this evolution from the ethics of the clan (I may start even from a pack of wolves) to the tribe, to the nation and hopefully to the family of nations. Religion did not, and still does not stop the human being from the most non-ethical deeds to his fellow human beings who do not share his religious dogma.*

Sub chapter 3. The Feeling Brain

> "Unfortunately, here we tend to think in terms of our own experience with 'reasoned' decisions. Animals, however, can neither reason nor engage in long-term planning like humans do (see next section); to reach a goal (hunting, feeding, finding a mate, building a shelter) they follow instructions based both on instincts (genetically acquired information) and experience (learned information)."

I agree that there is a difference in the range of long term planning between humans and animals but I disagree with the distinction in "reasoning" faculties. The principal of reasoning is the same, i.e. *if...then* and *adding* one *if...then* experience to the other.

Sub chapter 4. The Information-generating Human Brain

> "So far we have dealt with the functions of the brains of higher vertebrates in general. From the neurophysiologic and neuroanatomical points of view the human brain is not particularly different from that of a chimpanzee. It does have a cortex with more neurons and some of the cortico-cortical fascicles have more fibers, but this difference is of barely a factor of 2 or 3. More significant is the total number of

> synapses in the adult human brain. Is the difference in the resulting information-processing capabilities only one of degree but not one of substance?"

As a geologist, investigating the evolution of the intelligence of the bio-world I claim that "the difference in the resulting information-processing capabilities" is only one of degree!!!

> "More specifically, the most fundamentally distinct operation that the human, and only the human, brain can perform, is to recall stored information as images or representations, manipulate them, and restore modified or amended versions thereof without any concurrent external sensory input—this is information generation par excellence (Roederer, 1978). In other words, the human brain has internal control over its own feedback information flow."

But again what is information? Is it *only* "spatio-temporal distribution of neural activity and the spatial distribution of synapses" *and nothing more?*

> "The capability of recalling information without any concurrent input had vast consequences for human evolution."

I would say just the opposite: human evolution brought the "capability of recalling information without any concurrent input".

> "Quite generally, the human thinking process involves the creation of new images, i.e., spatio-temporal distributions of neural activity that do not correspond to any previously sensed or experienced information input."

Is it only spatio-temporal distributions of neural activity or something more? (See my former remarks.)

> "In parallel with this development came the ability to encode complex mental images into simple acoustic signals and the emergence of human language. This was of such decisive importance for the development of human intelligence"

Again! …"development of human intelligence" brought "emergence of human language".

> "In other words, in human beings the informational processes schematically depicted in Fig. 1 have an integral representation at a higher level, and a top-down propagation allows those higher-level patterns to influence the neural activity distribution at the lower levels."

I will say it is action along the Space-Time-Information coordinates.

5. The Believing and Knowing Brain

> "Finally, the 'binary' mode of the affective response controlled by the limbic system transcended into the behavior of human society since its very beginning, and is responsible for the mutually antagonistic pairs of good/bad, heaven/hell, peace/war, compromise/winner-takes-all, etc. On the other hand, we also referred to the capability of overruling the dictates of the limbic system; this makes it possible to suppress behavior judged detrimental to society (e.g., limit sexual drive and predatory instincts) or to oneself (e.g., hygiene, diets). All this leads to the roots of moral codes and the moral content of religions."

One should not ignore in the first place the "moral codes" of animal societies, whether a pack of wolves, a herd of elephants or a family of chimpanzee. I think that the same process, which Prof. Roederer finds in the brain of the human being, is functioning also in the brain of the animals, yet on a lower degree of complexity. In the second place the increasing degree of complexity due to evolution from animals to hominids, continues from *Homo erectus* to the Neanderthals to *H. sapiens*, etc. This is emphasized by the code of ethics existing in the clan, progressing to that of the nation, etc.?

> "In this case we have the chain 'thoughts → information → beliefs. The belief in divine beings who govern uncontrollable aspects of environmental events, human life and the after-death may have been the first 'practical application' in the spiritual realm. In particular, evidence of attempts to predict what happens after death is found in the earliest burial sites, which included food and artifacts for the deceased's soul. Even today, the continuity of our self throughout and

> beyond death is perhaps the most basic subject of every religion on earth. As intelligence evolved and early hominids became more and more preoccupied with this kind of after-death prediction-making, peace of mind (controlled by the limbic system) became an important factor for survival."

I am afraid I disagree with this conceptual model. As I see it religion started when, due to evolution of the hominids, the reaction to instinctive "existence angst" (from the hostile environment as well as a random non-forecasted future) turned from instinctive reaction to conscious. This caused the primitive human being to create an image of a creator which is ready to accept a donation or a sacrifice and protect the donor and his family. At some time a certain clan in the Middle East transferred the image of the creator from the dimensions of space-time to the dimension of information, and attributed to him the ethical laws, which evolved as a result of the struggle for the survival of human societies, evolving from family to nation.

> "Finally, let us turn to the emergence of science. The brain works with models that are only approximations of the so-called 'reality' outside. These models are built during learning experiences and amended in later experiences, in which errors can be corrected and details can be expanded."

On what dimension are these models built?

> "...and (iii) the mathematical idealization of regularities, symmetries and trends in cause-and-effect relationships in the form of laws of nature."

Is nature a mathematician, or maybe mathematics (formalization of steps along the coordinates of the dimension of information) is intrinsic in the universe?

Josef Svoboda

Professor Roederer, as his short biography reveals, has an amazing breadth of interests and an admirable creative drive. This is meant as a heartfelt compliment. He is also an organist. The pipe organ is also my favoured instrument, unfortunately only to listen to. There is another common, yet equally unequal liaison: both of us have been involved in Arctic studies, and both of us branched away from our original disciplines into the speculative realm of evolution.

The present chapter is based on Dr. Roederer's recent book: *Information and its Role in Nature.* I would have certainly greatly benefited from reading this book, but time-wise this was not possible. Thus, my comments are based solely on reading the chapter alone. Being a biologist by training, not everything was completely new to me. However, this is a substantial text and one is able to learn a lot about the brain's neurophysiology. The brain is a *central processor* requiring a *substantial supply of energy* for its information-processing activity. It is a complex neural network of billions of cells and trillions of synapses. A comparison with a present computer is utterly inadequate. It has a memory function and *hologic mode* of information processing. Lots of brain activity is going on at the same time. Some activities and responses are *pre-wired* and *genetically inherited* (instincts), many must be learned and re-learned during the life of an individual. Yet there is an established hierarchy. An organizing agent is overseeing what information is coming from the rest of the body, its sensors and senses, what is glowing in the cortex, and in the various parts and subdivisions of the brain. Admirably, in spite of the brain's prevalent firing inferno, consuming energy and oxygen, organisms as a whole steer well and act 'reasonably', even intelligently. We, ecologists, say that is for the purpose of survival.

And then there is a *human* brain. Especially that of *Homo sapiens sapiens.* New performance abilities, new qualities emerge: language, reasoning, self consciousness, moral imperative, etc. Anatomically, "*the human brain is not particularly different from that of a chimpanzee*" (Subchapter 4). Consequently, the author asks, is there a difference in degree or of substance? I read the section carefully looking for the answer. There is a rich list of differences which set us apart from the highest animals but I couldn't find a clear answer. Or have I missed it in the quantity of information about the brain processes?

Teilhard de Chardin wrote about the process of "cerebralisation" of higher animal forms but mainly its unprecedented rate in early ancestors of modern Man (de Chardin, 1964). It took several million years for the 450cc cranium of the *Australopithecus* to expanded to 850 - 1000cc in *Homo habilis and H. erectus*, respectively, but only the last 150 000 years or less for the brain case to pop up to the present 1450cc in modern humans. While the evolution of the ape's ancestors into *Homo erectus* happened "gradually", fitting the concept of a "change in degree", the last transformation into modern Man was explosive. This was clearly out of the range of anticipated rate of the evolutionary change, leading to questions: why so fast and why only humans?

Man's monophyletic origin 'out of Africa' is being upheld by most anthropologists now. Prior to the last great exodus 50 - 60 thousand years ago, there had been several migrations of ancient groups, formerly considered as ancestors of modern humans. They all vanished. The new pioneering research in variation in human genetics (mitochondrial DNA and Y chromosome) proposes that the *present* human population with all its races, anatomical difference, languages and cultures has a common origin. It is leading the genetic trail to a small group of individuals, if not to a single pair of parents, called symbolically Adam and Eve, to a cradle of mankind in the East-Central and South Africa (Wade, 2006).

As a consequence of the increased brain size and modification of life style (or vice versa?), the proportionality of the evolving human body has also dramatically modified. Unprecedented changes in behaviour followed. Was it an alteration in the environmental conditions which would account for such a thorough metamorphosis of a single species? Or was something more specific and direct involved? While a theory of aliens' invasion has been considered as a valid hypothesis (Crick, 1981), Divine intervention is not.

Was any noticeable evolutionary progress found of our closest family cousins, the gorillas and chimpanzees? They also had the opportunity to climb down the trees and adapt to the savannah or other suitable habitats. Not even a trace of transient, perhaps unsuccessful groups originating from the present apes has been found. These otherwise amazing primates seem to represent a dead end of macro-evolution while humans have been, and still are, moving forward, now with the help of self-engineering. Here I am inclined to accept that the relatively recent rapid development of the brain size, its structure and function (propelling fast-track civilization) has been *in substance* rather than in degree only. The author claims that *"The evolutionary expansion of the higher-level processing network of the human brain did not take place in order to satisfy the intricate requirements for scientific thought,"* (I fully agree) *"but rather, was shaped by an explosive expansion of requirements...to assure survival in the rapidly changing environment..."* (p. 178). Here, I bag to disagree: The abilities of our brain by far eclipse demands for physical survival (cf. the higher Primates). In other words, the evolutionary advancement of the brain cannot be explained as a result of ecological pressure only, since it leaped ahead beyond the practical needs.

Emergence of Religion

As pointed out, the brain is structured to regulate the body functions and behaviour of the individual. Through the external stimuli delivered by senses, the brain also interacts with the outer environment. The incoming information is sorted and stored in certain regions of the brain. The young individual is learning to live safely in its neighbourhood. Its brain is learning. The favourable information (e.g. food discovery) is remembered and associated with positive experiences. Information about a possible danger, harm or threat (e.g. thunderstorms and lighting) will also be remembered. In combination with inherited primordial experiences of fear and horror this may trigger unreal images and phantasms. Such reactions have been observed already in animal behavioural studies. In early- and not so early humans, it may have provoked imagery of surrealistic beings and forces. Amplified and enforced by shamanistic rituals this led to animism, natural religious experiences and practices. I don't see anything pathological in this evolutionary sequence. Yet, there is a thin line between imagination and hallucination. The early humans did not have the luxury of psychoanalysis. Everything was real, seen or perceived. In other words, over time, the brain/mind had to sort out what is real and what is imaginary. Modern neurophysiology has located brain areas preoccupied with speech, language (left hemisphere), and with musical, artistic and intuitive abilities (right hemisphere). The 'pre-disposed' sections of the brain, now found 'dedicated' to religious imagery and experiences probably evolved from the primeval experiences of fear and submission. Theologians may have their own explanation for the origin of religion.

Emergence of Science

Modern science with its *systematic* approach and continuous *re-examination* of results is an amazing invention for generating new pieces of knowledge. Nevertheless, it is a human invention, a method of discovery, in line with other ground-breaking discoveries which have changed the world. Cultural Revolution some 10 000 + years ago was the most significant achievement of the human race ever and profoundly affected the life style of modern humanity. Nomadic groups became sedentary, cities were built and social hierarchy with specialized classes established. Soon *religion, arts and science* began to flourish, stemming from two close related mental sources, intuition and rationality. Science has not developed separately. It has been an integral part of Man's cultural history.

Science-prone Brain?

Where to start? With Babylonian, Egyptian and Mayan astronomers? With Euclid, Pythagoras, Archimedes, or unfairly as late as with Copernicus, Galileo, Newton…? What about philosophy and its protagonists? Where to start? With pre-Socratics, Plato, Aristotle, or with medieval philosophers, Avicenna, Averroes, Aquinas, Maimonides, Descartes? Western science branched from philosophy, that from theology, and theology developed from ancient religious beliefs. Oriental philosophy and science predates our Western culture by millennia! Maybe this is the meaning of the author's notion about the emergence of science, that *"Much later, long past the latest stage of human brain evolution*, (emphasis added) *came (i) systematic measurements... (ii) sharing of information... (iii) the mathematical idealization..."* (p.178). All that was achieved within the last few thousand years. In an 'Augenblick', an evolutionary instant. After the disintegration of Roman Empire, the onset of modern science had been brewing for centuries in European monasteries and the first universities established *by them*. Refinement of modern science, in spite of its exponential advancement, is a progression in *degree not in substance*. This is similar to the industrial revolution, which replaced individual shoemakers and blacksmiths. Certainly, modern science is here due to the latest stage of human brain evolution, capable to 'work with models'. Already the builders of pyramids would highly qualify as scientists, since they definitely worked with models. Brains of Europeans of the 15-16th century, practitioners of modern science, were not more capable to learn, think and imagine than brains of Assyrian, Egyptian and Hellenistic master-builders, astronomers and mathematicians, and their common folks. The biological evolution of the brain outpaced the progress and regress of various civilizations. However, not all societies took advantage of this ability. If the brain capability were the required condition as, in fact, it is, *modern* science could have as well originated within the Mayan or Inca civilization. Sadly, their cultural environment was just not conducive for it. These master astronomers and mathematicians built perfectly designed temple-pyramids, irrigation facilities and agriculture, yet, curiously, they did not invent a wheel and axle! All because of their culturally biased narrow focus. The present progeny of the same group of people study at the universities achieving the highest degrees. Could something similar happen to us at the present culture of our science? That we may be missing something over which the future generations will be shaking their heads? Consequently, we have to extend our mental probe to the very roots of Cultural Revolution, and possibly beyond.

On page 178 the author also writes: Understanding of scientific facts, like existence of self-organizing systems, Darwinian evolution, the operating modes of the brain and the strange behaviour of the quantum world is nearly impossible for someone who is not scientifically trained..." I agree entirely. However, why there is need to add: "*... and thus at the mercy of naïve intuition — on which most religions are built."* Indeed, the advanced scientific understanding is based on proper training, but so it is in philosophy, performing arts (e.g. playing the pipe organ), classical composition, gymnastics and chess, to name a few. There are more scientists living at present than ever lived. Many more, giving the chance and means, will be trained in the future. Still, most people won't become scientists. This, however, not because of brain deficiencies but because of other priorities. Let's be objective. Narrow religious beliefs (in general) may negatively interfere with accepting certain scientific theories. On the other side, they may also lead to acceptance of the most phantasmagorical sci-fi as science. Critical believers may find in their religion a great support for their science and *vice versa.* Do I need to cite some renowned names?

Brain as a Person...

Brain, brain, brain! "*Neural expression of human intelligence*". Is that all there is? The narcissistic self-built tool processing its own products, thoughts and feelings? Brain, inventing means of its own satisfaction: the faith, religion, culture, science? No mention of self, mind, unique personality! I = brain. Intellect = Intelligence. Are we just a biological version of a futuristic quantum computer? Consciousness = emanation of brain's *hologic mode?* A small glitch and it crashes as my PC sometimes does?! All this seems so real and persuasive, yet at the same time *the same brain resists* to consider itself the ultimate singularity. Brain's schizophrenia? Brain, able to get intoxicated, even to commit suicide?! Hardly! It might not have been the author's primary objective but something is missing in his chapter: the search for self and its meaning, and meaning of everything else. Everything is connected: the brain to the body, an organism to the supporting environment, this to the geo-climatic conditions on Earth, those to the size of the planet and its distance from Sun, Sun to its origin and position within the Galaxy, that to its own origin at the beginning of time at Big Bang. Sliding down the trail is relatively easy and *comprehensible*. More difficult and less comprehensible is climbing up the ladder of cosmic succession. The trunk of the universal evolution has produced a crown of countless branches, many with dead ends. A relatively very few celestial bodies may have progressed up to living creatures. Extremely exceptionally (certainly at this planet and potentially at some other worlds), life may have advanced to organisms with a central nervous (or analogical) system - a 'brain'. As far as we know, the human brain is the pinnacle of the evolution of the Biocosmos. It is the extreme anti-pole of the cosmic free fall into ultimate oblivion (Svoboda, 2006). Yet the brain, as such, is not the ultimate objective of evolution. It is not a parallel of the selfish gene! *The whole organism is*. It is not that the *brain* has eyes, hands, legs, etc., but that the *organism* has eyes, hands, legs, etc., and also a heart and a brain. With all its sophistication, the brain is an accessory to the whole organism, as is the organism to the ecosystem, and this to the biosphere…

A question of the meaning of all of this is legitimate and clearly leads us beyond science. Although I was irked by the authors repeated reference to religion as a negative force in the advancement of science, his references to illusions, dreams and hallucinations, etc., as foundations of myths and at least some religions (and I agree with some of it), I found a solace in his closing observation (p 179): *"Scientific intuition is thus condemned to be based on what the brain machinery has originally evolved for, namely creating biologically useful neural images from sensory exposure to the classical macroscopic world – which is also the world as viewed by religion!" "We are motivated to know and to believe, and we are rewarded by knowing and believing. The coordinated interaction between cortical and limbic functions in fact constitute the neural basis* (wiring?) - *and they are the "antenna" through which human beings relate with their God and the universe."* The ultimate question still remains: Who is "we"?

Recently, *The New York Times* published an article discussing the likelihood of the emergence of multiple universes. In the endless cycle of big bangs, some theorists concluded, *it is easier to make fragments of one, like planets… or a 'naked brain' floating in space. From the standpoint of energy and probability… brains would appear far more frequently than real full-fledged universes.* Alan

Guth discusses the 'Boltzmann brain paradox' quite seriously and so do other renowned cosmologists. Since one or more of the myriad of brains might be identical with our own, some "freaky observers" consider the Big Brain Theory as a vehicle of reincarnation. Here the loop of the absurd is back firing. Reincarnation is an issue ordinary science is not supposed to touch (Overbye, 2008).

References

Crick, F. H. C. (1981). *Life Itself: Its Nature and Origin.* Simon and Schuster, New York.

de Chardin, P. T. (1964). *The Future of Man.* Collins. London. pp. 319.

Overbye, D. (2008). *Big Brain Theory: Have Cosmologists Lost Theirs?* The New York Times. January 15, 2008.

Svoboda, J. (2006). Life as an unfolding Biocosmos. In Seckbach J. (ed.): *Life as We know It.* pp. 434-444. Springer, Dordrecht.

Wade, N. (2006). *Before the Dawn. Recovering the Lost History of Our Ancestors*. The Penguin Press. New York. pp. 312.

Juan G. Roederer

Most regrettably I must decline my participation in a written discussion of our respective contributions, due to an absolute lack of time.

Still, I would like to thank the above discussants very much for interesting comments on my paper. Concerning their specific ideas about the concept of information, I strongly encourage them delve into the recent literature on the subject, especially that written by biologists (especially Küppers pioneering book "*Information and the Origin of Life*", MIT Press, 1989) as well as by physicists. In the hope of not sounding too pretentious, among the latter I would include my recent book "*Information and its Role in Nature*" (Springer, 2005), on which, after all, my DINA article was based.

Without getting into the details of the above comments (it is precisely for this that I would not have the necessary time!) I attach a brief summary of ideas about information expressed in my book, which I had prepared earlier for a similar purpose.

SUMMARY ABOUT THE CONCEPT OF INFORMATION

Let me explain my approach in the book "*Information and its Role in Nature*" (Springer-Verlag, 2005). It is, in a nutshell, the following:

I consider as an epistemological primitive the process of "*Interaction between two bodies in isolation of all others*" (like Ernst Mach's approach to Newtonian mechanics) and take it from there.

I divide all such interactions into the following two major groups (1a and 1b considered as belonging to one group):

1a. Interactions between two "simple" bodies (classical mass points; elementary particles, nuclei, atoms in the quantum domain)

1b. Interactions between two natural *complex* systems—interactions which can be described as a *linear superposition* of the mutual physical interactions between the component parts (e.g., gravitationally interacting macroscopic bodies, colliding rigid bodies, neighboring fluid parcels, chemical bonds between two molecules, etc)

2. Interactions between complex systems which do not fit into category 2), because forms and/or *patterns* (in space and/or time) play the determining role, rather than linear superpositions of fields, forces and energy (an enzyme reading the DNA template, a microorganism moving along a concentration gradient, an insect "in orbit" around a light source, the measurement of a physical quantity, etc.)

The concept of "information" does not appear as an active, controlling agent in the physical interaction processes 1a) and 1b); it only appears there, when an *observer* intervenes (as in a *measurement*—Shannon information) or when he/she describes and models them (as in the formulation of *physical laws*—algorithmic information). Unfortunately, in teaching physics we often use an anthropomorphic jargon like: "the electromagnetic field "carries information" about the source", or "in a diffraction experiment the electron seems to "know" if the second slit is open or closed").

The simplest example of interaction of type 2 is an arrangement in which the presence of a specific pattern in complex system A (the emitter) leads to a specific, univocal change in complex system B (the recipient). *Pragmatic information is defined as "that which represents a univocal correspondence pattern → change"*. This is why we call this type of interactions "information-driven".

All natural information-driven interactions *have a purpose* (to cause a specific change that would not occur in absence of the triggering pattern, or which only could occur by chance), and require a "common code" (or "meaning") somehow embedded in the physical mechanism responsible for the specific change. No "purpose" can be identified for purely physical interactions (to say that the "purpose" of the gravitational field is to attract other bodies would be advocating an animistic or even "intelligent-design" viewpoint).

Natural information-driven interactions can only emerge through biological *evolution* or, on a shorter time-scale, through a *learning or adaptation process.* In other words, they involve *living matter*—moreover they represent the *defining property of life*. Any information-driven interaction between *inanimate* complex systems is ultimately life-generated or designed, involving at one stage or another purposeful actions by a living system (a nest providing shielded space, a thermostat regulating temperature, a guided missile hitting a target, etc.). There should be a "grey zone" of emergent information-driven interactions (transition from macromolecules to a virus?) which also represent the "grey zone" between non-life and life.

There is no such thing as a "measure of pragmatic information". Pragmatic information *cannot be quantified*—it represents a correspondence (which either exists or not, but cannot be assigned a magnitude). Shannon information, algorithmic information, Fisher information and others are all *quantitative measures* of uncertainty, disorder, expectancy, novelty, quality of information, number of binary steps to describe something, error distributions, noise, etc. But neither of the respective theories actually *defines* in a strictly objective manner the concept of "*information per se*" (that is, without any direct or indirect reference to, or invocation of, human thought processes and actions).

Even the mathematical quantities used in traditional information theories are coupled to highly subjective concepts related to how the human brain reacts to certain sensory input. In most cases they relate to how the neural cognitive state changes from *not-knowing to knowing* the final state of a system when there exist several a priori alternatives. I like to describe this transition as a "collapse" of the initial brain state into one of possible "basis states". And such a collapse is triggered by pragmatic information processing. This pragmatic information represents the link between a pattern (e.g., the position of a dial in an instrument, the dots on cast dice) with another pattern (of spatio-temporal neural activity distribution in the prefrontal lobes corresponding to the knowledge "it's *this* particular final state and not any other possible final state"). (I think this indeed represents a modern-day expression of one of Kant's fundamental dicta.)

Quite generally, animal brains *only* handle pragmatic information, in which one specific spatio-temporal pattern of neural activity is mapped or transformed into another neural pattern—in its most fundamental form, from a physically triggered sensory or somatic input pattern to the neurally triggered physical/chemical behavioral output pattern (stimulation of muscle and gland fibers). Only the *human* brain can generate the latter (or anything in-between like a "reverse internal stimulation" of sensory receiving areas in the process of imagining an object or sound) *in absence* of the former (in general, this represents the human thinking process).

Since an animal or human brain can only handle pragmatic information, it is an eminently classical information-processing device (Penrose notwithstanding—his ideas about the possible role of quantum information processing in the brain have been disproved on grounds of quantum decoherence times—somewhere between 10^{-13} and 10^{-20} s). Internally triggered brain images about a physical system as needed in scientific model-making, however abstract, are ultimately based on the memory of interactions with systems that can be perceived by the senses. This has a fundamental impact on the whole gamut of emotion-guided human behavior, and transcends even into the quantum domain where it has many consequences such as the proverbial quantum paradoxes. But it also impacts classical thermodynamics.

In fact, I believe that Schrödinger cats and Maxwell demons are creatures conceived through common misunderstandings of the concept of information. And, I claim, so is the question of the relationship of the observer to a quantum system under observation.

Take a qubit in a superposed state, which could be a single photon in a Mach-Zehnder interferometer after having passed the first half-silvered mirror. We try to picture in our mind the resulting system, but the very moment we imagine the photon actually running along *one* given path (or a Schrödinger cat in *one* of its possible final states), we have to be consistent and take into account the fact that what we are mentally representing is a system whose wave function has already collapsed. In other words, it is as if that particular "mental observation" was an actual measurement—*even* if only a Gedanken Experiment. It is the *study* of a quantum system, whether experimental or theoretical, that requires considering the observer or the thinker as part of the system. Note that the traditional "Copenhagen interpretation" of quantum mechanics only refers to the experimental (measuring) aspect.

Interestingly, we have rather similar situations in classical physics. As an example, take two identical ideal gases with a total of N indistinguishable molecules, at the same p, T, initially separated in equal-volume vessels A and B. Now let them mix. We picture *in our mind* the molecules of one gas expanding adiabatically into the other vessel and vice versa. So the entropy of the total system should increase by $kN \ln 2$ —but that's nonsense because in reality nothing happens macroscopically (this is the famous *Gibb's paradox*). Here, too, we have to be consistent: by just imagining molecules from gas A diffusing into the vessel B containing an identical gas, we are following them and tagging them *in our mind*, just as we have followed the photon going through one path and not the other. To obtain the correct final state of *indistinguishable* molecules, the tags *must be erased* in our Gedanken-experiment, implying an outside intervention which eliminates *one bit* worth of alternatives per molecule (that of "being from vessel A or from vessel B"). Erasing one bit of alternatives decreases the entropy by $\Delta s = -k \ln 2$, according to Shannon. For N molecules, this is precisely equal to minus the paradoxical increase of Clausius entropy.

As a corollary: Is what we today call "quantum information" *really* information? (I mean here objective, pragmatic information, *detached* from the observer, representing the univocal correspondence between a pattern and a specific elicited change). For two real numbers like the relative amplitude and phase of a qubit to represent information, one must be able to identify some specific *change* that they induce in some *classical* macroscopic system during a measurement. But only *one* classical bit can ever be extracted from this measurement: the two real numbers are inaccessible to any observer. We can subject the qubit to unitary transformations into other qubits, but the *accessible* information still will never exceed 1 bit. We may call those two real numbers "quantum information", and we obviously do—but it isn't information defined in a strictly objective way.

Take, for instance, quantum teleportation (see example in Roederer (2005)): how is it possible that the two real numbers defining the qubit to be teleported can be "encoded" in *only two* classical bits (00,01,10,or 11), which far away are then used to re-create the original qubit (which had been destroyed in the process long ago)? We are puzzled because of our urge to picture quantum processes with our eminently classical-world brains: if we want to consider those two real numbers "information", we are *forced* to imagine the infinite bits that describe them as being transmitted through the system with infinite velocity backwards and forwards in time. But *nothing that transcends into the classical world* in the teleportation process violates any physical laws—neither classical nor quantum nor relativistic! Quantum mechanics is not incomplete, as Einstein insisted—our brains are (fortunately)!

Finally, a word about the boundary between the "quantum world" and the "classical world". This limit is a relative one. It can be defined only for a given observable or set of observables and the state in which a given system is found. Indeed, there are states in which a system would behave classically with respect to that set of observables: the common *eigenstates* of the observables in question (provided they exist). The very essence of a quantum system is the possibility of being in *superposed* eigenstates. If by means of an interaction with another quantum system we can transform a superposed state into another state, and later *recover* the original superposed state by means of yet another appropriate transformation, we have a "true" quantum system. But if, because of unavoidable and/or unpredictable external influences, such reversible transformations are *de facto* impossible (this is called *decoherence*), the system in question belongs to the classical domain. The device to kill a Schrödinger cat is an example.

Biography

Dr. Seth Shostak is a Senior Astronomer at the SETI (Search for Extraterrestrial Intelligence) Institute, in Mountain View, California. He has an undergraduate degree in physics from Princeton University, and a doctorate in astronomy from the California Institute of Technology. For much of his career, Dr. Shostak conducted radio astronomy research on galaxies, and has published approximately fifty papers in professional journals. He also founded and ran a company producing computer animation for TV.

Seth has written several hundred popular magazine and Web articles on various topics in astronomy, technology, film, and television. He now lectures on astronomy and other subjects at the California Academy of Sciences, and for the last five years, has been a Distinguished Speaker for the American Institute of Aeronautics and Astronautics. He has edited and contributed to a half dozen books. His first popular tome, *Sharing the Universe: Perspectives on Extraterrestrial Life*, appeared in March, 1998. His most recent books are *Life in the Universe* (2002, textbook with Jeff Bennett and Bruce Jakosky) and *Cosmic Company* (2003, with Alex Barnett).

40
Intelligent Design and the Science of Space

SETH SHOSTAK

SETI Institute, 515 N. Whisman Road
Mountain View, CA 94043, USA
seth@seti.org

1. Introduction

Nearly one-and-a-half centuries after Darwin's On the Origin of Species hit the book stores, and eighty years after Tennessee's infamous Scopes trial hit the headlines, we now have the following enlightened situation: dozens of states in the USA are embroiled in legislative controversies that would either water down the teaching of evolution, or push creationism into the science classroom.

Creationism, and its latest forensic wrinkle, intelligent design, maintain that the fundamental underpinning of biology's edifice — the theory of evolution — is merely a bit of flimsy scaffolding. Creationism's supporters claim that evolution is an unproven idea of small intellectual heft. The process of descent with modification, and the inference that all species share a common ancestor — well, those are just fanciful notions. The human genome that was so recently sequenced is not, in the view of those promoting intelligent design, a structure incorporating the history of nearly four billion years of life's experience, but a separate and deliberately planned blueprint just for us.

This might strike you as odd, especially when you note that if we (and just about any other creature you care to name) have been deliberately engineered, then the designer merits only a middling grade. Our bodies are fragile, vulnerable, and replete with flawed and useless parts, which bespeak their modification from earlier forms. The skeletal structure of your back might work well if you ambled around on all fours, but for an upright creature, vertebra are prone to pain and malfunction. An upright stance also exposes some of our most sensitive anatomy, as well as our internal organs, to easy injury. Our retinas have their nerve cell "wiring" spread across the surface facing the lens, a mistake as silly as loading film into your camera backwards. Your appendix and coccyx would have been fine and really functional for some other creature, but not for you.

However, even setting aside the doubtful merits of intelligent design, should you care whether it's taught or not? Should you get your knickers in a knot because of a recent Gallup poll revealing that 40% of Americans believe that creationism, not evolution, ought to be taught in the schools? If a lot of people have trouble accepting the fact that their forefathers were short, furry, beetle-browed and dim, should you insist on rubbing it in? Alas, you should.

2. Dangerous Consequences

For one thing, the world needs more science literacy, not less. If we wish to continue living well, our youth must be conversant with science and its logic, not merely passive consumers of its spin-off technology. This is a plaint obvious, long-standing, and overworked.

But keep in mind that the failure to educate the public in the ways of science has already cost us money and more. Silicone implants were deep-sixed by junk science, and as a result medical device manufacturers had difficulty getting a material useful for helping heart patients. Global warming continues to be debated — at least in the United States — and the public seems incapable of evaluating the merits of the case, other than judging on the basis of the local weather. An example on a smaller scale is the recurring, pathetic sideshow of befuddled juries trying to understand the statistics of DNA evidence.

The public's frequent failure to grasp what science is about gums up modern society's machinery. And creationism adds real and dangerous grit. In the 1980s, paleontologist Stephen Jay Gould wrote in *Natural History* magazine about the famous case of Baby Fae, an infant born with a severe heart defect. The baby was given a baboon heart transplant at the Loma Linda University Medical Center in California, and died shortly thereafter. It doesn't seem to have crossed the surgeon's mind that Baby Fae would be immunologically doomed by the evolutionary distance between baboons and humans. "I don't believe in evolution," he admitted, showing that even the technically trained can do damage when their personal philosophies don't mesh with scientific evidence.

Clearly, any philosophy that denies evolution will have immediate and ominous consequences for biology, medicine, and our prospects for the type of genetic engineering that could obliterate much disease. But even astronomy and space exploration will be affected, and none-too-prettily. If the complexity of our bodies is basically unfathomable because of the numinous workings of a designer, why should anyone believe that we can understand cosmic mysteries such as the formation of planets, black holes, and galaxies? Will we ever be able to grasp why the universe is so finely tuned for life, so kindly disposed for our existence? Do we give up on that puzzle, too?

Much of space exploration is motivated by the desire to find life beyond Earth. So what happens if we trip across microbes on Mars, or extremophiles on Europa? We would want to understand the history of that life, too, and evolution is the most valuable tool in our kit bag. Without it we would be like geologists stripped of plate tectonics. It's also possible that worlds seed one another, possibly even across interstellar distances. Could we have distant cosmic cousins? Not if you believe in intelligent design. According to that benighted philosophy, we don't even have non-human cousins here on Earth. Would people be keen to embrace a view that deprives them even of the possibility of being biologically connected to life on other worlds?

3. A Bad Idea

Creationism, and its subtler incarnation, intelligent design, are attempts to turn the data of science into "Exhibit A," a collection of physical evidence that proves God's existence. Scientists who have assembled and studied these data long and hard argue otherwise: that they speak to a process that is autonomous and completely natural — biological evolution. The ironic result is that the creationists, in some cases scientists themselves, are forced to challenge evolution in order to be able to invoke the data that spawned it.

This results in wobbly logic. Even if the creationists prove that the evolution scenario is wrong, it wouldn't prove that their idea is right. If they wish to argue from observation, then they will need to come up with predictions and experiments that could test their hypotheses — at least if they hope to convince scientists. As it is, the creationists are on thin ice: they are willing to admit that species can adapt (a phenomenon you will speedily witness if you hang out with fruit flies or finches). But they balk at the idea that one species can eventually become another. Speciation is God's job, they claim. However, if we accept this, then the separation between God and nature seems only one of degree, and God is relegated to filling in the gaps.

This impoverishment of God's function bespeaks bad theology as well as bad science and, frankly, resorting to such an untenable argument is a poor defense of faith. The universe's purpose, as opposed to its structure and workings, is a profound mystery. Indeed, it's a mystery that science cannot solve. You would think that this grandest of puzzles — the meaning of life, as Monty Python would say — would be turf enough for theology. But it seems that Galileo still rankles in some quarters. Astronomy has been troublesome before, and might be so again.

Consequently, those of us who are keen to extend our ken to the stars should pay heed. The public is being fed bad juju. And now some villagers with their pitchforks and lanterns are storming the citadel of science.

Dialogue

Taner Edis

I share Seth Shostak's concerns. This is not unusual; most scientists I know are, at least in the back of their minds, worried about the deep-seated popularity of creationism. So clearly I do not have much critical that I can say.

I do, however, have a question. Like many scientists who speak out publicly against creationism, Shostak says that it is "bad theology as well as bad science." Is this really a good idea?

Now, academic theologians are much less likely than, say, televangelists, to attack evolution. Being able to point to theologians who accept evolution is very useful in the political and cultural struggle over science education. I'd love it if larger numbers of ordinary religious people would be influenced by sophisticated theologians rather than by anti-intellectual preachers. But should we really be going around making pronouncements on what is good theology and bad theology?

One immediate problem is that if we come across as saying that "good" theology does not overtly interfere with science, that's a pretty self-serving definition. It's hard enough to convince people to agree with scientists about nature; it's hardly likely that too many will listen to scientists telling them what proper religion is. In a public debate where scientists have to defend elite forms of knowledge against populist religiosity, declaring that we're the ones who know what is good and bad theology is a dubious idea. Would this not reinforce perceptions of high-handedness?

A deeper problem is that there's little common ground that would allow us to come to an agreement on what is good and bad theology. Communities of faith decide what they believe among themselves, usually with little regard to what academics say. Good theology in one tradition or sect is rank heresy or a gross theological error in another. Especially in the typical situation where theological commitments come down to acts of faith, it becomes difficult to see declarations of "bad theology" as being much more than arbitrary expressions of disapproval.

It's tempting to try and stamp out creationism by any rhetorical means necessary. Often I find myself trying to reassure a student that accepting evolution does not mean abandoning their religious heritage. One way to get this across is to say that there are more liberal theological options. But I also feel uncomfortable with telling them what's good or bad theology. It strikes me as none of my business.

Florence Raulin Cerceau
Dr. Shostak's paper was very interesting and I was very happy to have the view of one of the most distinguished members of the SETI Institute. I completely agree with Dr. Shostak when he says that we must be careful to educate the public about evolution only with the help of scientific facts and data. I think Dr. Shostak has himself worked a lot along these lines, which can be regarded as a very worthy function. In my opinion it is very important indeed to popularize to a very large public any scientific notion of whatever difficulty.

For my part, I was involved for six years in the preparation of the *Grande Galerie de l'Evolution* (a huge permanent exhibition exclusively devoted to biological evolution in the National Museum of Natural History, Paris-France) and I saw how difficult it was to display to the public still debated topics. I also saw the numerous obstacles we had to overcome in order to demonstrate the theory of evolution. Of course, at the place where J.B. Lamarck gave the first lectures on *transformism*, the ground was already cleared.... This experience led me to understand that we can (very often) make the distinction between dogma and science thanks to the history of science. If the public can grasp the advances of the ideas and hypotheses put forward by scientists in the course of the centuries, then this public can be in a position to understand how a theory (such as the one of biological evolution) has been built. In that case, a notion such as creationism could not only be pushed aside thanks to the demonstration of scientific facts but also with the help of the analysis of the progress of the knowledge: really, are creationists able to use an historical thought process to demonstrate their assertions? I'm afraid they are not.

Despite the fact that I'm in agreement with Dr. Shostak's paper, I have a very short comment to make dealing with section 3 *A Bad Idea*: perhaps it is dangerous to speak about the "universe's purpose" in this context. It could lead to some misunderstanding and I am not sure we can assert that "it's a mystery that science cannot solve"....

Biography

Dr. John Baumgardner is currently associate professor of geophysics at the Institute for Creation Research in Santee, California. He obtained his Ph.D. in geophysics and space physics from the University of California, Los Angeles, in 1983 and continued his research in the Theoretical Division of the Los Alamos National Laboratory in Los Alamos, New Mexico, from 1984 until his retirement in 2004. He joined the Institute for Creation Research in 2005. Dr. Baumgardner's primary scientific interest is planetary mantle dynamics.

41
Language, Complexity, and Design

JOHN BAUMGARDNER

Institute for Creation Research
10946 Woodside Avenue North, Santee, California 92071 USA
jrbaumgardner@cox.net

Abstract: The phenomenon of language represents a notable counter-example to the materialist claim that there are no non-material realities. Language involves the assignment of meaning to arbitrary symbols to form a vocabulary and the application of a set of rules to join elements from the vocabulary together to generate more complex meaning structures. Although such structures are normally associated with a material medium, the meaning they convey is independent of the medium. Language is not an incidental aspect of our world. Not only is it a fundamental aspect of our conscious experience as humans and central to our social interactions with others, it also plays a significant role in the ways the material world is structured around us. Microprocessors, with language-based programming, for example, make possible increasingly complex communication, transportation, manufacturing, and commercial networks. At a much higher level of complexity, biological systems rely on language-based specification and control. As mathematical expressions are linguistic, so are the laws of chemistry and physics. This implies that the material realm itself has linguistic underpinnings. Because language is non-material and matter itself displays no apparent language generating capability, how language arises is obviously an important question. That human beings possess language capacity points to an answer.

1. Introduction

Language is such a spontaneous aspect of human experience that we rarely pause to consider its fundamental nature or how its fits with the rest of reality. In academic circles, although most might be aware that language theory is a scholarly subject treated in linguistics and philosophy departments, generally speaking, the issue of the fundamental nature of language is not a topic that garners much attention. Yet the phenomenon of language plays a profoundly important role in our world. Just what is language? Language involves the assignment of meaning to arbitrary symbols to form a vocabulary and the application of a set of rules to join elements from the vocabulary together to generate more complex meaning structures. Hence language deals with a non-material stuff called 'meaning.' Although language structures are normally associated with some sort of material medium, the meaning these structures convey is independent of the medium. Language thus is a non-material phenomenon and represents a notable counter-example to the materialist claim that there are no non-material realities.

Language ability is at the heart of what it means to be human. Language allows us as human beings to create incredibly complex social interrelationships. Language enables human institutions of government, industry, commerce, education, science, to name but a few. Electronic communication, including the Internet, is facilitating human linguistic interchange at astonishing new levels, transforming the manner in which we conduct our business, obtain our news, manage our finances, and add to our common understanding of our world. When we consider the number of books published each year, the number of professional journals and other periodicals produced on a regular basis, the amount of mail processed each day, not to mention all the telephone and email communication, the radio and television programming, and other types of electronic transmissions of linguistic data, it is clear that language is not an incidental aspect of our world but instead represents a prominent category of reality in which we all actively participate.

Not only are there thousands of relatively distinct natural human languages, but we have also devised languages for our machines. Computer software represents language within the scope of the definition just given, namely, the assignment of meaning to a set of arbitrary symbols to form a vocabulary and a specification of a set of rules by which elements of the vocabulary may be joined together to form more complex meaning structures. Such software enables machines to specify, monitor, and control highly complex manufacturing processes, communications networks, power grids, and transportation systems. In these applications we observe an amazing interplay between linguistic specifications and physical, material phenomena. Apart from software that microprocessors utilize to monitor and control, and in most cases, even design, humanly engineered systems of the current complexity simply are not imaginable. In the biological realm we observe a similar connection between linguistic information and the specification and control of a diverse array of processes, except the level of complexity is dramatically greater. In biological systems the structural specification is at the level of the individual atom — approaching the ultimate in nanotechnology. In both cases, we witness the intertwining of linguistic specifications and physical, material processes.

2. Language — Non-Material in Its Essence

For many people it is intuitively obvious that language, which has to do with conveyance of meaning, is ultimately a non-material phenomenon. But because of the strong influence of materialist philosophy for the past two centuries in the Western world, some today are uneasy with the notion that there might be entities which are indeed real but at the same time also non-material. Notably, Albert Einstein recognized the fundamental difference between the realm of matter and the realm of abstract entities. In this context he states, "We have the habit of combining certain concepts and conceptual relations so definitely with certain sense experiences that we do not become conscious of the gulf — logically unbridgeable — which separates the world of sensory experiences from the world of concepts and propositions" (Einstein, 1944). The world of sensory experiences to which Einstein refers is the material realm, and the world of concepts and propositions includes the realm of language. Linguists, giving due credit to Einstein for this observation, refer to the 'logically unbridgeable' separation of these two realms as the 'Einstein gulf' (Oller, 1989, Oller et al., 2006). Einstein correctly recognized that the 'world of concepts and propositions', which includes the laws of physics, belongs to a category of reality distinct from that of matter, despite the fact that we so commonly associate the two. Let us explore this issue a bit further.

A person who struggles with the proposition that there could be something real that is not at the same time material can correctly point out that wherever we observe an expression of language, it is associated with a material carrier. Human speech, for example, is normally generated by the human voice box as acoustic oscillations in air. If a microphone is present, the acoustic oscillations can be converted into electrical signals and relayed and/or recorded in a variety of ways, and in every case, a material medium is involved. But the crucial question here is whether or not the linguistic content of the message depends in any way on the material carrier. To be sure, the message can be degraded or even lost altogether as a result of defects and interruptions in the medium. However, to the extent that the medium is able to record/transmit the message reliably, the meaning the message conveys does not depend on the medium. If handled reliably, the message remains the same whether it is carried by acoustic waves through the air, transmitted electronically, faxed, emailed, encrypted, sent through the mail on a CD, or relayed via smoke signals. The meaning remains unaltered whether it is recorded with paper and ink, on a magnetic disk, on a plastic CD, or chiseled on a rock. The meaning remains unaltered if the message is switched from one medium to another, so long as the conversion is performed reliably. A linguistic message therefore possesses an identity and a reality that is independent of its carrier. The essence of a linguistic message is the meaning it conveys. Indeed, meaning, which encompasses the 'world of concepts and propositions', referred to by Einstein, is a reality distinct from matter, separated from it by a 'logically unbridgeable' gulf.

But just what is this stuff that language encodes which we are here calling 'meaning'? The ancient Greeks seemed to have had a partial grasp of this issue. A notable feature of classical Greek philosophy, primarily due to Plato, was a realm of ideal entities. As to location, this realm existed outside and independent of the human mind but could be apprehended by the mind. An example of an entity from this realm of ideals is the circle. We can grasp with our minds the concept of a perfect circle. Yet even the best circle we might attempt to construct, if inspected closely enough, will be found to deviate from this ideal. Plato argued that most, if not all, features of our changing physical world have unchanging ideal counterparts in this realm of forms (Plato, 360 BC). This latter realm includes not only geometric entities such as points, lines and circles, not only concrete entities like squirrels and ships, but also more abstract entities like justice and beauty and love.

Plato's notion of semantic abstractions, generalized to the proposition that meaning of every sort — not merely of what might be considered ideal — belongs to an extra-material realm of abstract entities is today taken very seriously among many linguists and philosophers of language (Oller et al., 2000). Advocates of 'entity theories' understand meanings to be individual 'things' or entities that are language-independent. Some understand these entities to be mental entities, which is to say that meanings of linguistic expressions are ultimately ideas in the human mind. An early advocate of this view was John Locke (Locke, 1490). Such theories of meaning are known as ideational theories. Others, however, understand meanings of linguistic expressions to correspond to abstract propositions that are not only language-independent but also have existence independent of the human mind. These theories are known as propositional theories. Bertrand Russell in 1919 argued in favor of this way of understanding (Russell, 1956). More recently, among the many philosophers who today defend this view, William Lycan, comments:

> "Like ideas, these abstract items [propositions] are "language-independent" in that they are not tied to any particular natural language. But unlike ideas, they are also people-independent. Mental entities depend upon the minds in which they inhere; a mental state has to be somebody's mental state, a state of some particular person's mind at a particular time. Propositions are entirely general and, if you like, eternal [by which he means, time independent]" (Lycan, 2000).

Other philosophers have sought other understandings of the ultimate nature of the content of linguistic expressions. Most restrict their scope to human language and emphasize human sociology and human psychology. One class of such theories is known as "use" theories. An example is the view set forth by Ludwig Wittgenstein who argued that words and sentences are like game tokens employed by individual human beings to make moves in the context of the rule-based society in which they find themselves (Wittgenstein, 1953). According to this theory, 'meaning' is not an abstract entity. Rather, meaning corresponds to the 'use' the expression has in a certain range of social contexts. Such theories obviously are inadequate outside the scope of natural human language, so for our purposes we mention them only in passing.

It is fitting in this context to consider computer languages that rely heavily on logical statements such as equality and inequality (i.e., greater than or less than or equal to), if-then conditions, arithmetic prescriptions, and assignment specifications. The propositional theory of meaning describes elegantly how these languages operate. The proposition of equality, for example, is precisely defined, and it retains this precise meaning as the software executes within the circuitry of the computer processor. Although logical equality as an idea is readily grasped by the human mind, it certainly appears to be a proposition that has reality and retains its force and content beyond the human mind in the inanimate world.

In concluding this section, the notion that the "world of concepts and propositions" is separate from the realm of matter, as Einstein proposed, seems to be testable and reasonably easy to establish as correct. While this "world of concepts and propositions" is distinct from that of matter, it nevertheless is capable of exerting powerful organizing influences on the material realm. This is especially evident as we consider just how it is that humans, especially in the past two centuries, have become so effective in altering their physical surroundings. Toward that end let us now consider the connection between language and complexity.

3. Language and Complexity

To most people, the meaning conveyed by the term complexity is more qualitative and relative than quantitative and absolute. One of the main reasons is that, generally speaking, complexity is difficult to quantify. Most people would readily judge a bicycle with brakes and gears and drive chain to have greater complexity than a child's tricycle. Similarly they would judge a motorcycle with an internal combustion engine and perhaps a transmission to have greater complexity than a bicycle. However, most people, including most scientists, have no criteria from which to derive quantitative values for the complexity of a physical object or system. If we seek to assess humanly engineered systems such as computers and automobiles and aircraft in regard to their complexity, we would observe that they commonly have large numbers of different components organized to perform a diversity of functions. But even if we might have available all the design drawings and other specifications for manufacturing each of the component parts and all the specifications for assembling the system, it would still be a daunting task to find a rigorous means by which to quantify the system's overall complexity.

Nevertheless, a theoretical measure of complexity does exist, a measure known as algorithmic or Kolmogorov complexity (Li and Vitanyi, 1997). In simple terms it corresponds to the minimum number of yes-no questions required to characterize a process or structure or system. Since the answer to a yes-no question can be represented by one binary bit, the Kolmogorov complexity is equal to the number of bits in the minimum length bit string needed to characterize the process or structure or system. This measure of complexity is useful, not so much in quantifying the complexities of actual systems, but rather in gaining further insight into the nature of complexity itself. It also provides a means for connecting linguistic specification with the structure of material systems, since a sequence of questions with their answers is a linguistic entity.

The interpretation of Kolmogorov complexity as the minimum number of yes-no questions needed to describe a structure highlights the fact that there is direct correspondence between structure in the material realm and language. Language, in this case as a set of yes-no questions and their answers, can fully characterize a material structure in all its complex details in a given context. Conversely, the linguistic description provides a set of specifications adequate for realizing the material structure in that same context. In other words, the features of a material structure can be translated into language and vice versa. Of course, the minimum length of the linguistic description is generally never achieved. That is not the issue here. The point is a simple, even obvious, one, namely, that language has the power to characterize structure in the material realm and, in the other direction, that linguistic expressions indeed can and do specify material structures.

As an illustration, using computer aided design (CAD) software I can design my dream house to a very high level of detail. Any additional details I can also specify linguistically. All these specifications I can record in electronic form on, say, a DVD. Potentially, after purchasing a suitable parcel of land and making appropriate financial arrangements, I could give this DVD to a contractor, leave for my five-month vacation in the Mediterranean, and return to find the house of my dreams, to an incredible degree of fidelity to what I had specified (assuming my contractor was competent and trustworthy) fully realized as part of this material world. The specifications consisted of nothing beyond marks melted into the surface of the plastic DVD encoding non-material linguistic data. But that linguistic data was adequate to specify the details of my house, including the placement of all the electrical outlets and recessed lights, the routing of all the pipes and locations of all the plumbing fixtures, choices for floor coverings and countertops, and even the wood and finish to be used for the banister on the front right staircase to the second level.

Containing the realm of ideal entities advocated by Plato, the more general realm of descriptions that language is capable of conveying is much larger and richer. In fact there appears to be no limit to the variety and complexity and detail that such linguistic descriptions can capture. Like the Greek realm of ideals, the realm of linguistic meaning is non-material, and, as argued by many contemporary philosophers, must also be independent of the human mind and independent of time. Nevertheless, it is very real and plays a central role in what it means to be a human being. It is what our thoughts are made of and is the primary medium by which we relate to others. It enables us to create complex social and economic structures, it enables us to do science and understand how the material world operates, it enables us to build machines and perform amazing feats of engineering. It enables us to articulate the issues we are probing in this very article and empowers us to explore how the realm of linguistic descriptions intersects the realm of matter.

Although we have briefly considered the *sufficiency* of language to characterize complex structures, at this point let us address the *necessity* of language for the realization of complex systems. In this regard it is useful to note that Kolmogorov complexity, measured in bits, has a close connection with Shannon self-information, also usually expressed in bits. Shannon defined the self-information I of a message m by $I(m) = -log_2p(m)$, where $p(m)$ is the probability of message m being chosen out of all possible choices in the message space M (Shannon, 1948). This means that if the message m carries an amount of self-information I bits, then the probability of that message in its context M is 2^{-I}. There is a close connection between Shannon information and Kolmogorov complexity (Gruenwald and Vitanyi, 2004), but reviewing this connection is beyond the scope of this article. Nevertheless, both Shannon self-information and Kolmogorov complexity can be interpreted as the length of a linguistic string. Since the Kolmogorov measure represents the minimum string length, to the extent that the contexts are similar, the probability which can be associated with the linguistic description implied by the Kolmogorov measure is similar to that implied by Shannon self-information of that same linguistic message. In other words, the probability in the context of similar linguistic descriptions of the linguistic description implied by a Kolmogorov complexity K is 2^{-K}.

The simple relationship between the amount of self-information in a string of linguistic symbols and the probability of the string in its context means that messages of several hundred bits in length can specify states that random searches in the material realm could never find. The reason is simple: there are not enough 'rolls of the die' available. For example, there is an upper bound on the number of atomic collisions that could have ever occurred during cosmological history. Suppose we let an atomic interaction with some other atom count as a 'roll of the die'. Let the reciprocal of the light transit time across the diameter of a free hydrogen atom serve as the frequency that each atom reacts with a neighbor. Let every atom in the cosmos react with other atoms at this rate for a period of time equal to the estimated age of the cosmos.

An atomic diameter for hydrogen of 10^{-10} m and a light speed of 3×10^8 m/s yield a maximum frequency of 3×10^{18} reactions atom^{-1} s^{-1}. To be generous, let us round this to 10^{20} reactions atom^{-1} s^{-1}. A generous estimate for the number of atoms in the observable universe is 10^{80}. Since we do not know the actual size of the universe, let us use 10^{100} as our estimate for the number of atoms, a number 10^{20} times larger. Let us assume 15×10^9 years, or 5×10^{17} s, for cosmological age. To be generous, let us round that number to 10^{20} s. Multiplying these numbers together, we get an upper bound of 10^{140} on the number of atomic collisions that could have ever occurred in cosmological history. Surely this represents an upper bound on the number of tries in any conceivable random search process. If we limit our search process to the matter near the surface of an earth-like planet, or even a single earth-like planet, the upper bound, of course, is reduced considerably. Yet a specific message with self-information of 1000 bits has a probability of $2^{-1000} = 10^{-301}$. In a lottery, chances are essentially nil that any material random search process would ever stumble upon a winning bit sequence of this length.

This lack of enough 'rolls of the die' has noteworthy implications in regard to the sort of complexity we observe in the biological world. Proteins with genuine biological function appear to be exceedingly rare within the sequence space of similar candidates. Hubert Yockey has estimated that for a candidate protein to have any significant biological function in the context of organisms on earth, half the amino acid sites, on average, must contain the correct amino acid (Yockey, 1978, 1992). Of course, several amino acids can substitute at some of the sites and the protein will continue to display biological function. At other sites, two or three substitutions are allowed for function to occur. But in most proteins there is a so-called conserved region, in which any substitution renders the protein non-functional. Yockey's research yields the rule of thumb that, on average, half the sites must be specified exactly for a protein to have biological function, while the other sites, on average, can accommodate any amino acid. While this implies that there are gigantic numbers of possible variants which display function, it also means that the *fraction* of viable possibilities is minuscule.

To illustrate, consider a protein with 400 sites. According to this rule of thumb, 200 sites must be specified exactly, while at the other sites any amino acid will do. This implies that, while there are $20^{200} = 10^{260}$ different configurations that indeed have the set of 200 crucial sites correctly specified, there are 20^{200} *times* this number that do not. Applying any sort of random search process to find even one functional configuration when the odds are 1 in $20^{200} = 10^{260}$ is hopeless. On the other hand, a relatively modest length linguistic string can specify all 400 sites perfectly. Using the coding that occurs in biological DNA, in which three nucleotide letters from a four-letter alphabet specify an amino acid, one requires a string only 1200 letters in length to specify the precise sequence.

Certainly, in the case of the most complex systems in the cosmos of which we are aware, namely, living organisms on our planet, linguistic specification appears to be essential. Simple organisms like bacteria have on the order of 1000 proteins, and more complex ones like mammals have on the order of a hundred times more. Linguistic coding in DNA specifies not only the very special protein sequences but also how these proteins are expressed in the diversity of processes involved in the organism's development, metabolism, self-repair, and reproduction. We observe that for complex systems produced by human engineering today, essentially all rely on both internal and external linguistic specification for their realization and function. Observation therefore leads us to conclude that the link between complexity and linguistic reality is more than incidental. Indeed, it is difficult to imagine how such systems could possibly be realized apart from prior linguistic description/specification.

4. Language — From Whence Does It Come?

Language, as we have seen, involves selecting and combining elements from a non-material realm of meaning entities. This is accomplished first by assigning specific meanings to a set of arbitrary symbols to form a vocabulary. The symbols possess no intrinsic meaning in themselves. The symbols are merely bearers of meaning. In the case of spoken human language, the symbol set consists of sounds produced by the human voice. Specific meanings are assigned to specific sounds. Among English speakers, the sound 'dog' has associated with it the meaning of a distinctive type of animal. But in other languages entirely different sounds such as 'perro', 'Hund', and 'chien' carry the same meaning. The sound itself is arbitrary; it is the meaning assigned to it that gives it significance. In addition to a vocabulary, languages also have rules by which elements from the vocabulary may be joined together. These rules allow for complexity of meaning far beyond what individual vocabulary words can convey. In fact, there is no limit to the complexity of meaning that is possible. Novels may weave together a hundred thousand words of human language to develop a single integrated story that no single word could possibly express. Computer programs combine hundreds of thousands of lines of code together to represent and explore the possible modes of dynamical behavior of a wide diversity of physical systems. So a pertinent issue is how such linguistic representations originate.

Let us begin by considering the properties of matter itself. We have already noted Einstein's observation that there is a logically unbridgeable gulf between the realm of matter and the realm of propositions and concepts, that is, language and meaning. But despite the fact that matter and language appear to belong to separate categories of reality, could matter, nevertheless, somehow possess an ability to generate language? As we survey the laws of chemistry and physics, including quantum mechanics, we simply find no clue that matter has any inherent tendency or ability to assign meaning to arbitrary symbols. None. The implication is that we must take Einstein's observation seriously and look for a non-material source for language. Indeed, most people do not consider this conclusion that surprising or profound. Most people appreciate the clear distinction between the realm of meaning and the realm of matter. They are not at all surprised by the inability of matter to assign meaning to arbitrary symbols and generate coherent sequences of such meaning-bearing symbols that obey a set of grammatical rules. Most people would naturally respond that it requires an intelligent being to generate linguistic messages. In fact, we humans do it with little effort, and do so at an early age, even in the absence of a great deal of coaching or encouragement from others (Chomsky, 1972).

Therefore in addressing the question of the origin of meaning-bearing language structures, the fact that humans generate linguistic expressions so readily points to an obvious place to focus our attention. Human language ability seems to be almost synonymous with intellect or intelligence or mind, or at least a crucial component of them (Oller et al., 2006). Yet just what the human mind is and how it relates to the neural activity are issues still shrouded in deep mystery. Techniques such as functional magnetic resonance imaging (fMRI) are indeed revealing that consistent patterns of neural activity in various regions of the brain are associated with various kinds of cognitive processes (Haynes and Rees, 2005). But the proposition that mental activity correlates with physiological phenomena in the brain does not represent anything new or surprising.

Moreover, the correlations discovered thus far provide no essential new understanding of the ultimate nature of human mental activity. Even the direction of causality has not as yet been established.

Some have suggested that mind and consciousness represent 'emergent behavior', that is, these phenomena 'emerge' from the connections between neurons in the human brain. Experimental evidence to date in support of this proposal is scant. Hardware and software simulations of very large neural networks have not yielded encouraging results. Most people find it difficult to imagine how a network of electrical components could ever give rise to anything akin to human consciousness and self-awareness. Especially in light of the non-material nature of language, the possibility that humans possess a distinct non-material component wherein resides language ability, intellect, and consciousness should be considered and not dismissed. Language, as we have seen, is a powerful category of reality, and it demands an adequate cause. It is not implausible for a non-material effect (in this case, language) to require a non-material cause. If we take cues from our own human perception and interpret our language ability and consciousness as manifestations of a non-material aspect of our being, then we have such a cause, at least as a working model. Which view ultimately will prove to be correct is not obvious at this point, that is, whether mind and consciousness are expressions of matter (i.e., connections of neurons in the brain) or of a real but non-material aspect of the human makeup.

The linguistic specifications that underlie all biological systems, from viruses to humans, certainly demand an adequate cause. The reality that DNA carries linguistic messages — with an alphabet, words, and grammar — represents the 'elephant in the living room' issue regarding the origin of linguistic structures. But recent investigations show the complexity of these linguistic structures to be beyond anything imagined just a few years before. In June 2007 the ENCODE (for ENCyclopedia Of DNA Elements) consortium, organized by the National Human Genome Research Institute (NHGRI) under the U. S. National Institutes of Health (NIH), reported results of its exhaustive, four-year pilot project to build a parts-list of all biologically functional elements in 1 percent of the human genome (ENCODE Project Consortium, 2007). The primary strategy for accomplishing this was to investigate the manner in which the DNA is transcribed to functional segments of RNA, similar to the copying paragraphs or articles from an encyclopedia. The results of this pilot effort are astonishing. They include the discovery that as much as 93% of *both* complementary strands of the DNA helix is transcribed into functional RNA. Moreover, many of these RNA transcripts overlap one another on a given segment of DNA. This implies, as this pilot report affirms, that in effect there is no 'junk' DNA and that the entire human genome has functionality. The fact that both DNA strands are transcribed into functional RNA messages, that a given segment of DNA can be part of multiple RNA messages, and that multiple splicings of multiple RNA messages seems to be common, means that the coding cannot be the product of random processes. This sort of multi-layered coded data compression scheme, even if the algorithms for doing it could be unraveled, would bring the world's most powerful computer to its knees to accomplish. The upshot of this work is that a much larger fraction of the genome than previously demonstrated is functional and that there appears to be a new, yet to be decoded, level of linguistic structure carried within the DNA nucleotide sequences.

Such incredible instantiations of non-material linguistic specification as we observe in the genomes of living organisms obviously demand a sufficient cause. Humans, as we have seen, display amazing linguistic facility. Not only do we have innate ability to acquire and utilize spoken language, but with some modest level of effort we can become proficient in written language. We can even develop computer languages and complex software programs to operate complex networks of machines. To account for the linguistic features of DNA, if we take our cue from our first-hand experience as human beings, it is but a small step to extrapolate from our own mental processes and facility with language to conceive of a being with similar attributes but with vastly greater intellectual capabilities. For many people this is an obvious extrapolation. Actually, there do not seem to be many alternatives.

5. Language and Design

Several have sought to find a means for identifying objects that are truly the result of what has been termed intelligent design (Dembski, 1998). In light of the foregoing discussion, it would seem that a simple and reliable indicator of intelligent design is the presence of linguistic specification associated with the object. For example, if it bears a serial number and has an instruction manual packaged with it, chances are high that the object is the result of intelligent design. If it relies on built-in coded software for its operation, it is almost certainly the product of intelligent design. Although reliable association with linguistic specification is not a necessary condition, it appears to be a sufficient one.

6. Language and Matter

What about matter itself? Does matter bear any evidence of linguistic specification? The rise of modern science is primarily a response to the discovery that the natural world behaves in ways that mathematics can successfully describe. What is today termed scientific understanding, to a large measure, corresponds to a conceptual and mathematical description of a physical process or phenomenon. These mathematical descriptions, as they are demonstrated to be reliable by repeated observation and application, become, of course, what we refer to as the laws of nature. Yet mathematical descriptions are linguistic entities. Mathematics involves the assignment of abstract meaning to an arbitrary set of symbols to form a vocabulary and the use of a set of rules to construct more complex meaning structures from the vocabulary elements. The laws of nature therefore correspond to a linguistic description of the character and behavior of the material world. This implies that matter, at a very fundamental level, has linguistic underpinnings.

7. Conclusions

Language plays a profound role in our world. It is central to our experience as humans. It enables us to create unimaginably complex machines and networks. It specifies biological systems to the level of the individual atom. Even matter itself, in the laws of chemistry and physics, appears to have linguistic underpinnings. Yet language, in that it encapsulates meaning, belongs to a realm separate from that matter and energy. Language nevertheless is real and exerts a powerful influence on the realm of material objects. The materialist axiom that matter/energy subsumes all that is real is therefore highly suspect. Almost all language that human beings normally experience is the product of the human mind. Yet whether human mental activity represents an 'emergent' property of interconnected neurons or is an expression of a non-material component of the human makeup at this point is not simple to resolve. On the other hand, language structures in complexity on a scale beyond what humans can imagine underlie biological systems. Since language facility is almost synonymous with intellect, these biological language structures imply an intellect that makes human intellect insignificant by comparison. If one is inclined to accept the biblical account of history as reliable, then the being who revealed himself to Abraham and Moses, who also, according to the account, created the cosmos, the earth, and all the living things it contains, is the logical candidate for the originator of the laws of nature as well as the biological genomes. According to the account, in making humans in his own image, he bestowed on them consciousness/mind/emotions/will, including language ability, similar in kind to his own. The observations we have considered here are in harmony with these claims.

References

Chomsky, N.A. (1972). *Language and Mind.* Harcourt, New York.

Dembski, W.A. (1998). *The Design Inference.* Cambridge University Press, Cambridge.

Einstein, A. (1944). Remarks on Bertrand Russell's theory of knowledge, In: P. A. Schilpp (ed.) *The Philosophy of Bertrand Russell.* Tudor Publishing, New York, p. 290.

ENCODE Project Consortium. (2007). Identification and analysis of functional elements in 1% of the human genome by the ENCODE pilot project. Nature **447**: 799-816; New Findings Challenge Established Views on Human Genome, NIH News, June 13, 2007 (http://www.genome.gov/25521554)

Gruenwald. P. and Vitanyi, P. (2004). Shannon information and Kolmogorov complexity. http://homepages/cwi.n./~paulv/papers/info.pdf.

Haynes, J.D. and Rees, G. (2005). Predicting the orientation of invisible stimuli from activity in human primary visual cortex. Nature Neuroscience, **8(5)**:686-691.

Li, M. and Vitanyi, P.M.B. (1997). *An Introduction to Kolmogorov Complexity and Its Applications*. Springer-Verlag, New York, 2nd edition.

Locke, J. (1490). Essay concerning human understanding. *Encyclopedia Britannica*, Chicago.

Lycan, W.G. (2000). *Philosophy of Language: A Contemporary Introduction.* Routledge, London and New York, p. 80.

Oller. J.W., Jr. (ed.) (1989). *Language and Experience: Classic Pragmatism.* University Press of America, Lanham, MD, p. 25.

Oller, J.W., Jr., Kim, K., & Choe, Y. (2000). Applying general sign theory to testing language (verbal) and nonverbal abilities. Language Testing, **17(4)**:377-396.

Oller, J.W., Jr., Oller, S.D., and Badon, L.C. (2006). *Milestones: Normal speech and language development across the lifespan.* Plural Publishing, Inc., San Diego, CA.

Plato (360 BC). Phaedo.

Rigoutsos, I., Huynh, T., Miranda, K., Tsirigos, A., MeHardy, A., and Platt, D. (2006). Short blocks from the noncoding parts of the human genome have instances within nearly all known genes and relate to biological processes. Proc. Natl. Acad. Sci. USA **103**:6605-6610.

Russell, B. (1956). On propositions: what they are and how they mean, In: R. Marsh (ed.) *Logic and Knowledge.* Allen & Unwin, London.

Shannon, C.E. (1948). A mathematical theory of communication. Bell System Tech. J. **27**:379-423, 623-656.

Wittgenstein, L. (1953). *Philosophical Investigations*, Anscombe GEM (trans.), Basil Blackwell, Oxford.

Yockey, H.P. (1978). A calculation of the probability of spontaneous biogenesis by information theory. J. Theor. Biol. **67**:377-398.

Yockey, H.P. (1992). *Information Theory and Molecular Biology.* Cambridge University Press, Cambridge.

Dialogue

Tom Barbalet

Natural languages and computer languages are fundamentally different. The philosophers of language quoted in the article — bar perhaps Kant speaking through Einstein — were instrumental in the philosophical groundwork for creating contemporary computer languages. Natural languages and abstract philosophical explorations of language are not the same thing.

The way to understand natural language versus computer language, in an applied context, is to look at the history of natural languages in contrast to the history of computer languages.

To save space, Wikipedia says most of what I want to say:

http://en.wikipedia.org/wiki/History_of_the_English_language

In stark contrast:

http://en.wikipedia.org/wiki/C_Language#History

or:

http://en.wikipedia.org/wiki/Java_%28programming_language%29#History

Questions of particular interest include:

* How many people were involved in the creation of English over what kind of timeframe versus C or Java?
* What was the reason for creating English versus C or Java?
* How has English been used versus C or Java?

Kolmogorov complexity assumes that there is a knowable minimum number of binary representations. This is clearly not applicable to the real-world.

The discussion of probability fails to acknowledge conditional probability. The author is calculating the probability of things occurring instantaneously. This in fact is a paradox because the author (in making non-conditional instantaneous probability claims) is arguing that one of his premises is completely improbable.

The author appeals to mass-ignorance of artificial intelligence rather than to the contemporary work of artificial intelligence developers. I agree with the author that the general public is sadly ignorant of the work being done in this field currently. I would encourage the author to participate in this community as a means of exploring his natural language and computer language hypothesis with practitioners whose work crosses this divide in an applied manner.

I'm not clear whether the article is complete. Sections 5-7 appear to be sketches rather than completed sections.

John Baumgardner
I would like to respond to Tom Barbalet's remarks on my chapter. While Tom raises the issue of the differences between natural languages and computer languages, particularly in an applied context, the focus in my chapter was on the ultimate *essence* of language, an essence that is characteristic of natural languages as well as computer languages. To reiterate, language, at its fundamental level, involves (1) meaning assignments to a set of otherwise arbitrary and meaningless symbols to form a vocabulary and (2) a rule set that specifies how elements from the vocabulary may be conjoined to form more complex meaning structures. Language, then, is a vehicle for conveying meaning. I argued that the meaning structures that languages convey are abstract and non-material while at the same time very real and a vital aspect of the world in which we live. My discussion was an attempt to capture the essence of what language *is* at its most basic level. This description is independent of how a given language may have originated. It has nothing to do with how many people may have been involved in creating it or the time scale for its generation or the purpose of the language or its specific application. Rather, it has to do with its ultimate ontology.

It is conceivable that Tom might respond, but of what practical use is such a description? My answer is that the insight provided by this description could radically alter one's approach to investigating topics such as artificial intelligence. It is my conclusion that human intelligence is intimately connected with the human ability to originate meaningful propositions and almost automatically to encode these into language. Humans also, of course, have the ability to decode language expressions of others and extract their meaning. If the arguments in my chapter are valid, then meaning-bearing language expressions represent an extremely important objective category of non-material reality. This implies that classical materialism, which posits that there are no objective non-material realities, cannot be valid. It suggests that materialist attempts to account for the human mind and human language ability solely in terms of matter and its built-in laws are almost certainly ill-founded. It also suggests that approaches to artificial intelligence that are ultimately materialist in their presuppositions are unlikely to be fruitful.

In regard to Kolmogorov complexity, Kolmogorov's limit is real; it exists. While in general it is not feasible from a computational standpoint to find the actual minimum length abstract representation of an arbitrary complex structure, that fact is not relevant to the point I was making. Restated, the point is that highly complex material structures can have relatively compact linguistic descriptions. If a complex material structure has a linguistic description, then its Kolmogorov complexity is no greater than the number of bits associated with its linguistic description.

With respect to my discussion of probability, my main point was that for complex structures with Kolmogorov complexity of only a few thousand bits, the dimensionality of the implied search space for a random search is so gigantic that no random scheme has any rational hope of success. The class of searches I had in mind was sequential, not instantaneous, in nature. I did, after all, include an allusion to time in my discussion. In regard to acknowledging conditional probability, while I did not discuss possible search strategies in which decisions at each step are conditioned on the outcomes from previous steps, for such schemes to be considered 'random' in the usual sense, the impact of such conditioned decision-making must be small. If someone is inclined to quibble about this, I invite them to offer from the natural world a contrary example, that is, a search scheme which can be seriously described as random while at the same time including significant levels of conditional decision-making.

As to my level of contact with the artificial intelligence community, I admit my interaction to date has been rather limited. During my tenure at Los Alamos National Laboratory, I had occasional discussions with colleagues at the Lab and also at the Santa Fe Institute who had interest in AI and related topics. For example, I want to acknowledge and thank Wojciech Zurek for introducing me to the idea that Kolmogorov complexity is a simple and elegant complexity measure for any system.

Tom Barbalet

> "It is conceivable that Tom might respond, but of what practical use is such a description? My answer is that the insight provided by this description could radically alter one's approach to investigating topics such as artificial intelligence."

The essence of language described by the author comes from the philosophers that created the paradigms (or the logic behind the paradigms) in topics such as classical artificial intelligence. There is nothing novel in these philosophers' approach to the essence of language. In fact, this approach is so problematic it creates the kind of popular scorn of artificial intelligence that the author references in his text.

Unless the essence methodology can encapsulate the application in most contexts, then it will not be useful in an applied context. This is a truism.

The applied comparison between natural languages and computer languages was designed to show, aside from the term "language", there is little in common between natural languages and computer languages even with appeals to essences.

> "In regard to Kolmogorov complexity, Kolmogorov's limit is real; it exists."

My understanding from reading on Kolmogorov is that he, himself, said his limit was only applicable to particular uses. Moreover his conjecture contains paradoxes of recursion even in the context of his agreed use. See:

http://en.wikipedia.org/wiki/Kolmogorov_complexity#Uncomputability_of_Kolmogorov_complexity

Kolmogorov's limit cannot be used in a real context unless the author wants to create a new methodology based on Kolmogorov's work that answers these problems.

I've read Wojciech Zurek's bio and he appears to be a brilliant fellow. I can't see Kolmogorov's limit being used to describe things like Wojciech Zurek, the domestic house cat or the English language. There don't seem to be a lot of binary decisions in the domestic house cat or the English language. I've never met Zurek, but I assume there were few binary decisions with him too.

Re: Probability. Conditional probability is really important in this context. The discussion here is not with regards to sorting conditional probability space, it is that if "condition A" occurs it could completely eliminate the possibility that "condition B" could occur at a future time. That can be more than a binary elimination if "condition B" is far more likely to occur if "condition A" didn't occur. There is no "allusion of time", there is time. If there is time, then there are events which occur at different times and can impact events in the future.

I can see the need to argue down conditional probability if you are making Kolmogorov complexity claims about things in the real world. As with the essence of language versus language, I fear this is the essence of reality versus reality.

John Baumgardner
Following are some brief comments in response to those by Tom Barbalet.

I stand by the simple and basic definition of language I presented in my article and reiterated in my first response to Tom. In his second response, Tom indicates that the concept of language I offered is problematic for workers in artificial intelligence like himself and that the concept has even generated popular scorn for his field. He does not elaborate much on these claims, but why this indeed ought to be the case should be fairly clear.

The reason is the following. If language, consistent with my definition, is merely the vehicle that conveys something more fundamental, namely, meaning, then the agency — let's call it intelligence — that produces linguistic expressions is actually doing more than that. Such an agency is first originating meaning and then selecting words from the vocabulary and observing rules from the grammar to package that meaning into a consequent linguistic expression. However, the 'applied' approach to artificial intelligence to which Tom is alluding is limited to searching for rules or algorithms that manipulate language elements, with the hope that somehow more 'intelligence' will 'emerge' than that which was folded in by the human algorithm designers. On the other hand, if the concept of intelligence which I am suggesting if correct, then the crucial essence of intelligent agency is the generation of meaning, while the packaging of that meaning into a linguistic expression is, by comparison, a secondary issue. This implies that the 'applied' approach to artificial intelligence, by restricting itself to mere manipulation of language elements, is of necessity missing the essential aspect of genuine intelligence. Not desiring to be unkind, I meekly observe that the degree of success of this 'applied' enterprise more or less speaks for itself.

Let me offer a final remark about Kolmogorov complexity. While there is no debate that as an exact quantity it is non-computable, nevertheless, from a thought experiment standpoint it is an elegant way to grasp how the complexity of an object might conceivably be quantified, regardless of whether it is Wojciech Zurek, his cat, or a piece of his writing in his native Polish language.

Tom Barbalet

To be clear, I made no allusion with regards to any applied approach to intelligence — artificial or otherwise. My interest was to understand the author's constructions in a context which could map onto very real claims the author was making. I am not a worker in artificial intelligence. As I have said, the author would be best served by engaging the artificial intelligence community directly with his thesis.

I rest with two claims the author makes:

> "Let me offer a final remark about Kolmogorov complexity. While there is no debate that as an exact quantity it is non-computable, nevertheless, from a thought experiment standpoint it is an elegant way to grasp how the complexity of an object might conceivably be quantified, regardless of whether it is Wojciech Zurek, his cat, or a piece of his writing in his native Polish language."

and:

> "In regard to Kolmogorov complexity, Kolmogorov's limit is real; it exists."

I remain unconvinced.

John Baumgardner
Just one concluding comment. Tom, if you have more satisfying alternative to Kolmogorov's measure for quantifying complexity, I am definitely eager to learn about it. It is just that I am not currently aware of such an alternative.

Steve Grand
This is an interesting and challenging article. I think it's worth pointing out that many of the things you refer to as using language are *describable* in linguistic terms but don't actually use arbitrary symbols manipulated by a grammar. A computer does not physically use symbols at all, but it embodies a mechanism that is *logically equivalent to* a language and thus can be described by humans in terms of symbols and grammar. That logical equivalence makes it powerful, and I think a lot of the processes we're talking about here have the elegant creative power of a language but aren't actually languages — including DNA. It makes for a good metaphor, but one shouldn't take it too literally.

You say: "The materialist axiom that matter/energy subsumes all that is real is therefore highly suspect." I agree, but one doesn't have to become a dualist of any kind, especially a Cartesian one, as a result, since it seems to me that matter is actually a form of organization (one name for the "other" aspect of reality that you seem to be talking about) and hence organization subsumes matter.

Photons, free particles and atoms are different organizations of the electromagnetic field. Each is a differently shaped disturbance that happens to have a property of self-maintenance. The vast majority of ways you can distort the electromagnetic field don't have this property but some do (and there would be some disturbances that had this property in just about any kind of universe). If you shake the field up enough, those that have a trick that makes them stable or capable of propagating themselves will arise and remain in existence, while the others arise but then dissipate (or turn into the persistent forms, just as an arbitrary splash in water turns into smooth ripples). People think of matter as "solid" because a side-effect of the way that atoms maintain themselves is that they exert forces on each other, which manifest themselves as gravitation, charge and volume. They impinge on our senses, which are also made from matter. Hence the materialist mistake of thinking that matter is somehow real, while other forms of organization are not. But atoms are really patterns in time and space, and it is *patterns of these patterns* (i.e. the relationships between atoms) that constitute everything else in the universe. All such patterns are equally real.

Organisms are self-maintaining disturbances in the electromagnetic field too. They piggyback on the properties made available by the existence of atoms. Organisms manage to persist by growing, adapting, replicating and predictively countering change (intelligence). An organism is not made from the same atoms from moment to moment; it is not a material entity but a stable pattern of organization that persists over time despite the flux of its components, and which manifests itself through the movements of atoms. A mob, a culture or a religion manifests itself through the movement of people. It is a stable, spatiotemporal pattern made of other stable, spatiotemporal patterns, which in turn are made of other patterns.

Persistence creates a ratchet: once something stable appears it tends to stay, and its existence makes new forms of persistence possible. Once atoms existed, molecules became possible and chemistry came into existence. Organisms are higher-order patterns of chemistry and, in this case at least, the existence of a "language" (DNA) massively increased the probability that new, more stable (fitter) forms would arise. Things that persist, persist; things that don't, don't. The universe pulls itself up by its bootstraps and is an endlessly creative place.

Now, a language is also an organization — describable as a set of symbols and a set of rules for the way the symbols interact. Meaning is equivalent to the language's functionality — languages express themselves through the movements of atoms too. Ideas are communicated by writing or speaking (the movements of atoms) to others, who change their behavior as a result (more moving atoms). So a human language in action is also a pattern of disturbance in the electromagnetic field.

Languages can generate new forms, whether they be new particle types, species or ideas. A language-like system is a sign of elegance — a minimal system that can create maximal results. So grammars are very effective at producing things that survive (organisms from the rules of DNA; ideas from the rules of Chinese). If a language gives rise to something that is persistent, will the language itself become more persistent?

It stands to reason that the DNA mechanism of heredity has every chance of becoming more persistent over time. It produces new, fitter kinds of organism and each carries the lexicon for the language with it. Any improvements to the language will thus be propagated widely and selected for. Books carry human languages with them too — a book contains ideas, which may be perceived as good and hence spread around. The book also contains examples of the language that it was constructed from, and so, if the ideas are successful, the language they're couched in will be successful too. Languages survive if they result in good literature. And they evolve by natural selection, including replication through the generations of native speakers, with small, local variations and mutations.

So, why shouldn't languages be emergent?

> "...it would seem that a simple and reliable indicator of intelligent design is the presence of linguistic specification associated with the object. For example, if it bears a serial number and has an instruction manual packaged with it, chances are high that the object is the result of intelligent design."

On the contrary, the existence of a "language" through which to describe a copyable blueprint is the key to Darwinian natural selection. You point out yourself that DNA enables life to solve the problem of how to find the right few proteins out of 10^{260} possibilities, turning randomness into a powerful design force (see my comments on Gerald Schroeder's chapter). There's every reason to suppose that the language of DNA (in its entirety, including punctuation, folding rules, repair mechanisms, etc.) has itself evolved. Grammars are powerful devices for engendering creativity, whether at the hands of an intelligence or not. Without languages, emergence wouldn't have much chance. And since languages themselves evolve (just as English has become more expressive over time), we'd expect languages to emerge, too. I suggest that the existence of languages (in the broadest sense, whether they operate using symbols or are just describable in symbolic terms) is actually supportive of a mechanistic interpretation of the universe.

John Baumgardner

Steve, let me congratulate you on your analysis of my article as an articulate expression of the emergent materialist perspective. Indeed, true to that perspective, you seek to categorize language as merely another type of pattern that emerges spontaneously out of the built-in laws of matter.

In passing, I note that you are not a physicist. Few, if any, physicists, at least of those I know, would suggest that the electromagnetic field is the only force, much less, as you seem to be suggesting, constitutes all there is. For example, without the so-called strong interaction, sometimes referred to as the strong nuclear force, the electromagnetic interaction would blow the atomic nucleus, because of all its like-charged protons, to smithereens. Thus, without the strong interaction, there would be no atoms or molecules or people. Although there has been a diligent quest to find it, so far no one thus far has produced a grand unified theory that accounts for the four fundamental interactions in terms of only one. So, given this present state of affairs, it is not a good idea to claim that all reality is simply "different organizations of the electromagnetic field." That is why I substituted 'the built-in laws of matter' in my summary of your worldview outlook above.

Now with regard to your basic claim that ultimately that language is no more than "a pattern of disturbance in the electromagnetic field," my response (wearing my physicist hat) is, to ask you to show me, in terms of physical law, that is, the laws of physics and chemistry, just how that occurs. As I discussed in my article, there is no hint at all in the laws of physics discovered thus far that matter/energy has any ability whatever to generate linguistic expressions. Absolutely none. And the reason from a physics standpoint is fairly obvious. It is that the substance which linguistic expressions convey, namely, meaning, is non-material. It is just that basic.

In terms of the most complex organizations of matter we know of, namely, living organisms, as I pointed out in my article, their organization depends critically upon massive language-mediated specification. And this specification emphatically cannot be deduced from, nor reduced to, the fundamental laws of physics. Moreover (and I did not discuss this in my article), genomic information, contrary to popular understanding, does not improve as it is transmitted from generation to generation; instead, it inexorably degrades. The reason is fairly simple. Natural selection has a threshold, below which it cannot 'see' a mutation. As realized by Kimura in the late 1970's, the vast majority of mutations (both favorable and deleterious) fall below this threshold and are therefore un-selectable. These low impact mutations simply drift and accumulate, untouched by selection. Because favorable mutations are so rare compared with deleterious ones, the net result is inexorable genomic deterioration, or genetic entropy[1], as it has been called.

---

[1] Sanford, J. (2005) *Genetic Entropy and the Mystery of the Genome*. Lima, NY: Elim Publications.

Emergent materialism, I conclude, represents a grand leap of faith. It is a leap of faith because it is in conflict with reality. The cosmos is not a huge self-organizing system that is ratcheting itself upward toward higher and higher levels of complexity and organization. The second law implies that everything is on a trajectory in the opposite direction, that is, toward dissolution and decay. Even the proton has a finite lifetime. Genomes to not improve and ratchet up to better ones; they likewise obey a law of deterioration with time.

The cosmos is not self-existent and self-organizing as emergent materialism asserts. Rather, the cosmos was designed and brought into existence by God. The *Bible* reveals that God has glorious purposes in view which will be realized, but that He has subjected this present created order "to futility" because of the abuse of volition on the part of beings He created. Emergent materialism is a manifestation of this corruption of the volitionary capacity — of this rebellion against God and His beneficent ways. Crudely expressed, it represents a spitting in God's face. Steve, I realize that you may not have thought through the implications of your worldview, but the implications are so significant I would be irresponsible in not urging you to do so.

Steve Grand

Thanks for your concern John, but I'm satisfied that I've thought the implications through well enough that if I turn out to be spitting in the face of God I'll be content to accept my punishment. My thoughts are my own and hard-won; they're not received wisdom. In fact by upbringing I ought to be on your side. But I have experimented with self-organization all my working life and have ample first-hand experience of its power, ubiquity and sufficiency. What I don't have is anything like the space I'd need to address your claims.

I'd like to pick up on a few important things, though. For one thing I accept that electromagnetism isn't the only force — I was using it as a simplification for brevity. My point was simply that particles and atoms seem to be resonant states in a multidimensional field, not little lumps sitting upon space. They're dynamic phenomena and hence are processes more than things (to use macroscopic analogies). So are minds, organisms and societies.

Of *course* there is no hint in the laws of physics of the emergence of language. It's hard enough to deduce the existence of vastly simpler phenomena from those laws even when we know the one leads to the other. This is a failure of our imagination and of our mathematical tools, nothing more. John Conway's "Life" game offers a simple example of phenomena that arise deterministically from a substrate which nobody would deduce (indeed *could* deduce, even in theory, without actually playing out the game) was capable of such things.

The laws of physics, just like the laws of Conway's game, are not in themselves enough to explain anything at all. They are merely functions waiting for data. You also need spatial distribution, aka organization. It is the spatial organization of ON and OFF cells in Conway's Life that interacts with the laws to produce complex phenomena requiring a higher level of description (including propagating and self-maintaining phenomena that are loosely analogous to matter). Different organizations lead to very different results, and you don't need a designer to obtain interesting ones. Likewise, it is the spatiotemporal organization of particles, not the laws alone, that is crucial to the emergence of higher-order structures in our universe, including molecules, minds and languages. Physics deals with averages and hence takes little account of organization.

The universe *as a whole* obeys the second law of thermodynamics, but regions of it are perfectly capable of moving in the opposite direction, like eddies in a stream. This is no violation of the law — they achieve a local drop in entropy at the cost of a larger rise elsewhere. Living systems are localized negative entropy. In any case, organization is not synonymous with order. Homogeneous stacks of electronic components make for a very ordered, low entropy state, but by increasing their disorder you can build a radio receiver, which does things that none of its parts can do. A zero entropy state is the simplest, most boring thing there is, but disorder brings complexity.

On that topic, genetic drift is not a degradation. Genes are not like perfect crystals that crumble into dust; they are dust that just turns into a different kind of dust. Entropy therefore has little to do with it (and in any case, gene repeats are a compensatory source of entropy reduction). One man's junk is another man's regulatory gene.

Emergence is neither a leap of faith nor in conflict with reality — it is demonstrable in ways that religious explanations are not. Belief in a god is nothing *but* a leap of faith, almost by definition. It solves no problems at all — it just pushes them back to a level where we can glibly ignore them.

John Baumgardner

Steve, certainly cellular automata demonstrate that just a few rules implemented on a lattice can give rise to many diverse and even unexpected patterns. Although there was excitement about CA methods in the 1980's and early 1990's, especially at the laboratory where I was working at that time, almost none of the optimistic forecasts concerning applications to the real world ever came to fruition. The reason is simple. The rules by which the real physical world operates are so much more numerous and complex, that one should not expect the trivially simple rules of CA systems to be able to reproduce the more complex behavior of the physical world. From your comments I am fairly sure you concur with this analysis.

To me the issue is, however, how can you rationally appeal to the behavior of cellular automata as justification to declare that "the laws of physics … are not in themselves enough to explain anything at all," that somehow physicists do not connect the laws of physics with particles so as to predict and understand their mutual interactions, and that "physics deals with averages and hence takes little account of organization." Most professionals in the hard sciences, and physicists in particular, would respond in utter amazement to such claims. To me this reveals that you are allowing something akin to mysticism or faith, instead of science, ultimately to drive your thinking on these matters.

On the issue of entropy, I am emphatically not conflating order with complexity. On the other hand, you seem to have a fundamental misunderstanding of the relationship between disorder and the sort of complexity we observe in both living systems and man-made systems like computers. The absolutely essential ingredient in these sorts of complex systems is specification. As I stress in my article, the level of specification required in these sorts of systems requires the compactness and power that only linguistic coding can provide. Without such specification in the physical world, one can get ordered crystals and transient eddies in a stream, but most of the time the result is randomness and noise. My verdict is that you seem to have missed a huge aspect of reality in the world around us, namely, its ubiquitous non-material linguistic specification.

Genetic entropy, a new term which I doubt you have encountered before, is only indirectly connected with thermodynamic entropy and has little to do with genetic drift. Genetic entropy is a direct consequence of natural selection's inability to 'see' the overwhelming majority of mutations, the vast majority of which are deleterious. There is then no effective mechanism to remove the errors in the genetic specifications, errors that accumulate inexorably, from one generation to the next. When the specifications become sufficiently garbled, the complex system fails and the 'pattern' disappears.

Finally, God has left us a vast amount of objective evidence revealing His reality. This evidence is so unambiguous in the linguistic specifications in living systems that any scientist living today has no serious justification for discounting God's reality. This evidence, of course, is in addition to the written record attesting unmistakably to God's disclosure of Himself in the context of objective human history, a record preserved so diligently and carefully by the Hebrew scribes. True faith in God involves a thoughtful and reasoned response to objective and trustworthy evidence, of which there is no real scarcity.

Steve Grand
Ok, I give in. Basically you're just insisting that everything you disagree with is simply a demonstration of my ignorance. Either that or you're willfully misinterpreting what I'm saying. My expertise is not on trial here, so I think it would be pointless trying to defend myself. I'll continue developing artificial systems that unquestionably improve their 'linguistic' specification without need of a designer, and let others make up their own mind what these tell us about reality. If Judgment Day turns up in the meantime I guess I'll take the problem up with God.

John Baumgardner
Steve, I sincerely do not think I was attributing our differences as ignorance on your part. Summarizing our interchange, I would describe my comments as questioning your view that CA behavior really does account for all you believe it does in the physical world. I was mostly giving you my reasoning why this sort of extrapolation from CA behavior to me seems to fall short. I definitely do not question or minimize your expertise in your field. I was simple raising questions as to the way you apply of your findings beyond their immediate context. I did detect that your work does indeed involve a conscious philosophical component for you. You alluded to the fact that you had some sort of religious input when you were young but that you had since left that behind. I'm curious about what sort of background that might have been, if you feel free to comment on it. My history is rather the opposite. My father was an agnostic college professor, so I had very little in the way of spiritual or religious influence growing up. It was not until I was in graduate school that I came to an awareness that there are indeed spiritual realities. In any case, thanks for taking time to dialogue.

Steve Grand
Forgive me if I misinterpreted your objections but you just seemed to be telling me baldy that I was wrong — "you seem to have a fundamental misunderstanding" or "you seem to have missed a huge aspect of reality", as if I'm just being dimwitted. Meanwhile your own assertions are stated to be unambiguous and obvious. I felt you were treating me like a student.

As it happens I am unqualified. I received the OBE (an order of knighthood) for my A-life [artificial life] work, I'm shortly to receive an honorary doctorate and I've held honorary fellowships in psychology, biomimetics and artificial life, but I'm not a professional academic because I missed out on the chance to obtain paper qualifications in early adulthood. But I do have a solid international reputation and I've devoted thirty years to building self-organizing and adaptive systems in the areas of physical simulation, artificial life and computational neuroscience. I'm a cyberneticist by trade, and I don't think I'm as naïve as you seem to imply.

The background I was referring to is that my father was a lay Baptist preacher and an electronics design engineer. His own scientific and systems-engineering experience lay uneasily with his Christianity, which was acquired (as is often the case) in his early soul-searching twenties, when Billy Graham came over to the UK. His own solution to this tension was to believe in a prime mover god, not an interfering one, and focus on the moral teachings of Christianity rather than metaphysical talk of a Second Coming or Heaven and Hell. His interest in science has never wavered. He was also a very independent thinker, who taught me to question everything and assume nothing. So although he has never referred to ID specifically, that was very much the environment I was raised in. It was not a dogmatic environment that I should either swallow whole or rebel against, so I think it left me with an open mind. But over the decades, from first-hand observations, I've become increasingly impressed by the self-organizing power of the universe and increasingly of the view that believing in a god creates far more problems than it solves (as well as having dangerous social and moral consequences).

Since we're still talking, I'll try to respond to your questions:

Firstly I wasn't extrapolating from the behavior of CAs at all — I was just using them as an illustration of the importance of organization in addition to laws. The lack of applications for CAs, incidentally, is irrelevant. What I was trying to say about Physics is that historically it derived its greatest triumphs, and hence its overarching paradigm of thought, from understanding linear systems, which are reducible (by definition) to totals and averages. In gases, for instance, you can safely reduce all the individual molecular trajectories and kinetic energies to a few aggregate values — temperature, pressure and volume. Abstracting and quantifying the world like this is the starting point for Physics. But you just can't do this with important aspects of crowds, collections of neurons or large networks of chemical reactions. Hence physics is essentially a science of quantities, while sociology and biology are sciences of organization (at least, they are when they aren't suffering from physics envy!).

Take a radio receiver and a radio transmitter as an illustration. Each has radically different properties and these are emergent — they are not properties of any of the components. But you can build a receiver or a transmitter from precisely the same set of transistors, resistors, capacitors and chokes. There is NO relevant quantitative difference between the two circuits — they have the same mass, composition, complexity, etc. They only differ qualitatively, in the spatial arrangement of their parts. Physics is primarily a quantitative science, and although it gave us the components themselves (for which I'm very grateful) it doesn't really provide the intellectual tools for thinking about the properties of specific networks of those components, especially when the relationships are nonlinear.

My main point was that laws are not enough to describe the behavior of the universe. Laws are like functions that operate on data; without the data they do nothing. But for complex nonlinear systems the data are spatial and so the functions must be applied in parallel or a specific sequence — you can't just slot an aggregate of all the values into a single equation. Crucially important properties of the system will be missed if you do that. And it is the properties of systems that form the properties of parts at a higher level of description: an organism is a real thing in and of itself, with properties and laws of its own; it is not just the atoms of which it is made (in fact it's not ***even*** the atoms, since they are replaced over and over again). The universe has a hierarchy of form, in which each level has a genuine right to be called real. Each level obeys the fundamental laws of physics, but cannot be inferred from them alone.

I'm not suggesting that something metaphysical is required in addition to the laws, just something qualitative — the set of ***relationships*** between component entities. Oddly, I thought this was a concept that united us — you too seemed to be talking about the need for both aspects of reality. Language, as I understood your thesis, was one aspect of organization — a kind of meta-organization. Where we differed, I thought, was in how we interpreted the significance of that.

I imagine you'd call electronic circuit design a linguistic process. We draw circuit diagrams using a grammar, made from a set of symbols and rules about how to interpret the spatial relationships between the symbols as electrical connections between physical objects. As it happens, most (but not all) electronic circuits are designed by an intelligence, and that intelligence uses the grammar to specify the functional relationships. In other words, the circuit diagram is not just a description of the circuit but the specification that creates it.

Where we differ is that you think the existence of such a grammar implies the existence of a designer, whereas I don't. I did say your article was challenging – normally people just suffer from the fallacy of imagination failure; they simply can't envision how something complex (the eye, say) could have come about by any other means so they invoke a creator. That's not a good enough argument. But you are saying something more sophisticated: you're not merely baffled by the existence of complex "phenotypes" — you are baffled by the existence of "genotypes". Now it could be the same fallacy — it could simply be that you are unable to conceive of how or why a "linguistic" system should have arisen spontaneously. Or you could be right that there is a reason ***in principle*** why that couldn't be so.

Simply telling me that it is self-evident that languages imply an author won't do. You have to give me reasons in principle. Likewise, I can't just say that languages self-evidently emerge all by themselves and their meanings are implicit, rather than imposed from the outside. I have to give you an existence proof. I was hoping we'd get to the point where we were agreed on the contextual stuff like the importance of organization to the properties of systems, and the existence of a hierarchy of form in nature (and hence at least the potential for real things to come into existence through emergence, as properties of systems). Then we could discuss mechanisms through which language-like meta-organizations could or could not come into existence without a mind to impose them. Could we ever reach that point, do you think? Or are we doomed to stay bogged down with the contextual stuff?

John Baumgardner

Steve, I appreciate very much your filling in some helpful details concerning your background and perspectives and for responding so generously to my questions. Yes, I do find that we have a substantial unity of outlook on many of the important issues. For example, I am essentially in full agreement with your two paragraphs:

> "My main point was that laws are not enough to describe the behavior of the universe. Laws are like functions that operate on data; without the data they do nothing. But for complex nonlinear systems the data are spatial and so the functions must be applied in parallel or a specific sequence – you can't just slot an aggregate of all the values into a single equation. Crucially important properties of the system will be missed if you do that. And it is the properties of systems that form the properties of parts at a higher level of description: an organism is a real thing in and of itself, with properties and laws of its own; it is not just the atoms of which it is made (in fact it's not ***even*** the atoms, since they are replaced over and over again). The universe has a hierarchy of form, in which each level has a genuine right to be called real. Each level obeys the fundamental laws of physics, but cannot be inferred from them alone."

> "I'm not suggesting that something metaphysical is required in addition to the laws, just something qualitative — the set of ***relationships*** between component entities. Oddly, I thought this was a concept that united us — you too seemed to be talking about the need for both aspects of reality. Language, as I understood your thesis, was one aspect of organization — a kind of meta-organization. Where we differed, I thought, was in how we interpreted the significance of that."

Let me respond a bit to the very good issues you raise in your last full paragraph:

> "Simply telling me that it is self-evident that languages imply an author won't do. You have to give me reasons in principle. Likewise, I can't just say that languages self-evidently emerge all by themselves and their meanings are implicit, rather than imposed from the outside. I have to give you an existence proof. I was hoping we'd get to the point where we were agreed on the contextual stuff like the importance of

> organization to the properties of systems, and the existence of a hierarchy of form in nature (and hence at least the potential for real things to come into existence through emergence, as properties of systems). Then we could discuss mechanisms through which language-like meta-organizations could or could not come into existence without a mind to impose them. Could we ever reach that point, do you think? Or are we doomed to stay bogged down with the contextual stuff?"

Yes, I think we agree reasonably well on the "contextual stuff' and the important issue indeed has to do with "mechanisms through which language-like meta-organizations could or could not come into existence without a mind to impose them." For me, as I have already intimated, an important issue is bridging the gap between what, for lack of better vocabulary, I will refer to as your cybernetic realm as against the realm of atoms and molecules and electromagnetic energy. I'm eager to hear what you might have to offer on that front.

Steve Grand
John, Ok, this is way too long but I decided to try and elaborate a bit more — brevity breeds too much misunderstanding.

> "For me, as I have already intimated, an important issue is bridging the gap between what, for lack of better vocabulary, I will refer to as your cybernetic realm as against the realm of atoms and molecules and electromagnetic energy."

That's largely why I was sticking my neck out about the idea that atoms and molecules are themselves primarily forms of organization. In my mind at least it helps to resolve a few things. Let me stick my neck out again for a while…

Hard-line materialism is our intuitive default position, don't you think? We grow up thinking that atoms are real because they're tangible and directly sensible, while other things (hurricanes, minds, ideas or whatever) are somehow of lower status and "not really real". But we recognize that this isn't enough to explain everything (especially ourselves). Atoms don't seem to have any properties that hint at the presence of "mind stuff", yet we know that we are real — *cogito ergo sum* — so we look for another kind of reality to contain minds, and that leads us to dualism. We want our minds to be "as real as matter". At the root is our innate assumption that atoms are fundamentally different from other forms in nature.

It's a kind of materialist snobbery. While this is the case we can't, as you say, bridge the gap, so we have to look elsewhere.

John Baumgardner
Wait just a minute! Atoms are not directly sensible. But certainly hurricanes are. I can honestly say that I never assigned hurricanes some substandard status of tangibility than, say, my car or my computer and entertained the idea that they are somehow "not really real". Such an outlook to me seems utterly bizarre! Similarly, I must say, I have never questioned the reality of my own mind nor my experience as a thinking, choosing person. I may have wondered, like Descartes, just how my mind works relative to my body, but I have never questioned its reality.

Steve Grand
But suppose atoms are part of the "cybernetic realm" too. That way there is no gap to be bridged. Instead of trying to reify minds and souls as some kind of "stuff", we can try to de-reify matter and see it as made of the same NON-stuff as minds and crowds and organisms. We can view everything from electrons to cultures as levels in a hierarchy of self-maintaining patterns. If even matter is organization, then we can use things we know about emergence at one level to help us understand it at another — there's nothing particularly special about the level we call matter.

John Baumgardner
The philosophical perspective you are presenting here perhaps gives me some better insight into some things you have been saying previously. Before, I had understood your outlook to be that of an emergent materialist, namely, that the patterns and organization we observe simply 'emerge' out of the innate properties of matter. But here you seem to be going beyond that outlook and 'de-reifying' matter itself to speculate that 'organization' or 'cybernetics' ultimately constitutes all there actually is. That in essence is very close to the pantheist view.

Steve Grand
I realize I don't have a leg to stand on here, because I'm not a physicist, but it seems to me as an intelligent layman that a number of lines of evidence point this way, towards the view that matter is no more real than minds, once we clear our heads of our innate assumption that atoms are little lumps of stuff superimposed on space. Even many physicists still seem to visualize matter this

way. Take the number of books where eminent physicists say how baffling it is that "light is sometimes a particle and sometimes a wave". Well of course light is neither a wave NOR a particle – it's just light. Both waves and particles are analogies from the macro world, and like all analogies they break down eventually. Sometimes one description is more apt than the other, but that's decidedly not the same thing as saying that light changes from one form of behavior to another according to circumstances.

This supposed wave/particle duality is, at least, one hint that matter is widely extended in space and is like a **distortion** of something — a little like a wave, yet also a little like a bump with a sharp peak in the centre — something like a Gabor wavelet, maybe, and certainly quite like a soliton — but emphatically NOT a little hard-edged lump floating in space. There are wavelet models that explain the two-slit experiment, quantum tunneling and the like quite satisfactorily without recourse to any quantum weirdness (unfortunately I can't give references, but I've seen some good simulation videos on the web).

Another hint comes from electron orbital patterns. Do these look even remotely like little moons orbiting a planet? No, they look distinctly like resonance modes. If you've ever played with a Chladni plate you'll have seen how the plate can support various resonant modes. Only one mode can exist at a time, and if the energy levels rise so that a second mode becomes more stable than the present one, there's a quantum jump from the one to the other, just like in an atom. I was at CERN once, and I asked a particle physicist whether he really thought that atoms **contain** electrons and other particles or whether they are a state of existence that simply **corresponds to** a whole number of particles (i.e. if you disrupt an atom, its energy will be dissipated as a set of simpler states that take the form of electrons, etc., but there are no such entities actually "inside" the atom). Unfortunately he didn't seem to understand the difference but I would have liked to hear some evidence that the latter is closer to the truth. Either way, atoms seem to me to be stable distortions of something, akin to a set of localized resonant states in it. Trapped electrons in quantum wells certainly act very much like distributed standing waves, and an atom's nucleus could potentially act like a well (although this begs the question of how the nucleus itself remains stable).

String theory also seems to point this way. Everyone draws strings like they're little loops of spaghetti, but the descriptions sound more like the way standing waves form around the perimeter of a bell. I can't get good answers from physicists about this stuff because their mental images seem quite weak and they switch straight to math, where they feel more comfortable but I don't. I'm a visual thinker and as a cyberneticist I think in terms of feedback loops and attractors in phase space, which are visual concepts.

Anyway, it doesn't seem improbable to me that the fundamental elements of matter are actually spatial distortions of something (some kind of multidimensional field) that happen to have the property of stability, self-propagation or self-maintenance. This doesn't seem in conflict with any evidence that I know of. As distortions they correspond to regions of potential energy and therefore exert force, which is why they seem to have "substance". But as distortions of something they are really examples of organization; of self-maintaining patterns; of systems.

John Baumgardner
Here you seem to be backing away a bit from that step of 'de-reifying' matter entirely.

Steve Grand
I can't even begin to imagine what the underlying "something" is like, but I guess the buck stops there and I'd have to say that this level is "physical" and not itself a form of organization, although even here things aren't quite that straightforward, since even space and time seem to be a product of *relationships*, rather than innate. But my point is, if matter is a reorganization of something in time and space then some such organizations will have the property of persistence while others won't. If you "bunch up" this "something" in a zillion different ways, most of them will dissipate immediately, but a few will hang around for a long time because they have a mechanism for persistence, just as drawing a bow across a violin string creates a mass of random vibrations which rapidly coalesce into a few resonances. And this, it seems to me, is the most fundamental principle in the universe: things that persist, persist, while things that don't, don't.

Everything we see around us exists because it persists, no matter whether it arose spontaneously or was put there by God. Some products survive while others fail; some weather patterns persist for weeks while others collapse in hours; some belief systems outlive others; some organisms are fit for a change in their environment while others die out. Wherever you look, you find that everything has some clever trick that maintains its coherence for extended periods. Obviously those that don't have a clever trick just disappear, so we're left with those that do.

We can identify these clever tricks and see how they work. We can't quite do that for electrons yet but with hindsight we can see how most successful patterns in the universe, from galaxies down to weather systems, cells and molecules, manage to persist. This is a remarkable fact, I think, and it needs to be taken notice of. Nothing exists (for long) that doesn't exhibit self-preservation. The most obvious corollary to this is that persistence is a selection mechanism. In any system where a wide variety of forms can come into existence, the system will tend to "discover" forms that are persistent and weed out those that aren't.

Many types of clever trick have been discovered so far, and each trick might be used at several levels in the hierarchy of form. Some things persist because they're inherently stable. Others copy themselves through space, so that their own collapse creates an identical copy close by (e.g. a collapse in the electric vector creates a magnetic force, which ends up with nothing to sustain it, so it collapses and creates an electric force…). Some copy themselves wholesale, which is how the forms of living things (species) persist through time and how colonies of living cells come to form collectives called organisms. At the same time, organisms also persist by converting their environment into substances they need, by adapting to environmental stress, and even by predicting stress and preventing it through the use of intelligence.

Any form that manages to persist creates new conditions and new opportunities that can be exploited by higher levels. Atoms make molecules possible; molecules make cells possible; cells make minds possible; minds make cultures and ideas possible…. A highly ordered, energetic universe has few properties to exploit. It's only as it cools and disorder increases (i.e. the number of different relationships or possible "circuits" increases) that new opportunities arise. Once there were no atoms, now there are; once there was no life, now there is.

John Baumgardner
Molecules may make cells possible, but will cells ever emerge from molecular interactions alone, apart from some additional specifications that transcend innate molecular behavior? I claim there are good reasons to conclude the answer to this question is no. Similarly, cells may (or may not) make minds possible, but do intellectual and volitional processes and other aspects of mind

emerge from solely from cell properties and dynamics? I believe an honest answer to this question is that, from a scientific standpoint, we simply do not know.

Steve Grand

It's worth re-emphasizing that an organism is NOT a material thing. It is a spatiotemporal PATTERN in a material thing. The actual material is constantly flowing through it and being replaced — we are not made of the same stuff we were made of as children. Clouds are the same — they're made of different water vapor from moment to moment. So is a wave on water. All these things are patterns in something. And the somethings that they are patterns in are themselves patterns in something else, all the way down.

John Baumgardner

To me it is more accurate to describe an organism as a dynamic (as opposed to static) material thing, whose inner workings are choreographed by a mind-bogglingly complex non-material linguistic instruction set. Clouds and waves on water, on the other hand, do not carry such explicit linguistic instructions. The instuctions (still nevertheless linguistic in character since we can identify many of them and express them using mathematical formalism) responsible for their behavior seem to be innate to their atoms.

Steve Grand

Mere stability is one kind of selection ratchet — a random splash in a pool will settle down until you're left with nothing more than ripples and vortices. Water "discovers" these stable forms by selecting them out of thousands of contenders. The basic laws of physics enable some distortions of the fundamental fields to persist when others don't (and just like splashes in water, patterns that are almost stable probably turn into stable ones), so in any big bang scenario, the universe can incrementally discover and/or create particles, atoms, etc. from the resulting melee. However, as ID-ers are fond of pointing out, that sort of dice throwing alone couldn't possibly account for the evolution of living systems — they're far too improbable. But natural selection and genetics provide a far more powerful ratchet that can search design space very efficiently. Leaving aside any ID objections about whether evolution is actually enough to account for what we find in the natural world, we DO know that it works — it's easy to demonstrate this in simulation. Darwinian evolution wasn't even possible until molecular networks and (possibly) a template mechanism existed. So a whole new ratchet came into existence once certain other patterns had been discovered. Therefore, not only do the forms themselves get more complex as the universe discovers them (particles to atoms to molecules) but new means by which forms may be discovered also emerge. The emergence of intelligence perhaps created a third level of ratchet, with the capacity of foresight — we humans can design things.

John Baumgardner
"But natural selection and genetics provide a far more powerful ratchet that can search design space very efficiently." I challenge that statement head-on! As have already stated in previous remarks, natural selection, as it actually operates in the natural world, is utterly helpless to affect the destiny of the vast majority of mutations because of noise from many sources. Population geneticists over the past 60 years have duped the rest of the professional biology community into believing the neo-Darwinian theory works when it does not. It is an embarrassing sham. It only works under highly artificial circumstances that are never satisfied in the real biological world. Neither does evolution work in numerical simulation experiments unless biologically unjustifiable assumptions are folded in. I can show you quite clearly why evolution does not work through numerical simulation if you are interested.

Steve Grand
You may well flatly disagree that evolution works, or intelligence emerged, or any number of other things about what I've said. What I propose asking next is, if the patterns themselves emerge and the ratchets that are able to discover/design them also emerge, and language-like systems are forms of organization with unusually potent design power, then will they also emerge? Have they? But given all the steps up to this point I guess I should hand the talking stick back to you first!

John Baumgardner
Let me get to what I am convinced is the heart of the issue, namely, God's reality. You are choosing to work within a framework which has God absent. That automatically leads to most of the logical difficulties with which we have grappling. Most of these problems disappear if the God of the *Bible* is truly part of the picture. For starters, something like Cartesian dualism indeed is true. Moreover, there is a perfectly reasonable explanation for the linguistic specification which we infer for things like atoms and electomagnetic fields and which we observe explicitly at such an astonishing level in living systems. As I expressed earlier, in my assessment, the evidence for God's reality is

overwhelming. And the evidence is objective and subject to rational evaluation and test. In particular, there is the historical individual Jesus, who inquired of Peter, "Who do you say I am?" I have concluded this is the most important question any human being can be called upon to answer. Can you honestly say that you have looked into the evidence in any sort of serious way and can give a thoughtful answer to this question? If not, I urge you, for your own sake, to do so.

Biography

Dr. Arie S. Issar is a Professor Emeritus at the J. Blaustein Institute for Desert Research, and the Geological Department of Ben Gurion University of the Negev. He founded, and was the head of, the Water Resources Center of the Institute for Desert Research since 1975 until his retirement in 1998. It was partly his inspiration, which saw the founding of the Institute for Water Sciences and Technologies.

He received his Ph.D. at the Hebrew University of Jerusalem in 1961. From 1980-1998 he was the holder of the Alain Poher Chair in Hydrogeology of Arid Zones. In 1985 he was awarded the Ernest D. Bergmann Prize for Special Scientific Contributions to the Development of the Negev Desert. In 2003, he received the Prize of the President of the International Association of Hydrogeologists for Outstanding International Contributions toward the Advancement of Hydrogeology. In 2005 he was awarded Honorary Member of the Israeli Association for Water Resources.

Prof. Issar's current research focuses on the impact of climate change on the hydrological cycle and socio-economic systems; and developing conceptual models in order to mitigate the negative impact of global change on the water resources of the Middle East, including long term policies of progressive development of the fossil aquifers of the region. He has published about one hundred papers, co-edited five books, and wrote five books in the fields of geology and hydrogeology.

In addition to his research in the above mentioned fields he is investigating the evolution of intelligence in the bio-world and has published a book with Late Professor Robert G. Colodny of Pittsburgh University, (1995), *From Primeval Chaos to Infinite Intelligence–Information as a Dimension and Entropy as a field of Force*, Ashgate-Avebury, UK.

42

Evolution of Intelligence in the Bio-World

ARIE S. ISSAR

Prof. Emeritus Ben-Gurion University of the Negev
Sede Boker Campus, Israel
issar@bgu.ac.il

1. Introduction

The conceptual model, herewith presented, evolved in connection with the author's involvement in the research of the geology of the Quaternary (Issar, 1968). In this research the evolution of artifacts, developed by hominids, beside other fossil remains, were basic tools for building a spatial-temporal, i.e. stratigraphic-geological model.

While investigating the evolution of the flint tools, from primitive pebble tools, which were just pebbles etched at one end to become sharp and pointed, to the evolved flint arrow heads, a number of questions arose:

- As the rising spatial-complexity of tools expressed progress in the intelligence of hominids, the question arises whether this trend of evolution could be explained in the framework of the conventional Darwinian to Neo-Darwinian paradigms, i.e. evolution through the process of random mutations filtered by the constraints of the hostile environment?
- While success in the process of Darwinian selection, i.e. survival of the fittest, can be measured according to the number of branchings off of similar forms of life and their number and distribution in space, how can the survival of the more intelligent be measured? In other words, who are more

successful from the point of view of the Darwinian paradigm, the hominids or the beetles?

- The general question which follows is how should the evolution of intelligence be quantified? In other words, on what dimension and by which units can intelligence be measured, when IQ tests are not feasible?

Generalizing these questions, it can be said that the Darwinian paradigm explained beautifully the evolution of forms, i.e. spatial changes and spread out along the space dimension, and as time passed, i.e. the time dimension. Now, at a certain point on the dimension of time the hominids branched off from the primates and started to produce tools. These tools became more and more sophisticated as time progressed and as the hominids multiplied and spread over the globe. The scales on which this progress on the tree of evolution is described, by conventional measures, are either temporal, i.e. the time passed since the first pebble-tools were produced, or spatial, i.e. the features of the hominids and their tools, as well as their geographical range. All these data are against space and time dimensions and the question is whether, once progress of intelligence became dominant in the evolutionary process, an additional dimension should not be added on which this progress can be presented?

Once this is answered the question would be whether the Darwinian paradigm is sufficient to explain the evolution of intelligence of the hominids? The reason for posing this question is that while the strategy of survival in Darwinian conceptual models was by adaptation to the environment, with the hominids a new strategy was introduced, which was changing of the environment, after intelligently understanding its constraints, in order to fit hominid requirements.

While the questions started with relation to hominids, in due time the author came to generalize this question to the entire bio-world, after he came upon the results of the research carried out by the psychologist Morton E. Bitterman (1965), who found that the evolution of intelligent behavior in the bio-world correlates with the place of the species on the evolutionary tree. In other words the increase of intelligence is parallel with the appearance of new forms of life on the geological timetable. Thus vertebrates are more intelligent than invertebrates, saurians than fishes, mammals than saurians, etc.

Bitterman investigated the level of intelligence by "the ability to develop a new way of reaction when an entirely new situation comes up." The question to address at this stage was whether there exists an abstract scale on which intelligence can be measured, except by the time needed to learn to push a button or find food in a maze? In other words, does another coordinate or set of coordinates exist, beyond space-time, on which the intelligence revealed by ability of a certain creature to behave more intelligently than another creature could be measured? Moreover once experience is gained and turned into instinctive behavior or abstract knowledge on what scale can this be presented, in addition to the spatial-temporal scales.

2. An Additional Dimension is Required

At this stage the author came upon an article by Lord Brain (1968) who, as a result of his research on the human mind, came to the conclusion: "...that there are living events, which we will fail to understand until we can describe them also in terms, which may involve additional coordinates." This conclusion was reached because of the difficulty of explaining by conventional physical models the ability of the brain to construct logical algorithms on the basis of physical data, fed into it by the senses.

This ability of the human mind to describe natural events, in logical and mathematical terms, spawns another basic question: How can it be that mathematics is, on one hand, the brainchild of the human being and, on the other hand, is intrinsic in the material framework of the universe?

Albert Einstein formulated this question in his lecture before the Prussian Academy of Sciences (Einstein 1921), namely: "How can it be that mathematics, being after all a product of human thought which is independent of experience, is so admirably appropriate to the objects of reality?"

His answer to this enigma was to accept as a fact that mathematics is intrinsic both in nature and in the human mind. Yet, while the human mind can build wonderful logical structures with the aid of this mysterious tool, these structures are not factual if not crosschecked by empirical observations. In another instance he admits facing a mystery: "One may say: 'the eternal mystery of the world is its comprehensibility'." (Einstein, 1936, p. 292).

Eugene Wigner (1960) brings this enigma to the level of the absurd in the title of his paper "The Unreasonable Effectiveness of Mathematics in the Physical Science". His claim is that "The miracle of the appropriateness of the language of mathematics to the formulation of the laws of physics is a wonderful gift which we neither understand nor deserve." This submissive conclusion differs from that of Sir James Jeans (1943) who pointed out that although "Our observational knowledge of the outer world is limited by the aperture of our senses and these form blinkers, which prevent our seeing beyond space and time...Still the events we see in space and time may have their origin outside space and time."

John Archibald Wheeler (1986) elaborates on this issue of comprehension of nature by saying that "If we're ever going to find an element of nature that explains space and time, we surely have to find something that is deeper than space and time — something that itself has no localization in space and time."

These questions and partial answers caused the author to ask the question whether mathematics is not this "seeing beyond space and time", namely a way of movement along a dimensional entity beyond Space-Time, yet intrinsic in a multi-dimensional general continuum.

This engendered a new question: how can a basic dimension be identified and then defined? Research, into the history of the theory of the dimensions of space and time, taught the author that when it comes to the basic dimensions of space and time, their existence cannot be substantiated, except by using their own attributes, or by proving that their non-existence hinders the explanation of observed phenomena. This brought up the premonition that maybe intelligence itself is a dimension. Exploring this direction of thought, the author, by a process of generalization came to the conclusion that the phenomenon of intelligence is a structure on a more abstract and general dimension, which is that of Information (Issar 1995).

3. Information as a Dimension

This dimension is that along which inherited information (DNA) and acquired knowledge are built (acquired instincts, memory). In other words the dimension along which the reader progressed since being an infant, in addition to his progress along the dimension of time, i.e. age, and along the dimension of space, i.e. size.

Analyzing the term 'more knowledge' will point to a process of the acquisition of information from the environment, its processing by collecting, i.e. adding impressions and an 'if-then' logical procedure. This procedure proceeds by imprinting the conclusion in the memory as data. These conclusions are reworked again, once new conclusions arrive, by the logical process of "if-then" to form composite structures of information, i.e. knowledge. This process can be described as movement from a certain point before the information was acquired to that of more knowledge, afterwards. It is argued that this movement is along a dimensional entity, like Space and Time. Thus instead of saying the readers aged, have grown up bodily and acquired more information, one has to say: they moved forwards along Space-Time-Information dimensions.

When it comes to the definition of the dimension of Information, being a basic dimension like space and time, it can only be defined by itself. Thus in order to avoid the philosophical vicious circle of arguments in which classical philosophers got entangled when trying to define Space and Time, the operational definition has to be adopted. Thus space is defined as: 'all that is measured by a ruler' and time is defined as: 'all that is measured by a clock'. The suggested definition of the dimension of information is 'all that is measured by a computer'. As for degrees of freedom of movement along the various dimensions: Space will be referred to as: a dimension with six degrees of freedom (x-y-z forwards and backwards), Time as that with one degree of freedom, namely from the past (through the present, which is the finite smallest portion of time i.e. instant) to future (t_1-t_2-t_3), while the dimension of Information has four degrees of freedom namely addition and subtraction (i.e. $1+1+1=3$ and $3-1-1=1$) and induction and deduction (i.e. 'if-then' and vice versa). Thus any piece of information is composed of notions, which make up by logical induction and deduction concepts, ideas, theories, etc. These can be computerized into bits making sentences, which can be generalized into paragraphs, etc. The magnitude of any theoretical conjecture can then be measured, according to the number of bits and logical sentences, without referring to its 'meaning'.

Elaborating on these definitions we can say the following: once Information is regarded as a dimension, in correlation with Space-Time and taking into account that the smallest finite entities, without direction, on Space and Time are point and instant, then their equivalent on the dimension of information is a notion (or a bit on a computer). A concept (coded as a word) is a few notions put together equivalent and reflecting a certain event on Space-Time. The different arrangements of a few notions (or bits) make a different word presenting a concept. The structuring of a few concepts one on the others, by logical steps, makes a new sentence presenting an idea. The structuring of a few ideas, again by logical steps, makes a theory or a world-view. All these are information-events, or images, which reflect on the dimension of information Space-Time events.

Thus intelligence can be defined as information arranged according to information, namely, as constructs on the dimension of information according to mathematical-logical laws. These are complex sets of notions--bits built by adding and by inductions and deductions, which are, themselves, built on inductions and deductions. Thus, ordered-complexity is increased by the combination of structures (along the arrow starting from notions, to concepts, to sentences, to ideas, to theories, etc.) along the dimension of Information according to mathematical-logical steps. When these laws are not obeyed by empirical observations the world-view is distorted and has to be restructured.

We call events theoretical or conceptual when we refer only to their structure along the dimension of Information (i.e. their measures on Space dimension and Time dimensions in the brain are so small that they can be neglected). Reality is structures composed of ordered sets of quanta of point-instant-notions, built along Space-Time-Information dimensions. Truism is tested by the conformity of the structure built along the dimension of Information with reality, namely 'Space-Time-Information'.

The size of an information-event, i.e. idea (equivalent to a body on the space dimension, and duration on time dimension), is measured by the number of steps taken along the dimension of information. The level of ordered-complexity, or organization of a theory, is a function of the number of logical steps (i.e. acts of addition of notions and their organization into sentences, ideas, etc.) combined to build the more general theory. The more elegant a theory is, the more economical it is in the number of steps and thus in the number of notions or bits of information which were used to reach the same goal (i.e. the straight line between the starting concept and the last conclusion).

It is suggested to adopt a practical physical unit of measurement of consciousness, which will be: the quantity of energy, which has to be invested by an ideal Turing machine to construct a unit logical sentence, like 1+1=2.

A computer program, as well as its software can be described as a map on the dimension of Information. The increase in the complexity of the programs is done by adding structures built on previous structures. The difference between the artificial intelligence of a computer and natural intelligence, namely consciousness of the bio-world, is that artificial intelligence is confined to the stage or level to which it was programmed by an intelligent conscious programmer. It can thus explore areas on the programmed map but not multiply its complexity. The brain on the other hand has the capability to "increase" the number of structures built on previous mental structures, in other words, building new structures as logical conclusions from previous structures, put together.

Evolution of intelligence in the bio-world, i.e. from the first living organism to *Homo sapiens* (which is inter-related with the evolution of the form, i.e. configuration on Space-Time coordinates) can be described as ascending the ladder of ordered-complexity along the dimension of Information. The level of intelligence of the individual is defined as the faculty to produce (and act upon) a program on a higher level of order than that imprinted in his DNA or brain as an instinct. In other words instinctive behavior is the facility to act only according to the program imprinted in the DNA or brain of the individual or genus.

4. Evolution as 'Take off' Against Entropy, Along Space-Time-Information Dimensions

Once the general coordinate system, i.e. Space-Time-Information along which evolution of the bio-world took place was described, and its direction towards ascending ordered-complexity was pointed out, then two questions arise: Why survival? Why ascending evolution?

These questions are sequential as, in the first place, the Darwinian paradigm takes it for granted that survival, in our case staying alive, is the goal of the process of evolution, which starts with random mutations and continues with the selection by the environment of the most adaptable forms. In the second place, this paradigm assumes that this process will result in more evolved forms of higher ordered-complexity, which have better chances to survive. Now these two 'truths', which are taken for granted, are in contrast with a basic physical law, which is the 'The Second Law of Thermodynamics' (SLT). The physicist Erwin

Schrödinger (1944), in his book *What is Life*, noted that life is a violation of the equilibrium in nature dictated by the SLT as it involves systems of higher order, with a higher state of dis-equilibrium with the environment. Thus in order not to violate this law any organism should either die or must, by its metabolic activities, produce greater disorder in the environment. Accordingly "the most important aspect of metabolism is that it represents the cell's way of dealing with all the entropy that it cannot help but produce as it builds its internal order." Schroedinger described this process as 'feeding on negative-entropy.'

In order to explain the term of negative entropy Schroedinger refers to Boltzmann-Maxwell's formulation of The Second Law of Thermodynamics by the equation of: Entropy = k log D (Where k is the Boltzmann constant ($3.2983.10^{-24}$cal/°C) and D the degree of disorder of the system). A living organism behaves, on the other hand, according to the "negative" equation: $-(\text{entropy}) = k \log(1/D)$.

This brings us to Shannon and Weaver's 'Information Theory' (1949). They found that the increase in disorder in the contents of messages sent through any system of communication could be quantified by the above-mentioned Boltzman-Maxwell's equation. The main difference is that in the SLT the constant k is an absolute value, while in a communication system k changes according to the different system and message. Shannon and Weaver, however, referred to 'information' in more general terms than referred to in the present article, namely to the "'information carrying or storage capacity' of any system of communication, like telegraph, telephone and radio." Yet, one cannot avoid the general conclusion that the loss of meaning of a certain message, either just an arrangement of numbers, or a chapter from the *Bible*, is through a process similar to that which determines the increase of entropy in a thermodynamic system. This means that Shannon and Weaver's formulation of the SLT, is it's generalizing to any system containing information. In the terms proposed in the present article it is suggested to see this formulation as referring to structures built along the dimension of Information.

Going back to Schrödinger's definition of life in terms of feeding on negative entropy in order not to die due to the law of "positive" entropy, one may find an analogy between this type of "climb" against, or "fall" into entropy and the movement of a body against the pull of gravity or its fall by following it. Thus when a mass is climbing up, or falls on, the slope of a Space-Time cone, its movement must abide to the laws of gravity dictated by the geometry of Space-

Time. This may bring the observer to conclude that movement in the field of gravity is "designed". In the case of an organism evolving into a more ordered-complex system, its movement must abide the laws of geometry of Space-Time-Information dimensions, which may also bring an observer to conclude that it is according to a design.

Indeed such a design and even an Intelligent Designer were claimed to exist in nature by Michael Behe (1996), who argues that his biochemical investigations on the scale of the cellular world reveal "irreducibly complex systems", which could not function if they were missing just one of their many parts. Thus, a plan, or an algorithm written by an Intelligent Designer, which was aiming to arrive at a certain target function, in this case *Homo sapiens sapiens*, decided the course of the evolution of the bio-world, and not just sheer random play mutations, and their selection according to fitness as claimed by the Darwinian paradigm of evolution.

To the author of the present paper, who is a secular geologist, the finding of such a design is a verification of his theory of a Space-Time-Information continuum deciding the geometry of our universe. This decides the logical pattern of the stages of evolution of the living world, which is movement along Space-Time-Information coordinates feeding on negative entropy.

In conclusion, the appearance of life and its evolution towards more ordered-complex and intelligent forms were inevitable once we take into consideration movement, pushed by the investment of energy (mainly solar) along a universal arrow of negative-entropy. Due to the fact that this movement is along the lattice of Space-Time-Information Dimensions, a spatial-temporal-logical design is inevitable.

5. Pre-Biotic Evolution of Intelligence

Another case in which a design was alleged, this time of an "intelligent observer", applies to what is termed the "Strong Anthropic Principle" which states that "... the Universe (and hence the fundamental parameters on which it depends) must be such as to admit the creation of observers within it at some stage..." (Carter 1974). It is claimed here that instead of speaking of an "intelligent observer", the dimension of Information along which intelligent observation takes place should be introduced. The physical meaning of such a rephrasing is that, at the instant of creation, presumably the Big Bang, the dimension of information came into existence together with space-time. Thus the geometry of the universe is dictated by this multi-dimensional coordinate system, which dictates the geometry along which the universe should spread out after the Big-Bang.

6. Summary

The author of the present article, a geologist, claims that the evolution of intelligence in the bio-world follows a design imprinted already in the pre-biotic universe. This design was dictated by the nature of the dimensions along which the universe has evolved since the Big Bang. Those dimensions, in addition to Space-Time, include also the dimension of Information.

This dimension is that along which inherited information (DNA) and acquired knowledge are built (acquired instincts, memory). In other words, the dimension along which the reader progressed since being an infant, in addition to his progress along the dimension of time, i.e. age and along the dimension of space, i.e. size. The dimension of Information has four degrees of freedom, which comprise the basic logical (mathematical) steps, i.e. 'addition' and 'deduction', 'if-then', both directions.

In order to proceed on the dimension of Information living forms must invest energy. This is because it is a movement opposing the arrow of universal entropy, operating along the coordinates of Space-Time-Information continuum.

References

Behe, M., (1996). *Darwin's Black Box – The Biochemical Challenge to Evolution*, Touchstone, New York.

Bitterman M. E. (1965 a) *Phyletic differences in learning. Amer. Psychol.* **20**:396-410.

Brain, L. (1968) In: L.R. Smythies (ed.) *Brain and Mind.* Routledge and Kegan Paul, London.

Carter B. (1974) Large Number Coincidences and the Anthropic Principle, In: M.S. Longair (ed.) *Confrontation of Cosmological Theories with Observational Data.* D. Reidel, Boston.

Einstein, A. (1921) Geometry and Experience, In: A. Einstein (1954) *Ideas and Opinions*. Crown Publishers, Inc., New York.

Einstein, A. (1936) Physics and Reality, The Journal of Franklin Institute, Vol **221** 3. In: A. Einstein (1954) Ideas and Opinions. Crown Publishers, Inc., New York.

Heisenberg, W. (1971) *Physics and Beyond – Encounters and Conversations*. Harper & Row. New York London.

Issar, A. (1968). Geology of the Central Coastal Plain of Israel. Israel J. Earth-Sci. **17**: 16-29.

Issar, A. S. (1995) *From Primeval Chaos to Infinite Intelligence – Entropy as a Field of Force and Information as a Dimension*. Ashgate-Avebury, UK.

Jeans, J. (1943) *The Background of Science*. Cambridge University Press.

Schrödinger, E., (1944) *What is Life*? Cambridge University Press.

Shannon, C.E. and Weaver, W. (1949) *The Mathematical Theory of Communication*. University of Illinois, Urbana.

Wheeler, J.A. (1986) In: P.C.W. Davies and J. R. Brown (eds.) *The Ghost in the Atom: A Discussion of the Mysteries of Quantum Physics*. Cambridge University Press.

Wigner E. (1960) The Unreasonable Effectiveness of Mathematics in the Physical Science, In: *Communications in Pure and Applied Mathematics*, Vol. **13**, No. I: John Wiley & Sons, Inc., New York.

Dialogue

Josef Svoboda
When I wrote my contribution to this book titled *To Creation via Evolution*, I tried to trace the winding path of human thought on this topic from the early times to the present. It has been a kind of examination 'for my own sake', hopefully useful to others as well. Since this is a rather non-controversial essay, its only intriguing point might be the revelation of the recent unexpected roll-reversal on the part of some evolutionists and creationist thinkers, as expressed in the subtitle of my paper. As for myself, I have been a convinced advocate of evolution yet also a deep believer in the Creator. For me there is no contradiction. The Prime Agent of my faith is independent. He is above and beyond, before and after his creation. His act of creation is immanent, allowing the process to unravel spontaneously with his act of will. He also maintains the created reality in existence and operating through 'rules' we have kept discovering as 'laws of nature'. His possession of the reality and presence in it are absolute. Not only here and there, not now and then, but as expressed succinctly a long time ago: the Creator's possession of reality is total, instantaneous and perfect *(Realitatis tota simul et perfecta possessio!)*. In other words, God need not be busy following every detail along the arrow of time. The extant reality was prescient by him, and predetermined and encoded from the beginning. The Creator cannot be 'caught in action'; neither can he be cheaply dismissed as a human invention and figment of our imagination. 'There is more to the extant reality than meets the eye.' One convincing reason to postulate God's primal involvement is the undisputable fact that evolution of the universe is a clearly directional and meaningful process. This from the moment of its conception and formation of the physical universe — to the emergence of intelligent entities, the humans being the present evolutionary pinnacle at this planet, 13.7 Billion years later. True, this may not be a 'proof' of God's existence but a strong impetus for it. The mystifying enigma of human reasoning is that what is obvious to some is unconvincing, even foolish to others. The staunchest advocates of chance may not see the transparent directionality of the evolutionary progression.

I found the chapter of Prof. Issar: Evolution of Intelligence in the Bio-world, interesting, plausible and challenging. <u>Interesting</u> that it deals with a topic I spent some time to ponder about myself, plausible, that by including a new factor of 'information' into the evolutionary mechanism, the author sees evolution as a non-random process, and challenging that by introducing a new variable, *information,* one runs into a difficulty to 'quantify the unquantifiable'. Please, allow me to elaborate.

A seamless garment of evolution exists from Big Bang to Man and possibly beyond, yet the process can be defined by three 'quantum leaps': emergence of our physical universe (out of nothing?.. something?), emergence of life, and emergence of consciousness (some speak of a cosmic self-awakening). However, the rate of change in time of the inanimate Cosmos and that of the enlivened Biocosmos are diametrically opposite. While at first all had 'happened' within a few seconds and minutes after the Universe's conception (the rest being a rather modest rearrangement of created matter in the expanding space and time), bio-evolution had an extremely slow start, at this planet lingering for four billions of years at a very rudimental level until the onset of the Cambrian era when life's own 'Biological Big Bang' took place. Since then bio-evolution has been expanding exponentially (cf. Svoboda, 2006). Yes, living beings, organisms from amoeba to man, have been feeding on negative entropy. The puzzling question is how this could happen? What made the non-living matter restructure itself to form a thermodynamically open system (a living proto-cell), able to exchange matter and energy with the surroundings to its great benefit?

Professor Issar has the answer. The catalyst, enabling the restructuring of inanimate matter into an energetically functioning living system is information! This triggers a subsequent question: did life pulled itself up from the inanimate brew by its own bootstraps (a physically impossible task) or did it require a 'crane' (help from outside, unacceptable as a scientific hypothesis)? Considering the myriads of chemical reactions running simultaneously even in the simplest cell, to accept the bootstrap theory is a tall order! As if the processes were guided by some cognitive agent. Maxwell's Demon in action! By using mere information, the demon manipulates the trapdoor between two adjacent gaseous systems, decreasing the entropy in one and increasing it in the other, thus violating the second law of thermodynamics. This process, however, to acquire information to manipulate the flow of the gas molecules, plus energy to operate the trap door which must be continually supplied to the demon, costs energy (Jacques Monod: *Chance and Necessity*, 1974).

Professor Issar recognizes the problem and clearly opts for the 'Crane scenario' without explicitly naming it:

> "Due to the fact this movement (evolution towards more ordered complex and intelligent forms) is along the lattice of Space-Time-Information Dimensions, a spatial-temporal-logical design is inevitable."

The author of this commentary complies with the above conclusion without reservation.

For the sake of argument I'll try to tackle some of the "technical" aspects of the paper.

<u>Information as a Dimension</u> Originally the concept of a dimension was a spatial one. Time as a dimension was added later with the Einstein's idea of space-time continuum, and it is already of a different nature (one speaks about apples and oranges in the same sentence). Modern physicists kept adding further 'dimensions' to satisfy the demand of their equations about the nature of matter and ultrastructure of the Universe. The concept itself became less definable and the multidimensional reality unimaginable even to those who introduced the additional dimensions. However, this has been deemed acceptable as long as each conceived dimension represents a new quantifiable aspect of the subject of concern.

Professor Issar assigned information the status of a dimension and built it into the Space-Time-Information equation. In reality, Information is the 5th dimension of living systems. As we accepted earlier above, this dimension is an entity, navigating the organism from conception to maturity, is responsible for a large portion of its preconscious behavior (instincts), and conscious activity such as analyzing this paper and trying to get the zest of it. Compared to space-time the 5th dimension is much more complex, multi-layered (hierarchical) and sophisticated. So, for practical purposes '*an operational definition has to be adopted*'. While space and time can be built of units (point, instant), the basic unit of information is 'notion' (whim, idea, a bit on computer). Here comes a problem. Would a greater number of bits signify greater information? Hardly so. How then to quantify the unquantifiable? Another point: in living systems, even in simple bacteria, the information's dimension (variable or factor) exceeds by orders of magnitude their space-time dimensions upon which it operates. In a human being it is immeasurable, approaching the infinite. Thus the mental construct of Space-Time-Information (individual) would be almost entirely defined by Information, since the biological blueprint is almost identical for all living organisms and decipherable. Maybe it is so. In fact, it must be so, when we accept the notion of information coming extraneously from design.

Prof. Issar asks: Why ascending evolution? To build a more complex system? What for? Structuralization of matter in time is subservient to a purpose, to enable conditions for life to emerge. Individual organisms (and their species) are subservient even to a higher purpose, defined by their role in the ecosystem. The purpose of the largest ecosystem, the Biosphere, is to become a breeding ground for Intelligence and self-conscious mind. The purpose of mind? It is up to each individual to answer this ultimate question.

References

Monod, J. 1974. *Chance and Necessity*. Collins — Fontana Books. Glasgow. 188pp.

Svoboda, J. 2006. Life as an Unfolding Biocosmos. In. J. Seckbach (ed.) *Life as We Know It*. 431-444. Springer. New York.

Biographies

Sandra Ramos-Amézquita is currently studying Sociology at the Department of Social and Political Sciences at the Universidad Nacional Autónoma de México, (UNAM). She recently earned a scholarship to study at the University of Ballarat, Australia. Her interests are mainly interdisciplinary studies; she is trying to associate Sociology with aspects of Biology specially connected to the origin of life. Ms. Ramos is also interested in daily life sociological relationships.

Ninel Valderrama-Negrón is currently studying History at the Department of Philosophy and Humanities at the Universidad Nacional Autónoma de México (UNAM). Her interest is mainly history of Mexico, but she is trying to incorporate the various aspects that include the human studies to fully understand our society. She is especially interest in the history of science and the origin of life.

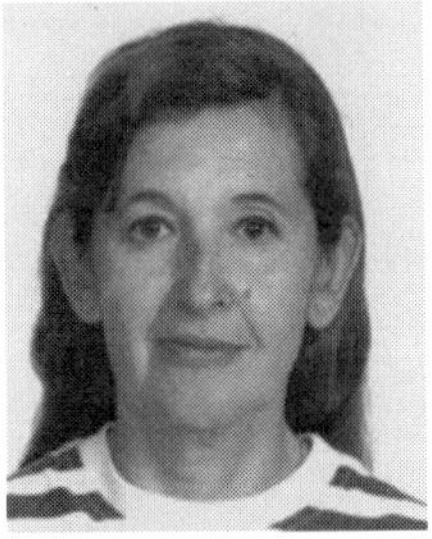

Dr. Alicia Negrón-Mendoza was born in Mexico City. She studied chemistry at the Universidad Nacional Autónoma de México; later she got her Ph.D. under the direction of Professor Ponnamperuma. She is working at the Instituto de Ciencias Nucleares, UNAM in the Chemistry Department. Her research field is related to radiation chemistry applied to problems in chemical evolution.

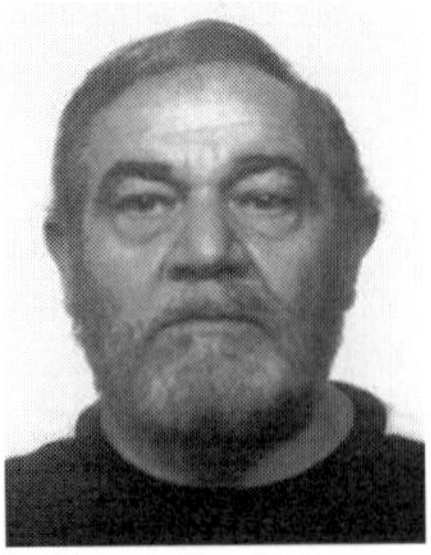

Dr. Sergio Ramos-Bernal is a researcher at "Instituto de Ciencias Nucleares", at the Universidad Nacional Autónoma de México (UNAM). He obtained his Ph.D. in 1974 at the University of Manchester, England. Dr. Ramos-Bernal carries out research in the field of solid state physics connected with problems of prebiotic processes.

43

Creation and Evolution in Harmonic Juxtaposition in Nahua Culture[1]

SANDRA RAMOS-AMÉZQUITA[1], NINEL VALDERRAMA-NEGRÓN[2], ALICIA NEGRÓN-MENDOZA[3,a], AND SERGIO RAMOS-BERNAL[3,b]

[1]*Facultad de Ciencias Políticas y Sociales, Universidad Nacional Autónoma de México, UNAM, México, sra.sandra@gmail.com*
[2]*Facultad de Filosofía y Letras, UNAM, México, ninelvn@gmail.com*
[3]*Instituto de Ciencias Nucleares, UNAM, A.P. 70-543, México, D.F. Mexico*
[a]*negron@nucleares.unam.mx* [b]*ramos@nucleares.unam.mx*

1. Introduction

Five hundred years ago, two very distinct ways of thinking collided upon what was improperly called the "Discovery of the Americas". An exhausted and old Europe faced startled the wonders of an even older world. Ignorance again played a very important role in giving birth to fear and violence; and the possibility of learning and understanding the ways of the 'other' was drowned in terms such as: "barbaric, blood-drinker and inhuman". Eurocentric visions on art, education and spirituality have flooded the world, and have made judgements without a clear understanding of other cultures or the natural surroundings of other populations. In spite of the differences that mark us, however, one thing is clear, and that is that human kind has been puzzled by unknowns that cannot be solved, and most societies have initially approached

[1] The reference notes were translated by the authors.

these issues by creating myths that answer precisely to their needs and surroundings. This and no other is the case of the question of the origin of life in most cultures around the world.

In Western cultures, the term "creationism" and its philosophy is closely linked to the historic development of the religions, showing its earlier roots in Judaism. Its primary bibliographical source is the ancient Hebrew text of the "Genesis". The term itself, "creationism", was not born until the late 19th century as a reaction to the emerging "evolutionist" ideas. Creationists see themselves as the philosophical and religious offspring of the traditions portrayed in the sacred scriptures. Most people in Europe, the Middle East and the Islamic world areas (Jews, Christians and Muslims) believe in the eternal existence of a Supreme Being and its influence on the creation of the world as we know it. Such a belief is based upon the veracity of the sacred scriptures, namely the Old and New Testament and the Koran, which have not necessarily been subjected to systematic or scientific inquiry. Islamic scholars have preserved the ancient Greek texts and developed their ideas based on them. This eventually led the world into what was known as the Renaissance during which a questioning of the biblical cosmology took place. Centuries later, the Enlightenment brought a variety of scientific and philosophical movements that challenged traditional viewpoints in Europe and the Americas. What later became known as Natural History arose with the clear objective of understanding God's plan. However, contradictions were found, and interpreted as science supporting evolution. In the mean time, clerical naturalists sought explanations compatible with their interpretations of the biblical texts. These explanations would latter become creationist arguments.

While the concept of an Ancient Earth was widely accepted, Darwin's theory of natural selection directly challenged the notion of God's immediate involvement in creating species. In response, Creationism arose as a movement that intended to reassert the literal accuracy of the sacred texts, particularly Genesis.

On the other hand, the Nahua culture placed much of its attention and curiosity around the unknowns of the origin of the universe. Their quest for finding answers to such a question is portrayed in the concept of *Omeyoalt*, in which a dynamic view on how the world evolves is developed through a structure of thought where the concepts of space and time are intertwined. This relationship is constructed much in the same manner as the Relativity theory, which has proven to be a step ahead from the Cartesian static world. Within this universe, every object has not only a place but also a moment in time. The sophistication of their belief system was congruent to both mundane and religious activities. The sacrificial practice, far from being barbaric, for them appeared to be flooded with their cosmologic notions. This characterized a highly vital world conceived by Nahuas, and allowed them to understand the universe as in constant movement. Such notion permits the understanding of the formation of life without the need of a new myth. Nahuas did not conceive absolutes, but instead proposed a point of reference that made sense to their interpretation. The introduction of dynamic parameters gave Nahuas an intuition that allowed them to understand life's formation in a manner that can be interpreted as an evolutionary conception of the universe. In the Nahua vision of the world an implicit concept of evolution coexist with that of creation, not in a controversial manner, but in harmonic juxtaposition.

It is important to introduce the reader to this culture in order to comprehend a different way of thinking, and to avoid the usual judgment that marks an ancient and highly developed culture as barbaric and violent. If we can succeed in this, then maybe we can incorporate some of their notions to provide new dynamism to old discussions such as the controversy between creation and evolution.

In Mesoamerican cultures, the origin of the universe and life was fundamental and are common subjects of research that reached its highest philosophical form in the "codices" and prehispanic art. In these documents, men expressed their thoughts, concerns and fears of the world, nature and life. For these cultures, understanding the cosmological vision and general ideas of the macro and micro cosmos has been one of the primary philosophical issues. In this research, we analyzed the concept of the origin of life from the Nahua perspective.

1.1. *Who are the Nahuas?*

The Nahua culture flourished in the central Plateau of Mexico, which was a favorable location for agriculture between the 14th century and the beginning of the 16th century AD (Durán, 1984). The Nahuas are a multiethnic culture that shares the same language, social structure, tradition, and philosophical cosmogony[2]. In ancient Mexico, the most known of these groups were the

---

[2] Nowadays, Nahuas are ethnic groups of México, and some of their philosophical ideas remain. For that reason their ideas are placed in present tense.

Aztecs (from 1325 to 1521 AD) because they were the last inheritors of this tradition, who were conquered by the Spaniards in 1521 AD.

The historical background characterizes Nahuas as a structured theocratic society that relied on natural cycles for agriculture. For Nahuas, the need of establishing a concrete order was of primary importance. In this research, we outline this fundamental idea to explain their concept of the origin of life. Nourishment shaped their line of thought creating a cosmogony based on agriculture, therefore, the physical cycles that ruled agriculture established a specific cosmic arrangement. The natural progressive pattern of the agricultural cycles defined their philosophy and daily life. Human activities such as harvest became intrinsically linked to religious festivities and rituals.

The high social and technological development allowed Nahua culture to have labor divisions, these divisions created the elite which held specialized jobs. A group named *tlamatini*, which means the one who has the knowledge, existed within this social division. The *tlamatini* were civil employees of the central power who made the function of a selective recovery of the past to justify the dominant sector. For these groups, the maintenance of hierarchized social structures was necessary, so for that reason, the meaning of order has a fundamental importance. The *tlamatini* looked for answers about the concept of life by empirical observation of the world. Many of these teachings became what we know today as myths, which are literary representations filled with metaphors and allegorical images.

2. The Understanding of the Myth

The function of myth is to organize the past and transmit it in a coherent structure. This mythical thought tries to explain the surroundings. Therefore, the most important feature of a myth is that it constitutes an alternative form of thinking. The historical and social background of the society shapes their myths (Soustelle, 1982). This background makes the difference between one social myth and another. A myth is filled with the values, beliefs, and social rules of a culture, so it can be said that it is a synthesis of all the Nahua's ideas.

3. Nahua's Cosmogonic Myth

As many cultures, Nahuas developed different characteristics due to their historical and geographical circumstances. For this culture, the first attempt to understand the world was to organize it in a specific order. This order is necessary to explain the existence of life and the universe. The cosmogonic myth gives a coherent disposition of an order, which guarantees that each element will be placed in a specific location. This disposition allows the universe to have a certain stability. The establishment of an order is manifested in all aspects of their life. The world of order begins with the cosmic one and reproduces itself in a sociopolitical order (Florescano, 1999, 2001). In this frame, the religious, political, and social activities are a reflection of the cosmic construction. "*Every creation is a repetition of the origin of the cosmos and all of it is in this way transformed in a sacred space ruled by the same initial force*" (Florescano, 2001).

3.1. *Order*

To understand fully the Nahuas cosmogony, it is important to keep in mind the concept of order. We understand order as the disposition of things in the specific place they belong in the human mind. The order allows us to have a schema to understand and comprehend the universe. For Nahuas, the concept of order gives structure and meaning to their world.

3.1.1. *Omeyoalt or "the unity of the multiple"*

Human beings, in their attempt to comprehend their world, have created several philosophical ideas that have helped structure and organize it. Their most important concern is to understand life's daily events. To accomplish this, man defines and shapes his perceived world, so it becomes the object of his studies. This perceived world is what Western philosophical tradition has named "the universe" or "reality". The term *Omeyoalt*, meaning "the endower of life", is the equivalent concept of universe or totality in the Nahua culture; because it is considered "the unity of the whole", since it summarizes the entire Nahua reality.

Omeyoalt has been defined by a series of elements that are organized through dualities. The dualities or oppositions are the divisions that form *Omeyoalt.* The entire representation of reality is captured within this term. Due to these dual characteristics, *Omeyoalt* is considered the creator of all that exists. For Nahuas, the entire reality was understood as a ramification from its wholeness. *Omeyoalt* is a metaphysical totality — beyond time and space — that incorporates everything. *Omeyoalt* embodies the whole, but at the same time divides it. In addition, *Omeyoalt* can be in all space at the same time, so it is a multipresent concept. Therefore, the Nahuas explained the maintenance of the world by this feature. *Omeyoalt*, as the origin of the universe, is also linked with the creation of the human being. *"In him, [Omeyoalt] we find the supreme explanation of cosmic generation and conception that gave origin to the universe and that constitutes the same being of Omeyoalt"* (León Portilla, 1956).

3.1.2. *Space and time divisions*

Once the meaning granted to the wholeness is understood, the ramification or parts that originate from it need to be described. The two basic concepts used by Nahuas to divide and give meaning to reality are space and time:

Space

The Nahuas visualized space as a division of Cartesian coordinates. The main division represents a universe pictured with horizontal and vertical planes. With these segmentations, Nahuas were able to place an object in their corresponding location. Within this main division (vertical and horizontal sections), we find subdivisions arranged in the following disposition:

The vertical plane is formed by an upper and lower level, also known as *supraworld* and *underworld.* These divisions established the beginning of the geometric order, since they are the first concrete representation of duality.

The *supraworld* is stratified in nine or thirteen hierarchical levels. Each level is characterized with a cosmic element, either profane or divine. "*The observation of cosmic bodies and their corresponding movement gave a foundation to establish different heavens, which are collected in the Vatican Codex A3738*" (Matos Moctezuma, 1998). The arrangement of these bodies in a confined level is not random, rather it has its origin in astronomical observation, which is to say, an empirically based-knowledge.

The *underworld* is formed with nine levels. The last one is known as *Mitlan*, the place where the dead reside (Westheim, 1985). This world is closely related to Mother Earth. It is characterized as a dark and humid space, also related to feminine features, such as fertility. In addition, the concept of life in the Nahua's cosmogony is always linked to death since life/death belongs to the same cycle (López-Austin, 1994).

> "… the nine steps that an individual that dies of a natural death leads to the return to the womb [Mother Earth] from which life was created […] The *Mexicas* [Aztecs], as well as other cultures, knew that menstruation stopped in nine times [nine months] signal that the woman is pregnant. These events culminated with the birth of a child. The interior of the womb was a dark place without windows; a similar description was assigned to the *Mictlan*. If we summarize our proposal, we can see that the dark, the water, and the position of the individual [fetal position] come from an anatomical and physiological knowledge that related life and death in a substantial manner: the road to life, that stops the bleeding in nine times, will be the backward road to become part of the big Mother Earth womb, which is the Earth" (Matos Moctezuma, 1998).

Horizontal plane. The horizontal division separates the universe into four sections that part from the center. Although there are several interpretations, most authors agree that this division is a symbolization of the four seasons. Our interpretation considers them as metaphors and symbols of natural phenomenon that where transformed into abstract terms. As mentioned previously, agriculture is the basic means of maintenance in the Nahua society. The observation of the movement of the sun based in the seasons of the year establishes the cardinal points.

Time

The concept of time is a necessary element in the mind's organization of the universe. The western idea of time is represented with a timeline. Time is always present for man to walk on it. This time has been, and will be there in a division of past, present, and future. The Nahua concept of time is different from the present linear concept because is linked with the existence of man, because man does not move through time, instead man has the ability to construct it. The Nahua's idea of time encompasses three kinds of time: original, cyclic, and circular. Although these types of time have their own characteristics, they all share the fact that time cannot be seen as a division of the past, present, and future. In the Nahua's time, all is part of the same unity, because the past is seen as an affiliation of the present, and the future is seen as destiny. Time is conceived with an inclusive preservation of the whole, including the past. Time is a unity that continuously grows, as it becomes part of the future.

1) The original time is considered sacred time and seen as the initial moment. In this concept, the arrangement of the cosmos is done, because it is seen as an archetypical time taken as a point of reference for the general disposition of the universe. The over-all significance of the original time can be found in its wearing out due to the cyclic time that would follow.

2) Cyclic time is based on the continuous phenomenon of creation, destruction, and re-creation. This type of time destroys the original time, but subsequently rebuilds it in five different occasions that constitute five geological eras. While this type of time is not perpetual, the duality of creation/destruction allows it to prevail and regenerate itself.

3) Circular time is so broad that the Nahuas conceived it as perpetual. This type of time incorporates the original and the cyclic time into its circular schema. This time is of *long durée*, so it includes the beginning (original time) and it is more similar to global re-creation than an end itself. "*The Mesoamerican societies [...] thought that in the infinite cyclic sequence (cyclic time) these would repeat themselves again after a long period of time*"[3] (Florescano, 2001).

In addition, time is seen as a continuous process of absolute innovation. The original moment is significant, but not as a creation; it is seen as a foundation or support of the universe. For the Nahuas, time is not important as an evolving series of events, rather as an organizing element to understand the sacred order of the universe.

---

[3] In the Florentine Codex, we can trace some ideas of the circular time "*Again it will be like this, again the things will be like this, in a time in a place. The things that were done long ago will be done again, again it will be as if was in past times. Them, who now live, will live again, and will be.*"

4. "Myth of the Suns" or the Explanation of the Eras of Time

The Nahua creation myth, known as the myth of the suns (Sahagún, 1989), has different versions. In the following we present the general ideas and their corresponding interpretation. The myth of the suns tries to unveil the concept of movement composed with dialectic characteristics[4] to find stability. *Omeyoalt*, the wholeness, produces a duality or ramifications that are symbolized as its "suns", named *Quetzalcoalt* and *Tezcatlipoca*[5] (Sejourné, 1957). They engage in a constant struggle that results in cosmic movement. From this struggle different cosmic eras, known as "suns" are created. Each era is produced when a certain kind of stability is reached due to a temporary supremacy of one of *Omeyoalt's* suns. This rivalry promotes the cyclical dialectic movement of creation/destruction. This way we can see that the balance and unbalance of the suns complements the dichotomy creation/destruction, which is the main component of cyclic time.

There are five eras[6] that are the concrete materialization of the cyclic type of time. Each one is constituted with a particular element (water, earth, wind, and fire), but each era will eventually be destroyed when an alteration of the equilibrium causes the excess of the particular element that originates it.

Therefore, the ideological aspects that convey this myth are important for the general understanding of the Nahua cosmogony. "*In the myth of the suns, we can find a series of elements that concern the Nahua's concept of the universe, its origin, its evolution, the dialectic of divine forces, and the place of the human individual in it*" (Moreno de los Arcos, 1953).

---

[4] We understand dialectic characteristic as a fluid transition from one state to another.

[5] At the same time, there is another ramification of *Tecaztlipoca* into three individuals symbolized by the blue, black, and red color. Together, the three *Tezcatlipocas* and *Quetzalcoalt* (symbolized by white) are the representations of the four sections of the universe stated in the horizontal division.

[6] The different worlds or eras were named "Sun of water", "Sun of earth", "Sun of fire", "Sun of wind", and finally the "Fifth sun" or *Ollin* that signifies "Sun of movement". Each "sun" is characterized by certain types of vegetation, animals, and a distinct kind of human being.

Among the many interpretations given to the myth of the suns, León-Portilla's contribution (León-Portilla, 1956) is remarkable because he is able to identify certain cosmological categories within the myth of the suns. We agree with the following cosmological categories:

The logical need for a universal foundation

The universal foundation is intrinsically linked with the original time, since it is described as a prototype. For the Nahuas, it is necessary to find a universal support that functions as the origin and maintenance of the world. In this framework, the idea of a universal foundation corresponds to their concept of truth, because it is perceived as the root or support of their ideas.

The serialization of the world in ages or cycles

Time is a necessary element to structure the universe conceived by the Nahuas. The concept of serialization of eras is related with movement, as said before it is a cyclic movement.

Distribution of universal space into four sections

Distribution of space is a mental organization, when space is divided and structured it acquires a certain meaning. Nahuas found this meaning by the astronomical observance of natural phenomenon, and they divided space according to its movement. When the universe obtains a specific meaning, men can shape, transform, and adapt it to the culture's social beliefs. An example is the horizontal distribution that is based on the four seasons, but in the Nahua culture it stands as the four sections of the world.

The action of "fighting" as a model to conceive the cosmic phenomenon

The establishment of the eras or worlds called "suns" represents the equilibrium of the whole system. The struggle represents the imbalance of the parts, which generates destruction in the cyclic phenomenon of creation/destruction. Nevertheless, this destruction is necessary for the continuation of cosmic occurrences. Inside Nahua's cosmogony, the world is made up by the complement of its parts seen as dual concepts (resulting from linguistic constructions); hence, for equilibrium to occur it needs a previous imbalance and vice versa. The concept of creation/destruction generates a movement that sustains life.

4.1. *The Movement Inside the Myth of the "suns"*

Once the main features that form the myth of the "suns" are understood, we can focus on the internal movement within the myth. The creation of the "suns" and their following disappearance reflect a materialized version of the Nahua's idea of movement. This movement is characterized by cyclic, spiral, and uprising characteristics. Due to these features, the Nahua universe is not a static place, rather it is dynamic. This dynamism can be found in the dichotomy of creation/destruction of the worlds, that establishes the idea of an internal equilibrium or imbalance. The Nahua's line of thought is based on dual concepts, so each equilibrium will bring an imbalance resulting in a movement that is generated by the continuity of this world arrangement. Hence, it can be established that the origin of the universe is a product of this movement, which is the cause for the creation of the worlds. Therefore, the existence of the life generated in each world is due to this phenomenon.

4.2. *The Nahua's Idea of "Evolution"?*

Within the features of movement in the myth of the "suns", we can find the idea of progress. Movement is defined by León Portilla (1983b) as an eternal cycle that evolves in a rising spiral. It has been mentioned that Nahuas conceived the universe as dynamic, meaning that each age cannot be static. On the contrary, each age has an internal movement. The evolutionary progress from one age to another relates to the concept of dynamic order. In the description of the ages or "suns", we can see that the Nahua's idea of life has the notion of gradual evolution. The food generated in each world evolves into a better kind of nourishment in each era.

> "[...] it must be added, that although the indigenous text does not mention evolution in the creation of better nourishments, this absence is replaced partially by the old testimony of *la Historia de los mexicanos por sus pinturas*, that successively assigned for each of the ages the following forms of maintenance: first, acorns of oak, then 'water corn', next *cincocopi* that is 'something very similar to corn', and finally the genuine corn" (León Portilla, 1983).

This idea demonstrates the progressive improvement of the Nahua's food supply through pictographic representations. As Alfonso Caso mentioned, this evolution also can be seen from a biological perspective:

> "*...the different plants quoted as the humanity's food supply progressively resembled the ideal nourishment of the Mesoamerican Indian*" (Caso 1953).

Biological research has discovered that *cincocopi*, as it is stated in the codex as "something similar to corn", is the predecessor of corn. *Cincocopi* is one of the plants that due to hybridization gave birth to the present corn plant. This way of conceiving the evolution of a plant demonstrates the great achievements reached by the Nahua culture.

5. "The Fifth Sun" or Our Present Time

The Fifth Sun, the time of man, is the name of our present age. It corresponds to the final part of a cyclic time engulfed within a broader circular time. This "sun" represents a total re-creation of the universe, so the foundation of the Earth and life that we now observe is exclusively formed in this age. The Fifth Sun or *Ollin* is the synthesis of the previous four "suns". This synthesis provokes the equilibrium of the elements in this age. The phenomenon of creation/destruction allows the existence of a Fifth Sun, which corresponds to the center of the schema. However, this age, like the previous cycles, is not perpetual; it will be destroyed by earthquakes, so this way, the element "movement" that created it also will be the cause of its destruction. Following the patterns of circular time, this system will start again; since the Fifth Sun is the center of the schema there is no place for a sixth "sun".

The myth tells us that the birth of this "sun" was due to the god's sacrifice in a city named *Teotihuacán*. As a consequence of this sacrifice, an internal movement in this "sun" is produced. The movement of the "sun" becomes a fundamental element for the creation of life.

> "*Inside the cycle, the creation of the sun appears as the primary cause of light, action, and life.... this movement starts a specific order, since the cosmogonic creation clearly establishes that from the sun*

movement the cosmos starts to function in a regular pattern" (Florescano, 2001).

This regular pattern is conceived as an agricultural cycle ruled by the sun movement or seasons.

5.1. *The Profane Influence in the Movement of the Sun*

The movement is represented in the Fifth Sun in the form of energy. Energy flows between the ramifications of *Omeyoalt* and the profane beings. This communication by an energy flow is the symbolic representation of the Nahua's movement of life. The movement of life is materialized with human blood.

6. The Importance of the Human Body in the Nahua Cosmogony

The medium used by a human to interact with the outer world is his body. In this sense, the human body occupies an important function in the human cognitive process. A human being uses his senses to grasp reality, so the understanding of the world starts initially from him. In many cultures, a human being's cosmogony will always be a reflection of man because he is an anthropocentric being. The appearance assigned to the human body is projected in their perception of macro cosmos. Hence, to fully understand the Nahua cosmology it is necessary to take into account the Nahua's physiology studies.

In their cosmogony, the study of the human body is essential. There was an important line of thought that conceived the cosmos from a corporal body projection, and in the same manner, explained human physiology in correspondence with the general processes of the universe (López Austin, 1989).

The Nahua culture reached a high degree of civilization, which developed a division of labor that produced the specialization to certain jobs. One specialized job was the study of the human body that produced, as a consequence, Nahua medicine. The Nahua's contact with the anatomy of the human body was direct because they performed many rituals in which the human body had a primary position. For example, certain sacrifices where performed to extract human hearts. This example expresses the importance of the human body inside their cosmogony. It is important to be aware that these types of ceremonies were in agreement with their cosmogony and the way that they understand their reality.

6.1. *The Significance of the Heart*

The clearest example to show how the Nahuas use certain material means (their body) to endow a specific meaning to the world that surrounds them is the heart. The heart's divisions and its vital action is extremely bond to the way the Nahuas understood the world and structure their cosmogony. The fundamental concepts to develop this hypothesis are the anatomical division of the organ and its functioning. We consider these aspects to be projected in the cosmic vertical and horizontal division. In other words, the importance of the heart in this culture resides as a mode of spaciousness (anatomic division) and movement (associated with the functioning process of the heart).

Space

a) Vertical Division. Nahua medicine identified three main organs: the mind, the liver, and the heart. The mind was known as the first conscience nucleus and was associated with logic processes. The liver was related to emotions and was known as the emotion nucleus. The heart, located in the middle of these two organs, was assigned to both logic and emotional processes.

> "[...] it is a division that goes from the rational (up) to the emotional (down), with a considerable emphasis on center, the confluence where the most valuable functions of human life reside. Even the highest thoughts and emotions that related to the endurance of human life occurred in the heart, and not in the liver or the head" (López Austin, 1989).

The main vertical division in the Nahua's cosmogony (*supraworld* and *underworld*) can be related to the anatomical distribution of these three main organs. The mind corresponds to the *supraworld*, the liver to the *underworld*, and the heart to the Earth's surface. The heart is seen as the synthesis of both levels because it is the point of equilibrium between these two human essences. The *supraworld* was divided into thirteen celestial levels that were associated with the thirteen body orifices conceived by the Nahuas. In these orifices, the essences of the gods penetrated human beings. Nahuas also conceived the *supraworld* as a space with masculine characteristics (López Austin, 1989). The nine levels of the *underworld* were related to the nine months that correspond to pregnancy. The underworld was seen as a space with feminine characteristics because it is visualized as a humid windowless cave similar to a womb. The concept of life was understood as a codependent cycle to death, to such an extent that the former could not work without the latter, because both were part of a cycle that gave continuity to existence (Matos Moctezuma, 1998). The Nahuas based their cosmogony on dichotomies; therefore, the masculine and feminine divisions are extremely important to understand their concept of the origin of universe and life.

b) Horizontal Division. This study focuses on the notion that it is necessary for humans to structure their universe to comprehend reality. This postulate can be studied through the conductive and cognitive processes that occur in human beings. First, the individual grasps a concrete reality, and then he makes the information abstract and processes it. Later, the information obtained is sent to the outside world with new meaning and in a concrete manner (Kovack, 1965). This process can be explained with the following model of understanding:

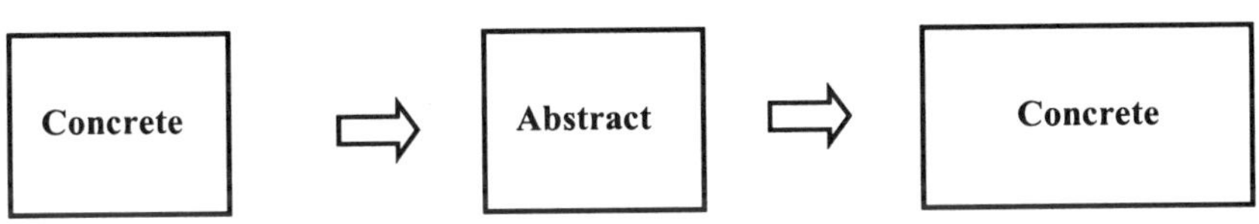

or

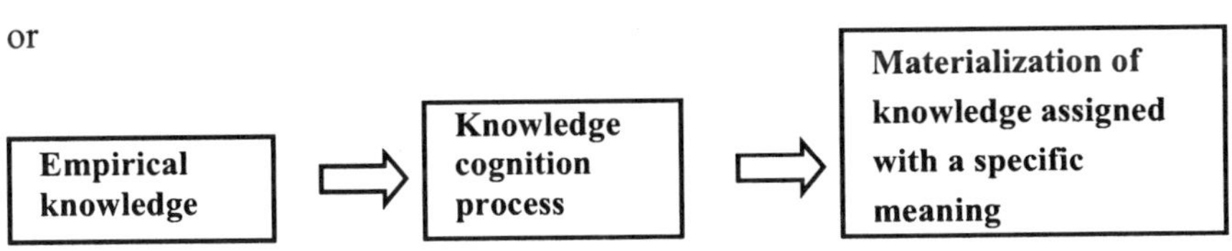

According to the last postulate, the horizontal division of the universe into four sections and the anatomical divisions of the heart are related to some extent. As has been established, the horizontal division of the universe is an abstraction used for the movement of the sun taken from the cardinal points. Thanks to this knowledge, the Nahuas had greater control over agricultural cycles. When the Nahuas completely understood the internal behavior of these cycles, a cultural link was established between the material cycles and ideological abstractions. Therefore, the Nahua cosmogony of the four sections of the universe is essentially an abstract representation of the agricultural cycles.

Surprisingly, the heart's anatomy is very similar to the horizontal division of their universe. The heart has four chambers: two upper thin-walled atria and two lower thick-walled ventricles. The heart is also divided into a right and left side with the septum.

Based on the meaning the Nahuas gave the heart, it is proposed that Nahuas found an intimate relationship between the model of organization of the universe and the anatomic heart divisions. The heart division of atriums and ventricles is similar to the four chambers division found in the universe. When the Nahuas discovered that the anatomy of the heart had the same division as their universe, they used it as an element to generate their cosmologic ideology.

Time

As mentioned previously, time is related to the concept of movement. Several authors agree that the Nahua word for heart, *yollolt*, is etymologically linked with the word *ollin*, which means movement. This linguistic approximation suggests a close relationship between the movement of the heart and the cosmic interpretation of the universe. The heart needs order, equilibrium, and movement to function well. Those characteristics also correspond to the needs for a correct functioning in the Nahua universe.

We must remember that the heart operates by a constant flow of blood divided into two sections. The heart mechanism does not allow the two types of blood to combine, for that reason the heart is called a double pump. The internal movement of the heart is called the cardiac cycle. In this cycle, the relaxation named diastole corresponds to the entrance of the blood. The blood is the pushed out of the heart by a contraction of the atria, which is called systole. The familiar beat of the heart is produced by the contraction/expansion of the organ.

Figure 1: The *Coatlicue*.

There are several interconnected ideas between the heart's functioning and their cosmic representation. Within these aspects, Ruben Bonifaz Nuño (2005) has proposed the thesis that the *Coatlicue*, one of the most famous Aztec sculptures (Fernández, 1959) (Figure 1), represents the condensation of the matter before disposition of the current universe. The Nahuas think that the universe (as indeed everything) came from *Omeyoatl* (symbol of the condensation of matter or "the unity of the whole"). These ideas share a certain parallelism with western ideas, which propose that the universe was generated from the explosion of condensed matter (the Big Bang theory). For that reason, Bonifaz Nuño considers the *Coatlicue* to be the representation of a "Mexican Big Bang". Circular time is essential, because it allows the movement of contraction/expansion of the universe. Everything will return to its initial status, condensation of matter. Therefore, *Omeyoalt* can be artistically represented in the *Coatlicue*. This idea grounded itself in the artistic forms of the sculpture, which has been described as the engulfment of the whole in it. As Bonifaz Nuño said, the horizontal and vertical lines provoke a feeling of intertwined folds which suggest that a great amount of matter is about to explode. In addition, it must be said that Nahuas consider art to be the only road that leads us to the truth.

Nahuas correlated the origin of the universe by a parallelism with thier own bodies, in particular, with their hearts. The intervals of the cycles, like circular time (contraction/expansion), in their conceptions of the origin of the universe, are similar to the cardiac cycle. Symbolic knowledge finds new ways of understanding this world in abstraction forms that have an evolutionary pattern (Luckacs, 1969). In this way, the idea of the heart is the initial motor that allows a new form of understanding the agricultural cycles, and it is simultaneously an affirmation of their ideology. See Figure 2. It can be observed that knowledge evolves in a spiral pattern; it always keeps the original schema of world comprehension, which is based in the agricultural phenomenon.

Human sacrifices are part of a very sophisticated structure of thoughts that are linking with their conception of space, time and surroundings. Their importance is because they reaffirmed concrete ideology expressed in heart anatomy and, at the same time, the culture's economic needs. In summary, the Nahua understand the world using an evolving self-constructed model.

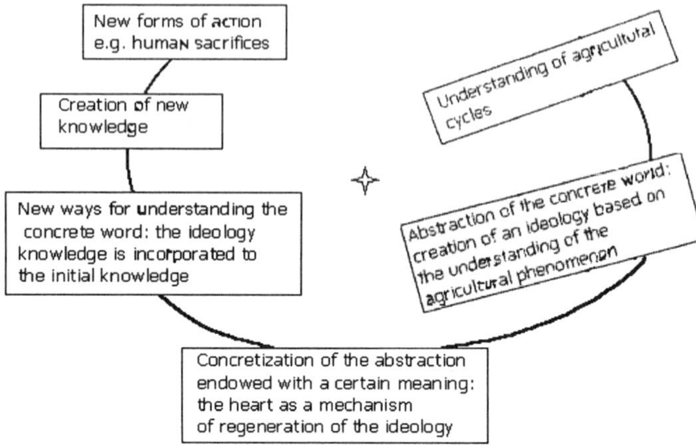

Figure 2: The creation of new knowledge.

On the other hand, there is also a close relationship between the different types of mythical times and the human body:

1) Original time: The relationship between the human body and the original time can be seen in the birth of a human being. For the Nahuas, life was intrinsically linked with death. One could not exist without the other. As mentioned previously, original time is an archetype. Within this time, there is a dual division of the universe into feminine and masculine characteristics. Both parts are necessary for the origin of life in a human being. Before a baby is born, this time is conceived as archetypical. Original time is fragmented with birth and the beginning of cyclic time, also related to death, because the beginning of life and time is the route that will end in death.

2) Cyclic time: The last part of cyclic time corresponds to the time of man that belongs to the Fifth Sun. The time of man is measured based on the material degradation of the human body placed in a cycle of life/death.

3) Circular time: The circular time of *long durèe* associated with the existence of the universe. The contraction/expansion phenomenon related with this time is associated with the cardiac cycle, where the same phenomenon occurs. This phenomenon is the inner movement of the heart, but it is also the movement of the universe, therefore, in its most essential meaning is the movement of life or, in a Nahua word, *ollin yolliztli*.

The heart has a fundamental place in Nahua culture. It is seen as the synthesis, center, and equilibrium where life resides. The heart represents the starting point in the origin of life, because it is the place where energy is in harmony and equilibrium. These characteristics are not only found in the human body, but also transmit its importance to the Nahua universe. The center of human beings and the universe provides enough energy that makes possible the movement interventions that are observed with the different types of time.

In other words, the Nahuas divided the body into two main sections, which are represented as horizontal and vertical planes. The vertical plane divided the body into upper and lower subsections, taking the heart as a reference. The horizontal plane was divided between right and left hemisphere. The horizontal division, the four sections of the universe can be traced in the anatomical division of the heart into its four separated chambers. Therefore, there is a correspondence between the anatomical divisions and the cosmic ones represented in myth. In this way, there is an identical equivalence between the micro and macrocosmos.

7. Materialization of the Cosmogony

The importance of artistic representation inside the Nahua culture comes from their concept of truth. Art is the materialized way to approach the root that supports the entire cosmic system. There are many artistic examples that manifest several aspects of the Nahua cosmogony; the example shown in Figure 3 is very illustrative of how the Nahuas placed their cosmogony. The urban planning of the cities is the most tangible demonstration of their cosmic arrangement taken to a profane level. The urbanization of Mexico-*Tenochtitlan* (capital of the Aztec empire) is example of the form in which the Nahuas impelled a pre-established order in all scopes of human activity. This organization of the city was a reflection of the cosmic displacement in a sociopolitical arrangement. The Nahua cosmogony is composed of two parts that represent symbols that were opposites, but in the Nahua's thought they were conceived as dichotomies. The main temple in the city is considered as the reference point; the vertical division was divided into three sections: *The supraworld*, formed by the body of the temple; the earth surface, represented by the base of the temple; and the *underworld*, that should be underneath the temple. The horizontal division divided the universe into four sections that were seen as main roads that divided the city into four different districts.

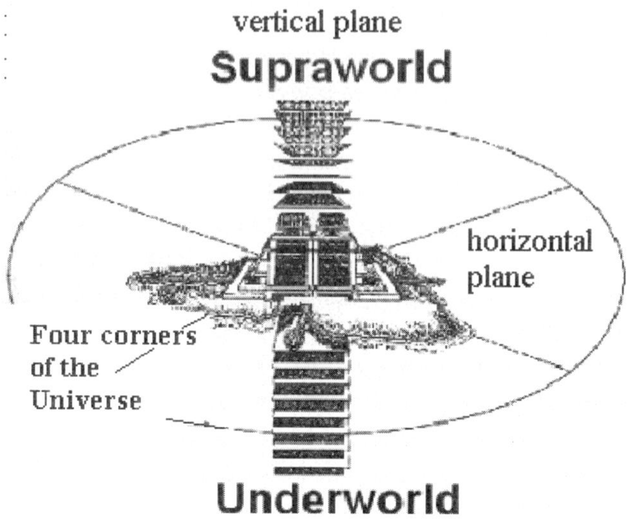

Figure 3: The urban planning of *México Tenochtitlan.*

8. Conclusions

Human kind has been puzzled by unknowns that cannot be solved, and most societies have initially approached these issues by creating myths that answer precisely to their needs and surroundings. This and no other is the case of the question of the origin of life in most cultures around the world.

Nahuas named their universe *Omeyoalt.* When they assigned this name they could grasp its vital space, after this, they established a series of parameters and meanings. Language creates structures of thought that allow man to be able to name his space and even to create it. The term *Omeyoalt* is used by the Nahuas to name the total reality or the unity of the whole. In this way, *Omeyoalt* is used to create parameters. The most concrete use of these parameters is: first, *Omeyoalt* as the concept of truth; second, *Omeyoalt* as space and circular time; and third, *Omeyoalt* as a method. The concept of truth can be understood because *Omeyoalt* constitutes the unity of the whole, so in this way, is the foundation or support in which the totality concentrates. The second, *Omeyoalt* is seen as space. After naming it, the Nahuas organize the total reality by dividing *Omeyoalt* into ramifications or divisions. The horizontal and vertical divisions are used as coordinates that limit and transform their space.

Surprisingly, the construction of the Nahua space is similar to a Cartesian plane, in which every element is organized without leaving anything out of its construction. The creation of certain "coordinates" is a useful mean to enclose elements and processes inside this space. The creation of a common space by the horizontal and vertical divisions transforms this space into a specific parameter.

In their quest for finding answers to the relation between time and space, is portrayed in the concept of Omeyoath, in which a dynamic view on how the world evolves is developed through a structure of thought where the concepts of space and time are intertwined. This relationship is constructed much in the same manner as the Relativity theory, which has proven to be a step ahead from the Cartesian static world. Nahuas not only created a plane to place objects, they also tried to unveil the movement inside this system. Nahuas created three different types of time: the original time is a parameter of movement, the cyclic time is the motor of this movement, and the circular time allows the reproduction of a cosmogonic idea about life in an evolving manner. However, *Omeyoalt* symbolizes the circular time because this time is so broad that it includes the others. *Omeyoalt* is the circular time because it allows the reproduction of movement in the system. In this way, *Omeyoalt* is a fluid stream of energy that allows the existence of life. Vitality of the universe is produced by its movement. In the Nahua culture, the space and time concepts divide and organize their reality.

The key elements to understand the Nahua cosmogony are: duality, movement, and interdependence. The Nahua characterize their universe by transforming it into pairs in a dual form. As seen before, movement is linked with the idea of time. Time is the main element from which life is generated and developed. This concept of time is represented with a rising and spiral movement. The Nahua universe operates with interdependence. Although it divides into pairs, the Nahuas acknowledge that everything cannot work without the sum of the parts. *Omeyoalt* is the unity of the whole, so in this sense, the Nahua "reality" needs to be an interdependent force. The unity of the whole makes the every element linked and united with the others.

In this way, the first dual division we can trace is related with the anatomic division and the cycle of birth and death. The horizontal division corresponds to the economic needs of maintenance (agricultural cycles). From their concrete reality, the Nahuas created abstractions with a broader meaning. This study concludes that the heart is the concrete form were all elements of the cosmogony synthesized and converged. The interdependence of the system makes that the microcosmos; in this case, the heart is reflected in the macrocosmos.

An important contribution of Nahuas is the certainty that human beings cannot have a complete and detailed knowledge of the broader and complex reality. For that reason, they established a point of reference to investigate the universe that surrounds them. A support or "truth" is necessary as a reference to scrutinize life. For the reasons exposited above, Nahuas do not believe that there was a single explanation of life and its origin. For the Nahuas, the method to comprehend the universe was an aesthetic path. Art was the method because the Nahuas conceive that nothing can be absolute. There were only points of reference for their interpretations. Art was used as a mean of addressing knowledge because it was considered as universal language since it is capable of projecting emotions. Art tries to realize of the most inner questions without giving a particular answer.

In the Nahua vision of the world an implicit concept of evolution coexists with that of creation, not in a controversial manner, but in harmonic juxtaposition.

Finally, this research concludes that the inquiry about life is always present in human beings, although, it was establish that the methods or the approaches differ with culture and historical background. Why are the methods of understanding life so different one from another? What makes man believe there is a characteristic type of origin of life? Should we expand our horizon? Should leaning be possible from another perspective?

9. Acknowledgements

We would like to specially thank the following people for their time, effort, patience, and valuable contributions to our study: Adriana Ramirez, Alejandro Ramos, Amoneeta Beckstein, Ana Lopez, Antonieta Lopez, Ma. Concepción Delgado, Cristina García Barragán, Dolores Nieto, Ena Reséndiz, and Gabriela Soní.

References

Bonifaz Nuño, R. (2005) *Cosmogonía Antigua de México*, UNAM, México.

Caso, A. (1953) *El Pueblo Sol*, Fondo de Cultura Economica, México.

Durán, Fray D. (1984) *Historia de las Indias de Nueva España e islas de Tierra*, Porrua, México.

Fernández, J. (1959) *Coatlicue, estética el arte indígena antiguo*, Centro de Estudios Filosóficos, México.

Florescano, E. (1999) *Memoria Indigena*, Taurus, México, D.F., Mexico.

Florescano, E. (2001) *Memoria Méxicana*, Taurus, México, D.F., Mexico.

Kovack, K. (1965) *Dialéctica de lo concreto*, Grijalbo, Mexico, D.F., México.

Leon Portilla, M. (1956) *La Filosofía Nahuatl*, Instituto de Investigaciones Historicas, UNAM, México.

Leon Portilla, M. (1983) *Los Antiguos Mexicanos*, Fondo de Cultura Economica, México.

Lopez Austin, A. (1989) *Cuerpo Humano e Ideología* - Instituto de Investigaciones Antropologicas, UNAM México.

Lopez Austin, A. (1994) *Tamoachan y Tlalocan*, Fondo de Cultura Economica, México.

Luckacs, G. (1969) *¿Qué es el marxismo ortodoxo?*, Historia conciencia de Clase, Grijalbo, México, p.10-80.

Matos Moctezuma, E. (1998) *Vida y Muerte en el Templo Mayor*, Fondo de Cultura Economica, México.

Moreno de los Arcos, R. (1953) *La Leyenda de los Soles*, Estudios de Cultura Nahualt **3,** 131-147.

Sahagún, Fray B. (1989) *Historia General de las Cosas de la Nueva España*, Porrúa, México.

Sejourné, L. (1957) *Pensamiento y religión en el México Antiguo,* Fondo de Cultura Economica, México.

Soustelle, J. (1982) *El universo de los aztecas*, Fondo de Cultura Economica, México.

Westheim, P. (1985) *La Calavera*, Fondo de Cultura Economica, México.

Biography

Dr. Arnie Gotfryd is Chief Environmental Scientist with the Gotfryd Group of consultants in Toronto, conducting environmental impact assessments for government, business, and individuals. He earned Canada's first doctorate in Applied Ecology at the University of Toronto (1985). His scientific work centers on urban wildlife habitats, assessment of observer bias in multivariate ecological analyses, and the health impacts of electromagnetic fields from power lines. He is a popular speaker, author and educator, especially on the interplay of science and faith.

He designed and taught an accredited, undergraduate course called Faith and Science in the Faculty of Arts and Science at the University of Toronto. The course won a Templeton Award of Excellence and was the most popular course at U of T's New College for many years.

Dr. Gotfryd has lectured in over 20 cities on 3 continents and has written 3 books and over 150 articles in peer-reviewed literature, popular science, magazines, and web-based media. His weekly e-newsletter, *Convergence, Exploring the Interplay of Science and Faith* reaches 3,000 subscribers. His book *Mind Over Matter, the Lubavitcher Rebbe on Science, Technology and Medicine* has been acclaimed by the likes of Nobel Physicist Arno Penzias and *Time Magazine*'s scholar of the century, Adin Steinsaltz.

44

Evolution
Myths and Facts

ARNIE GOTFRYD

Gotfryd Group Landscape Architects and Environmental Scientists
3727 Chesswood Rd., Toronto, ON M3J 2P6, Canada
arnie@gotfryd.com

1. Introduction

It is commonly believed that Darwin's Theory of Evolution has disproved the Biblical account of creation in general, and individual creation of species in particular. The result is that many deny the truth of *Torah* because they are convinced that science has vindicated evolution over revelation. However, both the creationists and the evolutionists tend to labor under gross misconceptions about Darwinism and its status within science. The purpose of this essay is to assess Darwin's evolutionary concept so that the rational person, whether layman or scientist, can determine for him/herself if it is tenable as an explanation of species origin.

2. Evolution Myths and Facts

To talk meaningfully about *species* evolving, we need a working definition of species. For practical purposes, we will use a common working definition, even though it only strictly applies to sexually reproducing organisms, and that is: A species is comprised of individuals capable of interbreeding to produce fertile offspring. Thus horses are a species and donkeys are a species but the product of their interbreeding, mules, are not a species for they are not capable of producing fertile offspring.

If we define *evolution* as simply a change in species over time, any student of biology must agree that species do evolve, for it is an often observed fact that many species do change over time. If, on the other hand, we define *evolution* in the Darwinian sense — as a process of *random mutation* and *natural selection* by which all living beings have arisen by chance from single-celled organisms over 100's of millions of years — we may not be on equally firm ground from a scientific perspective.

To explain: *Random mutation* refers to tiny, unpredictable changes in the hereditary qualities of a living being that get passed down to the next generation. If this slight change improves the chances of survival of the offspring, the next generation of that species will have slightly more of this new quality. This is the concept of survival of the fittest, which is equivalent to the term *natural selection*. Darwin argued that the cumulative effect of these small hereditary changes coupled with natural selection should eventually lead to new species and suggested that all species have come into existence from common ancestry in this way.

Introductory classes in Darwinian evolution universally cite the famous example of the peppered moth. Individuals of this British species are either black or white. From the mid-1950's until recently, the accepted story was as follows: The white form was believed to be predominant until the industrial revolution when the black variant gained a selective advantage and became far more common. This was thought to be the case because the soot in the atmosphere from the burning of coal darkened the tree trunks where the moths would rest, rendering the white moths easy prey while the black moths remained well camouflaged. Once industrial processes became more efficient and the trees became lighter in color, the white variety returned to dominance while the frequency of the black variety declined (Kettlewell, 1955).

However, this entire view has been challenged in academia (Wells, 1999), and an investigative journalist, Judith Hooper, has published a widely acclaimed book, "*Of Moths and Men*" (2002) on this very subject. The validity of these arguments is demonstrated by the credentials of the giants of biology who praise the book, including the legendary biologist Ernst Mayr and distinguished geoscientist, Lynn Margulis. But even if we assume that the old peppered moth scenario is correct, there are issues with extrapolating that speciation could occur in this way. First, the peppered moth population started out with both varieties. We cannot use the fact of polymorphism (multiple forms) to prove they arose through mutation. Thus at best we may have a case of natural selection only and hence not a Darwinian evolution. Second, no speciation has occurred here — only changes in the frequency of one variety over the other. And third, even if one of the variants had gone extinct, it would not exemplify Evolution, but rather a kind of *Devolution* since there would be a loss, not a gain in hereditary information.

In any case, microevolution, or *subspecific* evolution, does occur in many species including dogs. All familiar breeds of dogs are actually one species, *Canis familiaris*, whether it's a chihuahua or great dane. That's why breeders are careful about with whom their thoroughbreds mate; they don't want a mutt. But with all the thousands of years of microevolution of dogs through selective breeding, no speciation has occurred, and each type of dog is capable of producing fertile offspring with another. With plants, we see the same phenomenon. For instance, one might think that cabbage, kohlrabi, brussels sprouts, cauliflower, kale and broccoli are different species but in fact it is not so. If you allow them to interpollinate, within a few generations, all the produce will look like cabbage. That's because they are all one species, *Brassica oleracea*, and the changes that growers achieve through selective breeding are microevolutionary, and not speciation.

In some cases, we do find speciation occurring, but never through adding bits of hereditary information. We have seen in both lab and field how some plant and animal species have developed or split into two such that the populations can no longer interbreed to produce fertile offspring (e.g., Callaghan, 1987). However, in none of these cases did this come about through random mutation or any other gradual addition of hereditary information. On the contrary, any genetic additions have been non-evolutionary, through hybridization, polyploidy, etc. Moreover, in most cases cited, the new species come about not through an addition but rather through a *reduction* in the amount of hereditary information. Hence none of the new species lend any credence to the idea that life gradually evolved from simple to complex forms (Spetner, 1996).

In short, in the several centuries that we have been making detailed biological observations, and in thousands of years of selectively breeding plants and animals, we have not seen any Darwinian evolution in the lab, farm or field. That does not mean it could not happen; it just means that we have no direct evidence of it ever having happened.

So what is the scientific status (e.g., Popper, 1965) of Darwinism, or macroevolution, i.e., the idea that all living species evolved from a common origin through random mutation and natural selection? Can we say that it is a fact? Well, if we define 'fact' as that which has been empirically observed, then no. Can we say it is a theory? Well, if a theory is an idea which generates falsifiable hypotheses that can be tested through experimentation, then once again the answer is no. The normative use in science of the term 'theory' involves the necessity to be able to disprove it through experimental observations. We cannot call macroevolution a scientific theory because we cannot go back in time to make the necessary observations that would either support or refute it.

So if macroevolution is not a scientific fact and not a scientific theory, then what is it? It is certainly not rationally compelling in the sense of *deductive reasoning* where one uses syllogisms of the type that A implies B and B implies C and therefore A implies C. These types of proofs are strong logical proofs characteristic of philosophy and mathematics, but not the natural sciences. Science progresses using *inductive reasoning*, that is, rational inferences from what is known or observed to what is not known, or what has not been observed. But within scientific inference, there are stronger and weaker methods (Gotfryd et al., 2003). When one infers from the known to the unknown, it is more reliable to use *interpolation* rather than *extrapolation*. That is, if one has measured a variable quantity at two points, one will be more secure in estimating the situation at some intermediate condition between the measurements than in some state that is beyond the range of observation.

For instance, consider the relationship of temperature and density in water. If we know the density of water at 4°C and 99°C, and then try to predict some other values at other temperatures, we will be tremendously better off interpolating the density between these two temperatures than extrapolating even one or a few degrees outside this range. After all, with one more degree of heat, the water vaporizes and the density crashes, while at the other end, cooler water becomes less dense instead of more dense, an anomaly in all of nature. Besides, just a few degrees cooler yet yields a solid, ice, which unlike any other solid form is actually less dense than its liquid form.

Evolution is based on the weaker inferential method of extrapolation and not the stronger method of interpolation. We scientists have been studying organisms in the lab, field, and fossil record for only two or three centuries, and yet we attempt to make conclusions over 100's of millions of years. These are not modest extrapolations, but very big ones indeed. Within inferences based on extrapolation, we again have two types: *forward* and *backward*. When we extrapolate forward from a known present to an unknown future, our inferences are much more secure than when we use the same means to infer backwards into an unknown past, and especially a distant past. To exemplify *forward extrapolation*, imagine we have two numbers, 2 and 3, which will interact and produce some result. Depending on whether we add, subtract, multiply, divide, take roots or exponents, we will get a small range of possible results based on extrapolation forward from known conditions. If however we *end* with the numbers 2 and 3, and try to *extrapolate backward*, i.e., to determine which numbers have combined and in what way to yield these two numbers, we will be confronted by a truly infinite number of possibilities. Clearly backward extrapolation is a far more uncertain and variable method than forward extrapolation.

Of course any uncertainty over a short period of time will be greatly magnified over a long period. In science, we calculate uncertainty using *confidence intervals*. This is the likelihood and margin of error we attach to our estimates. The farther into time we guesstimate, the larger these confidence intervals become, but not in a linear or gradual way. Indeed it is the tendency of confidence intervals to widen geometrically with linear increase in time. In other words, errors multiply. For example, if doubling the time gives four times the uncertainty; tripling the time will result in nine times the uncertainty, and so on.

All this applies even when environmental conditions are constant. But what happens when the *uniformitarian principle* is violated, i.e., when conditions have been variable over the purported period of study? For example, if we have two substances that when mixed together produce a third, we cannot assume that the rate of production is always the same. It is possible that some *catalyst* has been present in the environment that changes the rate of reaction. Modern chemistry has discovered many such catalysts that can increase reaction rates by thousands of times, even though they are only present in minute amounts. All of the fossil and rock dating techniques rely on the uniformitarian principle and yet every worker in the field believes that it most certainly has been violated in very significant ways, rendering calculations unfathomably vague.

The most common of these methods is carbon dating. This involves comparing the relative amounts of two forms (*isotopes*) of carbon in the fossilized remains.

The idea is that while the organism was alive it had a known amount of each type of carbon but that once it has died, the amount of one type decreases at a known rate through a process of radioactive decay. This would allow the scientist to calculate the age of the fossil. One of the problems with this is that the relative amounts originally in the living organism depend on such environmental factors as temperature, humidity, radiation, and magnetic fields, solar flux, and ambient levels of organic combustion, all of which have been subject to change to an unknown degree in the distant past. Consequently experts continually revise their opinions and frequently disagree about dates with high and low estimates varying by as much as 20 times and more (reviewed in Hanoka, 1987).

Rocks are dated in a similar way using elements other than carbon, and these dates are even more variable. Volcanic rock from 25-50 year-old lava flows of known origin have been analyzed in commercial laboratories with results typically overestimated by a factor of 100,000 (Snelling, 1999). In fact the very same rock dated with different elements, samarium and potassium, have given results that vary by one billion years (Chandler, 1997). Considering that the lower age estimate was 0.7 billion years, the margin of error was even more than the estimated age!

Another issue is that Darwin's Theory of Evolution makes fairly specific predictions about what the *fossil record* should reveal about the history of life on earth. The fossil record is presumed to be like a vertical time line with more recent organisms near the surface and more ancient ones deeper down. In his 1859 book, *Origin of Species*, Darwin predicted that the fossil record will show that 1) species appear gradually, 2) change constantly, 3) disappear gradually, and 4) missing links between major types will be filled in. After some century and a half of digging up fossils all over the world, we now know that all of Darwin's predictions have been refuted: 1) Species appear suddenly, 2) show no significant change, 3) disappear suddenly, and 4) the missing link problem gets more acute instead of more resolved with time. Under these conditions, Darwin himself would have dropped evolution as an explanation for the origin and diversity of life.

Evolutionists have themselves noted these glaring flaws in Darwinian theory and have sought to deal with them in the manner of Stephen J. Gould who has suggested that speciation is a sudden and dramatic event which therefore does not show up in the fossil record (Gould and Eldridge, 1977). Gould himself states that "the fossil record with its abrupt transitions offers no support for gradual change," and then proposed that "macroevolution proceeds by the rare success of these hopeful monsters, not by continuous small changes within populations" (Gould, 1977) It sounds nice, but from a scientific standpoint, the fatal objection to his *punctuated equilibrium* notion is the absolute lack of any conceivable mechanism by which the necessary genetic and organic changes could occur. Gould and Eldridge (*op. cit.*) admitted as much, saying, "No theory of evolutionary mechanisms can be generated directly from paleontological data... we cannot generate new mechanisms."

In addition to all the above, are the unanswered challenges to macroevolution posed by *information theory* and *molecular genetics* (Lewin, 1980; Spetner, 1964, 1968, 1997) in such prestigious journals such as *Science* and the *Journal of Theoretical Biology*. For example, Spetner's calculations, published over 35 years ago, show that billions of years are insufficient to evolve even one new species, yet somehow not one scientist has ever even attempted to refute his arguments in a scientific journal. Spetner (1997) calculates the likelihood of one species evolving from another at no better than $1:10^{2738}$. This is comparable to the probability of every person on the planet entering a daily lottery with over 6 billion tickets and the same person winning every day for a year. At this ratio, even one speciation event would be an impossibility in the eyes of the rational man. How much more so if we were to recreate such an unlikelihood for each and every one of billions of speciation events purported to have happened over the history of the planet.

Many, if not most, leading scientists agree. Royal Society astronomer Sir Frederick Hoyle (1981) says that a tornado generating a jet in a junkyard is more likely than one species evolving from another. Nobel Prize-winning Chemist Harold Urey (1962), famous for his leading role in recreating the building blocks of life from inorganic matter, said "All of us who study the origin of life find that the more we look into it, the more we feel that it is too complex to have evolved anywhere." And Francis Crick (1981), Nobel laureate, father of modern genetics, and discoverer of DNA's helical structure, said, "An honest man, armed with all the knowledge available to us now, could only state that in some sense, the origin of life appears at the moment to be almost a miracle."

Above and beyond all the probabilistic arguments is the biochemical challenge to evolution. When Darwin proposed his theory, no scientist could imagine in his wildest dreams the incredible chemical intricacies underlying every biological process. This posed a new problem for the Darwinists: *irreducible complexity*. This means that if any one of dozens of key elements of a biochemical process would be missing, the entire process would simply shut down. Just as the dysfunction of one small screw could destroy a jetliner, so too one missing chemical can terminate an essential life process such as photosynthesis, respiration, blood clotting, or reproduction (Behe, 1996). This is an impossible outcome for Darwinian evolution. Macroevolution requires a progression of one beneficial mutation after another, with each generation becoming more fit and more developed than the previous one, until more complex organisms evolve from simpler ones. But if an irreducibly complex system of, say, 10 elements is to evolve, then element 1 has to add some fitness, element 2 has to add some fitness, and so on until all the parts are in place. The problem with the complex system is that elements 1, 2, 3… and 9 do not add any survivorship to the species, and there is no natural selection favoring those intermediate stages. On the contrary, they will be selected *against*. Thus irreducibly complex systems cannot evolve into existence, and therefore higher life forms cannot evolve from simpler ones.

3. Summary

The notion that the diversity of life arose through random mutation and natural selection is neither an empirical fact nor a scientific theory, but rather a groundless conjecture based on weak, inferential methods of backward extrapolation through eons of unobserved time over unknown conditions and having known and uncontrollable systematic errors. According to Darwin's own criteria in *Origin of Species*, he himself would have rejected evolution based on today's knowledge of the fossil record. Even the most modern formulations of Darwinian evolution have been shown to be impossible, based on unchallenged statistical models of molecular genetics, as well as the irreducible biochemical complexity of all physiological processes.

All this does not prove that the *Torah* is true or that the Biblical story of creation is true. What it does show is that accepting Darwinian evolution requires a leap of faith that may be more radical and less substantiated than to believe that G-d created the world in six days and on the seventh day He rested.

References

Behe, Michael J. (2006) (2nd revised ed'n) *Darwin's Black Box: The Chemical Challenge to Evolution*. Free Press, NY. 352pp.

Callaghan, C.A. (1987) Instances of observed speciation. *The American Biology Teacher*. **49**:34-36.

Chandler B., (1997) The geophysics of God, in: *US News & World Report* (1997 June 16) **122**: 55-58.

Crick, F.H. (1981) *Life Itself, Its Origin and Nature*. Simon & Schuster, NY. p.88.

Darwin, C. (1859 [new ed. 1998]) *The Origin of Species*. Gramercy Books, London. Chapters 6 and 9.

Gotfryd, A., H. Branover and J. Ginsburgh. (2003) Ch. 2, Theories of Evolution in: *Mind Over Matter: The Lubavitcher Rebbe on Science, Technology and Medicine*. Shamir Books. NY, Jerusalem.

Gould, S.J. (1977) The return of hopeful monsters. *Natural History* **86**:23.

Gould, S.J. and N. Eldredge. (1977) Punctuated Equilibrium: the tempo and mode of evolution reconsidered. *Paleobiology* **3**:115.

Hanoka, Y. (1987) Some thoughts on the age of the world. *Wellsprings*. **4**(1):4-6.

Hooper, J. (2002) *Of Moths and Men: An Evolutionary Tale*. W.W.Norton, NY.

Hoyle, F. and C. Wickramasinghe. (1981) *Evolution from Space*. Dent. London.

Kettlewell, H.B.D. (1955) Selection experiments on industrial melanism in the Lepidoptera. *Heredity*, **9**:323-342.

Lewin, R. (1980) Evolutionary theory under fire: an historic conference in Chicago challenges the four-decade long dominance of the Modern Synthesis. *Science* **210**:883-887.

Popper, K.R. (1965) *Conjectures and Refutations*. Harper and Row, NY.

Snelling, A. (1999) Radioactive 'dating' failure. *Creation* **22**(1):18–21.

Spetner, L.M. (1964) Natural selection: an information-transmission mechanism for evolution. *J. Theor. Biol.* **7**: 412-419.

Spetner, L.M. (1968) Information transmission in evolution. *IEEE Trans on Info Theory IT* **14**(1):1-6.

Spetner, L.M. (1997) *Not By Chance: Shattering the Modern Theory of Evolution*. Judaica Press, NY.

Urey, H. (1962) Interview. *Christian Science Monitor* (Jan. 4, 1962) p.4.

Wells, J. (1999) Second thoughts about peppered moths. *The Scientist*, **13**(11): 13.

Dialogue

Doron Aurbach

I read with interest your article, which provides the usual claims against evolution theory, including the irreducible complexity of living systems and the fossils record that suffers from so many missing links, and does not show evidence to the intermediate species that would be expected to live in the past in light of the Darwinistic theory. However, leading evolutionists such as Profs. F. Ayala or R. Dawkins sound very self confident about Darwin's theory. A latest document issued by the National Academy of Science of the US entitled "Science, evolution & creationism", tries to convince the world that evolution theory becomes stronger and stronger, based on recent findings. You can see the full document at:

Ayala, F.J., B. Alberts, M.R. Berenbaum, B. Carvellas, M.T. Clegg, G.B. Dalrymple, R.M. Hazen, T.M. Horn, N.A. Moran, G.S. Omenn, R.T. Pennock, P.H. Raven, B.A. Schaal, N.D. Tyson & H. Wichman (2008). *Science, Evolution, and Creationism* [http://www.nap.edu/catalog/11876.html], Washington, D.C.: National Academies Press.

Please respond to the main claims of this document, that intends to present the most updated propaganda, the last word of the evolutionists. To what extent do the anti-evolution claims in your article remain valid, after one reviews the most updated claims of the evolutionists, as reflected by the above document?

Arnie Gotfryd

I have read the NAS document and it is essentially a polemic piece which does not address the concerns raised in my article. There are plenty of facts and figures and plenty of interesting arguments but anyone who reads both documents objectively will admit that the NAS piece contains no refutations or disproofs of any of the issues I raised, whatsoever.

Part 9

Conclusion

Biographies

Professor Julian Chela-Flores was born in Caracas, República Bolivariana de Venezuela and studied in the University of London, England, where he obtained his Ph.D. in quantum mechanics (1969). He was a researcher at the Venezuelan Institute for Scientific Research (IVIC) and Professor at Simon Bolivar University (USB), Caracas until his retirement in 1990. During his USB tenure he was Dean of Research for six years. He is a Fellow of The Latin American Academy of Sciences, The Academy of Sciences of the Developing World, the Academy of Creative Endeavors (Moscow) and a Corresponding Member of the Venezuelan "Academia de Fisica, Matematicas y Ciencias Naturales". His current positions are Staff Associate of the Abdus Salam International Center for Theoretical Physics (ICTP), Trieste, Research Associate, Dublin Institute for Advanced Studies (DIAS) and Profesor-Titular, Institute of Advanced Studies (IDEA), Caracas. His particular area of expertise is astrobiology, in which he is the author of numerous papers. He organized a series of Conferences on Chemical Evolution and the Origin of Life from 1992 till 2003. In 2001 he published the book: *The New Science of Astrobiology From Genesis of the Living Cell to Evolution of Intelligent Behavior in the Universe.*

Professor Joseph Seckbach is the initiator and chief editor of **Cellular Origins, Life in Extreme Habitats and Astrobiology (COLE)** series. www.springer.com/sereis/5775. He is the author of several chapters in this series. Dr. Seckbach earned his Ph.D. from the University of Chicago, Chicago, IL (1965). Recently he spent three months in Ludwig Maximilians University in Munich with a DAAD fellowship from the German service of exchange academicians, where several forward steps of this volume have been performed.

Among his publications are books, scientific articles concerning plant ferritin (phytoferritin), cellular evolution, acidothermophilic algae (mainly on *Cyanidium caldarium*), and life in extreme environments. He also edited and translated several popular books. Dr. Seckbach is the co-author (with R. Ikan) of the Chemistry Lexicon (1991, 1999) and other volumes, such as the *Proceedings of Endocytobiology VII Conference* (Freiburg, Germany, 1998) and the *Proceedings of Algae and Extreme Environments Meeting* (Trebon, Czech Republic, 2000) http://www.schweizerbart.de/pubs/books/bo/novahedwig-051012300-desc.ht). His recent interest is in the field of enigmatic microorganisms and life in extreme environments.

45

Divine Action and Evolution by Natural Selection

A Possible and Necessary Dialogue•

JULIAN CHELA-FLORES[1] AND JOSEPH SECKBACH[2]

[1]*The Abdus Salam International Centre for Theoretical Physics, Strada Costiera 11; 34014 Trieste, Italy and Instituto de Estudios Avanzados, Caracas 1015A, Republica Bolivariana de Venezuela*

[2]*Hebrew University of Jerusalém, P.O.Box 1132, Efrat, 90435, Israel*

1. Introduction

This volume deals with the question of faith and reason, specifically concerning the emergence and evolution of life on Earth, one of the important topics of the new science of astrobiology. The same subject is the main aspect of Genesis, the first book of the *Bible*. From the point of view of faith this common objective — to reflect on the emergence of life on Earth from two different cultural points of view — has led in the past to a fecund dialogue between faith and reason. In this work we inquire how to interpret the holy books of the main monotheistic religions against the background of modern science.

• The topics covered in this chapter have been expanded in a forthcoming book by one of the authors (JCF): *A Second Genesis: Stepping Stones Towards the Intelligibility of Nature*. World Scientific Publishers, Singapore.

We face the dialogue between science and religion with the conviction that it will be fruitful, because the search for the truth is a common objective of both of these aspects of culture. We are also convinced of the unity of knowledge and consider that a fractured unity is due to the current excessive specialization that creates a barrier to the eventual integration of science and religion in our society. The lack of unity also interferes with an appropriate communication of science by scientists. Too much specialization has been forced upon us by professional commitments of both science and the humanities. Consequently, contemporary culture is far from the bigger picture that is relevant for a unified knowledge. Avoiding a fractured culture is of vital importance for discussing the question of faith and reason. This is especially relevant for the question of divine action and evolution by natural selection.

Since classical times significant changes have been introduced in our understanding of the constraints that philosophy and natural theology impose on matters related to life on Earth. A prominent question has always been the special position that has been assigned to humans in the universe. Neoplatonism is an ancient philosophical doctrine that was defended by Plotinus (c. 205-270 AD), who was a Roman citizen born in Egypt. He is considered to be its founder and an expositor of the doctrines of Plato. This philosophic school also based his doctrines on Pre-Socratic literature. Platonism achieved a brief moment of medieval renewal advocating a special position for man in the universe. However, science, since its renaissance with Copernicus and Galileo, introduced experimental and observational methods, a different approach from Neoplatonism to the position of man in the universe.

2. Lord Russell's Frontiers of Philosophy, Science and Theology

Science has made considerable progress since Galileo's introduction of the experimental approach for understanding nature. Since Galileo science does not appeal to philosophical and theological conclusions in the formulations of new theories. All conclusions that are inferred from hypothesis based on experimental observations are provisional. They are subject to improvement by evidence from new experiments and by observation. This aspect is its most significant feature. There appears to exist a misunderstanding in relation to a universe evolving for 13.7 billion years since the Big Bang. According to the new science of astrobiology, life emerged some 3 to 4 billion years before the present by an evolutionary process of random mutations and by the mechanism of natural selection. By these means the earliest microorganisms evolved at the end of a process of chemical evolution. The US National Academy of Sciences has reviewed these issues in a very relevant document for scientists who are also believers (NAS, 1999):

> "Many religious persons, including many scientists, hold that God created the universe and the various processes driving physical and biological evolution and that these processes then resulted in the creation of galaxies, our solar system, and life on Earth. This belief, which sometimes is termed 'theistic evolution,' is not in disagreement with scientific explanations of evolution. Indeed, it reflects the remarkable and inspiring character of the physical universe revealed by cosmology, paleontology, molecular biology, and many other scientific disciplines."

Religion has come a long way in its natural-theologic approach to deeper philosophical questions, always basing its considerations on revelation and tradition. Both scientists and humanists are well represented in this book with their worthy reflections. It is understandable that in addressing the phenomenon of the origin and evolution of life in the universe, the ensuing dialogue has not always excluded mutual contradictions. These discussions are necessary to delineate clearly the frontiers of each contribution. In this respect we should not forget Galileo's statement that the Book of Scripture and the Book of Nature speak of the same God. In a modest effort towards aiming at the unity of knowledge we have included many contributions in the preceding pages. We have brought together multiple points of view. Heterogeneous in their approaches from either humanities or science, the chapters have been brought together, even if some of them may have ignored the strict frontiers of either theology or science.

In the present context, one of the most distinguished philosophers of modern times, Bertrand Russell (1872-1970), explained this point particularly well. Lord Russell was an English logician and philosopher, and Nobel Prize winner, whose seminal work in mathematical logic was published in the early Twentieth Century. Russell collaborated with Alfred North Whitehead on *Principia Mathematica* (1910-13).

Russell's interests ranged over a wide spectrum including philosophy, mathematics, science, ethics, sociology, education, and history. And most significantly from the point of view of this book, his interests also extended to the area of religion. He had a personal gift for explaining philosophical and scientific arguments, an unusual ability that led to his 1950 Nobel Prize in literature (Russell, 1991). His main point is that philosophy is something intermediate between theology and science. Like theology it consists of speculations on matters to which definite knowledge has, so far, been unascertainable. But Russell makes it clear that like science, philosophy appeals to reason, rather than tradition, or revelation. He underlines that philosophy is like a "no-man's land" that is approachable from science, and also from theology. What is still lacking is a clear demarcation of the frontiers of science. In other words, the boundaries of the three areas demarcated by Russell are not always maintained. To a large extent our disagreements in this book reflect present-day uncertainties in these boundaries.

3. Saint Augustine's Reading of the Holy Books

The other important topic of faith and reason that is pertinent to our arguments concerns a certain caution of theologians with respect to the questions on which we have dwelt at some length. In the fourth century CE, Saint Augustine of Hippo (354-430) discussed faith and reason in *The City of God.* Augustine raised the point of possible conflicts that may arise from a literal reading of the *Bible* (Augustine, 1984). The *Bible* is not a single book, but rather a library with texts of different nature, but uniformly displaying a deep theological content revealing aspects of God and his divine action. Today, bearing in mind the significant progress in the space sciences, especially in astronomy, Augustine's point can be clearly illustrated with a quotation from the Book of Daniel.

The Book of Daniel is a book in both the Hebrew *Bible* (*Tanakh*) and the *Christian Old Testament.* The book may have been written during Daniel's lifetime in the sixth century BCE or it may have come to us from a version written later. An inspiring poetical analogy with the stars in the firmament comes from chapter 12: 3:

> *And those who are wise shall shine like the brightness of the firmament; and those who turn many to righteousness, like the stars forever and ever.*

According to Augustine's view (and ours), this quotation should be taken strictly as an analogy, and not as scientific information on the longevity of stars, since we know today from a considerable amount of data on supernovas that stars are not eternal, but once their nuclear fuel runs out several scenarios are possible, including supernova explosions. Not only is the message in the Book of Daniel valid today, but it still retains its usefulness after well over two millennia since the prophet's words were written. Daniel's main point is not one of accuracy in astronomy, but rather it is a point that emphasizes the importance of good living and good behavior.

To such questions pertaining to the philosophical area of ethics, no answer can be found in the laboratory, and the prophet's comments are relevant even today. A number of additional discussions on the question of faith and reason are available (Barbour, 1995; Chela-Flores, 1998; Coyne, 1998; Haught, 1998, 2005; McMullin, 2000; Peacock, 1988; Polkinghorne, 1996; Russell, 1995). Besides, previous books of ours have also contributed to these discussions (Seckbach, 2004, 2006; Seckbach et al. 2004). Still today the subject of faith and reason has captured the attention of the general public. We illustrate this assertion with the wide publicity given to the frontier of science and religion in the special issues of leading newspaper and magazines (*Editorial*, 2005; *News*, 2006; *Spiegel*, 2005; *Time Magazine*, 2005; 2006).

4. Einstein's Views on Faith and Reason

Albert Einstein, who together with Charles Darwin, Isaac Newton, and Nicholas Copernicus, is among a handful of scientists who have radically transformed our view of the world, considered that a conflict arises when a religious community insists on the absolute truthfulness of all statements recorded in the *Bible* (Einstein, 1950). He had an impersonal Spinozan view of God, as in the following citation:

> "Spinoza believed that everything that exists is God (a doctrine known as pantheism). But he opposed the view that God is no more than the sum of what exists, since He had infinite qualities. Only two, thought and extension can be perceived by human intelligence. Hence God must also exist in dimensions far beyond those of the visible world (Spinoza, 2002)."

The literal interpretation of the *Bible* means an intervention on the part of religion into the sphere of science. Einstein often spoke about trying to understand how the Lord shaped the universe. Scientists have an analogous faith that the world is comprehensible to reason in terms of a mathematical description of the physical sciences. Einstein concludes this part of his reflections in *Out of My Later Years* with the statement that he cannot conceive of a genuine scientist without that profound faith. It is at this stage that Einstein inserted his often-quoted image:

> "But science can only be created by those who are thoroughly imbued with the aspiration towards truth and understanding. This source of feeling, however, springs form the sphere of religion.... I cannot conceive of a genuine scientist without that profound faith. The situation may be expressed by an image: Science without religion is lame, religion without science is blind."

5. Astrobiology and Rational Bases for Evolution beyond Darwin

The new science of astrobiology covers research in the field of biological aspects of the subjects of the origin, evolution, and distribution of life in the universe. Astrobiology is currently in a period of fast development due to the many space missions that are indeed already in operation or in planning stages (Chela-Flores, 2004).

Among its pioneers a place of honor is undoubtedly assigned to Stanley Miller (1930-2007), and other distinguished organic chemists, including Sidney Fox (1912-1998), John Oro (1923-2004) and Cyril Ponnamperuma (1923-1994). They were pioneers in demonstrating the feasibility of the synthesis of amino acids (which are the building blocks of proteins) in an atmosphere that simulated prebiotic conditions that may have reigned on Earth soon after its formation over four billion years before the present. Miller was only a second-year graduate student at the University of Chicago, when he published a remarkable paper in 1953 on the generation of amino acids. It was a simple experiment that attempted to reproduce conditions similar to those on the early Earth, when life first originated. As the subject of his doctoral thesis Miller demonstrated experimentally that amino acids, the building blocks of the proteins, could be formed without the intervention of man in environmental conditions, which we have called prebiotic — similar to those that presumably were valid at the earliest stages in the evolution of the Earth itself. The corresponding geologic period was the Archean. Miller's work was an important step in the growth of the subject of chemical evolution. The 50th anniversary of the 1953 paper was celebrated with a scientific event in Trieste (Seckbach et al., 2004, where there is further information on the relevance of this singular contribution to the development of astrobiology).

However, even though no single living cell has been formed as yet in the test tube, chemical evolution continues to be a solid scientific pursuit that has been reviewed extensively over the last decade in the Trieste and Caracas series of conferences (Ponnamperuma and Chela-Flores, 1993, 1995; Chela-Flores et al, 2000; 2001; Seckbach et al. 2004).

Going beyond Darwinism strictly within the life sciences, including astrobiology is not new. One illustration is provided by adaptive radiation. This is a process in which one species gives rise to multiple species that exploit different habitats that they may occupy. This is an evolutionary process driven by mutation and natural selection. A well-known example of adaptive radiation as the result of an environmental change is the relatively rapid spread and development of mammalian species after the extinction of the dinosaurs. The speed of the adaptation is measured in time scales that are familiar to geologists. For example, the periods of time that are relevant for adaptive radiation are measured in millions of years. Darwin understood this phenomenon. By studying the work of his contemporaries he brought this phenomenon to the attention of science in his seminal work *The Origin of Species by Means of Natural Selection or the Preservation of Favored Races in the Struggle for Life* (Darwin, 1859).

What is new to us, at the present time when we are approaching the year 2009, the second centenary of Charles Darwin's birth and the 150th anniversary of the publication of *The Origin of Species*, is a phenomenon observed in, for example, fish. A small population of Trinidadian guppies was scooped from a waterfall pool, where predators were abundant. Later they were released upstream in a pool in the presence of only one enemy species. The guppies adjusted to the new environment with few predators by growing bigger, living longer, and having fewer and bigger offspring. Although natural selection was assumed by Darwin to be a slow process, it was found that natural selection is able to act speedily, not in the span of millions of years as in the case of the mammalian radiation that took place about 65 million years ago (Reznick et al., 1997). On the contrary, the guppies adapted to their new environment in a mere 4 years, which is a rate of change some 10,000 to 10 million times faster than the average rates determined from the fossil record (cf. Morell, 1997).

Keeping within the boundaries of science, in the future we could expect that Darwinian theory may eventually have to be improved. However, the evident avenue to follow for scientific progress would be to replace the theory by another one that may explain some of the new data better. One significant example is the following: The rate of change of species is assumed in *The Origin of Species* to be gradual. An alternative approach is that evolution may take place in bursts after long periods when little change takes place. This latter alternative that has been put forward with some support from the fossil record has been called the theory of evolution by punctuated equilibria (cf., for references, Eldredge, 2006). However 'bursts' in this context are short periods of time in a geological context. Darwin, on the other hand, was confronted with evidence suggesting that species are fairly stable entities with distinct beginnings and ends. Darwin's discovery of natural selection led him to believe that evolution must be slow and gradual.

To sum up, if neither Darwinism, nor punctuated equilibrium, eventually were shown to satisfy all the data that might be forthcoming, especially, if the current missions of the space agencies identified a non-terrestrial microorganism, then science will undoubtedly show its traditional strength by challenging biologists to produce a more refined theory for the evolution of life in the Solar System to incorporate such new discoveries. More generally, the question of the universality of biology would have to be discussed (Chela-Flores, 2007).

6. Towards an Understanding of Darwinism within Natural Theology

Religion has had great difficulty in assimilating the real significance of Darwinism. From the point of view of theology the difficulty focuses on how to reconcile evolution with the idea of divine action. It is possible to look at the natural world for explanations with scientific ideas that by the very definition of science must be provisional, namely capable of being discarded by evidence from experiments and observation. This approach to science is in sharp contrast to intelligent design that will be considered in the next section.

There are various other efforts along the lines of trying to understand the significance of Darwinism in the context of natural theology. In another chapter of this book we have brought to the attention of the reader examples of such theologies. For this purpose the concept of kenosis is taken to mean self-emptying and voluntary sacrifice on behalf of others, based on genuine and freely given love for others, and resulting in generosity and respect that flow from it. This approach to natural theology explains the world in terms of Darwinism. It focuses on features of process thought (see: Glossary, Chela-Flores, 2008). This philosophical system is considered to be particularly helpful in the task of constructing an evolutionary theology that may throw some further insights on Darwinism (Haught, 2005).

Another approach along these lines points out that theologians already have the concept of God's continuous creation with which to explore the implications of modern science for religious belief (Coyne, 2005):

> "God is working with the universe. The universe has a certain vitality of its own like a child does. You discipline a child but you try to preserve and enrich the individual character of the child and its own passion for life. A parent must allow the child to grow into adulthood, to come to make its own choices, to go on its own way in life. In such wisdom does God deal with the universe" (Coyne, this volume).

Along the above lines it is possible that evolution may also provide a way in which the tradition of natural theology may undergo a renewal. Instead of focusing on design without a designer, which can be accounted for scientifically in terms of Darwinism (Ayala, 1998), a revived natural theology may take place if we interpret correctly the origin and evolution of life on Earth.

7. Religion, Science and Creationism

In spite of the significant opinions expressed by the theologian Saint Augustine, the philosopher Lord Russell and the scientist Albert Einstein that we have mentioned in this chapter, the literal biblical interpretation of the emergence of life on Earth by a Divine Designer has emerged in recent times. This isolated approach to the emergence of natural events has been called Intelligent Design (ID). The advocates of ID, as opposed to the science of biology, dispute the idea that natural selection fully explains the complexity of life. This represents an unprecedented metaphysical basis for denying the validity of natural selection without providing any significant new data. Besides, ID proponents say that life is so intricate that only a powerful guiding force, or intelligent designer, could have created it. (In fact, science is neutral with respect to this theological statement.) ID does not identify the designer, but ID is one interpretation for God and divine action in the universe that lies well beyond the realm of science. Most significantly, ID also lies well beyond natural theology, as natural theology is, strictly speaking, the body of knowledge about religion that can be obtained by human *reason* alone, *without* appealing to revelation. It is especially relevant to highlight the fact that ID is in no way whatsoever the only possible approach to gaining insights into divine action in a theological context (see the above section: "Towards an understanding of Darwinism within natural theology).

The confrontation between Darwinism and ID has returned to the school curricula, where it has once again caused an embroiled hostility analogous to the 1925 Scopes evolution trial. It was also called the "monkey trial" as a reference to the second aspect of the theory of evolution, besides natural selection, namely the descent from a common ancestor, which, at the time of the publication of *The Origin of Species*, was emphasizing the evolution of primates (and the popular press singled out "monkeys"). The general dialogue at the time did not refer to a last common ancestor that astrobiology has traced back to a microbial organism (by careful arguments from molecular biology). The scientific and religious questions that the Scopes trial raised, unfortunately have not disappeared, as we are constantly reminded by leading journals (News, 2006).

However, we believe that there should not be a real conflict between the holy books and science. There are at least two reasons for an apparent contradiction between these two aspects of culture. Firstly, disagreements may arise from a lack of appreciation of the natural frontier of science, which is strictly an academic activity in which theory is subject to refutation only by appropriate experiments, or accurate observations. Secondly, an apparent conflict between the holy books and science may arise by an analogous lack of appreciation of the frontier of natural theology, as an academic activity in which the body of knowledge about religion can be obtained by human reason alone, without appealing either to revelation or to tradition. Judaism and other monotheistic religions may accept natural selection (cf., the chapters of Halperin and the one of Chela-Flores in this volume).

In this case one interpretation of the holy books is that God may have used the mechanism of evolution to create everything from microorganisms to human beings. Jewish traditions also maintain that God renews and guides his original creation as stated in the Jewish morning prayers (cf. Psalm 136: 4–9; Nehemiah 9: 6).

8. Concluding Remarks

Science and theology have raised similar questions from their very beginnings. It is sufficient to remember that stellar evolution predicts that the lifetime of the Sun will continue for a few billion years, while the science of anthropology suggests that humans are a fairly recent addition to an Earth biota, whose origins must be traced back to microorganisms that emerged on the early Earth from 3 to 4 billion years before the present. We expect that theology will continue its independent search for a deeper understanding of divine action. Science will be confronted with ever increasing mysteries, as our scientific instruments become better and more accurate to allow confrontation of theory with experiment. The questions on how the universe and life in it were formed will reveal new insights. Some questions escape the scope of science. We expect that future developments in philosophy and theology will gradually give us better insights into the question that was raised by the philosopher Gottfried Wilhelm Leibniz:

> "*Why is there something rather than nothing?*" (Leibniz, 1714).

We are convinced that in the future the question of the proper place of Darwinism in science, and its teaching in colleges, will be settled to the satisfaction of society in general. This will take place without excluding the questions of faith that many scientists accept, including both authors of this chapter. It is of paramount importance for non-scientists to understand that we do not consider science as the ultimate truth, or a cultural sector that has the monopoly on wisdom. Science is an activity whose main strength is that it invites arguments to improve itself. But such improvements are not implemented by acts of faith, or by authority. Progress is achieved with accurate measurements and new observations. This, in turn, will lead to more comprehensive theories that are themselves open to eventual improvement.

Acknowledgement

We thank Fern Seckbach for proofreading and presenting constructive points for this chapter.

References

Augustine of Hippo (1984). *Concerning the City of God against the Pagans,* Augustine of Hippo (1984). *Concerning the City of God against the Pagans,* Penguin Classics, London (cf. Book XVII, p. 4).

Ayala, F.J. (1998). Darwin's Devolution: Design without Designer, in: R.J. Russell, W. R. Stoeger, S.J. and F.J. Ayala, (eds.) *Evolutionary and Molecular Biology: Scientific Perspectives on Divine Action***,** Vatican City State/Berkeley, California: Vatican Observatory and the Center for Theology and the Natural Sciences. pp. 101-116.

Barbour, I.G. (1995). Ways of relating science and theology, in: R.J. Russell, W.R. Stoeger, SJ and G.V. Coyne, S.J (eds.), *Physics, Philosophy and Theology. A common quest for understanding,* 2nd. ed., Vatican Observatory Foundation, Vatican City State, pp. 21-48.

Chela-Flores, J. (1998). The Phenomenon of the Eukaryotic Cell, in: R.J. Russell. W. R. Stoeger, S.J. and F. J. Ayala, (eds.) *Evolutionary and Molecular Biology: Scientific Perspectives on Divine Action*, Vatican City State/Berkeley, California: Vatican Observatory and CTNS, pp. 79-99. http://www.ictp.trieste.it/~chelaf/ss20.html

Chela-Flores, J. (2004). *The New Science of Astrobiology From Genesis of the Living Cell to Evolution of Intelligent Behavior in the Universe.* Series: Cellular Origin, Life in Extreme Habitats and Astrobiology, vol. 3 Kluwer Academic Publishers: Dordrecht, The Netherlands, 251 p., softcover edition of the 2001 book, ISBN: 1-4020-2229-8 http://www.springeronline.com/sgw/cda/frontpage/0,11855,5-198-72-33595256-0,00.html

Chela-Flores, J. (2007). *Testing the universality of biology. International Journal of Astrobiology*, **6** (3): 241-248. (Cambridge University Press), http://www.ictp.it/~chelaf/ss116.html.

Chela-Flores, J. (2008). *Astrobiological reflections on faith and reason The issues of agnosticism, relativism and natural selection.* (In this volume.)

Chela-Flores, J., Lemarchand, G.A. and Oro, J. (2000). *Astrobiology: Origins from the Big Bang to Civilisation*. Kluwer Academic Publishers: Dordrecht, The Netherlands.

Chela-Flores, J, Owen, T. and Raulin, F. (2001). *The First Steps of Life in the Universe.* Kluwer Academic Publishers: Dordrecht, The Netherlands.

Coyne, G.V.S.J. (1998) The concept of matter and materialism in the origin and evolution of life, in: J. Chela-Flores, and F. Raulin (eds.), *Chemical Evolution: Exobiology. Matter, Energy, and Information in the Origin and Evolution of Life in the Universe*, Kluwer Academic Publishers, Dordrecht, pp. 71-80.

Coyne, G. V. SJ. (2005). God's chance creation, *The Tablet*, 6 August. http://www.thetablet.co.uk/cgi-bin/register.cgi/tablet-01063

Darwin, C. (1859). *The Origin of Species by Means of Natural Selection or the Preservation of Favored Races in the Struggle for Life,* John Murray/Penguin Books, London, 1968, p. 455.

Editorial (2005). Keeping religion out of science class, *Nature* **436**, 753.

Einstein, A. (1950). *Out of My Later Years*. Philosophical Library, New York, p. 25.

Eldredge, N. (2006). *Darwin Discovering the Tree of Life*. Norton, New York.

Haught, J.F. (1998). Darwin's gift to theology, in: R.J. Russell, W.R. Stoeger, SJ and F.J. Ayala, (eds.) *Evolutionary and Molecular Biology: Scientific Perspectives on Divine Action*, Vatican City State/Berkeley, California: Vatican Observatory and CTNS, pp. 393-418.

Haught, J.F. (2005). The Boyle Lecture 2003: Darwin, design and the promise of nature, *Science & Christian Belief*, **17** (1), 5-20.

Leibniz, G. (1714). *The Principles of Nature and Grace, Based on Reason*, in: L. Loemker (ed.) Philosophical Papers and Letters, 1969, Dordrecht: D. Reidel, pp. 636-42.

McMullin, E. (2000). Life and Intelligence far From Earth: Formulating Theological Issues, in: S. Dick (ed) *In Many Worlds*: *The New Universe, Extraterrestrial Life and the Theological Implications*, Philadelphia: Templeton Foundation Press.

Morell, V. (1997). Evolution: predator-free guppies take an evolutionary leap forward, *Science* Vol. **275**. no. 5308, p. 1880.

NAS (1999). *Science and Creationism: A View from the National Academy of Sciences*, Second Edition, National Academy of Sciences, pp. 35.

News (2006). Anti-evolutionists raise their profile in Europe, *Nature* **444,** 406-407.

Russell, B. (1991). *History of Western Philosophy and its Connection with Political and Social Circumstances from the Earliest Times to the Present Day*, Routledge, London, p. 3.

Russell, B. (1995). *Religion and Science*, New York: Dover.

Peacocke, A. (1988). Biological evolution — a positive theological appraisal, in: R.J. Russell, W.R. Stoeger and F.J. Ayala (eds.) *Evolutionary and Molecular Biology: Scientific Perspectives on Divine Action.* Vatican City State/Berkeley, California: Vatican Observatory and the Center for Theology and the Natural Sciences, pp. 357-376.

Polkinghorne, J. (1996). *Scientists as Theologians.* SPCK: London.

Ponnamperuma, C. and Chela-Flores, J. (eds.) (1993). *Chemical Evolution: Origin of Life* A. Deepak Publishing, Vol. **135**: Hampton, Virginia, USA.

Ponnamperuma, C. and Chela-Flores, J. (eds.) (1995). *Chemical Evolution: The Structure and Model of the First Cell.* Kluwer Academic Publishers: Dordrecht, The Netherlands.

Reznick, D.N., Shaw, F.H., Rodd, F.H., and Shaw, R.G. (1997). Evaluation of the Rate of Evolution in Natural Populations of Guppies (Poecilia reticulata), *Science* 275 1934-1937.

Seckbach, J. (ed.) (2004). *Origins: Genesis, Evolution and the Biodiversity of Life*, in the series of: *Cellular Origin, Life in Extreme Habitats and Astrobiology*, 6, Kluwer, Dordrecht, The Netherlands.

Seckbach, J. (ed.) (2006). *Life as we know it.* In the series of: Cellular Origins, Life in Extreme Habitats and Astrobiology, Springer: Dordrecht, The Netherlands.

Seckbach, J., Chela-Flores, J., Owen, T. and Raulin F. (eds.) (2004). *Life in the Universe. From Miller experiment to the Search for Life in other Worlds.* Kluwer, Dordrecht.

Spiegel Magazine (2005). *Darwinism Completely Refutes Intelligent Design*, December 26,
http://www.spiegel.de/international/spiegel/0,1518,392319,00.html.

Spinoza (2002). *Ethics.* Everyman, London.

Time (2005). Evolution wars, August 15, **166** (7) 26-35.

Time (2006). God vs. science, 13 November 2006. **168** (20) 48-55.

Index of Species

Index of Keywords

Listed below are the first occurrences of keywords as provided by authors as well as terms from the dialogues.

Index of Authors and Dialoguers

Index of Names